ELECTRICAL WIRING
Residential

BASED ON THE 2011
NATIONAL ELECTRICAL CODE®

ELECTRICAL WIRING
Residential

17TH EDITION

RAY C. MULLIN

PHIL SIMMONS

DELMAR
CENGAGE Learning™   Australia Canada Mexico Singapore Spain United Kingdom United States

Electrical Wiring Residential, 17th Edition
Ray C. Mullin, Phil Simmons

Vice President, Career and Professional
Editorial: Dave Garza

Director of Learning Solutions: Sandy Clark

Acquisitions Editor: Stacy Masucci

Managing Editor: Larry Main

Senior Product Manager: John Fisher

Editorial Assistant: Andrea Timpano

Vice President, Career and Professional
Marketing: Jennifer Baker

Marketing Director: Deborah Yarnell

Marketing Manager: Kathryn Hall

Associate Marketing Manager: Scott A. Chrysler

Production Director: Wendy Troeger

Production Manager: Mark Bernard

Content Project Manager: Barbara LeFleur

Senior Art Director: David Arsenault

Technology Project Manager: Christopher
Catalina

Cover Images:

   Electronic schematic: © Shane White/
   iStockphoto

   Lightbulb illustration: © Joseph Villanova

   Mansion @ twilight: © Dan Eckert/
   iStockphoto

Compositor: MPS Limited, a Macmillan
Company

Library of Congress Control Number: 2010932235

ISBN-13: 978-1-4354-9825-9
ISBN-10: 1-4354-9825-9

Delmar
5 Maxwell Drive
Clifton Park, NY 12065-2919
USA

Cengage Learning is a leading provider of customized learning solutions with office locations around the globe, including Singapore, the United Kingdom, Australia, Mexico, Brazil and Japan. Locate your local office at: **international. cengage.com/region**

Cengage Learning products are represented in Canada by Nelson Education, Ltd.

For your lifelong learning solutions, visit **delmar.cengage.com**

Visit our corporate website at **cengage.com**.

Printed in the United States of America
1 2 3 4 5 6 7 14 13 12 11 10

Contents

CHAPTER 3

Determining the Required Number of Lighting Outlets, Receptacle Outlets, and Small-Appliance Branch Circuits

CHAPTER 4

Conductor Sizes and Types, Wiring Methods, Wire Connections, Voltage Drop, Neutral Conductor Sizing for Services

CHAPTER 5

Conductor Identification, Switch Control of Lighting Circuits, Bonding/Grounding of Wiring Devices, Induction Heating

CHAPTER 6

Ground-Fault Circuit Interrupters, Arc-Fault Circuit Interrupters, Surge Protective Devices, Immersion Detection Circuit Interrupters, and Appliance Leakage Current Interrupters

CHAPTER 19

Special-Purpose Outlets—Water Pump, Water Heater 421

CHAPTER 20

Special-Purpose Outlets for Ranges, Counter-Mounted Cooking Unit ▲G, and Wall-Mounted Oven ▲F 448

CHAPTER 21

Special-Purpose Outlets—Food Waste Disposer ▲H, Dishwasher ▲I

CHAPTER 22

Special-Purpose Outlets for the Bathroom Ceiling Heat/Vent/Lights ▲K ▲J, the Attic Fan ▲L, and the Hydromassage Tub ▲A

CHAPTER 23

Special-Purpose Outlets—Electric Heating ▲M, Air Conditioning ▲N

CHAPTER 27

CHAPTER 28

CHAPTER 29

CHAPTER 30

CHAPTER 31

CHAPTER 32

CHAPTER 33

Sheet 1 of 10	Basement Plan
Sheet 2 of 10	Floor Plan
Sheet 3 of 10	South (Front) Elevation; Window Schedule, Door Schedule
Sheet 4 of 10	East (Right) Elevation; Interior Elevations
Sheet 5 of 10	North (Rear) Elevation; Plot Plan
Sheet 6 of 10	West (Left) Elevation; Schedule of Special-Purpose Outlets
Sheet 7 of 10	Section A-A
Sheet 8 of 10	Electrical Basement Plan
Sheet 9 of 10	Electrical First Floor Plan
Sheet 10 of 10	Code Requirements for Swimming Pool Wiring
Blank Floor Plan	First Floor
Blank Floor Plan	Basement

Foreword

THE IMPORTANCE OF PROPER TRAINING

Now that I have retired after 48 years in the electrical industry, it has become even more evident that a good solid education about the world of electricity is of utmost importance.

Accurate materials and training are the two sides of the electrical safety coin. This coin is spent every day by various persons intrinsically involved in the electrical construction industry. Unfortunately, some spend it less wisely than others. Usually, the unwise spenders are those who rush to career, having neglected to acquire accurate materials and to focus on training themselves to a high level of proficiency.

Ray Mullin, coauthor of this book, *Electrical Wiring—Residential*, has often stated, "The cost of education is small when compared to the price paid for ignorance." All too often, we, the citizens, pay the price for others' ignorance—ignorance of the codes, ignorance of proper wiring methods, ignorance of proper installation procedures, ignorance of design requirements, ignorance of product evaluations. This price becomes dear when our friends and family lose health or life or when our homes are destroyed.

It is exciting to see that Phil Simmons has joined with Ray as coauthor of the 17th edition of *Electrical Wiring—Residential*. Phil has served the electrical industry with distinction for many years. His ability to express complex electrical issues clearly and to illustrate them accurately is unparalleled among his peers.

Fortunately, accurate materials are so easy to obtain. Ray Mullin and Phil Simmons are both technical writers who have paid their dues in the electrical industry. Each has put in many years as an apprentice, a journeyman, and then as a master electrician before beginning to write about his trade. Phil was additionally a professional in the electrical inspection arena and managed the International Association of Electrical Inspectors (IAEI) for several years. Both have served or are serving on *NEC* Code Making Panels. *Electrical Wiring—Residential* contains accurate, up-to-date information about all aspects of residential wiring.

When installers and inspectors don't keep abreast of installation procedures and code requirements, things like cables across scuttle access to attics, improper spacing of receptacle outlets, improper short-circuit and ground-fault protection, and improper grounding of electrical systems, phone system, and CATV systems can lead to hazardous situations causing electrical shocks and fire. Not just anybody can install or inspect safe electrical systems. Trained professionals can, but even they must be constantly improving their knowledge and skills.

Because Ray Mullin and Phil Simmons care about the electrical safety coin, they have striven to provide the most accurate information possible. It is up to each of us, however, to focus on the training. Some training can be acquired simply by reading the best books in our trade; some training can come through the online programs available; and other training, through participation in classes and seminars. In each instance, though, motivation and desire come from within—to know everything involved in our trade, to be totally proficient, to focus continually on improvement. As we seek both accurate information and training, we learn to spend the coin of safety to benefit others as well as ourselves. I commend you for acquiring *Electrical Wiring—Residential*; now I challenge you to make it part of yourself. I challenge you to spend the electrical safety coin wisely.

James W. Carpenter
Former CEO and Executive Director,
International Association of Electrical Inspectors
Past Chair of the NEC Technical Correlating Committee

Preface

INTENDED USE AND LEVEL

STOP . . . Don't read any further . . . yet. Take a moment to familiarize yourself with how to use this text to get the most benefit from it. Think of it as a three-legged stool. One leg is this text, the second leg is the 2011 edition of the *National Electrical Code®*, and the third leg is the set of Plans that are in the packet in the inside back cover. If any one of the legs is missing, the stool will collapse. Stated another way, you will not get as much out of this course. When you have completed all of the chapters in *Electrical Wiring—Residential,* you will have virtually wired a typical house according to the requirements of the 2011 *National Electrical Code.* An accomplishment you can be proud of!

The *NEC®* defines a "qualified person" as *One who has skills and knowledge related to the construction and operation of the electrical equipment and installations and has received safety training to recognize and avoid the hazards involved.**

Electrical Wiring—Residential is intended for use in residential wiring courses at high schools, two-year and four-year colleges, as well as apprenticeship training programs. This comprehensive book guides readers, room by room, through the wiring of a typical residence and builds a foundation of knowledge by starting with the basic requirements of the *National Electrical Code (NEC)*, then continuing on to the more advanced wiring methods. Each *Code* rule is presented through text, illustrations, examples, and wiring diagrams. In addition, an accompanying set of plans at the back of the book guides the reader through the wiring process by applying concepts learned in each chapter to an actual residential building in order to understand and meet the requirements set forth by the *NEC*.

An Important Note about Safety

In the educational field, it is pretty much a given that "Society will pay for education . . . one way or another." Proper training of a skilled trade is much better than hit-or-miss learning. Having to do the job over, having a house burn down, or having someone get electrocuted because of improper wiring is costly!

It really doesn't take any longer to do it right the first time than to have to do it over. You probably have heard the phrase "Measure twice . . . cut once. Measure once . . . cut twice." How true!

Electrical wiring is a skilled trade. Wiring should not be done by anyone not familiar with the hazards involved. It is a highly technical skill that requires much training. This

---

**National Electrical Code®* and *NEC®* are registered trademarks of the National Fire Protection Association, Inc., Quincy, MA 02169.

material provides all of the electrical codes and standards information needed to approach house wiring in a safe manner. In fact, *Electrical Wiring—Residential* has been adopted as the core text by the major electrical apprenticeship programs across the country. Their residential curriculum program directors and committee members made this text their top choice for their residential wiring training.

Electrical Wiring—Residential will provide you with the know-how so you can wire houses that "Meet Code."

Electrical Wiring—Residential has become an integral part of approved (accredited) training programs by an increasing number of states that require residential electricians to have a residential license if they are going to wire homes and small apartments.

The *NEC* has one thing in mind—safety! There is too much at stake to do less than what the *NEC* requires. Anything less is unacceptable! The *NEC* in *90.1(A)* makes it pretty clear. It states that *The purpose of this Code is the practical safeguarding of persons and property from hazards arising from the use of electricity.* *

Do not work on live circuits! Always de-energize the system before working on it! There is no compromise when it comes to safety! Many injuries and deaths have occurred when individuals worked on live equipment. The question is always: "Would the injury or death have occurred had the power been shut off?" The answer is "No!"

All mandatory safety-related work practices are found in the Federal Regulation Occupational Safety and Health Administration (OSHA), Title 29, Subpart S—Electrical, Sections 1910.331 through 1910.360.

SUBJECT AND APPROACH

The 17th edition of *Electrical Wiring—Residential* is based on the 2011 *National Electrical Code* (*NEC*). The *NEC* is used as the basic standard for the layout and construction of residential electrical systems. In this text, thorough explanations are provided of *Code* requirements as they relate to

*Reprinted with permission from NFPA 70-2011.

residential wiring. To gain the greatest benefit from this edition, the student must use the *NEC* on a continuing basis.

It is extremely difficult to learn the *NEC* by merely reading it. This text brings together the rules of the *NEC* and the wiring of an actual house. You will study the rules from the *NEC* and apply those rules to a true-to-life house wiring installation.

Take a moment to look at the Table of Contents. It is immediately apparent that you will not learn such things as how to drill a hole, tape a splice, fish a cable through a wall, use tools, or repair broken plaster around a box. These things you already know or are learning on the job. The emphasis of this text is to teach you how to wire a house that "Meets Code." Doing it right the first time is far better than having to do it over because the electrical inspector turned down your job.

The first seven chapters in this book concentrate on basic electrical code requirements that apply to house wiring. This includes safety when working with electricity; construction symbols, plans, and specifications; wiring methods; conductor sizing; circuit layout; wiring diagrams; numerous ways to connect switches and receptacles; how to wire recessed luminaires; ground-fault circuit interrupters (GFCIs); arc-fault circuit interrupters (AFCIs); and surge suppressors.

The remaining chapters are devoted to the wiring of an actual house—room by room, circuit by circuit. All of these circuits are taken into account when calculating the size of the main service. Because proper grounding is a key safety issue, the subject is covered in detail.

You will also learn about security systems, fire and smoke alarms, low-voltage remote-control wiring, swimming pools, standby generators, and you will be introduced to structured wiring for home automation.

You will find this text unique in that you will use the text, an actual set of plans and specifications, and the *NEC*—all at the same time. The text is perfect for learning house wiring and makes an excellent reference source for looking up specific topics relating to house wiring. The blueprints serve as the basis for the wiring schematics, cable layouts, and discussions provided in the text. Each chapter dealing with a specific type of wiring is referenced to the appropriate plan sheet. All wiring systems are described in detail—lighting, appliance, heating, service entrance, and so on.

The house selected for this edition is scaled for current construction practices and costs. Note, however, that the wiring, luminaires, appliances, number of outlets, number of circuits, and track lighting are not all commonly found in a home of this size. The wiring may incorporate more features than are absolutely necessary. This was done to present as many features and *Code* issues as possible, to give the student more experience in wiring a residence. Also included are many recommendations that are above and beyond the basic *NEC* requirements.

Note: The *NEC* (NFPA 70) becomes mandatory only after it has been adopted by a city, county, state, or other governing body. Until officially adopted, the *NEC* is merely advisory in nature. State and local electrical codes may contain modifications of the *NEC* to meet local requirements. In some cases, local codes will adopt certain more stringent regulations than those found in the *NEC*. For example, the *NEC* recognizes nonmetallic-sheathed cable as an acceptable wiring method for house wiring. Yet, the city of Chicago and surrounding counties do not permit nonmetallic-sheathed cable for house wiring. In these areas, all house wiring is done with electrical metallic tubing (thinwall).

There are also instances where a governing body has legislated action that waives specific *NEC* requirements, feeling that the *NEC* was too restrictive on that particular issue. Such instances are very rare. The instructor is encouraged to furnish students with any local variations from the *NEC* that would affect this residential installation in a specific locality.

THE ELECTRICAL TRADE—TRAINING PROGRAMS

As you study *Electrical Wiring—Residential*, study with a purpose—to become the best residential wireman possible.

There will always be a need for skilled electricians! Qualified electricians almost always have work. It takes many hours of classroom and on-the-job training to become a skilled electrician. The best way to learn the electrical trade is through a training program approved by the U.S. Department of Labor (http://www.dol.gov). Many times an apprenticeship program is called "Earn while You Learn." These programs offer the related classroom training and the advantage of working on the job with skilled journeymen electricians. Completion of a Registered Apprenticeship program generally leads to higher pay, job security, higher quality of life, recognition across the country, and the opportunity for college credit and future degrees.

As a rule, these training programs require 144 to 180 hours of classroom-related technical training and 2000 hours of on-the-job training per year. Some programs have day classes and some have night classes. An electrical apprenticeship training program might run four to five years. The end-result—becoming a full-fledged licensed journeyman electrician capable of doing residential, commercial, and industrial electrical work. A residential electrician training program might run two to three years, with the training limited to the wiring of single- and multifamily dwellings. The end result—receiving a license limited to residential wiring.

To get into an apprenticeship program, the individual usually must have a high school education, with at least 1 year of high school algebra; be at least 18 years old; must be physically in shape to perform the work electricians are called upon to do (e.g., climbing, lifting, work in inclement weather); and, most importantly, be drug free. There generally is a qualifying aptitude test to make sure the applicant has the ability to take on the responsibility of a rigid apprenticeship training program. In some areas, passing the high school equivalency General Education Development (GED) test is acceptable in place of a high school diploma.

What does it take to make a good apprentice and journeyman electrician? In no particular order: commitment to master the electrical field, willingness to study and understand the training material, strong math skills, ability to think clearly and logically to analyze and solve problems, ability to work indoors and outdoors, comfortable working with your head and hands, good mechanical skills, ability to communicate and work with others, good verbal skills, ability to follow directions, strong work and personal ethics, and being a self-starter.

Following completion of an apprenticeship program, continuing education courses are available to keep the journeyman up to date on codes and other related topics and skills.

Journeymen electricians who have an interest in teaching apprentices will usually have to take instructor training courses. In certain programs, satisfactory completion of the required courses can lead to an associate degree. Others will go on to become crew leaders, supervisors, and contractors.

There are some areas where a "pre-apprenticeship" program is offered. To learn more about the careers possible in the electrical field, chat with your instructor; your local high school's guidance counselor; your vocational, technical, and adult education schools; and electricians and electrical contractors. Go online and search for electrical apprenticeship programs.

Your future is in your hands.

Some very important two-letter words that you should remember are

IF IT IS TO BE, IT IS UP TO ME!

Job Titles

Most building codes and standards contain definitions for the various levels of competency of workers in the electrical industry. Here are some examples of typical definitions:

Apprentice shall mean a person who is required to be registered, who is in compliance with the provisions of this article, and who is working at the trade in the employment of a registered electrical contractor and is under the direct supervision of a licensed master electrician, journeyman electrician, or residential wireman.

Residential Wireman shall mean a person having the necessary qualifications, training, experience, and technical knowledge to wire for and install electrical apparatus and equipment for wiring one-, two-, three-, and four-family dwellings. A residential wireman is sometimes referred to as a *Class B Electrician*.

Journeyman Electrician shall mean a person having the necessary qualifications, training, experience, and technical knowledge to wire for, install, and repair electrical apparatus and equipment for light, heat, power, and other purposes, in accordance with standard rules and regulations governing such work.

Master Electrician means a person having the necessary qualifications, training, experience, and technical knowledge to properly plan, lay out, and supervise the installation and repair of wiring apparatus and equipment for electric light, heat, power, and other purposes, in accordance with standard codes and regulations governing such work, such as the *NEC*.

Electrical Contractor means any person, firm, partnership, corporation, association, or combination thereof who undertakes or offers to undertake for another the planning, laying out, supervising and installing, or the making of additions, alterations, and repairs in the installation of wiring apparatus and equipment for electrical light, heat, and power.

Most electrical inspectors across the country are members of the International Association of Electrical Inspectors (IAEI). This organization publishes one of the finest technical bimonthly magazines devoted entirely to the *NEC* and related topics, and it is open to individuals who are not electrical inspectors. Electrical instructors, vo-tech students, apprentices, electricians, consulting engineers, contractors, and distributors are encouraged to join the IAEI so they can stay up to date on all *NEC* issues, changes, and interpretations. An application form that explains the benefits of membership in the IAEI can be found in the Appendix of this text.

NEW TO THIS EDITION

Continuing in the tradition of previous editions, this edition thoroughly explains how *Code* changes affect house wiring installations. New and revised full-color illustrations supplement the explanations to ensure that electricians understand the new *Code* requirements. New photos reflect the latest wiring materials and components available on the market. Revised review questions test student understanding of the new content. New tables that summarize *Code* requirements offer a quick reference tool for students. Other reference aids are the tables reprinted directly from the 2011 edition of the *NEC*. The extensive revisions for the seventeenth edition make

Electrical Wiring—Residential the most up to date and well organized guide to house wiring. Coverage of the *NEC* has been expanded to well over 1000 *Code* references.

This text focuses on the technical skills required to perform electrical installations. It covers such topics as calculating conductor sizes, calculating voltage drop, determining appliance circuit requirements, sizing service, connecting electric appliances, grounding service and equipment, installing recessed luminaires (fixtures), and much more. These are critical skills that can make the difference between an installation that "Meets *Code*" and one that does not. The electrician must understand the reasons for following *Code* regulations to achieve an installation that is essentially free from hazard to life and property.

Note: Symbols have been added to indicate changes in the 2011 *National Electrical Code* vs. the 2008 *National Electrical Code.*▶◀

This text might be called "Work in Progress." The authors stay in touch with the latest residential wiring trends and the *National Electrical Code.* Because the *NEC* is revised every three years, this text follows the same cycle. *Electrical Wiring—Residential* has been carefully reviewed to editorially simplify, streamline, and improve its readability. Many diagrams have been simplified. Some units were reorganized so the *Code* requirements for the various applications are more uniform.

Much rewriting was done. The 2011 *NEC* contains many editorial changes as well as renumbering and relocation of numerous *Code* references. All of these have been addressed in this edition of *Electrical Wiring—Residential.* Some text has been condensed and reformatted for ease in reading. Many diagrams have been simplified for clarity.

The Objectives have been fine-tuned for easier readability.

- There were 5016 Proposals submitted to make changes to the 2011 *NEC* with 2910 Comments submitted relating to the Proposals. The end result was the publishing of the 2011 edition of the *National Electrical Code.*

- In *Electrical Wiring—Residential,* all *Code* requirements have been updated to the 2011 edition of the *NEC.* These have been revised throughout the text, wiring diagrams, and illustrations.

- Illustrations have been enhanced for improving clarity and ease in understanding.

- Emphasis given to making the wiring of the residence conform to energy saving standards. In other words, the residence in *Electrical Wiring—Residential* is "Green."

- One of the most far-reaching new requirements in the 2011 *National Electrical Code* is that the grounded circuit conductor must be brought to every switch location. This new requirement has been addressed in *Electrical Wiring—Residential,* with all wiring diagram revised accordingly. This means that more 3-wire and 4-wire cable, and possibly larger boxes will be required.

- Major requirement when replacing receptacles in existing installations where AFCIs are now required. The replacement must be AFCI protected. See *NEC 406.4(D)(4).*

- Major requirement when replacing receptacles in existing installations where tamper-resistant receptacles are now required. The replacement must be of the tamper-resistant type. See *NEC 406.4(D)(5).*

- Major requirement when replacing receptacles in existing installations where weather-resistant receptacles are now required. The replacement must be of the weather-resistant type. See *NEC 406.4(D)(6).*

- Permission now given to install special types of receptacles in countertops. See *NEC 210.52(C)(5).*

- All of the wiring diagrams have been updated to show the latest system of electrical symbols. This is based on the National Electrical Contractors Association's National Electrical Installation Standard.

- Major revisions of many diagrams and figures have made to improve the clarity and ease of understanding the *Code* requirements.

- Many new full-color illustrations have been added.

- Because of concern and confusion over how to cope with the heat generated in confined areas such as circular raceways like EMT, RMC, and IMC, the *2011 NEC* calls attention to the difference between circular raceways and other wireways such as surface metal raceways, auxiliary gutters, and the like. The new term *circular raceways* has been addressed in this text.

An important new chapter has been added on residential utility-interactive solar photovoltaic systems. Many of these systems are being installed and being fully informed of the many unique *NEC* requirements is vital to successful installations.

SUPPLEMENT PACKAGE

An *Instructor's Guide* contains answers to all Review questions included in the book and a blank service-entrance calculation form; also available in electronic format on the accompanying *Instructor Resources CD*.
(Order #: 1-4354-9824-0)

An *Instructor Resources CD* provides instructors with valuable classroom materials on CD-ROM:

- *PowerPoint Presentations* outline the important concepts covered in each chapter. Extensively illustrated with photos, tables, and diagrams from the book, the presentations enhance classroom instruction. PowerPoint presentations also allow instructors to tailor the course to meet the needs of their individual class.

- A *Testbank* in the latest *ExamView* format provides instructors with approximately 1500 test questions to evaluate students as they work through each chapter in the book. Answers and book page references are also provided. Completely editable, instructors may also wish to delete or add questions to meet the needs of their individual class.

- An *Image Gallery* contains nearly all the images in the book and can be used to enhance the PowerPoint presentation, or to create transparency masters and handouts.

- *Electronic Instructor's Guide* in Microsoft Word enables instructors to view and print answers to review questions contained in the book.

- *Electronic Blueprints* provide an online version of the drawings that are included at the back of the book, allowing instructors to project and reference in classroom presentations.

- *Video Clips* from the accompanying video series visually highlight important concepts presented in the book

(Order #: 1-4354-9823-2)

A *Lab Manual* provides over twenty exercises to aid students in learning both basic and complex wiring circuits. Each lab consists of a hands-on wiring exercise as well as *NEC* drill problems. Students are also required to draw an electrical layout of wiring booths to familiarize themselves with electrical symbols. Allowing for instructor verification and student self-assessment, this manual is essential to applying important wiring concepts.
(Order #: 1-4354-9822-4)

Electrical Wiring Residential DVD Series: Available DVD, *Electrical Wiring—Residential Video Series* correlates directly with the *Electrical Wiring—Residential* book. Shot on various construction sites and enhanced with quality animations, each video is devoted to a specific, specialized topic. In addition to high-quality animations, questions have also been incorporated into each tape at strategic points to promote discussion aimed at enhancing viewers' understanding and preparing them for successful attainment of all learning objectives. Each video is approximately 20 minutes in length, making it easy to incorporate into a comprehensive electrician education and training program.

The complete set includes: Video #1: Introduction to Electrical Installation, Video #2: Planning for Circuit Installation, Video #3: Ground-Fault Circuit Interrupters, Video #4: Lighting by the Room, Video #5: Special Circuits, Video #6: Special-Purpose Outlets, Video #7: Miscellaneous and Custom Installations, Video #8: Service Entrances.
(DVD 1, Videos 1-4 Order #: 1-4354-9530-6) DVD 2, Videos 5-8 Order #: 1-4354-9525-X)

Blackboard supplement features include chapter objectives, practice tests, glossary, and links to relevant websites. *(Order #: 1435498216).*

Web Tutor supplement features include chapter objectives, practice tests, glossary, and links to relevant websites. *(Order #: 1435498208).*

Also Available from Delmar Cengage Learning

Videos. The *House Wiring DVD Series* is an integrated part of the *Residential Construction Academy House Wiring* package. The series contains a set of eight 20-minute videos that provide step-by-step instruction for wiring a house. All the essential information is covered in this series, beginning with the important process of reviewing the plans and following through to the final phase of testing and troubleshooting. Need-to-know *NEC* articles are highlighted, and Electrician's Tips and Safety Tips offer practical advice from the experts.

The complete set includes the following: Video #1: Safety and Safe Practices, Video #2: Hardware, Video #3: Tools, Video #4: Initial Review of Plans, Video #5: Rough-In, Video #6: Service Entrance, Video #7: Trim-Out, Video #8: Testing & Troubleshooting.

For more information, visit http://www .residentialconstructionacademy.com. ***Visit us at www.cengage.com/community/electrical, now LIVE for the 2011 Code cycle!***

This newly designed Web site provides information on other learning materials offered by Delmar, as well as industry links, career profiles, job opportunities, and more! To access additional course materials including CourseMate, please visit www.cengagebrain.com. At the CengageBrain .com home page, search for the ISBN of your title (from the back cover of your book), using the search box at the top of the page. This will take you to the product page where these resources can be found.

⎍⎍⎍ ABOUT THE AUTHORS

This text was coauthored by Ray C. Mullin and Phil Simmons.

Mr. Mullin is a former electrical instructor for the Wisconsin Schools of Vocational, Technical, and Adult Education. He is a former member of the International Brotherhood of Electrical Workers. He is a member of the International Association of Electrical Inspectors, the Institute of Electrical and Electronic Engineers, and the National Fire Protection Association, Electrical Section, and has served on Code-Making Panel 4 of the *National Electrical Code*.

Mr. Mullin completed his apprenticeship training and worked as a journeyman and supervisor for residential, commercial, and industrial installations. He has taught both day and night electrical apprentice and journeyman courses, has conducted engineering seminars, and has conducted many technical *Code* workshops and seminars at International Association of Electrical Inspectors Chapter and Section meetings, and has served on their code panels.

He has written many technical articles that have appeared in electrical trade publications. He has served as a consultant to electrical equipment manufacturers regarding conformance of their products to industry standards, and on legal issues relative to personal injury lawsuits resulting from the misuse of electricity and electrical equipment. He has served as an expert witness.

Mr. Mullin presents his knowledge and experience in this text in a clear-cut manner that is easy to understand. This presentation will help students to fully understand the essentials required to pass the residential licensing examinations and to perform residential wiring that "Meets *Code*."

Mr. Mullin is the author of *House Wiring with the NEC*—a text that focuses entirely on the *National Electrical Code* requirements for house wiring. He is coauthor of *Electrical Wiring—Commercial*, *Illustrated Electrical Calculations*, and *The Smart House*. He contributed technical material for Delmar's *Electrical Grounding and Bonding* and to the International Association of Electrical Inspectors' texts *Soares' Book On Grounding* and *Ferm's Fast Finder*.

He served on the Executive Board of the Western Section of the International Association of Electrical Inspectors and on their Code Clearing Committee, and, in the past, served as Secretary/Treasurer of the Indiana Chapter of the IAEI.

Mr. Mullin is past Chairman of the Electrical Commission in his hometown.

Mr. Mullin is past Director, Technical Liaison for a major electrical manufacturer. In this position, he was deeply involved in electrical codes and standards as well as contributing and developing technical training material for use by this company's field engineering personnel.

Mr. Mullin attended the University of Wisconsin, Colorado State University, and the Milwaukee School of Engineering.

Phil Simmons is self-employed as Simmons Electrical Services. Services provided include consulting on the *National Electrical Code* and other codes; writing, editing, illustrating and producing technical publications; and inspection of complex electrical installations. He develops training programs related to electrical codes and safety and has been a presenter on these subjects at numerous seminars and conferences for universities, the NFPA, IAEI, Department of Defense, and private clients. Phil also provides plan review of electrical construction documents. He has consulted on several lawsuits concerning electrical shocks, burn injuries, and electrocutions.

Mr. Simmons is coauthor and illustrator of *Electrical Wiring—Residential* (17th edition), coauthor and illustrator of *Electrical Wiring—Commercial* (14th edition) and author and illustrator of *Electrical Grounding and Bonding* (3rd edition),

all published by Delmar Cengage Learning. While at IAEI, Phil was author and illustrator of several books, including the *Soares' Book on Grounding of Electrical Systems* (five editions), *Analysis of the NEC* (three editions), and *Electrical Systems in One- and Two-Family Dwellings* (three editions). Phil wrote and illustrated the National Electrical Installation Standard (NEIS) on *Types AC and MC Cables* for the National Electrical Contractors Association.

Phil presently serves NFPA on Code Making Panel-5 of the National Electrical Code Committee (grounding and bonding). He previously served on the *NEC* CMP-1 (*Articles 90, 100,* and *110*), as Chair of CMP-19 (articles on agricultural buildings and mobile and manufactured buildings), and member of CMP-17 (health care facilities). He served six years on the NFPA Standards Council, as NFPA Electrical Section President, and on the NEC Technical Correlating Committee.

Phil began his electrical career in a light-industrial plant. He is a master electrician in the state of Washington and was owner and manager of Simmons Electric Inc., an electrical contracting company. He is a licensed journeyman electrician in Montana and Alaska. Phil passed the certification examinations for Electrical Inspector General, Electrical Plan Review, and Electrical Inspector One- and Two-Family.

He previously served as Chief Electrical Inspector for the State of Washington from 1984 to 1990 as well as an Electrical Inspector Supervisor, Electrical Plans Examiner, and field Electrical Inspector. While employed with the State, Phil performed plan review and inspection of health care facilities, including hospitals, nursing homes, and boarding homes.

Phil served the International Association of Electrical Inspectors as Executive Director from 1990 to 1995 and as Education, Codes, and Standards Coordinator from 1995 through June 1999. He was International President in 1987 and has served on local and regional committees.

He served Underwriters Laboratories as a Corporate Member and on the Electrical Council from 1985 to 2000. He served on the UL Board of

Directors from 1991 to 1995 and is a retired member of the International Brotherhood of Electrical Workers.

IMPORTANT NOTE

Every effort has been made to be technically correct, but there is always the possibility of typographical errors. If changes in the *NEC* do occur after the printing of this text, these changes will be incorporated in the next printing.

The National Fire Protection Association has a standard procedure to introduce changes between *Code* cycles after the actual *NEC* is printed. These are called "Tentative Interim Amendments," or TIAs. TIAs and typographical errors can be downloaded from the NFPA Web site, http://www.nfpa.org, to make your copy of the *Code* current.

Acknowledgments

Ray Mullin wishes to again thank his wife, Helen, for her understanding and support while he devoted unlimited time attending meetings and working many hours on revising this edition of *Electrical Wiring—Residential*. Major revisions such as this take somewhere between 1000 and 1500 hours! Patience is a virtue!

Phil Simmons once again wants to express his appreciation to his wife Della for her generosity in allowing him to devote so much time and effort to updating this book as well as *Electrical Wiring Commercial* and *Electrical Grounding and Bonding* to the new *NEC* during the year. Time after time she picked up the ball and ran with it on projects Phil would customarily attend to.

As always, the team at Cengage Delmar Learning has done an outstanding job in bringing this edition to press. Their drive, dedication, and attention to minute details ensure that this text, without question, is the country's leading text on house wiring. They sure know how to keep the pressure on!

Special thanks to our good friend Jimmy Carpenter, former Executive Director of the International Association of Electrical Inspectors, for his inspiring Foreword to this text regarding the "Importance of Proper Training."

We would also like to thank many individuals in the electrical industry across the country. They are our friends and are outstanding *Code* experts. These include Madeline Borthick, David Dini, Paul Dobrowsky, Joe Ellwanger, Ken Haden, David Hittinger, Mike Johnston, Robert Kosky, Richard Loyd, Bill Neitzel, Cliff Rediger, Clarence Tibbs, Charlie Trout, Gordon Stewart, Ray Weber, David Williams, the electrical staff at NFPA headquarters who responded timely to our questions on particular *Code* issues, the many Code-Making Panel members who clarified specific *Code* proposals for the 2011 *NEC*, and the other reviewers for their excellent suggestions that helped make this edition the most comprehensive and *Code*-compliant ever. Special thanks again to Robert Boiko for his technical input on water heaters and their safety related controls.

We wish we could name all our friends in the electrical industry, but there are so many, it would take many pages to include all of their names. Thanks to all of you for your input. We apologize if we missed anyone.

The authors gratefully acknowledge the contribution of the chapter on Residential Utility Interactive Photovoltaic Systems by Pete Jackson, electrical inspector for the City of Bakersfield, CA.

The coauthors and publisher would like to thank the following reviewers for their contributions:

William Dunakin
Independent Electrical Contractors
West Hartford, CT

Orville Lake
Augusta Technical College
Augusta, GA

Bill F. Neitzel
Madison, WI

Michael Ross
San Jacinto College
Pasadena, TX

General Information for Electrical Installations

OBJECTIVES

After studying this chapter, you should be able to

- understand the basic safety rules for working on electrical systems.

- access the Internet to obtain a virtual unlimited source of safety and technical related information.

- become familiar with important electrical codes, safety codes, and building codes such as NFPA 70, 70A, 70B, 70E, 73, OSHA, NIOSH, ADA, NRTL, and the ICC.

- learn about licensing, permits, plans, specifications, symbols, and notations.

- understand the role of the electrical inspector and the International Association of Electrical Inspectors.

- understand the metric system of measurement.

SAFETY IN THE WORKPLACE

Electricity is great when it is doing what it is intended to do, and that is to stay in its intended path and doing the work intended. But electricity out of its intended path can be dangerous, often resulting in fire, serious injury, or death.

Before getting into residential wiring and the *National Electrical Code* (*NEC*), we need to discuss on-the-job safety. Safety is not a joke! Electricians working on new construction, remodel work, maintenance, and repair work find that electricity is part of the work environment. Electricity is all around us, just waiting for the opportunity to get out of control. Repeat these words: **Safety First . . . Safety Last . . . Safety Always!**

Working on switches, receptacles, luminaires, or appliances with the power turned on is dangerous. Turn off the power! In addition, check with a voltmeter to be sure the power is off.

The voltage level in a home is 120 volts between one "hot" conductor and the "neutral" conductor or grounded surface. Between the two "hot" conductors (line-to-line), the voltage is 240 volts.

An electrical shock is received when electrical current passes through the body. From basic electrical theory, you learned that line voltage appears across an open in a series circuit. Getting caught "in series" with a 120-volt circuit will give you a 120-volt shock. For example, open-circuit voltage between the two terminals of a single-pole switch on a lighting circuit is 120 volts when the switch is in the "OFF" position and the lamp(s) are in place. See Figure 1-1. Likewise, getting caught "in series" with a 240-volt circuit will give you a 240-volt shock.

Working on equipment with the power turned on can result in death or serious injury, either as a direct result of electricity (electrocution or burns) or from an indirect secondary reaction, such as falling off a ladder or jerking away from the "hot" conductor into moving parts of equipment such as the turning blades of a fan. For example: A workman was seriously injured while working a "live" circuit that supplied a piece of equipment. He accidentally came into contact with a "hot" terminal, and reflex action caused him to pull his hand back into a turning pulley. The pulley cut deeply into his wrist, resulting in a tremendous loss of blood.

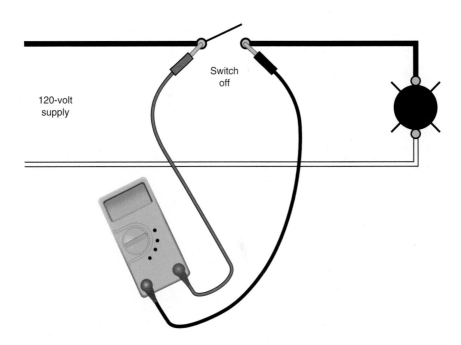

FIGURE 1-1 The voltage across the two terminals of the single-pole switch is 120 volts. (*Delmar/Cengage Learning*)

Dropping a metal tool onto live parts, allowing metal shavings from a drilling operation to fall onto live parts of electrical equipment, cutting into a "live" conductor and a "neutral" conductor at the same time, or touching the "live" wire and the "neutral" conductor or a grounded surface at the same time can cause injury directly or indirectly.

A short circuit or ground fault can result in a devastating arc flash that can cause serious injury or death. The heat of an electrical arc has been determined to be hotter than the sun. Tiny hot "balls" of copper can fly into your eye or onto your skin. Figure 1-2 shows a disconnect switch that has been locked and tagged.

Lockout/tagout (sometimes called LOTO) is the physical restraint of all hazardous energy sources that supply power to a piece of equipment. It simply means putting a padlock on the switch and applying a warning tag on the switch.

Dirt, debris, and moisture can also set the stage for equipment failure and personal injury. Neatness and cleanliness in the workplace are a must.

What about Low-Voltage Systems?

Although circuits of less than 50 volts generally are considered harmless, don't get too smug when working on so-called low voltage. Low-voltage circuits are not necessarily low hazard. A slight tingle might cause a reflex. A capacitor that is discharging can give you quite a jolt, causing you to jump or pull back.

In commercial work, such as telephone systems with large battery banks, there is extreme danger even though the voltage is "low." Think of a 12-volt car battery. If you drop a wrench across the battery terminals, you will immediately see a tremendous and dangerous arc flash.

It is the *current* that is the harmful component of an electrical circuit. *Voltage* pushes the current through the circuit. If you're not careful, you might become part of the circuit.

Higher voltages can push greater currents through the body. Higher voltages like 240, 480, and 600 volts can cause severe skin burns and possibly out-of-sight injuries such as internal bleeding and/or destruction of tissues, nerves, and muscles.

FIGURE 1-2   A typical disconnect switch with a lock and a tag attached to it. In the OSHA, ANSI, and NFPA standards, this is referred to as the *lockout/tagout* procedure. (*Delmar/Cengage Learning*)

It's the Law!

Not only is it a good idea to use proper safety measures as you work on and around electrical equipment, it is **required** by law. Electricians and electrical contractors need to be aware of these regulations. Practicing safety is a habit—like putting on your seat belt as soon as you get into your car.

The *NEC* is full of requirements that are safety related. For example, *430.102(B)* requires that a disconnecting means be located in sight from the motor location and the driven machinery location. This section also has "in-sight" and "lock-off" requirements, discussed in detail in Chapter 19.

The *NEC* defines a qualified person as: *One who has skills and knowledge related to the construction and operation of the electrical equipment and installations and has received safety training to recognize and avoid the hazards involved.* Merely telling someone or being told to "be careful" does not meet the definition of "proper training" and does not make the person qualified. An individual "qualified" in one skill might very well be "unqualified" in other skills.

According to *NFPA 70E, Electrical Safety in the Workplace,* circuits and conductors are not considered to be in an electrically safe condition until all sources of energy are removed, the disconnecting means is under lockout/tagout, and the absence of voltage is verified by an approved voltage tester.

The U.S. Department of Labor **Occupational Safety and Health Administration (OSHA)** regulations (Standards–29 CFR) is the law! This entire standard relates to safety in the workplace for general industry. Specifically, Part 1910, Subpart S, involves electrical safety requirements. The letters *CFR* stand for *Code of Federal Regulations.*

Key topics in the standard are: electric utilization systems, wiring design and protection, wiring methods, components and equipment for general use, specific purpose equipment and installations, hazardous (classified) locations, special systems, training, selection and use of work practices, use of equipment, selection and use of work practices, use of equipment, safeguards for personnel protection, and definitions (a mirror image of definitions found in the *NEC*).

A direct quote from *1910.333(a)(1)* states that *Live parts to which an employee may be exposed shall be de-energized before the employee works on or near them, unless the employer can demonstrate that de-energizing introduces additional or increased hazards or is infeasible due to equipment design or operational limitations. Live parts that operate at less than 50 volts to ground need not be de-energized if there will be no increased exposure to electrical burns or to explosion due to electric arcs.*

OSHA 1910.333(c)(2) states that "Only qualified persons may work on electric circuit parts or equipment that have not been de-energized under the procedures of paragraph (b) of this section. Such persons shall be capable of working safely on energized circuits and shall be familiar with the proper use of special precautionary techniques, personal protective equipment, insulating and shielding materials, and insulated tools."

OSHA 1910.399 defines a *qualified person* as "One familiar with the construction and operation of the equipment and the hazards involved," almost the same definition as that of the *NEC*.

For the most part, turning the power off then locking and tagging the disconnecting means is the safest. As the OSHA regulations state, "A lock and a tag shall be placed on each disconnecting means used to de-energize circuits and equipment."

Part 1926 in the OSHA regulation (Standards–29 CFR) deals with *Safety and Health Regulations for Construction.* Here we find the rules for anyone involved in the construction industry, not just electrical. A few of the topics are: medical services and first aid, safety training and education, recording and reporting injuries, housekeeping, personal protective equipment, means of egress, head protection, hearing protection, eye and face protection, ladders, scaffolds, rigging, hand and power tools, electrical requirements (a repeat of Part 1910, Subpart S), fall protection, and required signs and tags.

PERSONAL PROTECTIVE EQUIPMENT

Safety courses refer to personal protective equipment (PPE). These include such items as rubber gloves, insulating shoes and boots (footwear suitable for electrical work is marked with the letters "EH"), face shields, safety glasses, hard hats, ear protectors, Nomex™, and similar products. OSHA 1910.132(f)(1) requires that "The employer shall provide training to each employee who is required by this section to use PPE."

Working on electrical equipment while wearing rings and other jewelry is not acceptable. OSHA states that "Conductive articles of jewelry and clothing (such as watch bands, bracelets, rings, key chains, necklaces, metalized aprons, cloth with conductive thread, or metal headgear) may not be worn if they might contact exposed energized parts. However, such articles may be worn if they are rendered nonconductive by covering, wrapping, or other insulating means."

⌁ ARC FLASH

Don't get too complacent when working on electrical equipment.

A major short circuit or ground fault at the main service panel or at the meter cabinet or base can deliver a lot of energy. On large electrical installations, an arc flash (also referred to as an arc blast) can generate temperatures of 35,000°F (19,427°C). This is hotter than the surface on the sun. This amount of heat will instantly melt copper, aluminum, and steel. The blast will blow hot particles of metal and hot gases all over, resulting in personal injury, fatality, and/or fire. An arc flash also creates a tremendous air pressure wave that can cause serious hearing damage and/or memory loss due to the concussion. The blast might blow the victim away from the arc source.

Don't be fooled by the size of the service. Typical residential services are 100, 150, and 200 amperes. Larger services are found on large homes. Electricians seem to feel out of harm's way when working on residential electrical systems and seem to be more cautious when working on commercial and industrial electrical systems. A fault at a small main service panel can be just as dangerous as a fault on a large service. The available fault current at the main service disconnect for all practical purposes is determined by the kVA rating and impedance of the transformer. Other major limiting factors for fault current are the size, type, and length of the service-entrance conductors. Available fault current can easily reach 22,000 amperes, as is evident by panels that have a 22,000/10,000-ampere series rating.

Short-circuit calculations are discussed in Chapter 28.

Don't be fooled into thinking that if you cause a fault on the load side of the main disconnect that that main breaker will trip off and protect you from an arc flash. An arc flash will release the energy that the system is capable of delivering for as long as it takes the main circuit breaker to open. How much current (energy) the main breaker will "let through" is dependent on the available fault current and the breaker's opening time.

Although not required for house wiring, *NEC 110.16* requires that electrical equipment, such as switchboards, panelboards, industrial control panels, meter socket enclosures, and motor control centers, that are in other than dwelling units, and are likely to require examination, adjustment, servicing, or maintenance while energized shall be field marked to warn qualified persons of potential electric arc flash hazards. The marking shall be located so as to be clearly visible to qualified persons before examination, adjustment, servicing, or maintenance of the equipment. More information on this subject is found in *NFPA 70E* and in the ANSI Standard Z535.4, *Product Safety Signs and Labels*.

> ⌁ **TIP:** When turning a standard disconnect switch "ON," don't stand in front of the switch. Instead, stand to one side. For example, if the handle of the switch is on the right, then stand to the right of the switch, using your left hand to operate the handle of the switch, and turn your head away from the switch. That way, if an arc flash occurs when you turn the disconnect switch "ON," you will not be standing in front of the switch. You will not have the switch's door fly into your face, and the molten metal particles resulting from the arc flash will fly past you. ●

Classifying Electrical Injuries

OSHA recognizes the four main types of electrical injuries as:

- Electrical shock (touching "live" line-to-line or line-to ground conductors) (ground-fault circuit interrupters are discussed in Chapter 6)

- Electrocution (death due to severe electrical shock)

- Burns (from an arc flash)

- Falls (an electrical shock might cause you to lose your balance, pull back, jump, or fall off a ladder)

What to Do If You Are Involved with a Possible Electrocution

The following is taken in part from the OSHA, NIOSH, NSC regulations, and the American Heart Association recommendations. These are steps that should be taken in the event of a possible electrocution (cardiac arrest). You need to refer to the actual cardiopulmonary resuscitation (CPR) instructions for complete and detailed requirements, and to take CPR training.

- First of all, you must recognize that an emergency exists. Timing is everything. The time between the accident and arrival of paramedics is crucial. Call 911 immediately. Don't delay.

- Don't touch the person if he or she is still in contact with the live circuit.

- Shut off the power.

- Stay with the person while someone else contacts the paramedics, who have training in the basics of life support. In most localities, telephoning 911 will get you to the paramedics.

- Have the caller verify that the call was made and that help is on the way.

- Don't move the person.

- Check for bleeding; stop the bleeding if it occurs.

- If the person is unconscious, check for breathing.

- The *ABCs* of CPR are: *a*irway must be clear; *b*reathing is a must, either by the victim or the rescuer; and *c*irculation (check pulse).

- Perform CPR if the victim is not breathing—within 4 minutes is critical. If the brain is deprived of oxygen for more than 4 minutes, brain damage will occur. If it is deprived of oxygen for more than 10 minutes, the survival rate is 1 in 100. CPR keeps oxygenated blood flowing to the brain and heart.

- Defibrillation may be necessary to reestablish a normal heartbeat. Ventricular fibrillation is common with electric shock, which causes the heartbeat to be uneven and unable to properly pump blood.

- By now, the trained paramedics should have arrived to apply advanced care.

- When it comes to an electrical shock, *timing is everything!*

Electrical Equipment

For safety, it is very important that electrical equipment be "listed" by a Nationally Recognized Testing Laboratory (NRTL). Concepts about listed electrical equipment are found in *NEC 90.7*.

NEC 110.3(B) ***Installation and Use:*** states that *"Listed or labeled equipment shall be installed and used in accordance with any instructions included in the listing or labeling."*

OSHA rules state that "All electrical products installed in the work place shall be listed, labeled, or otherwise determined to be safe by a Nationally Recognized Testing Laboratory (NRTL)."

Who Is Responsible for Safety?

You are!

The electrical inspector inspects electrical installations for compliance to the *NEC*. He or she is not really involved with on-the-job safety.

For on-the-job safety, OSHA puts the burden of responsibility on the employer. OSHA can impose large fines for noncompliance with its safety rules. But since it's your own safety that we are discussing, you share the responsibility by applying safe work practices, using the proper tools and PPE equipment the contractor furnishes, and installing "Listed" electrical equipment. Be alert to what's going on around you! Do a good job of housekeeping!

Tools

Using the proper tools for a job is vital to on-the-job safety.

OSHA Standard 1926.302 specifically covers the requirements for hand and power tools. The American National Standards Institute (ANSI) also has standards relating to tools.

If you want to learn more about tools, visit the Web site of the Hand Tools Institute at www.hti.org. The institute has a number of excellent safety education materials available. Of particular interest is its 90-plus-page publication *Guide to Hand Tools* that includes topics for selecting, proper use, maintaining, hazards involved, as well as special emphasis on eye protection using all types of hand tools.

Electrical Power Tools

You will be using portable electric power tools on the job. Electricity is usually in the form of temporary power, covered by *Article 590* of the *NEC.*

NEC 590.6(A) requires that *"All 125–volt, single-phase, 15-, 20-, and 30-ampere receptacle outlets that are not a part of the permanent wiring of the building or structure and that are in use by personnel shall have ground-fault circuit interrupter protection for personnel."*

Because this requirement is often ignored or defeated on job sites, you should carry and use as part of your tool collection a portable GFCI of the types shown in Figure 1-3—an inexpensive investment that will protect you against possible electrocution. Remember, *"The future is not in the hands of fate, but in ourselves."*

Digital Multimeters

Some statistics show that more injuries occur from using electrical meters than from electric shock.

For safety, electricians should use quality digital multimeters that are *category rated.* The International Electrotechnical Commission (IEC) Standard 1010 for *Low Voltage Test, Measurement, and Control Equipment* rates the ability of a meter to withstand voltage transients (surges or spikes). This standard is very similar to UL Standard 3111. When lightning strikes a high line, or when utilities are performing switching operations, or when a capacitor is discharging, a circuit can "see" voltage transients that greatly exceed the withstand rating of the digital multimeter. The meter could explode, causing an arc flash (a fireball) that in all probability will result in personal injury. A properly selected category-rated digital multimeter is able to withstand the spike without creating an arc blast. The leads of the meter are also able to handle high transient voltages.

Digital multimeters also are category rated based on the location of the equipment to be tested, because the closer the equipment is to the power source, the greater the danger from transient voltages.

FIGURE 1-3 Two types of portable plug-in cord sets that have built-in GFCI protection. (*Delmar/Cengage Learning*)

Cat IV multimeters are used where the available fault current is high, such as a service entrance, a service main panel, service drops, and the house meter.

Cat III multimeters are used for permanently installed loads such as in switchgear, distribution panels, motors, bus bars, feeders, short branch circuits, and appliance outlets where branch-circuit conductors are large and the distance is short.

Cat II multimeters are used on residential branch circuits for testing loads that are plugged into receptacles.

Cat I multimeters are used where the current levels are very low, such as electronic equipment. Note that the lower the category rating, the lower is the meter's ability to withstand voltage transients. If you will be using the multimeter in all of the above situations, select the higher category rating.

Category-rated digital multimeters also contain fuses that protect against faults that happen when the meter is accidentally used to check voltage while it is inadvertently set in the current reading position.

To learn more about meters, visit the Web site of Fluke Corporation, http://www.fluke.com, for a wealth of technical information about the use of meters and other electrical and electronic measuring instruments.

Ladders

To learn more about ladders, visit the Web site of Werner Ladder Company, http://www.wernerladder .com. You can download their pamphlet entitled *Ladder Safety Tips*. You will learn about the right and wrong ways to use a ladder such as: Never work on a step ladder in which the spreaders are not fully locked into position; the 4:1 ratio, which means that the base of an extension ladder should be set back (S) one-fourth the length (L) of where the upper part of the ladder is supported (S = ¼ L); the duty ratings, such as do not stand higher than the second step from the top for step ladders, and do not stand higher than the fourth rung from the top for extension ladders; plus many more safety tips.

Ladders are labeled with their duty rating. Medium-duty commercial (Type II—225#), heavy-duty industrial (Type I—250#), and extra-heavy-duty (Type IA—300#) ladders bear an OSHA compliance label. Light-duty household (Type III—200#) ladders do not bear an OSHA logo.

Hazardous Chemicals

You will find more and more hazardous chemicals on the job. What do you do if you get a spilled chemical on your skin or in your eyes, or if you breathe the fumes?

Every manufacturer of this type of product is required to publish and make available a comprehensive data sheet called the **m**aterial **s**afety **d**ata **s**heet (MSDS). There are supposedly over 1.5 million of these data sheets. They contain product identification, ingredients, physical data, fire and explosion hazard data, health-hazard data, reactive data, spill or leak procedures, protection information, and special precautions.

The least you can do is to be aware that this information is available. Apprenticeship programs include some training about MSDS.

You can learn more about MSDS by checking any search engine for the letters MSDS.

TRAINING

If you want to learn more, visit manufacturers' Web sites. For example, Bussmann's Web site is http:// www.bussmann.com. It is easy to use and has a computer program for making arc-flash and fault-current calculations. This Web site also has a technical publication *Selecting Protective Devices*, or bulletin SPD, a 268-plus-page publication about overcurrent protection selection, application, *NEC*, and safety. You can order a hard copy of the publication from them.

The OSHA Training Institute offers outreach training programs of interest to electricians, contractors, and instructors. The basic safety courses for general construction safety and health are the OSHA 10-hour and OSHA 30-hour courses. Instructors interested in becoming an outreach trainer for the 10- and 30-hour courses must complete the OSHA 500 course entitled "Trainer Course in Occupational Safety and Health Standards for the Construction Industry." To become an outreach trainer, a test must be passed. Before the end of 4 years, outreach trainers must take the OSHA 502 update course for the construction industry or the OSHA 502 update course "Update for Construction Industry Outreach Trainers." Completion cards are issued on completion of these courses.

Other courses, publications, "free loan" videos, schedules of upcoming safety training seminars, and other important information relating to safety on the job are available from OSHA for electricians, contractors, and trainers.

Visit the OSHA Web site at http://www.OSHA .gov for everything there is to know about OSHA safety requirements in the workplace. The OSHA Web site is a virtual gold mine of information relating to safety on the job.

Another valuable source of safety information is the National Institute for Occupational Safety and Health (NIOSH), a division of the Department of Health and Human Services Centers for Disease Control and Prevention (CDC). Check out its Web site at http://www.cdc.gov/niosh. NIOSH offers an excellent downloadable 80-plus-page manual on *Electrical Safety*.

The National Safety Council has a vast amount of information relative to all aspects of safety. Check out its Web site at http://www.nsc.org.

The Consumer Product Safety Commission offers many safety publications for downloading. Visit its Web site at http://www.cpsc.gov, click on Library, click on CPSC Publications, click on By General Category, and then click on Electrical Safety. Here you will see a list of CPSC publications about GFCIs, AFCIs, metal ladder hazards, home wiring hazards, repairing aluminum wiring, and others.

The *National Fire Protection Association* offers many publications, videos, and a training course relating to safety. Browse its Web site at http://www.nfpa.org.

NFPA 70E Standard for Electrical Safety in the Workplace and NFPA 70B Recommended Practices for Electrical Equipment Maintenance present much of the same text regarding electrical safety as does the OSHA regulation.

Accredited apprenticeship training programs incorporate safety training as an integral part of their curriculum.

SAFETY CANNOT BE COMPROMISED!

It is impossible to put a dollar value on a life.

Don't take chances! Use the right tools! Turn off the power. Follow a lockout/tagout procedure. Mark the tag with a description of exactly what that particular disconnect controls.

How many times have we heard "The person would not have been injured (or electrocuted) had he turned the power OFF"? How many more times can we say it? **Turn OFF the power before working on the circuit!**

Visit the Web sites of the various organizations mentioned earlier. The Web site list can also be found in the back of this text. These organizations have a wealth of information about on-the-job safety educational material and safety training courses.

Check out the Web site of the Electrical Safety Foundation International (ESFI) at http://www.electrical-safety.org. This organization has

a tremendous amount of down-to-earth, simple-to-understand electrical safety material. Some of their educational material is free; other items are priced. Certain items are downloadable. The bottom line is to reduce deaths and injuries from preventable electrical accidents.

LICENSING AND PERMITS

Most communities, counties, and/or states require electricians and electrical contractors to be licensed. This usually means they have taken and passed a test. To maintain a valid license, many states require electricians and electrical contractors to attend and satisfactorily complete approved continuing education courses consisting of a specified number of classroom hours over a given period of time. Quite often, a community will have a "Residential Only" license for electricians and contractors that limits their activity to house wiring.

Permits are a means for a community to permanently record electrical work to be done and who is doing the work, and to schedule inspections during and after the "rough-in" stage and in the "final" stages of construction. Usually, permits must be issued prior to starting an electrical project. In most cases, homeowners are allowed to do electrical work in their own home where they live but not in other properties they might own.

Figure 1-4 is a simple application for an electrical permit form. Some permit application forms are much more detailed.

If you are not familiar with licensing and permit requirements in your area, it makes sense to check this out with your local electrical inspector or building department before starting an electrical project. Not to do so could prove to be very costly. Many questions can be answered: you will find out what tests, if any, must be taken; which permits are needed; which electrical code is enforced in your community; minimum size electrical service; and so on. Generally, the electrical permit is taken out by an electrical contractor who is licensed and registered as an electrical contractor in the jurisdictional area.

For new construction or for a main electrical service change, you will also need to contact the electric utility.

APPLICATION FOR ELECTRICAL PERMIT
VILLAGE OF ANYWHERE, USA   1-234-567-8900, EXT. 1234

Date _____   Permit No. _____

Owner_____   Job Address_____

Telephone No._____   Job Start Date _____

CONTRACTOR INFORMATION AND SIGNATURE

Electrical Contractor_____   Tel. No. _____

Address_____ City_____ State_____ Zip _____

Registration No._____ City of Registration _____

Supervising Electrician (Please Print Name)_____

Supervising Electrician's Signature _____

Insurance Bond_____ Village Business License _____

SERVICE INSPECTIONS OR REVISIONS

Existing Service Size: Amps_____ Volts_____ No. of Circuits_____ No. Added _____

New Service Size: Amps_____ Volts_____ No. of Circuits_____

Type: Overhead_____ Underground_____

Service Installation Fees: 100-200 amps $50_____ over 200 amps $75_____

MOTORS AND AIR-CONDITIONING EQUIPMENT

No. of Motors: up to 1 HP @ $50_____   Over 1–10 HP @ $50_____

                11–25 HP @ $25_____   Over 25 HP @ $25_____

Air Conditioner/Heat Pump:

No. of Tons_____ ($20 for first ton, $5 for each additional ton)   _____

Furnace (electric):   kW_____ Amps_____ ($25)   _____

Dryer (electric):   kW_____ Amps_____ ($10)   _____

Range, oven, cooktop (electric): Total kW_____ Amps_____ ($10 each)   _____

Water Heater (electric):   kW_____ Amps_____ ($10)   _____

TYPE OF MISCELLANEOUS ELECTRIC WORK

_____

_____

Minimum Inspection Fee: $40.00   _____

Escrow Deposit, if applicable   _____

TOTAL DUE   _____

FIGURE 1-4   A typical Application for Electrical Permit. (*Delmar/Cengage Learning*)

Temporary Wiring

There is an ever-present electrical shock hazard on construction sites. The *NEC* addresses this in *Article 590*. This is covered in Chapter 6.

Construction Terms

Electrical Wiring—Residential covers all aspects of typical residential wiring, with focus on the *NEC*. Electricians work with others on construction sites.

Knowing construction terms and symbols is therefore a key element to getting along with the other workers. A rather complete dictionary of construction terms can be found on http://www.constructionplace.com/glossary.asp. Architectural symbols are found in the Appendix of this text.

PLANS

An architect or electrical engineer prepares a set of drawings that shows the necessary instructions and details needed by the skilled workers who are to build the structure. These are referred to as **plans**, **prints**, **blueprints**, **drawings**, **construction drawings**, or **working drawings**. The sizes, quantities, and locations of the materials required and the construction features of the structural members are shown at a glance. These details of construction must be studied and interpreted by each skilled construction craftsperson—masons, carpenters, electricians, and others—before the actual work is started.

The electrician must be able to (1) convert the 2-dimensional plans into an actual electrical installation and (2) visualize the many different views of the plans and coordinate them into a 3-dimensional picture, as shown in Figure 1-5.

The ability to visualize an accurate 3-dimensional picture requires a thorough knowledge of blueprint reading. Because all of the skilled trades use a common set of plans, the electrician must be able to interpret the lines and symbols that refer to the electrical installation and also those used by the other construction trades. The electrician must know the structural makeup of the building and the construction materials to be used

Plans might be black line prints, which are simply photocopies; diazo prints, referred to as white line prints; or computer-aided drawings (CAD), which are most commonly used today. The need for the draftsman has for the most part become history. We now need CAD technicians and engineers.

Today's high-tech computers and wide-frame printers allow for storage and backup of clear and concise drawings, standardization of lines in symbols, ease of making revisions, the ability to zoom in and out when viewed on a monitor, simplicity of printing (black or color), and in many cases the ability to view the structure in 2D or 3D, turn a drawing

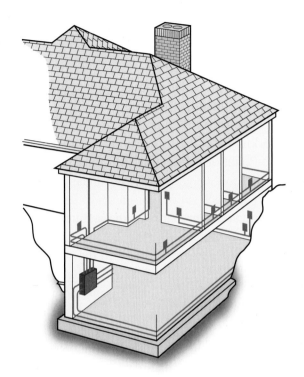

FIGURE 1-5   Three-dimensional view of house wiring. (*Delmar/Cengage Learning*)

to any angle, and view the more complicated plans in layers. The larger the job, the greater the need for more detail.

Although blueprints for the most part have gone the way of the dinosaur, they are still around. Electricians that are from an older generation are familiar with blueprints. Those of you coming into the trade will probably not see blueprints on the jobs you work on.

Blueprints are created by running yellowish, light-sensitive paper and transparent Mylar with black images on it through a blueprint machine, where the two sheets are exposed to a bright light for a short period of time. The light-sensitive paper turns white in all places except where the black images are. Where the black images are, the light-sensitive paper turns blue. That is why it is called a blueprint.

A tear-out/fold-out set of full-size plans for the residence referred to throughout this text is included in the back of the book.

A few of the more common sizes for blueprints are

- Size C (17 inches by 22 inches),
- Size B (22 inches by 34 inches), and
- Size E (34 inches by 44 inches).

SPECIFICATIONS

The **specifications** (specs, for short) for the electrical work indicated on the plans for the residence discussed throughout this text are found in the back of the book, before the Appendix.

Working drawings are usually complex because of the amount of information that must be included. To prevent confusing detail, it is standard practice to include with each set of plans a set of detailed written specifications prepared by the architect.

These specifications provide general information to be used by all trades involved in the construction. In addition, specialized information is given for the individual trades. The specifications include information on the sizes, the type, and the desired quality of the standard parts to be used in the structure.

Typical specifications include a section on "General Clauses and Conditions," which is applicable to all trades involved in the construction. This section is followed by detailed requirements for the various trades—excavating, masonry, carpentry, plumbing, heating, electrical work, painting, and others.

In the electrical specifications, the listing of standard electrical parts and supplies frequently includes the manufacturers' names and the catalog numbers of the specified items. Such information ensures that these items will be of the correct size, type, and electrical rating, and that the quality will meet a certain standard. To allow for the possibility that the contractor will not always be able to obtain the specified item, the phrase "or equivalent" is usually added after the manufacturer's name and catalog number.

The specifications are also useful to the electrical contractor in that all of the items needed for a specific job are grouped together and the type or size of each item is indicated. This information allows the contractor to prepare an accurate cost estimate without having to find all of the data on the plans.

If there is a difference between the plans and specifications, the specifications will take preference. The electrical contractor should discuss the matter with the homeowner, architect, and engineer. The cost of the installation might vary considerably because of the difference(s), so obtain any changes to the plans and/or specifications in writing.

SYMBOLS AND NOTATIONS

The architect uses **symbols** and **notations** to simplify the drawing and presentation of information concerning electrical devices, appliances, and equipment. For example, an electric range outlet symbol is shown in Figure 1-6.

Most symbols have a standard interpretation throughout the country as adopted by ANSI. Symbols are described in detail in Chapter 2.

A notation will generally be found on the plans (blueprints) next to a specific symbol calling attention to a variation, type, size, quantity, or other necessary information. In reality, a symbol might be considered to be a notation because symbols represent words, phrases, numbers, quantities, and so on.

Another method of using notations to avoid cluttering up a blueprint is to provide a system of symbols that refer to a specific table. For example, the written sentences on plans could be included in a table referred to by a notation. Figure 2-10 on page 35 is an example of how this could be done. The special symbols that refer to the table would have been shown on the actual plan.

NATIONAL ELECTRICAL CODE (NEC)

The *NEC* is the electrical *Code* standard recognized by everyone in the electrical industry.

The first sentence in the *NEC* is found in *90.1*. This sentence lays the foundation for all electrical installations. It states, *The purpose of this Code is the practical safeguarding of persons and property from hazards arising from the use of electricity*. It goes on to state that the *NEC contains provisions that are considered necessary for safety* and that *compliance therewith and proper maintenance will result in an installation that is essentially free from*

FIGURE 1-6 An electric range outlet symbol. (*Delmar/Cengage Learning*)

*hazard but not necessarily efficient, convenient, or adequate for good service or future expansion of electrical use.**

As you study this text, you will learn how to "Meet Code" for wiring a typical house, not "Beat Code."

As *NEC* requirements are discussed throughout this text, the sheer number of *Code* references can become mind-boggling. To simplify using this text as a reference, in addition to the conventional "subject" index at the back of the text, there is a Cross Index, making it easy for you to pinpoint specific *Code* sections and articles found in this text.

The *NEC* is published by the National Fire Protection Association and is referred to as *NFPA 70*. The *NEC* was first published in 1897. It is revised every 3 years so as to be as up to date as possible. The *NEC* does not become law until adopted by official action of the legislative body of a city, municipality, county, or state. Because of the ever-present danger of fire or shock hazard through some failure of the electrical system, the electrician and the electrical contractor must use listed materials and must perform all work in accordance with recognized standards.

One- and Two-Family Dwelling Electrical Code

NFPA 70A is the *National Electrical Code Requirements for One- and Two-Family Dwellings.* It is a condensed version of the *NEC* that contains excerpts directly from the *NEC NFPA 70*. It includes requirements and tables that relate only to house wiring. It does not include *Code* requirements for other types of construction.

Other Electrical Codes

In addition to the *NEC*, you must also consider local and state electrical codes.

Code Arrangement

It is important to understand the arrangement of the *NEC*.

• *Introduction*	
Article 90	An introduction to the *NEC*. Explains what is and what is not covered in the *NEC*, the arrangement of the *NEC*, who has the authority to enforce the *Code*, what mandatory rules are, what permissive rules are, and the basics of metric vs. inch-pound measurements found throughout the *NEC*.
• *Chapter 1*	General *Article 100: Definitions Article 110: Requirements for Electrical Installations* Applies to all electrical installations.
• *Chapter 2*	*Wiring and Protection Articles 200–299* Applies to all electrical installations.
• *Chapter 3*	*Wiring Methods Articles 300–399* Applies to all electrical installations.
• *Chapter 4*	*Equipment for General Use Articles 400–499* Applies to all electrical installations.
• *Chapter 5*	*Special Occupancies Articles 500–599*
• *Chapter 6*	*Special Equipment Articles 600–699*
• *Chapter 7*	*Special Conditions Articles 700–799*
• *Chapter 8*	*Communications Systems Articles 800–899* The articles in *Chapter 8* are not subject to the requirements of *Articles 1* through *7* unless specifically stated.
• *Chapter 9*	Tables showing dimensional data for raceways and conductors, resistance and reactance values of conductors, Class 2 and Class 3 circuit power limitations.
• *Informative Annex A*	Provides a comprehensive list of product safety standards from Underwriters Laboratories (UL).
• *Informative Annex B*	Provides data for determining conductor ampacity where there is engineering supervision.

---

*Reprinted with permission from NFPA 70-2011.

(continues)

• *Informative Annex C*	Has tables showing the maximum number of conductors permitted in various types of raceways.
• *Informative Annex D*	Examples of load calculations.
• *Informative Annex E*	Types of building construction.
• *Informative Annex F*	*Availability and Reliability of Critical Operations Power Systems*
• *Informative Annex G*	*Supervisory Control and Data Acquisitions (SCADA)*
• *Informative Annex H*	*Administration and Enforcement A* Comprehensive suggested typical electrical ordinance that could be adopted on a local level.
• *Informative Annex I*	*Recommended Tightening Torque Tables from UL Standard 486A-B*
• *Index*	The alphabetical index for the *NEC*.

Note: *Annexes* and *Informational Notes* are informational only. They are not mandatory.

Easy to Find Things!

Finding things in the current edition of the *NEC* is much easier than in past editions. All articles in *Chapter 3* of the *NEC* cover wiring methods. In each article, the same section number has been assigned for a particular requirement. Here are a few examples:

- *Scope* is found in *XXX.1 such as 320.1, 330.1 and 344.1*

- *Definitions* (if present in the *Article*) are found in *XXX.02*

- *Permitted Uses* are found in *XXX.10.*

- *Uses Not Permitted* are found in *XXX.12.*

- *Securing and Supporting* is found in *XXX.30.*

This is referred to as "the parallel numbering system."

How to Spot Changes in the 2011 *NEC*

Highlighted text makes it easy to quickly spot where changes in the *NEC* have been made from the previous edition.

Who Writes the *Code*?

For each *Code* cycle, the NFPA solicits proposals from anyone interested in electrical safety. Anyone may submit a proposal to change the *NEC* using the Proposal Form found in the back of the *NEC*. A form may also be downloaded from www.nfpa.org. Proposals received are then given to a specific Code-Making Panel (CMP) to accept, reject, accept in part, accept in principle, or accept in principle in part. These actions are published in the Report on Proposals (ROP) and may be downloaded from www.nfpa.org. Individuals may send in their comments on these actions using the Comment Form found in the ROP. The CMPs meet again to review and take action on the comments received. These actions are published in the Report on Comments (ROC) and also may be downloaded from www.nfpa.org. Final action (voting) on proposals and comments is taken at the NFPA annual meeting.

However, before the *NEC* is published, if there is disagreement on any specific Code requirement adopted through the above process, NFPA will consider an *Appeal* that is reviewed and acted upon by the NFPA Standards Council about 6 weeks after the annual meeting. After an *Appeal* is acted upon by the Standards Council, should there still be controversy, another step not often used in the *Code* adoption process is a petition that is reviewed and acted upon by the NFPA Board of Directors.

Individuals who serve on CMPs are electrical inspectors, electrical contractors, electrical engineers, individuals from utilities, manufacturers, testing laboratories, the Consumer Product Safety Commission, insurance companies, and similar organizations. CMP members are appointed by the NFPA. The CMPs have 10 to 20 principal members, plus a similar number of alternate members. All of the CMPs have a good balance of representation.

Which Edition of the *NEC* to Use

This text is based on the 2011 edition of the *NEC*. Some municipalities, cities, counties, and states have not yet adopted this edition and may continue using older editions. Check with your local electrical inspector to find out which edition of the *Code* is in force.

Copies of the latest edition of the *NEC NFPA 70* may be ordered from the following:

Cengage Learning
P.O. Box 8007
Clifton Park, NY 12065
Phone: 800-998-7498
http://www.delmar.cengage.com

National Fire Protection Association
1 Batterymarch Park
Quincy, MA 02269-9101
Phone: 617-770-3000
http://www.nfpa.org

International Association of Electrical Inspectors
901 Waterfall Way, Suite 602
Richardson, TX 75080-7702
Phone: 800-786-4234
http://www.iaei.org

Electrical Inspection Code for Existing Dwellings

This code is published by the NFPA and is referred to as *NFPA 73*. It is a brief nine-page code that provides requirements for evaluating installed electrical systems within and associated with existing dwellings to identify safety, fire, and shock hazards—such as improper installations, overheating, physical deterioration, and abuse. This code lists most of the electrical things to inspect in an existing dwelling that could result in a fire or shock hazard if not corrected. It only points out things to look for that are visible. It does not get into examining concealed wiring that would require removal of permanent parts of the structure. It also does not get into calculations, location requirements, and complex topics as does the *NEC NFPA 70*.

This code can be an extremely useful guide for electricians doing remodel work and for electrical inspectors wanting to bring an existing dwelling to a reasonably safe condition. Many localities require that when a home changes ownership, the wiring must be brought up to some minimum standard, but not necessarily as extensively as would be the case for new construction.

Homes for the Physically Challenged

The *NEC* does not address the wiring of homes for the physically or mentally challenged or disabilities associated with the elderly.

Each installation for the physically challenged must be based on the specific need(s) of the individual(s) who will occupy the home. Some physically challenged people are bedridden, require the mobility of a wheelchair, have trouble reaching, have trouble bending, and so forth. There are no hard and fast rules that *must* be followed—only many suggestions to consider.

Some of these follow:

- Install more ceiling luminaires instead of switching receptacles. Cords and lamps are obstacles.

- Install luminaires having more than one bulb.

- Go "overboard" in the amount of lighting for all rooms, entrances, stairways, stairwells, closets, pantries, bathrooms, and so on.

- Use higher wattage bulbs—not to exceed the wattage permitted in the specific fixture.

- Consider installing luminaires in certain areas (such as bathrooms and hallways) to be controlled by motion detectors.

- Consider installing exhaust fans in certain areas (such as laundries, showers, and bathrooms) that turn on automatically when the humidity reaches a predetermined value.

- Consider the height of switches and thermostats, usually 42 in. (1.0 m) or lower, instead of the standard 46–52 in. (1.15–1.3 m).

- Consider rocker-type switches instead of toggle type.

- Install pilot light switches.

- Consider "jumbo" switches.

- Be sure stairways and stairwells are well lit.

- Consider stair tread lighting.

- Position lighting switches so as not to be over stairways or ramps.

- Locate switches and receptacles to be readily and easily accessible—not behind doors or other hard-to-reach places.

- Consider installing wall receptacle outlets higher 24–27 in. (600–675 mm) than normal 12 in. (300 mm) height.

- Install lighted doorbell buttons.

- Chimes: Consider adding a strategically located "dedicated" lamp(s) that will turn on when doorbell buttons are pushed. The wiring diagram for this is found in Figure 25-25.

- Telephones: Consider adding visible light(s) strategically located that will flash at the same time the telephone is ringing.

- Consider installing receptacle outlets and switches on the face of kitchen cabinets. Wall outlets and switches can be impossible for the physically challenged person to reach.

- Consider fire, smoke, and security systems directly connected to a central office for fast response to emergencies that do not depend on the disabled to initiate the call.

- Consider installing the fuse box or breakerpanel on the first floor instead of in the basement.

- Consider the advanced home systems concept of remote control of lighting, receptacles, appliances, television, telephones, and so on. The control features can make life much easier for a disabled person.

When involved with multifamily dwellings, check with your local building authority. They will have copies of the Americans with Disabilities Act (ADA) from the U.S. Department of Justice and the Fair Housing Act from the U.S. Department of Housing and Urban Development (HUD) that basically require that all units must have accessible light switches, electrical outlets, thermostats, and other environmental controls.

The ANSI publication *ANSI A117.1–2003, Standard for Accessible and Useable Buildings and Facilities*, contains many suggestions and considerations for buildings and facilities for the physically challenged. This and other standards can be ordered from

American National Standards Institute
25 West 43rd Street
New York, NY 10036
Tel: 212-642-4900
Fax: 212-398-0023

The same standard is available from the International Code Council (ICC) under the publication number *ICC/ANSI A117.1–2003*.

Another virtually unlimited source of standards is

Global Engineering Documents
15 Inverness Way East
Englewood, CO 80112
Phone: 800-624-3974 ext. 1950
 303-792-2181 ext. 1950
Fax: 303-792-2192
http://www.global.ihs.com
e-mail: globalcustomerservice@ihs.com

BUILDING CODES

The majority of the building departments across the country have for the most part adopted the *NEC* rather than attempting to develop their own electrical codes. As you study this text, you will note numerous references to the *NEC*, the electrical inspector, or the authority having jurisdiction.

An authority's level of knowledge varies. The electrical inspector may be full-time or part-time and may also have responsibility for other trades, such as plumbing or heating. The heads of the building departments in many communities are typically called the Building Commissioners or Directors of Development. Regardless of title, they are responsible for ensuring that the building codes in their communities are followed.

International Code Council (ICC)

Rather than writing their own codes, most communities for the most part adopt the codes of the International Code Council (ICC). Over the years, the ICC has developed comprehensive and coordinated model construction codes. These include building, mechanical, plumbing, fire, energy conservation, existing building, fuel gas, sewage disposal, property maintenance, zoning, residential, and electrical codes. The ICC provides technical, educational, and informational products. Its address is

ICC Headquarters
500 New Jersey Avenue, NW, 6th Floor
Washington, DC 20001
Phone: 888-422-7233
http://www.iccsafe.org

The *ICC International Residential Code*, Chapters 33–42, contains electrical provisions that were written and produced under the guidance of the NFPA. The material in these chapters is copyrighted by the NFPA. These provisions are similar to the *NEC* other than the layout and numbering system.

Electrical Wiring—Residential is based on the *NEC*.

Because local electrical codes may differ from the *NEC*, you should check with the local inspection authority to determine which edition of the *NEC* is enforced, and what, if any, local requirements or amendments take precedence over the *NEC*.

American National Standards Institute

The **American National Standards Institute (ANSI)** is an organization that coordinates the efforts and results of the various standards-developing organizations, such as those mentioned in previous paragraphs. Through this process, ANSI approves standards that then become recognized as American National Standards. One will find much similarity between the technical information found in ANSI standards, the UL standards, the International Electronic and Electrical Engineers standards, and the *NEC*.

International Association of Electrical Inspectors

The International Association of Electrical Inspectors (IAEI) is a nonprofit organization. The IAEI membership consists of electrical inspectors, building officials, electricians, engineers, contractors, and manufacturers throughout the United States and Canada. The major goal of the IAEI is to improve the understanding of the *NEC*. Representatives of this organization serve as members of the various CMPs of the *NEC* and share equally with other members in the task of reviewing and revising the *NEC*.

The IAEI publishes a bimonthly magazine—*The IAEI News*. It is devoted entirely to electrical code topics. Anyone in the electrical industry is welcome to join the IAEI. An application form is found after the Appendix of this text. Its address is

International Association of Electrical Inspectors
901 Waterfall Way, Suite 602
Richardson, TX 75080-7702
Phone: 800-786-IAEI
http://www.iaei.org

Code Definitions

The electrical industry uses many words (terms) that are unique to the electrical trade. These terms need clear definitions to enable the electrician to understand completely the meaning intended by the *Code*.

Article 100 of the *NEC* is a "dictionary" of these terms. *Article 90* also provides further clarification of terms used in the *NEC*. Here are a few examples:

Ampacity: The maximum current, in amperes, that a conductor can carry continuously under the conditions of use without exceeding its temperature rating.

Approved: Acceptable to the authority having jurisdiction.

Authority Having Jurisdiction (AHJ): An organization, office, or individual responsible for enforcing the requirements of a code or standard, or for approving equipment, materials, an installation, or a procedure. See *NEC 90.4*.

Dwelling Unit: A single unit, providing complete and independent living facilities for one or more provisions for living, sleeping, cooking, and sanitation.

Informational Note: Informational Notes are found throughout the *Code. Informational Notes* are "explanatory" in nature, in that they refer to other sections of the *Code*. They also describe things where further description is necessary. See *90.5. Informational Notes* are not to be enforced.

Listed: Equipment, materials, or services included in a list published by an organization that is acceptable to the authority having jurisdiction and concerned with evaluation of products or services, that maintains periodic inspection of production of listed equipment or materials or periodic evaluation of services, and whose listing states that either the equipment, material, or services meets appropriate designated standards or has been tested and found suitable for a specified purpose.

Shall: Indicates a mandatory rule, *90.5.* As you study the *NEC*, think of the word "shall" as meaning "must." Some examples found in the *NEC* where the word "shall" is used in combination with other words are: *shall be, shall have, shall not, shall be permitted, shall not be permitted*, and *shall not be required.*

Refer to Key Terms in the Appendix of this text.

Read *110.3(B)* Carefully!

One of the most far-reaching *NEC* rules is *110.3(B).* This section states that *Listed or labeled equipment shall be installed and used in accordance with any instructions included in the listing or labeling.* This means that an entire electrical system and all of the system's electrical equipment must be *installed* and *used* in accordance with the *NEC* and the numerous standards to which the electrical equipment has been tested.

METRICS (SI) AND THE *NEC*

The United States is the last major developed country in the world not using the metric system of weights and measures as the primary system. For most of our lifetime, we have used the English system of weights and measures, also referred to as inch-pound and U.S. Customary. But this is changing!

Manufacturers are now showing both inch-pound and metric units in their catalogs. By law, plans and specifications for new governmental construction and renovation projects have used the metric system since January 1, 1994.

You may not feel comfortable with metric measurements, but metric measurements are here to stay. You might just as well become familiar with the metric system.

All measurements in the *NEC* are shown in both inch-pound and metric values.

For more information about metrics, refer to the "Metric System of Measurement" section found in the Appendix of this text.

Trade Sizes

A unique situation exists. Strange as it may seem, what electricians have been referring to for years has not been correct!

Raceway sizes have always been an approximation. For example, there has never been a ½-in. raceway. Measurements taken from the *NEC* for a few types of raceways are shown in Table 1-1.

You can readily see that the cross-sectional areas, critical when determining conductor fill, are different. It makes sense to refer to conduit, raceway, and tubing sizes as *trade sizes.* The *NEC* in *90.9(C)(1)* states that *where the actual measured size of a product is not the same as the nominal size, trade size designators shall be used rather than dimensions.* This edition of *Electrical Wiring— Residential* uses the term *trade size* when referring to conduits, raceways, and tubing. For example, a ½-in. EMT is referred to as trade size ½ EMT. EMT is also referred to in the trade as "thinwall."

The *NEC* also uses the term *metric designator.* A trade size ½ EMT is shown as *metric designator 16 (½).* A trade size 1 EMT is shown as *metric designator 27 (1).* The numbers 16 and 27 are the metric designator values; the (½) and (1) are the trade sizes. The metric designator is the raceway's inside diameter—in rounded-off millimeters (mm). Table 1-2 shows some of the more common sizes of conduits, raceways, and tubing. A complete listing is found in *NEC Table 300.1(C).*

For ease in understanding, this text uses only the term *trade size* when referring to conduit and raceway sizes.

TABLE 1-1

Comparison of trade size vs. actual inside diameters.

Trade Size	Inside Diameter (I.D.)
½ Electrical Metallic Tubing	0.622 in.
½ Electrical Nonmetallic Tubing	0.560 in.
½ Flexible Metal Conduit	0.635 in.
½ Rigid Metal Conduit	0.632 in.
½ Intermediate Metal Conduit	0.660 in.

TABLE 1-2

Trade sizes of raceways and their metric designator identification.

METRIC DESIGNATOR AND TRADE SIZE

Metric Designator	Trade Size
12	⅜
16	½
21	¾
27	1
35	1¼
41	1½
53	2
63	2½
78	3

Conduit knockouts in boxes do not measure up to what we call them. Table 1-3 shows some examples.

Outlet boxes and device boxes use their nominal measurement as their *trade size*. For example, a 4 in. × 4 in. × 1½ in. does not have an internal cubic-inch volume of 4 in. × 4 in. × 1½ in. = 24 in³. *Table 314.16(A)* shows this size box as having a 21-cubic-in. volume. This table shows *trade sizes* in two columns—millimeters and inches.

In this text, a square outlet box is referred to as trade size 4 × 4 × 1½. Similarly, a single-gang device box would be referred to as a trade size 3 × 2 × 3 box.

Trade sizes for construction material will not change. A 2 × 4 is really a *name*, not an actual dimension. A 2 × 4 will keep its name forever. This is its *trade size*.

In this text, most measurements directly related to the *NEC* are given in both inch-pound and metric

TABLE 1-3

Comparison of knockout trade size vs. actual measurement.

Trade Size Knockout	Actual Measurement
½	⅞ in.
¾	1³⁄₃₂ in.
1	1⅜ in.

units. In many instances, only the inch-pound units are shown. This is particularly true for the examples of raceway and box fill calculations, load calculations for square foot areas, and on the plans (drawings).

Because the *NEC* rounded off most metric conversion values, a calculation using metrics results in a different answer when compared to the same calculation done using inch-pounds. For example, load calculations for a residence are based on 3 volt-amperes per square foot or 33 volt-amperes per square meter.

For a 40 ft × 50 ft dwelling: 3 VA × 40 ft × 50 ft = 6000 volt-amperes.

In metrics, using the rounded off values in the *NEC*: 33 VA × 12 m × 15 m = 5940 volt-amperes.

The difference is small, but, nevertheless, there is a difference.

To show calculations in both units throughout this text would be very difficult to understand and would take up too much space. Calculations in either metrics or inch-pounds are in compliance with the *NEC, 90.9(D)*. In *90.9(C)(3)* we find that metric units are not required if the industry practice is to use inch-pound units.

It is interesting to note that the examples in *Chapter 9* of the *NEC* use inch-pound units, not metrics.

Guide to Metric Usage

The metric system is a "base-10" or "decimal" system in that values can be easily multiplied or divided by 10 or "powers of 10." The metric system as we know it today is known as the International System of Units (SI), derived from the French term "le Système International d'Unités."

In the metric system, the units increase or decrease in multiples of 10, 100, 1000, and so on. For instance, one megawatt (1,000,000 watts) is 1000 times greater than one kilowatt (1000 watts).

By assigning a name to a measurement, such as a *watt*, the name becomes the unit. Adding a prefix to the unit, such as *kilo*, forms the new name *kilowatt*, meaning 1000 watts. Refer to Table 1-4 for prefixes used in the metric system.

The prefixes used most commonly are *centi*, *kilo*, and *milli*. Consider that the basic unit is a meter (one). Therefore, a centimeter is 0.01 meter, a kilometer is 1000 meters, and a millimeter is 0.001 meter.

TABLE 1-4

Metric prefixes, symbols, multipliers, powers, and values.

Prefix	Symbol	Multiplier	Scientific Notation (Powers of Ten)	Value
tera	T	1 000 000 000 000	10^{12}	one trillion (1 000 000 000 000/1)
giga	G	1 000 000 000	10^{9}	one billion (1 000 000 000/1)
mega	M	1 000 000	10^{6}	one million (1 000 000/1)
kilo	k	1 000	10^{3}	one thousand (1 000/1)
hecto	h	100	10^{2}	one hundred (100/1)
deka	da	10	10^{1}	ten (10/1)
unit		1	—	one (1)
deci	d	0.1	10^{-1}	one tenth (1/10)
centi	c	0.01	10^{-2}	one hundredth (1/100)
milli	m	0.001	10^{-3}	one thousandth (1/1 000)
micro	m	0.000 001	10^{-6}	one millionth (1/1 000 000)
nano	n	0.000 000 001	10^{-9}	one billionth (1/1 000 000 000)
pico	p	0.000 000 000 001	10^{-12}	one trillionth (1/1 000 000 000 000)

Some common measurements of length and equivalents are shown in Table 1-5.

Electricians will find it useful to refer to the conversion factors and their abbreviations shown in Table 1-6.

Refer to the Appendix of this text for a comprehensive metric conversion table. The table includes information on how to "round off" numbers for practical use on the job.

TABLE 1-5

Some common measurements of length and their equivalents.

one inch	=	2.54	centimeters
	=	25.4	millimeters
	=	0.025 4	meter
one foot	=	12	inches
	=	0.304 8	meter
	=	30.48	centimeters
	=	304.8	millimeters
one yard	=	3	feet
	=	36	inches
	=	0.914 4	meter
	=	914.4	millimeters
one meter	=	100	centimeters
	=	1,000	millimeters
	=	1.093	yards
	=	3.281	feet
	=	39.370	inches

TABLE 1-6

Useful conversions and their abbreviations.

inches (in) × 0.0254	= meters (m)
inches (in) × 0.254	= decimeters (dm)
inches (in) × 2.54	= centimeters (cm)
centimeters (cm) × 0.393 7	= inches (in)
inches (in) × 25.4	= millimeters (mm)
millimeters (mm) × 0.039 37	= inches (in)
feet (ft) × 0.304 8	= meters (m)
meters (m) × 3.280 8	= feet (ft)
square inches (in²) × 6.452	= square centimeters (cm²)
square centimeters (cm²) × 0.155	= square inches (in²)
square feet (ft²) × 0.093	= square meters (m²)
square meters (m²) × 10.764	= square feet (ft²)
square yards (yd²) × 0.836 1	= square meters (m²)
square meters (m²) × 1.196	= square yards (yd²)
kilometers (km) × 1 000	= meters (m)
kilometers (km) × 0.621	= miles (mi)
miles (mi) × 1.609	= kilometers (km)

NATIONALLY RECOGNIZED TESTING LABORATORIES (NRTL)

How does one know if a product is safe to use?

Manufacturers, consumers, regulatory authorities, and others recognize the importance of independent, "third-party" testing of products in an effort to reduce safety risks. Unless you have all of the necessary test equipment and knowledge of how to properly test a product for safety, the surest way is to accept the findings of a third-party testing agency.

Nationally Recognized Testing Laboratories (NRTL) have the knowledge, wherewithal, and test equipment to test and evaluate products for safety.

A NRTL will perform tests on a product based on a specific nationally recognized safety standard. After the product has been tested and found to comply with the safety standard, the product is considered to be free from reasonably foreseeable risk of fire, electric shock, and related hazards. The product is then "listed" and will have a listing marking (label).

As you work with electrical products, make sure the product has a listing marked on it from a NRTL. If the product is too small to have a listing mark on the product itself, then look for the mark on the carton the product came in.

The following laboratories do a considerable amount of testing and listing of electrical equipment:

Underwriters Laboratories, Inc. (UL®)

Underwriters Laboratories, Inc. (UL), founded in 1894, is a highly qualified, nationally recognized testing laboratory with several testing laboratories in the United States and service locations in several other countries. UL develops product safety standards, and performs tests to these standards. Most reputable manufacturers of electrical equipment submit their products to UL, where the equipment is subjected to numerous tests. These tests determine whether the product can perform safely under normal and abnormal conditions to meet published standards. After UL tests and evaluates a product, and determines that the product complies with the specific standard, the manufacturer is then permitted

FIGURE 1-7   The Underwriters Laboratories mark. (*Courtesy Underwriters Laboratories, Inc.*)

to *label* its product with the UL Mark (Figure 1-7). The products are then *listed* in a UL Directory.

UL Marking. The UL marking is required to be on the product! The UL marking will always consist of four elements—UL in a circle, the word "LISTED" in capital letters, the product identity, and a unique alphanumeric control or issue number. If the product is too small, or has a shape or is made of a material that will not accept the UL Mark on the product itself, the marking is permitted on the smallest unit carton or container that the product comes in. Marking on the carton or box is nice but does not ensure that the product is UL listed!

The letters will *always* be staggered: U_L. They will *not* be side by side: UL.

The listing mark shown in Figure 1-7 indicates the product is in compliance with the applicable product safety standards in the United States and in Canada. A listing mark with only UL in a circle indicates the product has been evaluated to only US standards.

When UL tests and lists products that comply to the requirements of a particular CSA standard, the UL logo shown in Figure 1-7 will appear with a "C" outside of and to the left of the circle. This means the product has been tested and evaluated for compliance *only* with Canadian requirements. Product standards are being harmonized in North America as a result of the North American Free Trade Act (NAFTA). Discussions are also going on with Mexico. When all of this is finalized, electrical equipment standards may be the same in the United States, Canada, and Mexico.

Additional efforts are being made to harmonize North American standards with those of Europe.

Counterfeit Products. Be on the lookout for counterfeit electrical products. These products have not been tested and listed by a recognized testing laboratory. They can present a real hazard to life and property. Most counterfeit products as of this writing supposedly come from China. Counterfeit electrical products might also be referred to as "black market products."

Look for unusual logos or wording. For example, the UL logo might be illustrated in an oval instead of a circle, or the UL logo might not be encircled with anything, or the wording might say *approved* instead of LISTED. UL doesn't approve anything! It tests the product. If the product meets UL standards, the product is then LISTED. The UL marking might be on the carton and not on the product. Marking on the carton or box is nice, but it can be considered advertising! Absence of the manufacturer's name should raise the caution flag.

Currently, there is federal legislation in progress that would make it a criminal offense to traffic in counterfeit products and counterfeit trademarks. The legislation makes it mandatory that the counterfeit products and any tools that make the products or markings be seized and destroyed.

To learn more about counterfeits, check out http://www.ul.com, then search on (type in) the word "counterfeit." Also, check out http://www.nema.org, then type in the word "counterfeit".

UL *does not approve* a product. Rather, UL *lists* those products that conform to a specific safety standard. A UL Listing Mark on a product means that representative samples of the product have been tested and evaluated to nationally recognized safety standards with regard to fire, electric shock, and related safety hazards.

Do not confuse a UL marking in a circle with the markings found on *recognized components*. Recognized components that have passed certain tests are marked with the letters *RU* printed backwards (Figure 1-8). By themselves, recognized components are not to be field-installed. They are intended for use in end-use products or systems that would ultimately be tested and listed, with the final assembly becoming a UL-listed product. Some examples of recognized

FIGURE 1-8 The *recognized components* mark. (*Delmar/Cengage Learning*)

components are relays, ballasts, insulating materials, special switches, and so on.

Useful UL publications are

Electrical Construction Equipment Directory (Green Book).

Electrical Appliance and Utilization Equipment Directory (Orange Book).

Guide Information for Electrical Equipment Directory (White Book).

It is extremely useful for an electrician, an electrical contractor, and/or an electrical inspector to refer to these directories when looking for specific requirements, permitted uses, limitations, and others for a certain product.

Many times the answer to a product-related question that cannot be found in the *NEC* can be found in one of these directories.

The *Green* and *Orange* Directories provide technical information about a particular product, and they list the names and addresses of manufacturers and the manufacturers' identification numbers. The *White Book* provides technical information regarding a product but does not show manufacturers' names and addresses.

The best companion to the *NEC* is probably the UL *White Book* (Figure 1-9).

These directories can be obtained by contacting:

Underwriters Laboratories, Inc.
333 Pfingsten Road
Northbrook, Illinois 60062-2096
Phone: 847-272-8800
http://www.ul.com

Underwriters Laboratories, Inc., and the Canadian Standards Association (CSA) have worked out an

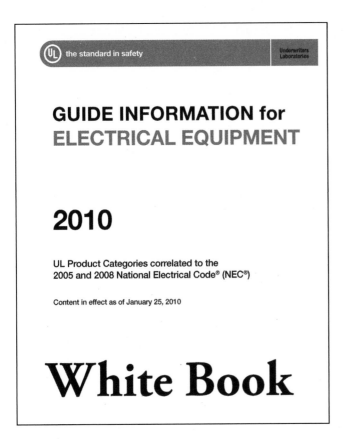

FIGURE 1-9   The Underwriters Laboratories *White Book.* (*Courtesy Underwriters Laboratories*)

FIGURE 1-10   The listing mark of Canadian Standards Association. (*Courtesy Canadian Standards Association*)

agreement whereby either agency can test, evaluate, and list equipment for the other agency. For example, UL might test and list an air conditioner unit to the requirements of UL Standard 1995 (Heating and Cooling Equipment) because the Canadian Standard C22.2 No. 236-M90 is a mirror image of UL Standard 1995. One by one, the UL and CSA Standards are becoming similar.

CSA International

In ways, Canadian Standards Association International is the Canadian counterpart of Underwriters Laboratories, Inc., in the United States. CSA International is the source of the *Canadian Electrical Code (CEC)* and of the Canadian Standards for the testing, evaluation, and listing of electrical equipment in Canada. The *Canadian Electrical Code* is quite different from the *NEC*. A Canadian version of *Electrical Wiring—Residential* is available in Canada.

Figure 1-10 is a representation of the CSA listing mark. Like the UL listing mark, the appearance

of "C" and "US" at the approximately 4:00 and 8:00 o'clock positions indicates what nation's standards the equipment has been evaluated.

The *Canadian Electrical Code* and CSA standards can be obtained by contacting

CSA International
178 Rexdale Boulevard
Toronto, Ontario, CANADA
M9W 1R3
Phone: 416-747-4044; 800-463-6727
Fax: 416-747-2510
http://www.csa-international.org

Intertek Testing Services

ITS is a nationally recognized testing laboratory. Its Intertek ETL division provides testing, evaluation, labeling, listing, and follow-up service for the safety testing of electrical products. This is done in conformance to nationally recognized safety standards.

The ETL listing mark, like the UL and CSA listing marks, indicates which nation's product safety standards the equipment has been found to be in compliance with. See Figure 1-11.

FIGURE 1-11   The listing mark of Intertek Testing. (*Courtesy Intertek*)

Information can be obtained by contacting

Intertek Testing Services, NA, Inc.
ETL SEMKO
3933 US Route 11
Cortland, NY 13045
Phone: 607-753-6711
http://www.intertek-etlsemko.com

National Electrical Manufacturers Association

The National Electrical Manufacturers Association (NEMA) represents nearly 600 manufacturers of electrical products. NEMA develops electrical equipment standards, which, in many instances, are very similar to UL and other consensus standards. NEMA has representatives on the CMPs for the *NEC*. Additional information can be obtained by contacting

National Electrical Manufacturers Association
1300 North 17th Street, Suite 1847
Rosslyn, VA 22209
Phone: 703-841–3200
http://www.nema.org

World Wide Web Sites

A comprehensive list of World Wide Web sites for manufacturers, organizations, and inspection and testing agencies is provided in the back of this text.

REVIEW

Note: Refer to the *NEC* or the plans in the back of this text where necessary.

1. What is the purpose of specifications? _____

   _____

   _____

2. Refer to the specifications in the back of this text following Chapter 32.
   a. What electrical codes must be conformed to?

      _____

      _____

      _____

   b. What is the wiring method to be used in the workshop?

      _____

   c. What size conductors are to be used for

      the lighting branch circuits? _____ AWG

      the small-appliance branch circuits? _____ AWG

   d. What section of the specifications tells us the size of the service entrance to be installed for this residence?

      _____

3. What is done to prevent a plan from becoming confusing because of too much detail?

   _____

   _____

   _____

4. Name three requirements contained in the specifications regarding material.

   a. _____   c. _____

   b. _____

5. The specifications state that all work shall be done _____

   _____

6. What phrase is used when a substitution is permitted for a specific item? _____

   _____

7. What is the purpose of an electrical symbol? _____

   _____

8. What is a notation? _____

   _____

9. Where are notations found? _____

10. List at least 12 electrical notations found on the plans for this residence. Refer to the plans at the back of the text. _____

    _____

    _____

    _____

    _____

    _____

11. What three parties must be satisfied with the completed electrical installation?

    a. _____   c. _____

    b. _____

12. What code sets standards for electrical installation work? _____

    _____

13. What authority enforces the requirements set by the *NEC*? _____

    _____

14. Does the *NEC* provide minimum or maximum standards? _____

15. What do the letters *UL* signify?

    _____

16. What section of the *NEC* states that all listed or labeled equipment shall be installed or used in accordance with any instructions included in the listing or labeling?

    _____

17. When the word "shall" appears in a *NEC* reference, it means that it (must)(may) be done. (Underline the correct word.)

18. What is the purpose of the *NEC*?

    _____

    _____

    _____

19. Does compliance with the *NEC* always result in an electrical installation that is adequate, safe, and efficient? Why? _____

_____

_____

_____

20. Name two nationally recognized testing laboratories. _____

_____

21. a. Do Underwriters Laboratories and the other recognized testing laboratories "approve" products? _____

    b. What do these testing laboratories do? _____

_____

_____

_____

22. a. Has the *NEC* been officially adopted by the community in which you live? _____

    b. By the state in which you live? _____

    c. If your answer is YES to (a) or (b), are there amendments to the *NEC*? _____

    d. If your answer is YES to (c), list some of the more important amendments.

_____

_____

_____

_____

23. Does the *NEC* make suggestions about how to wire a house that will be occupied by handicapped persons? _____

24. A junction box on a piece of European equipment is marked 200 cm³. Convert this to cubic inches. _____

_____

25. Convert 4500 watts to Btu/hour. _____

_____

26. You will learn in Chapter 3 that residential lighting loads are based on 3 volt-amperes per ft² (33 volt-amperes per m²). Determine the minimum lighting load required for an area of 186 m². Do calculations for both feet squared and meters squared so you can see the difference in answers. To convert meters squared to feet squared, refer to Table 1-6. _____

_____

_____

_____

SAFETY RELATED QUESTIONS

27. What federal organization dictates requirements for work-related safety issues? Circle the correct answer.
    a. *NEC*
    b. OSHA
    c. IAEI

28. What can you do to reduce or eliminate the possibility of receiving an electric shock when working on electrical circuits? Circle the correct answer.
    a. Turn off the power on the circuit you are working on.
    b. Turn off the power on the circuit you are working on, tag it, and lock the disconnect in the "OFF" position.
    c. Turn off the power on the circuit you are working on, then tell everyone else working with you not to turn the power back on.

29. What is the *NEC* definition of a *qualified person*?

    _____

30. Are you a *qualified person*? Explain your answer.

    _____

    _____

31. Are low-voltage systems totally safe? Explain.

    _____

    _____

32. What NFPA standard specifically covers safety in the workplace? _____

33. What do the letters PPE stand for?

    _____

34. Explain briefly what an "arc flash" is.

    _____

    _____

35. Where might you obtain information about on-the-job safety and safety training? ____

    _____

    _____

    _____

36. a. Is it a safe practice to use a step ladder in which the spreaders have not been fully locked into position? _____

    b. If the ladder collapses, who is at fault? _____

CHAPTER

2

Electrical Symbols and Outlets

OBJECTIVES

After studying this chapter, you should be able to

- identify and explain the electrical outlet symbols used in the plans of the single-family dwelling.

- discuss the types of outlets, boxes, luminaires, and switches used in the residence.

- explain the methods of mounting the various electrical devices used in the residence.

- understand the preferred way to position receptacles in wall boxes.

- understand issues involved in remodel work.

- understand how to determine the maximum number of conductors permitted in a given size box.

- understand the concept of fire-resistance rating of walls and ceilings.

ELECTRICAL SYMBOLS

Electrical symbols used on an architectural plan show the location and type of electrical device required. A typical electrical installation as taken from a plan is shown in Figure 2-1.

The *NEC* has many words unique to the electrical trade. For example:

- A **device** is *A unit of an electrical system that carries or controls electric energy as its principle function.*

- An **outlet** is *A point on the wiring system at which current is taken to supply utilization equipment.*

- A **receptacle** is *A contact device installed at the outlet for the connection of an attachment plug.*

- A **receptacle outlet** is *An outlet where one or more receptacles are installed,* Figure 2-2.

- A **lighting outlet** is *An outlet intended for the direct connection of a lampholder or luminaire.* See Figure 2-3.

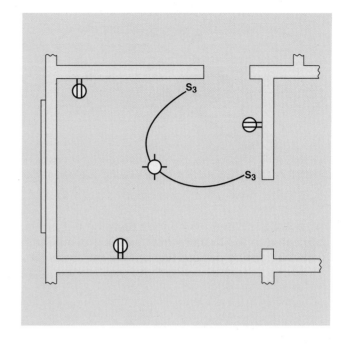

FIGURE 2-1   Use of electrical symbols and notations on a floor plan. (*Delmar/Cengage Learning*)

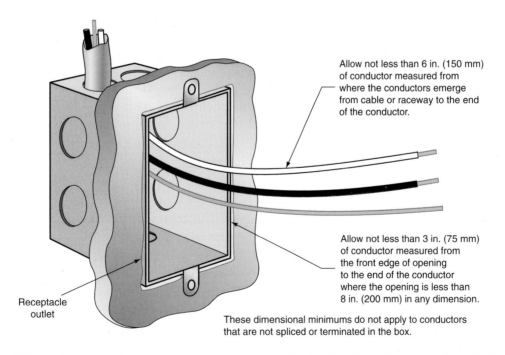

Allow not less than 6 in. (150 mm) of conductor measured from where the conductors emerge from cable or raceway to the end of the conductor.

Allow not less than 3 in. (75 mm) of conductor measured from the front edge of opening to the end of the conductor where the opening is less than 8 in. (200 mm) in any dimension.

These dimensional minimums do not apply to conductors that are not spliced or terminated in the box.

Receptacle outlet

FIGURE 2-2   Where a receptacle is connected to the branch-circuit wires, the outlet is called a *receptacle outlet. NEC 300.14* states the minimum length of conductors at an outlet, junction, or switch point for splices or the connection of luminaires or devices. Leaving the conductors too long can cause crowding of the wires in the box, leading to possible short circuits and/or ground faults. (*Delmar/Cengage Learning*)

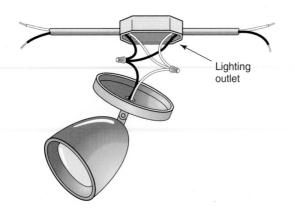

FIGURE 2-3 When a luminaire (lighting fixture) is connected to the branch-circuit wires, the outlet is called a *lighting outlet*. (*Delmar/Cengage Learning*)

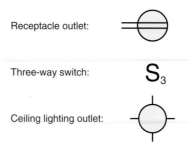

FIGURE 2-4 Some common electrical symbols. (*Delmar/Cengage Learning*)

- A **split-wired receptacle** is electrician's jargon, not an official *NEC* definition. Electricians are very creative in their use of terms. Other terms for the use of these receptacles include *split-receptacle*, *split-wired*, *split-switched*, *switched-receptacle*, and *half-switched*.

To convert a conventional duplex receptacle into a split-wired receptacle, simply remove the tab between the two ungrounded conductor terminals (brass colored). The receptacle can then be used where one receptacle is "hot" at all times, and the other receptacle is switch controlled. Another common application is to connect each receptacle of the duplex to a separate branch circuit.

By definition, toggle switches, receptacles circuit breakers, fuses, and occupancy sensors are *devices* because they carry or control current as their principle function.

The term *opening* is widely used by electricians and electrical contractors when estimating the cost of an installation. The term *opening* covers all lighting outlets, receptacle outlets, junction boxes, switches, etc. The electrician and/or electrical contractor will estimate a job at "X dollars per lighting outlet," "X dollars per switch," "X dollars per receptacle outlet," and so on. These estimates will include the time and material needed to complete the job. Each type of electrical *opening* is represented on the electrical plans as a symbol. The electrical openings in Figure 2-1 are shown by the symbols in Figure 2-4.

ANSI recently published a totally revised standard entitled *Symbols for Electrical Construction Drawings*. This was the first revision in over 25 years. Figures 2-5, 2-6, 2-7, 2-8, 2-9, and 2-10 show the electrical symbols most commonly found on architectural and electrical plans. Because some items may have more than one symbol, it is important to check the plans and specifications of any job you are working on for a symbol schedule to make sure you have interpreted the symbols correctly.

The curved lines in Figure 2-1 run from the outlet to the switch or switches that control the outlet. Curved lines are used to differentiate the electrical circuitry from the building construction drawing lines. Several drawings show a curved line between 3-way and 4-way switches to indicate the control points. Some receptacles have one-half of the duplex switched. A curved line runs from the receptacle to the switch or control point. Other curved lines show the connection of luminaires or the circuiting for receptacles.

A study of the plans for the single-family dwelling shows that many different electrical symbols are used to represent the electrical devices and equipment used in the building.

Be very careful when interpreting symbols because of the similarity between them, Figure 2-11. For example, the left symbol can be used for a split-wired or a ½-switched duplex receptacle. Several of the rooms in this dwelling have duplex receptacles with half of the receptacle switched and the other half remaining hot, or energized, all the time. Table or floor lamps can be plugged into the receptacle that is switched to allow control of room lighting from near the door. The receptacle remains hot

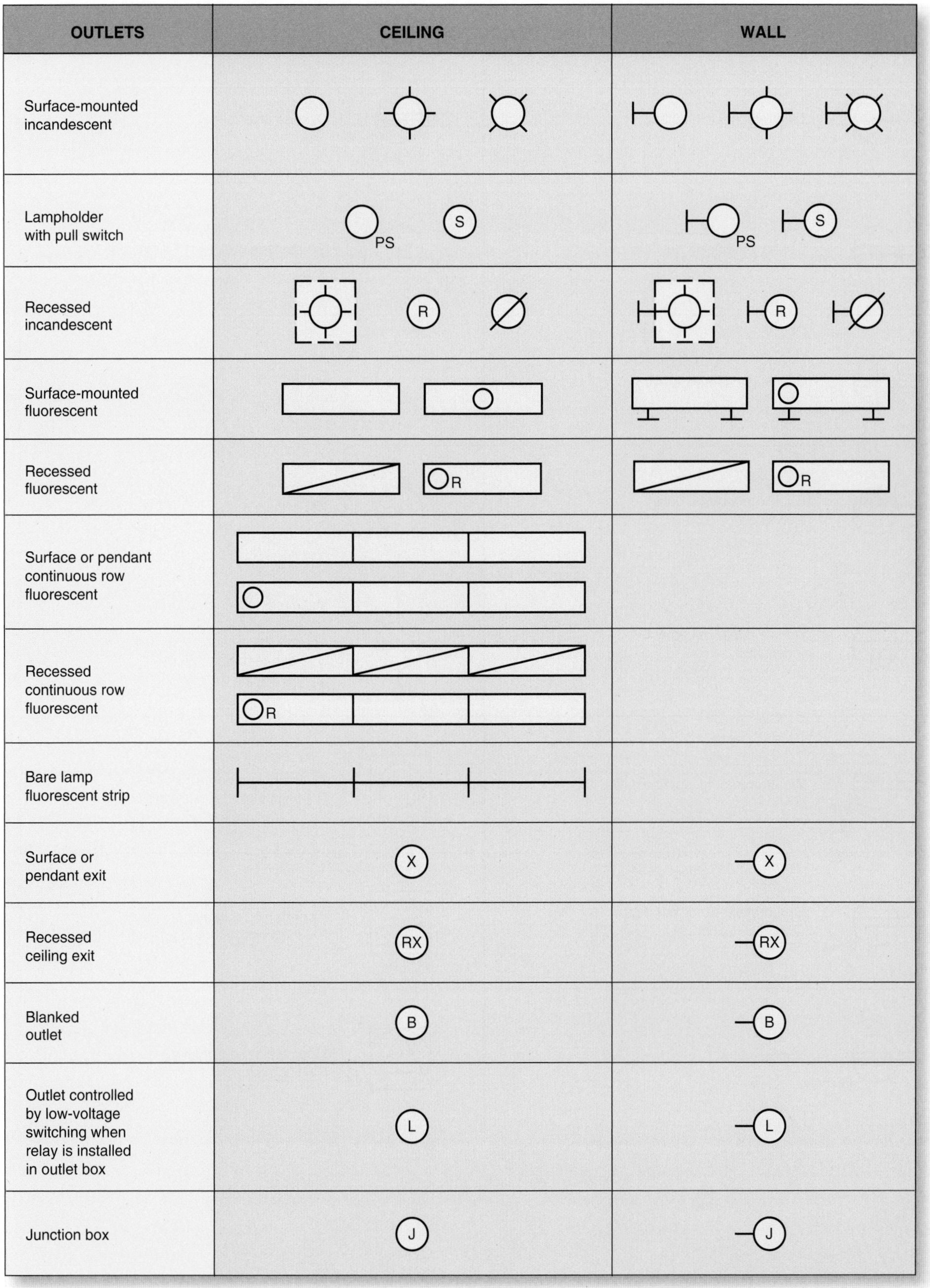
OUTLETS
CEILING
WALL
Surface-mounted incandescent
Lampholder with pull switch
PS
S
PS
S
Recessed incandescent
R
R
Surface-mounted fluorescent
Recessed fluorescent
R
R
Surface or pendant continuous row fluorescent
Recessed continuous row fluorescent
R
Bare lamp fluorescent strip
Surface or pendant exit
X
X
Recessed ceiling exit
RX
RX
Blanked outlet
B
B
Outlet controlled by low-voltage switching when relay is installed in outlet box
L
L
Junction box
J
J

FIGURE 2-5 Lighting outlet symbols. (*Delmar/Cengage Learning*)

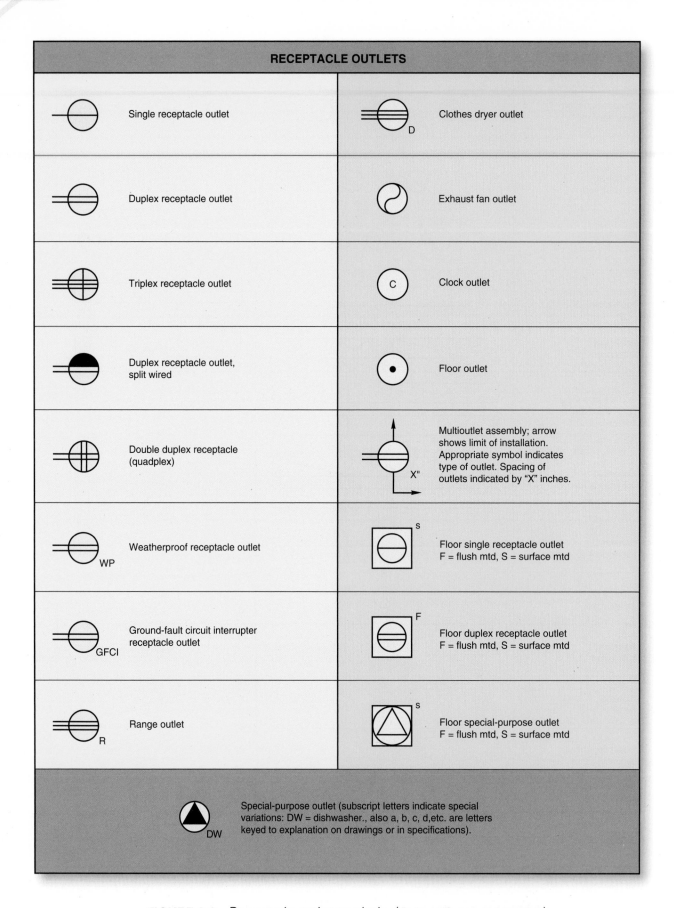

RECEPTACLE OUTLETS	
Single receptacle outlet	Clothes dryer outlet
Duplex receptacle outlet	Exhaust fan outlet
Triplex receptacle outlet	Clock outlet
Duplex receptacle outlet, split wired	Floor outlet
Double duplex receptacle (quadplex)	Multioutlet assembly; arrow shows limit of installation. Appropriate symbol indicates type of outlet. Spacing of outlets indicated by "X" inches.
Weatherproof receptacle outlet	Floor single receptacle outlet F = flush mtd, S = surface mtd
Ground-fault circuit interrupter receptacle outlet	Floor duplex receptacle outlet F = flush mtd, S = surface mtd
Range outlet	Floor special-purpose outlet F = flush mtd, S = surface mtd

Special-purpose outlet (subscript letters indicate special variations: DW = dishwasher., also a, b, c, d,etc. are letters keyed to explanation on drawings or in specifications).

FIGURE 2-6 Receptacle outlet symbols. (*Delmar/Cengage Learning*)

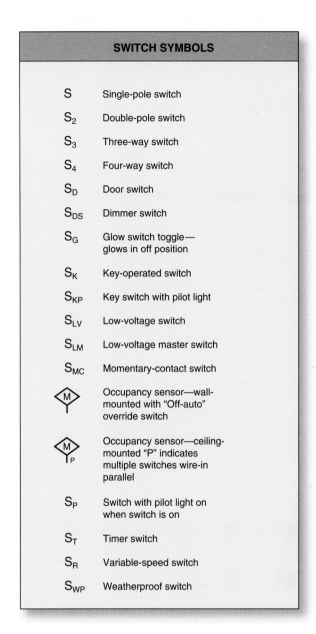

FIGURE 2-7 Switch symbols.
(*Delmar/Cengage Learning*)

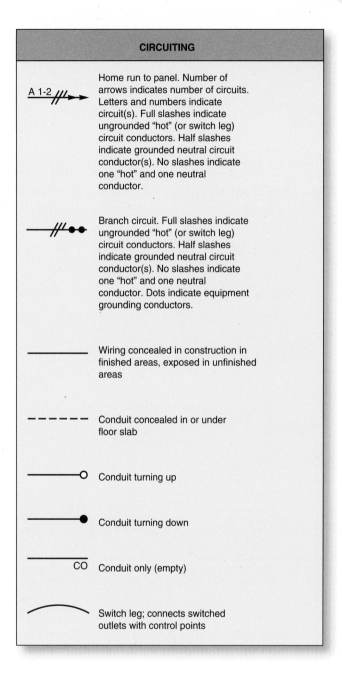

FIGURE 2-8 Circuiting symbols.
(*Delmar/Cengage Learning*)

all the time for any appliance or fixture that is to remain energized at all times. In some areas these receptacles are referred to as switched receptacles or split-wired receptacles. The concept of split-wired receptacles includes using a branch circuit with two hot or energized conductors and a shared neutral.

For both applications mentioned here, the tab is removed on the side of the duplex receptacle so that each receptacle can be controlled separately or have both ungrounded conductors connected to it.

This switching arrangement satisfies *NEC 210.70(A)(1), Exception No. 1*, which permits

switch-controlled receptacles to serve as the lighting outlet in habitable rooms.

The notation for the receptacle, shown at the right in Figure 2-11, indicates the receptacle is to be in an enclosure suitable for a weatherproof location.

When preparing electrical plans for commercial and industrial buildings, most architects and electrical engineers use symbols approved by ANSI wherever

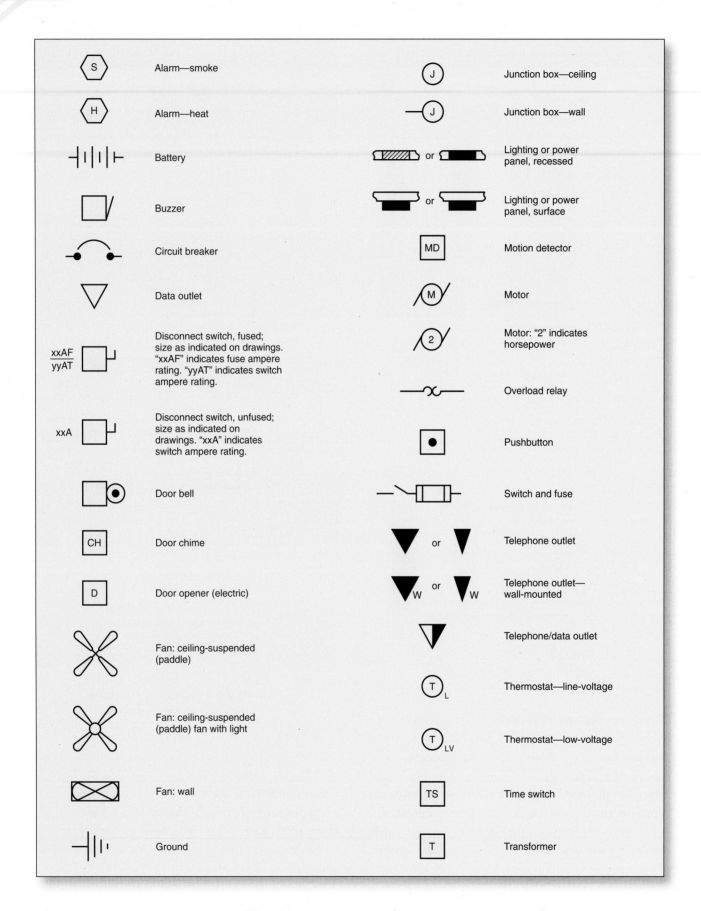

S	Alarm—smoke	J	Junction box—ceiling
H	Alarm—heat	J	Junction box—wall
	Battery	or	Lighting or power panel, recessed
	Buzzer	or	Lighting or power panel, surface
	Circuit breaker	MD	Motion detector
	Data outlet	M	Motor
xxAF / yyAT	Disconnect switch, fused; size as indicated on drawings. "xxAF" indicates fuse ampere rating. "yyAT" indicates switch ampere rating.	2	Motor: "2" indicates horsepower
xxA	Disconnect switch, unfused; size as indicated on drawings. "xxA" indicates switch ampere rating.		Overload relay
	Door bell		Pushbutton
CH	Door chime		Switch and fuse
D	Door opener (electric)	or	Telephone outlet
	Fan: ceiling-suspended (paddle)	W or W	Telephone outlet—wall-mounted
	Fan: ceiling-suspended (paddle) fan with light		Telephone/data outlet
	Fan: wall	T L	Thermostat—line-voltage
		T LV	Thermostat—low-voltage
		TS	Time switch
	Ground	T	Transformer

FIGURE 2-9 Miscellaneous symbols. (*Delmar/Cengage Learning*)

SYMBOL	NOTATION
1	Plugmold entire length of workbench. Outlets 18 in. (457mm) O.C. install 48 in. (1.2m) to center from floor. GFCI protected.
2	Track lighting. Provide 5 lampholders.
3	Two 32-watt rapid start fluorescent lamps in valance. Control with dimmer switch.

FIGURE 2-10   Example of how certain notations might be added to a symbol when a symbol itself does not fully explain its meaning. The architect or engineer has a choice of explaining fully the meaning directly on the plan if there is sufficient room; if insufficient room, then a notation could be used. (*Delmar/Cengage Learning*)

Split-wired receptacle outlet   Weatherproof receptacle outlet

FIGURE 2-11   Variations in significance of outlet symbols. (*Delmar/Cengage Learning*)

possible. For residential projects, usually only the basic symbols are used, showing the location of receptacles, switches, appliances, and luminaires. Specific circuitry, as shown in Figure 2-8, is generally left up to the electrician.

Later in this text, you will find suggested *cable layouts* for all of the branch circuits in this residence. These *cable layouts* show the required number of conductors between outlets and switches.

LUMINAIRES AND OUTLETS

The term *luminaire* is found in the *NEC*. In the United States, we have used the term *lighting fixture* for years. But because the *NEC* is an international code, we must become familiar with some new terms such as *luminaire*. By definition, a *luminaire* is *A complete lighting unit consisting of a light source such as a lamp or lamps, together with the parts designed to position the light source and connect it to the power supply. It may also include parts to protect the light source, ballast, or distribute the light. A lampholder itself is not a luminaire.*

Simply stated for our purposes, a luminaire and a lighting fixture are one and the same.

The location of lighting outlets is determined by the amount and type of illumination required to provide the desired lighting effects. It is not the intent of this text to describe how proper and adequate lighting is determined. Rather, the text covers the proper methods of installing the circuits for such lighting.

Luminaire manufacturers publish catalogs that provide a tremendous amount of information regarding recommendations for residential lighting. These publications are available at electrical distributors, lighting showrooms, and home centers.

Architects often include a specific amount of money in the specifications, referred to as a *fixture allowance*. Electrical contractors include this amount in their bid, and the choice of luminaires is then left to the homeowners. If the owner selects luminaires whose total cost exceeds the fixture allowance, the owner is expected to pay the difference between the actual cost and the fixture allowance. If the luminaires are not selected before the rough-in stage of wiring the house, an electrician will usually install outlet boxes such as those illustrated in Figure 2-20.

How luminaires are handled varies by job and geographical area you are working in. Sometimes the recessed luminaire housings are included in the base bid, and the trims are included in the fixture allowance. Always read the specifications and study the plans carefully for important symbols and notations that might reference luminaires.

Surface-mounted luminaires are secured to an outlet box, a raised plaster (adapter) ring, or a device box using the mounting hardware (straps and No. 8-32 or No. 6-32 screws) furnished with the luminaire. Some outlet box/bar hanger assemblies have a fixture stud on which to fasten the luminaire. See Figure 2-12.

Symbols Used to Represent Lighting Outlets and Switches

Typical lighting outlet symbols are shown in Figure 2-12. Typical switch symbols are shown in Figure 2-13.

OUTLET, DEVICE, AND JUNCTION BOXES

Article 314 contains much detail about outlet boxes, device boxes, pull boxes, junction boxes, and conduit bodies (LB, LR, LL, C, etc.), all of which are used to provide the necessary space for making electrical connections, contain wiring, and support luminaires, and so on. *Article 314* is quite lengthy. For the moment, we will focus on some of the basic *Code* requirements for outlet boxes, device boxes, and junction boxes. Additional *Code* requirements are covered on an "as needed" basis in this chapter and in later chapters.

Screw sizes for typical electrical boxes are as follows:

Device boxes and faceplates	No. 6-32
Outlet boxes	No. 8-32
Ceiling fan boxes (two sets of holes)	No. 10-32 and No. 8-32

The numbers 6, 8, and 10 refer to the diameter of the screw, with the smaller number being a smaller diameter. The larger number, 32, refers to the number of threads per inch or the "pitch" of the thread. For these screws and similar screws, a coarse and fine thread is established according to the Unified Thread Standard (UTS). "Coarse" and "fine" do not refer to the quality of the thread but to the number of threads per inch. Dimensions and threads for these screws are shown in the following table:

UTS Number	Diameter	Coarse Pitch	Fine Pitch
6	0.1380	32	40
8	0.1640	32	36
10	0.1900	24	32

Electricians commonly carry a multitap tool to clean out threaded holes that may have become damaged or filled with paint or other contaminates.

Some Basic Code Requirements

- *NEC 300.11* and *314.23:* Boxes must be securely mounted and fastened in place.

- *NEC 300.15:* Where conduit, electrical-metallic tubing, nonmetallic-sheathed cable, Type AC cable, or other cables are installed, a box or conduit body must be installed at each conductor splice connection point, outlet, switch point, junction point, or pull point. This requirement is illustrated in Figure 2-14. Exceptions to this are found in *300.15(A)* through *(M).* Typical exceptions are surface-mounted strip luminaires, as shown in Figure 10-2. Most of these have a back plate that has a trade size ½ knockout intended for the direct entry of a cable without the need of a box behind the luminaire.

- *NEC 300.15(C):* Where cables enter or exit from conduit or tubing that is used to provide cable support or protection against physical damage, a fitting shall be provided on the end(s) of the conduit or tubing to protect the cable from abrasion. Figure 2-15 shows a transition from a cable to a raceway.

- *NEC 314.16:* Boxes must be large enough for all of the enclosed conductors and wiring devices. Installing boxes too small for the number of conductors, wiring devices, luminaire studs, and splices is a real problem! This topic is covered in detail later.

- *NEC 314.17(C):* The exception allows multiple cables to be run through a single knockout opening in a nonmetallic box. This permission is not given for metallic boxes. See Figure 2-16.

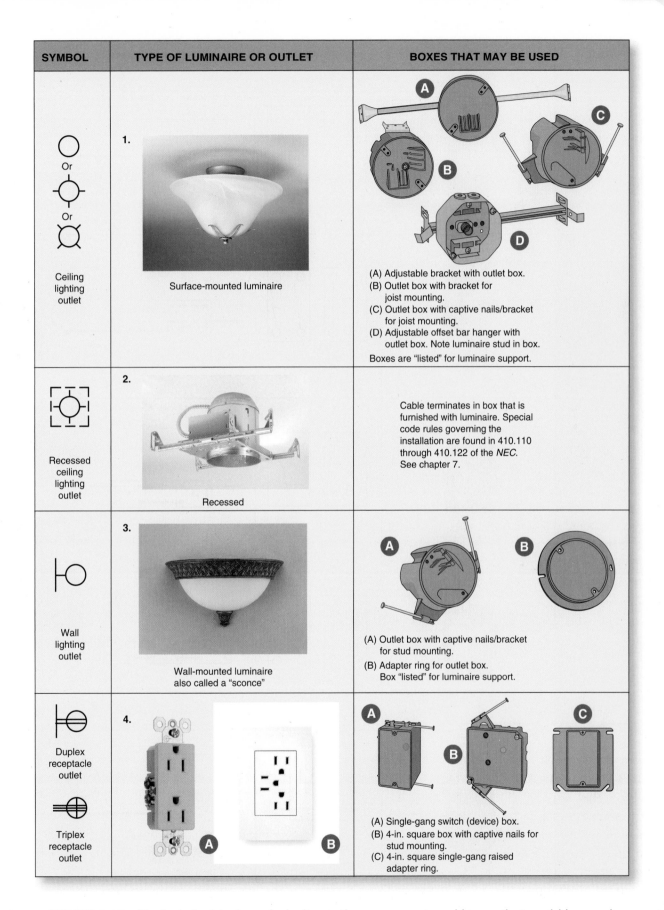

SYMBOL	TYPE OF LUMINAIRE OR OUTLET	BOXES THAT MAY BE USED
Ceiling lighting outlet	**1.** Surface-mounted luminaire	(A) Adjustable bracket with outlet box. (B) Outlet box with bracket for joist mounting. (C) Outlet box with captive nails/bracket for joist mounting. (D) Adjustable offset bar hanger with outlet box. Note luminaire stud in box. Boxes are "listed" for luminaire support.
Recessed ceiling lighting outlet	**2.** Recessed	Cable terminates in box that is furnished with luminaire. Special code rules governing the installation are found in 410.110 through 410.122 of the *NEC*. See chapter 7.
Wall lighting outlet	**3.** Wall-mounted luminaire also called a "sconce"	(A) Outlet box with captive nails/bracket for stud mounting. (B) Adapter ring for outlet box. Box "listed" for luminaire support.
Duplex receptacle outlet / Triplex receptacle outlet	**4.**	(A) Single-gang switch (device) box. (B) 4-in. square box with captive nails for stud mounting. (C) 4-in. square single-gang raised adapter ring.

FIGURE 2-12   Typical electrical symbols, items they represent, and boxes that could be used.
(Photos courtesy of Progress Lighting [1, 3], Halo/Cooper Lighting [2], Leviton Manufacturing Company [4A & 4B])

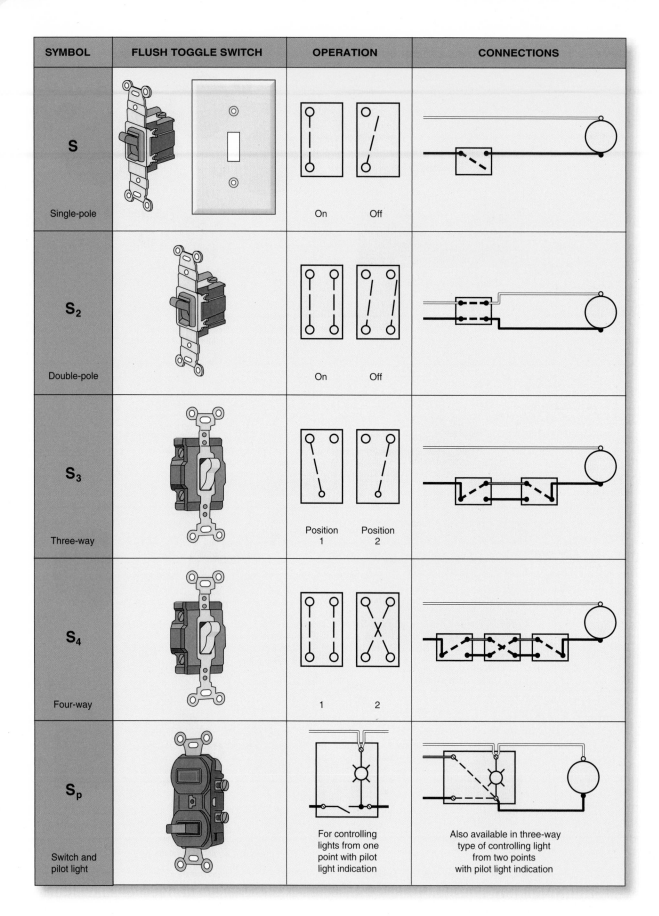

SYMBOL	FLUSH TOGGLE SWITCH	OPERATION	CONNECTIONS
S Single-pole		On Off	
S₂ Double-pole		On Off	
S₃ Three-way		Position 1 Position 2	
S₄ Four-way		1 2	
Sₚ Switch and pilot light		For controlling lights from one point with pilot light indication	Also available in three-way type of controlling light from two points with pilot light indication

FIGURE 2-13 Standard switches and symbols. (*Delmar/Cengage Learning*)

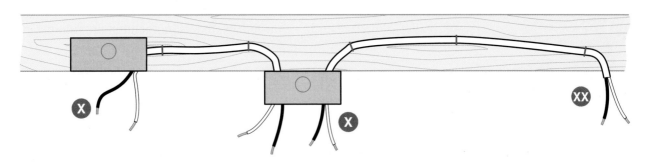

FIGURE 2-14   A box (or fitting) must be installed wherever there are splices, outlets, switches, or other junction points as indicated by the "X." Unless listed for use without a box, such as a typical surface-mounted strip luminaire that has a back plate with a trade size ½ knockout, point XX is a *Code* violation, *300.15*. (*Delmar/Cengage Learning*)

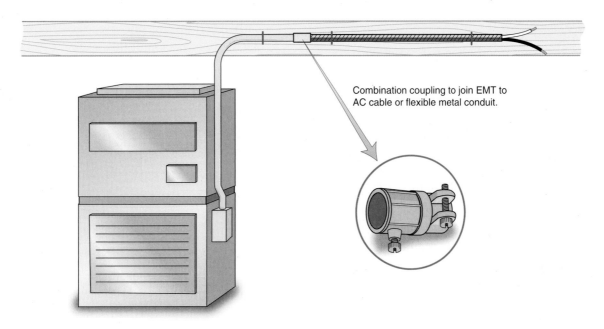

Combination coupling to join EMT to AC cable or flexible metal conduit.

FIGURE 2-15   *NEC 300.15(F)* permits a transition to be made from one wiring method to another wiring method. In this case, the armor of the Type AC cable is removed, allowing sufficient length of the conductors to be run through the conduit. A proper fitting must be used at the transition point, and the fitting must be accessible after installation. The conductors in Type AC cable often have THHN insulation. (*Delmar/Cengage Learning*)

- *NEC 314.20:* In walls or ceilings where the surface material is noncombustible, boxes, plaster rings, domed covers, extension rings, or listed extenders shall be mounted so that they will be set back not more than ¼ in. (6 mm). In walls or ceilings where the surface material is combustible, boxes, plaster rings, domed covers, extension rings, or listed extenders shall be set flush with the surface.

- *NEC 314.22:* A surface extension from a box of a concealed wiring system is made by mounting and mechanically securing an extension ring over the concealed box. An exception to this is that a surface extension is permitted to be made

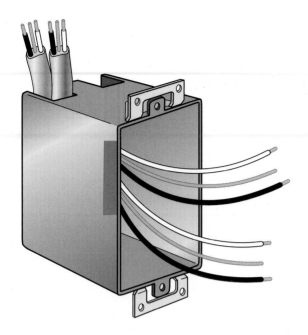

FIGURE 2-16 Multiple cables may run through a single knockout opening in a nonmetallic box, *314.17(C), Exception*. (*Delmar/Cengage Learning*)

from the cover of a concealed box where the cover is designed so it is unlikely to fall off, or be removed if its securing means becomes loose. The wiring method must be flexible and arranged so that any required grounding continuity is independent of the connection between the box and cover. This is shown in Figure 2-17.

- *NEC 314.24:* Outlet and device boxes shall be deep enough so that conductors will not be damaged when installing the wiring device or equipment into the box. Merely calculating and providing the proper volume of a box is not always enough. The volume calculation might prove adequate, yet the size (depth) of the wiring device or equipment might be such that conductors behind it may possibly be damaged.

- *NEC 314.25:* In completed installations, all boxes shall have a cover, faceplate, or luminaire canopy. Do not leave any electrical boxes uncovered.

- *NEC 314.27(A):* Boxes used at lighting outlets must be designed or installed so that a luminaire

may be attached to it. This is typically accomplished by the manufacturer providing tapped holes in boxes.

- *NEC 314.27(A)* and *(B)*: Listed outlet boxes are suitable for hanging luminaires that weigh 50 lb. (23 kg) or less. Luminaires that weigh more than 50 lb. (23 kg) must be supported independently of the outlet box, or the outlet box must be "listed" for the weight to be supported. Wall-mounted luminaires of not more than 6 lb. (3 kg) may be supported by a box, or by a plaster ring secured to a box, by not less than two No. 6-32 or larger screws, as in Figure 2-18.

- *NEC 314.27(D):* Standard outlet boxes must not be used as the sole support for ceiling-suspended (paddle) fans unless they are specifically listed for this purpose. See Chapter 9 for a detailed discussion of this issue.

- *NEC 314.29:* Conduit bodies, junction, pull, and outlet boxes must be installed so the wiring contained in them is accessible without removing any part of the building. Never install outlet boxes, junction boxes, pull boxes, or conduit bodies where they will be inaccessible behind or above permanent finished walls or ceilings. Should anything ever go wrong, it would be a nightmare, if not impossible, to troubleshoot and locate hidden splices. Wiring located above "lay-in" ceilings is considered accessible because it is easy to remove a panel(s) to gain access to the space above the ceiling. The *NEC* definition of *accessible* is *Capable of being removed or exposed without damaging the building structure or finish, or not permanently closed in by the structure or finish of the building.**

The same accessibility requirement is true for underground wiring. Boxes must be accessible without excavating sidewalks, paving, earth, or other material used to establish the finished grade.

The exception to *314.29* permits listed boxes to be covered with *gravel, light aggregate, or non-cohesive granulated soil* (not clay) if their location is clearly marked.

*Reprinted with permission from NFPA 70-2011.

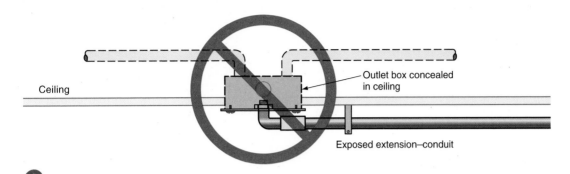

A

Violation. It is a violation of the *NEC* to make a rigid extension from a cover that is attached to a concealed outlet box, *314.22*. It would be difficult to gain access to the connections inside the box.

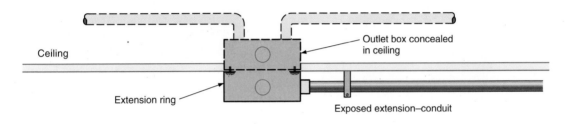

B

This meets *Code*. An extension ring must be mounted over and mechanically secured to the concealed outlet box, *314.22*.

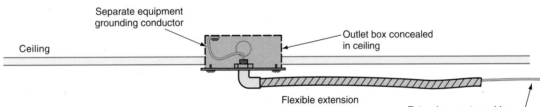

C

This meets *Code*. It is permitted to make a surface extension from a cover fastened to a concealed box where the extension wiring method is flexible and where a separate grounding conductor is provided so that the grounding path is not dependent upon the screws that are used to fasten the cover to the box. See *314.22*, exception.

It would be a *Code* violation to make the extension if a flexible wiring method was used but had no provisions for a separate equipment grounding conductor.

FIGURE 2-17 How to make an extension from a flush-mounted box. (*Delmar/Cengage Learning*)

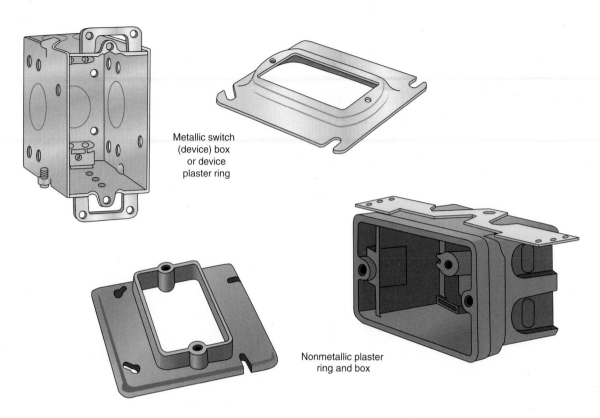

Metallic switch
(device) box
or device
plaster ring

Nonmetallic plaster
ring and box

FIGURE 2-18 Wall-mounted luminaires (lighting fixtures) that do not weigh more than 6 lb. (3 kg) are permitted to be secured to these types of boxes and plaster rings by not less than two No. 6-32 or larger screws, *314.27(A), Exception.* For standard outlet boxes, the maximum weight is 50 lb. (23 kg), *314.27(B).* (*Delmar/Cengage Learning*)

• *NEC 314.71:* Pull and junction boxes must provide adequate space and dimensions for the installation of conductors, and they shall comply with the specific requirements of this section.

• *NEC 410.22:* This is another repeat of the requirement that boxes must be covered. The "cover" could be a blank cover, a luminaire canopy, a lampholder, a switch or receptacle, or similar device with faceplate.

• *NEC 410.36(A):* Here we find permission to use outlet boxes and fittings to support luminaires provided the requirements found in *314.27(A)* and *(B)* are met.

What to Do Where Recessed Luminaires Will Be Installed. This topic is covered in detail in Chapter 7.

GROUNDING/BONDING

As you study this text, you will come across many terms unique to the electrical industry. For now, let us explain the meaning of *grounding* and *bonding*.

What Does *Grounded (Grounding)* Mean?

The definition of *grounded (grounding)* found in *NEC Article 100* is *Connected to ground or to a conductive body that extends the ground connection.*

What Does *Bonded (Bonding)* Mean?

The definition of *bonded (bonding)* found in *NEC Article 100* is *Connected to establish electrical continuity and conductivity.*

Take a look at Figures 27-17(A), 27-17(B), 27-26, and 27-27 to see the meanings of these terms. Other terms are explained and illustrated as they appear.

NONMETALLIC OUTLET AND DEVICE BOXES

Today, most house wiring is done using nonmetallic boxes, as permitted by *314.3*, along with nonmetallic sheathed cable Type NM. This is usually because of economics. Although not often done, *NEC 314.3 (Exceptions)* allows metal raceways and metal-jacketed cables (BX) to be used with nonmetallic boxes provided all metal raceways or cables entering the box are bonded together to maintain the integrity of the grounding path to other equipment in the installation.

The house wiring system usually is formed by a number of specific circuits. Each circuit consists of runs of cable from outlet to outlet or from box to box. The residence plans show many branch circuits for general lighting, appliances, electric heating, and other requirements. The specific *Code* rules for each of these circuits are covered in later chapters.

GANGED SWITCH (DEVICE) BOXES

Multiple-Gang Boxes

Where more than one wiring device is to be installed at one location, multiple-gang boxes are used. Nonmetallic boxes for wiring devices are available in 2-gang, 3-gang, and 4-gang.

Some designs of metal device boxes can be ganged together by removing and discarding one side from both the first and third boxes, and both sides from the second (center) box.

Nonmetallic multiple-gang boxes are shown in Figure 2-19.

Box Fill

Figure 2-20 shows the dimensions of typical metal boxes and the number of conductors permitted in a given size box. This figure replicates *Table 314.16(A)* of the *NEC*.

Nonmetallic boxes are marked with their cubic-inch volume. *Table 314.16(B)* shows the cubic-inch volume allowance required for different size conductors.

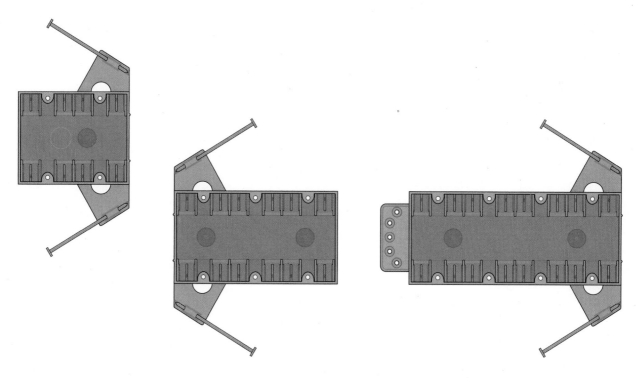

FIGURE 2-19   Examples of multiple-gang device boxes. (*Delmar/Cengage Learning*)

QUIK-CHEK BOX SELECTION GUIDE
FOR BOXES GENERALLY USED FOR RESIDENTIAL WIRING

DEVICE BOXES

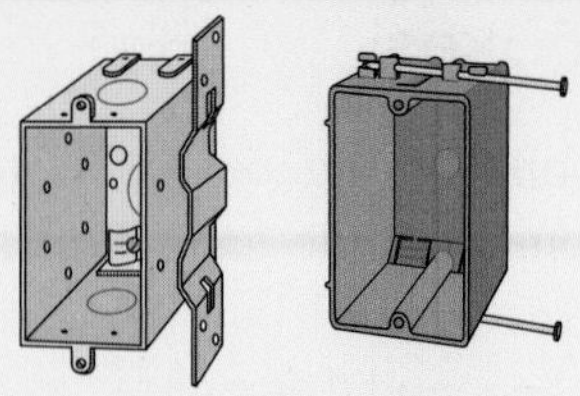

Wire size	3x2x1½ (7.5 in.³)	3x2x2 (10 in.³)	3x2x2¼ (10.5 in.³)	3x2x2½ (12.5 in.³)	3x2x2¾ (14 in.³)	3x2x3 (16 in.³)	3x2x3½ (18 in.³)
14 AWG	3	5	5	6	7	8	9
12 AWG	3	4	4	5	6	7	8

The number of conductors is "per gang"

SQUARE BOXES

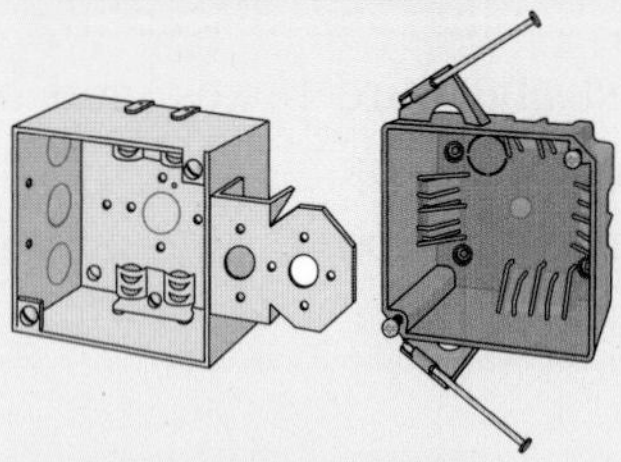

Wire size	4x4x1½ (21 in.³)	4x4x2⅛ (30.3 in.³)
14 AWG	10	15
12 AWG	9	13

OCTAGON BOXES

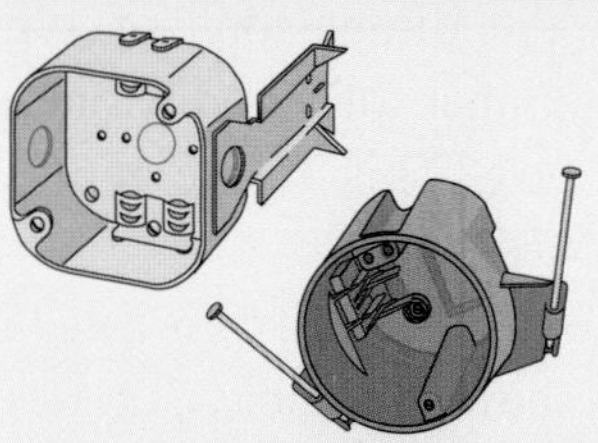

Wire size	4x1½ (15.5 in.³)	4x2⅛ (21.5 in.³)
14 AWG	7	10
12 AWG	6	9

HANDY BOXES

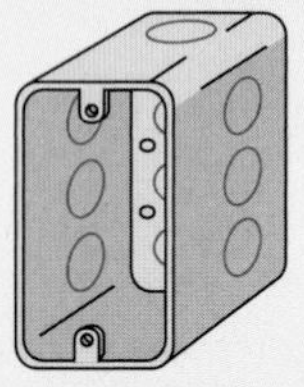

Wire size	4x2⅛x1½ (10.3 in.³)	4x2⅛x1⅞ (13 in.³)	4x2x2⅛ (14.5 in.³)
14 AWG	5	6	7
12 AWG	4	5	6

RAISED COVERS

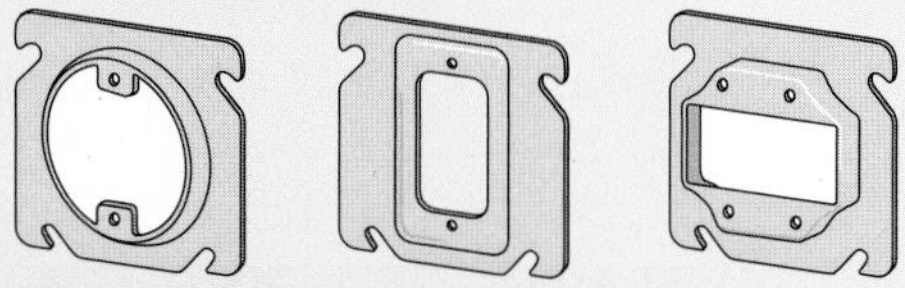

Where raised covers are marked with their volume in cubic inches, that volume may be added to the box volume to determine maximum number of conductors in the combined box and raised cover. Nonmetallic raised covers are available.

MASONRY BOXES

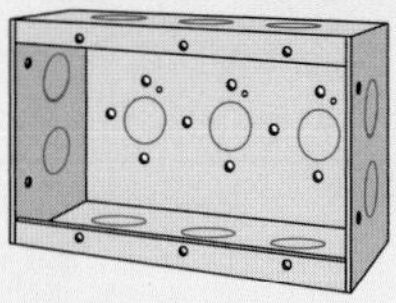

3-gang box

Wire size	3¾x2x2½ (14 in.³)	3¾x2x3½ (21 in.³)
14 AWG	7	10
12 AWG	6	9

Masonry boxes have lots of room. Available up to 6 gang. The conductor fill is for each gang. Refer to *table 314.16 (A)* and *(B)*.

Note: Be sure to make deductions from the above maximum number of conductors permitted for wiring devices, cable clamps, fixture studs, and grounding conductors. The cubic-inch (in.³) volume is taken directly from *table 314.16(A)* of the *NEC*. Nonmetallic boxes are marked with their cubic-inch capacity.

FIGURE 2-20 Quik-chek box selection guide. (*Delmar/Cengage Learning*)

BOX MOUNTING

NEC 314.20 states that where noncombustible surface material (tile, gypsum, plaster, and like material) is used on walls or ceilings, boxes, plaster rings, domed covers, extension rings, or listed extenders must be mounted so that they will be set back not more than ¼ in. (6 mm) from the face of the surface material. Where combustible surface material (wood paneling) is used on walls or ceilings, boxes, plaster rings, domed covers, extension rings, or listed extenders must be set flush with the surface material, Figure 2-21.

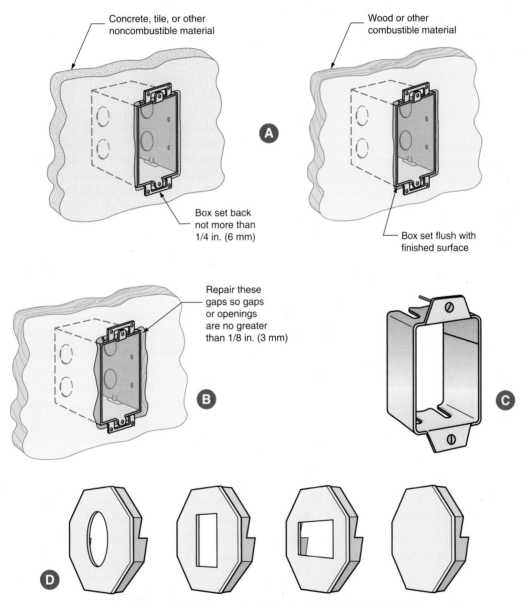

Concrete, tile, or other noncombustible material

Box set back not more than 1/4 in. (6 mm)

Wood or other combustible material

Box set flush with finished surface

Repair these gaps so gaps or openings are no greater than 1/8 in. (3 mm)

An assortment of plastic mounting blocks to attain neat appearance when mounting lighting fixtures and receptacles where outdoor siding is lapped.

FIGURE 2-21 Box position requirements in walls and ceilings constructed of various materials. If the box is not flush with the finished surface (as a result of improper installation) or if paneling, drywall, tile, or a mirror is added, then install a box extender, sometimes referred to as a "goof ring." One type is shown in (C). Box extenders that are ¼ in., ³/₈ in., ½ in., or ¾ in. deep are also available for 1- and 2-gang boxes. Refer to *314.20* and *314.21*. For setback requirements for panelboard cabinets, see *312.3* and *312.4*. (*Delmar/Cengage Learning*)

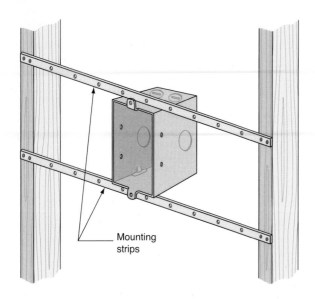

FIGURE 2-22 Switch (device) boxes installed between studs using metal switch box supports, often referred to as "Kruse strips." If wood strips are used, they must have a cross-sectional dimension of not less than 1 in. × 2 in. (25 mm × 50 mm), *314.23(B)(2)*. (*Delmar/Cengage Learning*)

Listed box that complies with the nail-positioning rules.

Listed box that has external nails for securing the box to wooden studs.

FIGURE 2-23 Using nails or screws to install a sectional switch box, *314.23(B)(1)*. UL listed boxes will have nail mounting holes positioned to meet the requirements of the *NEC*. A better choice is to use boxes that have external brackets or external nail mounting fixtures. (*Photos courtesy of Thomas and Betts*)

These requirements hold back, retard, and slow down the spread of fire caused by arcing at a loose connection, a short circuit, or a ground fault that might occur within the box. This significantly reduces the supply of oxygen needed to keep a fire burning. Closing the gaps around boxes further reduces the spread of fire.

Ganged sectional switch (device) boxes can be installed using a pair of metal mounting strips. These strips are also used to install a switch box between wall studs, Figure 2-22. When an outlet box is to be mounted at a specific location between joists, as for ceiling-mounted luminaires, an adjustable bar hanger is used, as in Figure 2-12.

NEC 314.23(B)(1) requires that mounting nails or screws through a box shall not be more than ¼ in. (6 mm) from the back or ends of the box, as shown in Figure 2-23. This ensures that the nails or screws will not interfere with the wiring devices in the box. Furthermore, you are not allowed to use screws through a box unless the threads are protected by an approved means to avoid abrasion of conductor insulation.

Why bother with these *Code* requirements that take up valuable wiring space and could lead to conductor damage? It is much better to use boxes that have external mounting means.

Most residences are constructed with wood framing. However, metal studs and joists are being used more and more. Figure 2-24 shows a few types of boxes for use with metal framing members.

The first one is a single-gang nonmetallic device box that can be riveted or screwed to the metal stud. The second is a 4-in. square steel box with nonmetallic sheathed cable (Romex) clamps. The third is a 4-in. square steel box with armored cable (BX) clamps. The second and third boxes "snap" onto the metal stud, and do not require drilling the metal studs and fastening the box with nuts and bolts or screws. See Chapter 4 for *Code* requirements about running cables through metal framing members.

Another type of device used for fastening electrical boxes to steel framing members is shown in Figure 2-25. The bracket is screwed to the steel framing members, then the box or boxes are attached to the bracket.

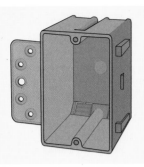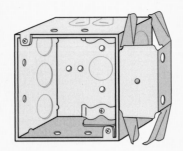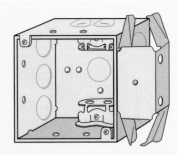

FIGURE 2-24 Types of boxes that can be used with steel framing construction. Note the side clamps that "snap" directly onto the steel framing members. (*Delmar/Cengage Learning*)

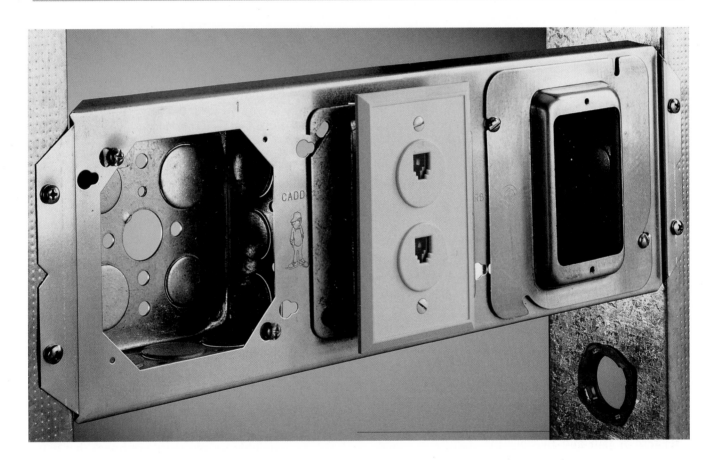

FIGURE 2-25 A bracket for use on steel framing members. The bracket spans the space between two studs. Note that more than one box can be mounted on this bracket. (*Courtesy of ERICO, Inc.*)

Damage Control!

One of the most common problems electricians face is what happens to conductors in boxes during the construction stage of a job. It's a major problem. In an attempt to minimize conductor damage, proposals similar to the following have been made to the *NEC:*

Conductors, inside electrical boxes, subject to physical damage from router bits, sheet rock saws, and knives, and nonconductive coatings, such as drywall mud, paint, lacquer and enamel, shall be protected during the construction process by means of a rigid cover, plate, or insert of a thickness and strength as to prohibit penetration by the above mentioned items.

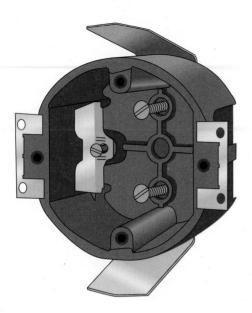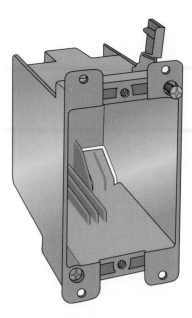

Note: After inserting the box into the hole cut in the drywall, when screws are turned, "ears" flip upward behind the drywall, tightly locking the box in place.

FIGURE 2-26 Nonmetallic boxes are commonly used for remodel work. When boxes are inserted into the hole cut in the drywall, they securely lock into place. This type of box is not for use in new work. (*Delmar/Cengage Learning*)

In the meantime, do everything possible to prevent conductors from being damaged in electrical boxes. Pushing the conductors deep into the box and using deeper boxes can reduce damage to conductors. Installing temporary metal covers on boxes will prevent damage to conductors as well as prevent unwanted material such as sheetrock finishing compounds and paint from entering the boxes.

Remodel (Old Work)

When installing switches, receptacles, and lighting outlets in remodel work where the walls and ceilings (paneling, drywall, sheetrock and plaster, or lath and plaster) are already in place, first make sure you will be able to run cables to where the switches, receptacles, and lighting outlets are to be installed. After making sure you can get the cables through the concealed spaces in the walls and ceilings, you can then proceed to cut openings the size of the specific type of box to be installed at each outlet location. Cables are then "fished" through the stud and joist spaces behind or above the finished walls and ceilings to where you have cut the openings. Then boxes having plaster ears and snap-in brackets, Figure 2-26, can be inserted from the front, through the holes already

cut at the locations where the wall or ceiling boxes are to be installed. After the boxes are snapped into place through the hole from the front, they become securely locked into place.

Another popular method of fastening wall and ceiling boxes in existing walls and ceilings is to use boxes that have plaster ears. Insert the box through the hole from the front. Then position a metal support into the hole on each side of the box. Be careful not to let the metal support fall into the hole. Next, bend the metal support over and into the box (Figure 2-27). Two metal supports are used. These metal switch box supports are known as Madison Hold-Its.[†]

Also available for old work are boxes that have a screw-type support on each side of the box, as in Figure 2-28. This type of box has plaster ears. It is inserted through the hole cut in the wall or ceiling, and then the screws are tightened. This pulls up a metal support tightly behind the wall or ceiling, firmly holding the box in place. The action is very similar to that of Molly screw anchors.

Because these types of remodel boxes are supported by the drywall or plaster, as opposed to new

[†]Hold-It™ is a brand name trademark owned by Madison Equipment Co., Cleveland, OH.

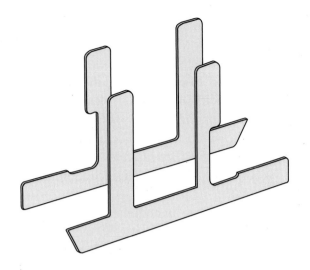

FIGURE 2-27 Metal device box supports referred to as Madison Hold-Its. (*Delmar/Cengage Learning*)

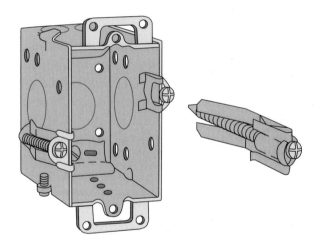

FIGURE 2-28 These boxes are commonly used in remodel work. The box is installed from the front, then the screws are tightened. The box is adequately supported. (*Delmar/Cengage Learning*)

work where the boxes are supported by the framing members, be careful not to hang heavy luminaires from them. Be sure that there is proper support if a ceiling fan is to be installed. This is covered in Chapter 9.

Spread of Fire

To protect lives and property, fire must be contained and not be allowed to spread. In addition to

the *NEC*, you must also become familiar with building codes.

Building codes such as the International Code Council (ICC) *International Residential Code* generally do not require fire-resistance-rated walls and ceilings in one- and two-family buildings. Some communities have amendments to the basic building codes, such as requiring ⅝-in. (16-mm) gypsum wallboard [instead of the normal ½-in. (13-mm)] on the ceilings and on the walls between the garage and living areas, and requiring *draft stopping* where ducts, cables, and piping run through bottom (sole) and top plates. Draft stopping is generally done by the insulation contractor. Townhouses, condominiums, and other multi-occupancy buildings do have fire-resistance rating requirements. To avoid costly mistakes, check with your local building officials to determine exactly what the requirements are!

In walls, partitions, and ceilings that are fire-resistance-rated, special consideration must be given when installing wall and ceiling electrical boxes "back to back" or in the same stud or joist space of common walls or ceilings, as might be found in multifamily buildings. Electrical wall boxes installed back to back or installed in the same stud or joist space defeat the fire-resistance rating of the wall or ceiling.

When discussing fire-rated assemblies, the term *membrane penetration* refers to penetrating one side of the fire-rated assembly. The term *through penetration* refers to penetrating entirely through the fire-rated assembly.

What Is Fire-Resistance Rating?

A building assembly's (walls, ceilings, partitions, gypsum, drywall, plaster, etc.) fire-resistance rating refers to the period of time the assembly will serve as a barrier to the spread of fire. It is an indication of how long it can hold back a fully developed blaze before it spreads to adjacent areas of the building and how long the assembly can function structurally after being exposed to a fire.

While undergoing specific tests, passage of flames and/or structural collapse are the determining factors for establishing fire-resistance ratings.

To be truly fire-resistance-rated, the construction and materials must conform to very rigid requirements, as set forth in the UL Standards.

Ratings of fire-resistance materials are expressed in hours. Terms such as a 1-hour fire-resistance rating, 2-hour fire-resistance rating, and so on, are used.

UL Standard 263 *Fire Tests of Building Construction and Materials* and NFPA 251 cover fire-resistance ratings. In UL Standard 263, and in all building codes, you will find requirements such as "the surface area of individual metallic outlet or switch boxes shall not exceed 16 square inches" (i.e., a 4-in. square box is 4 in. × 4 in. = 16 in.²); "the aggregate surface area of the boxes shall not exceed 100 in.² per 100 ft² of wall surface"; "boxes located on opposite sides of walls or partitions shall be separated by a minimum horizontal distance of 24 in. (600 mm)"; "the metallic outlet or switch boxes shall be securely fastened to the studs and the opening in the wallboard facing shall be cut so that the clearance between the box and the wallboard does not exceed ¹⁄₈ in. (3 mm)"; and "the boxes shall be installed in compliance with the *NEC.*"

The UL *Fire Resistance Directory* covers fire-resistant materials and assemblies and contains a rather detailed listing of *Outlet Boxes and Fittings Classified for Fire Resistance (QBWY)*, showing all manufacturers' product part numbers for *non-metallic boxes* to be installed in walls, partitions, and ceilings that meet the fire-resistance standard. This category also covers special-purpose boxes for installation in floors.

When using nonmetallic boxes in fire-resistance rated walls, the restrictions are more stringent than for metallic boxes. Nonmetallic boxes shall be marked with

- UL in a circle (or other listing mark from an NRTL) located in the base of the box

- the hour rating (1 hour or 2 hours)

- the letters F, W, and/or C where F = floor, W = wall, and C = ceiling

Nonmetallic boxes are UL Classified for use in specific fire-resistant rated assemblies. The Classification data shows spacing restrictions and installation details.

Molded fiberglass outlet and device boxes have become available and are listed for use in fire-resistance-rated partitions without the use of putty pads, mineral wool batts, or fiberglass batts, provided a minimum horizontal distance of 3 in.

(75 mm) is maintained between boxes and the boxes are not back to back. See Figure 2-31(B).

If you are dealing with fire-resistance-rated construction, be sure to carefully read the box manufacturer's instructions and "Listing" details.

For both metallic and nonmetallic electric outlet boxes, the *minimum* distance between boxes on opposite sides of a fire-rated wall or partition is 24 in. (600 mm). There is an exception, which is discussed later.

For both metallic and nonmetallic electric outlet boxes, the maximum gap between the box and the wall material is ¹⁄₈ in. (3 mm). This holds true for fire-resistance-rated walls and non-fire-resistance-rated walls.

To maintain the hourly fire rating of a fire-rated wall that contains electrical outlet and switch boxes, the UL *Fire Resistance Directory* lists an *intumescent* (expands when heated) fire-resistant material that comes in "pads."

These "putty pads" are moldable and can be used to wrap metal electrical boxes or inserted on the inside back wall of metal electrical boxes. When exposed to fire, the material expands and chars, sealing off the opening to prevent the spread of flames and limit the heat transfer from the fire-resistance-rated side of the wall to the non-fire-resistance-rated side of the wall. It is not to be used on nonmetallic boxes.

The ¹⁄₈-in. (3-mm) thickness provides a 1-hour fire-resistance rating. A 2-hour rating is obtained with ¼-in. (6-mm) thickness.

When this material is properly installed, the 24-in. (600-mm) separation between boxes is not required. However, the electrical outlet boxes *must not* be installed back to back. Using this material meets the requirements of *NEC 300.21* of the *NEC* and other building codes.

The ICC *International Residential Code* states that "fireblocking" be provided to cut off all concealed draft openings (both vertical and horizontal) and to form an effective fire barrier between stories, and between a top story and the roof space. This includes openings around vents, pipes, ducts, chimneys, and fireplaces at ceiling and floor levels. Oxygen supports fire. Firestopping cuts off or significantly reduces the flow of oxygen. Most communities that have adopted the ICC code require

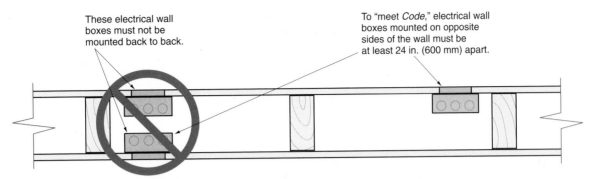

These electrical wall boxes must not be mounted back to back.

To "meet *Code*," electrical wall boxes mounted on opposite sides of the wall must be at least 24 in. (600 mm) apart.

This side of wall has a 1-hour fire-resistance rating.

FIGURE 2-29 Building codes prohibit electrical wall boxes to be mounted back to back in fire-resistance-rated walls unless specific installation procedures, as illustrated in Figures 2-30, 2-31, and 2-32, are followed. One of these requirements is that electrical wall boxes must be kept at least 24 in. (600-mm) apart. This is to maintain the integrity of the fire-resistance rating of the wall. (*Delmar/Cengage Learning*)

firestopping around electrical conduits. In residential work, this is generally done by the insulation installer, before the drywall installer closes up the walls.

Be careful when roughing in wall boxes where wood paneling is to be installed. Some communities require that ½-in. (13-mm) gypsum board be installed on the studs behind finished wood paneling to establish a fire-resistance rating. Remember that where the finish wall or ceiling surface is combustible material such as wood paneling, the front edge of the box, plaster ring, domed cover, extension ring, or listed extender shall be set flush with the wall surface. Refer to *NEC 314.20* and Figure 2-21.

The previous discussion makes it clear that the electrician must become familiar with certain building codes in addition to the *NEC*. Knowing the *NEC* is not enough!

It is extremely important to check with the electrical inspector and/or building official for complete clarification on the matter of fire-resistance requirements before you find yourself cited with serious violations and lawsuits regarding noncompliance with fire codes. Changes can be costly and will delay completion of the project. Improper wiring is hazardous. The *NEC* addresses this issue in *300.21*.

Figures 2-29, 2-30, 2-31, and 2-32 illustrate the basics for installing electrical outlet and switch

This side of wall is non-fire-resistance-rated.

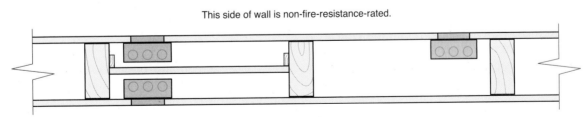

This side of wall has a 1-hour fire-resistance rating.

FIGURE 2-30 Building codes permit boxes to be mounted back to back or in the same wall cavity when the fire-resistance rating is maintained. In this detail, gypsum board has been installed in the wall cavity between the boxes, creating a new fire-resistance barrier. Verify that this is acceptable to the building authority in your area. (*Delmar/Cengage Learning*)

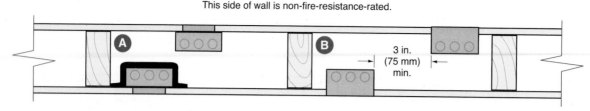

This side of wall is non-fire-resistance-rated.

3 in. (75 mm) min.

This side of wall has a 1-hour fire-resistance rating.

FIGURE 2-31   Building codes permit electrical outlet and device boxes to be mounted in the same stud space under certain conditions. (A) Here we see "Wall Opening Protective Material" packed around one outlet box. (B) Here we see molded fiberglass boxes in the same stud space with a minimum horizontal clearance of 3 in. (75 mm). These boxes are listed for this application. In both (A) and (B), the boxes are not permitted to be back to back. (*Delmar/Cengage Learning*)

FIGURE 2-32   This illustration shows an electrical outlet box that has been covered with "Listed" fire-resistance "putty pad" to retain the wall's fire-resistance rating per building codes. (*Delmar/Cengage Learning*)

boxes in fire-resistance-rated partitions. This is an extremely important issue. If in doubt, check with your local electrical inspector and/or building official.

BOXES FOR CONDUIT WIRING

Some local electrical ordinances require a metal raceway rather than cable wiring. Conduit wiring is discussed in Chapter 18. Examples of conduit fill (how many wires are permitted in a given size conduit) are presented.

When a metal raceway is installed in a residence, it is quite common to use 4-in. square boxes trimmed with suitable plaster covers, Figure 2-33(A). There are sufficient knockouts in the top, bottom, sides, and back of the box to permit a number of EMTs to run to the box. Plenty of room is available for the conductors and wiring devices. Note how easily these 4-in. square outlet boxes can be mounted back to back by installing a small fitting between the boxes. This is illustrated in Figure 2-33(B) and 2-33(C). These boxes are available in 1½ in. (38 mm) and 2⅛ in. (54 mm) depths.

Four-inch square outlet boxes can be trimmed with 1-gang or 2-gang plaster rings where wiring devices will be installed. Where luminaires will be installed, a plaster ring having a round opening should be installed.

Any and all unused openings in electrical equipment must be closed! This includes unused knockouts in boxes, spaces for circuit breakers in a panel, meter socket enclosures, and all other similar

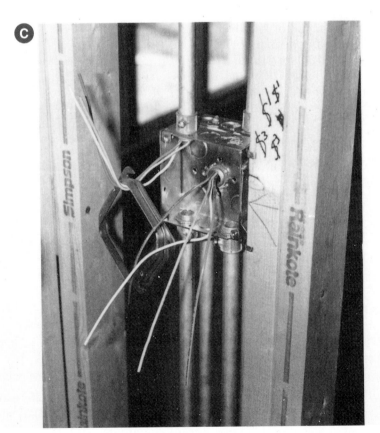

FIGURE 2-33 (A) shows electrical metallic tubing run to a 4-in. square box. (B) and (C) show 4-in. square boxes attached together for "back-to-back" installations. When mounting boxes back to back, be sure to consider possible transfer of sound between the rooms. IMPORTANT: See text ("Spread of Fire") for details of fire-resistance ratings and the restrictions related to installing electrical boxes back to back. (*Delmar/Cengage Learning*)

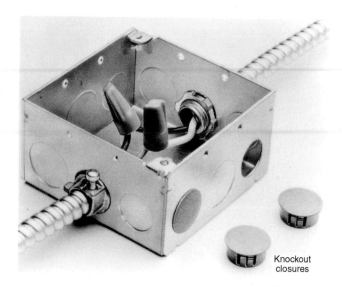

Knockout
closures

FIGURE 2-34 Unused openings in boxes must be closed according to *110.12(A)* and *408.7*. This is done to contain electrical short-circuit problems inside the box or panel and to keep rodents out. Note that the conductors *do not* meet the 3-in. (75-mm) and 6-in. (150-mm) conductor length requirements of *300.14*. This is a good example of a *Code violation*. See Figure 2-2 for proper installation regarding conductor length at boxes. (*Delmar/Cengage Learning*)

electrical equipment. The closing material shall provide substantially the same protection as the wall of the equipment. See *110.12(A)*, *408.7*, and Figure 2-34.

Openings that are intended for mounting purposes, for the operation of the equipment, or as part of the design of listed equipment do not have to be closed.

Close the Gap around Boxes!

NEC 314.21 requires that openings around electrical boxes shall be repaired so that *there will be no gaps or open space greater than* ¹/₈ *in. (3 mm) at the edge of the box or fitting*. This is to minimize the spread of fire. The UL *Fire Resistance Directory* mirrors the *NEC* rule by stating that "the outlet or switch boxes shall be securely fastened to the studs and the opening in the wallboard facing shall be cut so that the clearance between the box and the wallboard does not exceed ¹/₈ in." See Figure 2-21(B).

This is easier said than done. The electrician installs the wiring (rough-in), followed by an inspection by the electrical inspector. Next comes the drywall/plaster/panel installer, who many times cuts out the box openings much larger than the outlet or switch box. In most cases, the walls and ceilings are painted or wallpapered before the electrician returns to install the receptacles, switches, and luminaires. For gaps or openings greater than ¹/₈ in. (3 mm) around the electrical boxes, should the electrician repair the gap with patching plaster, possibly damaging or marring the finished wall? Whose responsibility is it? Is it the electrician's? Is it the drywall/plaster/panel installer's?

The electrician should check with the general contractor to clarify who is to be responsible for seeing to it that gaps or openings around electrical outlet and switch boxes do not exceed ¹/₈ in. (3 mm). *NEC 314.21* puts the responsibility on the electrician.

YOKE

We use the term *yoke*. The *NEC* uses the term *yoke*. Yet, the *NEC* does not define a *yoke*. What is a yoke? It is the metal or polycarbonate mounting strap on which a wiring device or devices are attached. The yoke in turn is attached to a device box or device cover with No. 6-32 screws. One yoke may have one, two, or three wiring devices attached to it.

Visit an electrical distributor or home center. You will find many combination wiring devices on a single yoke or strap. Selecting one of these combination devices may eliminate the need to use combinations of interchangeable wiring devices.

SPECIAL-PURPOSE OUTLETS

Special-purpose outlets are usually indicated on the plans. These outlets are described by a notation and are usually detailed in the specifications. The plans indicate special-purpose outlets by a triangle inside a circle with subscript letters. In some cases, a subscript number is added to the letter.

When a special-purpose outlet is indicated on the plans or in the specifications, the electrician must check for special requirements. Such a requirement may be a separate circuit, a special 240-volt circuit, a special grounding or polarized receptacle, or other preparation.

A set of plans and specifications for this residence is found at the end of this text. The specifications include a *Schedule of Special Purpose Outlets.*

NUMBER OF CONDUCTORS IN BOX

NEC 314.16 dictates that outlet boxes, switch boxes, and device boxes should be large enough to provide ample room for the wires in that box, without having to jam or crowd the wires into the box. Jamming the

Table 314.16(A) Metal Boxes

Box Trade Size			Minimum Volume		Maximum Number of Conductors* (arranged by AWG size)						
mm	in.		cm³	in.³	18	16	14	12	10	8	6
100 × 32	(4 × 1¼)	round/octagonal	205	12.5	8	7	6	5	5	5	2
100 × 38	(4 × 1½)	round/octagonal	254	15.5	10	8	7	6	6	5	3
100 × 54	(4 × 2⅛)	round/octagonal	353	21.5	14	12	10	9	8	7	4
100 × 32	(4× 1¼)	square	295	18.0	12	10	9	8	7	6	3
100 × 38	(4 × 1½)	square	344	21.0	14	12	10	9	8	7	4
100 × 54	(4 × 2⅛)	square	497	30.3	20	17	15	13	12	10	6
120 × 32	(4¹¹⁄₁₆ × 1¼)	square	418	25.5	17	14	12	11	10	8	5
120 × 38	(4¹¹⁄₁₆ × 1½)	square	484	29.5	19	16	14	13	11	9	5
120 × 54	(4¹¹⁄₁₆ × 2⅛)	square	689	42.0	28	24	21	18	16	14	8
75 × 50 × 38	(3 × 2 × 1½)	device	123	7.5	5	4	3	3	3	2	1
75 × 50 × 50	(3 × 2 × 2)	device	164	10.0	6	5	5	4	4	3	2
75× 50 × 57	(3× 2 × 2¼)	device	172	10.5	7	6	5	4	4	3	2
75 × 50 × 65	(3 × 2 × 2½)	device	205	12.5	8	7	6	5	5	4	2
75 × 50 × 70	(3 × 2 × 2¾)	device	230	14.0	9	8	7	6	5	4	2
75 × 50 × 90	(3 × 2 × 3½)	device	295	18.0	12	10	9	8	7	6	3
100 × 54 × 38	(4 × 2⅛ × 1½)	device	169	10.3	6	5	5	4	4	3	2
100 × 54 × 48	(4 × 2⅛ × 1⅞)	device	213	13.0	8	7	6	5	5	4	2
100 × 54 × 54	(4 × 2⅛ × 2⅛)	device	238	14.5	9	8	7	6	5	4	2
95 × 50 × 65	(3¾ × 2 × 2½)	masonry box/gang	230	14.0	9	8	7	6	5	4	2
95 × 50 × 90	(3¾ × 2 × 3½)	masonry box/gang	344	21.0	14	12	10	9	8	7	4
min. 44.5 depth	FS — single cover/gang (1¾)		221	13.5	9	7	6	6	5	4	2
min. 60.3 depth	FD — single cover/gang (2⅜)		295	18.0	12	10	9	8	7	6	3
min. 44.5 depth	FS — multiple cover/gang (1¾)		295	18.0	12	10	9	8	7	6	3
min. 60.3 depth	FD — multiple cover/gang (2⅜)		395	24.0	16	13	12	10	9	8	4

*Where no volume allowances are required by 314.16(B)(2) through (B)(5).

Reprinted with permission from NFPA 70-2011.

Table 314.16(B) Volume Allowance Required per Conductor

Size of Conductor (AWG)	Free Space Within Box for Each Conductor	
	cm³	in.³
18	24.6	1.50
16	28.7	1.75
14	32.8	2.00
12	36.9	2.25
10	41.0	2.50
8	49.2	3.00
6	81.9	5.00

Reprinted with permission from NFPA 70-2011.

wires into the box can not only damage the insulation on the wires, but can result in heat buildup in the box that can further damage the insulation on the wires. The *Code* specifies the maximum number of conductors allowed in standard metal outlet, device, and junction boxes; see *Tables 314.16(A)* and *(B)*. Nonmetallic boxes are marked by the manufacturer with their cubic-inch capacity.

When conductors are the same size, the proper metal box size can be selected by referring to *Table 314.16(A)*. When conductors are of different sizes and nonmetallic boxes are installed, refer to *Table 314.16(B)* and use the cubic-inch volume for the particular size of wire being used.

Table 314.16(A) and *Table 314.16(B)* do not take into consideration the space taken by luminaire studs, cable clamps, hickeys, switches, pilot lights, or receptacles that may be in the box. These require additional space. Table 2-1 shows the additional allowances that must be considered for different situations.

SELECTING THE CORRECT SIZE BOX

Selecting the correct size box depends on the mix of conductors and wiring devices that will be contained in the particular box. Here are some examples.

When Using a Nonmetallic Box Where All Conductors Are the Same Size. A nonmetallic box is marked as having a volume of 22.8 in.[3] The box contains no luminaire stud or cable clamps. How many 14 AWG conductors are permitted in this box?

From *Table 314.16(B)*, the volume requirement for a 14 AWG conductor is 2.00 in.[3] Therefore, the maximum number of 14 AWG conductors permitted in this box is

$$\frac{22.8}{2} = 11.4 \text{ (Round down to 11 conductors.)}$$

When Using a Metal Device Box Where All Conductors Are the Same Size. A box contains one wiring device and two internal cable clamps. The wiring method is two 12/2 NM w/ground cables.

TABLE 2-1

Quick checklist for possibilities to be considered when determining proper size boxes. See *314.16(B)(1)* through *314.16(B)(5)*.

• If box contains no fittings, devices, fixture studs, cable clamps, hickeys, switches, receptacles, or equipment grounding conductors:	• Refer directly to *Table 314.16(A)* or *Table 314.16(B)*.
• **Clamps.** If box contains one or more internal cable clamps:	• Add a single-volume based on the largest conductor in the box.
• **Support Fittings.** If box contains one or more luminaire studs or hickeys:	• Add a single-volume for each type based on the largest conductor in the box.
• **Device or Equipment.** If box contains one or more wiring devices on a yoke:	• Add a double-volume for each yoke based on the largest conductor connected to a device on that yoke. Some large wiring devices, such as a 30-ampere dryer receptacle, require a 2-gang box.
• **Equipment Grounding Conductors.** If a box contains one or more equipment grounding conductors:	• Add a single-volume based on the largest equipment grounding conductor in the box.
• **Isolated Equipment Grounding Conductor.** If a box contains one or more additional "isolated" (insulated) equipment grounding conductors as permitted by 250.146(D) for "noise" reduction:	• Add a single-volume based on the largest equipment grounding conductor in the box.
• For conductors less than 12 in. (300 mm) long that are looped or coiled in the box without being spliced:	• Add a single-volume for each conductor that is looped or coiled through the box.
• For conductors 12 in. (300 mm) or longer that are looped or coiled in the box without being spliced:	• Add a double-volume for each conductor that is looped or coiled through the box.
• For conductors that originated outside of the box and terminate inside the box:	• Add a single-volume for each conductor that originates outside the box and terminates inside the box.
• If no part of the conductor leaves the box, as with a "jumper" wire used to connect three wiring devices on one yoke, or pigtails:	• Don't count this (these). No additional volume required.
• For small equipment grounding conductors or not more than four conductors smaller than 14 AWG that originate from a luminaire canopy or similar canopy (like a fan) and terminate in the box:	• Don't count this (these). No additional volume required.
• For small fittings, such as locknuts, bushings, and wire connectors:	• Don't count this (these). No additional volume required.

four 12 AWG conductors	4
one wiring device	2
two cable clamps (only count one)	1
two 12 AWG equipment grounding conductors (only count one)	<u>1</u>
Total	8

Referring to *Table 314.16(A)* and *Table 314.16(B)*, a $3 \times 2 \times 3\frac{1}{2}$ device box is suitable for this example.

When Using a Metal or Nonmetallic Box Where the Conductors Are Different Sizes. What is the minimum cubic-inch volume required for a box that will contain one internal cable clamp, one switch, and one receptacle mounted on one yoke? One 14/2 Type AC and one 12/2 Type AC cable are used. There are no separate equipment grounding conductors involved as the bonding wire in the armored cables are laid back over the armor and do not take up any space within the box.

two 14 AWG conductors @ 2 in.³ per conductor	= 4.00 in.³
two 12 AWG conductors @ 2.25 in.³ per conductor	= 4.50 in.³
two cable clamp (only count one) @ 2.25 in.³	= 2.25 in.³
one switch and one receptacle on one yoke @ 2.25 in.³ × 2	= <u>4.50 in.³</u>
Total	15.25 in.³

Select a box having a minimum volume of 15.25 in.³ of space. The cubic-inch volume might be marked on the box; otherwise refer to the second column of *Table 314.16(A)* entitled *Minimum Volume*.

The *Code* in *314.16(A)(2)* requires that all boxes *other* than those listed in *Table 314.16(A)* be durably and legibly marked by the manufacturer with their cubic-inch capacity. Nonmetallic boxes have their cubic-inch volume marked on the box. When sectional boxes are ganged together, the calculated volume is the total cubic-inch volume of the assembled boxes.

When Installing a Box That Will Have a Raised Cover Attached to It. When marked with their cubic-inch volume, the *additional* space provided by plaster rings, domed covers, raised covers, and extension rings is permitted to be used when determining the overall volume. This is illustrated in Figure 2-35.

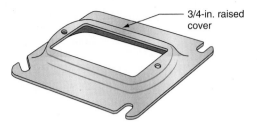

Additional wiring space provided by the raised cover is 2 in. x 3 in. x 3/4 in. = 4 1/2 in.³

FIGURE 2-35 Raised cover. Raised covers are sometimes called plaster rings. (*Delmar/Cengage Learning*)

How many 12 AWG conductors are permitted in this box and raised plaster ring? Refer to Figure 2-36 and *Table 314.16(A)* and *Table 314.16(B)*.

$$\frac{25.5 \text{ in.}^3 \text{ of total space}}{2.25 \text{ in.}^3 \text{ per 12 AWG conductor}} = \text{Eleven 12 AWG conductors}$$

This calculation actually resulted in 11.33. The conductor fill numbers in *Table 314.16(A)* were created by dropping any excess above the whole number after the calculations were made. Following the same procedure, this conductor fill calculation dropped off the excess.

Electricians and electrical inspectors have become very aware of the fact that GFCI receptacles, dimmers, and timers take up a lot more space than regular wiring devices. Therefore, it is a good practice to install boxes that will provide plenty of room for the wires, instead of pushing, jamming, and crowding the wires into the box.

Figure 2-37 shows that transformer leads 18 AWG or larger must be counted when selecting a proper size box.

The minimum required size of equipment grounding conductors is shown in *Table 250.122*. The equipment grounding conductors are the same size as the circuit conductors in cables having 14-AWG, 12-AWG, and 10-AWG circuit conductors. Thus, for normal house wiring, box sizes can be calculated using *Table 314.16(A)*. Refer to Figure 2-38.

Figure 2-20, a quick-check box selection guide, shows some of the most popular types of boxes used in residential wiring. This guide also shows the box's cubic-inch volume.

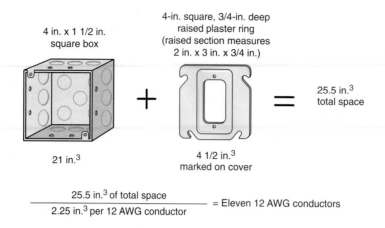

FIGURE 2-36 When marked with their cubic-inch volume, the volume of the cover may be added to the volume of the box to determine the total cubic-inch volume. (*Delmar/Cengage Learning*)

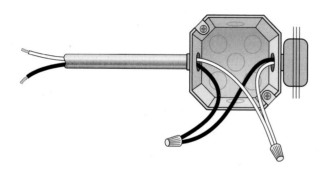

FIGURE 2-37 Transformer leads 18 AWG or larger must be counted when selecting the proper size box, *314.16(B)(1)*. In the example shown, the box contains four conductors. (*Delmar/Cengage Learning*)

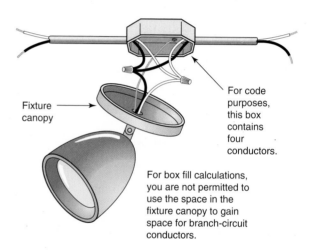

FIGURE 2-38 Four or fewer luminaire conductors, smaller than 14 AWG, and/or small equipment grounding conductors that originate in the luminaire and terminate in the box, need not be counted when calculating box fill, *314.16(B)(1), Exception*. (*Delmar/Cengage Learning*)

Figure 2-39 illustrates typical wiring using cable and conduit. Note in this figure how the conductor count is determined.

Using electrical metallic tubing (EMT) makes it possible to "loop" conductors through the box, which counts as one conductor only. Looping conductors through a box is not possible when using cable as the wiring method.

The basic rules for box fill are summarized in Table 2-1. Boxes that are intended to support ceiling fans are discussed in detail in Chapter 9.

Depth of Box—Watch Out!

Determining the cubic volume of a box is not the end of the story. Outlet and device boxes shall be deep enough so that conductors will not be damaged when installing the wiring device or equipment into the box. Merely calculating and providing the proper volume of a box is not always enough. The volume calculation might prove adequate, yet the size (depth) of the wiring

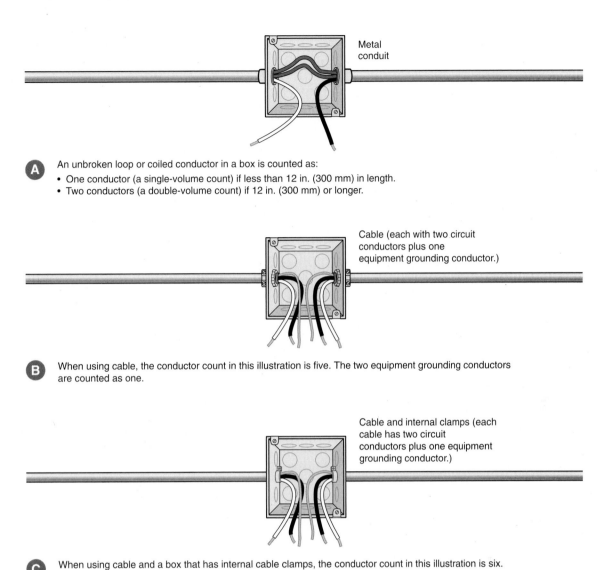

A An unbroken loop or coiled conductor in a box is counted as:
  • One conductor (a single-volume count) if less than 12 in. (300 mm) in length.
  • Two conductors (a double-volume count) if 12 in. (300 mm) or longer.

Metal conduit

Cable (each with two circuit conductors plus one equipment grounding conductor.)

B When using cable, the conductor count in this illustration is five. The two equipment grounding conductors are counted as one.

Cable and internal clamps (each cable has two circuit conductors plus one equipment grounding conductor.)

C When using cable and a box that has internal cable clamps, the conductor count in this illustration is six. The two equipment grounding conductors are counted as one. The two cable clamps are counted as one.

FIGURE 2-39   Example of conductor count for both metal conduit and cable installations. (*Delmar/Cengage Learning*)

device or equipment might be such that conductors behind it may possibly be damaged. See *NEC 314.24*.

Width of Box—Watch Out!

Large wiring devices, such as a 30-ampere, 3-pole, 4-wire dryer or a 50-ampere-range receptacle will not fit into a single-gang box that is 2 in. (50.8 mm) wide. These receptacles measure 2.10 in. (53.3 mm) in width. Likewise, a 50-ampere, 3-pole, 4-wire receptacle measures 2.75 in. (69.9 mm) in width. Consider using a 4-in. (101.6-mm) square box with a 2-gang plaster device ring for flush mounting or a 2-gang raised cover for surface mounting, or use a 2-gang device box. The center-to-center mounting holes of both the 30-ampere and the 50-ampere receptacles are 1.81 in. (46.0 mm) apart, exactly matching the center-to-center holes of a 2-gang plaster ring, raised cover, or 2-gang device box, so mounting the devices is not a problem. See *NEC 314.16(B)(4)*.

Means for Connecting Equipment Grounding Conductors in a Box

NEC 314.4 requires that all metal boxes be grounded and bonded. The specific requirements for specific applications are found in *Article 250, Parts I, IV, V, VI, VII,* and *X.*

NEC 314.40(D) requires that *a means shall be provided in each metal box for the connection of* **equipment grounding conductors.** Generally, these are tapped No. 10-32 holes in the box that are marked GR, GRN, GRND, or similar identification. The screws used for terminating the equipment grounding conductor must be hexagonal and must be green. Electrical distributors sell green, hexagonal No. 10-32 screws for this purpose.

HEIGHT OF RECEPTACLE OUTLETS

There are no hard-and-fast rules for locating most outlets. A number of conditions determine the proper height for a switch box. For example, the height of the kitchen counter backsplash determines where the switches and receptacle outlets are located between the kitchen countertop and the cabinets.

The residence featured in this text is electrically heated, which is discussed in Chapter 23. The type of electric heat could be

- electric furnace (as in this text).
- electric resistance heating buried in ceiling plaster or "sandwiched" between two layers of drywall material.
- electric baseboard heaters.
- heat pump.

Let us consider the electric baseboard heaters. In most cases, the height of these electric baseboard units from the top of the unit to the finished floor seldom exceeds 6 in. (150 mm). The important issue here is that the manufacturer's receptacle accessories may have to be used to conform to the receptacle spacing requirements, as covered in *210.52.*

Electrical receptacle outlets are not permitted to be located above an electric baseboard heating unit. Refer to the section "Location of Electric Baseboard Heaters in Relation to Receptacle Outlets," in Chapter 23.

It is common practice among electricians to consult the plans and specifications to determine the proper heights and clearances for the installation of electrical devices. The electrician then has these dimensions verified by the architect, electrical engineer, designer, or homeowner. This practice avoids unnecessary and costly changes in the locations of outlets and switches as the building progresses.

POSITIONING OF RECEPTACLES

Although no *NEC* rules exist on positioning receptacles, there is always the possibility of a metal wall plate coming loose and falling downward onto the blades of an attachment plug cap that is loosely plugged into the receptacle. This can create sparks that could result in a fire, burns, and potential shock. Positioning the *equipment grounding* conductor slot or the *grounded* conductor slot on top will minimize these hazards.

The equipment grounding blade on a 3-wire attachment plug cap is longer than the other blades. Thus, when inserted into a receptacle, in agreement with *250.124(A)*, the equipment grounding blade is first to make and last to break.

There is also the argument that with the equipment grounding slot to the bottom, if a plug cap comes loose and starts to fall out, the longer equipment grounding blade would be the last connection to be disconnected. See Figures 2-40, 2-41, 2-42, and 2-43.

Recommended

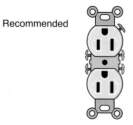

FIGURE 2-40 Grounding hole to the top. A loose metal faceplate could fall onto the grounding blade of the attachment plug cap, but no sparks would fly. (*Delmar/Cengage Learning*)

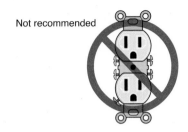

FIGURE 2-41   Grounding hole to the bottom. A loose metal faceplate could fall onto both the grounded neutral and "hot" blades. Sparks would fly. (*Delmar/Cengage Learning*)

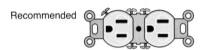

FIGURE 2-42   Grounded neutral blades on top. A loose metal faceplate could fall onto these grounded neutral blades. No sparks would fly. (*Delmar/Cengage Learning*)

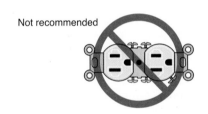

FIGURE 2-43   "Hot" terminal on top. A loose metal faceplate could fall onto these live blades. If this were a split-wired receptacle fed by a 3-wire, 120/240-volt circuit, the short would be across the 240-volt line. Sparks would fly. (*Delmar/Cengage Learning*)

TYPICAL HEIGHTS FOR SWITCHES AND OUTLETS

There are no hard-and-fast *Code* rules regarding the heights for switches and receptacles; local preferences prevail. Typical heights for switches and outlets are shown in Table 2-2. These dimensions usually are satisfactory. However, the electrician must check the blueprints, specifications, and details for measurements that may affect the location of a particular outlet or switch. The cabinet spacing, available space between the countertop and the cabinet, and the tile height may influence the location of the outlet or switch. For example, if the top of the wall tile is exactly 4 ft (1.2m) from the finished floor line, a wall switch should not be mounted 4 ft (1.2 m) to center. This is considered poor workmanship. The switch should be located entirely within the tile area or entirely out of the tile area, as in Figure 2-44. This situation requires the full cooperation of all craftspersons involved in the construction job.

Faceplates

Faceplates for switches and receptacles shall be installed so as to completely cover the wall opening and seat against the wall surface, *404.9(A)* and *406.5*.

Faceplates should be level! No matter how good the wiring inside the wall is, no matter how good the connections and splices are, no matter how proper the equipment grounding is, the only thing the homeowner will see is the wiring devices and the faceplates. That establishes the impression the homeowner has of the overall wiring installation. Because of the elongated slots in the yoke of a wiring device, it is rather easy to level a wiring device and faceplate on a single-gang box even if the box is not level. This is not true for multigang boxes. It is extremely important to make sure multi-gang boxes are level. If a multigang box is not level, it will be virtually impossible to make the wiring devices and faceplate level.

Another nice detail that can improve the appearance is to align the slots in all of the faceplate mounting screws in the same direction, as in Figure 2-45. This is not a *Code* rule—just a nice little finishing touch.

TABLE 2-2

Typical heights for switches, receptacles, and wall luminaire outlets. Local height preferences prevail.

SWITCHES	
Regular	46 in. (1.15 m). Some homeowners want wall switches mounted 30–36 in. (750–900 mm) so as to be easy for children to reach and easy for adults to reach when carrying something. The choice is yours!
Between counter and kitchen cabinets. Depends on backsplash.	44–46 in. (1–1.15 m)
RECEPTACLE OUTLETS	
Regular	12 in. (300 mm)
Between counter and kitchen cabinets. Depends on backsplash.	44–46 in. (1–1.15 m)
In garages	46 in. (1.15 m) Minimum 18 in. (450 mm)
In unfinished basements	46 in. (1.15 m)
In finished basements	12 in. (300 mm)
Outdoors (above grade or deck)	18 in. (450 mm)
WALL LUMINAIRE OUTLETS	
Outdoor entrance bracket luminaires	72 in. (1.8 m). If luminaire is "upward," mount wall box lower. If luminaire is "downward," mount wall box higher. A good appearance for typical outdoor wall lanterns is when the center of wall box is approximately 10 in. (0.254 m) below the top of the door. The final choice is yours.
Inside wall brackets	5 ft (1.5 m)
Side of medicine cabinet or mirror above medicine cabinet or mirror	You need to know the measurement of the medicine cabinet. Check the rough-in opening of medicine cabinet measurement of mirror. Mount electrical wall box approx. 6 in. (150 mm) to center above rough-in opening or mirror. Medicine cabinets that come complete with luminaires have a wiring compartment with a conduit knockout(s) in which the supply cable or conduit is secured using the appropriate fitting. Many strip luminaires have a backplate with a conduit knockout in which the supply cable or conduit is secured using the appropriate fitting. Where to bring in the cable or conduit takes careful planning.

Note: All dimensions are from finished floor to center of the electrical box. If possible, try to mount wall boxes for luminaires based on the type of luminaire to be installed. Verify all dimensions before "roughing in." If wiring for physically handicapped, the above heights may need to be lowered in the case of switches and raised in the case of receptacles.

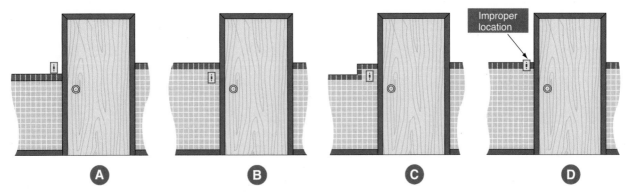

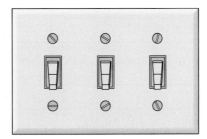

Where a wall is partially tiled, a switch or convenience outlet must be located entirely out of the tile area (A) or entirely within the tile area (B), (C).

The faceplate in (D) does not "hug" the wall properly. This installation is considered unacceptable by most electricians and is a violation of the last sentence of *NEC 404.9(A)*.

FIGURE 2-44   Locating an outlet on a tiled wall. (*Delmar/Cengage Learning*)

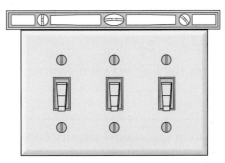

Align screw–slots
like this,

Not like this.

FIGURE 2-45   Aligning the slots of the faceplate. Mounting screws in the same direction makes the installation look neater. Make sure the faceplate is level. (*Delmar/Cengage Learning*)

REVIEW

Note: For assistance in finding the answers to these review questions, refer to the *NEC*, the plans in the back of this text, and the specifications found in the Appendix.

⌐∿⌐ PART 1—ELECTRICAL FEATURES

1. What does a plan show about electrical outlets? _____

_____

_____

2. What is an outlet? _____

_____

_____

3. Match the following switch types with the proper symbol. _____
    a. single-pole   **S$_p$**
    b. three-way   **S$_4$**
    c. four-way   **S**
    d. single-pole with pilot light   **S$_3$**

4. The plans show curved lines running between switches and various outlets. What do these lines indicate? _____

_____

5. Why are the lines referred to in Question 4 usually curved?_____

6. a. What are junction boxes used for? _____

    b. Are junction boxes normally used in wiring the first floor? Explain. _____

    _____

    _____

    c. Are junction boxes normally used to wire exposed portions of the basement? Explain. _____

    _____

7. How are standard sectional switch (device) boxes mounted? _____

_____

_____

_____

8. a. What is an offset bar hanger? _____

    _____

    _____

    b. What types of boxes may be used with offset bar hangers? _____

    _____

    _____

9. What methods may be used to mount luminaires to an outlet box fastened to an offset bar hanger? _____

   _____

   _____

10. What advantage does a 4-in. octagon box have over a 3¼-in. octagon box?

    _____

    _____

11. What is the size of the opening of a switch (device) box for a single device?

    _____

12. The space between a door casing and a window casing is 3½ in. (88.9 mm). Two switches are to be installed at this location. What type of switches will be used?

    _____

    _____

13. Three switches are mounted in a 3-gang switch (device) box. The wall plate for this assembly is called a _____ plate.

14. The mounting holes in a device (switch) box are tapped for No. 6-32 screws. The mounting holes in an outlet box are tapped for No. 8-32 screws. The mounting holes in metal boxes for attaching equipment grounding conductors are tapped for _____ screws.

15. a. How high above the finished floor in the living room are switches located?

       _____

    b. How high above the garage floor are switches located?

       _____

16. a. How high above the finished floor in the living room are receptacles located?

       _____

    b. How high above the garage floor are receptacles located?

       _____

17. Outdoor receptacle outlets in this dwelling are located _____ in. above grade.

18. In the spaces provided, draw the correct symbol for each of the descriptions listed in (a) through (r).

a. _____	Lighting panel	i. _____	Weatherproof outlet
b. _____	Clock outlet	j. _____	Special-purpose outlet
c. _____	Duplex outlet	k. _____	Fan outlet
d. _____	Outside telephone	l. _____	Range outlet
e. _____	Single-pole switch	m. _____	Power panel
f. _____	Motor	n. _____	3-way switch
g. _____	Duplex outlet, split-wired	o. _____	Push button
		p. _____	Thermostat
h. _____	Lampholder with pull switch	q. _____	Electric door opener
		r. _____	Multi-outlet assembly

19. The front edge of a box installed in a combustible wall must be _____ with the finished surface.

20. List the maximum number of 12 AWG conductors permitted in a

    a. 4 in. × 1½ in. octagon box. _____

    b. 4¹¹/₁₆ in. × 1½ in. square box. _____

    c. 3 in. × 2 in. × 3½ in. device box. _____

21. When a switch (device) box is nailed to a stud, and the nail runs through the box, the nails must not interfere with the wiring space. To accomplish this, keep the nail:

    a. halfway between the front and rear of the box.

    b. a maximum of ¼ in. (6 mm) from the front edge of the box.

    c. a maximum of ¼ in. (6 mm) from the rear of the box.

22. Hanging a ceiling luminaire directly from a plastic outlet box is permitted only if

    _____

    _____

    _____

    _____

23. It is necessary to count luminaire wires when counting the permitted number of conductors in a box according to *314.16*. True or false? _____

24. *Table 314.16(A)* allows a maximum 10 ten wires in a certain box. However, the box will have two cable clamps and one fixture stud in it. What is the maximum number of wires allowed in this box? _____

    _____

25. When laying out a job, the electrician will usually make a layout of the circuit, taking into consideration the best way to run the cables and/or conduits and how to make up the electrical connections. Doing this ahead of time, the electrician determines exactly how many conductors will be fed into each box. With experience, the electrician will probably select two or three sizes and types of boxes that will provide adequate space to meet *Code*. *Table 314.16(A)* of the *Code* shows the maximum number of conductors permitted in a given size box. In addition to counting the number of conductors that will be in the box, what is the additional volume that must be provided for the following items? Enter *single* or *double* volume allowance in the blank provided.

    a. one or more internal cable clamps: _____ volume allowance.

    b. for a fixture stud: _____ volume allowance.

    c. for one or more wiring devices on one yoke: _____ volume allowance.

    d. for one or more equipment grounding conductors: _____ volume allowance.

26. Is it permissible to install a receptacle outlet above an electric baseboard heater?

    _____

    _____

27. What is the maximum weight of a luminaire permitted to be hung directly from an outlet box in a ceiling? _____ lb.

28. Two 12/2 and two 14/2 nonmetallic-sheathed AWG cables enter a box. Each cable has an equipment grounding conductor. The 12 AWG conductors are connected to a receptacle. Two of the 14 AWG conductors are connected to a toggle switch. The other two 14 AWG conductors are spliced together because they serve as a switch loop. The box contains two cable clamps. Calculate the minimum cubic-inch volume required for the box.

29. Using the same number and size of conductors as in question 28, but using electrical metallic tubing, calculate the minimum cubic-inch volume required for the box. There will be no separate equipment grounding conductors, nor will there be any clamps in the box.

30. To allow for adequate conductor length at electrical outlet and device boxes to make up connections, *300.14* requires that not less than [3 in. (75 mm), 6 in. (150 mm), 9 in. (225 mm)] of conductor length be provided. This length is measured from where the conductor emerges from the cable or raceway to the end of the conductor. For box openings having any dimension less than 8 in. (200 mm), the minimum length of conductor measured from the box opening in the wall to the end of the conductor is [3 in. (75 mm), 6 in. (150 mm), 9 in. (225 mm)]. Circle the correct answers.

31. When wiring a residence, what must be considered when installing wall boxes on both sides of a common partition that separates the garage and a habitable room? _____

    _____

    _____

32. Does the *NEC* allow metal raceways to be used with nonmetallic boxes?

    Yes _____ No _____ , *NEC* _____

PART 2—STRUCTURAL FEATURES

1. To what scale is the basement plan drawn? _____

2. What is the size of the footing for the steel Lally columns in the basement? Refer to Plan 1 of 10. (Lally is the trademark used for a concrete-filled steel cylinder used as a supporting member for wood or steel girders and beams. The Lally column was named after John Lally, born in Ireland in 1859. The term *lolly* is also used.) _____

    _____

3. To what kind of material will the front porch lighting bracket luminaire be attached?

_____

4. Give the size, spacing, and direction of the ceiling joists in the workshop. _____

_____

5. What is the size of the lot on which this residence is located? _____

_____

6. The front of the house is facing which compass direction?_____

7. How far is the front garage wall from the curb? _____

8. How far is the side garage wall from the property lot line?_____

9. How many steel Lally columns are in the basement and what size are they? _____

_____

10. What is the purpose of the I-beams that rest on top of the steel Lally columns?_____

_____

11. To make sure that switch boxes and outlet boxes are set properly in the garage walls and ceilings, we need to know the thickness of the gypsum wallboard used in these locations. In the garage, what is the thickness and type of the gypsum wallboard?

    a. On the "warm walls" of the garage?_____

    b. On the "cold walls" of the garage? _____

    c. On the ceiling of the garage? _____

12. Where is access to the attic provided? _____

13. Give the thickness of the outer basement walls. _____

14. What material is indicated for the foundation walls?_____

15. Where are the smoke detectors located in the basement? _____

    What is the ceiling height in the basement workshop from the bottom of the joists to the floor? _____

    _____

16. Give the size and type of the front door.

    _____

17. What is the stud size for the partitions between the bathrooms in the bedroom area where substantial plumbing is to be installed?_____

    _____

18. Who is to furnish the range hood? _____

    _____

19. Who is to install the range hood?_____

    _____

Determining the Required Number of Lighting Outlets, Receptacle Outlets, and Small-Appliance Branch Circuits

OBJECTIVES

After studying this chapter, you should be able to

- understand the *NEC* requirements for calculating branch-circuit sizing and loading.

- understand the term *volt-amperes per square foot.*

- calculate the occupied floor area of a residence.

- determine the minimum number of lighting and small-appliance branch circuits.

- know where receptacle outlets and lighting outlets are required.

It is standard practice in the design and planning of dwelling units to permit the electrician to plan the circuits. Thus, the residence plans do not include layouts for the various branch circuits. The electrician may follow the general guidelines established by the architect. However, any wiring systems designed and installed by the electrician must conform to the *NEC*, as well as local and state code requirements.

This chapter focuses on lighting branch circuits and small-appliance branch circuits. The circuits supplying the electric range, oven, clothes dryer, and other specific circuitry not considered to be a lighting branch circuit or a small-appliance branch circuit are covered later in this text. Refer to the Index for the specific circuit being examined.

BASICS OF WIRE SIZING AND LOADING

The *NEC* establishes some very important fundamentals that weave their way through the decision-making process for an electrical installation. They are presented here in brief form, and are covered in detail as required throughout this text.

The *NEC* defines a **branch circuit** as *The circuit conductors between the final overcurrent device protecting the circuit and the outlet(s).** See Figure 3-1. In the residence discussed in this text, the wiring to wall outlets, the dryer, the range, and so on, are all examples of a branch circuit.

The *NEC* defines a **feeder** as *All circuit conductors between the service equipment, the source of a separately derived system, or other power supply source and the final branch-circuit overcurrent device.** In the residence discussed in this text, the wiring between Main Panel A and Subpanel B is a feeder.

The ampacity (current-carrying capacity) of a conductor must not be less than the rating of the **overcurrent device** protecting that conductor, *NEC 210.19* and *NEC 210.20*. A common exception to this is a motor branch circuit, where it is quite common to have overcurrent devices (fuses or breakers) sized larger than the ampacity of the conductor. Motors and motor circuits are covered specifically in *NEC Article 430*. The **ampere** rating of the branch-circuit overcurrent protective device (fuse or circuit breaker) determines the rating of the branch circuit. For example, if a 20-ampere conductor is protected by a 15-ampere fuse, the circuit is considered to be a 15-ampere branch circuit, *NEC 210.3*.

Standard branch circuits that serve more than one receptacle outlet or more than one lighting outlet are rated 15, 20, 30, 40, and 50 amperes. A branch circuit that supplies an individual load can be of any ampere rating, *NEC 210.3*.

If the ampacity of the conductor does not match up with a standard rating of a fuse or breaker, the next higher standard size overcurrent device may be used, provided the overcurrent device does not exceed 800 amperes, *NEC 240.4(B)*. This deviation is not permitted if the circuit supplies receptacles where "plug-connected" appliances, and so on, could be used, because too many "plug-in" loads could result in an overload condition, *NEC 240.4(B)(1)*. You may go to the next standard size overcurrent device only when the circuit supplies other than

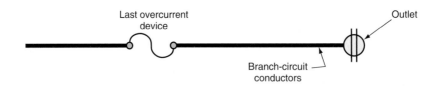

FIGURE 3-1 The branch circuit is that part of the wiring that runs from the final overcurrent device to the outlet. The rating of the overcurrent device, not the conductor size, determines the rating of the branch circuit. (*Delmar/Cengage Learning*)

*Reprinted with permission from NFPA 70-2011.

receptacles for cord-and-plug-connected portable loads.

For instance, when a conductor having an allowable ampacity of 25 amperes [see *NEC Table 310.15(B)(16)*] 14 AWG Type THHN is derated to 70%:

$$25 \times 0.70 = 17.5 \text{ amperes}$$

It is permitted to use a 20-ampere overcurrent device if the circuit supplies *only* fixed lighting outlets or other fixed loads.

If the previous example were to supply receptacle outlets, then the rating of the overcurrent device would have to be dropped to 15 amperes; otherwise it is possible to overload the conductors by plugging in more load than the conductors can safely carry.

The allowable ampacity of conductors commonly used in residential occupancies is found in *NEC Table 310.15(B)(16)*. This includes Type NM cable. It is required to be manufactured with 90°C insulated conductors. Typically, the insulation is Type THHN. As a result, the cable is limited to use in dry locations. *See NEC 334.10(A)(1)*. *NEC 334.80* allows the 90°C ampacity to be used for derating purposes so long as the final ampacity is selected from the 60°C column of *NEC Table 310.15(B)(16)*.

The ampacities in *Table 310.15(B)(16)* are subject to **correction factors** that must be applied if high ambient temperatures are encountered—for example, in attics; see *NEC Table 310.15(B)(2)(a)*.

Conductor ampacities are also subject to a **derating factor** if more than three current-carrying conductors are installed in a single raceway or cable; see *NEC Table 310.15(B)(3)(a)*. See Chapter 18 for complete coverage of correction and derating factors.

Most general-use receptacle outlets in a residence are included in the general lighting load calculations, *NEC 220.14(J)*.

Receptacle outlets connected to the 20-ampere small-appliance branch circuits in the kitchen, dining room, laundry, and workshop are not considered part of the general lighting load. Additional load values must be added into the calculations for these receptacle outlets. This is discussed later in this text.

The minimum lighting load for dwellings is 3 volt-amperes per square foot. See *NEC 220.12* and *Table 220.12*.

Continuous Loads

The *NEC* defines **continuous load** in *Article 100* as *a load where the maximum current is expected to continue for three hours or more*.* Continuous loads shall not exceed 80% of the rating of the branch circuit. General lighting outlets and receptacle outlets in residences are not considered to be continuous loads.

Certain loads in homes are considered to be continuous and must be treated accordingly. Examples are electric water heaters (*422.13*), central electric heating [*424.3(B)*], snow-melting cables (*426.4*), and air-conditioning equipment (*440.32*). For these loads, the branch-circuit rating, the conductors, and the overcurrent device shall not be less than 125% of the rating of the equipment. Mathematically, sizing the conductors and overcurrent device at 125% of the load is the same as loading the conductors and overcurrent device to 80%.

For example, an electric furnace with a nameplate rating of 40 amperes would require the supply conductors and overcurrent protection to be not less than

$$40 \times 1.25 = 50 \text{ amperes}$$

VOLTAGE

All calculations throughout this text use voltage values of 120 volts and 240 volts. This complies with the requirements of *NEC 220.5*, which lists the various voltage values to use for different electrical systems. This is repeated in the second paragraph of *Annex D* of the *NEC*.

The word *nominal* means "in name only, not a fact." For example, in our calculations, we use 120 volts even though the actual voltage might be 110, 115, or 117 volts. This provides us with a uniformity that without it would lead to misleading and confusing results in our calculations.

Most public service commissions require that the voltage for residential services be held to ±5% of the nominal voltage as measured at the service point.

CALCULATING LOADS

When wiring a house, it is all but impossible to know which appliances, lighting, heating, and other loads will be turned on at the same time. Different families lead different lifestyles. There is tremendous

---

*Reprinted with permission from NFPA 70-2011.

diversity. There is a big difference between "connected load" and "actual load." Who knows what will be plugged into a wall receptacle, now or in the future? It's a guess at best. Over the years, the *NEC* has developed procedures for calculating loads in typical one- and two-family homes.

The rules for doing the calculations are found in *Article 220.* For lighting and receptacles, the computations are based on volt-amperes per square foot. For the small-appliance circuits in kitchens and dining rooms, the basis is 1500 volt-amperes per circuit. For large appliances such as dryers, electric ranges, ovens, cooktops, water heaters, air conditioners, heat pumps, and so on, which are not all used continuously or at the same time, there are demand factors to be used in the calculations. Following the requirements in the *NEC*, the various calculations roll together in steps that result in the proper sizing of branch circuits, feeders, and service equipment.

As you work your way through this text, examples are provided for just about every kind and type of load found in a typical residence.

Inch-pounds versus Metrics When Calculating Loads

As discussed in Chapter 1, converting inch-pound measurements to metric measurements and vice versa results in odd fractional results. Adding further to this problem are the values rounded off when the *NEC* Code-Making Panels did the metric conversions. When square feet are converted to square meters and the unit loads are calculated for each, the end results are different—close, but nevertheless different. To show both calculations would be confusing as well as space consuming. Many of the measurements in this text are shown in both inch-pound units and metric units. Load calculations throughout this text use inch-pound values only, which is in agreement with the *Examples* given in *Annex D* of the 2011 *NEC*.

CALCULATING FLOOR AREA

The general lighting load for a dwelling is based on the square footage of the dwelling. Here is how this is done.

First Floor Area

To estimate the total load for a dwelling, the occupied floor area of the dwelling must be calculated. Note in the residence plans that the first floor area has an irregular shape. In this case, the simplest method of calculating the occupied floor area is to determine the total floor area using the outside dimensions of the dwelling. Then, the areas of the following spaces are subtracted from the total area: open porches, garages, or other unfinished or unused spaces if they are not adaptable for future use, *NEC 220.12.*

Many open porches, terraces, patios, and similar areas are commonly used as recreation and entertainment areas. Adequate lighting and receptacle outlets should be provided for these areas.

For practicality, the author has chosen to round up dimensions for the determining of total square footage and to round down dimensions for those areas (garage, porch, and portions of the inset at the front of the house) not to be included in the computation of the general lighting load. This produces a slightly larger result as opposed to being on the conservative side. Don't be miserly with your measurements; rather, be generous. Figure 3-2 shows the procedure for calculating the total square footage of this residence.

Basement Area

Although the *NEC* in *220.12* tells us that *unused or unfinished spaces not adaptable for future use* do not have to be included in calculating the square footage of a dwelling, it makes sense to include some of these spaces.

Nearly all basements in homes today certainly could be considered as being adaptable for future use. A crawl space and most attics would not normally be considered as being adaptable for future use. This is a judgment call based on a close examination of the Plans and Specifications. In this residence, more than half of the total basement area is finished off as a recreation room, which certainly is considered a living area. The workshop area also is intended to be used.

To simplify the calculation for this residence, we will consider the entire basement as usable space and figure the basement square footage area as being the same as the area of the first floor.

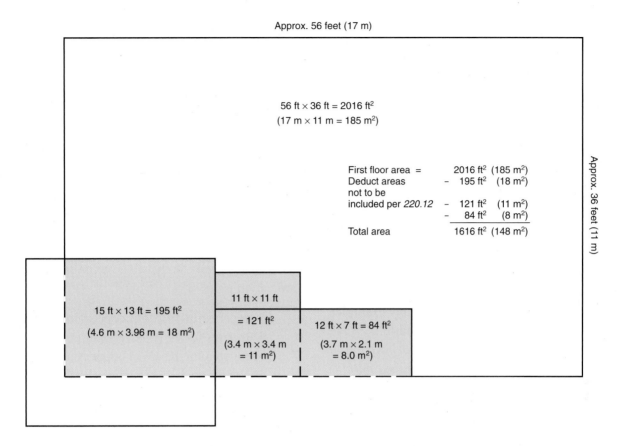

Approx. 56 feet (17 m)

56 ft × 36 ft = 2016 ft²
(17 m × 11 m = 185 m²)

First floor area =   2016 ft² (185 m²)
Deduct areas   −   195 ft²   (18 m²)
not to be
included per *220.12*   −   121 ft²   (11 m²)
                        −   84 ft²   (8 m²)
Total area   1616 ft² (148 m²)

Approx. 36 feet (11 m)

15 ft × 13 ft = 195 ft²
(4.6 m × 3.96 m = 18 m²)

11 ft × 11 ft
= 121 ft²
(3.4 m × 3.4 m
= 11 m²)

12 ft × 7 ft = 84 ft²
(3.7 m × 2.1 m
= 8.0 m²)

FIGURE 3-2   Determining the first floor square footage area. The basement area is the same as the first floor. See text for explanation of how the dimensions of the residence were rounded off to make the calculations much simpler and much more practical for the electrician and electrical inspector to calculate. (*Delmar/Cengage Learning*)

The combined occupied area of the dwelling is found by adding the first floor and basement areas together:

First Floor	1616 ft² (149 m²)
Basement	1616 ft² (149 m²)
Total	3232 ft² (298 m²)

DETERMINING THE MINIMUM NUMBER OF LIGHTING BRANCH CIRCUITS

Table 220.12 of the *NEC* tells us that the minimum load requirement for dwelling units is 3 volt-amperes (VA) per square ft (0.093 m²) of occupied area.

To determine the minimum number of 15-ampere lighting branch circuits required for a residence, here are the simple steps:

Step 1: $= \dfrac{3 \text{ volt-amperes} \times \text{square ft}}{120 \text{ volts}} = \text{amperes}$

Step 2: $\dfrac{\text{amperes}}{15} = \dfrac{\text{minimum number of 15-amperes}}{\text{lighting branch circuits}}$

This equates to a minimum of

- one 15-ampere lighting branch circuit for every 600 ft² (55.8 m²).

- one 20-ampere lighting branch circuit for every 800 ft² (74.4 m²).

Note: 20-ampere branch circuits are not commonly used for lighting in residential installations. In this residence, let's calculate the load for the total occupied area of 3232 ft² (298 m²):

$$3232 \times 3 = 9696 \text{ volt-amperes}$$

$$\text{Amperes} = \frac{\text{volt-amperes}}{\text{volts}} = \frac{9696}{120} = 80.8 \text{ amperes}$$

The minimum number of 15-ampere lighting branch circuits is

$$\frac{80.0 \text{ amperes}}{15 \text{ amperes}} = \frac{5.4 \text{ branch circuit}}{\text{(round up to 6)}}$$

Because we cannot have a fraction of a branch circuit, we rounded up to a minimum of six 15-ampere lighting branch circuits.

We get the same answer using 600 ft² (55.8 m²) for each 15-ampere lighting branch circuit as follows:

$$\frac{3232 \text{ ft}^2}{600} = \frac{5.4 \text{ branch circuit}}{\text{(round up to 6)}}$$

No matter which method we use, a minimum of six 15-ampere lighting branch circuits is required.

In *Annex D* of the *NEC*, we find many examples of load calculations. Here, we also find that *except where the calculations result in a major fraction of an ampere (0.5 or larger), such fractions are permitted to be dropped.* **CAUTION: This permission is for load calculations such as in Step 1. This permission to drop fractions is *not* to be used in Step 2 because it is impossible to have a fraction of a branch circuit. That is why we rounded up from 5.4 to arrive at a minimum of six 15-ampere branch circuits.**

Load calculations for small-appliance branch circuits and other major appliance branch circuits are in addition to the general lighting loads. These calculations are discussed throughout this text as they appear.

The *NEC* in *210.11(B)* states that the calculated load *shall be evenly proportioned among multioutlet branch circuits within the panelboard(s).*

Table 3-1 is the schedule of the 15-ampere lighting and 20-ampere small-appliance branch circuits in this residence.

240.4(D) tells us that the maximum overcurrent protection is

15 amperes for 14 AWG copper conductors.
20 amperes for 12 AWG copper conductors.
30 amperes for 10 AWG copper conductors.

In conformance to *210.3*, a branch circuit is rated according to the rating or setting of the overcurrent device. For example, a lighting branch circuit is rated 15 amperes even if 12 AWG copper conductors are installed.

TABLE 3-1

Summary of 15-ampere lighting and 20-ampere receptacle circuits.

15-Ampere Circuits		20-Ampere Circuits	
A14	Bathrooms, hall ltg.	A18	Workbench Receptacles
A15	Front entry/porch	A20	Workshop Receptacles, Window Wall
A16	Front Bedroom Ltg. & Outdoor Receptacle	A22	Master Bath Receptacle
A17	Workshop Ltg.	A23	Hall Bath Receptacle
A19	Master Bedroom Ltg. & Outdoor Receptacle		
A21	Study/bedroom Ltg.		
B7	Kitchen Ltg.	B13	Kitchen North & East Wall Receptacles
B9	Wet Bar Ltg. & Receptacles	B15	Kitchen Receptacles, West Countertop
B10	Laundry, Rear Entry, Powder Room, Attic Ltg.	B16	Kitchen Receptacles, South Countertop, & Island
B11	Recreation Room Receptacles	B18	Washer Receptacle
B12	Recreation Room Ltg.	B20	Laundry Room Receptacles & Outdoor Receptacle
B14	Garage, Post Light	B21	Powder Room Receptacle
B17	Living Room & Outdoor Receptacle	B22	Refrigerator

Notes: Panelboard "A" is the main panel located in the workshop.
Panelboard "B" is the subpanel located in the corner of the recreation room.

How Many Outlets per Circuit

This is covered in detail in Chapter 8, where you will find *Estimating Loads for Outlets, How Many Outlets per Circuit, Circuit Loading Rules of Thumb,* and the *80% Rule.*

TRACK LIGHTING LOADS

Track lighting loads are considered to be part of the general lighting load for residential installations, based on 3 volt-amperes per square foot as previously discussed. There is no need to add more wattage (volt-amperes) to the load calculations for track lighting in homes.

As required by *410.151(B)*, the connected load on a lighting track shall not exceed the rating of the track. Also, be sure that a lighting track is supplied by a branch circuit having a rating not more than that of the track. There is more on track lighting in Chapters 12 and 13.

DETERMINING THE NUMBER OF SMALL-APPLIANCE BRANCH CIRCUITS

NEC 220.52(A) and *(B)* state that an additional load of not less than 1500 volt-amperes shall be included for each two-wire small-appliance circuit and each laundry circuit.

The receptacle for the refrigerator in the kitchen is permitted to be supplied by an individual 15- or 20-ampere branch circuit rather than connecting it to one of the two required 20-ampere small-appliance circuits. This permission is found in *210.52(B)(1), Exception No. 2.*

The *NEC* defines an *individual branch circuit* as *A branch circuit that supplies only one utilization equipment.*

Providing an individual branch circuit for the refrigerator diverts the refrigerator load from the receptacle branch circuits that serve the countertop areas. Because this individual branch circuit does not serve countertops, it is not considered a small-appliance branch circuit. An additional 1500 volt-amperes *does not* have to be added into the load

calculations, *220.52(A), Exception No. 2.* There is nothing in the *Code* that prohibits adding another 1500 volt-amperes to the load calculations for this additional branch circuit, as this would result in more capacity in the electrical system.

NEC 210.11(C)(1) requires that two or more 20-ampere small-appliance circuits must be provided for supplying receptacle outlets, as specified in *210.52(B)*.

NEC 210.52(B)(1) states that the 20-ampere appliance circuits shall serve only the receptacle outlets in the kitchen, pantry, dining room, breakfast room, or similar rooms of dwelling units. *NEC 210.52(B)(2)* also requires that these 20-ampere small-appliance circuits shall not supply other outlets.

Exceptions in *210.52(B)(2)* allow clock receptacles and receptacles that are installed to plug in the cords of gas-fired ranges, ovens, or counter-mounted cooking units in order to power up their timers and ignition systems, Figure 3-3.

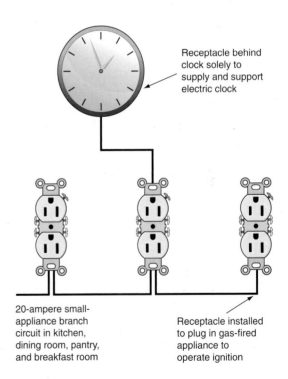

Receptacle behind clock solely to supply and support electric clock

20-ampere small-appliance branch circuit in kitchen, dining room, pantry, and breakfast room

Receptacle installed to plug in gas-fired appliance to operate ignition

FIGURE 3-3 A receptacle installed solely to supply and support an electric clock, or a receptacle that supplies power for clocks, clock timers, and electric ignitions for gas ranges, ovens, and cooktops, may be connected to a 20-ampere small-appliance circuit or to a general-purpose lighting circuit, *210.52(B)(2), Exceptions.*
(Delmar/Cengage Learning)

NEC 210.52(B)(3) requires that not less than two 20-ampere small-appliance circuits must supply the receptacle outlets serving the countertop areas in kitchens.

Food waste disposers and/or dishwashers shall not be connected to these required 20-ampere small-appliance circuits. They must be supplied by other circuits, as is discussed later. Generally, 12 AWG conductors are used to connect these appliances. The use of this larger conductor, rather than a 14 AWG conductor, helps improve appliance performance and lessens the danger of overloading the circuits or of experiencing excessive voltage drop.

A complete list of the branch circuits for this residence is found in Chapter 27. The circuit directories for Main Panel A and Subpanel B show clearly the number of lighting, appliance, and special-purpose circuits provided.

RECEPTACLE OUTLET BRANCH-CIRCUIT RATINGS

The receptacles in the kitchen, breakfast area, workshop, powder room, bathrooms, and laundry room are connected to 20-ampere branch circuits. All other receptacles are connected to 15-ampere branch circuits.

SUMMARY OF WHERE RECEPTACLE AND LIGHTING OUTLETS MUST BE INSTALLED IN RESIDENCES

The following is a recap of the *NEC* requirements for receptacle and lighting outlets. We will look at these *Code* rules, room by room, later in this text.

Remember, the *NEC* provides *minimum* requirements. There is nothing wrong with installing more receptacles than the *NEC minimum.* You will never have enough receptacles! Think about it. A duplex receptacle can't handle computers, cable modems, small UPS devices, printers, stereos, VCRs, VDRs, CD players, CD burners, shredders, television sets, radios, fax machines, answering machines, typewriters, adding machines, pencil sharpeners, lamps, telephones, and others.

Consider spacing receptacles closer than the minimum. Consider quadplex receptacles or two

duplex receptacles at each receptacle outlet location and consider surge suppression receptacles or surge protection multiple plug-in strips. See Chapter 6 for a discussion of surge protection.

> **TIP:** Some electricians install a receptacle below the wall switch in bedrooms, living rooms, family rooms, and similar rooms. As is usually the case, the required receptacles will be behind furniture, making it difficult to reach for plugging in a vacuum cleaner or other temporarily used appliance. A receptacle below the wall switch will be very easy to get to. A real plus! ●

See Figure 3-4 for an explanation of *receptacle* and *receptacle outlet.*

Receptacle Outlets (125 Volts, Single-Phase, 15 and 20 Amperes)

Ground-Fault Circuit-Interrupter (GFCI). GFCI protection is required in specific locations in a home. GFCIs are discussed in Chapter 6.

Arc-Fault Circuit-Interrupter (AFCI). AFCI protection is required in specific locations in a home. AFCIs are discussed in Chapter 6.

The following is a list of *Code* requirements where receptacles are required in new homes:

- Receptacles are located throughout a home for the convenience of plugging "things" in. The following required locations for receptacles are in addition to receptacles that are wall switch controlled, are part of a luminaire or appliance, or if located inside of a cabinet, or located more than 5½ ft (1.7 m) above the floor. A duplex receptacle that has one receptacle switched and one receptacle "hot" continuously meets the receptacle requirements of *210.52* and the switching requirements of *210.70.* See Figures 3-9, 3-10, and 8-10.

- Wall receptacles must be placed so that no point measured horizontally along the floor line in any wall space is more than 6 ft (1.8 m) from a receptacle outlet. Fixed room dividers and

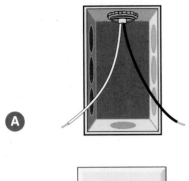

A receptacle outlet where one or more receptacles will be installed.

A

A receptacle outlet with a single receptacle. (one contact device)

B

A receptacle outlet with a multiple (duplex) receptacle. This is two receptacles. (two contact devices)

C

A receptacle outlet with two multiple (duplex) receptacles. This is four receptacles. (four contact devices)

D

FIGURE 3-4   The branch-circuit wiring and the box where one or more receptacles will be installed is defined by the *Code* as a *receptacle outlet.* The receptacle itself, whether a single or multiple device (one, two, or three receptacles) on one strap (yoke) is defined as a *receptacle.* For house wiring, receptacle loads are considered to be part of the general lighting load calculations, *220.14(J).*
(*Delmar/Cengage Learning*)

railings are considered to be walls for the purpose of this requirement. See *210.52(A)(1).*

- Any wall space 24 in. (600 mm) or more in width in kitchens, family rooms, dining rooms, living rooms, parlors, libraries, dens, sunrooms, bedrooms, recreation rooms, or similar room or area of dwelling units must have a receptacle outlet, *210.52(A)(2)(1).*

- Nonsliding fixed glass panels on exterior walls are considered to be wall space and are to be figured in when applying the rule that no point along the floor line is more than 6 ft (1.8 m) from a receptacle. Sliding panels on exterior walls are not considered to be a wall, *210.52(A)(2).*

- Receptacle outlets located in the floor more than 18 in. (450 mm) from the wall are not to be counted as meeting the required number of wall receptacle outlets, *210.52(A)(3).* An example of this would be a floor receptacle installed under a dining room table for plugging in warming trays, coffee pots, or similar appliances.

- Hallways 10 ft (3.0 m) or longer in homes must have at least one receptacle outlet, *210.52(H).* See Figure 3-5.

- ▶Foyers that have an area that is greater than 60 ft² (5.6 m²) shall have a receptacle(s) located in each wall space as defined in *210.52(I).*◀

- Circuits supplying wall receptacle outlets for lighting loads are generally 15-ampere circuits but are permitted by the *Code* to be 20 amperes.

- Outdoors: One receptacle is required in front and another in back, *210.52(E).* This is discussed in detail later in this chapter.

- Receptacle(s) for HVAC equipment, *210.63.* This is discussed in detail later in this chapter.

- Outdoor receptacles are required to be GFCI protected, *210.8(A)(3).* Exempt are receptacles that are not readily accessible, are supplied by a dedicated branch circuit, and serve snow-melting or deicing equipment; see *426.28.* Ground-fault protection of equipment is required for fixed outdoor electric deicing and snow-melting equipment. This equipment protection will typically trip or open at 30 mA or greater leakage current rather than at from 4 to 6 mA for personnel protection.

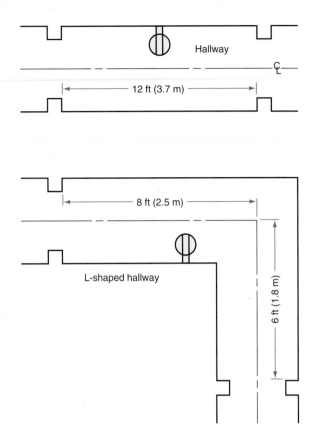

FIGURE 3-5 Determination of hallway length. Hallways 10 ft (3.0 m) or longer must have at least one receptacle outlet, *210.52(H)*. (*Delmar/Cengage Learning*)

Basements

- At least one receptacle outlet must be installed in addition to the laundry outlet, *210.52(G)*.

- Receptacles must be supplied by a 15- or 20-ampere circuit.

- In finished basements, the required number, spacing, and location of receptacles, lighting outlets, and switches shall be in conformance to *210.52(A)* and *210.70*.

- In finished basements, GFCI protection for receptacles is *not* required unless the receptacles are near a sink.

- ▶Any receptacle within 6 ft (1.8 m) from the outside edge of a sink shall be GFCI protected. There are no exceptions to this requirement.◀

- Unfinished basements, or unfinished portions of basements that are not generally "lived in,"

such as work areas and storage areas, are not subject to the spacing requirements for receptacles as specified in *210.52(A)*.

- In each separate unfinished basement area that is not intended to be "lived in," such as storage areas or work areas, at least one 125-volt, single-phase, 15- or 20-ampere receptacle must be installed, *210.52(G)*. This receptacle(s) must be GFCI protected, *210.8(A)(5)*. This includes receptacles that are an integral part of porcelain or plastic pull chain or keyless lampholders.

- See the "Laundry Area" section in this chapter for receptacle outlet requirements in laundry areas.

- Refer to *210.8(A)*, *210.52(A)*, *210.52(G)*, and *210.70*.

Small-Appliance Receptacle Outlets

- At least two 20-ampere small-appliance circuits must supply the receptacles that serve the countertop surfaces in the kitchen, *210.11(C)(1)* and *210.52(B)(3)*. Two small-appliance branch circuits is the minimum. It is advisable to run additional 20-ampere branch circuits to these areas because of the heavy concentration of appliances.

- These two 20-ampere small-appliance circuits may also supply the receptacles in the dining area, pantry, breakfast room, and the refrigerator receptacle, *210.52(B)(1)*. *Exception No. 2* permits connecting the refrigerator receptacle to an individual 15- or 20-ampere branch circuit.

What is meant by the term *individual branch circuit*? The *NEC* defines an *individual branch circuit* as *A branch circuit that supplies only one utilization equipment*. To meet this definition, either a single receptacle or a duplex receptacle with only the refrigerator plugged in is permitted to be installed for the refrigerator. Keep in mind the receptacle meets the definition of a device and not utilization equipment. The refrigerator meets the definition of utilization equipment. It's doubtful that a duplex receptacle behind a refrigerator would be used for anything other than the refrigerator. The same logic would apply in other situations such as a receptacle located behind a gas range. It's up to the electrical inspector.

A single receptacle supplied by an individual branch circuit must have an ampere rating not less than the rating of the branch circuit, *NEC 210.21(B)(1)*. If the branch-circuit rating is 20 amperes, the single receptacle would have to be rated 20 amperes.

—∿— **TIP:** There are too many nuisance power outages when circuit breakers trip or fuses open because too many appliances are plugged in and used at the same time in the kitchen. Many large combination refrigerator/freezers today draw quite a bit of current. It is better to be safe than sorry. Connect the refrigerator receptacle to an individual 15- or 20-ampere branch circuit. Being behind the refrigerator, the receptacle does not have to be GFCI protected. Only receptacles that serve the countertops are required to be GFCI protected, *210.8(A)(6).* ●

See Chapter 12 for more details about receptacles for refrigerators.

- Space countertop receptacles in kitchens according to *210.52(C)*.

- GFCI protection shall be provided for all receptacles in kitchens that serve countertop areas, *210.8(A)(6)*. GFCI protection may be GFCI receptacles or GFCI circuit breakers, as discussed in detail in Chapter 6. ▶The GFCI device must be readily accessible for monthly testing.◀

- These two or more 20-ampere small-appliance circuits shall not supply other outlets, *210.52(B)(2)*. Exceptions to this are (1) a clock receptacle and (2) a receptacle(s) for plugging in a gas-fired range, oven, or counter-mounted cooktop, which are permitted to be supplied by one of the 20-ampere small-appliance circuits.

- Allow 1500 volt-amperes for each 20-ampere small-appliance branch circuit when computing loads for determining the size of the service-entrance equipment, *220.52(A)*.

- Refer to *210.8(A)(6)*, *210.11(C)(1)*, *210.52(B)(1)*, *(2)*, and *(3)*, *210.52(C)*, and *220.52(A)*.

Countertops in Kitchens

See Figure 3-6.

- The following requirements pertain to countertops in kitchens, pantries, breakfast rooms, dining rooms, and similar areas; see *210.52(C)*.

- Receptacles that serve countertops in kitchens must be GFCI protected. This includes a receptacle located inside an "appliance garage."

- Receptacles are not permitted to be installed "face up" in countertops.

- For wall countertop surfaces, receptacles shall be installed above all countertop spaces 12 in. (300 mm) or wider.

- Receptacles shall be installed not more than 20 in. (500 mm) above a countertop.

- ▶*NEC 210.52(C)(5)* states that receptacle outlet assemblies listed for the application are permitted to be installed in countertops.◀ See Figure 3-6.

- Receptacles above wall countertop spaces shall be positioned so that no point along the wall line is more than 24 in. (600 mm), measured horizontally, from a receptacle outlet in that space. Example: A section of a countertop measures 5 ft 3 in. (1.6 m) along the wall line. The minimum number of receptacles is 5 ft 3 in. ÷ 4 ft = 1+. Therefore, install two receptacles in this space.

- Receptacles may be installed below countertops only:
  - where the construction is for the physically impaired, or
  - for island and peninsula countertop spaces where the countertop is flat across its entire surface and there is no way to mount a receptacle within 20 in. (500 mm) above the countertop.

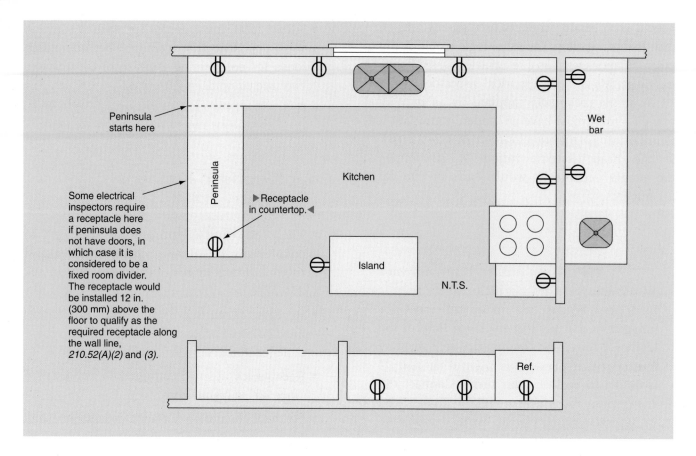

FIGURE 3-6 Floor plan showing placement of receptacles that serve countertop surfaces in kitchens and wet bar areas in conformance to *210.52(C)*. All 120-volt receptacles that serve kitchen countertops must be GFCI protected, *210.8(A)(6)*. Receptacles that are located behind the refrigerator, under the sink for plugging in a food waste disposer, or inside the cabinets for plugging in a microwave oven do not have to be GFCI protected because they do not serve the countertop surfaces. ▶Receptacles located in areas other than kitchens require GFCI protection if installed within 6 ft (1.8 m) of the outside edge of the sink.◀ *NEC 210.8 (A)(7)*. Drawing is not to scale. See text for detailed explanation. ▶*NEC 210.52(C)(5)* states that receptacle outlet assemblies listed for the application are permitted to be installed in countertops.◀ (*Delmar/Cengage Learning*)

For either of these two conditions:
— Receptacles shall be mounted not more than 12 in. (300 mm) below the countertop.
— No receptacles shall be installed below a countertop that extends more than 6 in. (150 mm) beyond its base cabinet. The logic to this rule is the popular use of stools where the countertop extends beyond the base cabinet, increasing the possibility of someone pulling coffee pots and similar appliances off of the countertop.

• Peninsulas and islands that have a long dimension of 24 in. (600 mm) or greater and a short dimension of 12 in. (300 mm) or greater shall have at least one receptacle.

▶More than one receptacle might be required if the countertop, peninsula, or island has a range, counter-mounted cooking unit, or sink that creates separate counter spaces.◀ A peninsula starts where it connects to the wall base cabinet.

• ▶If a countertop, peninsula, or island has a range, counter-mounted cooking unit, or sink that creates a counter space less than 12 in. (300 mm) between the appliance and the wall or edge of the countertop, separate countertop spaces on either side of these appliances have been created. Receptacles shall be installed to comply with the required number of receptacles in each of these spaces.◀

- Receptacles that do not serve countertops, such as behind a refrigerator, under the sink for plugging in a food waste disposer or dishwasher, or behind or above an appliance that is fastened in place, are not considered part of the receptacle outlet requirement for serving countertops. These receptacles are not required to be GFCI protected.

- Receptacles inside an "appliance garage" are *in addition* to the receptacles required to serve countertops. They must be GFCI protected. Although permitted to be connected to one of the two required 20-ampere small-appliance circuits, consider running a separate 20-ampere branch circuit.

- A receptacle located inside an upper cabinet that serves a "hang below cabinet" microwave oven should be connected to an individual branch circuit. This is discussed in Chapter 20.

- ▶Receptacles are generally not required on the wall behind a range, counter-mounted cooking unit, or sink. However, a receptacle is required
  - if the distance from the wall to the range, counter-mounted cooking unit, or sink is 12 in. (300 mm) or greater. In other words, if the space is 12 in. (300 mm) or greater, it is counter space; if less than 12 in. (300 mm), it is not counter space.
  - if the range, counter-mounted cooking unit, or sink is corner mounted on an angle, and the distance from the corner to the range, counter-mounted cooking unit, or sink is 18 in. (450 mm) or greater. In other words, if the space is 18 in. (450 mm) or greater, it is counter space; if less than 18 in. (450 mm), it is not counter space.◀

  This is illustrated in *Figure 210.52* in the *NEC* and in Figures 3-7 and 3-8.

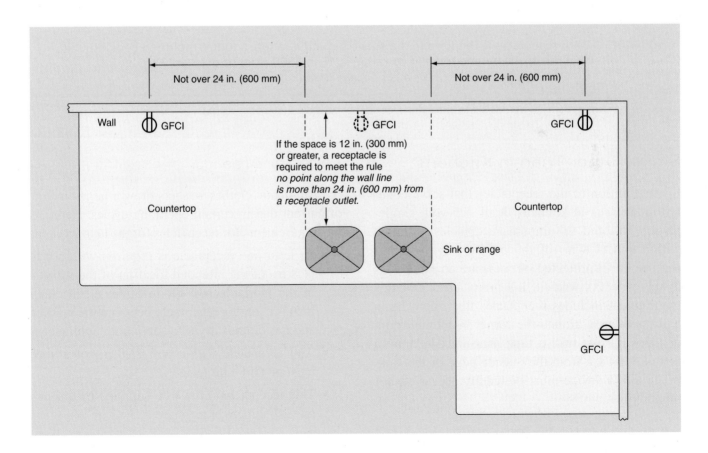

FIGURE 3-7 ▶A receptacle is required in the space behind a range, counter-mounted cooking unit, or sink if the space behind the sink or range is 12 in. (300 mm) or greater. *210.52(C)(1)*◀ *(Delmar/Cengage Learning)*

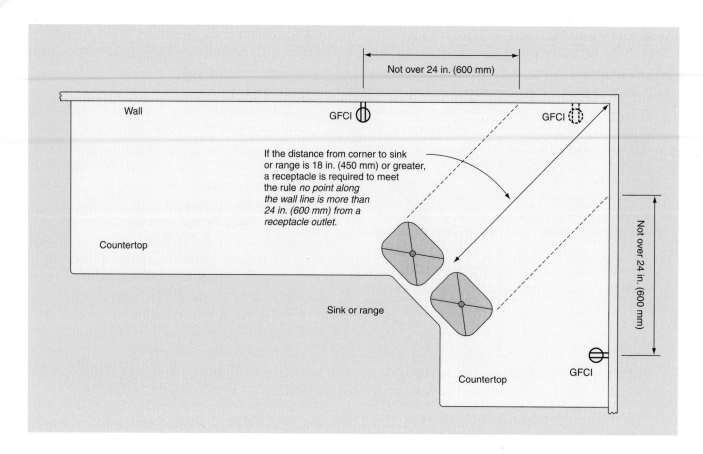

FIGURE 3-8 A receptacle is required in the space behind a range, counter-mounted cooking unit, or sink if the space behind the sink or range is 18 in. (300 mm) or greater, *210.52(C)(1)*. (*Delmar/Cengage Learning*)

Sinks, Other Than in Kitchens

In addition to the receptacles that serve countertop surfaces in kitchens, ▶all 125-volt, single-phase, 15- and 20-ampere receptacles installed within 6 ft (1.8 m) of the outside edge of sinks shall be GFCI protected. ◀There are no exceptions to this rule. As written, it's pretty clear that this requirement includes receptacles that might be in a different area, around the corner in another room or hallway. Just take a tape measure and draw a 6-ft (1.8-m) arc from the outside edge of the sink. ▶The GFCI device must be readily accessible for monthly testing.◀

- See Chapter 6 for details on GFCI protection.

- See *210.8(A)(6)* and *(7)*, *210.11(C)*, *210.52(C)*, *220.52(A)*, and *406.4(E)*.

Laundry Area

Automatic clothes washers draw a large amount of current during certain operating cycles. Here are the requirements for receptacles for the laundry area:

- At least one receptacle is required within 6 ft (1.8 m) of the intended location of the washer in the laundry area. Quite often, electricians install a duplex receptacle between the washer and gas clothes dryer to serve both appliances.

- Any receptacle within 6 ft (1.8 m) of a sink must be GFCI protected.

- This receptacle(s) must be supplied by a separate 20-ampere branch circuit.

- This branch circuit shall not supply any other outlets outside the laundry area.

- This branch circuit and receptacle(s) are in addition to other required receptacles and branch circuits.

- GFCI protection is required for receptacles in unfinished basements. ▶The GFCI device must be readily accessible for monthly testing.◀

- If you install a single receptacle on a 20-ampere branch circuit, it must be rated 20 amperes.

- If you install a duplex receptacle on a 20-ampere branch circuit, it may be rated 15 amperes or 20 amperes.

- Include 1500 volt-amperes for this laundry branch circuit when calculating the ampacity of the feeder or service-entrance conductors.

- Include 1500 volt-amperes for this laundry branch circuit when computing loads for determining the ampere rating of the feeder panelboard or service-entrance equipment.

- Refer to *210.8(A)(5), 210.8(A)(7), 210.11(C)(2), 210.21(B)(1), 210.21(B)(3), 210.50(C), 210.52(C)(5), 210.52(F)* and *(G), 220.14(J), 220.16(B),* and *220.52(B).*

Bathroom Receptacles

Here is a recap of the *NEC* requirements for receptacles in bathrooms:

- At least one receptacle must be installed in a bathroom.

- Receptacle(s) in bathrooms must be GFCI protected. ▶The GFCI device must be readily accessible for monthly testing.◀

- The receptacle(s) must be installed within 3 ft (900 mm) of the outside edge of each basin:
  — on the wall or partition adjacent to the basin.
  — on the side or face of the basin cabinet not more than 12 in. (300 mm) below the countertop.

- ▶*NEC 210.52(C)(5)* states that receptacle outlet assemblies listed for the application are permitted to be installed in countertops.◀

- The receptacle(s) must be supplied by at least one separate 20-ampere branch circuit and generally must not supply other loads such as lighting outlets.

- The 20-ampere branch circuit and receptacle is in addition to other required outlets and branch circuits.

- The separate 20-ampere branch circuit is permitted to supply more than one bathroom, although it is not recommended because of the use of high-wattage hair dryers. It is better to run a separate 20-ampere branch circuit to each bathroom.

- If the separate 20-ampere branch circuit supplies only one bathroom, other outlets in that bathroom are permitted to be connected to it. The other permitted loads are limited to not more than 80% of the rating of the branch circuit if they are cord-and-plug-connected, and 50% if they are fixed or fastened-in-place equipment.

- Instead of the "only one bathroom" exception, it is recommended that you run a separate 20-ampere branch circuit to each bathroom. That's what the plans for the residence call for.

- Don't be left in the dark. Do not connect bathroom lighting to this separate 20-ampere receptacle branch circuit.

- Receptacles are not permitted face up on counters or other work surfaces.

- A receptacle on a medicine cabinet or luminaire does not qualify as the required receptacle. It is virtually impossible to conform to the requirements that call for a separate circuit to be within 3 ft (900 mm) of the basin rim and be GFCI protected.

- No additional load need be added for this separate 20-ampere branch circuit when doing the load calculations. The separate branch-circuit requirement merely redirects the receptacle load off the lighting branch circuit.

- See *210.8(A)(1), 210.11(C)(3), 210.23(A)(1)* and *(2), 210.52(C)(5), 210.52(D), 220.14(J),* and *406.4.*

Outdoors

- At least one receptacle shall be installed in the front and another in the back of a one-family dwelling, *210.52(E)(1).* These receptacles shall

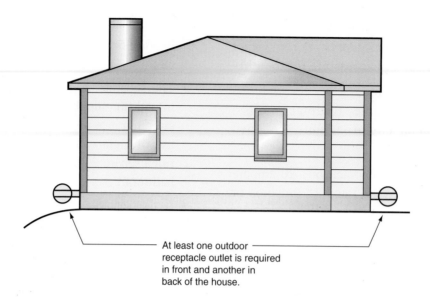

At least one outdoor receptacle outlet is required in front and another in back of the house.

FIGURE 3-9 Outdoor receptacle requirements for residences, *210.52(E),* require that for one-family dwellings, and each unit of a two-family dwelling that is at grade level, at least one receptacle outlet must be installed in the front and another in the back. These receptacle outlets must be accessible while standing at grade level and be mounted not more than 6½ ft (2.0 m) above grade. (*Delmar/Cengage Learning*)

be accessible while standing at grade level and not more than 6½ ft (2.0 m) above grade. See Figure 3-9. The same is true for each unit of a multifamily dwelling where the entrance/egress is at grade level, *210.52(E)(2).*

Grade level could be grass, sidewalk, or driveway. It is possible that a concrete or brick patio might be built at grade level.

In addition to the "front and back" receptacle requirement, let's take a look at receptacle requirements for balconies, decks, and porches.

- **Balconies, decks, and porches:** We don't want people running extension cords through doors or window openings and pinching the cord. In *210.52(E)(3),* we find these requirements for balconies, decks, and porches that are accessible from inside the dwelling:
  - At least one receptacle shall be installed within the perimeter of the balcony, deck, or porch.
  - The receptacle shall not be more than 6½ ft (2.0 m) above the surface of the balcony, deck, or porch.

Roughing-in receptacle(s) on the front of the house is usually a no-brainer. Plans clearly show the construction of the porch. In the residence in this text, there is one switched receptacle on the front porch, and another nonswitched receptacle on the outside of the front bedroom wall.

A rear deck/patio/porch is another story. They are generally built after the house is completed. It is easier said than done to figure out where to rough-in the required receptacle.

The electrician must consider: Where will the deck/patio be built? Is the deck/patio poured concrete or decorative brick/stone at grade level? What is the height if built above grade? Does it have rails or walls? Is it screened in? Sometimes the surest thing to do is

- rough-in the one receptacle on the rear of the house close to the means of entrance/egress such as a sliding door not over 6½ ft (2.0 m) above grade.
- rough-in a second receptacle somewhere else on the rear of the house outside the footprint of the intended deck/patio/porch.

For the more elaborate deck/patio/porch, consider installing a few additional receptacles around the perimeter of the deck/patio/porch. These would be wired in after the deck/patio/porch is completed.

- Additional receptacles may be installed higher, such as receptacles used for plugging in ice and snow-melting heating cables and/or decorative outdoor lighting.

- For servicing HVAC equipment, *210.63* requires that a 125-volt, single-phase, 15- or 20-ampere-rated receptacle be installed as follows:
  — Be at an accessible location.
  — Be on the same level as the HVAC equipment.
  — Be within 25 ft (7.5 m) of the HVAC equipment.
  — *Not* be connected to the load side of the equipment disconnecting means.

  Evaporative coolers are exempt from the above requirements. Why are evaporative coolers, often referred to as *swamp coolers*, exempt? Because they do not contain anywhere near as many complicated internal parts as a typical air conditioner or heat pump. There is less need for maintenance on this type of equipment. Less expensive than an air conditioner or a heat pump, they contain one motor that pumps water to the top of the unit. The water "falls" downward over porous filter pads where the water is cooled by evaporation. A second motor (fan) pulls in outside air and blows the air over the cooled water into the room. Evaporative coolers work great in hot, dry climates. The higher the humidity, the less efficient is an evaporative cooler.

- The outdoor receptacles required by *210.52(E)* (one in the front, one in the back) might or might not meet the requirements of *210.63*. If the HVAC equipment is located on a roof, in an attic, in a crawl space, or in a similar location, a receptacle must be installed in that location so as to be on the same level as the equipment.

- All outdoor receptacles must be GFCI protected, *210.8(A)(3)*. ▶The GFCI device must be readily accessible for monthly testing.◀ Exempt from this requirement are GFCI outdoor receptacle(s) that are not readily accessible and are supplied by a dedicated branch circuit installed for snow- and ice-melting equipment (e.g., heating cables).

- Ground-fault protection of equipment (GFPE) must be provided for "fixed" outdoor deicing and snow-melting equipment. GFPE can be a circuit breaker, a receptacle, or be self-contained in the equipment. GFPE devices for equipment protection trip at approximately 30 milliamperes; see *426.28*.

  When used for deicing and snow melting, mineral-insulated metal-sheathed cable embedded in noncombustible material such as concrete does not require ground-fault protection.

- See *210.8(A)(3)*, *210.52(E)*, *210.63*, and *426.28*.

Garages and Accessory Buildings

- ▶At least one receptacle outlet in addition to receptacles installed for specific equipment must be provided for each garage, and in each detached garage or accessory building with electric power.◀

- Receptacles must be supplied by a 15- or 20-ampere circuit.

- All receptacles in garages must be GFCI protected.

- ▶The GFCI device must be readily accessible for monthly testing.◀

- If the garage or accessory building is detached, the *Code* does not require electric power, but if electric power is provided, then at least one GFCI receptacle outlet must be installed.

  Receptacles are not required in an accessory building, but if installed, they must be GFCI protected if the floor is at or below grade level; see *210.8(A)(2)* and *210.52(G)*.

Receptacles in Other Locations. All of the required receptacles discussed previously are *in addition* to any receptacles that are part of luminaires or appliances, are located inside cabinets, or are located more than 5½ ft (1.7 m) above the floor, *210.52*.

Specific Loads. Appliances such as electric ranges, ovens, air conditioners, electric heat, heat pumps, water heaters, and similar loads are discussed later in this text. See the Index for their location.

Receptacle Definitions. The *NEC* in *Article 100* defines the following:

- **Receptacle:** *A receptacle is a contact device installed at the outlet for the connection of an attachment plug.*

- **Single receptacle:** *A single receptacle is a single contact device with no other contact device on the same yoke.*

- **Multiple receptacle:** *A multiple receptacle is two or more contact devices on the same yoke.*

See Figure 3-4 for an illustration of these definitions.

Receptacle and Branch-Circuit Ratings

Table 3-2 shows the requirements of *210.21(B)* and *Table 210.21(B)(3)* where the branch circuit supplies two or more receptacles or outlets in homes. The table also points out that where the branch circuit supplies a single receptacle, the receptacle must have a rating not less than the ampere rating of the branch circuit.

Arc-Fault Circuit Interrupters (AFCIs). See Chapter 6 for complete information about AFCI requirements in bedrooms.

Replacing Existing Receptacles. See Chapter 6 for detailed discussion and illustrations about replacing existing receptacles in older homes.

Lighting Outlets. *(210.70)* The *NEC* contains minimum requirements for providing lighting for dwellings.

Switched Lighting Outlets. Per *210.70*, lighting outlets in dwellings must be installed as shown in Figure 3-10. Here is a summary of these requirements.

Habitable Rooms. In every habitable room and in every bathroom, at least one wall switch–controlled lighting outlet must be installed. Switched receptacles meet the intent of this requirement except in

TABLE 3-2

Ampere rating for receptacles connected to 15- and 20-ampere branch circuits. See *210.21(B)* and *Table 210.21(B)(3)*.

Branch-Circuit Rating	Receptacle Rating
15-ampere lighting branch circuit: Circuit has two or more receptacles or outlets.	15-ampere maximum
20-ampere small-appliance branch circuit: Circuit has two or more receptacles.	15 or 20 amperes
Individual 15-ampere branch circuit: Circuit has only a single receptacle connected.	15 or 20 amperes
Individual 20-ampere branch circuit: Circuit has only a single receptacle connected.	20 amperes

kitchens and bathrooms. Quite often, ceiling luminaires are omitted in bedrooms, living rooms, and similar rooms. Instead, the room lighting is dependent on cord-and-plug-connected table, floor, wall, and/or swag lamps.

The NFPA defines a habitable room as: *A room in a residential occupancy used for living, sleeping, cooking, and eating, but excluding bath, storage and service area, and corridors.*

As illustrated in Figure 3-11, swag lamps are not hardwired. They are connected by weaving the cord through the chain.

Wiring split-wired receptacles is the key to providing convenient control of cord-and-plug-connected lighting. The required minimum spacing of receptacles is found in *210.52*, already discussed in this chapter. When laying out and roughing in the wiring for split-wired receptacles, remember that:

- the "always hot" portion of a split-wired receptacle qualifies for the required receptacle.

- the "switched" portion of a split-wired receptacle is not considered to be one of the required receptacles but, rather, is considered to be "in addition."

If plug-in swag lighting as illustrated in Figure 3-11 is used in a dining room, a switched receptacle connected to a general lighting branch circuit must be installed. In other words, you still must install the receptacles supplied by the

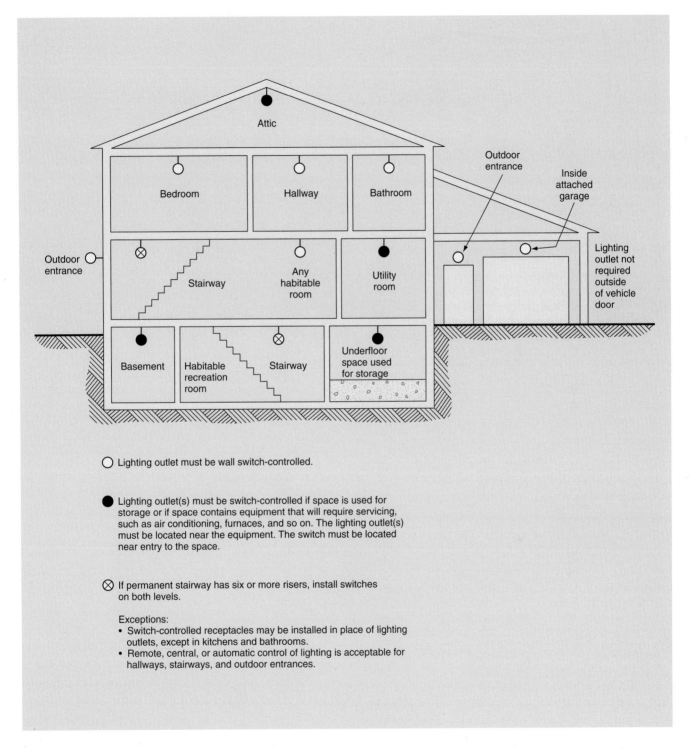

○ Lighting outlet must be wall switch-controlled.

● Lighting outlet(s) must be switch-controlled if space is used for storage or if space contains equipment that will require servicing, such as air conditioning, furnaces, and so on. The lighting outlet(s) must be located near the equipment. The switch must be located near entry to the space.

⊗ If permanent stairway has six or more risers, install switches on both levels.

Exceptions:
• Switch-controlled receptacles may be installed in place of lighting outlets, except in kitchens and bathrooms.
• Remote, central, or automatic control of lighting is acceptable for hallways, stairways, and outdoor entrances.

FIGURE 3-10   Lighting outlets required in a typical dwelling unit. (*Delmar/Cengage Learning*)

required 20-ampere small-appliance branch circuits, *210.52(B)(1), Exception No. 1,* Figure 3-12.

Where Do I Learn More about Split-Wired Receptacles?   In this residence, split-wired receptacles are used in the Living Room and the bedrooms. The wiring of split-wired receptacles is covered in Chapters 8, 12, 13, 14, and is shown on Blueprint sheet 9.

Occupancy motion sensors may be used if (1) they are in addition to the required wall switch, or (2) they have a manual override and are located where the wall switch would normally be located.

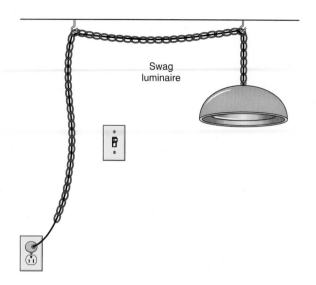

FIGURE 3-11 A receptacle controlled by a wall switch is acceptable as the required lighting outlet in rooms other than a kitchen or bathroom, *210.70(A)(1), Exception.* (*Delmar/Cengage Learning*)

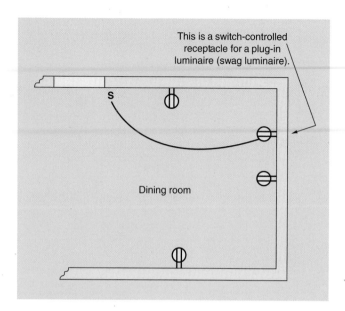

FIGURE 3-12 When a switch-controlled receptacle outlet is installed in rooms where the receptacle outlets normally would be 20-ampere small-appliance circuits (breakfast room, dining room, and so on), it must be in addition to the required small-appliance receptacle outlets, *210.52(B)(1), Exception No. 1.* (*Delmar/Cengage Learning*)

Additional Locations. At least one wall switch-controlled lighting outlet must be installed in hallways, stairways, attached garages, and detached garages with electric power, *210.70(A)(2)(a)*.

The *NEC* does *not* require that a detached garage have electric power, but if it does have electric power, then it must have at least one GFCI receptacle, and it must have wall switch–controlled lighting.

Some electrical inspectors consider a tool shed or other similar accessory building to be a storage or utility room that comes under the requirements of *210.70(A)(3)*, thus requiring a wall switch–controlled lighting outlet.

At least one wall switch–controlled lighting outlet must be installed to provide illumination on the exterior side of outdoor entrances or exits that have grade-level access to dwelling units, attached garages, and detached garages if they have electric power. A sliding glass door is considered to be an outdoor entrance. A vehicle door is not considered to be an outdoor entrance; see *210.70(A)(2)(b)*.

Although not necessarily good design practice,

- one switch could control the required outdoor lighting at more than one entrance.

- one switch could control both the required outdoor and indoor lighting of a garage.

- one outdoor luminaire, properly located, could meet the required outdoor lighting for both the house entrance and the garage entrance.

Stairways. Where there are six stair risers or more on an interior stairway, the *NEC* requires that wall switch–controlled stairway lighting must be provided on each floor level and any landing level that has an entryway, *210.70(A)(2)(c)*.

This switching requirement includes permanent stairways to basements and attics but does not include fold-down storable ladders commonly installed for access to residential attics.

The *NEC* does not tell us how to illuminate the stairway.

The *International Residential Code* is more specific about stairway lighting. It requires indoor stairways to have lighting that will illuminate the landings and treads, and that the lighting shall be located in the immediate vicinity of each landing at the top and bottom of the stairs. This *Code* also requires that the switching must be accessible at the top and bottom of the stairway, without using any step to reach a switch.

If You Don't Install Switches, What Else Is Acceptable?   Remote control, central control, motion sensors, photocells, and other automatic devices may be installed instead of wall switches for the control of lighting in hallways, stairways, and at outdoor entrances. For example, an outdoor post light controlled by a photocell does not require a separate switch.

This is permitted by the *Exception* to *210.70(A)(2)(a), (b)*, and *(c)*.

Clothes Closet Lighting.   The *NEC* does not require lighting in clothes closets, but some local codes do. Check this out in your locality. Lighting in clothes closets is particularly hazardous because of the possibility of broken light bulbs. See Chapter 8 for a detailed explanation of *Code* requirements for clothes closet lighting. The *NEC* defines a clothes closet as *A non-habitable room or space intended primarily for storage of garments and apparel.* A utility closet, broom closet, china closet, or "game" closet would not come under the stringent clothes closet requirements found in *410.16* unless it is obvious that the room or area is primarily for storage of garments and apparel.

Basements, Attics, Storage, and Other Equipment Spaces.   At least one switch-controlled lighting outlet must be installed in spaces that are used for storage or contain equipment that requires servicing, such as attics, underfloor spaces, utility rooms, and basements. Examples of equipment that may need servicing are furnaces, air conditioners, heat pumps, sump pumps, and so on. The control is permitted to be a wall switch or a lighting outlet that has an integral switch such as a pull chain switch. The lighting outlet must be at or near the equipment that may require servicing. The control must be at the usual point of entrance to the space.

The Number of Branch Circuits Required and How to Determine the Number of Lighting and Receptacle Outlets to Be Connected to One Branch Circuit

This chapter explains the *NEC* rules on where receptacle and lighting outlets are to be installed. The question of *how many* receptacle outlets and lighting outlets should be connected to each branch circuit is covered in detail in Chapter 8, where we begin the actual layout of the wiring for the residence discussed throughout this text.

Common Areas in Two-Family or Multifamily Dwellings

NEC 210.25 contains specific requirements for circuits that serve common areas of two-family or multifamily dwellings. Be careful when wiring a two-family or multifamily dwelling. For areas such as a common entrance, common basement, or common laundry area, the branch circuits for lighting, central fire/smoke/security alarms, communication systems, or any other electrical requirements that serve the common area *shall not* be served by circuits from any individual dwelling unit's power service. Common areas will require separate circuits, associated panels, and metering equipment. The reasoning for this requirement is that power could be lost in one occupant's electrical system, resulting in no lights or alarms for the other tenant(s) using the common shared area.

For example, a two-family dwelling would probably have two separate stairways to the basement. If only one stairway is provided to serve the basement of a two-family dwelling, then lighting would have to be provided and served from a branch circuit originating from each tenant's electrical system, so that each tenant would have control of the lighting. For a common entrance, two luminaires would be installed and connected to each tenant's electrical system.

Likewise, fire/smoke/security systems in a two-family dwelling shall be installed and connected to branch circuits originating from each tenant's electrical panel. This becomes an architectural design issue. This is not generally a problem in two-family dwellings but can become a problem in dwellings having more than two tenants. Rather than having to install a third watt-hour meter and associated electrical equipment on a two-family dwelling, the design of the structure can be such that there are no common areas, Figure 3-13. A fourplex will probably require a fifth watt-hour meter, four watt-hour meters for the tenants and one watt-hour meter to serve common area lighting and the fire/smoke/security systems.

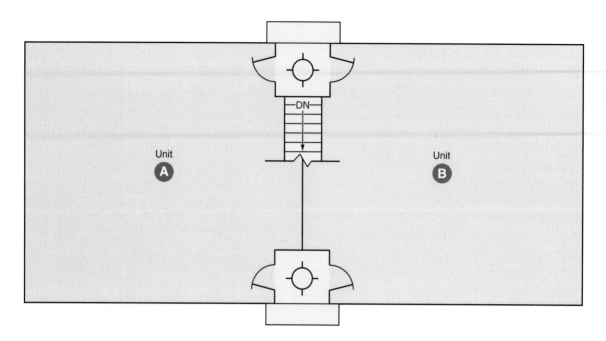

FIGURE 3-13   This outline of a two-family dwelling shows a common rear entry, a common stairway to the basement, and a common front entry. Note that each entry has one luminaire. These common area luminaires are not permitted to be connected to either of the tenants' electrical systems in the individual dwelling units, *210.25.* The choices are (1) the architect/designer must redo the layout of the building to eliminate the common entrances; (2) two luminaires must be provided in each entry, with each luminaire connected to the respective individual tenant's electrical circuit; or (3) a third meter and panel must be installed to supply the electrical loads in the common areas. (*Delmar/Cengage Learning*)

REVIEW

For the following questions, you may refer to the *NEC* to obtain the answers. To help you find these answers, references to *Code* sections are included after some of the questions. These references are not given in later reviews, where it is assumed that you are becoming familiar with the *Code*.

1. What is the meaning of calculated load? _____

   _____

2. How are branch circuits rated? See *NEC 210.3.* _____

   _____

3. How is the rating of the branch-circuit protective device affected when the conductors used are of a larger size than called for by the *Code*? See *NEC 210.3.* _____

   _____

4. What dimensions are used when measuring the area of a building? See *NEC 220.12.*

   _____

5. What spaces are not included in the floor area when computing the load in volt-amperes per square foot? See *NEC 220.12.* _____

   _____

6. What is the unit load per square foot for dwelling units? See *Table 220.12.* _____

   _____

7. According to *NEC 210.50(C)*, a laundry equipment outlet must be placed within _____ ft (m) of the intended location of the laundry equipment.

8. How is the total load in volt-amperes for lighting purposes determined? See *NEC 220.12.* _____

_____

_____

9. How is the total lighting load in amperes determined? _____

_____

10. How is the required number of branch circuits determined? _____

_____

11. What is the minimum number of 15-ampere lighting branch circuits required if the dwelling has an occupied area of 4000 ft² (368 m²)? Show all calculations.

12. How many lighting branch circuits are provided in this dwelling? _____

   a. What is the minimum load allowance for small-appliance circuits for dwellings? See *NEC 220.52(A).* _____

   b. An individual 15-ampere branch circuit is run to the receptacle outlet behind the refrigerator instead of connecting it to one of the two 20-ampere small-appliance branch circuits that are required in kitchens. For this separate circuit, an additional 1500 volt-amperes (shall) (does not have to) be added to the load calculations for dwellings. Circle the correct answer.

13. What is the smallest size wire that can be used in a branch circuit rated at 20 amperes?

_____

14. How is the load determined for outlets supplying specific appliances? See *NEC 220.14.* _____

_____

15. What type of circuits must be provided for receptacle outlets in the kitchen, pantry, dining room, and breakfast room? See *NEC 210.11(C)(1).* _____

_____

16. How is the minimum number of receptacle outlets determined for most occupied rooms? See *NEC 210.52(A).* _____

_____

_____

_____

_____

_____

_____

_____

17. In a single-family dwelling, what types of overload protection for circuits are used? See *NEC 240.6.* _____

_____

_____

_____

_____

18. The *NEC* in *Article 100* defines a *continuous load* as *a load where the maximum current is expected to continue for three hours or more.* According to *210.19(A)* and *210.20(A),* the branch-circuit conductors and overcurrent protection for a continuous load shall be sized at not less than (100%) (125%) (150%) of the continuous load. Circle the correct answer.

19. The minimum number of outdoor receptacles for a residence is _____, *NEC* _____. State the location. _____

20. The *Code* indicates the rooms in a dwelling that are required to have switched lighting outlets or switched receptacles. Write yes (switch required) or no (switch not required) for the following areas [see *NEC 210.70(A)*]:

   a. attic   _____

   b. stairway   _____

   c. crawl space (where
      used for storage)   _____

   d. hallway   _____

   e. bathroom   _____

   f. clothes closet   _____

21. Is a receptacle required in a bedroom on a 3-ft (900-mm) wall space behind the door? The door is normally left open. _____

22. The *Code* requires that at least one 20-ampere circuit feeding a receptacle outlet must be provided for the laundry. May this circuit supply other outlets? _____

23. In a basement, at least one receptacle outlet must be installed in addition to the receptacle outlet installed for the laundry equipment. This additional receptacle outlet and any other receptacle outlets in unfinished areas must be _____ protected.

24. *NEC 210.8(A)(3)* requires all receptacles installed outdoors to have GFCI protection. Explain the exception and describe the conditions of the exception.

_____

_____

_____

_____

25. Define a branch circuit. _____

_____

26. Although the *Code* contains many exceptions to the basic overcurrent protection requirements for conductors, in general, the rating of the branch-circuit overcurrent device must (not be less than) (not be more than) the ampacity of a conductor. Circle the correct answer.

27. The rating of a branch circuit is based on (circle the correct answer)

    a. the rating of the overcurrent device.

    b. the length of the circuit.

    c. the branch-circuit wire size.

28. a. A 25-ampere branch-circuit conductor is derated to 70%. It is important to provide proper overcurrent protection for these conductors. The derated conductor ampacity is _____

_____

    b. If the connected load is a "fixed" nonmotor, noncontinuous load, the branch-circuit overcurrent device may be sized at (20) (25) amperes. Circle the correct answer.

    c. If the above circuits supply receptacle outlets, the branch-circuit overcurrent device must be sized at not over (15) (20) (25) amperes. Circle the correct answer.

29. Small-appliance receptacle outlets are (included) (not included) in the 3-volt-ampere-per-square-foot calculations. Circle the correct answer.

30. If a homeowner wishes to have a switched receptacle outlet for a swag lamp in a dining room, may this switched receptacle outlet be considered to be one of the receptacle outlets required for the 20-ampere small-appliance circuits? _____

31. How many receptacle outlets are required on a 13-ft (4.0-m) wall space between two doors? Refer to *NEC 210.52(H)*. Draw the outlets in on the diagram.

32. Is a receptacle required in a hallway in a home when the hallway is 8 ft (2.5 m) long?

_____

33. A split-level home has one stairway that has six risers between two levels of the home. Which of the following choices meets *Code*? Choose the correct answer.

    a. Two three-way switches must be installed (one switch at the top of the stairs, one switch at the bottom of the stairs) to control the lighting for the stairway.

    b. One single-pole switch is permitted on either level to control the lighting for the stairway.

    c. No switches are necessary for the stairway lighting because the lighting on the upper level and the lower level provides enough illumination on the stairway.

34. A sliding glass door is installed in a family room that leads to an outdoor deck. The sliding door has one 4-ft (1.22-m) sliding section and one 4-ft (1.22-m) permanently mounted glass section. For inside the recreation room, what does the *Code* say about wall receptacle outlets relative to the receptacle's position near the sliding glass door?

_____

_____

_____

_____

_____

_____

_____

_____

35. The *Code* permits certain other receptacles to be connected to a 20-ampere small-appliance circuit. What are they? Which section of the *NEC* supports your answer?

_____

_____

_____

_____

_____

_____

_____

36. What is the minimum number of 20-ampere small-appliance circuits required by the *Code* to supply the receptacle outlets that serve the countertop surfaces in the kitchen? In which section of the *Code* is this requirement found? _____

37. Does the *Code* allow a 120-volt receptacle located behind a gas range to be supplied by a small-appliance circuit? (Yes) (No) Circle the correct answer.

38. When determining the location and number of receptacles required, fixed room dividers and railings (shall be) (need not be) considered to be wall space. Circle the correct answer.

39. No point along the wall line shall be more than 6 ft (1.8 m) from a receptacle outlet. This requirement (does apply) (does not apply) to unfinished residential basements. Circle the correct answer.

40. Receptacle outlets that serve countertop surfaces in kitchens are required to be _____ protected and shall be installed so that no point along the wall line above the base cabinets is more than [18 in. (450 mm)] [24 in. (600 mm)] [36 in. (900 mm)] from a receptacle. Circle the correct answer.

41. In kitchens, the *Code* requires that at least (one) (two) receptacle(s) be installed to serve the countertop surfaces on an island or peninsula. Circle the correct answer. Where in the *Code* is this requirement found? *NEC* _____

42. In the past, it was common practice to connect the lighting and receptacle(s) in a bathroom to the same circuit. Because of overloads caused by high-wattage grooming appliances, the *Code* now requires that these receptacle(s) be supplied by a separate 20-ampere branch circuit. Where in the *Code* is this requirement found? *NEC* _____

43. If a residence has two bathrooms, the *Code* states:

   a. The receptacles in each bathroom must be connected to a separate 20-ampere branch circuit. For two bathrooms, this would mean two circuits. (True) (False) Circle the correct answer.

   b. The receptacles in both bathrooms are permitted to be served by the same 20-ampere branch circuit. (True) (False) Circle the correct answer.

44. In your own words, explain the GFCI exemptions for receptacle outlets installed in unfinished basements. Give some examples.

   _____

   _____

   _____

   _____

45. *NEC* _____ of the *Code* prohibits connecting lighting and/or fire/smoke/security systems to an individual tenant's electrical source where common (shared) areas are present, in which case the lighting provided for both tenants is connected on one tenant's electrical source.

46. When installing weatherproof outdoor receptacles high up under the eaves of a house intended to be used for plugging in deicing and snow-melting cable for the rain gutters, must these receptacles be GFCI protected for personnel protection? (Yes) (No) Circle the correct answer.

47. An individual 20-ampere branch circuit supplies a receptacle for a refrigerator. The receptacle shall be or is permitted to be

   a. a duplex receptacle.

   b. a single receptacle.

   Explain why you chose your answer. _____

   _____

   _____

   _____

48. A balcony on the second floor of a new residence measures 4 ft (1.22 m) × 10 ft (3.05 m). The balcony is accessed by a sliding door. Is a receptacle required for this balcony? Circle the correct answer, and give the *NEC* section where the answer is found.

   Yes   No   *NEC* _____

49. a. What section of the *NEC* permits installing a receptacle in a kitchen countertop?

   _____

   b. A receptacle installed on a countertop is required to be a (conventional receptacle) (receptacle listed for the application). Circle the correct answer.

Conductor Sizes and Types, Wiring Methods, Wire Connections, Voltage Drop, Neutral Conductor Sizing for Services

OBJECTIVES

After studying this chapter, you should be able to

- determine the current-carrying capacity (ampacity) of conductors.

- understand overcurrent protection for conductors and maximum loading of branch circuits.

- understand aluminum conductors and the possible fire hazards if they are not properly installed.

- know the *NEC* installation requirements for all types of cables and raceways.

- understand the special ampacity ratings of service-entrance conductors.

- make voltage-drop calculations.

- learn an alternate cost and time-saving method permitted to bring nonmetallic-sheathed cables into the top of a surface-mounted panel.

CONDUCTORS

Throughout this text, all references to conductors are for copper conductors, unless otherwise stated.

Wire Size

The copper wire used in electrical installations is graded for size according to the American Wire Gauge (AWG) Standard. The wire diameter in the AWG standard is expressed as a whole number. The higher the AWG number, the smaller the wire. AWG sizes vary from fine, hairlike wire used in coils and small transformers to very large diameter wire required in industrial wiring to handle heavy loads.

The wire may be a single strand (solid conductor), or it may consist of many strands. Each strand of wire acts as a separate conducting unit. The wire size used for a circuit depends on the maximum current to be carried. The *NEC* in *Table 210.24*

shows that the minimum conductor size for branch-circuit wiring is 14 AWG. Smaller size conductors are permitted for bell wiring, thermostat wiring, communications wiring, intercom wiring, luminaire wires, and similar low-energy circuits.

Don't Get Confused!

In the past, conductor size was shown as, for example, "No. 12 AWG." This same conductor now appears in the *NEC* as "12 AWG." The "12" is a size, not a quantity.

Table 4-1 shows typical applications for different size conductors.

Solid and Stranded Conductors

For residential wiring, sizes 14, 12, and 10 AWG conductors are generally solid when the wiring method is nonmetallic-sheathed cable or armored cable.

TABLE 4-1

Conductor applications chart.

Conductor Size	Overcurrent Protection	Typical Applications (Check wattage and/or ampere rating of load to select the correct size conductors based on *Table 310.15(B)(16)*.)
20 AWG	Class 2 circuit transformers provide overcurrent protection; see *Article 725*.	Telephone wiring is usually 20 or 22 AWG.
18 AWG	Class 2 circuit transformers provide overcurrent protection; see *Article 725*. For motor control circuits, 7 amperes, see *Table 430.72(B)*.	Low-voltage wiring for thermostats, chimes, security, remote control, home automation systems, etc. For these types of installations, 18 or 20 AWG conductors can be used depending on the connected load and length of circuit.
16 AWG	Class 2 circuit transformers provide overcurrent protection; see *Article 725*. For motor control circuits, 10 amperes, see *Table 430.72(B)*.	Same applications as above. Good for long runs to minimize voltage drop.
14 AWG	15 amperes	Typical lighting branch circuits.
12 AWG	20 amperes	Small-appliance branch circuits for the receptacles in kitchens and dining rooms. Also laundry receptacles, bathroom and workshop receptacles. Often used as the "home run" for lighting branch circuits. Some water heaters.
10 AWG	30 amperes	Most clothes dryers, built-in ovens, cooktops, some central air conditioners, some water heaters, some heat pumps.
8 AWG	40 amperes	Ranges, ovens, heat pumps, some large clothes dryers, large central air conditioners, heat pumps.
6 AWG	50 amperes	Electric ranges, electric furnaces, heat pumps.
4 AWG	70 amperes	Electric furnaces, feeders to subpanels.
3 AWG and larger	100 amperes	Main service-entrance conductors, feeders to subpanels, electric furnaces.

In a raceway, the preference is to use stranded 10 AWG conductors because of the flexibility and ease of handling and pulling in the conductors.

NEC 310.106(C) requires that unless permitted elsewhere in the *Code*, conductors 8 AWG and larger must be stranded where installed in a raceway.

Ampacity

Ampacity means *the maximum current, in amperes, that a conductor can carry continuously under the conditions of use without exceeding its temperature rating.** This value depends on the conductor's cross-sectional area, whether the conductor is copper or aluminum, and the type of insulation around the conductor. Ampacity values, also referred to as current-carrying capacity, are found in *Article 310*. The most commonly used conductor ampacities are found in *Table 310.15(B)(16)*.

The allowable ampacity values in the tables are valid where there are no more than three current-carrying conductors in a raceway or cable and where the temperature does not exceed 86°F (30°C). These are considered the *conditions of use* for the conductors.

When there are more than three current-carrying conductors in a raceway or cable, the allowable ampacity values are adjusted according to the factors in *Table 310.15(B)(3)(a)*.

When the ambient temperature exceeds 86°F (30°C), the allowable ampacity values are corrected according to the factors found in *Table 310.15(B)(2(a)*.

If both conditions are present, then both penalties (adjustment and correction factors) must be applied. More on conductor ampacities, derating, adjusting, and correction factors is presented in Chapter 18.

Conductors must have an ampacity not less than the maximum load that they are supplying, as shown in Figure 4-1. All conductors of a specific branch circuit must have an ampacity of the branch circuit's rating, as shown in Figure 4-2. There are exceptions to this rule, such as taps for electric ranges (see Chapter 20).

Ampacity of Flexible Cords

Table 4-2 shows the allowable ampacities for some sizes of flexible cords. Refer to *Table 400.5(A) (1) and (A)(2)* in the *NEC* for other specific types and sizes of flexible cords.

*Reprinted with permission from NFPA 70-2011.

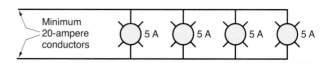

FIGURE 4-1 Branch-circuit conductors shall have an ampacity not less than the maximum load to be served, *210.19(A)*. See *210.23* for permissible loads. (*Delmar/Cengage Learning*)

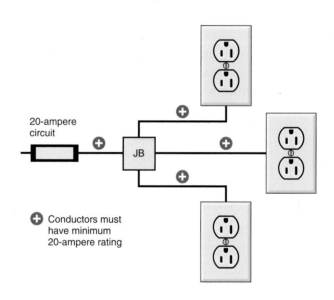

FIGURE 4-2 All conductors in this circuit supplying receptacles shall have an ampacity of not less than the rating of the branch circuit. In this example, 20-ampere conductors must be used; see *NEC 210.19(A)(2)* and *Table 210.24*. (*Delmar/Cengage Learning*)

Overcurrent protection requirements for extension cords are found in *NEC 240.5*.

Conductor Sizing

The diameter of wire is usually given in a unit called a mil. A *mil* is defined as one-thousandth of an inch (0.001 inch). Mils squared are known as circular mils.

Table 8 in *Chapter 9* of the *NEC* clearly shows that conductors are expressed in AWG numbers from 18 (1620 circular mils) through 4/0 (211,600 circular mils). Wire sizes larger than 4/0 are expressed in circular mils.

Large conductors, such as 500,000 circular mils, are generally expressed as 500 kcmil. Because the letter "k" designates 1000, the term *kcmil* means "thousand circular mils." This is much easier to express in both written and verbal terms.

TABLE 4-2		
Allowable ampacities flexible cords.		
Conductor Size AWG Copper	**Cords in Which Three Conductors Carry Current** Example: black, white, red, plus equipment grounding conductor	**Cords in Which Two Conductors Carry Current** Example: black, white, plus equipment grounding conductor
18	7	10
16	10	13
14	15	18
12	20	25
10	25	30
8	35	40
6	45	55
4	60	70

Allowable ampacities for some of the more common flexible cords used for residential applications. These values were taken from *Table 400.5(A)(1), NEC.* Refer to the *NEC* table for other specific types and sizes.

Older texts used the term *MCM*, which also means "thousand circular mils." The first letter "M" refers to the Roman numeral that represents 1000. Thus, 500 MCM means the same as 500 kcmil. Roman numerals are no longer used in the electrical industry for expressing conductor sizes.

Overcurrent Protection for Conductors

Conductors must be protected against overcurrent by fuses or circuit breakers rated not more than the ampacity of the conductors, *NEC 240.4*.

NEC 240.4(B) permits the use of the next higher standard overcurrent device rating, as shown in *240.6(A)*. This permission is granted *only* when the overcurrent device is rated 800 amperes or less. For example, from *Table 310.15(B)(16)*, a 6 AWG conductor with Type THWN insulation rated for 75°C has an allowable ampacity of 65 amperes. In *240.6(A)*, we find that the next higher standard rating for an overcurrent device is 70 amperes. Nonstandard ampere ratings of fuses and breakers are also permitted.

NEC 240.4(D) spells out the maximum overcurrent protection for small branch-circuit conductors. Table 4-3 illustrates the requirements of *240.4(D)*.

There are exceptions for motor branch circuits and HVAC equipment. Motor circuits are discussed in Chapter 19 and Chapter 22. HVAC circuits are discussed in Chapter 23.

TABLE 4-3	
Maximum overcurrent protection for 14, 12, and 10 AWG copper conductors.	
Conductor Size	**Maximum Overcurrent Protection**
14 AWG copper	15 amperes
12 AWG copper	20 amperes
10 AWG copper	30 amperes

Branch-Circuit Rating

The rating of the overcurrent device (OCD) determines the rating of a branch circuit, as shown in Figure 4-3.

⎍ PERMISSIBLE LOADS ON BRANCH CIRCUITS (210.23)

The *NEC* is very specific about the loads permitted on branch circuits. Here is a recap of these requirements.

- The load shall not exceed the branch-circuit rating.

- The branch circuit must be rated 15, 20, 30, 40, or 50 amperes when serving two or more outlets, *NEC 210.3*.

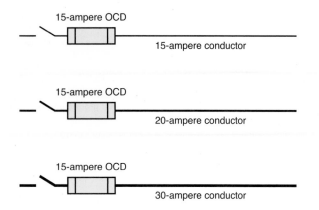

FIGURE 4-3   All three of these branch circuits are rated as 15-ampere circuits, even though larger conductors were used for some other reason, such as solving a voltage-drop problem. The rating of the overcurrent device (OCD) determines the rating of a branch circuit, *210.3.* (*Delmar/Cengage Learning*)

- An individual branch circuit may supply any size load.

- 15- and 20-ampere branch circuits

  a. may supply lighting, other equipment, or both types of loads.

  b. for cord-and-plug-connected equipment, shall not exceed 80% of the branch-circuit rating.

  c. for equipment fastened in place, shall not exceed 50% of the branch-circuit rating if the branch circuit also supplies lighting, other cord-and-plug-connected equipment, or both types of loads.

- The 20-ampere small-appliance circuits in homes shall not supply other loads, *NEC 210.11(C)(1)*.

- 30-ampere branch circuits may supply equipment such as dryers, cooktops, water heaters, and so forth. Cord-and-plug-connected equipment shall not exceed 80% of the branch-circuit rating.

- 40- and 50-ampere branch circuits may supply cooking equipment that is fastened in place, such as an electric range, as well as HVAC equipment.

- Over 50-ampere-rated branch circuits are for electric furnaces, large heat pumps, air-conditioning equipment, large double ovens, and similar large loads.

ALUMINUM CONDUCTORS

The conductivity of aluminum is not as great as that of copper for a given size. For example, checking *Table 310.15(B)(16)*, an 8 AWG Type THHN copper conductor has an allowable ampacity of 55 amperes. An 8 AWG Type THHN aluminum or copper-clad aluminum conductor has an ampacity of 45 amperes.

In *240.4(D)*, the maximum overcurrent protection for a 12 AWG copper conductor is 20 amperes but only 15 amperes for a 12 AWG aluminum or copper-clad aluminum conductor.

Aluminum conductors have a higher resistance compared to a copper conductor of the same size. When considering voltage drop, a conductor's resistance is a key ingredient. Voltage-drop calculations are discussed later on in this chapter.

Common Connection Problems

Some common problems associated with aluminum conductors when not properly connected may be summarized as follows:

- A corrosive action is set up when dissimilar wires come in contact with one another when moisture is present.

- The surface of aluminum oxidizes as soon as it is exposed to air. If this oxidized surface is not penetrated, a poor connection results. When installing aluminum conductors, particularly in large sizes, an inhibitor (antioxidant) is brushed onto the aluminum conductor, and then the conductor is scraped with a stiff brush where the connection is to be made. The process of scraping the conductor breaks through the oxidation, and the inhibitor keeps the air from coming into contact with the conductor. Thus, further oxidation is prevented. Aluminum connectors of the compression type usually have an inhibitor paste already factory installed inside of the connector.

- Aluminum wire expands and contracts to a greater degree than does copper wire for an equal load. This is referred to as *creep* or *cold flow*. This factor is another possible cause of a poor connection. Crimp connectors for aluminum conductors are usually longer than those for comparable copper conductors, thus resulting in greater contact surface of the conductor in the connector.

- Older technology aluminum such as Alloy 1350 experienced the above problems. The newer technology aluminum such as AA-8000 has much better conductivity, creep resistance, and strength. Still, extreme care must be taken when terminating aluminum conductors. The manufacturer of the aluminum conductor or connector will provide detailed installation instructions.

Proper Installation Procedures

Proper, trouble-free connections for aluminum conductors require terminals, lugs, and/or connectors that are suitable for the type of conductor being installed.

Terminals on receptacles and switches must be suitable for the conductors being attached. Table 4-4 shows how the electrician can identify these terminals. Listed connectors provide proper connection when properly installed. See *NEC 110.14, 404.14(C), and 406.2(C).*

Electrical Fires Caused by Improper Connections of Aluminum Conductors

Many electrical fires have been caused by improper connections and terminations of aluminum conductors. Most often, the problem is not the aluminum conductors. The problem is with the connections and terminations.

Many of these problems occurred in the meter base and at the main panel where aluminum service entrance conductors were installed years ago. In the course of time, the connections failed. Troubles also happened where aluminum conductors were terminated on receptacles and switches.

The Consumer Product Safety Commission (CPSC) (Washington, DC, 20207) publishes an excellent free booklet (#516) entitled *Repairing Aluminum Wiring* that explains what you can do with connections and terminations for the small aluminum wires that were installed years ago for the branch-circuit wiring of receptacles and switches.

The CPSC states that older homes wired with aluminum conductors are 55 times more likely to have one or more connections that will reach "Fire Hazard Conditions" than are homes wired with copper.

TABLE 4-4

Terminal identification markings and the acceptable types of conductors permitted to be connected to a specific type of terminal.

Type of Device	Marking on Terminal or Conductor	Connector Permitted
15- or 20-ampere receptacles and switches	CO/ALR	Aluminum, copper, copper-clad aluminum
15- and 20-ampere receptacles and switches	None	Copper, copper-clad aluminum
30-ampere and greater receptacles and switches	AL/CU	Aluminum, copper, copper-clad aluminum
30-ampere and greater receptacles and switches	None	Copper only
Screwless pressure terminal connectors of the push-in type	None	Copper or copper-clad aluminum
Wire connectors	AL	Aluminum
Wire connectors	AL/CU or CU/AL	Aluminum, copper, copper-clad aluminum
Wire connectors	CC	Copper-clad aluminum only
Wire connectors	CC/CU or CU/CC	Copper or copper-clad aluminum
Wire connectors	CU or CC/CU	Copper only
Any of the above devices	COPPER OR CU ONLY	Copper only

Although the *NEC*, notably in *110.5, 110.14, 240.4(D), 310.106(B), Table 310.15(B)(16),* and other sections, recognizes aluminum conductors in sizes 12 AWG and larger, there are some local electrical codes that do not allow aluminum conductors to be used in branch-circuit sizes. Check this out before using aluminum conductors.

Some of the signs of potential problems might be

- wall switches and receptacles that feel warm to the touch.

- the smell of burning plastic or rubber at switches, receptacle, main panel, or meter.

- flickering lights.

- lights getting bright or dim.

- circuits that don't work; some circuits are on, others are off.

- TV shifting to another channel for no apparent reason.

- appliances with electronic controls shutting off or changing settings for no apparent reason.

- security system sounding off for no apparent reason.

Wire Connections

Wire connectors are known in the trade by such names as *screw terminal, pressure terminal connector, wire connector, Wing-Nut, Wire-Nut, Scotchlok, Twister, split-bolt connector, pressure cable connector, solderless lug, soldering lug, solder lug,* and others.

All cartons and manufacturers' literature show the size, number, and types of conductors permitted.

Solder-type lugs and connectors are rarely, if ever, used today. In fact, connections that depend on solder are *not* permitted for connecting service-entrance conductors to service equipment, *230.81.* Nor is solder permitted for grounding and bonding connections, *250.8.* The labor costs and time spent make the use of the solder-type connections prohibitive.

Solderless connectors designed to establish *Code*-compliant electrical connections are shown in Figure 4-4.

As with the terminals on wiring devices (switches and receptacles), wire connectors must be marked

"AL" when they are to be used with aluminum conductors. This marking is found on the connector itself, or it appears on or in the shipping carton.

Connectors marked "AL/CU" are suitable for use with aluminum, copper, or copper-clad aluminum conductors. This marking is found on the connector itself, or it appears on or in the shipping carton.

Connectors not marked "AL" or "AL/CU" are for use with copper conductors only.

Unless specially stated on or in the shipping carton, or on the connector itself, conductors made of copper, aluminum, or copper-clad aluminum *may not* be used in combination in the same connector.

When combinations are permitted, the connector will be identified for the purpose and for the conditions when and where they may be used. The conditions usually are limited to dry locations only. There are some "twist-on" wire connectors that have recently been listed for use with combinations of copper and aluminum conductors.

The preceding data are found in *110.14* and in the Underwriters Laboratories (UL) Standards. When terminating large conductors such as service-entrance conductors, be sure to read any instructions that might be included with the equipment where these conductors will be terminated. Terminations for all electrical connections to devices and equipment shall be tightened to the torque as required by the manufacturer of such electrical device or equipment, *110.12(D)*.

Old Houses Wired with Aluminum Conductors

If you come across a house wired with aluminum conductors, it was probably wired back in the mid-sixties or early seventies. You will very likely find problems where the conductors are terminated on switches and receptacles. Take special care when changing out switches and receptacles or making splices. Use wiring devices that bear the AL/CU marking. You can also use special "pigtails" designed to make the connection between aluminum and copper conductors and terminals. One such connector is COPALUM—a compression (crimp) connector. Apply an antioxidant to the splice or connection as recommended by the manufacturer of the connector.

Crimp connectors used to splice and terminate 20 AWG to 500 kcmil aluminum-to-aluminum, aluminum-to-copper, or copper-to-copper conductors.

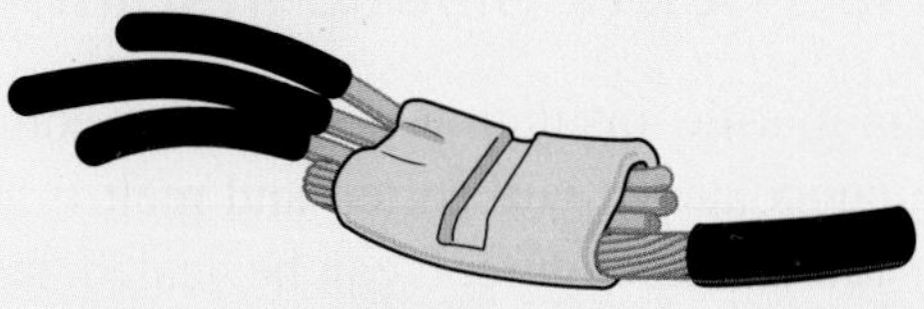

Properly crimp then tape

Connectors used to connect wires together on combinations of 18 AWG through 6 AWG conductors. They are twist-on, solderless, and tapeless.

*Wire-Nut,® Wing-Nut,® and Twister® are registered trademarks of Ideal Industries, Inc. Scotchlok® is a registered trademark of 3M.

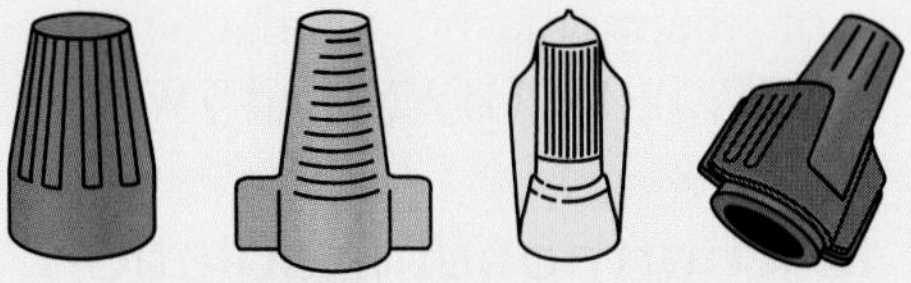

Wire connectors variously known as Wire-Nut,® Wing-Nut,® Twister,® and Scotchlok®

Connectors used to connect wires together in combinations of 16, 14, and 12 AWG conductors. They are crimped on with a special tool, then covered with a snap-on insulating cap.

Crimp-type wire connector and insulating cap

Solderless connectors are available in sizes 4 AWG through 500 kcmil conductors. They are used for one solid or one stranded conductor only, unless otherwise noted on the connector or on its shipping carton. The screw may be of the standard screwdriver slot type, or it may be for use with an allen wrench or socket wrench.

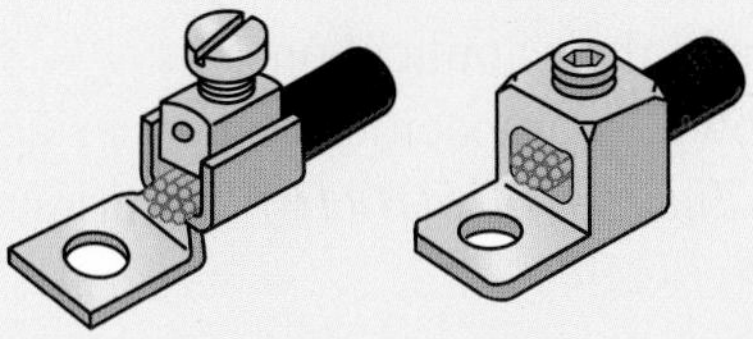

Solderless connectors

Compression connectors are used for 8 AWG through 1000 kcmil conductors. The wire is inserted into the end of the connector, then crimped on with a special compression tool.

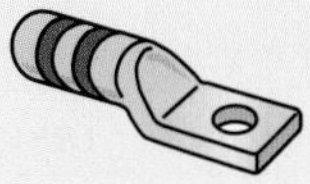

Compression connector

Split-bolt connectors are used for connecting two conductors together, or for tapping one conductor to another. They are available in sizes 10 AWG through 1000 kcmil. They are used for two solid and/or two stranded conductors only, unless otherwise noted on the connector or on Its shipping carton.

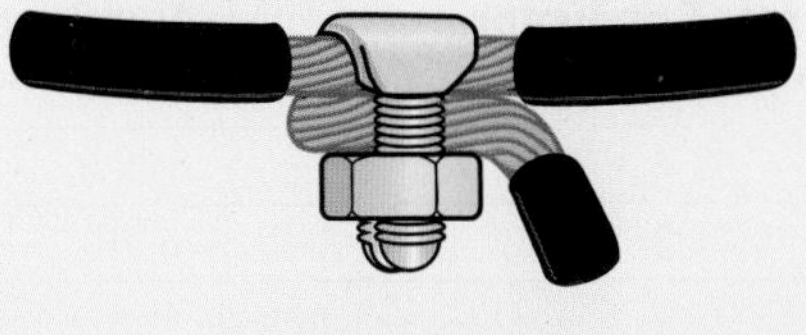

Split-bolt connector

FIGURE 4-4   Types of wire connectors. (*Delmar/Cengage Learning*)

CONDUCTOR INSULATION

The *NEC* requires generally that all conductors be insulated, *310.106(D)*. There are a few exceptions, such as the permission to use a bare neutral conductor for services, and bare equipment grounding conductors.

NEC Table 310.104(A) shows many types of conductors, their applications, and insulations. The conductors most commonly used fall into the thermoplastic and thermoset categories.

What Is Thermoplastic Insulation?

Thermoplastic insulation is the most common. It is like chocolate. It will soften and melt if heated above its rated temperature. It can be heated, melted, and reshaped. Thermoplastic insulation will stiffen at temperatures colder than 14°F (−10°C). Typical examples of thermoplastic insulation are Types THHN, THHW, THW, THWN, and TW.

What Is Thermoset Insulation?

Thermoset insulation can withstand higher and lower temperatures. It is like baking a cake. Once the ingredients have been mixed, heated, and formed, it can never be reheated and reshaped. If heated above its rated temperature, it will char and crack. Typical examples of thermoset insulation are Types RHH, RHW, XHH, and XHHW.

Table 310.104(A) lists the various conductor insulations and their applications.

The allowable ampacities of copper conductors are given in *Table 310.15(B)(16)* for various types of insulation. *Table 310.15(B)(7)* is a table of conductor ampacities that are greater than the ampacities found in *Table 310.15(B)(16)*. This table is permitted to be used *only* for dwelling unit 120/240-volt, 3-wire, single-phase service-entrance conductors, service-lateral conductors, and feeder conductors that serve as the main power feeder to a dwelling unit. This table is based on the tremendous known load diversity found in homes. These special ampacities must not be used for feeder conductors that do not serve as the main power feeder to a dwelling. An example of this would be the feeder from the Main Panel A to Panel B in the residence discussed later on in this text.

Table 4-5 shows the special ampacities discussed previously. This table is the same as *Table 310.15(B)(7)* in the *NEC*.

The insulation covering wires and cables used in house wiring is usually rated at 600 volts or less. Exceptions to this statement are low-voltage wiring and luminaire wiring.

TABLE 4-5

Special ampacity ratings for residential 120/240-volt, 3-wire, single-phase service-entrance conductors, service-lateral conductors, and feeder conductors that serve as the main power feeder to a dwelling unit.

CONDUCTOR TYPES RHH, RHW, RHW-2, THHN, THHW, THW, THW-2, THWN, THWN-2, XHHW, XHHW-2, SE, USE, USE-2.

Copper Conductor (AWG or kcmil)	Aluminum or Copper-Clad Aluminum Conductors (AWG or kcmil)	Service or Feeder Rating (Amperes)
4	2	100
3	1	110
2	1/0	125
1	2/0	150
1/0	3/0	175
2/0	4/0	200
3/0	250 kcmil	225
4/0	300 kcmil	250
250 kcmil	350 kcmil	300
350 kcmil	500 kcmil	350
400 kcmil	600 kcmil	400

Notes: Where the conductors will be exposed to temperatures in excess of 86°F (30°C), apply the correction factors found in *Table 310.15(B)(2)(a)*.

Conductors having a suffix "2," such as THWN-2, are permitted for continuous use at 90°C in wet or dry locations. Though these conductors cannot be used at their 90°C ampacities, the higher ampacities can be used for derating for elevated temperatures or for more than three current-carrying conductors installed without maintaining spacing.

Older Types of Conductor Insulation

Although Type THHN/THWN is the most common building wire used today, older types of conductor insulation might still be found, such as Types A, RH, RU, RUH, RUW, T, TW, and THW. Some of these are still shown in the *NEC* but may no longer be manufactured. Some are still manufactured but are difficult, if not impossible, to find at an electrical supply house. Others are available only as a special-order product.

⎓⤳⊣ WET, DAMP, DRY, AND SUNLIGHT LOCATIONS

Conductors are listed for specific locations. Be sure that the conductor you use is suitable for the location. Table 4-6 shows some of the more typical conductors used in house wiring. Here are some definitions taken directly from the *NEC:*

- **Damp Location.** *Locations protected from weather and not subject to saturation with water or other liquids, but subject to moderate degrees of moisture. Examples of such locations include partially protected locations under canopies, marquees, roofed open porches, and like locations; and interior locations subject to moderate degrees of moisture, such as some basements, some barns, and some cold-storage warehouses.** (See Figure 11-4.)

- **Dry Location.** *A location not normally subject to dampness or wetness. A location classified as dry may be temporarily subject to dampness or wetness, as in the case of a building under construction.**

- **Wet Location.** *Installations underground or in concrete slabs or masonry in direct contact with*

*Reprinted with permission from NFPA 70-2011.

TABLE 4-6

Typical conductors used for residential wiring.

Trade Name	Type Letter	Maximum Operating Temp.	Application Provisions	Insulation	AWG	Outer Covering
Heat-resistant thermoplastic	THHN	90°C	Dry and damp locations	Flame-retardant, heat-resistant thermoplastic	14–1000 kcmil	Nylon jacket or equivalent
Moisture- and heat-resistant thermoplastic	THHW	75°C 90°C	Wet location Dry location	Flame-retardant, moisture- and heat-resistant thermoplastic	14–1000 kcmil	None
Moisture- and heat-resistant thermoplastic	THWN Note: If marked THWN–2, okay for 90°C in dry or wet locations	75°C	Dry and wet locations	Flame-retardant, moisture- and heat-resistant thermoplastic	14–1000 kcmil	Nylon jacket or equivalent
Moisture- and heat-resistant thermoplastic	THW Note: If marked THW–2, okay for 90°C in dry or wet locations	75°C	Dry and wet locations	Flame-retardant, moisture- and heat-resistant thermoplastic	14–2000 kcmil	None
Underground feeder and branch-circuit single conductor or multiconductor cable. See *Article 339*	UF	60°C 75°C	See *Article 339* See *Article 339*	Moisture-resistant Moisture-resistant	14 through 4/0	Integral with insulation

the earth; in locations subject to saturation with water or other liquids, such as vehicle washing areas; and in unprotected locations exposed to weather. The interior of raceways installed in wet locations is considered a wet location. Insulated conductors and cables in these locations shall be listed for wet locations. See *NEC 300.5(B), 300.9, and 310.10(C)*. Also see Figure 11-4.

- **Locations Exposed to Direct Sunlight.** Insulated conductors and cables used where exposed to direct rays of the sun shall be listed, or listed and marked as being "sunlight resistant."

See *NEC 310.10(D), 310.15(A)(3), 310.104*, and *Table 310.104(A)* for additional information.

Conductor Insulation Temperature Ratings

Conductors are also rated as to the temperature their insulation system can withstand. Even though the *NEC* provides Fahrenheit temperatures for conductors, they are not used by the electrical industry. Only the Celsius temperatures are used in referring to the temperature rating of insulation. The temperature rating of conductor insulations commonly used in residential wiring is:

	Celsius	Fahrenheit
Type TW	60°	140°
Type THWN	75°	167°
Type THHN	90°	194°

Without question, conductors with Type THHN/THWN insulation are the most popular and most commonly used, particularly in the smaller sizes, because of their small diameter (easy to handle; more conductors of a given size permitted in a given size raceway) and their suitability for installation where high temperatures are encountered, such as attics, buried in insulation, and supplying recessed fixtures.

The temperature rating of nonmetallic-sheathed cable is 90°C. However, *NEC 334.80* states that the allowable ampacity is that of the 60°C column in *Table 310.15(B)(16)*. The 90°C ampacity is permitted to be used for derating purposes.

*Reprinted with permission from NFPA 70-2011.

Multiple-Type Designations

Some conductors are listed for more than one application. For example, a conductor marked THHN/THWN is rated 600 volts, 90°C when used in dry locations, *and* 600 volts, 75°C when used in wet locations. As always, read the surface marking on the conductor insulation, on the tag, on the reel, or on the carton. The *NEC* covers *Conductors for General Wiring* in *Article 310*.

The Weakest Link of the Chain

A conductor has two ends!

Selecting a conductor ampacity based solely on the allowable ampacity values found in *Table 310.15(B)(16)* is a very common mistake that can prove costly.

In any given circuit, we have an assortment of electrical components that have different maximum temperature ratings. Our job is to find out which component is the "weakest link" in the system and do our circuit design based on the lowest rated component in the circuit. Maximum temperature ratings are found in the *NEC* and in the UL Standards.

- **Circuit breakers:** UL 489. Generally marked "75°C Only" or "60°C/75°C."

- **Conductors:** *Table 310.104(A)*, *Table 310.15(B)(16)*, and UL 83.

- **Disconnect switches:** UL 98. Generally marked "75°C Only" or "60°C/75°C."

- **Panelboards:** UL 67. Panelboards are generally marked 75°C. The temperature rating of a panelboard has been established as an assembly with circuit breakers in place. Do not use the circuit breakers' temperature marking by itself. It is the temperature marking on the assembly that counts.

- **Receptacles and attachment-plug caps:** UL 498. Most 15- and 20-ampere branch circuits have receptacles that are marked with a wire size and not a temperature rating. As a result, the default temperature rating in *NEC 110.14(C)* applies. This results in a maximum temperature rating of 60°C. Some 30-ampere receptacles are rated 60°C and some are rated for 75°C. Larger 50-ampere receptacles are

TABLE 4-7

Types of conductor temperature ratings permitted for various terminal temperature ratings in conformance to the *NEC* and the UL Standards.

Termination Rating	Conductor Insulation Rating		
	60°C	75°C	90°C
60°C	Okay	Okay (at 60°C ampacity)	Okay (at 60°C ampacity)
75°C	No	Okay	Okay (at 75°C ampacity)
60/75°C	Okay	Okay (at 60°C or 75°C ampacity)	Okay (at 60°C or 75°C ampacity)
90°C	No	No	Okay (only if equipment has a 90°C rating)

rated 75°C. You will want to verify the marking on the device.

- **Snap switches:** The safety standard is UL 20. Most 15- and 20-ampere branch circuits have snap switches with a maximum terminal temperature rating of 60°C.

- **Wire splicing devices and connectors:** Wire-Nuts, Wing-Nuts, Twisters, and Scotchloks for copper conductors (UL 486) and for aluminum conductors (UL 486B). These types of connectors are commonly rated 105°C.

Table 4-7 indicates the correct applications for conductors and terminations.

Another key *NEC* requirement is found in *110.14(C)*, *"Termination Limitations,"* where we find the following:

For circuits rated 100 amperes or less, or marked for conductor sizes 14 through 1 AWG, unless otherwise identified, the terminals on wiring devices, switches, breakers, motor controllers, and other electrical equipment are based on the ampacity of 60°C conductors. It is acceptable to install conductors having a higher temperature rating, such as 90°C THHN, but you must use the 60°C ampacity values.

For circuits rated over 100 amperes, or marked for conductors larger than 1 AWG, the conductors

installed must have a minimum 75°C rating. The ampacity of the conductors is based on the 75°C values. It is acceptable to install conductors having a higher temperature rating, such as 90°C THHN, but you must use the 75°C ampacity values.

When using high-temperature insulated conductors where adjustment factors are to be applied, such as *derating* for more than three current-carrying conductors in a raceway or cable, or *correcting* where high temperatures are encountered, the final results of these adjustments (in amperes) must comply with the requirements of *110.14(C)*. Similar requirements are found in the UL Standards.

Examples of applying *derating factors* and *correcting factors* are found in Chapter 18.

Temperature ratings of conductors and cables are discussed later in this chapter.

In *240.4(D)*, there is a built-in code compliance limitation for applying small branch-circuit conductors. Here we find that the maximum overcurrent protection is 15 amperes for 14 AWG, 20 amperes for 12 AWG, and 30 amperes for 10 AWG.

From *Table 310.15(B)(16)* we find that

- 14 AWG copper conductors' allowable ampacity is

  15 amperes in the 60°C column
  20 amperes in the 75°C column
  25 amperes in the 90°C column

- 12 AWG copper conductors' allowable ampacity is

  20 amperes in the 60°C column
  25 amperes in the 75°C column
  30 amperes in the 90°C column

- 10 AWG copper conductors' allowable ampacity is

  30 amperes in the 60°C column
  35 amperes in the 75°C column
  40 amperes in the 90°C column

There are exceptions to this maximum overcurrent protection rule for motor, air-conditioner, and heat pump branch circuits. These exceptions are covered elsewhere in this text.

In summary, we cannot arbitrarily use the ampacity values for conductors as found in *Table 310.15(B)(16)*. Conductor ampacity is not a "stand-alone" issue. We must also consider the temperature limitations of the equipment, such as panelboards, receptacles, snap switches, receptacles, connectors, and so on. The lowest maximum temperature-rated device in an electrical system is the weakest link, and it is the weakest link on which we must base our ultimate conductor ampacity and conductor insulation decision.

Knob-and-Tube Wiring

Older homes were wired with open, individual, insulated conductors supported by porcelain *knobs* and *tubes*. In its day, knob-and-tube wiring served its purpose well. This text does not cover knob-and-tube wiring because this method is no longer used in new construction. On occasion, it is necessary to make some modifications to knob-and-tube wiring

when doing remodel work. The *NEC* covers this wiring method in *Article 394*.

Watch out when working on old knob-and-tube wiring. You will possibly find the old rubber insulation on the conductors dried out and brittle from the many years of heat generated by luminaires and from the extreme heat in attics, particularly when the conductors are completely buried in thermal insulation. The conductor insulation might fall apart when touched or moved. This is a potential fire hazard!

If you do find this situation, one fix is to slide insulation (heat shrink) tubing (available from electrical distributors) over the conductors to reinsulate them.

Figure 4-5 is a photo of typical knob-and-tube wiring from the 1930s.

FIGURE 4-5 Photograph of a typical knob-and-tube wiring from the 1930s. (*Delmar/Cengage Learning*)

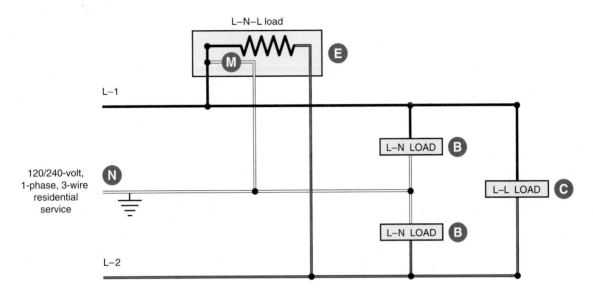

FIGURE 4-6 Diagram showing line-to-line and line-to-neutral loads. (*Delmar/Cengage Learning*)

Neutral Conductor Size

Figure 4-6 shows how *line-to-line* loads and *line-to-neutral* loads are connected. The grounded neutral (white) conductor Ⓝ for residential services and feeders is permitted to be smaller than the "hot" ungrounded phase (black, red) conductors L1 and L2 only when the neutral conductor is properly and adequately sized to carry the maximum unbalance loads calculated according to *NEC 215.2, 220.61,* and *230.42.*

NEC 215.2 relates to feeders and refers us to *Article 220* for calculation requirements.

NEC 230.42 relates to services and refers us to *Article 220* for calculation requirements.

Focusing on the neutral conductor, *220.61* states that *the feeder or service neutral load shall be the maximum unbalance of the load determined by this Article.*

NEC 220.61 further states that *the maximum unbalanced load shall be the maximum net calculated load between the neutral and any one ungrounded conductor.*

In a typical residence, we find a number of loads that carry little or no neutral current, such as an electric water heater, electric clothes dryer, electric oven and range, electric furnace, and air conditioner. Checking Figure 4-6, you can readily see that load (C) is connected to a straight 240-volt, 2-wire circuit and has no neutral current. Note that load (E) is connected to a 120/240-volt 3-wire circuit, and in this hookup the 120-volt motor (M) will result in

current flowing in the neutral conductor. Loads (B) are 120-volt loads that are connected line to neutral.

Thus we find logic in the *NEC*, which permits reducing the neutral conductor size on services and feeders where the calculations prove that the neutral conductor will be carrying less current than the "hot" phase conductors.

See Chapter 20 for sizing neutral conductors for electric range branch circuits and Chapter 29 for sizing neutral service-entrance conductors.

VOLTAGE DROP

Low voltage can cause lights to dim, some television pictures to "shrink," motors to run hot, electric heaters to not produce their rated heat output, and appliances to not operate properly.

Low voltage in a home can be caused by

- wire that is too small for the load being served.
- a circuit that is too long.
- poor connections at the terminals.
- conductors operating at high temperatures having higher resistance than when operating at lower temperatures.

A simple formula for calculating voltage drop on single-phase systems considers only the dc resistance of the conductors and the temperature of the conductor. See *Table 8, Chapter 9,* in the *NEC* for the dc

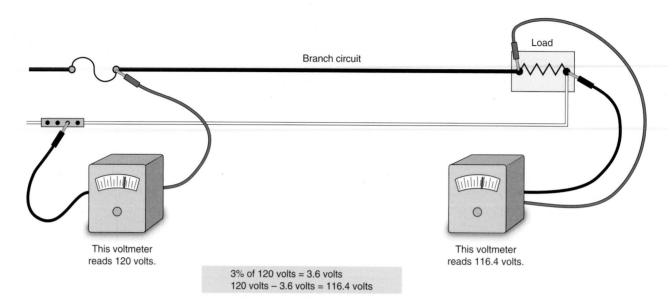

This voltmeter reads 120 volts.

This voltmeter reads 116.4 volts.

3% of 120 volts = 3.6 volts
120 volts − 3.6 volts = 116.4 volts

FIGURE 4-7 Maximum recommended voltage drop on a branch circuit is 3%, *210.19(A), Informational Note No. 4.* (*Delmar/Cengage Learning*)

resistance values. The more accurate formulas consider ac resistance, reactance, temperature, spacing, in metal conduit and in nonmetallic conduit, *NEC Table 9, Chapter 9.* Voltage drop is covered in great detail in *Electrical Wiring—Commercial* (Delmar/Cengage Learning). The simple voltage-drop formula is more accurate with smaller conductors and gets increasingly less accurate as conductor size increases. It is sufficiently accurate for voltage-drop calculations necessary for residential wiring.

To Find Voltage Drop in a Single-Phase Circuit

$$E_d = \frac{K \times I \times L \times 2}{CMA}$$

To Find Conductor Size for a Single-Phase Circuit

$$CMA = \frac{K \times I \times L \times 2}{E_d}$$

In the above formulae:

E_d = result of voltage drop calculation in volts.

K = approximate resistance in ohms per circular-mil foot at 75°C.
  • For uncoated copper wire, use 12 ohms.
  • For aluminum wire, use 20 ohms.

I = current in amperes flowing through the conductors.

L = length in feet from beginning of circuit to the load.

CMA = cross-sectional area of the conductors in circular mils.**

We use the factor of 2 for single-phase circuits because there is voltage drop in both conductors to and from the connected load.

To Find Voltage Drop and Conductor Size for a 3-Phase Circuit

Although residential wiring does not commonly use 3-phase systems, commercial and industrial systems do. To make voltage-drop and conductor sizing calculations for a 3-phase system, substitute the factor of 1.732 (the square root of 3) in place of the factor 2 in the voltage-drop formulae.

Code References to Voltage Drop

For branch circuits, refer to *NEC 210.19(A), Informational Note No. 4.* The recommended maximum voltage drop is 3%, as shown in Figure 4-7.

Note: From Table 4-8 in this text or from *Table 8, Chapter 9, NEC.*

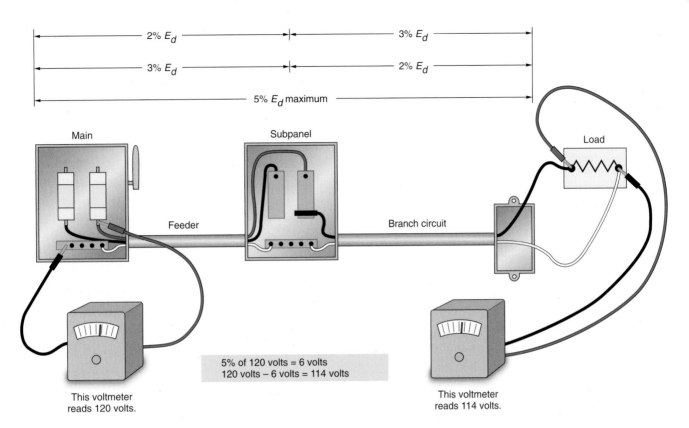

FIGURE 4-8   *NEC 210.19(A), Informational Note No. 4, and 215.2(A), Informational Note No. 2, of the Code* state that the total voltage drop from the beginning of a feeder to the farthest outlet on a branch circuit that does not exceed 5% will provide reasonable efficiency of operation. In this figure, if the voltage drop in the feeder is 3%, then do not exceed 2% voltage drop in the branch circuit. If the voltage drop in the feeder is 2%, then do not exceed 3% voltage drop in the branch circuit. (*Delmar/Cengage Learning*)

For feeders, refer to *NEC 215.2(A)*, *Informational Note No. 2*. The recommended maximum voltage drop is 3%. See Figure 4-8.

When both branch circuits and feeders are involved, the total voltage drop should not exceed 5%. This is shown in Figure 4-8.

According to *NEC 90.5(C)*, *Informational Notes* are explanatory only and are not to be enforced as *Code* requirement.

There is nothing in the *NEC* that dictates what the incoming voltage to a home must be. Incoming voltage is really determined by the electric utility and is affected by the type, size, and length of the service conductors, transformer(s), and primary lines. Electric utilities come under the jurisdiction of the *National Electrical Safety Code (NESC)*. A local public service commission, usually at the state level, mandates maximum over- and undervoltages. This is generally within the 5% range of nominal voltage.

Examples of Voltage-Drop Calculations

EXAMPLE 1

What is the approximate voltage drop on a 120-volt, single-phase circuit consisting of 14 AWG copper conductors where the load is 11 amperes and the distance of the circuit from the panel to actual load is 85 ft (25.91 m)?

Solution:

$$E_d = \frac{K \times I \times L \times 2}{CMA}$$

$$E_d = \frac{12 \times 11 \times 85 \times 2}{4110}$$

$$E_d = 5.46 \text{ volts drop}$$

Table 310.15(B)(16) (formerly Table 310.16) **Allowable Ampacities of Insulated Conductors Rated Up to and Including 2000 Volts, 60°C Through 90°C (140°F Through 194°F), Not More Than Three Current-Carrying Conductors in Raceway, Cable, or Earth (Directly Buried), Based on Ambient Temperature of 30°C (86°F)***

Size AWG or kcmil	Temperature Rating of Conductor [See Table 310.104(A).]						Size AWG or kcmil
	60°C (140°F)	75°C (167°F)	90°C (194°F)	60°C (140°F)	75°C (167°F)	90°C (194°F)	
	Types TW, UF	Types RHW, THHW, THW, THWN, XHHW, USE, ZW	Types TBS, SA, SIS, FEP, FEPB, MI, RHH, RHW-2, THHN, THHW, THW-2, THWN-2, USE-2, XHH, XHHW, XHHW-2, ZW-2	Types TW, UF	Types RHW, THHW, THW, THWN, XHHW, USE	Types TBS, SA, SIS, THHN, THHW, THW-2, THWN-2, RHH, RHW-2, USE-2, XHH, XHHW, XHHW-2, ZW-2	
	COPPER			ALUMINUM OR COPPER-CLAD ALUMINUM			
18	—	—	14	—	—	—	—
16	—	—	18	—	—	—	—
14**	15	20	25	—	—	—	—
12**	20	25	30	15	20	25	12**
10**	30	35	40	25	30	35	10**
8	40	50	55	35	40	45	8
6	55	65	75	40	50	55	6
4	70	85	95	55	65	75	4
3	85	100	115	65	75	85	3
2	95	115	130	75	90	100	2
1	110	130	145	85	100	115	1
1/0	125	150	170	100	120	135	1/0
2/0	145	175	195	115	135	150	2/0
3/0	165	200	225	130	155	175	3/0
4/0	195	230	260	150	180	205	4/0
250	215	255	290	170	205	230	250
300	240	285	320	195	230	260	300
350	260	310	350	210	250	280	350
400	280	335	380	225	270	305	400
500	320	380	430	260	310	350	500
600	350	420	475	285	340	385	600
700	385	460	520	315	375	425	700
750	400	475	535	320	385	435	750
800	410	490	555	330	395	445	800
900	435	520	585	355	425	480	900
1000	455	545	615	375	445	500	1000
1250	495	590	665	405	485	545	1250
1500	525	625	705	435	520	585	1500
1750	545	650	735	455	545	615	1750
2000	555	665	750	470	560	630	2000

*Refer to 310.15(B)(2) for the ampacity correction factors where the ambient temperature is other than 30°C (86°F).
**Refer to 240.4(D) for conductor overcurrent protection limitations.

Reprinted with permission from NFPA 70-2011.

Note: Refer to Table 4-8 for the CMA value.

This exceeds the voltage drop information in the *Code*, which is

$$3\% \text{ of } 120 \text{ volts} = 3.6 \text{ volts}$$

Let's try it again using 12 AWG conductors:

$$E_d = \frac{12 \times 11 \times 85 \times 2}{6530}$$

$$E_d = 3.44 \text{ volts drop}$$

This satisfies the information provided in the Code.

EXAMPLE 2

Find the wire size needed to keep the voltage drop to no more than 3% on a single-phase, 240-volt, air-conditioner circuit. The

TABLE 4-8

Circular mil area for many of the more common conductors.

Conductor Size, AWG	Cross-Sectional Area in Circular Mils
18	1620
16	2580
14	4110
12	6530
10	10,380
8	16,510
6	26,240
4	41,740
3	52,620
2	66,360
1	83,690
0	105,600
00	133,100
000	167,800
0000	211,600

nameplate reads: "Minimum Circuit Ampacity 40 Amperes." The circuit originates at the main panel located approximately 65 ft (19.81 m) from the air-conditioner unit. No neutral conductor is required.

Solution: Checking *Table 310.15(B)(16)*, the conductors could be 6 AWG Type TW copper or 8 AWG Type THHN copper.

The permitted voltage drop is

$$E_d = 240 \times 0.03 = 7.2 \text{ volts}$$

Let's see what voltage drop we might experience if we installed the 8 AWG Type THHN conductors:

$$E_d = \frac{12 \times 40 \times 65 \times 2}{16,510}$$

$$= 3.78 \text{ volts drop}$$

This is well below the recommended 7.2 volts drop and would make an acceptable installation.

A Word of Caution When Using High-Temperature Conductors

Note in *NEC Table 310.15(B)(16)* that conductor insulations fall into three classes of temperature ratings: 60°C, 75°C, and 90°C. For a given conductor size, we find that the allowable ampacity of a 90°C insulated conductor is greater than that of a 60°C insulated conductor. Therefore, be careful when selecting conductors based on their ability to withstand high temperatures. For example:

- 8 AWG THHN (90°C) copper has an allowable ampacity of 55 amperes.
- 8 AWG THW (75°C) copper has an allowable ampacity of 50 amperes.
- 8 AWG TW (60°C) copper has an allowable ampacity of 40 amperes.

We learned earlier in this chapter that according to *110.14(C)*, we use the 60°C column of *NEC Table 310.15(B)(16)* for circuits rated 100 amperes or less, and the 75°C column for circuits rated over 100 amperes. However, when the equipment is marked as being suitable for 75°C, we can take advantage of the higher ampacity of 75°C conductors. This oftentimes results in a smaller size conductor and a correspondingly smaller raceway.

But installing smaller conductors might result in excessive voltage drop.

Therefore, after selecting the proper size conductor for a given load, it is always a good idea to run a voltage-drop calculation to make sure the voltage drop is not excessive.

What Is the Advantage of High-Temperature Conductors?

They can withstand the high temperatures found in attics and high-temperature climates.

Another big advantage of the higher ampacities of high-temperature insulation is that the *NEC* permits the higher ampacity values to be used as the starting point when applying derating and correcting factors. Example calculations of this are found in Chapter 18.

Effect of Voltage Variation

Chapter 19 has more information on the effect voltage differences have on wattage output of appliances and on electric motors.

APPROXIMATE CONDUCTOR SIZE RELATIONSHIP

There is a definite relationship between the circular mil area and the resistance of conductors. The following rules explain this relationship.

Rule One. For wire sizes up through 0000, every third size doubles or halves in circular mil area.

Thus, a 1 AWG conductor is 2 times larger than a 4 AWG conductor (83,690 versus 41,470). Thus, a 0 wire is one-half the size of 0000 wire (105,600 versus 211,600).

Rule Two. For wire sizes up through 0000, every consecutive wire size is approximately 1.26 times larger or smaller than the preceding wire size.

Thus, a 3 AWG conductor is approximately 1.26 times larger than a 4 AWG conductor (41,740 × 1.26 = 52,592). A 2 AWG conductor is approximately 1.26 times smaller than a 1 AWG conductor (83,690 ÷ 1.26 = 66,420).

Try to fix in your mind that a 10 AWG conductor has a cross-sectional area of 10,380 circular mils and that it has a resistance of 1.2 ohms per 1000 ft (300 m). The resistance of aluminum wire is approximately 2 ohms per 1000 ft (300 m). By remembering these numbers, you will be able to perform voltage-drop calculations without having the wire tables readily available.

EXAMPLE

What is the approximate cross-sectional area, in circular mils, and resistance of a 6 AWG copper conductor?

Solution: Note in Table 4-9 that when the CMA of a wire is doubled, then its resistance is cut in half. Inversely, when a given wire size is reduced to one-half, its resistance doubles.

TABLE 4-9

"Every third size" and the 1.26 relationship between the circular-mil area (CMA) and the resistance in ohms per 1000 ft of copper conductors sizes 10 AWG through 6 AWG.

Wire Size (AWG)	CMA (in circular mils)	Ohms per 1000 ft
10	10,380	1.2
9		
8		
7	20,760	0.6
6	26,158	0.476

NONMETALLIC-SHEATHED CABLE (*ARTICLE 334*)

The *NEC* and UL Standard 719 describe the construction details of nonmetallic-sheathed cable. The following is a summary of these details.

Description

Most electricians still refer to nonmetallic-sheathed cable as *Romex*, a name chosen years ago by the Rome Wire and Cable Company. Romex is now a registered trademark of Southwire Company.

Nonmetallic-sheathed cable is a factory assembly of two or more insulated conductors having an outer sheath of moisture-resistant, flame-retardant, nonmetallic material. This cable is available with two or three current-carrying conductors. The conductors range in size from 14 AWG through 2 AWG for copper conductors, and from 12 AWG through 2 AWG for aluminum or copper-clad aluminum conductors. Two-wire cables contain one black conductor, one white conductor, and one bare equipment grounding conductor. Three-wire cables contain one black, one white, one red, and one bare equipment grounding conductor. Equipment grounding conductors are permitted to have green insulation, but bare equipment grounding conductors are the most common.

NEC 334.108 requires that nonmetallic-sheathed cable have an insulated or a bare equipment grounding conductor.

FIGURE 4-9   A nonmetallic-sheathed Type NM-B cable showing (A) black "ungrounded" (hot) conductor, (B) bare equipment "grounding" conductor, and (C) white "grounded" conductor. (*Courtesy of Southwire Company*)

Figure 4-9 clearly shows a bare equipment grounding conductor. Sometimes the bare equipment grounding conductor is wrapped with paper or fiberglass, which acts as a filler. The equipment grounding conductor is *not* permitted to be used as a current-carrying conductor.

Types of Nonmetallic-Sheathed Cable

UL 719 lists two types of nonmetallic-sheathed cable. The *NEC* shows three types of nonmetallic-sheathed cable.

- **Type NM-B** is the most common type of nonmetallic-sheathed cable in use today. Type NM-B cable has a flame-retardant, moisture-resistant, nonmetallic outer jacket. The conductors are rated 90°C. The ampacity is based on the 60°C column in *NEC Table 310.15(B)(16)*. The conductors in Type NM-B meet all of the requirements of THHN but do not have the identifying marking along the entire length of the individual conductors. In the past, Type NM cable contained conductors having Type TW insulation.

- **Type NMC-B** cable has a flame-retardant, moisture-resistant, fungus-resistant, and corrosion-resistant, nonmetallic outer jacket. The conductors are rated 90°C. The ampacity is based on the 60°C column in *NEC Table 310.15(B)(16)*. Type NMC-B cable is not commercially available. Underground feeder Type UF-B cable can be used as a substitute for NMC.

- **Type NMS-B** cable is a hybrid cable that contains the conventional insulated power conductors as well as telephone, coaxial, home entertainment, and signaling conductors all in one cable. This type of cable is used for home automation systems using the latest in digital technology.

Prior to the 2008 *NEC, Article 780* recognized this wiring method for closed loop and programmed power distribution. In the 2008 *NEC, Article 780* was deleted. The Type NMS-B cable had a moisture-resistant, flame-retardant, nonmetallic outer jacket. Type NMS-B cable is not commercially available. This type of home automation never did take off. It fell by the wayside. Today, we are seeing a tremendous amount of wireless home automation systems.

Although somewhat confusing, the UL standards use the suffix B, whereas the *NEC* refers to nonmetallic-sheathed cables as Type NM, NMC, and NMS. The conductors in these cables are rated 90°C. Types NM, NMC, and NMS cable, identified by the markings NM-B, NMC-B, and NMS-B, meet this 90°C requirement. See *334.112*.

As of the writing of this edition of *Electrical Wiring—Residential*, there are no UL listings for Type NMS and NMS-B cables.

Why the Suffix *B*?

The suffix *B* means that the conductors have 90°C insulation.

Older nonmetallic-sheathed cable contained 60°C-rated conductors. There were many problems with the insulation becoming brittle and breaking off due to the extremely high temperatures associated with recessed and surface-mounted luminaires: fires resulted. In hot attics, after applying correction factors, the adjusted allowable ampacity of the conductors could very well be zero. To solve this dilemma, since December 17, 1984, UL has required that the conductors be rated 90°C, and that the cable be identified with the suffix *B*. This suffix makes it easy to differentiate the newer 90°C from the old 60°C cable.

The ampacity of the conductors in nonmetallic-sheathed cable is based on that of 60°C conductors. The 90°C ampacity is permitted to be used for derating purposes, such as might be necessary in attics. The final derated ampacity must not exceed the ampacity for 60°C conductors. More specifics can be found in *NEC 334.80*.

Table 310.15(B)(16) shows the allowable ampacities for conductors.

Overcurrent Protection for Small Conductors

Overcurrent protection for small copper conductors, in conformance to *240.4(D)*, is as follows:

- 14 AWG   15 amperes
- 12 AWG   20 amperes
- 10 AWG   30 amperes

A typical nonmetallic-sheathed cable is illustrated in Figure 4-9.

Color Coding of Type NM-B Jacket

To make installations easier for the electrician and identification of the conductor size easier for the electrical inspector, color-coded nonmetallic-sheathed cable is available.

Conductor Size	Color of Jacket
14 AWG	White
12 AWG	Yellow
10 AWG	Orange
8 and 6 AWG	Black

New Types of NM-B Cable

At least one manufacturer offers a 4-conductor Type NM-B that can be used for two branch circuits.

This cable is used where sharing of the neutral is *not wanted*, as in a multiwire branch circuit, or for branch circuits supplying GFCIs and AFCIs, where sharing the neutral is *not permitted*.

This cable contains the following:

- One bare equipment grounding conductor
- Pair 1: one black and one white
- Pair 2: one black and one white (with a red stripe)

Because all four conductors in this cable are current-carrying conductors, the ampacity of the conductors must be adjusted (derated) according to *NEC 310.15(B)(3)(a)*.

Where Nonmetallic-Sheathed Cable May Be Used

Table 4-10 shows the uses permitted for Types NM-B, NMC-B, and NMS-B.

Equipment Grounding Conductors (*250.122*)

Equipment grounding conductor (EGC) sizing is based on the rating or setting of the overcurrent device protecting a given branch circuit or feeder. See *NEC Table 250.122*.

The EGC in a nonmetallic-sheathed cable is sized according to *NEC Table 250.122* and is usually bare. Table 4-11 is a short version of *Table 250.122*.

An EGC does not have to be larger than the ungrounded conductors of a given circuit. Why? Because in the event of a ground fault, the amount of ground-fault current returning on the EGC can never be more than the ground-fault current flowing on the ungrounded conductor of that circuit that caused the problem. It's a simple series circuit. What goes out—comes back! See *NEC 250.122(A)*.

Installation

Nonmetallic-sheathed cable is probably the least expensive of the various wiring methods. It is relatively lightweight and easy to install. It is widely

TABLE 4-10

Uses permitted for nonmetallic-sheathed cable.

	Type NM-B	Type NMC-B	Type NMS-B
May be used on circuits of 600 volts or less	Yes	Yes	Yes
May be run exposed or concealed in dry locations	Yes	Yes	Yes
May be installed exposed or concealed in damp and moist locations	No	Yes	No
Has flame-retardant and moisture-resistant outer covering	Yes	Yes	Yes
Has fungus-resistant and corrosion-resistant outer covering	No	Yes	No
May be used to wire one- and two-family dwellings or multifamily dwellings; see *334.10*	Yes	Yes	Yes
May be embedded in masonry, concrete, plaster, adobe, fill	No	No	No
May be installed or fished in the hollow voids of masonry blocks or tile walls where not exposed to excessive moisture or dampness	Yes	Yes	Yes
May be installed or fished in the hollow voids of masonry blocks or tile walls where exposed to excessive moisture	No	Yes	No
In outside walls of masonry block or tile	No	Yes	No
In inside wall of masonry block or tile	Yes	Yes	Yes
May be used as service-entrance cable	No	No	No
Must be protected against physical damage	Yes	Yes	Yes
May be run in shallow chase of masonry, concrete, or adobe if protected by a steel plate(s) at least 1/16 in. (1.6 mm) thick, then covered with plaster, adobe, or similar finish	No	Yes	No

TABLE 4-11

Minimum size equipment grounding conductors is based on the rating or setting of the overcurrent device protecting that particular branch circuit. This table is based on *Table 250.122* of the *NEC*.

MINIMUM SIZE (AWG) EQUIPMENT GROUNDING CONDUCTOR

Rating or Setting of Overcurrent Device	Copper	Aluminum or Copper-Clad Aluminum
15	14	12
20	12	10
30	10	8
40	10	8
60	10	8
100	8	6
200	6	4

used for dwelling unit installations on circuits of 600 volts or less. Figure 4-10 shows an example of a stripper used for stripping nonmetallic-sheathed cable. A razor knife also works fine with care. The installation of nonmetallic-sheathed cable must conform to the requirements of *NEC Article 334*. Refer to Figures 4-11, 4-12, 4-13, 4-14, 4-15, and 4-16.

- The cable must be strapped or stapled not more than 12 in. (300 mm) from a box or fitting.

- Do not staple flat nonmetallic-sheathed cables on edge, Figure 4-15.

- Unless run through horizontal holes or notches in studs and joists, the intervals between straps or staples must not exceed 4½ ft (1.4 m).

- Exposed runs of cables can be subjected to lots of abuse. A child may pull or stand on the cables and adults may use the cables to hang things on. *NEC 334.15* requires that cables be protected

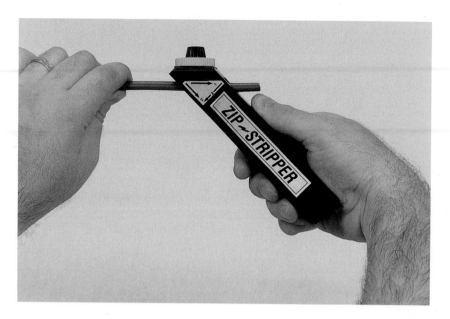

FIGURE 4-10 This stripper is used to remove the outer jacket from nonmetallic-sheathed cable. (*Courtesy of Seatek Co., Inc.*)

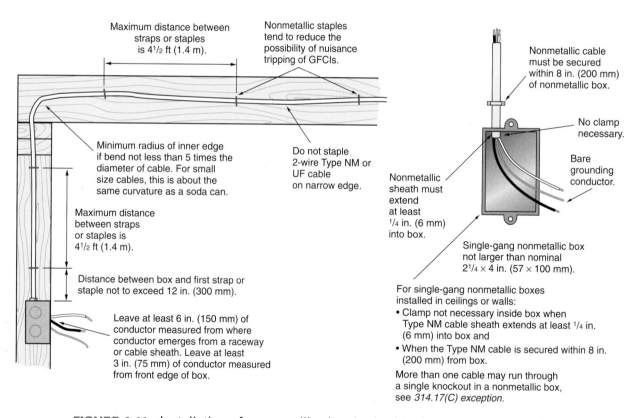

Maximum distance between straps or staples is 4½ ft (1.4 m).

Nonmetallic staples tend to reduce the possibility of nuisance tripping of GFCIs.

Minimum radius of inner edge if bend not less than 5 times the diameter of cable. For small size cables, this is about the same curvature as a soda can.

Do not staple 2-wire Type NM or UF cable on narrow edge.

Maximum distance between straps or staples is 4½ ft (1.4 m).

Distance between box and first strap or staple not to exceed 12 in. (300 mm).

Leave at least 6 in. (150 mm) of conductor measured from where conductor emerges from a raceway or cable sheath. Leave at least 3 in. (75 mm) of conductor measured from front edge of box.

Nonmetallic cable must be secured within 8 in. (200 mm) of nonmetallic box.

No clamp necessary.

Bare grounding conductor.

Nonmetallic sheath must extend at least ¼ in. (6 mm) into box.

Single-gang nonmetallic box not larger than nominal 2¼ × 4 in. (57 × 100 mm).

For single-gang nonmetallic boxes installed in ceilings or walls:
• Clamp not necessary inside box when Type NM cable sheath extends at least ¼ in. (6 mm) into box and
• When the Type NM cable is secured within 8 in. (200 mm) from box.

More than one cable may run through a single knockout in a nonmetallic box, see *314.17(C)* exception.

FIGURE 4-11 Installation of nonmetallic-sheathed cable. (*Delmar/Cengage Learning*)

against physical damage where necessary. For example, where exposed in a garage with open studs, run the cables on the sides of the studs. Where cables are run horizontally from or through stud to stud, protection is required. Some inspectors require protective boards over exposed horizontal cable runs, or sheet rock, or other wall finish up to the ceiling, or at least 7 ft

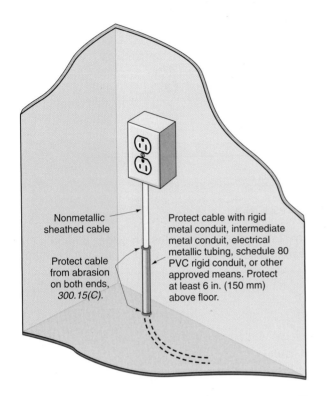

FIGURE 4-12 Installation of exposed nonmetallic-sheathed cable where passing through a floor; see *334.15(B)*. (*Delmar/Cengage Learning*)

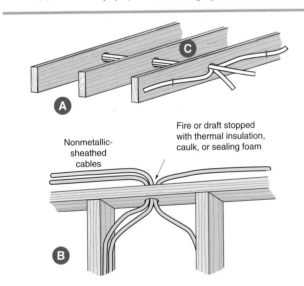

FIGURE 4-13 ▶The conductor ampacity in NM cables shall be adjusted according to *Table 310.15(B)(3)(a)*: (A) when NM cables are "bundled" or "stacked" for distances more than 24 in. (600 mm) without maintaining spacing; or (B) when more than two NM cables pass through wood framing members if sealed with thermal insulation, caulk or sealing foam, and there is no spacing between the cables; or (C) when more than two NM cables are installed in contact with thermal insulation. ◀ *See 310.15(B)(3)(a) and 334.80*. (*Delmar/Cengage Learning*)

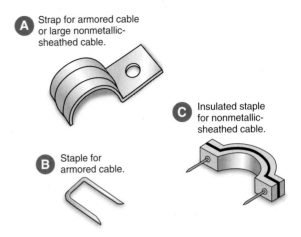

FIGURE 4-14 (A) One-hole strap used for conduit, EMT, large nonmetallic-sheathed cable and armored cable. (B) Metal staple for armored cable and nonmetallic-sheathed cable. (C) Insulated staple for nonmetallic-sheathed cable. Certain types of insulated and noninsulated staples can be applied with a staple gun. (*Delmar/Cengage Learning*)

(2.13 m) up. For protection of cables in attics, see Chapter 15.

- The 4½ ft (1.4 m) securing requirement is not needed where nonmetallic-sheathed cable is run *horizontally* through holes in wood or metal framing members (studs, joists, rafters, etc.). The cable is considered to be adequately supported by the framing members, *NEC 334.30(A)*.

- The inner edge of the bend shall have a minimum radius not less than 5 times the cable diameter, *NEC 334.24*.

- The cable must not be used in circuits of more than 600 volts.

- **CAUTION: Be very careful when installing staples. Driving them too hard can squeeze and damage the insulation on the conductors, causing short circuits and/or ground faults.**

- See Figure 4-11 for special conditions when using single-gang nonmetallic device boxes.

- Nonmetallic-sheathed cable must be protected where passing through a floor by at least 6 in.

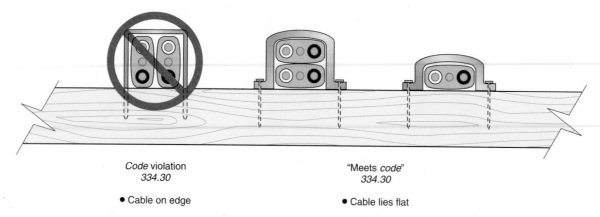

Code violation
334.30

● Cable on edge

"Meets *code*"
334.30

● Cable lies flat

FIGURE 4-15   It is a *Code* violation to staple flat nonmetallic-sheathed cables on edge, *334.30*.
(*Delmar/Cengage Learning*)

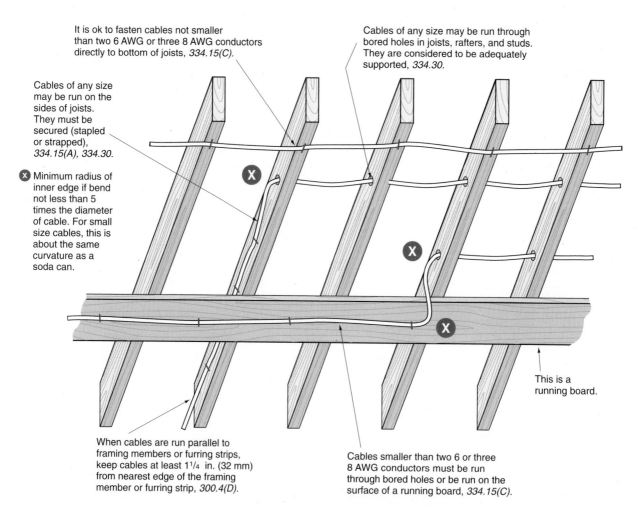

It is ok to fasten cables not smaller than two 6 AWG or three 8 AWG conductors directly to bottom of joists, *334.15(C)*.

Cables of any size may be run through bored holes in joists, rafters, and studs. They are considered to be adequately supported, *334.30*.

Cables of any size may be run on the sides of joists. They must be secured (stapled or strapped), *334.15(A)*, *334.30*.

Ⓧ Minimum radius of inner edge if bend not less than 5 times the diameter of cable. For small size cables, this is about the same curvature as a soda can.

This is a running board.

When cables are run parallel to framing members or furring strips, keep cables at least 1¼ in. (32 mm) from nearest edge of the framing member or furring strip, *300.4(D)*.

Cables smaller than two 6 or three 8 AWG conductors must be run through bored holes or be run on the surface of a running board, *334.15(C)*.

FIGURE 4-16   The requirements for running nonmetallic-sheathed cable in unfinished basements and crawl spaces. (*Delmar/Cengage Learning*)

(150 mm) of rigid metal conduit, intermediate metal conduit, electrical metallic tubing, Schedule 80 rigid PVC conduit, or other approved means, *NEC 334.15(B)*. A fitting (bushing or connector) must be used at both ends of the conduit to protect the cable from abrasion, Figure 4-12.

- When nonmetallic-sheathed cables are "bundled" or "stacked" for distances of more than 24 in. (600 mm) without maintaining spacing, their ampacities must be reduced (adjusted, derated) according to *Table 310.15(B)(3)(a)*. When cables are bundled together, the heat generated by the conductors cannot easily dissipate. See Figure 4-13(A).

- When more than two nonmetallic-sheathed cables are installed through the same opening in wood framing members that are to be fire- or draft-stopped with thermal insulation, caulk, or sealing foam, and there is no spacing between the cables, their conductor ampacities shall be adjusted according to *NEC Table 310.15(B)(3)(a)*. See *NEC 334.80* and Figure 4-13(B).

- When Type AC and Type MC cables are "bundled" or "stacked" without maintaining spacing, derating is not necessary if the following is true:

  — The conductors are 12 AWG copper.

  — The bundling or stacking is not more than 24 in. (600 mm).

  — There are no more than three current-carrying conductors in each cable.

  — There are no more than 20 current-carrying conductors in the bundle.
    When there are more than 20 current-carrying conductors in the bundle, for a length greater than 24 in. (600 mm) a 60% adjustment factor shall be applied. See *NEC 310.15(B)(3)(a)(4)* and *(B)(3)(a)(5)*.

- Type NM cable may be run in unsupported lengths of not more than 4½ ft (1.4 m) between the last point of support of the cable and a luminaire or other equipment within an accessible ceiling. For this last 4½ ft (1.4 m), the 12-in.

(300-mm) and 4½-ft (1.4-m) securing requirements mentioned above are not necessary if the nonmetallic-sheathed cable is used as a luminaire (fixture) whip to connect luminaires of other equipment; see *NEC 334.30*. Luminaire (fixture) whips are discussed in Chapter 7 and Chapter 17.

Figure 4-16 shows the installation of nonmetallic-sheathed cable in unfinished basements and crawl spaces, *334.15(C)*.

See Chapter 15 and Figures 15-13 and Figure 15-14 for additional text and diagrams covering the installation of cables in attics.

Figure 4-11 shows the *Code* requirements for securing nonmetallic-sheathed cable when using nonmetallic boxes.

The *Code* in *314.40(D)* requires that all metal boxes have provisions for the attachment of an equipment grounding conductor. Figure 4-17 shows a gang-type switch (device) box that is tapped for a screw by which the grounding conductor may be connected underneath. Figure 4-18 shows an outlet box, also with provisions for attaching grounding

Ground screw tapped hole

FIGURE 4-17 Gang-type switch (device) box.
(*Delmar/Cengage Learning*)

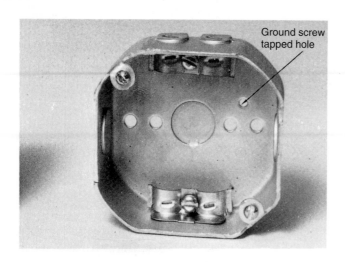

FIGURE 4-18   Outlet box.
(*Delmar/Cengage Learning*)

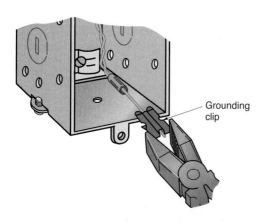

FIGURE 4-19   Method of attaching ground clip to a metal switch (device) box; see *250.148*.
(*Delmar/Cengage Learning*)

conductors. Figure 4-19 illustrates the use of a small grounding clip.

NEC 410.44 requires that there be a provision whereby the equipment grounding conductor can be attached to the exposed metal parts of luminaires.

Are Cables Permitted in Raceways?

Yes, but only if certain conditions are met.

NEC 300.15(C) permits using a listed conduit or tubing without a box where the conduit or tubing is used to protect a nonmetallic-sheathed cable against physical damage. Figure 4-20 shows an example of this in a basement where the wiring method in the ceiling is NM cable and a switch or receptacle is in a box mounted on the wall. An EMT is installed from the box to a point above the ceiling joists, where the EMT is secured to the side of the joists. The NM cable is then run inside the EMT, and an insulating fitting is installed on the upper end of the tubing to protect the cable against abrasion. The cable does not require securing inside the raceway, *334.30*. The finished job is neat.

The conductors inside a nonmetallic-sheathed cable meet all of the UL requirements for Type THHN conductors but are not surface marked in any manner.

NEC 334.15(B) requires protection of nonmetallic-sheathed cable from physical damage where necessary. This section lists different types of raceways permitted to be used for protection of the cable.

The raceway fill for a cable in a conduit or tubing is based on the allowable percentage fill values specified in *Table 1, Chapter 9*.

Alternate Method of Installing NM above Panel

Nonmetallic-sheathed cables are usually secured to a panel, box, or cabinet, with a cable clamp or connector. For surface-mounted enclosures only, *312.5(C)*, *Exception*, provides an alternate method for running nonmetallic-sheathed cables, as illustrated in Figure 4-21.

Be aware of a significant rule in *Note 9* to *NEC Table 1* of *Chapter 9*. This note requires that if cables with elliptical cross sections are installed in a raceway, the cross-sectional area of the cable is determined by assuming the greater dimension is the diameter of a circle. This rule is needed because some installers do not straighten the cables before installing them through the raceway but pull them out of the center of a carton, which results in the cables twisting inside the raceway.

Let's do a calculation using a 12-2 W/G Type NM cable that is ¼ in. × ½ in. in size. We will assume the ½ in. dimension is the diameter of a

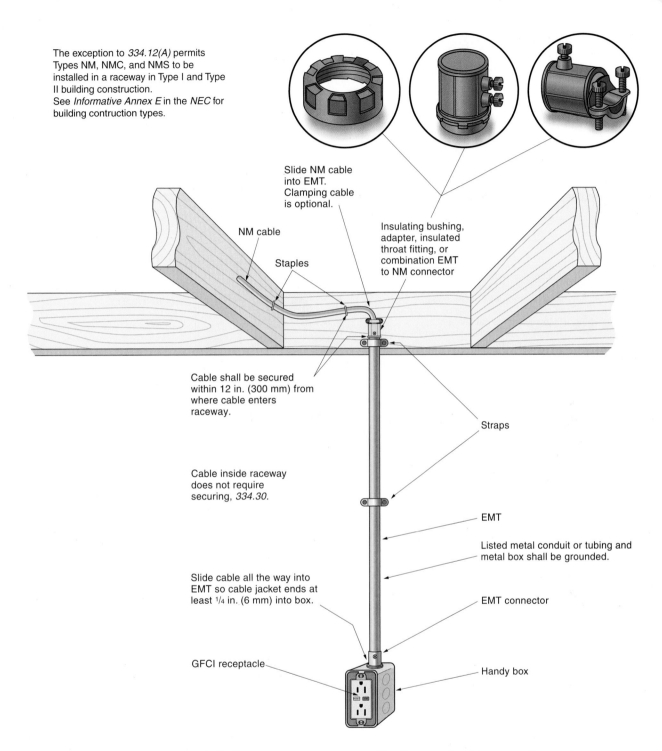

The exception to *334.12(A)* permits Types NM, NMC, and NMS to be installed in a raceway in Type I and Type II building construction.
See *Informative Annex E* in the *NEC* for building contruction types.

Slide NM cable into EMT. Clamping cable is optional.

NM cable

Staples

Insulating bushing, adapter, insulated throat fitting, or combination EMT to NM connector

Cable shall be secured within 12 in. (300 mm) from where cable enters raceway.

Straps

Cable inside raceway does not require securing, *334.30*.

EMT

Listed metal conduit or tubing and metal box shall be grounded.

Slide cable all the way into EMT so cable jacket ends at least ¼ in. (6 mm) into box.

EMT connector

GFCI receptacle

Handy box

FIGURE 4-20   An example of EMT protecting a nonmetallic-sheathed cable from physical damage where the cable is run down the wall of an unfinished basement. This also makes a neat installation. The EMT and the metal box must be grounded. Physical protection and grounding are required by *334.15(C)*. NEC *300.18(A), Exception*, also recognizes this type of installation.
(*Delmar/Cengage Learning*)

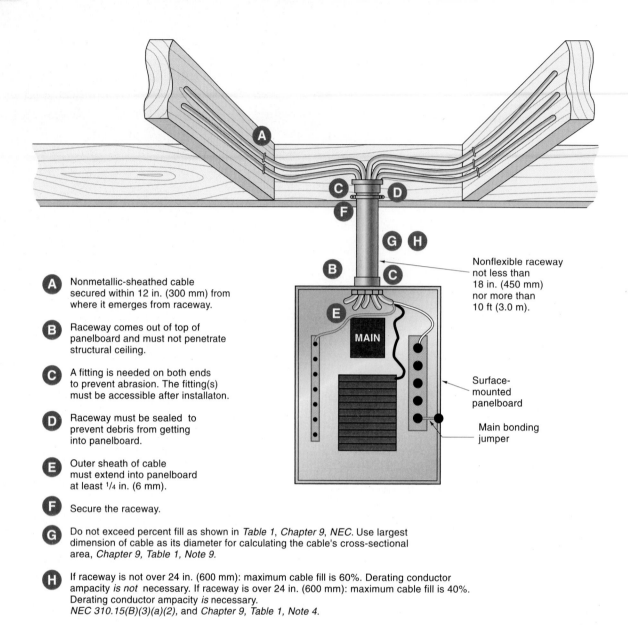

(A) Nonmetallic-sheathed cable secured within 12 in. (300 mm) from where it emerges from raceway.

(B) Raceway comes out of top of panelboard and must not penetrate structural ceiling.

(C) A fitting is needed on both ends to prevent abrasion. The fitting(s) must be accessible after installaton.

(D) Raceway must be sealed to prevent debris from getting into panelboard.

(E) Outer sheath of cable must extend into panelboard at least ¼ in. (6 mm).

(F) Secure the raceway.

(G) Do not exceed percent fill as shown in *Table 1, Chapter 9, NEC*. Use largest dimension of cable as its diameter for calculating the cable's cross-sectional area, *Chapter 9, Table 1, Note 9*.

(H) If raceway is not over 24 in. (600 mm): maximum cable fill is 60%. Derating conductor ampacity *is not* necessary. If raceway is over 24 in. (600 mm): maximum cable fill is 40%. Derating conductor ampacity *is* necessary. *NEC 310.15(B)(3)(a)(2)*, and *Chapter 9, Table 1, Note 4*.

Nonflexible raceway not less than 18 in. (450 mm) nor more than 10 ft (3.0 m).

Surface-mounted panelboard

Main bonding jumper

MAIN

FIGURE 4-21   An alternate method for running nonmetallic-sheathed cables into a surface-mounted panel. This is permitted for surface-mounted panels only, *312.5(C), Exception*. (*Delmar/Cengage Learning*)

circle and perform the calculation using the following formula:

$$A = \pi r^2$$
$$A = 3.14159 \times (0.25 \times 0.25)$$
$$A = 3.14159 \times 0.0625$$
$$A = 0.1963 \; Sq.In.$$

Let's look at the following table to determine how many of these cables we are permitted to install in various sizes of EMT and whether derating is required.

100% Area of EMT	Area of 12/2 WG Type NM Cable	Number Permitted with 60% Fill*	Number Permitted with 40% Fill**	Derating Required?
1 in. = 0.864	0.1963	2	1	No/No
1¼ = 1.496	0.1963	4	3	No/Yes
1½ = 2.036	0.1963	6	4	No/Yes
2 = 3.356	0.1963	10	6	No/Yes

* Maximum 24 in. (600 mm) long
** Longer than 24 in. (600 mm)

Be sure to comply with all the conditions stated in *NEC 312.5(C) Exception*. Also be aware Type NM cable is limited to installation in dry locations and thus the installation in EMT as described above is not permitted in damp or wet locations.

Identifying Nonmetallic-Sheathed Cables

One nice thing about nonmetallic-sheathed cable, particularly that with a white outer jacket, is that at panelboards and device and outlet boxes, you can use a permanent marking pen to mark the outer jacket with a few words indicating what the cable is for. Examples: *FEED TO ATTIC; FEED TO L.R.; FEED TO DINING ROOM RECEPTACLES; FEED TO PANEL; BC#6; TO OTHER 3-WAY SWITCH; SWITCH LEG TO CEILING LIGHT; TO OUTSIDE LIGHT*; and so on.

By marking the cables when they are installed, confusion when making the connections is reduced considerably.

Another great idea is to make a sleeve by cutting off a short 2 in. length of NM cable, pulling out the conductors, then marking an identifying statement on the sleeve. Slide the sleeve over the specific conductors inside the panel or box.

Multifamily Dwellings

Nonmetallic-sheathed cable is permitted to be installed in one- and two-family dwellings and in multifamily dwellings where the building construction is Type III, Type IV, or Type V; see *334.12(A)*. Definitions of construction types are found in building codes, the glossary of this text, and in *Annex E* of the *NEC*.

For years, the *NEC* restricted nonmetallic-sheathed cable to a maximum of three floors above finished grade. This limitation was removed in the *2002 NEC*.

ARMORED CABLE (TYPE AC) AND METAL-CLAD CABLE (TYPE MC)

Type AC and Type MC cables look very much the same (Figure 4-22), yet there are significant differences. Table 4-12 compares some of these differences.

To remove the outer metal armor, a rotary armor-cutting tool like that shown in Figure 4-24 may be used.

If a hacksaw is used, make an angle cut on one of the armor's convolutions as long as the

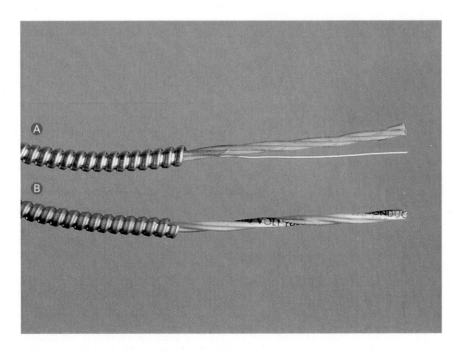

FIGURE 4-22 (A) Type AC cable (Note paper wrap) (B) Type MC cable
(*Courtesy of AFC Cable Systems, Inc., New Bedford, MA*)

TABLE 4-12

Comparisons of Type AC and Type MC cables.

Type AC (*Article 320*)	Type MC (*Article 330*)
Article 320: This article contains all of the installation requirements for the installation of Type AC armored cable.	*Article 330:* This article contains all of the installation requirements for the installation of Type MC metal-clad cable.
320-1. Definition: ▶ Type AC cable is a fabricated assembly of insulated conductors in a flexible interlocked metallic armor. ◀	***330.1. Definition:*** Type MC cable is a factory assembly of one or more insulated circuit conductors with or without optical fiber members and enclosed in an armor of interlocking metal tape or a smooth or corrugated metallic sheath.
Number of Conductors: Two to four current-carrying conductors plus bonding wire. It may also have a separate equipment grounding conductor.	**Number of Conductors:** Any number of current-carrying conductors. At least one manufacturer has a "Home Run Cable" that has conductors for more than one branch circuit, plus the equipment grounding conductors.
Conductor Size and Type: 14 AWG through 1 AWG for copper conductors. 12 AWG through 1 AWG for aluminum conductors will be marked "AL" on a tag on the carton or reel. 12 AWG through 1 AWG for copper-clad aluminum conductors will be marked "AL (CU-CLAD)" or "Cu-Clad Al" on a tag on the carton or reel.	**Conductor Size and Type:** 18 AWG through 2000 kcmil for copper conductors. 12 AWG through 2000 kcmil for aluminum or copper-clad aluminum conductors. 12 AWG through 2000 kcmil for copper-clad aluminum conductors will be marked "AL (CU-CLAD)" or "Cu-Clad Al" on a tag on the carton or reel.
Color Coding: Two-conductor: one black, one white. Three-conductor: one black, one white, one red. Four-conductor: one black, one white, one red, one blue. These are in addition to any equipment grounding and bonding conductors.	**Color Coding:** Two-conductor: one black, one white. Three-conductor: one black, one white, one red. Four-conductor: one black, one white, one red, one blue. These are in addition to any equipment grounding and bonding conductors.
Bonding & Grounding: Type AC cable shall provide an adequate path for fault current. Type AC cable has a bonding wire or strip. Bonding wire is in continuous direct contact with the metal armor. The bonding wire and armor act together to serve as the acceptable equipment ground. The bonding wire does not have to be terminated, just folded back over the armor. Figure 4-23 shows how the bonding strip is folded back over the armor. Never use the bonding strip as a grounded neutral conductor or as a separate equipment grounding conductor. It is an internal bonding strip—nothing more!	**Bonding & Grounding:** This has a separate green insulated equipment grounding conductor. It may have two equipment grounding conductors for isolated ground requirements, such as for computer wiring. Unless specifically listed for grounding, the jacket of MC cable is not to be used as an equipment grounding conductor. The sheath of smooth or corrugated tube Type MC is listed as acceptable as the required equipment grounding conductor. Some MC cables have a full size bare aluminum conductor in continuous, direct contact with the metal armor. This bare conductor and the metal armor act together to serve as the acceptable equipment ground. Cut off the bare conductor at the armor.
Insulation: Type ACTHH has 90°C thermoplastic insulation, by far the most common. Type ACTH has 75°C thermoplastic insulation. Type AC has thermo set insulation. Type ACHH has 90°C thermoset insulation. Conductors are individually wrapped with a flame-retardant fibrous cover: light brown Kraft paper. Author's comment: This is an easy way to distinguish Type AC from Type MC.	**Insulation:** Type MC has 90°C thermoplastic insulation. Some Type MC has thermoset insulation. Type MC cable is available with a PVC outer jacket suitable for direct burial. No individual wrap. Has a polyester (Mylar) tape over all conductors. Some constructions have the extra layer of Mylar tape over the individual conductors. Has no bare bond wire but may have a full-size aluminum grounding/bonding conductor. Author's comment: These are easy ways to distinguish Type AC from Type MC.
Armor: Galvanized steel or aluminum. Armored cable with aluminum armor is marked "Aluminum Armor."	**Armor:** Galvanized steel, aluminum, or copper. Three types: *Interlocked (most common).* Requires a separate equipment grounding conductor. The armor by itself is not an acceptable equipment grounding conductor. Steel or aluminum. *Smooth tube.* Does not require a separate equipment grounding conductor. Aluminum only. *Corrugated tube.* Does not require a separate equipment grounding conductor. Copper or aluminum.

(continues)

TABLE 4-12

(continued)

Type AC (*Article 320*)	Type MC (*Article 330*)
Covering over armor: None	**Covering over armor:** Available with PVC outer covering. Suitable for wet locations, for direct burial and concrete encasement if so covered.
Locations: Okay for dry locations. Not okay for damp or wet locations. Not okay for direct burial.	**Locations:** Okay for wet locations if so listed. Okay for direct burial if so listed. Okay to be installed in a raceway.
Insulating bushings: Required to protect conductor insulation from damage. Referred to as "antishorts," these keep any sharp metal edges of the armor from cutting the conductor insulation. Usually supplied in a bag with cable. Figure 4-23(A) shows an antishort being inserted. Figure 4-23(B) shows the recommended way of bending the bonding strip over the antishort to hold the antishort in place.	**Insulating bushings:** Recommended but not required to protect conductor insulation from damage. Usually in a bag supplied with cable.
Minimum radius of bends: 5× diameter of cable. For small size cables, this is about the same curvature as a soda can.	**Minimum radius of bends:** Smooth Sheath: • 10× for sizes not over ¾ in. (19 mm) in diameter. • 12× for sizes over ¾ in. (19 mm) and not over 1½ in. (38 mm) in diameter. • 15× for sizes over 1½ in. (38 mm) in diameter. Corrugated or interlocked: • 7× diameter of cable.
Not available with "super neutral conductor."	Available with "super neutral conductor" when required for computer branch-circuit wiring. For example, three 12 AWG and one 10 AWG neutral conductor.
Not available with fiber-optic cables.	Available with power conductors and fiber-optic cables in one cable.
Support: Not more than 4½ ft (1.4 m) apart. Not more than 12 in. (300 mm) from box or fitting. Support not needed where cable is fished through walls and ceilings. Cables run horizontally through holes in studs, joists, rafters, etc., are considered supported by the framing members. Support not needed when used as a fixture whip within an accessible ceiling for lengths not over 6 ft (1.8 m) from last point of support.	**Support:** Not more than 6 ft (1.8 m) apart. Not more than 12 in. (300 mm) from box or fitting for cables having four or fewer 10 AWG or smaller conductors. Support not needed where cable is fished through walls and ceilings. Cables run horizontally through holes in studs, joists, rafters, etc., are considered supported by the framing members. Support not needed when used as a fixture whip within an accessible ceiling for lengths not over 6 ft (1.8 m) from last point of support.
Voltage: 600 volts or less.	**Voltage:** 600 volts or less. Some MC has a voltage rating not to exceed 2000 volts.
Ampacity: Use the 60°C column of *Table 310.15(B)(16)*. Derating permitted from 90°C column.	**Ampacity:** Similar to wire pulled in raceways, determine in accordance with *310.15*.
Thermal insulation: If buried in thermal insulation, the conductors must be rated 90°C.	**Thermal insulation:** No mention in *NEC*. Type MC has 90°C thermoplastic insulation.
Connectors: Set screw-type connectors not permitted with aluminum armor. Listed connectors are suitable for grounding purposes.	**Connectors:** Set screw-type connectors not permitted with aluminum armor. Listed connectors are suitable for grounding purposes.
UL Standard 4. Because UL standards have evolved in numerical order, it is apparent that the armored cable standard is one of the first standards to be developed.	**UL Standard 1569.**

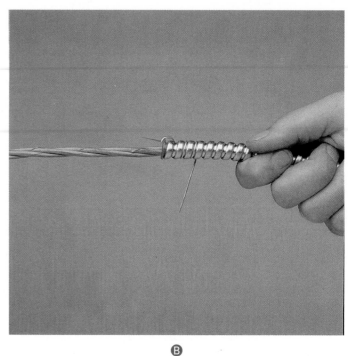

Ⓐ Ⓑ

FIGURE 4-23 Antishort bushing prevents cutting of the conductor insulation by the sharp metal armor. (*Courtesy of AFC Cable Systems, Inc., New Bedford, MA*)

conductors need to be, perhaps 10–12 in. Be careful not to cut too deep or you might cut into the conductor insulation. Bend the armor at the cut and snap it off. Then slide the 10–12 in. of armor off the end of the cable. This method is not recommended by cable manufacturers because of the risk of damage to the insulated conductors. See Figure 4-25.

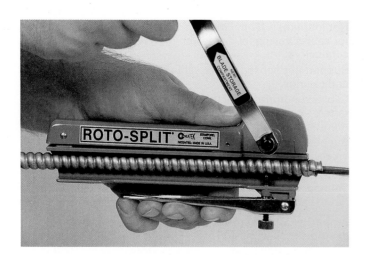

FIGURE 4-24 This is an example of a tool that precisely cuts the outer armor of armored cable, making it easy to remove the armor with a few turns of the handle. (*Courtesy of Seatek Co., Inc.*)

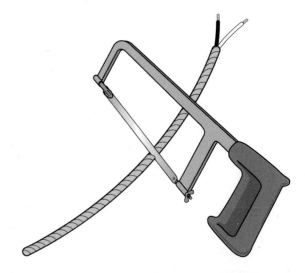

FIGURE 4-25 Using a hacksaw to cut through a raised convolution of the cable armor. This method is not recommended by manufacturers due to the risk of damaging the conductors by cutting too deeply. (*Delmar/Cengage Learning*)

Most electricians, when using armored cable, will make three cuts with their hacksaw. The middle cut is where the cable and conductors will be cut off completely. The first and third cuts are where the armor will be snapped off. This saves time because it actually prepares two ends of the cable.

Most electricians still refer to armored cable as BX, supposedly derived from an abbreviation of the Bronx in New York City, where armored cable was once manufactured. BX was a trademark owned by the General Electric Company, but over the years it has become a generic term.

Figure 4-22 shows armored cable with the armor removed to show the conductors that are wrapped with paper and the bare bonding wire.

Table 4-13 shows the uses and installation requirements for Type AC and Type MC cables for residential wiring.

INSTALLING CABLES THROUGH WOOD AND METAL FRAMING MEMBERS (300.4)

Wood Construction

When wiring a house, nonmetallic-sheathed cables are run through holes drilled in studs, joists, and other framing members. Typically, holes are drilled in the approximate centers of wood framing members. To avoid possible damage to the cables, make sure that the edge of the hole is at least 1¼ in. (32 mm) from the nearest edge of the framing member, *300.4(A)(1)*. Refer to Figure 4-26.

You have probably noticed that a 2 × 4 measures 1½ in. × 3½ in., and that a 2 × 3 measures 1½ in. × 2½ in. This can present a real problem!

TABLE 4-13		
Uses and installation in dwellings (Type AC and Type MC).		
	Type AC	**Type MC**
Branch circuits & feeders	Yes	Yes
Services	No	Yes
Exposed & concealed work in		
• dry locations.	Yes	Yes
• wet locations.	No	Yes (See *300.10(A)(12)* for specifics)
Where subject to physical damage	No	No
May be embedded in plaster finish on masonry walls or run through the hollow spaces of such walls if these locations are not considered damp or wet.	Yes	Yes
Run or fished in air voids of masonry block or tile walls where not exposed to excessive moisture or dampness. Masonry in direct contact with earth is considered to be a wet location.	Yes	Yes
In any raceway	No	Yes
Direct burial, embedded in concrete, or in cinder fill.	No	Yes, if identified for such use
In accessible attics or roof spaces, must be protected by guard strips at least as high as the cable when run across top of floor joists, or within 7 ft (2.1 m) of the floor or floor joists when the cable is run across the face of studs or rafters. If there is no permanent stairway or ladder, protection is needed only within 6 ft (1.8 m) from edge of scuttle hole or entrance to attic. No guard strips or running boards are necessary for cables run parallel to the sides of joists, rafters, or studs. See Chapter 15 for additional discussion and diagrams regarding physical protection of cables in attics.	Yes	Yes

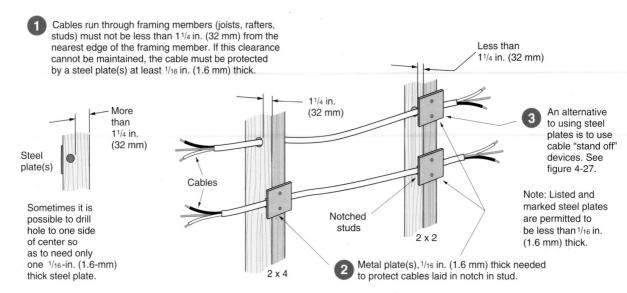

1 Cables run through framing members (joists, rafters, studs) must not be less than 1¼ in. (32 mm) from the nearest edge of the framing member. If this clearance cannot be maintained, the cable must be protected by a steel plate(s) at least ¹/₁₆ in. (1.6 mm) thick.

Less than 1¼ in. (32 mm)

More than 1¼ in. (32 mm)

Steel plate(s)

Sometimes it is possible to drill hole to one side of center so as to need only one ¹/₁₆-in. (1.6-mm) thick steel plate.

1¼ in. (32 mm)

Cables

Notched studs

2 x 2

2 x 4

3 An alternative to using steel plates is to use cable "stand off" devices. See figure 4-27.

Note: Listed and marked steel plates are permitted to be less than ¹/₁₆ in. (1.6 mm) thick.

2 Metal plate(s), ¹/₁₆ in. (1.6 mm) thick needed to protect cables laid in notch in stud.

FIGURE 4-26 Methods of protecting nonmetallic and armored cables so nails and screws that "miss" the studs will not damage the cables, *300.4(A)(1)* and *(2)*. Because of their strength, intermediate and rigid metal conduit, PVC rigid conduit, and electrical metallic tubing are exempt from this rule per the *Exceptions* found in *300.4.* See Figure 4-28 for alternative methods of protecting cables.
(*Delmar/Cengage Learning*)

Steel Plates

If the 1¼ in. (32 mm) from the edge of hole to the edge of framing member cannot be maintained, or if the cable is laid in a notch, a steel plate(s) at least ¹/₁₆ in. (1.6 mm) thick must be used to protect the cable from errant nails or drywall screws, *300.4(A)(1)*. In all cases where the *NEC* requires ¹/₁₆-in. (1.6-mm) steel plates, listed and marked steel plates are permitted to be less than ¹/₁₆ in. (1.6 mm) because they have passed tough mechanical tests. This permission is found in *300.4(A)(1) Exception 2, 300.4(A)(2) Exception 2, 300.4(B)(2) Exception, 300.4(D) Exception 3,* and *300.4(E) Exception 2.* See Figure 4-26.

Instead of using steel plates, Figure 4-27 illustrates devices that can be used to *stand off* the cables from a framing member.

Cables Installed Parallel to Framing Members or Furring Strips

▶ To protect nonmetallic-sheathed cables, armored cables, or nonmetallic raceways that are run parallel to framing members or furring strips from being damaged by drywall screws and nails,

you have two choices: (1) keep them more than 1¼ in. (32 mm) from the nearest edge of the framing member or furring strip, or (2) use steel plates. Refer to *300.4(D), 320.17,* and *334.17.* See Figures 4-16, 4-26, 4-27, and 4-28.◀

Metal Construction

Metal studs are being used more and more. They don't burn and termites don't eat them. They are stronger than wood and can be set 20 in. (500 mm) on center instead of the usual 16-in. (400-mm) spacing. Metal studs are 10 times more conductive than wood, so treatment of insulation must be addressed.

Where nonmetallic-sheathed cables pass through holes or slots in metal framing members, the cable must be protected by a listed bushing or listed grommet securely fastened in place prior to installing the cable. The bushing or grommet shall:

- remain in place during the wall finishing process,
- cover the complete opening, and
- be listed for the purpose of cable protection.

This information is found in *300.4(B)(1)* and *334.17.* See Figure 4-29. Bushings that snap into

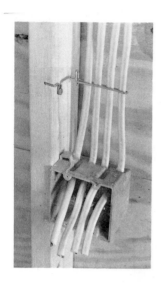

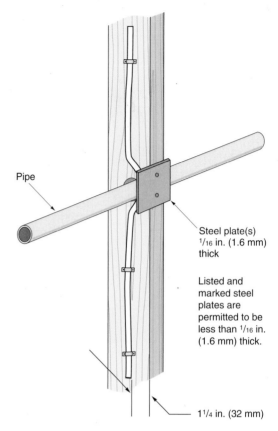

Pipe

Steel plate(s)
$^1/_{16}$ in. (1.6 mm) thick

Listed and marked steel plates are permitted to be less than $^1/_{16}$ in. (1.6 mm) thick.

$1^1/_4$ in. (32 mm)

FIGURE 4-28 Where it is necessary to cross pipes or other similar obstructions, making it impossible to maintain at least 1¼-in. (32-mm) clearance from the framing member to the cable, install a steel plate(s) at least $^1/_{16}$-in. (1.6 mm) thick to protect the cable from possible damage from nails or screws. See *300.4(A)(1)* and *(A)(2)*.
(*Delmar/Cengage Learning*)

FIGURE 4-27 Devices that can be used to meet the *Code* requirement to maintain a 1¼-in. (32-mm) clearance from framing members.
(*Delmar/Cengage Learning*)

a hole sized for the specific size bushing (½ in., ¾ in., etc.), or a two-piece snap-in bushing that fits just about any size precut hole in the metal framing members are acceptable to use, as shown in Figure 4-29. These types of bushings (insulators) are not required in metal framing members when the wiring method is armored cable, EMT, FMC, or electrical nonmetallic tubing.

When running nonmetallic-sheathed cable or electrical nonmetallic tubing through metal framing members where there is a likelihood that nails or

screws could be driven into the cables or tubing, a steel plate, steel sleeve, or steel clip at least $^1/_{16}$ in. (1.6 mm) thick must be installed to protect the cables and/or tubing. This is spelled out in *300.4(B)(2)*. See Figures 4-28 and 4-29.

CABLES IN SHALLOW GROOVES AND CHASES

NEC 300.4(F) addresses the wiring problems associated with cutting grooves into building material because there is no hollow space for the wiring. Nonmetallic-sheathed cable and some raceways must be protected when laid in a groove. Protection must be provided with a $^1/_{16}$-in. (1.6-mm) steel plate(s), sleeve, or equivalent, or the groove

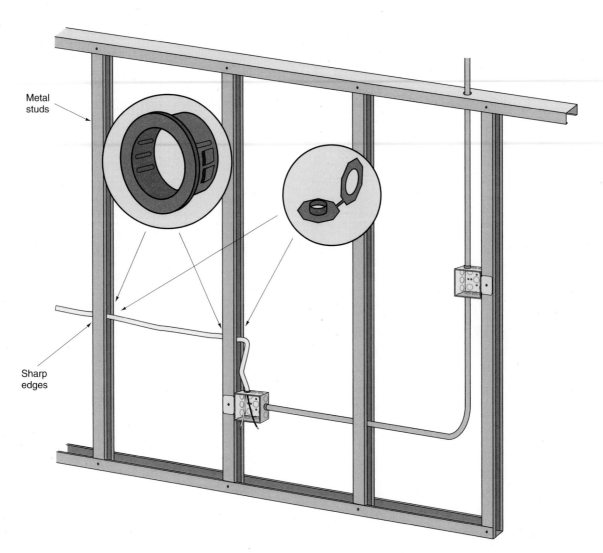

Metal studs

Sharp edges

FIGURE 4-29 *NEC 300.4(B)* and *334.17* require protection of nonmetallic-sheathed cable where it runs through holes in metal framing members. Shown are two types of listed bushings (grommets) that can be installed to protect nonmetallic-sheathed cable from abrasion. Electricians generally avoid using nonmetallic-sheathed cable through metal studs and joists. They use electrical metallic tubing, flexible metal conduit, armored cable (Type AC or MC), or electrical nonmetallic tubing. Where there is a likelihood that nails or screws might penetrate the nonmetallic-sheathed cable, ¹⁄₁₆-in. (1.6-mm) thick steel plates, sleeves, or clips must be installed to protect the cable or electrical nonmetallic tubing.
(*Delmar/Cengage Learning*)

must be deep enough to allow not less than 1¼-in. (32-mm) free space for the entire length of the groove. This situation can be encountered where styrofoam insulation building blocks are grooved to receive the electrical cables, then covered with wallboard, wood paneling, or other finished wall material. Another typical example is where solid wood planking is installed on top of wood beams. The top side of the planking is covered with roofing material, and the bottom side is exposed and serves as the finished ceiling. The only way to install wiring for ceiling luminaires and fans is to groove the planking in some manner, as shown in Figure 4-30.

This additional protection is not required when the raceway is intermediate metal conduit (*Article 342*), rigid metal conduit (*Article 344*), rigid PVC conduit (*Article 352*), or electrical metallic tubing (*Article 358*) (see Figure 4-29).

NEC 334.30(B)(1), and *300.4(D), Exception 2*, allow cables to be "fished" between outlet boxes or other access points without additional protection.

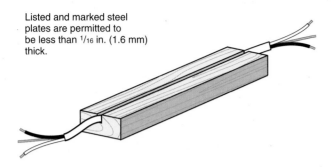

Listed and marked steel plates are permitted to be less than $^1/_{16}$ in. (1.6 mm) thick.

FIGURE 4-30 Example of how a solid wood framing member or Styrofoam insulating block might be grooved to receive the nonmetallic-sheathed cable. After the cable is laid in the groove, a steel plate(s) not less than $^1/_{16}$-in. (1.6-mm) thick must be installed over the groove to protect the cable from damage, such as from driven nails or screws. See *300.4(F)* and *334.15(B)*. (*Delmar/Cengage Learning*)

In the recreation room of the residence plans, precautions must be taken so there is adequate protection for the wiring. The walls in the recreation room are to be paneled. Therefore, if the carpenter uses 1 × 2 or 2 × 2 furring strips, as in Figure 4-31, nonmetallic-sheathed cable would require the additional $^1/_{16}$-in. (1.6-mm) steel plate(s) protection or the use of *stand off* devices similar to those illustrated in Figure 4-27.

Some building contractors attach 1-in. insulation to the walls, then construct a 2 × 4 wall in front of the basement structural foundation wall, leaving 1-in. spacing between the insulation and the back side of the 2 × 4 studs. This makes it easy to run cables or conduits behind the 2 × 4 studs and eliminates the need for the additional mechanical protection, as shown in Figure 4-32.

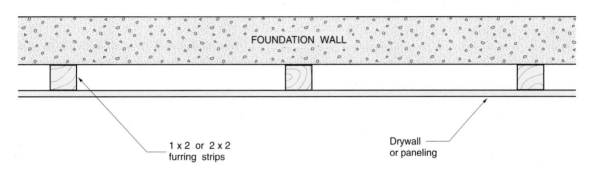

FOUNDATION WALL

1 x 2 or 2 x 2 furring strips

Drywall or paneling

FIGURE 4-31 When 1 × 2 or 2 × 2 furring strips are installed on the surface of a basement wall, additional protection, as shown in Figures 4-26, 4-27, 4-29, and 4-30, must be provided. It is impossible to maintain the minimum distance of 1¼ in. (32 mm) from the nearest edge of the framing member, in which case protection is required the entire length of the cable run, *300.4(D)*. (*Delmar/Cengage Learning*)

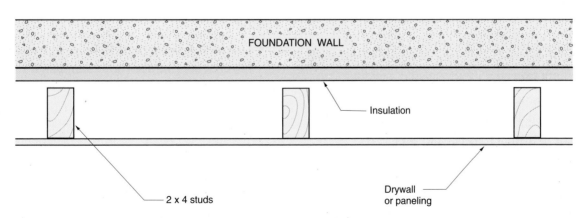

FOUNDATION WALL

Insulation

2 x 4 studs

Drywall or paneling

FIGURE 4-32 This sketch shows how to insulate a foundation wall and how to construct a wall that provides a space to run cables and/or raceways that will not require the additional protection as shown in Figures 4-26, 4-27, 4-29, and 4-30. (*Delmar/Cengage Learning*)

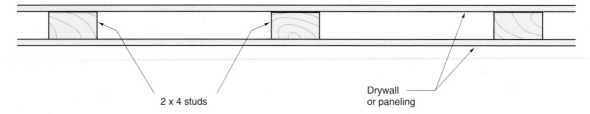

2 x 4 studs

Drywall
or paneling

FIGURE 4-33 When 2 × 4 studs are installed "flat," such as might be found in nonbearing partitions, cables running parallel to or through the studs must be protected against the possibility of having nails or screws driven through the cables. This means protecting the cable with ⅟₁₆-in. (1.6-mm) steel plates or equivalent for the entire length of the cable, or installing the wiring using intermediate metal conduit, rigid metal conduit, rigid PVC conduit, or EMT. See Figures 4-26, 4-27, 4-28, and 4-29. (*Delmar/Cengage Learning*)

Watch out for any wall partitions where 2 × 4 studs are installed "flat," as in Figure 4-33. In this case, the choice is to provide the required mechanical protection as required by *300.4*, or to install the wiring in EMT.

See Chapter 15 concerning the installation of cables in attics.

Building Codes

When complying with the *NEC*, be sure to also comply with the building code. Building codes are concerned with the weakening of framing members when drilled or notched. Meeting the requirements of the *NEC* can be in conflict with building codes and vice versa. Here is a recap of some of the key building code requirements found in the ICC *International Residential Code*.

Studs:

- Cuts or notches in exterior or bearing partitions shall not exceed 25% of the stud's width.
- Cuts or notches in nonbearing partitions shall not exceed 40% of the stud's width.
- Bored or drilled holes in any stud:
  - shall not be greater than 40% of the stud's width,
  - shall not be closer than ⅝ in. (15.9 mm) to the edge of the stud, and
  - shall not be located in the same section as a cut or notch.

Joists:

- Notches on the top or bottom of joists shall not exceed ⅟₆ the depth of the joist, shall not be longer than ⅓ the width of the joists, and shall not be in the middle third of the joist.
- Holes bored in joists shall not be within 2 in. (50 mm) of the top or bottom of the joist or any other hole, and the diameter of any such hole shall not exceed ⅓ the depth of the joist. If the joist is also notched, the hole shall not be closer than 2 in. (50 mm) to the notch.

Engineered Wood Products

Today, we find many homes constructed with engineered wood I-joists and beams. The strength of these factory-manufactured framing members conforms to various building codes. It is critical that certain precautions be taken relative to drilling and cutting these products so as not to weaken them. In general:

- Do not cut, notch, or drill holes in trusses, laminated veneer lumber, glue-laminated members, or I-joists unless specifically permitted by the manufacturer.
- Do not cut, drill, or notch flanges. Flanges are the top and bottom pieces.
- Do not cut holes too close to supports or to other holes.
- For multiple holes, the amount of web to be left between holes must be at least twice the diameter of the largest adjacent hole. The web is the piece between the flanges.

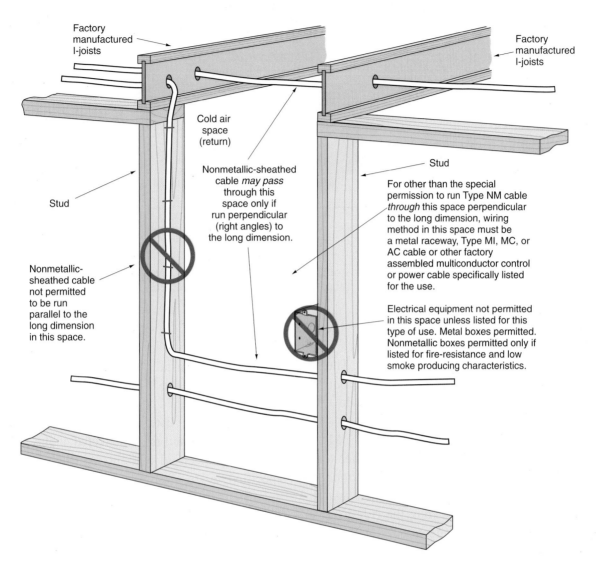

FIGURE 4-34   Nonmetallic-sheathed cable may pass through cold air return, joist, or stud spaces in dwellings, but only if it is run at right angles to the long dimension of the space, *300.22(C)*, *Exception*. (*Delmar/Cengage Learning*)

- Holes not over 1½ in. (38 mm) usually can be drilled anywhere in the web.
- Do not hammer on the web except to remove knockout prescored holes.
- Check and follow the manufacturer's instructions!

INSTALLATION OF CABLES THROUGH DUCTS

NEC 300.22 is extremely strict as to what types of wiring methods are permitted for installation of cables through ducts or plenum chambers. These stringent rules are for fire safety.

The *Code* requirements are somewhat relaxed by *300.22(C)*, *Exception*. In this exception, permission is given to run nonmetallic-sheathed and armored cables in joist and stud spaces (i.e., cold air returns), but only if they are run perpendicular to the long dimensions of such spaces, as illustrated in Figure 4-34.

CONNECTORS FOR INSTALLING NONMETALLIC-SHEATHED AND ARMORED CABLE

The connectors shown in Figure 4-35 are used to fasten nonmetallic-sheathed cable and armored cable to

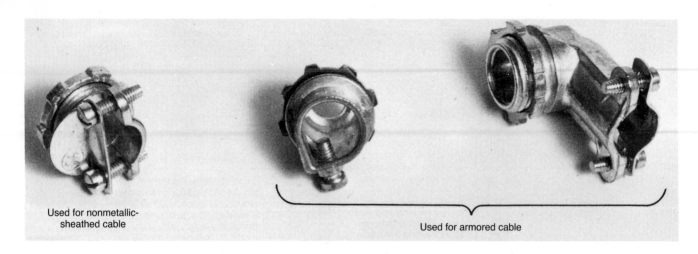

Used for nonmetallic-sheathed cable

Used for armored cable

FIGURE 4-35 Cable connectors. (*Delmar/Cengage Learning*)

the boxes and panels in which they terminate. These connectors clamp the cable securely to each outlet box. Many boxes have built-in clamps and do not require separate connectors.

The question continues to be asked: May more than one cable be inserted in one connector? Unless the UL listing of a specific cable connector indicates that the connector has been tested for use with more than one cable, the rule is *one cable, one connector.* Connectors suitable for more than one cable will be marked on the carton.

INTERMEDIATE METAL CONDUIT (*ARTICLE 342*), RIGID METAL CONDUIT (*ARTICLE 344*), RIGID PVC CONDUIT (*ARTICLE 352*), AND ELECTRICAL METALLIC TUBING (*ARTICLE 358*)

Some communities do not permit cable wiring. They require that the wiring method be a raceway—metallic or nonmetallic. You will have to check this out before starting a project.

The following comparison chart provides the highlights of the *NEC* requirements for different types of raceways that might be used in a residence. For more detailed information, refer to the

manufacturer's installation instructions, the specific *NEC* article, and the UL *White Book.*

Grounding. All listed fittings for listed metallic raceways and listed metal jacketed cables are tested to carry a specified amount of fault current for a specified length of time, making the fittings acceptable for grounding when properly installed.

In this residence, the wiring method for all of the exposed wiring in the workshop is to be electrical metallic tubing (EMT). Wiring with EMT is covered in detail in Chapter 18.

Table 4-14 provides a summary of the requirements for installing intermediate metal conduit (IMC), rigid metal conduit (RMC), rigid polyvinyl chloride conduit (PVC), and electrical metallic tubing (EMT).

FLEXIBLE CONNECTIONS

The installation of some equipment requires flexible connections, both to simplify the installation and to stop the transfer of vibrations. In residential wiring, flexible connections are used to connect equipment such as attic fans, food waste disposers, dishwashers, air conditioners, heat pumps, and recessed luminaires.

TABLE 4-14

Conduit and tubing.

	Intermediate Metal Conduit (IMC) *Article 342*	Rigid Metal Conduit (RMC) *Article 344*	Rigid PVC Conduit *Article 352*	Electrical Metallic Tubing (EMT) *Article 358*
Bends	Not more than the equivalent of four quarter bends between pull points. 360° total.	Not more than the equivalent of four quarter bends between pull points. 360° total.	Not more than the equivalent of four quarter bends between pull points. 360° total.	Not more than the equivalent of four quarter bends between pull points. 360° total.
	Manufactured bends available.	Manufactured bends available.	Manufactured bends available.	Manufactured bends available.
	Field bends made with benders.	Field bends made with benders.	Field bends made with benders that electrically heat the PVC to just the right temperature.	Field bends made with benders.
Bend radius	See *Table 2, Chapter 9, NEC.*	See *Table 2, Chapter 9, NEC.*	See *Table 2, Chapter 9, NEC.*	See *Table 2, Chapter 9, NEC.*
Bushings	Required where a conduit enters a box, fitting, or other enclosure unless the design of the box, fitting, or enclosure provides needed protection, *342.46.*	Required where a conduit enters a box, fitting, or other enclosure unless the design of the box, fitting, or enclosure provides needed protection, *344.46.*	Required where a conduit enters a box, fitting, or other enclosure unless the design of the box, fitting, or enclosure provides needed protection, *352.46*	Where 4 AWG or larger ungrounded conductors are installed, smoothly rounded insulating fittings or the equivalent must be used, *300.4(G)* and *312.6(C).* See Figure 27-30 and Figure 27-31.
	Where 4 AWG or larger ungrounded conductors are installed, smoothly rounded insulating fittings or the equivalent must be used, *300.4(G)* and *312.6(C).* See Figure 27-30 and Figure 27-31.	Where 4 AWG or larger ungrounded conductors are installed, smoothly rounded insulating fittings or the equivalent must be used, *300.4(G)* and *312.6(C).* See Figure 27-30 and Figure 27-31.	Where 4 AWG or larger ungrounded conductors are installed, smoothly rounded insulating fittings or the equivalent must be used, *300.4(G)* and *312.6(C).* See Figure 27-30 and Figure 27-31.	Not required where the PVC terminates in a hub or boss. See Figure 27-16.
	Not required where the RMC terminates in a hub or boss. See Figure 27-16.	Not required where the RMC terminates in a hub or boss. See Figure 27-16.	Not required where the PVC terminates in a hub or boss. See Figure 27-16.	
Cable permitted inside	Yes, if permitted by the specific cable *Code* article.	Yes, if permitted by the specific cable *Code* article.	Yes, if permitted by the specific cable *Code* article.	Yes, if permitted by the specific cable *Code* article.
Cinder fill	Yes, where subject to permanent moisture if protected on all sides by noncinder concrete not less than 2 in. (50 mm), or if at least 18 in. (450 mm) under the fill, or if it has corrosion protection.	Yes, where subject to permanent moisture if protected on all sides by noncinder concrete not less than 2 in. (50 mm), or if at least 18 in. (450 mm) under the fill, or if it has corrosion protection.	Yes.	Yes, where subject to permanent moisture if protected on all sides by noncinder concrete not less than 2 in. (50 mm), or if at least 18 in. (450 mm) under the fill, or if it has corrosion protection.
Concealed in walls, ceilings, and floors	Yes.	Yes.	Yes.	Yes.
Concrete above grade	Yes.	Steel: Yes. Aluminum: Yes.	Yes.	Steel: Yes. Aluminum: Yes, if it has supplemental corrosion protection.

(continues)

TABLE 4-14

(continued)

	Intermediate Metal Conduit (IMC) *Article 342*	Rigid Metal Conduit (RMC) *Article 344*	Rigid PVC Conduit *Article 352*	Electrical Metallic Tubing (EMT) *Article 358*
Concrete on grade	Yes.	Steel: Yes for galvanized. Steel: No for nongalvanized. Aluminum: Needs supple-mental corrosion protection.	Yes.	Steel: Yes Aluminum: Yes, if it has supplemental corrosion protection.
Concrete below grade	Yes.	Steel: Yes for galvanized. Steel: No for non-galvanized. Aluminum: Needs supplemental corrosion protection.		Steel: Yes, if it has supplemental corrosion protection. Aluminum: Yes, if it has supplemental corrosion protection.
Conductor fill	As determined by *Table 1, Chapter 9, NEC.* Inside interior diameter greater than RMC of same trade size. More room for conductors. See Chapter 18 for examples of conductor fill calculations.	As determined by *Table 1, Chapter 9, NEC.* See Chapter 18 for examples of conductor fill calculations.	As determined by *Table 1, Chapter 9, NEC.* See Chapter 18 for examples of conductor fill calculations.	As determined by *Table 1, Chapter 9, NEC.* See Chapter 18 for examples of conductor fill calculations.
Conductors, when to install?	After complete raceway system is installed, *300.18(A).*	After complete raceway system is installed, *300.18(A).*	After complete raceway system is installed, *300.18(A).*	After complete raceway system is installed, *300.18(A).*
Corrosive conditions	Yes, if protected by corrosion protection and judged suitable for the condition. Listing will indicate this.	Yes, if protected by corrosion protection and judged suitable for the condition. Listing will indicate this. Aluminum resists most corrosive atmospheres.	Yes. Like those found around swimming pool areas.	Yes, if protected by corrosion protection and judged suitable for the condition. Listing will indicate this.
Dry and damp locations	Yes.	Yes.	Yes.	Yes.
Expansion fittings for thermal expansion	Not required.	Not required.	Yes. Refer to *Table 352.44(A)* and *(B).* Without expansion fittings, RNC in direct sunlight will pull out of hubs of boxes, or will pull the boxes off the wall.	Not required.
Exposed	Yes.	Yes.	Yes.	Yes.
Fittings	Threaded and threadless.	Threaded and threadless compression. For aluminum conduit, aluminum threaded fittings recommended, but cadmium plated or galvanized fittings are satisfactory in most installations.	Attached with solvent-type cement.	Set screw and threadless compression.
Grounding	Is considered to be an equipment grounding conductor.	Is considered to be an equipment grounding conductor.	Where required, a separate equipment grounding conductor must be installed.	Is considered to be an equipment grounding conductor.

(continues)

TABLE 4-14

(continued)

	Intermediate Metal Conduit (IMC) *Article 342*	Rigid Metal Conduit (RMC) *Article 344*	Rigid PVC Conduit *Article 352*	Electrical Metallic Tubing (EMT) *Article 358*
Joints	Use listed couplings, connectors, and fittings.	Use listed couplings, connectors, and fittings.	Use proper cement.	Use listed couplings, connectors, and fittings.
Length	10 ft	10 ft	10 ft	10 ft
Listed	Yes.	Yes.	Yes.	Yes.
Material	Steel–galvanized. Is as strong as same trade size RMC.	Steel–galvanized. Also aluminum, red brass, and stainless steel.	Polyvinyl chloride (PVC). Schedule 40 (heavy wall thickness). Schedule 80 (extra-heavy wall thickness). O.D. same for both. I.D. of Schedule 40 is greater than that of Schedule 80.	Steel—galvanized. Also aluminum.
Rough edges	Must be properly reamed to remove rough edges.	Must be properly reamed to remove rough edges.	Must be properly reamed to remove rough edges.	Must be properly reamed to remove rough edges.
Support of conduit bodies	Yes.	Yes.	Yes. Conduit bodies not to contain device or support luminaires or other equipment.	Yes.
Securing	Maximum 3 ft (900 mm) from box, panel, cabinet, conduit body, etc.	Maximum 3 ft (900 mm) from box, panel, cabinet, conduit body, etc.	Maximum 3 ft (900 mm) from box, panel, cabinet, conduit body, etc.	Maximum 3 ft (900 mm) from box, panel, cabinet, conduit body, etc.
Supporting	At least every 10 ft (3 m). See *Table 344.30(B)(2)*. See Figure 4-36.	At least every 10 ft (3 m). See Table *344.30(B)(2)*. See Figure 4-36.	See *Table 352.30* for distance between supports. See Table 4-15 and Figure 4-36.	At least every 10 ft (3 m). See Figure 4-36.
Sizes	Trade sizes ½ through 4.	Trade sizes ½ through 6.	Trade sizes ½ through 6.	Steel: Trade sizes ½ through 4. Aluminum: Trade sizes 2 through 4.
Threaded	Yes. Same threads as RMC. Field cut threads shall be coated with an approved electrically conductive corrosion-resistant compound.	Yes. Field cut threads shall be coated with an approved electrically conductive corrosion-resistant compound.	No.	No.
Underground in soil.	Yes.	Steel: Yes for galvanized. Steel: No for non-galvanized. Aluminum: Needs supplemental corrosion protection	Yes.	Steel: Yes, if it has supplemental corrosion protection. Aluminum: Yes, if it has supplemental corrosion protection.
Weight	Medium. For example, trade size ½: 100 ft (30.5 m) approx. 62 lb. (28.1 kg).	Heavy. For example, trade size ½: 100 ft (30.5 m) approx. 82 lb. (37.2 kg). Aluminum is approx. ⅓ as heavy as steel. For example, trade size ½: 100 ft (30.5 m) approx. 28 lb. (12.7 kg).	Very light. For example, trade size ½ Schedule 40: 100 ft (30.5 m) approx. 17 lb. (7.7 kg); Schedule 80 approx. 21 lb. (9.53 kg).	Light. For example, trade size ½: 100 ft (30.5 m) approx. 30 lb. (13.5 kg).
Wet locations	Yes, if raceway, fittings, bolts, screws, straps, etc., are of corrosion-resistant material.	Yes, if raceway, fittings, bolts, screws, straps, etc., are of corrosion-resistant material.	Yes, if raceway, fittings, bolts, screws, straps, etc., are of corrosion-resistant material.	Yes, if raceway, fittings, bolts, screws, straps, etc., are of corrosion-resistant material.

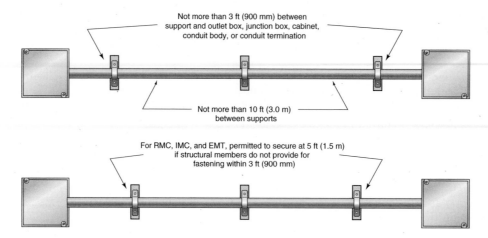

FIGURE 4-36 *Code* requirements for securing and supporting raceways are found in *342.30* (intermediate metal conduit), *344.30* (rigid metal conduit, *352.30* (rigid PVC conduit), and *358.30* (electrical metallic tubing). (*Delmar/Cengage Learning*)

TABLE 4-15

Maximum distance between supports for rigid PVC conduit.

Rigid PVC Conduit Trade Size	Maximum Spacing between Supports
½, ¾, 1	3 ft (900 mm)
1¼, 1½, 2	5 ft (1.5 m)

The following Table 4-16 shows the *NEC* requirements for three types of commonly used flexible conduit.

ELECTRICAL NONMETALLIC TUBING (ENT) (*ARTICLE 362*)

Electrical nonmetallic tubing (ENT) is covered in *Article 362* of the *NEC*. It is a pliable, lightweight, corrugated raceway made of PVC. It requires fittings

TABLE 4-16

Flexible connections.

	Flexible Metal Conduit (FMC) *Article 348*	Liquidtight Flexible Metal Conduit (LFMC) *Article 350*	Liquidtight Flexible Nonmetallic Conduit (LFNC) *Article 356*
	See Figure 4-37(A).	See Figure 4-37(B).	See Figure 4-37(C).
Bends	Not more than the equivalent of four quarter bends between pull points. 360° total.	Not more than the equivalent of four quarter bends between pull points. 360° total.	Not more than the equivalent of four quarter bends between pull points. 360° total.
Bend radius	See *Table 2, Chapter 9, NEC*.	See *Table 2, Chapter 9, NEC*.	See *Table 2, Chapter 9, NEC*.
Bushings	Where 4 AWG or larger ungrounded conductors are installed, smoothly rounded insulating fittings or the equivalent must be used, *300.4(G)* and *312.6(C)*.	Where 4 AWG or larger ungrounded conductors are installed, smoothly rounded insulating fittings or the equivalent must be used, *300.4(G)* and *312.6(C)*.	Where 4 AWG or larger ungrounded conductors are installed, smoothly rounded insulating fittings or the equivalent must be used, *300.4(G)* and *312.6(C)*.

(continues)

TABLE 4-16

(continued)

	Flexible Metal Conduit (FMC) *Article 348*	Liquidtight Flexible Metal Conduit (LFMC) *Article 350*	Liquidtight Flexible Nonmetallic Conduit (LFNC) *Article 356*
Cable permitted inside	Yes, if permitted by the specific cable *Code* article.	Yes, if permitted by the specific cable *Code* article.	Yes, if permitted by the specific cable *Code* article.
Concealed in walls, ceilings, and floors	Yes.	Yes.	Yes.
Concrete above grade	No.	Yes, if marked "Direct Burial."	Yes, if marked "Direct Burial."
Concrete on grade	No.	Yes, if marked "Direct Burial."	Yes, if marked "Direct Burial."
Concrete	No.	Yes, if marked "Direct Burial."	Yes, if marked "Direct Burial."
Conductor fill	As determined by *Tables 1 and 4, Chapter 9, NEC.* For trade size ³/₈, use *Table 348.22.*	As determined by *Table 1, Chapter 9, NEC.* For trade size ³/₈, use *Table 348.22.*	As determined by *Table 1, Chapter 9, NEC.* For trade size ³/₈, use *Table 348.22.*
Conductors, when to install?	After complete raceway system is installed, *300.18(A).*	After complete raceway system is installed, *300.18(A).*	After complete raceway system is installed, *300.18(A).*
Corrosive conditions	Yes, if protected by corrosion protection and judged suitable for the condition. Listing will indicate this.	Yes, if protected by corrosion protection and judged suitable for the condition. Listing will indicate this.	Yes. Like those found around swimming pool areas.
Dry and damp locations	Yes.	Yes.	Yes.
Exposed	Yes.	Yes.	Yes.
Fittings	Must be listed. Sizes larger than trade size ¾ will be marked GRND if acceptable for grounding. Do not conceal angle-type fittings. See Figure 4-38.	Must be listed. Sizes larger than trade size ¾ will be marked GRND if acceptable for grounding. Do not conceal angle-type fittings. See Figure 4-38.	Must be listed. Do not conceal angle-type fittings. See Figure 4-38.
Grounding	When terminated with listed fittings, it is considered to be an EGC • if it is not over 6 ft (1.8 m) long and the overcurrent device does not exceed 20 amperes. See Figure 4-39 and Figure 4-40. Longer than 6 ft (1.8 m), not acceptable as a grounding means. Install a separate equipment grounding conductor sized per *Table 250.122.* If used to connect equipment where flexibility is necessary to minimize the transmission of vibration from equipment or to provide flexibility for equipment that requires movement after installation, an equipment grounding conductor shall be installed.	When terminated with listed fittings, it is considered to be an EGC • in trade sizes ³/₈ and ½ not over 6 ft (1.8 m) long and the overcurrent device does not exceed 20 amperes. • In trade sizes ¾, 1, and 1¼ not over 6 ft (1.8 m) long and the overcurrent device does not exceed 60 amperes. Trade size 1½ and larger, not acceptable for grounding. See Figures 4-39, 4-40, 4-41, and 4-42. Longer than 6 ft (1.8 m), LFMC is not acceptable as a grounding means. Install a separate equipment grounding conductor sized per *Table 250.122.*	For grounding, install an equipment grounding conductor sized per *250.122.* Install per *250.134(B).* For bonding, install per *250.102.* See Figure 4-41.

(continues)

TABLE 4-16

(continued)

	Flexible Metal Conduit (FMC) *Article 348*	Liquidtight Flexible Metal Conduit (LFMC) *Article 350*	Liquidtight Flexible Nonmetallic Conduit (LFNC) *Article 356*
Grounding (Continued)	Sized per *250.122*. For grounding, install per *250.134(B)*. For bonding, install per *250.102*. See *250.118(5)*.	If used to connect equipment where flexibility is necessary to minimize the transmission of vibration from equipment or to provide flexibility for equipment that requires movement after installation, an equipment grounding conductor shall be installed. Size per *250.122*. For grounding, install per *250.134(B)*. For bonding, install per *250.102*. See *250.118(6)*.	
Length	No limit.	No limit.	Type LFNC-B permitted longer than 6 ft (1.8 m) if secured per *356.30*.
Listed	Must be listed. See UL 1.	Must be listed. See UL 360.	Must be listed. See UL 1660.
Material	Steel and aluminum. Must not be used where subject to physical damage.	Steel with nonmetallic outer jacket. Must not be used where subject to physical damage.	Nonmetallic. Three types. See *356.2* and UL *White Book*. Must not be used where subject to physical damage.
Rough edges	Must be properly trimmed to remove rough edges.	Must be properly trimmed to remove rough edges.	Must be properly trimmed to remove rough edges.
Securing and supporting	Maximum 12 in. (300 mm) from box, cabinet, or fitting. Maximum 4½ ft (1.4 m) between supports. Waived if fished through walls and ceilings—where flexibility is needed and the flex is not more than 3 ft (900 mm) from where it is terminated, or as a whip not over 6 ft (1.8 m). See Figures 4-43, 7-2, 7-6, and 7-10. Also waived when supported horizontally through framing members not more than 4½ ft (1.4 m) apart and secured within 12 in. (300 mm) of box, cabinet, or fitting.	Maximum 12 in. (300 mm) from box, cabinet, or fitting. Maximum 4½ ft (1.4 m) between supports. Waived if fished through walls and ceilings—where flexibility is needed and the flex is not more than 3 ft (900 mm) from where it is terminated, or as a whip not over 6 ft (1.8 m). See Figures 4-43, 7-2, 7-6, and 7-10. Also waived when supported horizontally through framing members not more than 4½ ft (1.4 m) apart and secured within 12 in. (300 mm) of box, cabinet, or fitting.	Maximum 12 in. (300 mm) from box, cabinet, or fitting. Maximum 4½ ft (1.4 m) between supports. Waived if fished through walls and ceilings—where flexibility is needed and the flex is not more than 3 ft (900 mm) from where it is terminated, or as a whip not over 6 ft (1.8 m). See Figures 4-43, 7-2, 7-6, and 7-10. Also waived when supported horizontally through framing members not more than 3 ft (900 mm) apart and secured within 12 in. (300 mm) of box, cabinet, or fitting.

(continues)

TABLE 4-16

(continued)

	Flexible Metal Conduit (FMC) *Article 348*	Liquidtight Flexible Metal Conduit (LFMC) *Article 350*	Liquidtight Flexible Nonmetallic Conduit (LFNC) *Article 356*
Sizes	Steel: Trade sizes ½ through 4. Aluminum: Trade sizes ½ through 3. Trade size ³/₈ okay for enclosing leads to a motor or as a luminaire (fixture) whip not longer than 6 ft (1.8 m).	Trade sizes ½ through 4. Trade size ³/₈ okay for enclosing leads to a motor or as a luminaire (fixture) whip not longer than 6 ft (1.8 m).	Trade sizes ½ through 4. Trade size ³/₈ okay for enclosing leads to a motor or as a luminaire (fixture) whip not longer than 6 ft (1.8 m).
In sunlight	Yes	Yes	Yes, if listed and marked "Outdoor."
Temperature	Limited to the temperature rating of the conductors.	Limited to 60°C unless marked otherwise. Also limited to temperature rating of the conductors.	Limited to 60°C unless marked otherwise. Also limited to temperature rating of the conductors. Can become brittle in extreme cold.
Underground in soil	No.	Yes, if marked "Direct Burial."	Yes, if marked "Direct Burial."
Uses	Where flexibility is required. Do not use where subject to physical damage.	Where flexibility is required. Do not use where subject to physical damage.	Where flexibility is required. Do not use where subject to physical damage.
Wet locations	No.	Yes.	Yes.

FIGURE 4-37 (A) Flexible metal conduit. (B) Liquidtight flexible metal conduit. (C) Liquidtight flexible nonmetallic conduit. (*Courtesy of Electri-Flex Co.*)

made specifically for the ENT. Figure 4–44 shows color-coded ENT and some fittings. Color-coded ENT can be used to differentiate power, communication, and fire protection signaling system.

Although not often used in typical house wiring, some electricians find ENT very practical for providing a raceway system for installing structured wiring conductors and cables. Structured wiring is discussed in Chapter 31.

SERVICE-ENTRANCE CABLE (*ARTICLE 338*)

Service-entrance cable is covered in *Article 338* of the *NEC*.

Service-entrance cable is defined as *a single conductor or multi-conductor assembly provided with or without an overall covering, primarily used for services.*

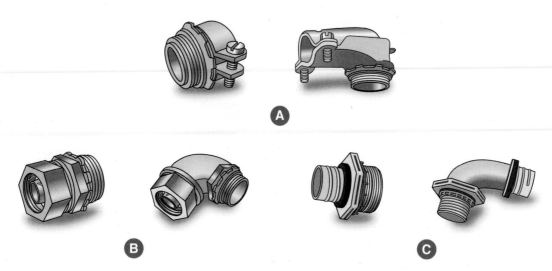

FIGURE 4-38 Various fittings for (A) flexible metal conduit, (B) liquidtight flexible metal conduit, and (C) liquidtight flexible nonmetallic conduit. (*Delmar/Cengage Learning*)

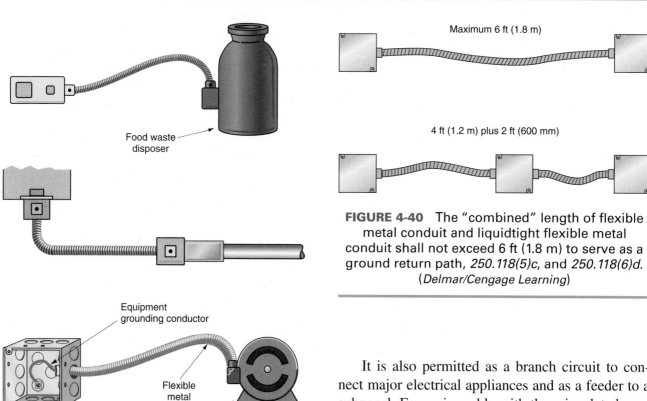

Food waste disposer

Equipment grounding conductor

Flexible metal conduit

FIGURE 4-39 If FMC or LFMC are serving as an equipment ground return path, the total length shall not exceed 6 ft (1.8 m). Refer to *250.118(5)c*, and *250.118(6)d*. (*Delmar/Cengage Learning*)

Maximum 6 ft (1.8 m)

4 ft (1.2 m) plus 2 ft (600 mm)

FIGURE 4-40 The "combined" length of flexible metal conduit and liquidtight flexible metal conduit shall not exceed 6 ft (1.8 m) to serve as a ground return path, *250.118(5)c*, and *250.118(6)d*. (*Delmar/Cengage Learning*)

Where permitted by local electrical codes, service-entrance cable is most often run from the utility's service drop (service point) to the meter base, then from the meter base to the main service panelboard.

It is also permitted as a branch circuit to connect major electrical appliances and as a feeder to a subpanel. Four-wire cable with three insulated conductor and a bare equipment grounding conductor is required for feeders to panelboards as well as to appliances such as electric ranges and dryers. For these situations, the choice of whether to use Type SE cable or nonmetallic-sheathed cable becomes a matter of choice, the major issue being the cost of the cable. Some Type SE cable with aluminum conductors may cost less than that for nonmetallic-sheathed cable with copper conductors. The labor for the installation of either type is the same.

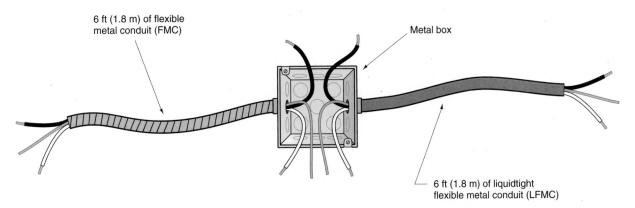

FIGURE 4-41 Use of liquidtight flexible metal conduit or flexible liquidtight nonmetallic conduit. (*Delmar/Cengage Learning*)

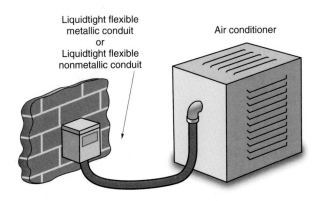

FIGURE 4-42 Some of the more common places where flexible connections could be used. (*Delmar/Cengage Learning*)

There are two styles of Type SE cable: SER and SEU. Table 4-17 shows the *NEC* installation requirements and the key differences in construction of these cables. See Table 4-17 and Figure 4-45.

Type UF Cable (*Article 340*)

Type UF is an underground feeder and branch-circuit cable. Type UF cable is discussed in Chapter 16.

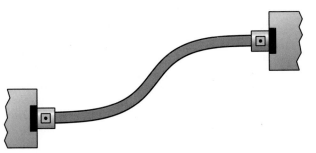

A Permitted as a grounding means if it is not over trade size 1¼, is not over 6 ft (1.8 m) long, and is connected by fittings listed for grounding. The 6 ft (1.8 m) length includes the total length of any and all flexible connections in the run. See Figure 4-41.

B Minimum trade size ½. Trade size ⅜ permitted as a luminaire (fixture) whip not over 6 ft (1.8 m) long.

C When used as the grounding means, the maximum overcurrent device is 20 amperes for trade sizes ⅜ and ½, and 60 amperes for trade sizes ¾, 1, and 1¼.

D Not suitable as a grounding means under any of the following conditions:
- Trade sizes 1½ and larger
- Trade sizes ⅜ and ½ when overcurrent device is greater than 20 amperes
- Trade sizes ⅜ and ½ when longer than 6 ft (1.8 m)
- Trade sizes ¾, 1, and 1¼ when overcurrent device is greater than 60 amperes
- Trade sizes ¾, 1, and 1¼ when longer than 6 ft (1.8 m)

For these conditions, install a separate Equipment grounding conductor sized per *Table 250.122.*

FIGURE 4-43 Grounding capabilities of liquidtight flexible metal conduit (LFMC). See *Article 350.* (*Delmar/Cengage Learning*)

FIGURE 4-44 Color-coded ENT, a coupling, a threaded connector, and a snap-in connector. (*Photos courtesy Carlon division of Thomas and Betts*)

TABLE 4-17

Service-entrance cables Type SE, SER, SEU, and USE comparisons.

Type	Covering	Insulation	Uses Permitted	Bends	Ampacity
SE Service Entrance Cable conforms to UL Standard 854: Service-Entrance Cables. See Figure 4-45.	Flame-retardant, moisture- and sunlight-resistant covering.	Type RHW, XHHW, or THWN insulation. Sizes 12 AWG and larger for copper. Sizes 10 AWG and larger for aluminum or copper-clad aluminum. If aluminum conductors, jacket will be marked "AL." If copper-clad aluminum conductors, jacket will be marked "AL (CU-CLAD)" or "Cu-Clad AL." If letter "H" on insulation, temperature rating 75°C, wet or dry locations.	600 volts or less. Above ground. Suitable for use where exposed to sun. Use "sill plate" where cable is bent to pass through a wall. See Figure 27-3. Permitted for interior wiring (feeder to a subpanel, ranges, dryers, cooktops, ovens) where all conductors are insulated. Must be installed according to the requirement in *Article 334* (Nonmetallic-sheathed cable).	Internal radius of bend not to exceed 5 times the diameter of the cable.	See: *Table 310.15(B) (16)* 230.42 for services. *215.2(A)(1)* for feeders. *310.15(B)(7)* for 120/240-volt, 3-wire single-phase dwelling services and feeders. ▶*338.10(B)* for branch circuits and feeders.◀ ▶*338.10(B)(4) (a)* for interior installations.◀ *338.10(B)(4) (b)* for exterior installations.

(continues)

TABLE 4-17

(continued)

Type	Covering	Insulation	Uses Permitted	Bends	Ampacity
SE (continued)		If letter "HH" on insulation, temperature rating 75°C wet/90°C dry locations. If marked with a "-2", suitable for wet or dry locations as high as 90°C. May have one conductor uninsulated. May have reduced neutral conductor as permitted by *220.61* and *310.15(B)(7)*.	A bare, uninsulated conductor is permitted • as the neutral conductor when used as service-entrance cable. • as an equipment grounding conduct or when used for interior wiring. For existing installations only, may have a bare neutral conductor. Must originate at service panel, *338.10(B)(2), Exception.* Also see *250.140.* Not permitted for new work. Shall not be used where subject to physical damage unless protected. Shall not be used underground whether in or not in a raceway.		
SER Conforms to UL Standard 854: Service-Entrance Cables. Subject to the same rules as Type SE.	Same as above.	Same as above. Available with two insulated phase conductors and one insulated neutral conductor. Also available with two insulated phase conductors, one insulated neutral conductor, and one bare equipment grounding conductor.	Same as above.	Same as above.	See: *Table 310.15(B) (16)* 230.42 for services. *215.2(A)(1)* for feeders. *310.15(B)(7)* for 120/240-volt, 3-wire single-phase dwelling services and feeders. ▶*338.10(B)* for branch circuits and feeders.◀ ▶*338.10(B)(4) (a)* for interior installations.◀ *338.10(B)(4) (b)* for exterior installations.

(continues)

TABLE 4-17

(continued)

Type	Covering	Insulation	Uses Permitted	Bends	Ampacity
SER (continued)		Available with reduced neutral conductor as permitted by *220.61* and *310.15(B)(7)*. Type XHHW-2 conductors. Rated 90°C, wet or dry.			
SEU Conforms to UL Standard 854: Service-Entrance Cables. Subject to the same rules as Type SE.	Same as above.	Type XHHW-2 conductors. Rated 90°C, wet or dry. Type THHN/THWN Easily identified by the bare wraparound neutral conductor. Available with two and three insulated conductors plus the bare wraparound neutral conductor. Available with reduced neutral conductor as permitted by *220.61* and *310.15(B)(7)*.	Same as above.	Same as above.	See: *Table 310.15(B)(16)* 230.42 for services. *215.2(A)(1)* for feeders. *310.15(B)(7)* for 120/240-volt, 3-wire single-phase dwelling services and feeders. ▶*338.10(B)* for branch circuits and feeders.◀ ▶*338.10(B)(4) (a)* for interior installations.◀ *338.10(B)(4) (b)* for exterior installations.
USE Service-Entrance Conductor. Individual conductors may be bundled in an assembly. Conforms to UL Standard 854: Service-Entrance Cables. USE cable is shown in Figure 4-46.	Moisture-resistant covering. Not required to have flame-retardant covering.	Insulation equivalent to Type RHW or XHHW. The temperature rating of the cable is the same as the temperature rating of the individual conductors. If this is not marked on the outer jacket, then the cable is rated 75°C. If Type USE-2, insulation equivalent to Type RHW-2 or XHHW-2. Rated 90°C, wet or dry. Cabled single conductor may have bare copper conductor with the assembly. May have reduced neutral conductor if multiple conductor assembly, as permitted by *220.61* and *310.15(B)(7)*.	Same as above except USE is also permitted for underground direct burial. Permitted to emerge above ground in meter bases and similar enclosures. Where cable emerges from the ground, protect according to the requirement found in *300.5(D)*. Not permitted for interior wiring.	Same as above.	See: *Table 310.15(B) (16)* 230.42 for services. *215.2(A)(1)* for feeders. *310.15(B)(7)* for 120/240-volt, 3-wire single-phase dwelling services and feeders. ▶*338.10(B)* for branch circuits and feeders.◀ ▶*338.10(B)(4) (a)* for interior installations.◀ *338.10(B)(4) (b)* for exterior installations.

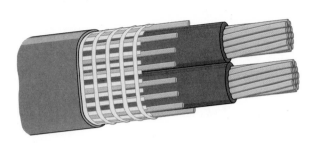

FIGURE 4-45 Type SEU service-entrance cable. The major characteristic is the uninsulated wraparound neutral conductor. See Table 4-17 for *NEC* requirements for the permitted uses of Type SE cables. (*Delmar/Cengage Learning*)

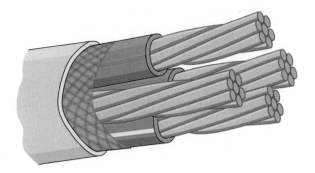

FIGURE 4-46 Type SER service-entrance cable. The major characteristics are the insulated neutral conductor and bare equipment grounding conductor. See Table 4-17 for *NEC* requirements for the permitted uses of Type SE cables. (*Delmar/Cengage Learning*)

*Reprinted with permission from NFPA 70-2011.

REVIEW

Note: Refer to the *Code* or the plans where necessary.

1. The largest size solid conductor permitted to be installed in a raceway is (10 AWG) (8 AWG) (6 AWG). Circle the correct answer.

2. What is the minimum branch-circuit wire size that may be installed in a dwelling?

   _____

3. What exceptions, if any, are there to the answer for Question 2? _____

   _____
   _____
   _____
   _____
   _____

4. What determines the ampacity of a wire? _____

   _____
   _____
   _____
   _____
   _____

5. What unit of measurement is used for the diameter of wires? _____

6. What unit of measurement is used for the cross-sectional area of wires? _____

   _____

7. What is the voltage rating of the conductors in Type NMC cable? _____

8. Indicate the allowable ampacity of the following Type THHN (copper) conductors. Refer to *Table 310.15(B)(16)*.

   a. 14 AWG _____ amperes

   b. 12 AWG _____ amperes

   c. 10 AWG _____ amperes

   d. 8 AWG _____ amperes

   e. 6 AWG _____ amperes

   f. 4 AWG _____ amperes

9. What is the maximum operating temperature of the following conductors? Give the answer in degrees Celsius.

   a. Type XHHW _____

   b. Type THWN _____

   c. Type THH _____

   d. Type TW _____

10. What are the colors of the conductors in nonmetallic-sheathed cable for

    a. 2-wire cable? _____ , _____

    b. 3-wire cable? _____ , _____ , _____

11. For nonmetallic-sheathed cable, may the uninsulated conductor be used for purposes other than grounding? _____

12. What size equipment grounding conductor is used with the following sizes of nonmetallic-sheathed cable?

    a. 14 AWG _____

    b. 12 AWG _____

    c. 10 AWG _____

    d. 8 AWG _____

13. Under what condition may nonmetallic-sheathed cable (Type NM) be fished in the hollow voids of masonry block walls? _____

    _____

    _____

14. a. What is the maximum distance permitted between straps on a cable installation?

    _____

    b. What is the maximum distance permitted between a box and the first strap in a cable installation? _____

    c. Does the *NEC* permit 2-wire Romex stapled on edge? _____

    d. When Type NMC cable is run through holes in studs and joists, must additional support be provided? _____

15. What is the difference between Type AC and Type ACT armored cables? _____

    _____

    _____

    _____

    _____

    _____

    _____

    _____

16. Type ACT armored cable may be bent to a radius of not less than _____ times the diameter of the cable.

17. When armored cable is used, what protection is provided at the cable ends?

    _____

    _____

    _____

    _____

18. What protection must be provided when installing a cable in a notched stud or joist, or when a cable is run through bored holes in a stud or joist where the distance is less than 1¼ in. (32 mm) from the edge of the framing member to the cable, or where the cable is run parallel to a stud or joist and the distance is less than 1¼ in. (32 mm) from the edge of the framing member to the cable? _____

    _____

    _____

    _____

    _____

19. For installing directly in a concrete slab, (armored cable) (nonmetallic-sheathed cable) (conduit) may be used. Circle the correct method of installation.

20. Circle the correct answer to the following statements:

    a. Type SE service-entrance cable is for aboveground use.   True   False

    b. Type USE service-entrance cable is suitable for direct burial in the ground.   True   False

    c. Type SE cable with an uninsulated neutral conductor is permitted for hooking up an electric range.   True   False

    d. Type SE cable with an insulated neutral conductor is permitted for hooking up an electric range.   True   False

    e. Are Types SE and USE service-entrance cables used in your area?   Yes   No

21. When running NMC through a bored hole in a stud, the nearest edge of the bored hole shall not be less than _____ in. from the face of the stud unless the cable is protected by a ¹/₁₆ in. (1.6 mm) metal steel plate(s).

22. Where is the main service-entrance panel located in this residence? _____

    _____

23. a. Is nonmetallic-sheathed cable permitted in your area for residential wiring?

    _____

    b. From what source is this information obtained? _____

    _____

24. Is it permitted to use flexible metal conduit over 6 ft (1.8 m) in length as a grounding means? (Yes) (No) Circle the correct answer.

25. Liquidtight flexible metal conduit may serve as a grounding means in sizes up to and including _____ in. where used with listed fittings.

26. The allowable current-carrying capacity (ampacity) of aluminum wire, or the maximum overcurrent protection in the case of 14 AWG, 12 AWG, and 10 AWG conductors, is less than that of copper wire for a given size, insulation, and temperature of 30°C. Refer to *Table 310.15(B)(16)* and *240.4(D)* and complete the following table. Important: Where the ampacity of the conductor does not match the rating of a standard fuse or circuit breaker as listed in *240.6(A), 240.4(B)* permits the selection of the next "higher" standard rating of fuse or circuit breaker if the next higher standard rating does not exceed 800 amperes. Nonstandard ampere ratings are also permitted.

WIRE (AWG-kcmil)	COPPER		ALUMINUM	
	Ampacity	Overcurrent Protection	Ampacity	Overcurrent Protection
12 AWG THHN				
10 AWG THHN				
3 AWG THW				
0000 THWN				
500 kcmil THWN				

27. It is permissible for an electrician to connect aluminum, copper, or copper-clad aluminum conductors together in the same connector. (True) (False) Circle the correct answer.

28. Terminals of switches and receptacles marked CO/ALR are suitable for use with _____, _____, and _____ conductors.

29. Wire connectors marked AL/CU are suitable for use with _____, _____, and _____ conductors.

30. A wire connector bearing no marking or reference to AL, CU, or ALR is suitable for use with (copper) (aluminum) conductors only. Circle the correct answer.

31. When Type NM or NMC cable is run through a floor, it must be protected by at least _____ in. (_____ mm) of _____

_____

_____

32. When nonmetallic-sheathed cables are bunched or bundled together for distances longer than 24 in. (600 mm), what happens to their current-carrying ability?

_____

33. In diagrams A and B, nonmetallic-sheathed cable is run through the cold air return. Which diagram "meets *Code*"? A _____ B _____ Check one.

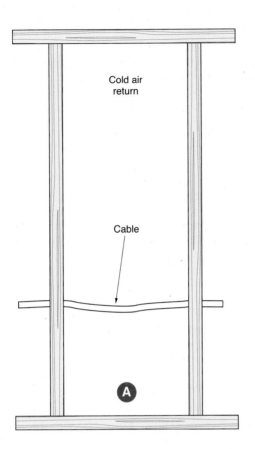

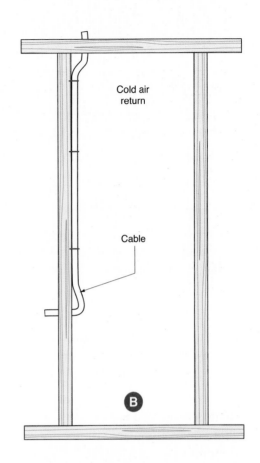

34. The marking on the outer jacket of a nonmetallic-sheathed cable indicates the letters NMC-B. What does the letter "B" signify? _____

_____

_____

35. A 120-volt branch circuit supplies a resistive heating load of 10 amperes. The distance from the panel to the heater is approximately 140 ft. Calculate the voltage drop using (a) 14 AWG, (b) 12 AWG, (c) 10 AWG, (d) 8 AWG copper conductors. *See 210.19(A), Informational Note No. 4* and *215.2(A)(3), Informational Note No. 2.*

36. In Question 35, it is desired to keep the voltage drop to 3% maximum. What minimum size wire would be installed to accomplish this 3% maximum voltage drop? See *210.19(A), Informational Note No. 4* and *215.2(A)(3), Informational Note No. 2.*

_____

_____

_____

_____

_____

37. *NEC 215.2, 220.61, 230.42,* and *310.15(B)(7)* state that the neutral conductor of a residential service or feeder can be smaller than the phase conductors, but only

_____

_____

_____

38. The allowable ampacity of a 4 AWG THHN from *Table 310.15(B)(16)* is 95 amperes. What is this conductor's ampacity if connected to a terminal listed for use with 60°C wire?_____

_____

39. If, because of some obstruction in a wall space, it is impossible to keep an NMC cable at least 1¼ in. (32 mm) from the edge of the stud, then it shall be protected by a metal plate at least _____ in. thick.

40. The recessed fluorescent luminaires installed in the ceiling of the recreation room of this residence are connected with trade size ⅜ flexible metal conduit. These flexible connections are commonly referred to as fixture whips. Does the flexible metal conduit provide adequate grounding for the luminaires, or must a separate equipment grounding conductor be installed? _____

_____

_____

_____

_____

41. Flexible liquidtight metal conduit will be used to connect the air-conditioner unit. It will be trade size ¾. Must a separate equipment grounding conductor be installed in this FLMC to ground the air conditioner properly? _____

_____

_____

_____

_____

42. What size overcurrent device protects the air-conditioning unit? _____

_____

43. May the 20-ampere small-appliance branch circuits in the kitchen of a residence also supply the lighting above the kitchen sink and under-cabinet lighting? Check the correct answer. Yes _____ No _____

44. A 30-ampere branch circuit is installed for an electric clothes dryer. What is the maximum ampere rating of a dryer permitted by the *NEC* to be cord-and-plug-connected to this circuit? Check the correct answer.

    a. _____ 30 amperes

    b. _____ 24 amperes

    c. _____ 15 amperes

45. In many areas, metal framing members are being used in residential construction. When using nonmetallic-sheathed cable, what must be used where the cable is run through the metal framing members? _____

46. Are set-screw-type connectors permitted to be used with armored cable that has an aluminum armor? _____

47. Most armored cable today has 90°C conductors. What is the correct designation for this type of armored cable? _____

48. If you saw two different types of SE cables, how would you distinguish them?

    _____

    _____

    _____

    _____

49. Circle the correct answer defining the type of location:

    a. A raceway installed underground   (Dry)   (Wet)

    b. A raceway installed in a concrete slab that is in direct
       contact with the earth   (Dry)   (Wet)

    c. A raceway installed in a concrete slab between the
       first and second floor   (Dry)   (Wet)

    d. A raceway installed on the outside of a building
       exposed to the weather   (Dry)   (Wet)

    e. The interior of a raceway installed on the outside of
       a building exposed to the weather   (Dry)   (Wet)

Conductor Identification, Switch Control of Lighting Circuits, Bonding/Grounding of Wiring Devices, Induction Heating

OBJECTIVES

After studying this chapter, you should be able to

- identify the grounded and ungrounded conductors in cable or conduit (color coding).
- identify the various types of toggle switches for lighting circuit control.
- select a switch with the proper rating for the specific installation conditions.
- describe the operation that each type of toggle switch performs in typical lighting circuit installations.
- determine when a neutral conductor must be added for switch boxes.
- demonstrate the correct wiring connections for each type of switch per *Code* requirements.
- understand the various ways to bond wiring devices to the outlet box.
- understand how to design circuits to avoid heating by induction.

WARNING! Improper wiring can result in fires, personal injuries, and electrocutions! All wiring must be done in conformance to the *NEC*. Anything less is unacceptable!

CONDUCTOR IDENTIFICATION (*ARTICLES 200* AND *210*)

Before making any electrical connections, an electrician must understand conductor color coding.

Ungrounded Conductor. An ungrounded conductor must have an outer finish that is a color other than green, white, gray, or three continuous white stripes. This conductor is commonly called the "hot" conductor. You will get a shock if this "hot" conductor and a grounded conductor or grounded surface such as a metal water pipe are touched at the same time.

Popular colors for cable wiring systems are black, red, and sometimes blue. Many other colors are available for conduit systems including blue, yellow, brown, orange, and purple.

Grounded Conductor. For alternating-current circuits, the *NEC* requires that the grounded (identified) conductor have an outer finish that is either continuous white or gray, or have an outer finish (not green) that has three continuous white stripes along the conductor's entire length. From a NEC definition standpoint, a *neutral conductor* connects to the *neutral point*. In multiwire circuits, the grounded conductor is also a *neutral* conductor.

Three Continuous White Stripes. In house wiring, you will probably never run across a conductor that has three continuous white stripes on other than green insulation. This method of identifying conductors is done by the cable manufacturer and would be found on larger size conductors and some that may be used by the electric utility. Some underground service conductors such as those used for a mobile home feeder may have two black conductors, one black conductor that has three white stripes, and one green conductor.

Grounded Neutral Conductor

For residential wiring, the grounded conductor with white insulation is commonly referred to as the *neutral*. However, this conductor is not always a neutral conductor, as you will learn.

For branch circuits and feeders, the grounded neutral conductor must be insulated, *310.106(D)*. For residential services, the grounded neutral conductor is permitted to be insulated or bare, *230.41*.

For residential wiring, the 120/240-volt electrical system is grounded by the electric utility at their transformer, and again by the electrician at the main service. This is in accordance with *250.20(B)(1)* requiring that the system be grounded if the maximum voltage to ground on the ungrounded conductors does not exceed 150 volts. *NEC 250.26(2)* requires that the neutral conductor be grounded for single-phase, 3-wire systems.

Electricians often use the term "neutral" whenever they refer to a white grounded circuit conductor. The correct definition is found in the *NEC*.

Neutral Conductor. *The conductor connected to the neutral point of a system that is intended to carry current under normal conditions.** See Figure 5-1.

Neutral Point. *The common point on a wye-connection in a polyphase system or midpoint on a single-phase, 3-wire system, or midpoint of a single-phase portion of a 3-phase delta system, or a midpoint of a 3-wire, direct current system.** See Figure 5-1.

By these definitions, the white grounded conductor in a 2-wire circuit is a neutral conductor. In a multiwire 3-wire branch circuit, the white grounded conductor is a neutral conductor. See Figure 5-1.

For those of you who will broaden your training into commercial and industrial wiring, you will find that there are electrical systems such as a 3-phase delta connected corner grounded system where the grounded conductor is a phase conductor and is not a neutral conductor. This is covered in detail in the *Electrical Wiring—Commercial* text.

A neutral conductor in residential wiring is always a grounded conductor—but a grounded conductor is not always a neutral conductor.

Although it may seem obvious to trained electricians, *NEC 200.2(B)* prohibits using a metal enclosure, a metal raceway, or metal cable armor to be part of a grounded conductor's path. Always connect the neutral conductor to the terminal bar for the neutral conductors rather than to an equipment grounding terminal bar.

---

*Reprinted with permission from NFPA 70-2011.

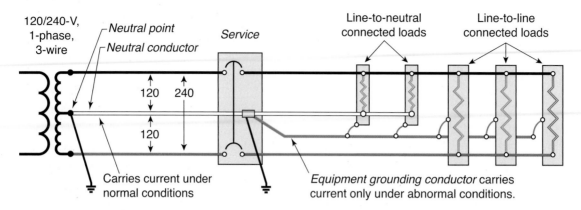

FIGURE 5-1 Definition of a neutral conductor and neutral point. (*Delmar/Cengage Learning*)

Multiwire Branch-Circuit

The *Code* defines a *multiwire branch circuit* as *A branch circuit consisting of two or more ungrounded conductors having a voltage between them, and a grounded conductor having equal voltage between it and each ungrounded conductor of the circuit and that is connected to the neutral or grounded conductor of the system.**

For additional information relative to the hazards involved with multiwire circuits, refer to Figures 17-5, 17-6, 17-7, 17-8, and 17-9 in Chapter 17.

Color Coding (Cable Wiring)

The conductors in nonmetallic-sheathed cable (Romex) are color coded with insulation as follows:

2-wire: one black ("hot" phase conductor)
one white (grounded "identified" conductor)
one bare, covered, or insulated (equipment grounding conductor)

3-wire: one black ("hot" phase conductor)
one white (grounded "identified" conductor)
one red ("hot" phase conductor)
one bare, covered, or insulated (equipment grounding conductor)

4-wire: one black ("hot" phase conductor)
one white (grounded "identified" conductor)

one red ("hot" phase conductor)
one blue ("hot" phase conductor)
one bare, covered, or insulated (equipment grounding conductor)

Four-wire nonmetallic-sheathed cable is also available with two ungrounded ("hot") and two neutral conductors. This cable is designed for wiring two 120-volt branch circuits without using a common neutral. This avoids the requirement of installing a tie handle on the circuit breakers or installing a 2-pole circuit breaker. This cable has the following insulated conductors:

- one black ("hot" phase conductor)
- one white with a black stripe (grounded "identified" conductor)
- one red ("hot" phase conductor)
- one white with a red stripe ("hot" phase conductor)
- one bare, covered, or insulated (equipment grounding conductor)

Manufacturers of Type MC cable also make a "home run cable." This cable is available with 6 or 8 12 AWG conductors and with 6, 8, 12, and 16 10 AWG conductors. The insulation is THHN/THWN, so derating [required by *NEC 310.15(B)(3)*] is started in the 90°C column of *NEC Table 310.15(B) (16)*. Table 5-1 illustrates application of derating for the number of current-carrying conductors in the home run cable. Notice that the number of current-carrying conductors can change depending on how connections are made.

---

*Reprinted with permission from NFPA 70-2011.

TABLE 5-1

Application of adjustment factors to home-run cable.

No. of Conductors	Connected	Number current-carrying conductors	Table 310.15(B)(16)	Factor from Table 310.15(B)(3)(a)	Adjusted Ampacity	Permitted Overcurrent Protection
12 AWG 6	Three 2-wire circuits	6	30	0.8	24	20
	Two 3-wire circuits	4	30	0.8	24	20
8	Four 2-wire circuits	8	40	0.8	24	20
	Two 3-wire circuits and one, 2-wire	6	30	0.8	24	20
10 AWG 6	Three 2-wire circuits	6	40	0.8	32	30
	Two 3-wire circuits	4	40	0.8	32	30
8	Four 2-wire circuits	8	40	0.8	32	30
	Two 3-wire circuits and one, 2-wire	6	40	0.8	32	30
12	Six 2-wire circuits	12	40	0.5	20	20
	Four 3-wire circuits	8	40	0.8	32	30
16	Eight 2-wire circuits	16	40	0.5	20	20
	Five 3-wire circuits	15	40	0.5	20	20

In every case, the allowable ampacity of the conductors after application of the adjustment factors permits a 20-ampere overcurrent device to be used. In some installations, home run cable can be used rather than installing a subpanel some distance away from the service equipment.

Color Coding (Raceway Wiring)

When the wiring method is a raceway such as EMT, the choice of colors for the ungrounded "hot" conductors is virtually unlimited, with the following restrictions:

> *Green or green with yellow stripe(s):* Reserved for use as an equipment grounding conductor only. Not permitted to be used as a grounded or ungrounded conductor. See *250.119* and *310.12(B).*

> *White or gray:* Reserved for use only as a grounded circuit conductor (neutral). See Table 5-2 for more information about identifying conductors.

What about Conductors with Gray Insulation?

The *NEC* recognizes a conductor with gray insulation to be used as a grounded conductor. See *200.6(A)* and *(B).*

The *NEC* prohibits a conductor with gray insulation to be used as an ungrounded "hot" conductor. See *200.6(A)* and *(B).*

For years, conductors with gray insulation were used in commercial and industrial installations as both grounded conductors and ungrounded conductors. Past editions of the *NEC* used the term *natural gray,* but no one really knew what it was. Was it "dirty white"? Was it "light gray"? Was it "dark gray"? Confusion was the end result. To bring an end to the confusion, all references to the term *natural gray* have been deleted in the *NEC.*

You might be saying to yourself, "No cable that I use for house wiring has a gray conductor in it." Right! But in commercial and industrial work, there are situations where two systems are installed in the same raceway, such as a 120/208-volt and a 277/480-volt circuit. This is an example of where you might want to use a white insulated conductor for the 120/208-volt system neutral conductor, and a gray insulated conductor for the 277/480-volt system neutral conductor.

Changing Colors When Conductors Are in a Cable

The basic rule in *200.6* is that grounded conductors must be white, gray, or have three continuous white stripes on other than green insulation.

TABLE 5-2

Changing colors of conductor insulation in a raceway or cable.

To Change This . . .	to This,	Do This
red, black, blue, etc.	an *equipment grounding conductor*	**Conductors 6 AWG or smaller:** No reidentification permitted for conductors 6 AWG and smaller. An equipment grounding conductor must be bare, covered, or insulated. If covered or insulated, the outer finish must be green or green with yellow stripes. The outer finish color must run the entire length of the conductor. **Conductors larger than 6 AWG:** Insulation on large conductors is usually black but may have other colors. Therefore, at time of installation, at both ends and at every place where the conductor is accessible except in a conduit body where there are no splices, do any one of the following: • strip off the insulation for the entire exposed length, or • at the termination, encircle the insulation with green color, or • at the termination, encircle the insulation with a few wraps of green tape, green adhesive labels, or green heat shrink tubing See *210.5(B), 250.119,* and *310.110(B).*
red, black, blue, etc.	a *grounded* conductor	**Conductors 6 AWG or smaller:** For the conductor's entire length, the insulation must be white, gray, or have three continuous white stripes on other than green insulation. Reidentification in the field is not permitted when the wiring method is a raceway. **Conductors larger than 6 AWG:** For the conductor's entire length, the insulation can be white, gray, or have three continuous white stripes on other than green insulation. If not white, gray, etc., reidentification at terminations to be white or gray paint, tape, or heat shrink tubing. This marking must encircle the conductor or insulation. Do this marking at time of installation. Identification must encircle the conductor. See *200.2, 200.6(A)* and *(B); 200.7(A), (B),* and *(C); 210.5(A);* and *310.110(A).*
white, gray	an *ungrounded* (hot) conductor	**Change permitted for cables but not for raceways:** Insulated conductors that are white, gray, or have three continuous white stripes are to be used only as grounded conductors in raceways. ▶Conductors that are part of a cable assembly having white or gray insulation are permitted to be reidentified for use as an ungrounded conductor by marking tape, painting or other effective means. Reidentification required at each location the conductor is visible, accessible and at terminations. For switch loops, the reidentified conductor is to be the supply to the switch but not the return.◀ See *200.7(A), (B),* and *(C);* and *310.110(C).*

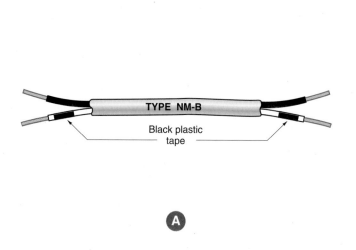

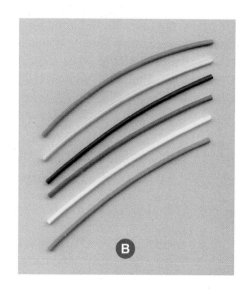

FIGURE 5-2   Shows a white conductor reidentified with black plastic electrical tape. (B) Shows an assortment of easy-to-use, 6-in.-long heat shrink tubing that is slid over the conductor to be reidentified, then shrunk tightly over the conductor by applying heat as specified in the instructions. (*Delmar/Cengage Learning*)

For cable wiring such as nonmetallic-sheathed cable or armored cable, *200.7(C)(1)* permits the white or gray conductor to be used for single-pole, 3-way, or 4-way switch loops. These sections require that when used for a switch "loop," the conductor that is white, gray, or marked with three continuous white stripes is to be used for the supply to the switch, and not as the return conductor from the switch to the switched outlet. Also, the conductor must be permanently reidentified to indicate its use by painting or other effective means, at its terminations and at each location where the conductor is visible and accessible. See Figure 5-2.

A new requirement for providing a neutral conductor at all switch boxes trumps the provisions in *200.7(C)(1)* for reidentifying the neutral conductor for switch loops. See the discussion later in this chapter as well as in Chapter 6.

How to Reidentify Conductor Insulation

Small conductors are available with insulation colors of black, blue, brown, gray, green, orange, pink, purple, red, tan, white, and yellow. There are a number of ways to reidentify a conductor's insulation. Plastic electrical tape, heat shrink tubing,

permanent felt tip marking pens, and fast-drying "touch-up" paint are generally considered acceptable ways to permanently reidentify conductor insulation. You might want to test the paint on a conductor to make sure that it does not have a harmful effect on the insulation, *NEC 310.10(G)*. Whichever method you choose to reidentify a conductor, be sure it encircles the insulation.

Increasingly, electricians are finding that permanent marking pens are a great way to reidentify conductor insulation. These pens are readily available with a wide felt tip ($5/8$ inch). Cut a notch or drill a small hole in the felt tip. Slide the notched marker up and down the conductor insulation, or slide the marker with the small hole over the conductor and slide it up and down. This makes the job of reidentifying a conductor very easy—and fast. Inspectors like this method.

Heat shrink tubing (UL recognized) with a shrink ratio of 2:1 is available at electrical distributors in packages of assorted colors and sizes, for example, 6 in. (150 mm) long. See Figure 5-2(B). This method is more labor intensive than using permanent marking pens or tape.

Some electrical inspectors have very strong feelings about what they will accept for reidentifying conductors. They might not consider plastic tape as being a "permanent" material. Check it out!

It is pretty obvious that the intent of the *Code* is to permanently reidentify a white conductor when it is used as a "hot" conductor. If not properly reidentified, someone someday will get a surprise. An accident waiting to happen—such "surprises" can prove to be fatal!

The wiring diagrams that follow are good examples of where permanent reidentification of the white conductor is necessary.

Changing Colors When Conductors Are in a Raceway

Because of the availability of so many insulation colors, it is rarely necessary to reidentify a conductor. If, however, it becomes necessary to change the actual color of a conductor's insulation, refer to Table 5-2.

CONNECTING WIRING DEVICES

Terminating and splicing conductors is a skill. The connections must be tight so as not to overheat under load.

NEC 110.3(B) requires us to follow the manufacturers' instructions regarding the listing and labeling of all electrical products. *NEC 110.14* provides clear requirements for making electrical connections.

Table 4-4 shows in detail the terminal identification markings found on wiring devices (switches and receptacles) indicating the types of conductors permitted to be connected to them.

Most commonly used wiring devices have screw terminals.

Some wiring devices have back-wiring holes only. For these, the conductor insulation is stripped off for the desired length, inserted into the hole, and held in place by the device's internal spring pressure mechanism. These wiring devices are referred to as having *push-in* terminals.

For convenience, some wiring devices have both screw terminals and back-wiring holes, as illustrated in Figures 5-3 and 5-4. These can be connected or hooked up in different ways. One way is if the conductor insulation is stripped off for the desired length and inserted into the back hole, and then the screw terminal is properly tightened. In another way, the conductor insulation is stripped

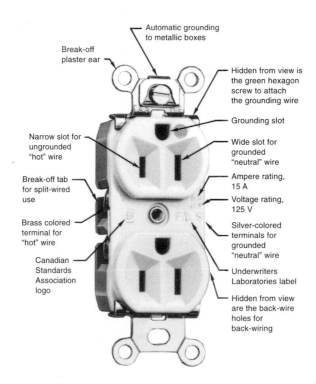

FIGURE 5-3 Grounding-type receptacle detailing various parts of the receptacle.
(Delmar/Cengage Learning)

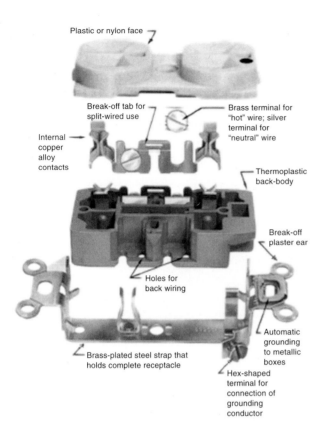

FIGURE 5-4 labels: Plastic or nylon face; Break-off tab for split-wired use; Brass terminal for "hot" wire; silver terminal for "neutral" wire; Internal copper alloy contacts; Thermoplastic back-body; Break-off plaster ear; Holes for back wiring; Automatic grounding to metallic boxes; Brass-plated steel strap that holds complete receptacle; Hex-shaped terminal for connection of grounding conductor

FIGURE 5-4   Exploded view of a grounding-type receptacle, showing all internal parts. *200.10(B)* requires that terminals on receptacles, plugs, and connectors that are intended for the ground-circuit conductor be substantially white (usually silver colored) or marked with the word "white" or the letter "W" adjacent to the terminal. The terminal for the ungrounded "hot" conductor is usually brass colored. (*Delmar/Cengage Learning*)

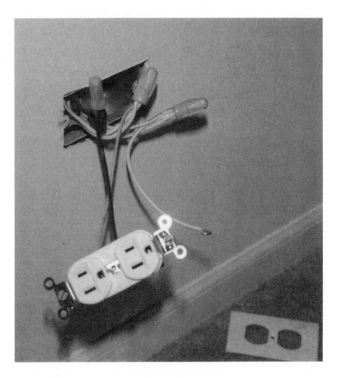

FIGURE 5-5   A receptacle connected with pigtails. This particular installation has one receptacle switched and one receptacle on continuously. In some areas this is referred to as a "split-wired receptacle," and as a "switched receptacle" in other areas. Six conductors enter the box because the wiring runs from receptacle to receptacle around the room. A 4-in. square outlet box trimmed with a single-gang plaster (drywall) cover is used. Removal of the receptacle does not affect the circuit to the other receptacles. (*Delmar/Cengage Learning*)

off for the desired length and terminated under the screw terminal, in which case the back-wiring holes are not used.

Using both the back-wiring terminals and the screw terminals, it would be possible to terminate as many as four conductors to the white terminal, and four conductors to the brass terminal. This is a violation of *110.3(B)*. UL does not test receptacles with so many conductors connected to the device. A much better way is shown in Figure 5-5, where all of the necessary splicing is done independent of the receptacle. Short lengths of wire called *pigtails* are included in the splice, and these pigtails in turn connect to the receptacle. Removal of the receptacle would not have any effect on the splice.

Installing Switches and Receptacles

Always make sure that the yokes of switches and receptacles are rigidly held in place so they do not twist. Not tightly securing wiring devices is unsafe as well as being an example of sloppy workmanship, *406.4(A)*.

Receptacle Configuration

Figure 5-6 illustrates some of the common receptacle configurations used in homes in conformance to *Table 210.21(B)(3)*. Range and dryer receptacles are discussed in Chapter 20.

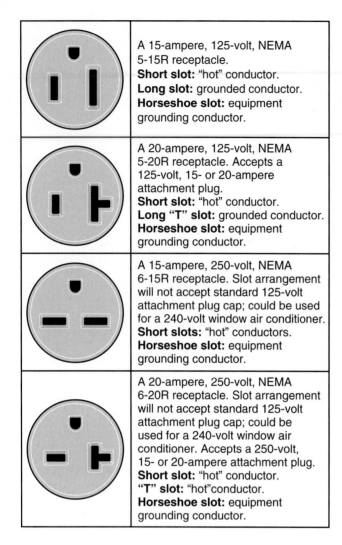

	A 15-ampere, 125-volt, NEMA 5-15R receptacle. **Short slot:** "hot" conductor. **Long slot:** grounded conductor. **Horseshoe slot:** equipment grounding conductor.
	A 20-ampere, 125-volt, NEMA 5-20R receptacle. Accepts a 125-volt, 15- or 20-ampere attachment plug. **Short slot:** "hot" conductor. **Long "T" slot:** grounded conductor. **Horseshoe slot:** equipment grounding conductor.
	A 15-ampere, 250-volt, NEMA 6-15R receptacle. Slot arrangement will not accept standard 125-volt attachment plug cap; could be used for a 240-volt window air conditioner. **Short slots:** "hot" conductors. **Horseshoe slot:** equipment grounding conductor.
	A 20-ampere, 250-volt, NEMA 6-20R receptacle. Slot arrangement will not accept standard 125-volt attachment plug cap; could be used for a 240-volt window air conditioner. Accepts a 250-volt, 15- or 20-ampere attachment plug. **Short slot:** "hot" conductor. **"T" slot:** "hot" conductor. **Horseshoe slot:** equipment grounding conductor.

FIGURE 5-6 Slot configuration of receptacles. Receptacles have the suffix "R" and plug caps have the suffix "P." (*Delmar/Cengage Learning*)

PUSH-IN TERMINATIONS

Be careful when using screwless push-in terminals.

Screwless push-in terminals on receptacles are listed by Underwriters Laboratories (UL) for use *only* with solid 14 AWG copper conductors. They *are not* to be used with

- aluminum or copper-clad aluminum conductors.
- stranded conductors.
- 12 AWG conductors. By design, the holes are large enough to take only a 14 AWG solid conductor.

Push-in terminals for 12 AWG solid copper conductors are still permitted on snap switches.

Figure 5-7 shows a typical tamper-resistant receptacle. These receptacles are required in most every location in a single-family dwelling. They were introduced into the *NEC* in an attempt to reduce the number of injuries caused

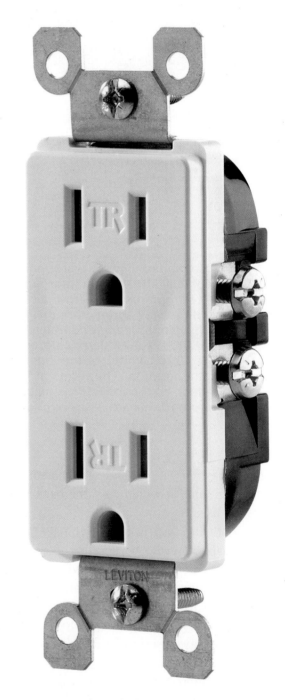

FIGURE 5-7 A typical duplex tamper-resistant receptacle. (*Courtesy of Leviton Manufacturing*)

by children who insert foreign objects into these receptacles. Electric shock and burn injuries commonly result when metallic foreign objects make contact with energized terminals in these receptacles.

The general rule in *406.12* is all nonlocking-type, 125-volt, 15- and 20-ampere receptacles are required to be tamper-resistant for new installations in dwelling units in all areas specified in *210.52*. If you take a look at that section, you will find it literally covers all locations in a dwelling including living areas, garages, unfinished basements, and accessory buildings. Scan through *210.52* and refresh your memory on the areas where tamper-resistant receptacles are required.

Several exceptions apply and exclude

1. receptacles located more than 5½ ft (1.7 m) above the floor.

2. receptacles that are part of a luminaire or appliance.

3. a single receptacle or a duplex receptacle for two appliances located within dedicated space for each appliance that in normal use is not easily moved from one place to another and that is cord-and-plug connected in accordance with *400.7(A)(6), (A)(7),* or *(A)(8).*

4. Nongrounding receptacles used for replacements as permitted in *406.4(D)(2)(a).*

Existing receptacles that are replaced in locations where tamper-resistant receptacles are now required must be replaced with tamper-resistant receptacles; see *406.4(D)(5).*

TOGGLE SWITCHES (ARTICLE 404)

UL refers to them as "snap switches." Manufacturers and electricians refer to them as "toggle switches." They are one and the same.

The most frequently used switch in lighting circuits is the flush toggle switch, Figure 5-8. When mounted in a flush switch box, the switch is concealed in the wall, with only the insulated handle or toggle protruding through the cover plate.

Figure 5-9 shows one of the ways switches can be weatherproofed.

Toggle Switch Ratings

UL lists toggle switches used for lighting circuits as *general-use snap switches.* The UL requirements are the same as *404.14(A)* and *(B)* of the *NEC.*

FIGURE 5-8   Toggle switches. (*Courtesy of Hubbell Lighting Outdoor & Industrial*)

FIGURE 5-9 Switch is protected by a weatherproof cover. (*Delmar/Cengage Learning*)

AC/DC General-Use Snap Switches [404.14(B)]

The requirements for general-use snap switches include the following:

- Alternating-current (ac) or direct-current (dc) circuits

- Resistive loads not to exceed the ampere rating of the switch at applied voltage

- Inductive loads not to exceed one-half the ampere rating of the switch at applied voltage

- Motor loads not to exceed the ampere rating of the switch at applied voltage only if the switch is marked in horsepower

- Tungsten filament lamp loads not to exceed the ampere rating of the switch at applied voltage when marked with the letter "T"

- For switches marked with a horsepower rating, a motor load not to exceed the rating of the switch at rated voltage.

Why a "T" Rating?

A tungsten filament lamp draws a very high momentary inrush current at the instant the circuit is energized. This is because the *cold resistance* of tungsten is very low. For instance, the cold resistance of a typical 100-watt lamp is approximately 9.5 ohms. This same lamp has a *hot resistance* of 144 ohms when operating at 100% of its rated voltage.

Normal operating current would be

$$I = \frac{E}{R} = \frac{120}{144} = 0.83 \text{ ampere}$$

The instantaneous inrush current could be as high as

$$I = \frac{E}{R} = \frac{170 \text{ (peak voltage)}}{9.5} = 17.9 \text{ amperes}$$

This instantaneous inrush current drops off to normal operating current in about 6 cycles (0.10 second). The contacts of T-rated switches are designed to handle these momentary high inrush currents. See Chapter 13 for more information pertaining to inrush currents.

The ac/dc general-use snap switch normally is not marked ac/dc. However, it is always marked with the current and voltage rating, such as 10A-125V or 5A-250V-T.

AC General-Use Snap Switches [404.14(A)]

This is the type most commonly used for house wiring projects. Alternating-current general-use snap switches are marked "ac only," in addition to identifying their current and voltage ratings. A typical switch marking is 15A, 120–277V ac. The 277-volt rating is required on 277/480-volt systems. Other requirements include these:

- Alternating-current (ac) circuits only are allowed.

- Resistive and inductive loads are not to exceed the ampere rating of the switch at the voltage involved. This includes electric-discharge lamps that involve ballasts, such as fluorescent lamps.

- Tungsten-filament lamp loads are not to exceed the ampere rating of the switch at 120 volts.

- Motor loads are not to exceed 80% of the ampere rating of the switch at rated voltage. A UL requirement is that the load shall not exceed 2 horsepower.

What Are the Switching Conductors Called?

In *200.7(C)(1)*, we find a few unique words that have to be explained. There are also local and regional expressions for these words. It is always best to use the terms found in the *NEC*.

- *Switch loop:* The conductors that run between (to and from) a switched outlet and the switch controlling that outlet.

- *Supply:* The conductor that runs from the source to the switch. It might also be referred to as the *feed*, *hot*, *hot leg*, or *supply leg*.

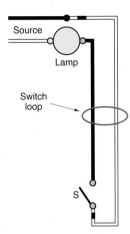

Source

Lamp

Switch loop

S

- *Return:* The conductor that runs from the switch back to the switched outlet. It might also be referred to as the *switch leg* or *switched leg*.

Conductor Color Coding for Switch Connections

The insulated conductors in nonmetallic-sheathed cable or armored cable that are commonly used for switching functions are typically black–white, black–white–red, or, in some cases, black–white–red–blue. We usually think of the black insulated conductor as the supply to a switch and a red insulated conductor as the return. However, when wiring with nonmetallic-sheathed cable or armored cable, we are "stuck" with the colors found in the cables. We must not use a white conductor as the return conductor from a switch, *200.7(C)(1)*. A color-coding scheme must be established for the supplies and returns. In the case of 3-way and 4-way switches, we also have "travelers"—the conductors that connect between 3-way and 4-way switches. The wiring diagrams in Figures 5-11 through 5-22 show the recommended color schemes when wiring with cable.

Do You Need the Grounded Circuit Conductor at the Switch? Maybe yes; maybe no.

Conventional switches operate on a mechanical principle and are simply connected "in series"

with the load. The contacts of the switch close (on) and open (off). This type of switch does not require a grounded circuit conductor to operate correctly.

However, a new requirement was added to the 2011 edition of the *NEC* in *404.2(C)*. The rule requires, ▶*Where switches control lighting loads supplied by a grounded general purpose branch circuit, the grounded circuit conductor for the controlled lighting circuit shall be provided at the switch location.**◀ This rule applies irrespective of the wiring method used or whether the switches are single-pole, 3-way, or 4-way.

Two exceptions are provided. The first applies to wiring methods such as EMT. It allows, ▶*(1) A grounded circuit conductor is not required if the conductors for the switches that control lighting loads enter the box through a raceway so long as the raceway has sufficient cross-sectional area to accommodate the extension of the grounded circuit conductor of the lighting circuit to the switch location even if the conductors in the raceway are required to be increased in size to comply with 310.15(B)(3)(a).**◀

A second exception applies to cable-type wiring methods and provides, ▶*A grounded circuit conductor is not required so long as the cable assemblies for switches controlling lighting loads enter the box through a framing cavity that is open at the top or bottom on the same floor level, or through a wall, floor, or ceiling that is unfinished on one side.**◀

The *Informational Note* tells us the purpose of the neutral conductor is to *complete a circuit path for electronic lighting control devices.*

The result of this rule requires a 3-wire cable to most single-pole switches if the source is at the lighting outlet—one supply conductor to the switch, one return conductor, and the neutral. This is shown in Figure 5-13.

For 3-way and 4-way switch loops, the number of conductors varies according to which end of the switches the supply is connected to. If the supply is at the furthest 3-way switch from the lighting outlet, three conductors are required between switches, as shown in Figure 5-21. This shows

---

*Reprinted with permission from NFPA 70-2011.

compliance with the new rule, as a neutral is available at each switch box. If the supply starts at the lighting outlet, the installation is more complicated and expensive. To provide a neutral at every switch location, a 3-wire cable is required from the lighting outlet to the first 3-way switch and a 4-wire cable between the 3-way switches. This is shown in Figure 5-22.

If the dwelling is wired with EMT, the electrician will have to be certain there is space for the neutral to be pulled into the EMT. The exception to the rule for EMT wiring includes the requirement to preplan for derating, because there are more than three current-carrying conductors in the raceway.

Electronic control devices such as dimmers, motion sensors, occupational sensors, photoelectric devices, timers, and fan speed controls must have a small voltage drop across them to operate. This type of switch (control) must have both the grounded and ungrounded circuit conductors connected to it.

Recently introduced to the marketplace are listed wiring devices such as occupancy sensors that have internal electronic circuitry that allow a maximum of 0.5 milliampere (mA) "leakage" to ground. This tiny amount of "leakage" current is enough to power up the electronic circuitry in the device, yet does not pose a shock hazard. This is referred to as low-level standby current. The maximum leakage current of 0.5 mA is the same value of leakage current permitted for electric appliances. These devices will have a white conductor that is connected to the neutral provided in the switch box.

Illuminated toggle switches, rocker switches, and illuminated faceplates do not require the grounded circuit conductor. These are merely "in series" with the load. Their operation is load "ON," pilot lamp "OFF"; or load "OFF," pilot lamp "ON." See Figure 5-25.

True pilot lights require the grounded circuit conductor. Their operation is load "ON," pilot lamp "ON"; or load "OFF," pilot lamp "OFF." See Figure 5-27.

What To Do?   Most decisions on how the switching run is installed are made on an installed-cost basis. The electrician or electrical contractor will need to look at the actual layout on the job to determine the most economical approach based on the cost of labor and materials. The most economical installation may be to always connect the supply at the switch. The two methods are shown in Figures 5-12 and 5-17.

Never use the equipment grounding conductor as a substitute for the circuit's grounded or neutral conductor. This could prove to be deadly.

When the wiring method is a raceway or electrical metallic tubing (EMT), the choices of insulation colors are virtually unlimited.

Always connect a white wire to the white (silverish) terminal or to the white wire of a lampholder or receptacle, *200.9* and *200.10*. Always connect the black switch leg conductor (or red in some cases) to the black wire (or dark brassy terminal) of a lampholder or receptacle.

Because the neutral is required at the switch box, the need to reidentify conductors as neutrals in 2-, 3-, and 4-wire cables used for switching functions has practically been eliminated.

Never use a green-colored insulation for a grounded or ungrounded conductor. Green is reserved for equipment grounding conductors. See *210.5* and *250.119*. See Table 5-2.

Induction Heating

Excessive heat damages conductor insulation. Care must be taken to avoid induction heating when running branch circuits, connecting switches, receptacles, and lighting outlets.

A conductor carrying an alternating current produces a magnetic field (flux) around the conductor. This magnetic field extends outside of the conductor. The greater the current, the stronger the magnetic field. In a 60-Hz (pronounced "60-hertz") circuit, the current and magnetic field reverses 120 times each second. If this single conductor is run through a steel raceway, a steel jacketed cable (i.e., Type AC or MC), a steel locknut, a steel bushing, or through a knockout in a steel box, the alternating magnetic field "induces" heat in the steel. Heat in the steel, in turn, can harm the insulation on the conductor. When all conductors of the same circuit are run through the same raceway, the magnetic fields around the conductors are equal and opposite, thereby canceling one another out.

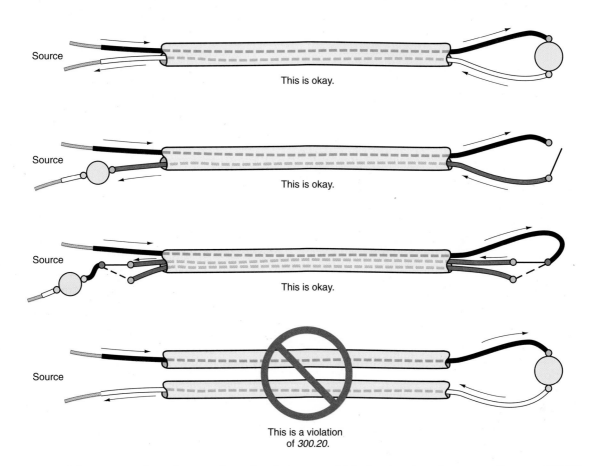

Source

This is okay.

Source

This is okay.

Source

This is okay.

Source

This is a violation
of *300.20.*

FIGURE 5-10   A few examples of arranging circuitry to avoid induction heating according to *300.20* when metal raceway or metal cables (BX) are used. Always ask yourself, "Is the same amount of current flowing in both directions in the metal raceway?" If the answer is *no*, there will be induction heating of the metal that can damage the insulation on the conductors. (*Delmar/Cengage Learning*)

To prevent induction heating and to keep the impedance of a circuit as low as possible where conductors are run in metal raceways, metal jacketed cables, and in metal enclosures, the installation must follow certain *NEC* requirements.

NEC 300.3(B) requires that *All conductors of the same circuit and, where used, the grounded conductor, all equipment grounding conductors, and bonding conductors shall be contained within the same raceway, auxiliary gutter, cable tray, trench, cable, or cord, unless otherwise permitted.** There are four exceptions: for paralleled installations, grounding conductors, nonferrous wiring methods, and certain enclosures.

NEC 300.20(A) requires that *Where conductors carrying alternating current are installed in ferrous metal enclosures or metal raceways, they shall be so arranged as to avoid heating the surrounding metal*

*by induction. To accomplish this, all phase conductors and, where used, the grounded conductor and all equipment grounding conductors shall be grouped together.**

Similar requirements are found in *300.3(B)* and *300.5.*

NEC 404.2(A) requires that *Three-way and four-way switches shall be so wired that all switching is done only in the ungrounded circuit conductor. Where in metal raceways or metal-armored cables, wiring between switches and outlets shall be in accordance with 300.20(A).** Switch loops do not require a grounded conductor.

Figures 5-10 and 5-11 illustrate the intent of the *NEC* to prevent induction heating.

*Reprinted with permission from NFPA 70-2011.

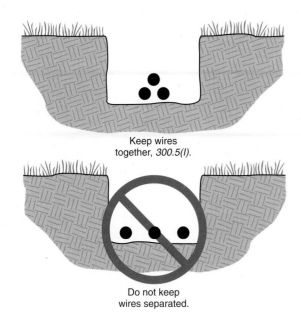

FIGURE 5-11   Arrangement of conductors in trenches. Refer to Chapter 16 for complete information pertaining to underground wiring. (*Delmar/Cengage Learning*)

Don't Do It!

Unusual switch connections may or may not conform to the *NEC*, depending on how the circuitry is run. Some electricians have found that they can save substantial lengths of 3-wire cable when hooking up 3-way and 4-way switches by using 2-wire cable.

Connections as shown in Figure 5-12 are in violation of *300.3(B)*, *300.20(A)*, and *404.3(C)* when metal boxes and/or steel jacketed Types AC and MC are used. If nonmetallic-sheathed cable and plastic boxes are used, this connection continues to be *Code* violation because there is no grounded conductor at each switch. If the framing members are steel, we have a *Code* violation because of the induction heating issue. The circuitry in Figure 5-12 results in greater impedance and greater voltage drop than if wired in a conventional manner. High impedance will cause the tripping time of a breaker or opening time of a fuse to increase. Another possible problem would be that of picking up "noise"

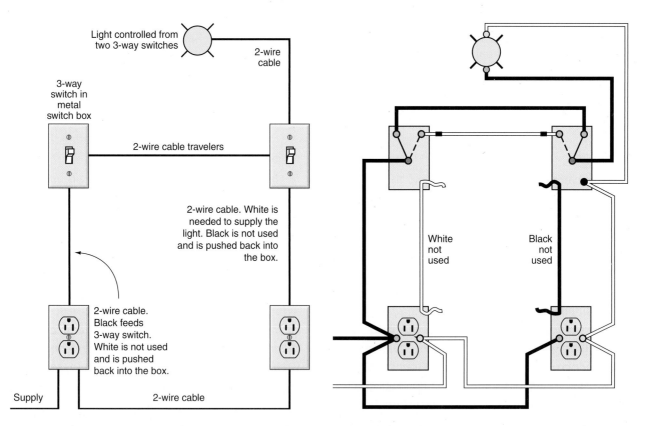

FIGURE 5-12   This connection is a violation of *300.3(B)*, *300.20(A)*, and *404.2(A)* when using steel jacketed cables, steel raceways, or steel outlet and device boxes. See text for a full explanation. This is not a recommended way of connecting 3-way and 4-way switches. (*Delmar/Cengage Learning*)

from the magnetic field surrounding the conductor. This could affect the operation of a computer or telephone located in the vicinity, particularly if non-twisted telephone cables are used. The circuitry is very confusing and is certainly an odd way to do a simple job.

Because of possible adverse interference created by magnetic fields, and confusion for someone who at some later date may be involved with that circuit for troubleshooting, repair, replacement, or remodel, the circuitry in Figure 5-12 is strongly discouraged. **Don't do it**.

Is It a Light? A Lamp? A Fixture?

Let's not get hung up on words! We usually say, "Turn on the light."

Light is "the radiant energy that is capable of exciting the retina and producing a visual sensation."

- A **lamp** is "a generic term for a man-made source of light."

- A bulb is "a glass envelope—the glass component part used in a bulb assembly."

- A **luminaire** is "a complete lighting unit consisting of a light source such as a lamp or lamps, together with the parts designed to position the light source and connect it to the power supply." The *NEC* uses the term *luminaire* as the internationally used term for *lighting fixture*.

Use whatever term that you feel comfortable with. Most of us will continue to say, "Turn on the light."

Switch Types and Connections

Switches are available in single-pole, 3-way, 4-way, double-pole, dimmer, occupancy sensor, timer, and speed control types. They are available in snap (toggle), push, rocker, slider, rotary, electronic, and with or without pilot lights. Faceplates are commonly available in white, brown, and ivory plastic, stainless steel, brass, chrome, and wood.

In all of the following wiring diagrams, the equipment grounding conductors are not shown, as this is a separate issue covered elsewhere in this text.

Single-Pole Switches. A single-pole switch is marked with "ON" and "OFF" positions. Used when the load is to be controlled from one switching point, a single-pole switch is connected in series with the ungrounded (hot) conductor.

Single-Pole Switch, Feed at Switch. In Figure 5-13, the 120-volt source enters the switch box where the white grounded conductor is spliced directly through to the lamp. The black conductor from the supply circuit connects to one terminal on the switch. The black conductor in the cable coming from the outlet box at the lamp is connected to the other terminal on the switch. Because a single-pole switch connection is simply a series circuit, it makes no difference which terminal is the supply and which terminal is the load.

This layout supplies a neutral conductor at the switch. That is why many electricians prefer to run the feed to the switch then up to the light.

Single-Pole Switch, Feed at Light. In Figure 5-14, the 120-volt source enters the outlet box where the luminaire will be connected. If a cable is used as the wiring method, a 3-wire cable is required. The conductor with black insulation is the supply to the switch and the conductor with red insulation is the return conductor. The conductor with white insulation is the neutral, so an electronic switch such as an occupancy sensor or timer can be connected properly.

Single-Pole Switch, Feed at Switch, Receptacle "ON" Continuously. Figure 5-15 shows the 120-volt source entering the switch box where the switch is located. The switch controls the lamp. The receptacle remains "hot" at all times. The white conductor in the switch box is spliced straight through to the outlet box. The black conductor in the switch box is spliced through to the outlet box and also feeds the switch. The red conductor in the 3-wire cable is the switch leg, connecting to the switch and to the lamp. A 2-wire cable from the outlet box carries the circuit to the receptacle. The circuitry is such that no reidentification of the white conductor is necessary.

Three-Way Switches. Three-way switches are used when a load is to be controlled from two

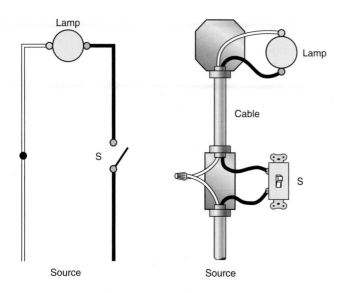

FIGURE 5-13 Single-pole switch in circuit with feed at switch. (*Delmar/Cengage Learning*)

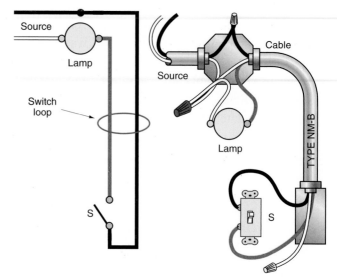

FIGURE 5-14 Single-pole switch in circuit with feed at light. (*Delmar/Cengage Learning*)

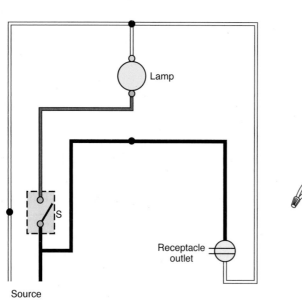

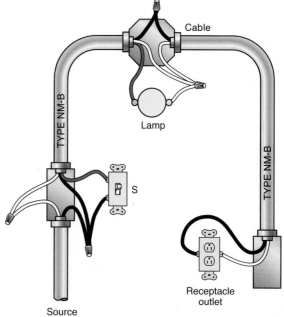

FIGURE 5-15 Lighting outlet controlled by single-pole switch with live receptacle outlet and feed at switch. (*Delmar/Cengage Learning*)

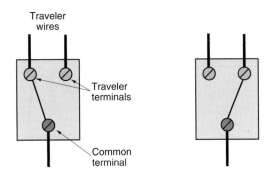

FIGURE 5-16 Two positions of a 3-way switch. (*Delmar/Cengage Learning*)

locations. The term "3-way" is very misleading. Three-way switches have three terminals. Maybe they should have been named "3-terminal" switches. One terminal is called the "common" and is darker in color than the other two terminals. The other two terminals are called "traveler terminals."

Figure 5-16 shows the two positions of a 3-way switch. Actually, a 3-way switch is a single-pole, double-throw switch.

There is no "ON/OFF" marking on a 3-way switch, as can be seen in Figure 5-17.

FIGURE 5-17 Toggle switch: 3-way snap switch. (*Courtesy of Legrand/Pass & Seymour*)

Three-Way Switch Control, Feed at Switch. Figure 5-18 shows the 120-volt source entering at the first switch location. A 3-wire cable is run between the two switches. A 2-wire cable is run from the second switch to the lamp. The white conductors in both switch boxes are spliced directly through to the lamp. The red and black conductors in the 3-wire cable are the "travelers." The black conductor in the 3-wire cable is reidentified with red tape, consistent with keeping both "travelers" red in color. The black conductor from the source is connected to the "common" terminal of the first switch. The black conductor in the cable coming from the outlet box at the lamp is connected to the "common" terminal of the second switch. The circuitry is such that no reidentification of the white conductor is necessary.

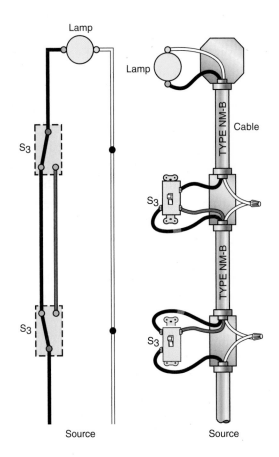

FIGURE 5-18 Circuit with 3-way switch control. The feed is at the first switch. The load is connected to the second switch. The black and red conductors are used for the travelers. The black conductor has been reidentified with red tape. (*Delmar/Cengage Learning*)

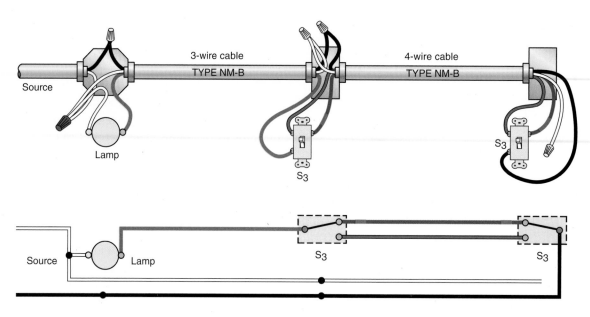

FIGURE 5-19 Circuit with 3-way switch control and feed at light. (*Delmar/Cengage Learning*)

Three-Way Switch Control, Feed at Light. Figure 5-19 shows the 120-volt source feeding into the outlet box for the luminaire. A 3-wire cable is run from the outlet box to the first switch location. The conductor with black insulation is the supply to the switch and is connected to the black conductor in the 4-wire cable. A 4-wire cable is run from the first switch to the second switch. The neutral is carried through the first box and is capped with a wire connector in the second switch box.

The conductor with red insulation of the 3-wire cable is the return conductor to the lighting outlet and is connected to the "common" terminal of the first switch. The neutral conductor is capped off for future use. The red conductor and the blue conductor with red marking tape in the 4-wire cable are the "travelers."

At the second switch location, the white conductor is capped off for future use. The black conductor of the 4-wire cable is connected to the common terminal of the 3-way switch. The blue conductor with red tape and the red conductor are the "travelers."

Three-Way Switch Control, Feed at Light (Alternate Connection). Figure 5-20 shows the 120-volt source entering at the outlet box. The white conductor from the 120-volt source at the outlet box is connected to the neutral that is required at both switch boxes and to a jumper for the luminaire. A 4-wire cable is run to both switch boxes.

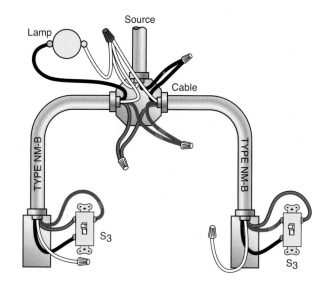

FIGURE 5-20 Circuit with a 3-way switch control with feed at the lamp. This is not the best way to wire multiple switches when wiring with cable. It is best to run the travelers between switches, not through the outlet box. Less confusion—fewer connections. (*Delmar/Cengage Learning*)

From the outlet box to the switch on the left, the black conductor in the 4-wire cable serves as the switch return to the lamp, and the red conductor and the blue conductor with the red marking tape are the "travelers." From the outlet box to the switch on the right, the black conductor serves as the supply to the 3-way switch, and the red conductor and the blue conductor with red marking tape are the "travelers." For uniformity, the conductor with red insulation and the conductor with blue insulation that has red marker tape used for "travelers" are identified and connected in the outlet box.

Four-Way Switches. Four-way switches are used where switch control is needed from more than two locations. The term *4-way* is misleading. Four-way switches have four terminals. Maybe 4-way switches should have been called "four-terminal" switches. A 4-way switch does not have "ON" and "OFF" markings. Four-way switches are connected to the "travelers" *between* two 3-way switches. One 3-way switch is fed by the source; the other 3-way switch is connected to the load. Make sure that the "travelers" are connected to the proper "pairs" of terminals on a 4-way switch; otherwise the switching will not work.

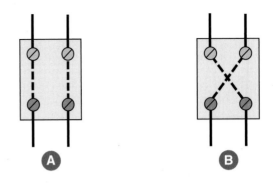

FIGURE 5-21 Two positions of a 4-way switch. (*Delmar/Cengage Learning*)

Figure 5-21 shows the internal switching of a 4-way switch.

Figure 5-22 shows a luminaire controlled from three switching points. Because the white conductor from the source is connected straight through all of the switch boxes and on to the lamp, it satisfies the requirement for a neutral conductor to be present in all switch boxes.

The black conductor from the source is connected to the common terminal on the first 3-way switch. The black conductor at the third switch is connected to the common terminal on the 3-way switch and supplies the luminaire. The black and

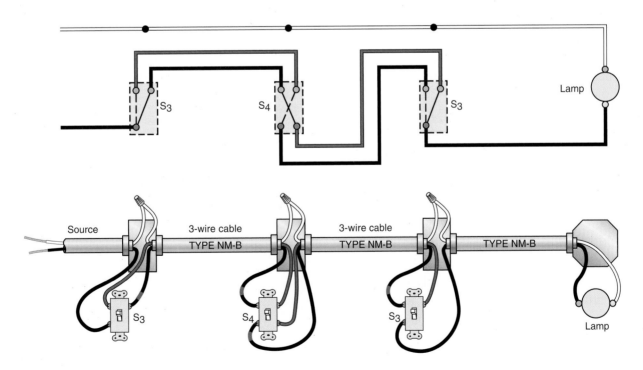

FIGURE 5-22 Circuit with switch control at three locations—feed at switch. (*Delmar/Cengage Learning*)

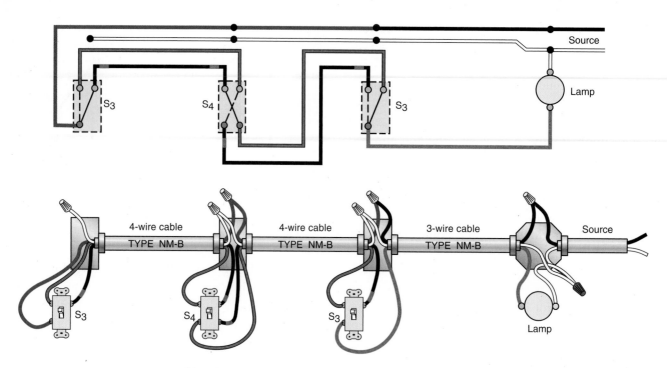

FIGURE 5-23   Circuit with switch control at three locations—feed at luminaire.
(*Delmar/Cengage Learning*)

red conductors in the 3-wire cables are the "travelers" and are connected to the proper "traveler" terminals of the 3-way and 4-way switches. The black conductors in the 3-wire cables have been reidentified with red tape, consistent with keeping both "travelers" red in color. At the lamp, the white conductor of the 2-wire cable connects to the white conductor (or silverish terminal) of the lampholder, and the black conductor connects to the black conductor (or dark brassy terminal) of the lampholder.

Figure 5-23 shows a lamp controlled from three switching points with two 3-way switches and one 4-way switch. The supply comes into the outlet box where the luminaire is connected. A 3-wire cable is run to the first switch box and 4-wire cables from the second to the third, and from the third to the fourth, box.

The conductor with black insulation is connected together at each switch box and is connected to the common terminal of the end 3-way switch. The conductor with white insulation is connected through each box and is available for future connection to an electronic switch or timer. The conductor with red insulation and the

conductor with blue insulation are the travelers. Red marking tape is placed on the conductor with blue insulation for consistent identification of traveler conductors.

Double-Pole Switches. Double-pole switches are not common in residential work. They can be used when two separate circuits must be controlled with one switch, as in Figure 5-24. If used for this purpose, they must be marked, "2-circuit." Otherwise,

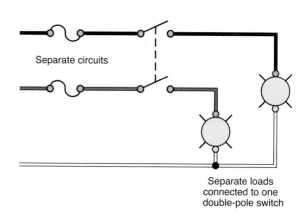

FIGURE 5-24   Application of a double-pole switch.
(*Delmar/Cengage Learning*)

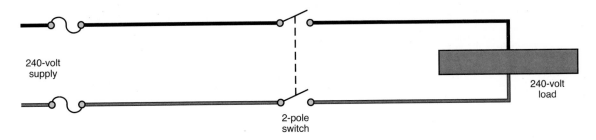

FIGURE 5-25 Double-pole (2-pole) disconnect switch. (*Delmar/Cengage Learning*)

2-pole switches must be used to switch two conductors of a single branch circuit.

They are commonly used to control 240-volt loads, such as electric heat, motors, electric clothes dryers, and similar 240-volt loads, as in Figure 5-25. When line-voltage thermostats for electric heat are double-pole, they switch both ungrounded "hot" conductors and are marked with "ON" and "OFF" positions.

Switches with Pilot Lights. There are instances when a pilot light is desired at the switch location. Figure 5-26 shows a "locator" type of switch where the lamp in the toggle glows when the switch is in the "OFF" position. *This is sometimes referred to as*

a "glow" or "lit-handle" switch. A resistor inside the switch is connected in series with the pilot light. Be careful when working on this type of switch because when the switch is in the "OFF" position, the load and the pilot light are actually connected in "series." Even though these switches are marked with an "OFF" position, they are not truly off. The circuit's path through the switch and connected load is still complete. There is a voltage drop (almost full open circuit voltage) across the pilot light/resistor and an infinitesimal voltage drop across the connected load. Although there is not enough voltage at the light to light it, there may be enough voltage and current present to cause a person to "flinch," which

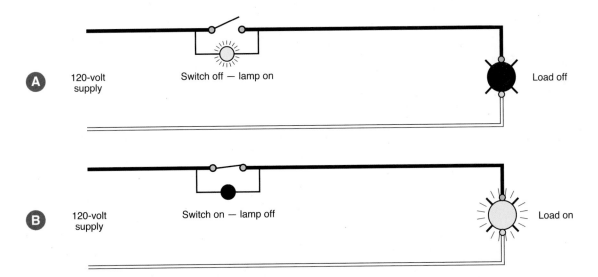

FIGURE 5-26 In (A) the switch is in the "OFF" position. The neon lamp in the handle of the switch glows and the load is off. This is a series circuit. The neon lamp has extremely high resistance. In a series circuit voltage divides proportionately to the resistance. Therefore, the neon lamp "sees" line voltage (120 volts) for all practical purposes and the load "sees" zero voltage. The neon lamp glows.
In (B) the switch is in the "ON" position. The neon lamp is shunted out and therefore has zero voltage across it. Thus, the neon lamp does not glow. Full voltage is supplied to the load. This type of switch might be referred to as a *locator* because it glows when the load is off and does not glow when the load is on.
(*Delmar/Cengage Learning*)

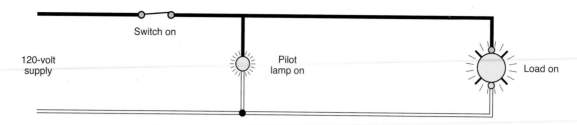

FIGURE 5-27 A true pilot light. The pilot lamp is an integral part of the switch. When the load is turned on, the pilot light is also on. When the load is turned off, the pilot light is also off. This switch has three terminals because it requires a grounded circuit conductor. (*Delmar/Cengage Learning*)

might result in the person jerking back and falling off a ladder.

Figure 5-27 shows a true pilot light that is "on" when the light is on.

COMBINATION WIRING DEVICES

In the space of one standard wiring device, it is possible to install combination wiring devices. Figure 5-28(A) shows three types of combination wiring devices. Other combinations are available. Figure 5-28(B) shows an interchangeable line of wiring devices, assembled on the metal yoke by the electrician. Switches, receptacles, and neon and incandescent pilot lights are available. Combination wiring devices offer a neater appearance as opposed to having 2-gang or 3-gang faceplates at a given location. Another common use is where wall space is limited, such as between a door casing and a window casing where there is not enough space for a multigang box and faceplate. Be sure the wall box has plenty of room for the number of cable clamps, wiring devices, and conductors.

Miscellaneous Connections. Figures 5-29 and 5-30 are diagrams that somewhat combine parts of the

FIGURE 5-28 Combination wiring devices. Many combinations are available on a single yoke: slider, rocker, toggle, 3-way plus two single poles, four single-pole switches, switches and pilot lights, switches and receptacles. (*Photos courtesy of Leviton Manufacturing Co., Inc.*)

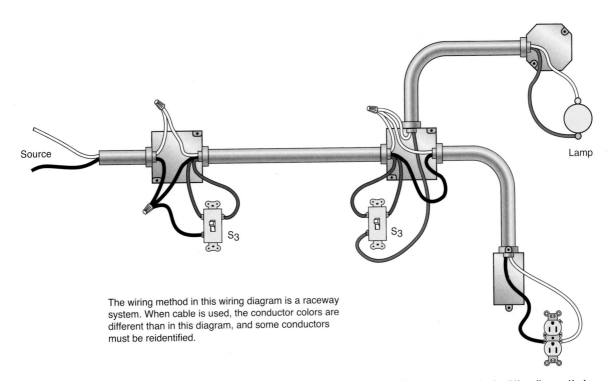

Source

Lamp

The wiring method in this wiring diagram is a raceway system. When cable is used, the conductor colors are different than in this diagram, and some conductors must be reidentified.

FIGURE 5-29   A lighting outlet controlled by two 3-way switches. The receptacle is "live" at all times. The wiring method is a raceway. When cable is used, the conductor colors are different from those in this wiring diagram, and some conductors must be reidentified. A typical example of this might be between a house and a detached garage or other outbuilding. (*Delmar/Cengage Learning*)

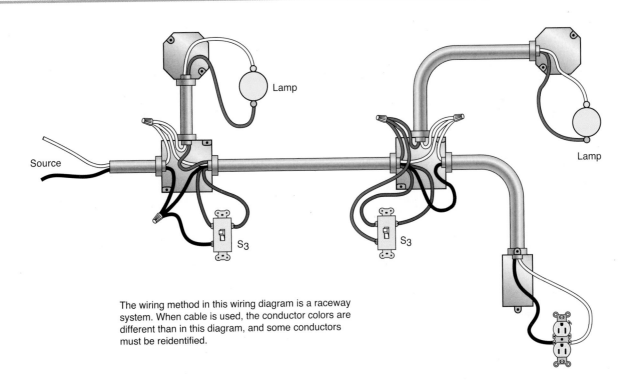

Source

Lamp

Lamp

The wiring method in this wiring diagram is a raceway system. When cable is used, the conductor colors are different than in this diagram, and some conductors must be reidentified.

FIGURE 5-30   Two lighting outlets controlled by two 3-way switches. The receptacle is "live" at all times. An example of this hookup might be between a house and a detached garage or other outbuilding, where one lighting outlet is on the house and the other lighting outlet is on or in the garage or outbuilding. (*Delmar/Cengage Learning*)

previous wiring diagrams. These connections become rather complicated using cable because of the limitation on the available insulation colors. These connections are more suited to raceway wiring.

Does "OFF" Really Mean "OFF?"

The answer is *yes*. When a switching device has a marked "OFF" position, it must completely disconnect the ungrounded conductor(s) to the load it serves when turned to the "OFF" position. UL Standards require this. Also see *404.15(B)*.

Electronic controls, such as some motion (occupancy) sensors, some dimmers, some fan speed controls, and X-10 transmitting and receiving devices (see Chapter 31), require a small voltage drop across the device to operate. When these devices are turned to what appears to be the "OFF" mode, there is still an ever-so-slight amount of voltage present at the load. The device is really not in a full "OFF" position. This can be hazardous to someone working on a luminaire, for example, thinking it is totally "OFF." See Figure 5-31.

You will probably find instructions furnished with an electronic switch (control) that the load must be a minimum of "X" watts. This is to provide the proper voltage drop across the switch (control).

Compact fluorescent lamps will not operate on an electronic motion sensor or electronic timer

because until voltage is applied to the lamp, there is no current draw. A fluorescent lamp is basically an open circuit until energized. The lamp cannot be energized because there will be no voltage across the electronic control to operate the electronic control.

Refer to Chapter 6 for additional information relating to the effects of an electrical shock.

Bonding and Grounding at Receptacles and Switches

A metal box is considered to be adequately grounded when the wiring method is armored cable, nonmetallic-sheathed cable with ground, or a metal raceway such as EMT, *250.118*. A separate equipment grounding conductor can also provide the required grounding, *250.134(B)*. Equipment grounding conductors may be bare or green.

Figure 5-32 illustrates a switch that has an equipment grounding conductor terminal. Using a switch of this type provides an easy means for providing the required grounding of a metal faceplate when using nonmetallic boxes.

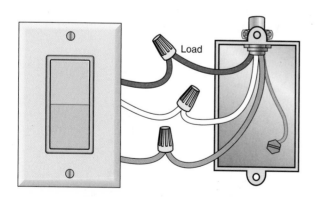

FIGURE 5-31 An X-10 switch. To operate properly, it requires the presence of the grounded circuit conductor. *Never* connect the white wire on the switch to an equipment grounding conductor in the box. This is hazardous and is a violation of *250.6(A)* and *250.24(A)(5)*.
(*Delmar/Cengage Learning*)

FIGURE 5-32 Single-pole toggle switch. Note that this switch has a terminal for the connection of an equipment ground conductor. Thus, when a metal face plate is installed, it will be grounded. See *404.9(A)* and *(B)*.
(*Courtesy of Legrand/Pass & Seymour*)

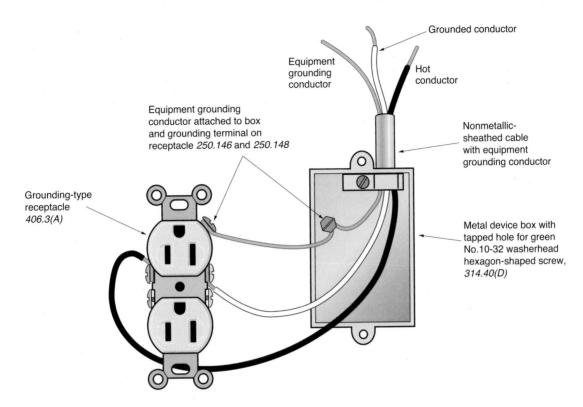

FIGURE 5-33   Connections for grounding-type receptacles. (*Delmar/Cengage Learning*)

Grounding and bonding of the equipment grounding conductor to a metal box, switch, or receptacle is important. Figures 5-32, 5-33, and 5-34 show how an equipment grounding conductor of a nonmetallic-sheathed cable can be attached to a metal box and to the equipment grounding screw of a receptacle. Most metal boxes have a No. 10-32 tapped hole for securing a green hexagon-shaped equipment grounding screw.

Other than for special "isolated ground" receptacles, the metal yoke of a receptacle is integrally

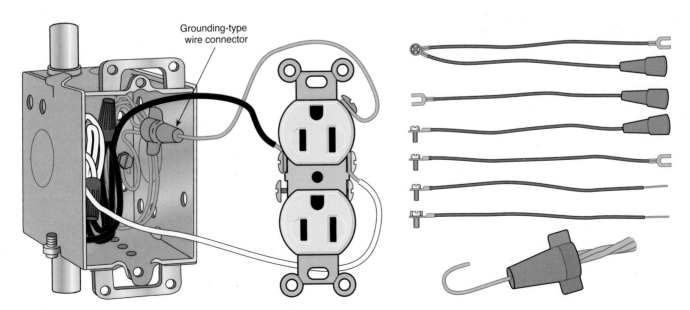

FIGURE 5-34   One method of connecting the grounding conductor using a special grounding-type wire connector. See *250.146* and *250.148*. (*Delmar/Cengage Learning*)

bonded to the equipment grounding terminal of the receptacle.

Figures 5-3 and 5-4 show a receptacle that has a special listed "self-grounding" feature for the No. 6-32 mounting screws that results in effective grounding of the metal yoke and equipment grounding terminal without the use of a separate equipment grounding conductor so long as a grounded metal box is used.

To ensure the continuity of the equipment grounding conductor path, *250.148* requires that where more than one equipment grounding conductor enters a box, they shall be spliced with devices "suitable for the use." Splices shall not depend on solder. The splicing must be done so that if a receptacle or other wiring device is removed, the continuity of the equipment grounding path shall not be interrupted. This is clearly shown in Figure 5-33. Splicing must be done in accordance with *110.14(B).*

Grounding Metal Faceplates and Switches

Assuming that an equipment grounding conductor has been carried to and properly connected at all metallic and nonmetallic boxes, we now need to be concerned with carrying that equipment grounding means to receptacles, switches, and metal faceplates.

Metal faceplates for receptacles shall be grounded, *406.5(B).* This is accomplished through the No. 6-32 screw that secures the faceplate to the metal yoke on the receptacle.

A similar requirement for the grounding of metal faceplates is found in *314.25(A).*

Snap switches and dimmer switches must provide a means to ground metal faceplates, *404.9(B).* This is a requirement whether metallic or nonmetallic faceplates are installed. The actual grounding of the faceplate is through the No. 6-32 screws that secure the faceplate to the metal yoke on the switch.

In *UL Standard 20,* we find the requirement that flush-type switches intended for mounting in a flush-device box shall provide a grounding means. In most cases this is a green hexagon-shaped screw connected to the metal yoke of the switch. This can

be seen on the switches illustrated in Figures 5-8, 5-17, and 5-32. The grounding means could also be a special self-grounding feature on the metal yoke, as illustrated in Figures 5-3 and 5-4, a bare copper wire, a copper wire with green insulation, or a copper wire with green insulation having one or more yellow stripes.

Connecting a bare equipment grounding conductor to the equipment grounding terminal on a switch complies with *406.5(B)* for the grounding of a metal faceplate. Where there are multiple switches, there is no need to attach an equipment grounding conductor on all of the switches. One connection is enough.

Figure 5-35 illustrates connecting equipment grounding conductors or jumpers to a metal box with a grounding clip.

If the equipment grounding path is accomplished through a self-grounding design of the metal yoke of a switch or receptacle, and the device is mounted in a metal device box that is grounded, you do not have to connect an equipment grounding conductor to the green hexagon-shaped grounding screw on the switch or receptacle.

Many wiring devices have small pieces of cardboard or plastic holding the No. 6-32 mounting screws from falling out of the yoke. Remove at least one of these small pieces of cardboard when direct metal-to-metal contact between the yoke of the wiring device and the box is needed, *250.146(A).* An example of this would be a receptacle secured to a handy box for surface wiring. If

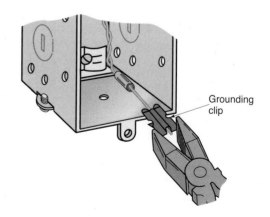

Grounding clip

FIGURE 5-35   Method of attaching grounding clip to switch (device) box. See *250.146* and *250.148.*
(*Delmar/Cengage Learning*)

there is no metal-to-metal contact, then an equipment grounding conductor must be connected to the grounding terminal on the wiring device or the wiring device must have the special self-grounding feature discussed previously. See Figures 5-33 and 5-34.

For surface-mounted boxes of the type shown in Figures 18-8 and 18-9, where cover-mounted receptacles are used, the metal-to-metal requirement mentioned earlier is not acceptable. Some other "listed" means for providing satisfactory ground continuity must be used, *250.146(A)*. This section makes it clear that there must be direct metal-to-metal contact between the metal box and the device yoke to be considered an acceptable means of grounding the receptacle equipment grounding terminal and the metal cover.

To make sure that there is direct metal-to-metal contact between the metal yoke of a receptacle and a surface-mounted metal box, *250.146(A)* accepts the automatic self-grounding device on the yoke of the receptacle, or the removal of at least one of the insulating washers. Automatic self-grounding devices on the yoke of a receptacle are clearly shown in Figures 5-3, 5-4, and 5-5.

In general, where no effective grounding means exists in the switch box, it is acceptable to replace a snap switch with another snap switch that does not have a provision for grounding.

However, in existing installations where no effective grounding means is present and where the snap switch is located within reach of the earth, a conductive floor (i.e., concrete, tile), or other conductive surfaces (i.e., metal pipes, metal ducts), use nonmetallic faceplates or provide GFCI protection for the wiring to that particular snap switch. See *404.9(B), Exception.*

CAUTION: *Never* make connections (jumpering) between the different colored terminals on receptacles. Lives have been lost because of this!

- Jumpering between the brass and silver terminals is a violation of *200.10* and *200.11*.

- Jumpering between the silver and green hexagon-shaped terminals is a violation of *250.24(A)(5)* and *250.142(B)*.

- Jumpering between the brass and green hexagon-shaped terminals is a violation of *250.126*.

COMMON *CODE* VIOLATION "TAPS"

As defined in *240.2*, a "tap" is a conductor that has overcurrent protection ahead of its point of supply that exceeds the ampacity of the conductor. In general, conductors shall be protected at their ampacities per *310.15*. Tap conductors are excluded from the basic rule, and this is found in *240.4(E)*. Some examples of "taps" are small luminaire wires. In some cases, taps can be cost-effective for hooking up electric ranges, counter-mounted cooktops, and wall-mounted ovens. This is discussed in Chapter 20.

Table 5-3 shows the maximum overcurrent protection for small conductors as permitted by *240.4(D)*.

Figure 5-36 illustrates a *Code* violation. The switch leg (loop) conductors are part of the branch-circuit wiring and are not to be considered a "tap." Although the connected load may be well within the ampacity of the switch leg, the switch leg must also be capable of handling short circuits and ground faults. The switch leg shown in Figure 5-36 must be the same size as the branch-circuit conductors, which are 12 AWG 20-ampere conductors protected by a 20-ampere circuit breaker.

Connecting Receptacles to 20-Ampere Branch Circuits

Another very common *Code* violation is using short 14 AWG pigtails to connect receptacles to 20-ampere small-appliance branch-circuit conductors in a kitchen or dining room. These short pigtails

TABLE 5-3

Maximum overcurrent protection for small conductors.

Conductor Size	Maximum Overcurrent Protection
14 AWG copper	15 amperes
12 AWG copper	20 amperes
10 AWG copper	30 amperes

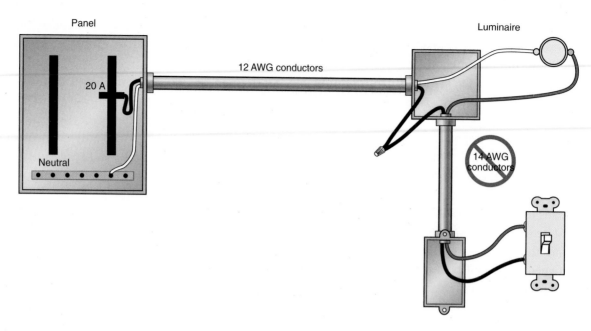

FIGURE 5-36 *Violation:* The electrician has run a 20-ampere branch circuit consisting of 12 AWG conductors to the outlet box where a luminaire will be installed. 14 AWG conductors are run as the switch legs between the lighting outlet and the switch. The switch legs are part of the branch circuit and must have an ampacity equal to that of the branch-circuit ampere rating, 20 amperes. Switch legs are not "taps." For the same reasons, it is a *Code* violation to use short pigtails of 14 AWG wire to connect receptacles to the 12 AWG conductors fed from a 20-ampere branch circuit. (*Delmar/Cengage Learning*)

must be rated 20 amperes because they are an *extension* of the 20-ampere branch-circuit wiring—they are not "taps." Figure 5-36 illustrates this type of connection.

Consider Using Larger Size Conductors for Home Runs

To minimize voltage drop, sometimes 12 AWG conductors are used for long "home runs" (50 ft or longer) even though the branch-circuit overcurrent device is rated 15 amperes. A "home run" is the branch-circuit wiring from the panel to the first outlet or junction box in that circuit. This is permitted by *210.3*, which states, *where conductors of higher ampacity are used for any reason, the ampere rating or setting of the specified overcurrent device shall determine the circuit rating.** In this case, 14 AWG conductors are permitted for the switch legs. But this leads to confusion later when someone removes the cover from the panel and wonders why 12 AWG branch-circuit conductors

are connected to 15-ampere circuit breakers. For house wiring, it is best to use the same size conductor throughout a given branch circuit.

Specifications for commercial and industrial installations oftentimes will call for a "home run" of 50 ft or more (≥ 15.2 m) to be 10 AWG conductors, even though the branch-circuit rating is 20 amperes.

Table 210.24 is a summary of branch-circuit requirements. This table lists branch-circuit ratings, branch-circuit conductor sizes, tap conductor sizes, overcurrent protection, outlet (lampholder and receptacle) ratings, maximum loads, and permissible loads.

⟿ TIMERS

Timers are unique in that they provide automatic control of electrical loads. Timers are also referred to as time clocks.

Timers are used where a load is to be controlled for specific "ON/OFF" times of the day or night. The capabilities of timers are endless. Some timers

*Reprinted with permission from NFPA 70-2011.

FD30MAC   FF30MH

FIGURE 5-37 Spring-loaded timers.
(*Courtesy of Invensys Climate Control*)

TK01

FIGURE 5-38 A 24-hour time clock.
(*Courtesy of Intermatic, Inc.*)

are astronomical in that the location of the timer is entered when installed. After that, the timer turns lights on and off based on the dusk-to-dawn time rather than by a photocell.

Figure 5-37 shows two types of spring-loaded timers that are connected in series with the load—the same as a standard wall switch. A typical application is in a bathroom for the control of an exhaust fan or an electric heater. These timers install in a standard single-gang device box. Because they take up quite a bit of space, make sure the wall box is large enough to meet the box-fill requirements of *314.16*. Electronic timers that do not have any moving parts are available as well. These timers require a neutral conductor for proper operation.

Figure 5-38 is a 24-hour time switch commonly used to control security lighting, decorative lighting, or energy management. This particular time switch has two adjustable and removable "pins" that control one "ON" and one "OFF" operations (additional "pins" can be added). There is an "override" switch for manual control.

Figure 5-39 is an electronic, 7-day astronomical time switch. It can be programmed to skip certain days, and up to 14 events can be programmed. There is an "override" switch for manual control.

ET8215CR

FIGURE 5-39 An electronic time switch with astronomic feature. (*Courtesy of Intermatic, Inc.*)

REVIEW

Note: Refer to the *Code* or the plans where necessary.

1. The identified grounded circuit conductor must be _____ or _____ in color.

2. Explain how lighting switches are rated. _____

   _____

   _____

   _____

   _____

   _____

3. A T-rated switch may be used to its _____ current capacity when controlling an incandescent lighting load.

4. What switch type and rating is required to control five 300-watt tungsten filament lamps on a 120-volt circuit? Show calculations. _____

   _____

   _____

   _____

   _____

   _____

5. List four types of lighting switches.

   a. _____   c. _____

   b. _____   d. _____

6. To control a lighting load from one control point, what type of switch would be used?

   _____

7. Single-pole switches are always connected to the _____ wire.

8. Complete the connections in the following arrangement so that both ceiling light outlets are controlled from the single-pole switch. Assume the installation is in cable.

9. Complete the connections for the diagram. Installation is cable.

120-volt
source

Lamp

Switch

10. A three-way switch may be compared to a _____ switch.

11. What type of switch is installed to control a luminaire from two different control points? How many switches are needed and what type are they? _____

_____

12. Complete the connections in the following arrangement so that the lamp may be controlled from either 3-way switch.

120-volt
source

S_3

Lamp

S_3

13. When connecting 4-way switches, care must be taken to connect the "travelers" to the _____ terminals.

14. Show the connections for a ceiling outlet that is to be controlled from any one of three switch locations. The 120-volt feed is at the light. Use colored pens or pencils. Assume the installation is in cable.

120-volt
source

Lamp

S_3

S_4

S_3

15. Match the following switch types with the correct number of terminals for each.

    Three-way switch   Two terminals

    Single-pole switch   Four terminals

    Four-way switch   Three terminals

16. When connecting single-pole, 3-way, and 4-way switches, they must be wired so that all switching is done in the _____ circuit conductor.

17. What section of the *Code* emphasizes the fact that all circuiting must be done so as to avoid the damaging effects of induction heating? _____

18. If you had to install an underground 3-wire feeder to a remote building using three individual conductors, which of the following installations "meets *Code*"? Circle the correct installation.

19. Is it always necessary to attach the bare equipment grounding conductor of a nonmetallic-sheathed cable to the green hexagon-shaped grounding screw on a receptacle? Explain.

    _____

    _____

    _____

    _____

    _____

    _____

20. List the methods by which an equipment grounding conductor is connected to a device box.

    _____

    _____

    _____

    _____

    _____

    _____

21. When two nonmetallic-sheathed cables (Romex) enter a box, is it permitted to bring both of the bare equipment grounding conductors directly to the grounding terminal of a receptacle, using the terminal as a splice point? _____

22. This installation is in electrical metallic tubing. Receptacle outlet "B" is only a few feet from switch "C." Receptacle outlet "A" is only a few feet from switch "D." The electrician saved a considerable amount of wire by picking up the supply (feed) for switch "C" from receptacle "B," and the switch leg (return) from switch "D." What is wrong with this installation?

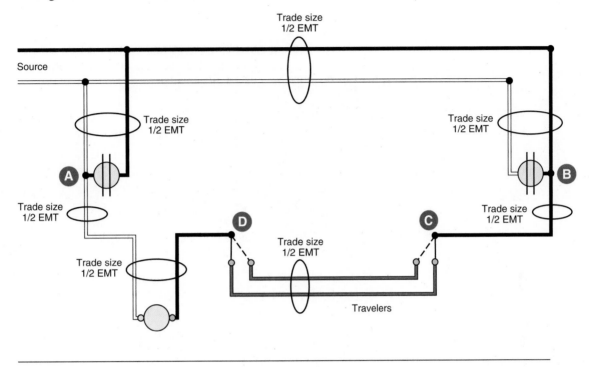

Trade size
1/2 EMT

Source

Trade size
1/2 EMT

Trade size
1/2 EMT

A

B

Trade size
1/2 EMT

Trade size
1/2 EMT

D

C

Trade size
1/2 EMT

Trade size
1/2 EMT

Travelers

_____

_____

_____

23. Define an *equipment grounding conductor.* _____

_____

_____

24. When metal toggle switchplates are used with nonmetallic boxes, the faceplate must be grounded according to *404.9(B)*. How is this accomplished?

_____

25. Does the *Code* permit the ampacity of switch legs to be less than the ampere rating of the branch circuit? _____

26. Does the *Code* permit connecting receptacles to a 20-ampere branch circuit using short lengths of 14 AWG conductors as the "pigtail" between the receptacle terminals and the 12 AWG branch-circuit conductors? _____

27. Do receptacles that have back-wired "push-in" terminals accept 12 AWG conductors? Give a brief explanation for your answer. _____

_____

_____

Ground-Fault Circuit Interrupters, Arc-Fault Circuit Interrupters, Surge Protective Devices, Immersion Detection Circuit Interrupters, and Appliance Leakage Current Interrupters

OBJECTIVES

After studying this chapter, you should be able to

- understand how GFCIs, AFCIs, IDCIs, and ALCIs operate.
- understand *NEC* requirements of where and how GFCIs and AFCIs are to be installed and connected.
- understand why AFCIs and GFCIs should not be used on a shared neutral branch circuit unless it is listed as such.
- select and install other special purpose devices including tamper-resistant and weather-resistant receptacles.
- understand the important *NEC* requirements for replacing existing receptacles.
- know the rules for providing GFCI protection on construction sites.
- understand the basics of surge protective devices.

ELECTRICAL SHOCK HAZARDS

There can be no compromise where human life is involved!

Many injuries have occurred and many lives have been lost because of electrical shock. Coming in contact with live wires or with an appliance or other equipment that is "hot" spells danger. Problems arise in equipment when there is a breakdown of insulation because of wear and tear, defective construction, or misuse of the equipment. Insulation failure can result when the "hot" ungrounded conductor in the appliance comes in contact with the metal frame of the appliance. If the equipment is not properly grounded, the potential for an electrical shock is present.

The shock hazard exists whenever the user can touch both the defective equipment and grounded surfaces such as grounded equipment or appliances, water pipes, stainless steel sinks, metal faucets, grounded metal luminaires, earth, concrete in contact with the earth, or water.

A severe shock can cause considerably more damage to the human body than is visible. A person may suffer internal hemorrhages, destruction of tissues, nerves, and muscles. Further injury can result from a fall, cuts, burns, or broken bones.

The effect of an electric current passing through a human body varies, depending on circuit characteristics (i.e., current, frequency [60 Hz is the worst], voltage, body contact resistance [open wounds are extremely hazardous as compared to a callused hand]), internal body resistance, the path of current, duration of contact, and environmental conditions (e.g., humidity).

Experts generally agree on values of 800 to 1000 ohms as typical body and contact resistance, but the resistance can vary from a few hundred ohms to many thousand ohms. Let's see how important the role of resistance is for a person coming in contact with a 120-volt circuit.

Take a dry hand, for example:

$$I = \frac{E}{R} = \frac{120 \text{ volts}}{100,000 \text{ ohms}} = \frac{0.0012 \text{ ampere}}{(1.2 \text{ milliamperes})}$$

This would probably be a little tingle.

Now take a wet hand and standing barefoot on the ground:

$$I = \frac{E}{R} = \frac{120 \text{ volts}}{1,000 \text{ ohms}} = \frac{0.12 \text{ ampere}}{(120 \text{ milliamperes})}$$

This would probably be fatal.

Figure 6-1 shows a time/current curve that indicates the amount of current that a normal, healthy adult can stand for a certain time. Table 6-1 shows expected sensations at various current levels in milliamperes.

The path of current is very important. Where and how is contact to the live conductor made? What are you touching with the other hand? Where and in what are you standing? The current flow might be finger to finger, hand to hand, head to hand, head to foot, or any other combination such as a hand holding a "hot" wire, a hand holding a pair of pliers, a hand holding an electric drill, a hand grasping a grounded metal pipe, two hands grasping a grounded metal pipe, a hand in water, standing in water, hand to foot, foot to foot, and so on. Hand to foot can be fatal because the current is probably passing through the heart and lungs.

Generally, voltages of less than 50 volts to ground are considered safe. Voltages 50 volts and greater are considered lethal. All ac systems (with few exceptions) of 50 volts to 1000 volts that supply premises wiring are required to be grounded, *250.20(B)*. This is covered in Chapter 1.

Some circuits of less than 50 volts must be grounded according to *250.20(A)*

- if the circuit is supplied by a transformer where the primary exceeds 150 volts to ground.

- if the circuit is supplied by a transformer where the primary supply is ungrounded.

- if the circuit is overhead wiring outside of a building.

CODE REQUIREMENTS FOR GROUND-FAULT CIRCUIT INTERRUPTERS (*210.8*)

- To protect against electric shocks, *210.8(A)* requires that ground-fault circuit interrupter (GFCI) protection be provided for specific

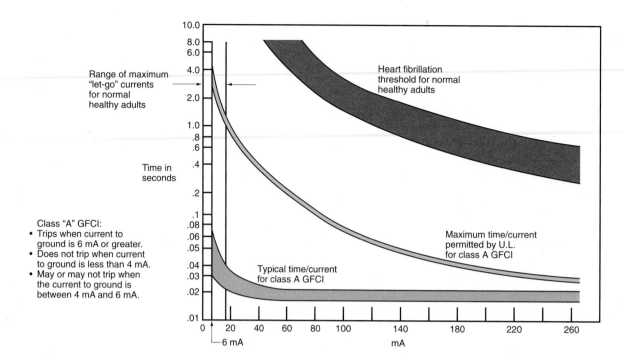

FIGURE 6-1   The time/current curve shows the tripping characteristics of a typical Class A GFCI. Note that if you follow the 6-mA line vertically to the crosshatched typical time/current curve, you will find that the GFCI will open in from approximately 0.035 second to just less than 0.1 second. One electrical cycle is $\frac{1}{60}$ of a second (0.0167 second). An air bag in an automobile inflates in approximately $\frac{1}{20}$ of a second (0.05 second). (*Delmar/Cengage Learning*)

TABLE 6-1

Effect of Electric Shock

	Current in Milliamperes @ 60 hertz	
	Men	**Women**
• Cannot be felt	0.4	0.3
• A little tingling—mild sensation	1.1	0.7
• Shock—not painful—can still let go	1.8	1.2
• Shock—painful—can still let go	9.0	6.0
• Shock—painful—just about to point where you can't let go—called "threshold"—you may be thrown clear	16.0	10.5
• Shock—painful—severe—can't let go—muscles immobilize—breathing stops	23.0	15.0
• Ventricular fibrillation (usually fatal)		
• Length of time: 0.03 sec.	1000	1000
• Length of time: 3.0 sec.	100	100

single-phase, 125-volt, 15- and 20-ampere receptacles in dwellings. GFCI protection can be provided with GFCI circuit breakers or with GFCI receptacles. The basic requirements for receptacle GFCI protection are as follows:

- **Bathrooms:** See Figure 10-7 for the definition of a bathroom.

- **Garages** (attached and detached) and in accessory buildings that are not habitable, such as sheds, workshops, storage buildings, or areas of similar use that have the floor at or below grade.

- **Outdoors:** An exception is that GFCI protection is not required for a receptacle not readily accessible and supplied by a dedicated branch circuit for snow melting or deicing equipment. Install these receptacles for snow melting or deicing equipment according to *426.28.*

- **Crawl spaces** that are at or below grade.

- **Unfinished basements:** An exception to this GFCI requirement is for a receptacle that serves a permanently installed fire alarm or burglar alarm system. See *760.41B* and *760.121(B)*. A receptacle on a porcelain or plastic lampholder must also be GFCI protected. Some individuals try to get around the requirement for GFCI protection in basements by calling the basement "finished," even though it is clearly evident that it is not "finished." If the basement is truly intended to be "finished," then the *Code* requirements for the spacing of receptacles, required lighting outlets, and switch controls must be followed.

- **Kitchens:** All receptacles that serve the countertop surfaces. This includes receptacles installed on islands and peninsulas, as in Figure 6-2. The key words here are *that serve the countertop surfaces.*

  GFCI protection is *not* required for receptacles that are obviously not intended to serve countertops, such as a receptacle installed

  — solely for a clock.

  — inside of an upper kitchen cabinet for plugging in a microwave oven that is fastened to the underside of the cabinet.

  — below the kitchen sink for plugging in a food-waste disposer or dishwasher.

  — behind a range or refrigerator.

- ▶**Sinks:** For other than kitchens as covered in *210.8(A)(6)*, any receptacle within 6 ft (1.8 m)

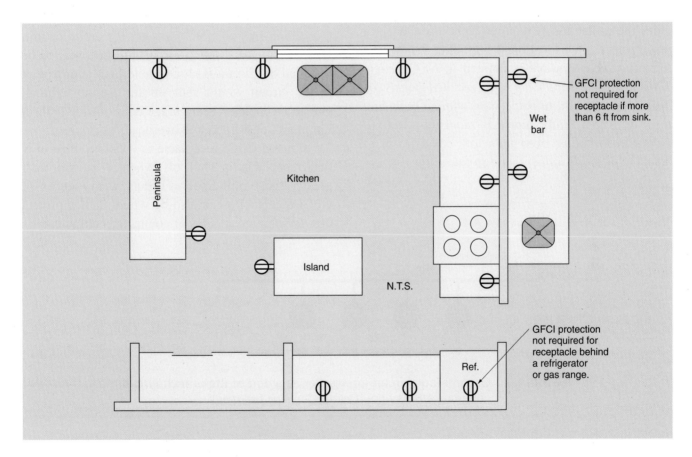

FIGURE 6-2   In kitchens, *all* 125-volt, single-phase, 15- and 20-ampere receptacles that serve countertop surfaces (walls, islands, or peninsulas) must be GFCI protected. The key words are *that serve countertop surfaces.* GFCI personnel protection is *not* required for the refrigerator receptacle, a receptacle under the sink for plugging in a food waste disposer, or a receptacle installed inside cabinets above the countertops for plugging in a microwave oven attached to the underside of the cabinets. For other than kitchens, all 125-volt, single-phase, 15- and 20-ampere receptacles within 6 ft (1.8 m) of a sink must be GFCI protected. See *210.8(A)(6)* and *210.8(A)(7)*. Drawing is not to scale. (*Delmar/Cengage Learning*)

of the outside edge of a sink. See Figure 6-2. This requirement is easy to understand. Just measure 6 ft (1.8 m) in any direction around a corner, even in another room or hallway.◄

- **Boathouses:** A boathouse for a boat is similar in concept to a garage for a car.

- **Swimming pools:** See Chapter 30.

- For the branch circuit supplying heated floors in bathrooms, and hydromassage bathtub, spa, and hot tub locations, see *424.44(G)*. This GFCI requirement is regardless of the type of flooring or heating cables used.

Readily Accessible Location

►The circuit breaker or GFCI receptacle that provides the protection is required to be installed in a readily accessible location.◄ See *NEC 210.8*. The term *readily accessible* is defined in *NEC Article 100* and means, *Capable of being reached quickly for operation, renewal, or inspections without requiring those to whom ready access is requisite (required or needed) to climb over or remove obstacles or to resort to portable ladders, and so forth.**

---

*Reprinted with permission from NFPA 70-2011.

This important new requirement ensures the device that provides the GFCI protection can be easily accessed to do the safety operation test that is required by the manufacturer. An operational test is required to be performed at least monthly. This is important as we rely on the GFCI device functioning correctly when an actual ground fault occurs. A record of the test should be made and maintained.

The "push-to-test" button on the circuit breaker or receptacle should be pressed to test the device. Several manufacturers make plug-in type GFCI testers as well.

GFCI Receptacles and Breakers

The *Code* requirements for ground-fault circuit protection can be met in many ways. Figure 6-3 illustrates a GFCI circuit breaker installed on a branch circuit. A ground fault in the range of 4 to 6 milliamperes or more shuts off the entire circuit. If, for example, a GFCI circuit breaker were to be installed on Circuit B14, a ground fault at any point in the circuit would shut off the garage lighting, garage receptacles, rear garage door outside bracket luminaire, post light, and overhead door opener.

When a GFCI receptacle is installed, then only that receptacle is shut off when a ground fault greater than 6 milliamperes occurs, as in Figure 6-4. GFCI receptacles can also be wired so they protect downstream receptacles or connected equipment.

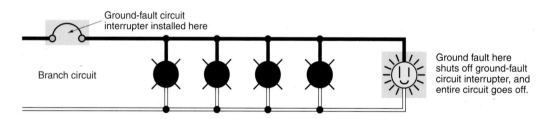

FIGURE 6-3   Ground-fault circuit interrupter as a part of the branch-circuit overcurrent device. (*Delmar/Cengage Learning*)

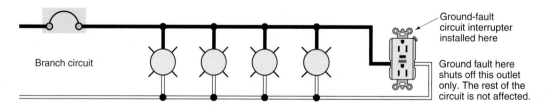

FIGURE 6-4   Ground-fault circuit interrupter as an integral part of a receptacle outlet. (*Delmar/Cengage Learning*)

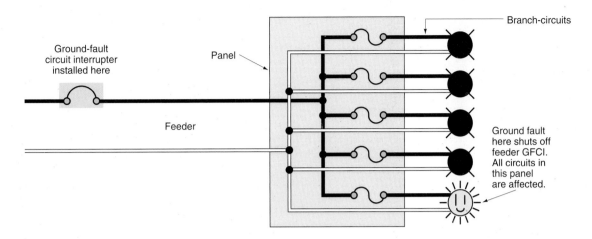

FIGURE 6-5   Ground-fault circuit interrupter as part of the feeder supplying a panel that serves a number of 15- and 20-ampere branch circuits. This is permitted by *215.9* but is rarely—if ever—used in residential wiring. (*Delmar/Cengage Learning*)

GFCI receptacles break both the ungrounded ("hot") and grounded conductors.

GFCI circuit breakers break only the ungrounded ("hot") conductor.

Figure 6-5 shows the effect of a GFCI installed as part of a feeder supplying 15- and 20-ampere receptacle branch circuits. This is rarely used in residential wiring.

Figure 6-6 is a pictorial view of how a ground-fault circuit interrupter operates. The GFCI device in the upper drawing does not operate because the current through the torroidal coil is in balance. As shown in the lower drawing, the GFCI will trip very quickly if the current returning to its source is 6 mA or more and is not permitted to trip if the "leakage current" is less than 4 mA. This leakage current is simulated by the push-to-test circuit internal to the device. See the additional information later in this chapter.

What Kinds of GFCIs Are Available?

- 120-volt receptacles
- Single-pole, 120-volt breakers
- Single-pole, 120-volt dual-function with both GFCI and AFCI features in one breaker
- Two-pole, 240-volt common trip breakers
- Two-pole, 120/240-volt common trip dual-function with both GFCI and AFCI features in one breaker

- Two-pole, 120/240-volt independent trip dual-function with both AFCI and GFCI features in one breaker. Suitable for use with a shared neutral on a multiwire branch circuit
- "Faceless." Strictly for GFCI protection. Have only the test and reset buttons. Do not have a receptacle. Mount in a single-device box. Commonly used to protect whirlpool tubs, etc.

Nuisance Tripping

Occasionally, a GFCI will trip for seemingly no reason. This could be "leakage" in extremely long runs of cable from a GFCI circuit breaker in a panel to the protected branch-circuit wiring. One manufacturer's literature specifies a maximum one-way length of 250 ft (76.2 m) for the branch circuit.

The allowable "leakage current" for listed cord-and-plug-connected appliances is so small that they should not cause nuisance tripping of GFCI devices. However, if nuisance tripping can be traced to an appliance, it might be a good idea to simply replace the GFCI circuit breaker or the GFCI receptacle before discarding the appliance. If the GFCI continues to nuisance trip, it probably is the appliance.

Nuisance tripping occasionally can be traced to moisture somewhere in the circuit wiring, receptacles, or lighting outlets.

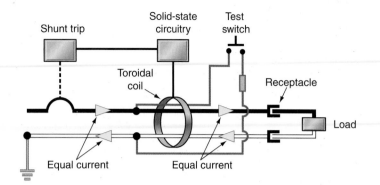

No current is induced in the toroidal coil since both circuit wires are carrying equal current. The contacts remain closed.

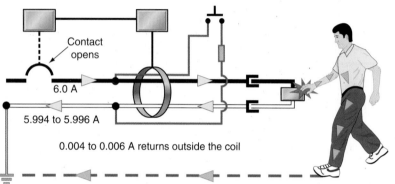

An imbalance of from 4 to 6 milliamperes in the coil will cause the contacts to open. The GFCI must open in approximately 25 milliseconds. Receptacle-type GFCIs have a switching contact in each circuit conductor.

FIGURE 6-6   Basic principle of how a ground-fault circuit interrupter operates. Receptacle-type GFCIs switch both the phase ("hot") and grounded conductors. Note that when the test button is pushed, the test current passes through the test button, the sensor, then back around (bypasses, outside of) the sensor, then back to the opposite circuit conductor. This is how the "unbalance" is created, then monitored by the electronic circuitry to signal the GFCI's contacts to open. Note in the upper drawing that because the load currents passing through the sensor are equal, no unbalance is present. (*Delmar/Cengage Learning*)

Some electricians are of the opinion that using nonmetallic staples (see Figure 4-14[C]) will reduce the possibility of nuisance tripping. No actual testing has been done to prove or disprove this opinion.

What a GFCI Does

A GFCI monitors the current balance between the ungrounded "hot" conductor and the grounded conductor. As soon as the current flowing through the "hot" conductor is in the range of 4 to 6 milliamperes more than the current flowing in the "return" grounded conductor, the GFCI senses this unbalance and trips (opens) the circuit off. The unbalance indicates that part of the current flowing in the circuit is being diverted to some path other than the normal return path along the grounded return conductor. If the "other" path is through a human body, as illustrated in Figure 6-6, the outcome could be fatal.

UL Standard No. 943 covers ground-fault circuit interrupters.

- Class "A" GFCI devices are the most common. They are designed to
  - trip when current to ground is 6 milliamperes ($^6/_{1000}$ of an ampere) or greater.
  - not trip when the current to ground is less than 4 milliamperes ($^4/_{1000}$ of an ampere).
  - may or may not trip when the current to ground is between 4 and 6 milliamperes.
  - will open very quickly, in approximately 25 milliseconds.

- Class "B" GFCI devices are pretty much obsolete. They were designed to trip on ground faults of 20 milliamperes ($^{20}/_{1000}$ of an ampere) or more. They were used only for underwater swimming pool lighting installed before the adoption of the 1965 *NEC*. For this application, Class "A" devices were too sensitive and would nuisance trip.

What a GFCI Does *Not* Do

- It *does not* protect against electrical shock when a person touches both circuit conductors at the same time (two "hot" wires, or one "hot" wire and one grounded neutral conductor) because the current flowing in both conductors is the same. Thus, there is no unbalance of current for the GFCI to sense and trip.

- It *does not* limit the magnitude of ground-fault current. It *does* limit the length of time that a ground fault will flow. The GFCI should trip in about 25 milliseconds. In other words, you will still receive a severe shock during the time it takes the GFCI device to trip "off." See Figure 6–1.

- It *does not* sense solid short circuits between the "hot" conductor and the grounded "neutral" conductor. The branch-circuit fuse or circuit breaker provides this protection.

- It *does not* sense solid short circuits between two "hot" conductors. The branch-circuit fuse or circuit breaker provides this protection.

- It *does not* sense and protect against the damaging effects of arcing faults, such as would occur with frayed extension cords. This protection is provided by an arc-fault circuit interrupter (AFCI) discussed later in this chapter.

- It *does not* provide overload protection for the branch-circuit wiring. It provides *ground-fault protection only*.

⎍⎍⎍ GROUND-FAULT CIRCUIT INTERRUPTERS IN RESIDENCE CIRCUITS

In this residence, GFCI personnel protection must be provided for receptacle outlets installed outdoors, in the bathrooms, in the garage, in the

FIGURE 6-7   (A) A ground-fault circuit interrupter and tamper-resistant receptacle (*courtesy of Leviton Manufacturing*), and (B) a ground-fault circuit interrupter circuit breaker (*Courtesy of Schneider Electric*). The switching mechanism of a GFCI receptacle opens both the ungrounded "hot" conductor and the grounded conductor. The switching mechanism of a GFCI circuit breaker opens the ungrounded "hot" conductor only.

workshop, specific receptacles in the basement area, and in the kitchen for those receptacles that serve countertops. ▶All 125-volt, single-phase, 15- and 20-ampere receptacles within 6 ft (1.8 m) of a sink must be GFCI protected.◀ See *210.8(A)(1)* through *210.8(A)(8)*. The GFCI protection can be provided by GFCI circuit breakers or by GFCI receptacles of the type shown in Figure 6-7. This is a design issue that confronts the electrician. The electrician must decide how to provide the GFCI personnel protection required by *210.8(A)*. These receptacle outlets are connected to the circuits listed in Table 6-2.

Swimming pools also have special requirements for GFCI protection. These requirements are covered in Chapter 30.

GFCIs operate properly only on grounded electrical systems, as is the case in all residential, condo, apartment, commercial, and industrial wiring. The GFCI will operate on a 2-wire circuit even though an equipment grounding conductor is not included with the circuit conductor. In this residence, equipment grounding conductors are in the cables. In the case of a metal raceway (EMT) or armored cable

TABLE 6-2

Circuit numbers and general locations of GFCI-protected receptacles. GFCI protection can be provided by using GFCI receptacles, GFCI circuit breakers, or GFCI/AFCI dual-function circuit breakers. Refer to panel schedules shown in Figures 27-13 and 27-16.

Circuits	GFCI Receptacle Locations
A15	Front-porch receptacle
A16	Outdoor receptacle on front of residence outside of Front Bedroom
A18	Workbench receptacles
A19	Outdoor receptacle on rear of residence outside of Master Bedroom
A20	Workshop receptacles on window wall
A22	Master Bath receptacle
A23	Bath (off hall) receptacle
B90	Wet-bar (Recreation Room) receptacles
B13	Kitchen receptacles
B14	Garage receptacles
B15	Kitchen receptacles
B16	Kitchen receptacles
B17	Outdoor receptacle outside of Living Room next to sliding door
B18	Laundry Room receptacle for clothes washer
B20	Laundry Room receptacles (2) and weatherproof receptacle (1) outside of laundry
B21	Powder Room receptacle

(Type AC or Type MC), the equipment grounding conductor is provided by the metal raceway or armored cable that contains a bonding strip.

Never ground a system neutral conductor except at the service equipment. A GFCI would be inoperative or trip unnecessarily.

Never connect the neutral conductor of one circuit to the neutral conductor of another circuit.

When a GFCI feeds an isolation transformer (separate primary winding and separate secondary winding), as might be used for swimming pool underwater luminaires, the GFCI *will not* detect any ground faults on the secondary of the transformer.

Ground-fault circuit interrupters may be installed on other circuits, in other locations, and even when rewiring existing installations where the *Code* does not specifically call for GFCI protection.

Do not connect fire and smoke alarms to a circuit that is protected by a GFCI breaker. A nuisance tripping of the GFCI would render the fire alarm and smoke detectors inoperative. This requirement is found in *NFPA 72, The National Fire Alarm Code.*

FEED-THROUGH GROUND-FAULT CIRCUIT INTERRUPTER

With rare exception, all GFCI receptacles on the market today have the "feed-through" feature.

The decision to use more GFCIs rather than trying to protect many receptacles through one GFCI becomes one of economy and practicality. GFCI receptacles are more expensive than regular receptacles. Here is where knowledge of material and labor costs comes into play. The decision must be made separately for each installation, keeping in mind that GFCI protection is a safety issue recognized and clearly stated in the *Code.* However, the actual circuit layout is left up to the electrician.

Figure 6-8 illustrates how a feed-through GFCI receptacle supplies many other receptacles. Should a ground fault occur anywhere on this circuit, all 11 receptacles lose power—not a good circuit layout. Attempting to locate the ground-fault problem, unless it's obvious, can be very time-consuming. The use of more GFCI receptacles is generally the more practical approach.

Figure 6-9 shows how a feed-through receptacle is connected into a circuit. Figure 6-10 shows a feed-through GFCI receptacle connected midway into a circuit. In both diagrams, the feed-through GFCI receptacle and all downstream outlets are ground-fault protected.

In UL Standard 943, we find that a GFCI receptacle may have black-and-white line wire leads, red-and-gray load wire leads, and a green lead for the equipment grounding conductor, as illustrated in Figure 6-9. GFCI receptacles may have ordinary screw terminals: brass color for the "hot" conductors, silver color for the grounded conductors, and a green hexagon-shaped screw for the equipment grounding conductor. The line and load connections are clearly marked.

Where the wiring method is a grounded method such as electrical metallic tubing, flexible metal conduit, or armored cable, the grounding terminal

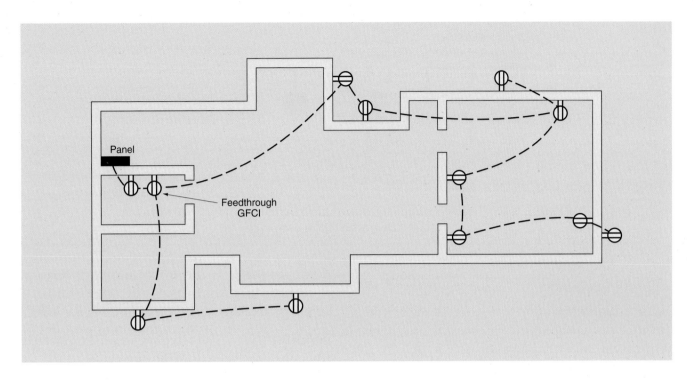

FIGURE 6-8   Illustration showing one GFCI feed-through receptacle protecting other receptacles. Although this might be an inexpensive way to protect many other downstream receptacles outlets that require GFCI protection, it could be considered as "putting too many eggs in one basket." The same situation would exist if a GFCI circuit breaker were to be installed in the panel. One major manufacturer of GFCI circuit breakers suggests that the maximum one-way length of a GFCI-protected branch circuit be limited to 250 ft (76.2 m). (*Delmar/Cengage Learning*)

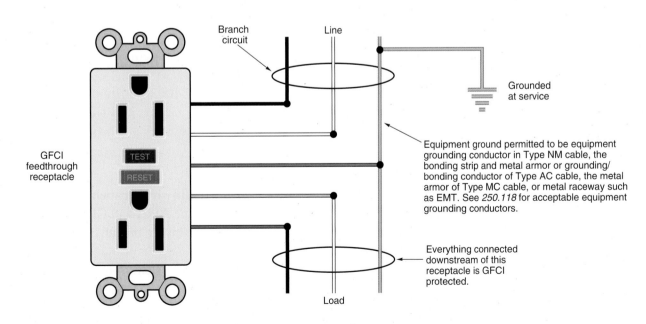

FIGURE 6-9   Connecting feed-through ground-fault circuit interrupter into circuit. (*Delmar/Cengage Learning*)

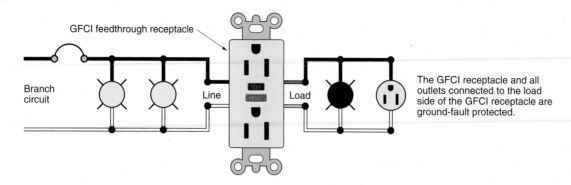

FIGURE 6-10 Feed-through GFCI receptacle installed midway in a branch circuit. (*Delmar/Cengage Learning*)

of the GFCI receptacle is considered adequately grounded by the metallic grounded wiring method, and a separate connection to the green lead or to the green hexagon-shaped screw is not necessary. This is discussed and illustrated in Chapter 5.

The wiring device manufacturer is required to provide a "safety yellow" adhesive label must over the GFCI receptacle load terminals, or be wrapped around the load wire leads. The label must be marked with wording like this: *"Attention! The load terminals under this label are for feeding additional receptacles. Miswiring can leave this outlet without ground-fault protection. Read instructions prior to wiring."*

The most important factors to be considered are the continuity of electrical power and the economy of the installation. The decision must be made separately for each installation.

WARNING: Be very careful when hooking up GFCI receptacles. It is extremely important that the LINE and LOAD connections are done correctly. On older style GFCI receptacles, it was possible to reverse the line and load connections. Under a ground-fault condition, this sort of misconnection would result in the GFCI receptacle itself still being "live" even though the GFCI mechanism has tripped "off." In the case of feed-through receptacles, a ground-fault condition shut off the circuit downstream, yet the feed-through GFCI receptacle itself was still energized.

It is easy to tell whether a GFCI receptacle has the line and load leads reversed. Push the test button. The GFCI will trip. If the GFCI receptacle is still "hot," it has been wired incorrectly.

The current UL Standard 943 requires that a GFCI receptacle must trip to the "OFF" position if the receptacle is miswired, such as a reversal of the line/load connections. In other words, if misconnected, the receptacle simply will not work!

WARNING: The circuitry in a GFCI detects any unbalance current in the circuit.

The circuitry in an AFCI detects various levels and types of arcing in the circuit.

Be extremely careful when hooking up GFCI and AFCI circuit breakers. AFCIs are discussed a little later in this chapter. A misconnection or misapplication could result in a fatal electrical shock! Before hooking up these devices, study the installation instructions (wiring diagrams, cautions, etc.).

Questions that have to be answered. Will all branch-circuit "home runs" be 2-wire? Will some "home runs" be 3-wire multiwire branch circuits in order to save "time and material"? You've got to think about this before you lay out your branch-circuit wiring. To "Meet *Code*," you must clearly understand the requirements in NEC 210.8 and 210.12 for GFCIs and AFCIs.

These NEC sections tell you where GFCI protection is required. These sections do not tell you how to provide the protection.

What about Multiwire Branch Circuits? It's plain and simple. Some AFCI and GFCI devices are "listed" for use on multiwire branch circuits—others are not.

Those that are not "listed" for use on multiwire branch circuits have clear instructions warning "Never share a neutral." Using two single-pole AFCIs or GFCIs on a multiwire branch circuit will not work because each needs to have the white grounded conductor from that branch circuit connected to that particular AFCI or GFCI.

As with any electrical equipment, follow the installation instructions (wiring diagrams).

Figure 6-11(A) shows a 3-wire, multiwire circuit. This particular GFCI is not "listed" as suitable for sharing a neutral conductor in a multiwire branch circuit. Trace the current flow and note that the GFCI will sense a current unbalance of 5 amperes and trip "OFF." In fact, as long as the unbalance exists, it will never be able to be turned "ON."

Figures 6-11(B) and 6-11(C) show a 2-wire circuit. Figure 6-11(B) is an improper connection of a feed-through GFCI receptacle. Note how the GFCI

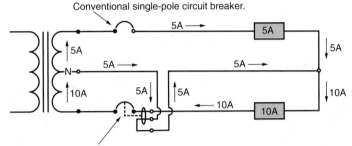

A **Improper** connection of single-pole GFCI circuit breaker on a multiwire branch circuit. The current flowing in and out of the GFCI circuit breaker is different (unbalanced) because the circuit is a multiwire circuit. This difference of current flow will cause the GFCI breaker to trip immediately.

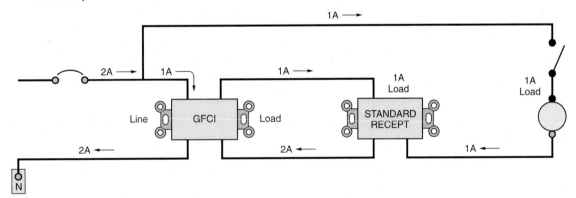

B **Improper** connection: The current flowing in and out of the GFCI receptacle is different (unbalanced). The GFCI receptacle will trip immediately.

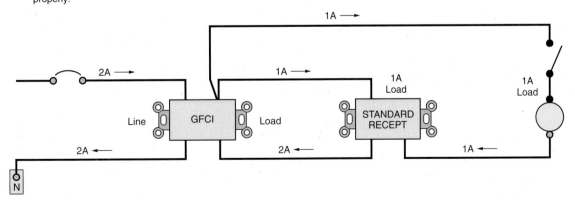

C **Proper** connection: The current flowing in and out of the GFCI receptacle is the same (balanced). The GFCI receptacle operates properly.

FIGURE 6-11   These diagrams show improper and proper connections of GFCI circuit breakers. (*Delmar/Cengage Learning*)

detects a current unbalance of 1 ampere and will trip "OFF." As long as the unbalance exists, the GFCI receptacle will continue to trip "OFF" until the misconnection is corrected. Figure 6-11(C) illustrates the proper connection of a GFCI receptacle.

Other Features for GFCI Receptacles

GFCIs have come a long way. All GFCI receptacles have a manual TEST button and a RESET button. In addition, look for GFCI receptacles that have:

- an LED that indicates that there is power to the receptacle.
- an LED that indicates that the device is not functioning properly.
- a built-in line/load reversal feature that prevents power to the receptacle face if the line/load connections are reversed.
- an "End-of-Life" feature that activates when the GFCI fails to respond to a manual test can be either:
  - a lockout feature so the GFCI cannot be reset if the internal GFCI circuit is not functioning properly, or
  - a "failure" light or alarm that indicates that the GFCI protection has failed yet the receptacle continues to supply power.

Nuisance Tripping

Occasionally, a GFCI will trip for seemingly no reason.

The maximum allowable "leakage" for listed cord-and-plug-connected appliances is 0.5 milliampere (mA). However, some product standards, such as for computers and televisions permit higher values of leakage current—some as high as 3.5 mA. If nuisance tripping can be traced to an appliance, the appliance should be checked by a qualified person.

Nuisance tripping might be traced to moisture somewhere in the branch-circuit wiring, wiring devices, or lighting outlets.

"Leakage" in extremely long runs of cable from a GFCI circuit breaker in a panel to the protected branch-circuit wiring could result in nuisance tripping. This

occurrence is very rare, but because this is a possibility, it is worth checking out. Of course, the problem could be a defective GFCI breaker or GFCI receptacle.

⌁ TESTING AND RECORDING OF TEST DATA FOR GFCI RECEPTACLES

Refer to Figure 6-7 and note that the GFCI receptacle has "T" (Test) and "R" (Reset) buttons. As shown in Figure 6-6, pushing the test button allows a small current to bypass the torroid coil in the device. If operating properly, the GFCI receptacle should trip to the "OFF" position. Pushing the reset button will restore power. The operation is similar for the GFCI circuit breakers.

Because of the fragile nature of the electronic circuitry in GFCIs, test them often, particularly in high lightning strike areas of the country. GFCIs and other delicate electronic circuitry are susceptible to transient voltage surges.

"Things" are changing so fast in the electronic world, it is hard to keep up with the changes. At least one manufacturer recently announced the availability of a self-testing GFCI receptacle. Its circuitry automatically generates a simulated ground-fault leakage current every 60 seconds. Instead of relying on the homeowners to do the recommended monthly test of the GFCI receptacles in their homes, the test is performed automatically.

Underwriters Laboratories (UL) requires that detailed installation and testing instructions be included in the packaging for GFCI receptacles or circuit breakers. Instructions are not only for the electrician but also for homeowners, and must be left in a conspicuous place so homeowners can familiarize themselves with the receptacle, its operation, and its need for testing. Figure 6-12 shows a chart homeowners can use to record monthly GFCI testing.

Arc-Fault Circuit Interrupters (AFCIs)

Arc-fault circuit interrupters are "the new kids on the street" compared to GFCIs. First came ground-fault circuit interrupters (GFCIs), then came

OCCUPANT'S TEST RECORD

TO TEST, depress the "TEST" button. The "RESET" button should extend. Should the "RESET" button not extend, the GFCI will not protect against electrical shock. Call a qualified electrician.

TO RESET, depress the "RESET" button firmly into the GFCI unit until an audible click is heard. If reset properly, the "RESET" button will be flush with the surface of the receptacle.

This label should be retained and placed in a conspicuous location to remind the occupants that for maximum protection against electrical shock, each GFCI should be tested monthly.

Year	Jan	Feb	Mar	Apr	May	Jun	Jul	Aug	Sep	Oct	Nov	Dec

FIGURE 6-12   Homeowner's testing chart for recording GFCI testing dates. UL requires that the instructions furnished with GFCIs be marked and that testing be performed upon installation and at least once each month as well as after severe lightning storms. (*Delmar/Cengage Learning*)

arc-fault circuit interrupters (AFCIs). As you will learn later in this chapter, dual-function circuit breakers are available that provide both AFCI and GFCI protection, making it very easy to "meet *Code*." The definition of *Arc-Fault Circuit Interrupter* has moved to *Article 100* for the 2011 *NEC*. It is defined as *A device intended to provide protection from the effects of arc faults by recognizing characteristics unique to arcing and by functioning to de-energize the circuit when an arc fault is detected.*

The requirements for AFCI protection have at the same time both broadened and become more specific. A new exception has been added and a new section added on branch circuit extensions or modifications.

NEC 210.12(A) states that in dwelling units, *All 120-volt, single-phase, 15- and 20-ampere branch circuits supplying outlets installed in dwelling unit family rooms, dining rooms, living rooms, parlors, libraries, dens, bedrooms, sun rooms, recreation rooms, closets, hallways, or similar rooms or areas shall be protected by a listed arc-fault circuit interrupter, combination-type, installed to provide protection of the branch circuit.**

What Is an Outlet?   This is an important term to understand for the correct application of the AFCI requirements. The word "Outlet" is defined in *NEC Article 100* as *A point on the wiring system at which current is taken to supply utilization equipment.** Several types of outlets that require AFCI protection are present in a dwelling unit. They include: lighting outlets, receptacle outlets, ceiling (paddle) fan outlets, and outlets for single station fire alarms. AFCI protection is required for all of these outlets that are installed in any of the rooms mentioned. Note that a lighting switch is not an outlet. The box where the luminaire connects is the outlet.

---

*Reprinted with permission from NFPA 70-2011.

Certain rooms and areas are exempt from the AFCI requirement:

- Attics—AFCI protection not required, but no problem if you do provide AFCI protection.
- Bathrooms—The receptacle(s) require GFCI protection. You could put the lighting on an AFCI-protected circuit, but that is not required.
- Garages—most of the receptacle(s) require GFCI protection. You could put the lighting on an AFCI-protected circuit, but that is not required.
- Kitchens—Receptacles serving countertops require GFCI protection. You could put the lighting on an AFCI-protected circuit, but again, that is not required.
- Laundry rooms—Receptacles within 6 ft (1.8 m) of a sink require GFCI protection, but other receptacles and the room lighting do not. AFCI protection is not required.
- Home Office—Careful here. If you take over a room that qualifies as a bedroom, the inspector is likely to require AFCI protection of the room.
- Outdoors—Outdoor receptacle(s) require GFCI protection, but not AFCI protection. Outdoor lighting is permitted to be supplied by a non-AFCI-protected branch circuit or an AFCI-protected branch circuit.
- Unfinished basements—Unfinished basement receptacle(s) require GFCI protection. You could put the lighting on an AFCI-protected circuit but are not required to do so.

Note the last few words in *210.12(A): installed to provide protection of the branch circuit.*

By definition, a branch circuit is *The circuit conductors between the final overcurrent device protecting the circuit and the outlet(s).** These are the branch-circuit conductors that connect to the circuit breaker (or fuse) in the panelboard. There are three exceptions.

►The first exception to *210.12(A)* is *If RMC, IMC, EMT, Type MC or steel armored Type AC cables meeting the requirements of 250.118 and metal outlet and junction boxes are installed for the portion of the branch circuit between the branch circuit overcurrent device and the first outlet, it shall be permitted to install an outlet branch circuit type AFCI at the first outlet to provide protection for the remaining portion of the branch circuit.*◄ This permits an outlet type AFCI receptacle to

be installed downstream (any distance) from the panel, but only if the wiring method is all metal to provide the desired level of protection against physical damage.

A new second exception has been added to *210.12(A)*. It provides that, ►*Where a listed metal or nonmetallic conduit or tubing is encased in not less than 50 mm (2 in.) of concrete for the portion of the branch circuit between the branch-circuit overcurrent device and the first outlet, it shall be permitted to install an outlet branch circuit AFCI at the first outlet to provide protection for the remaining portion of the branch circuit.*◄ This exception is sufficiently self-explanatory.

►The third exception to *210.12(A)* is that AFCI protection is not required for an individual branch circuit that supplies a fire alarm system installed in accordance with *760.41(B)* and *760.121(B)* so long as the wiring methods comply with *Exception 1.*◄ See Chapter 26 in this text. Fire alarm system requirements are found in NFPA 72.

A new *210.12(B)* was added to the 2011 NEC. It covers branch circuit extensions or modifications and reads, ►*In any of the areas specified in 210.12(A), where branch circuit wiring is modified, replaced or extended, the branch circuit shall be protected by*:

1. *A listed combination AFCI located at the origin of the branch circuit; or*
2. *A listed outlet branch circuit AFCI located at the first receptacle outlet of the existing branch circuit.**◄

This rule extends the AFCI requirements for areas of a dwelling where branch circuits are extended or modified, such as for room additions or remodels.

Don't Get Confused

Don't confuse an AFCI with a GFCI. These two devices protect against two different kinds of electrical hazards.

- An AFCI device is designed to sense and respond to an arcing fault that could develop into a fire. Some AFCI devices also provide personnel GFCI protection (6 mA). Some AFCI devices provide equipment ground-fault protection (GFP—30 mA). A typical application for equipment GFP in dwellings is for outdoor electric deicing and snow melting equipment, covered in *Article 426*.

*Reprinted with permission from NFPA 70-2011.

Always read the label and the instructions of an AFCI device to be sure you know where and how to connect it.

- A GFCI device is designed to sense and respond to a ground fault, protecting people from severe electrical shock that could lead to death.
- Dual-function AFCI/GFCI breakers provide arcing and ground-fault protection.

Smoke Alarms and AFCI Protection

As you will learn in Chapter 26, smoke alarms are required to have secondary battery backup protection. Single-station smoke alarms require AFCI protection in the rooms or areas specified even if interconnected so that all alarms sound if one detector goes into alarm. This arrangement does not constitute a fire alarm system, as there is no fire alarm control panel. Permanently installed fire alarm systems shall not be GFCI- or AFCI-protected, *760.41(B)* and *760.121(B)*.

Why AFCI Protection?

Electrical arcing is an "early" event in the history of a typical electrical fire. Arcing is considered one of the leading causes of electrical fires in homes. Think about frayed cords, cords under carpets, cords pinched under legs of furniture, and nails being driven through plaster into cables in the wall. The Consumer Product Safety Commission recently cited 150,000 residential electrical fires annually in the United States, 850 deaths, 6000 injuries, and more than $1.5 billion in property loss. These statistics are pretty convincing of the need for greater electrical protection. The heat of an arc is extremely high, known to reach 10,000°F (5538°C) or more. Hot particles of metal expelled by the arc are enough to ignite most surrounding combustible materials.

We have all been guilty of blaming electrical fires on overloaded circuits, overloaded extension cords, and too many extension cords plugged into a receptacle. The heat of an overloaded cord by itself is generally not hot enough to ignite surrounding combustible materials. An overload condition can cause the breakdown of the insulation on the conductors, allowing the conductors to touch one another. This sets the stage for an arcing condition between the conductors,

or between the "hot" conductor and ground. We now have the ingredients of an electrical fire.

For example, let's say that the cord on an electric iron becomes frayed and the conductor(s) break. The iron draws 10 amperes. The arcing of 10 amperes trying to bridge the gap in the broken conductor(s) develops the tremendous heat capable of igniting surrounding combustible material, yet 10 amperes flowing in the circuit will not trip a 15-ampere circuit breaker or open a 15-ampere fuse. This condition is not an overload and initially is not a short circuit. Conventional overcurrent devices "see" this as normal. They are not designed to detect and open under current levels below their trip setting. What is needed is a device that can detect the unique characteristics of an arcing condition—an AFCI.

Types of Arcs

Arcing is current flowing outside of the intended path. Arcing is a function of voltage and the amount of current flowing in the arc.

Taking the liberty of using familiar terms, we could say that arcs come in two varieties—*good* and *bad*.

Good arcs are not dangerous. For example, the slight arcing that occurs between the contacts of a switch when turned ON and OFF, or when a male attachment plug cap is inserted into a receptacle is not considered hazardous. This type of arcing creates a unique "signature."

Bad arcs are dangerous. Arcs develop a tremendous amount of heat (I^2R) at the point of arcing (sputtering). Arcing could be line to line, line to neutral, or line to ground. This type of arcing also creates a "signature."

To avoid nuisance tripping, the sophisticated electronic circuitry in an AFCI has the ability to recognize the "signature," and distinguish between *good* (harmless) and *bad* (dangerous) arcs. It makes a decision as to whether the arcing or glowing is dangerous or harmless.

Arcs are classified as follows:

- *Series Arcs:* When a conductor that is carrying current breaks or when a loose connection occurs, a "series" arcing fault is created. The current draw of the connected load is trying to jump across the opening created by the break or loose connection.

TABLE 6-3

Clearing times for listed AFCI combination-type circuit breakers when subjected to various values of arcing-fault current.

Ampere Rating of AFCI Circuit Breaker	VARIOUS TIMES TO CLEAR ARCING CURRENT IN AMPERES			
	5 Amperes	10 Amperes	Rated Current of the AFCI Circuit Breaker	150% of Rated Current of the AFCI Circuit Breaker
15 amperes	1 second	0.40 second	0.28 second	0.16 second
20 amperes	1 second	0.40 second	0.20 second	0.11 second
30 amperes	1 second	0.40 second	0.14 second	0.10 second

Table 6-3 gives you a good idea of how fast an AFCI circuit breaker trips for various values of arcing current as specified in UL Standard 1699. Remember, there are 60 cycles in 1 second.

- *Parallel Arcs:* When arcing occurs between the black ("hot") and white grounded (neutral) circuit conductors, or between the black ("hot") conductor and ground, it is referred to as a "parallel" arc. These conditions might be caused by a nail being driven through a cable, a staple driven too tightly, or a clamp or connector squeezing the cable too tightly. The current is traveling outside of its intended path. A broken or frayed cord could result in a "series" or "parallel" arc, or both.

Table 6-4 shows UL Standard 1699 specified current values for "parallel" arc testing of an AFCI device.

Types of AFCI Devices

The *NEC* tells us where to install AFCIs in homes. UL Standard 1699 specifies the sensing and tripping characteristics of AFCIs.

AFCIs are available in the following types:

1. *Combination AFCI:* This type of AFCI is mandatory in *210.12(A)*. It is installed in the

TABLE 6-4

Arcing current values for parallel arcing tests.

ARCING CURRENT IN AMPERES

75	100	150	200	300	500

For each of the above arcing current values, the total arcing time shall not exceed 8 half-cycles in 0.5 second. The 8 half-cycles of arcing could be consecutive or intermittent.

panel where the branch circuit originates. It provides arc-fault protection as specified in the UL Standard for both *Branch/feeder AFCIs* and *Outlet AFCIs.*

Note: Don't get confused. A *combination AFCI* means that the device has been tested and listed as meeting the requirements for both *Branch/feeder AFCIs* and *Outlet AFCIs.* When referring to a device that provides both AFCI and GFCI personnel protection, think of it as having *dual-function* characteristics. Don't use the word *combination!*

2. *Branch/feeder AFCI:* This type of AFCI circuit breaker is installed in a panel. It provides arc-fault protection for the branch-circuit wiring. It is not quite as sensitive as an *Outlet AFCI* although it does provide limited protection for extension cords plugged into receptacles on the circuit. The time permitted for the use of this type ended on January 1, 2008.

3. *Outlet AFCI:* This is an AFCI receptacle. It provides protection for cord sets that are plugged into the receptacle. It is more sensitive than a *Branch/feeder AFCI.* A feed-through AFCI receptacle installed at the first receptacle is permitted by *210.12(A), Exception 1 and 2.* There is no limitation on the length of the branch-circuit wiring to the first receptacle. However, the branch-circuit wiring method must comply with *Exception 1 or 2.* Everything downstream from the feed-through AFCI receptacle is AFCI protected.

4. *Portable AFCI:* A plug-in device with one or more receptacles. It provides AFCI protection

for cords that are plugged to it. This type of AFCI device is a "must" for workers using electrical tools on the job.

5. *Cord AFCI:* A plug-in device that provides AFCI protection to the power cord connected to it. It has no additional outlets. An example of this would be the AFCI device required to be an integral part of room air-conditioning equipment in accordance with *440.65.*

How Do You Know What Type It Is? The marking is found on the device. When installed in a panelboard, the marking must be visible after you remove the cover from the panelboard.

What Kinds of AFCIs Are Available?

- Single-pole, 120-volt breaker
- Single-pole, 120-volt dual-function with both AFCI and GFCI features in one breaker
- Two-pole, 240-volt, common trip breaker
- Two-pole, 120/240-volt breaker
- Two-pole, 120/240-volt common trip dual-function with both AFCI and GFCI features in one breaker
- Two-pole, 120/240-volt independent trip dual-function with both AFCI and GFCI features in one breaker; suitable for use with a shared neutral on a multiwire branch circuit
- 120-volt feed-through receptacles (As of the date this text is being written, no AFCI receptacles are in production for sale and installation.)

UL Standards for AFCIs and GFCIs

Type	Conforms to Requirements of	If Circuit Breaker Type, Also Conforms to Requirements of
AFCI	UL 1699	UL489
GFCI	UL 943	UL 489
AFCI/GFCI combination	UL 1699 and 943	UL 489

What Types of AFCI Devices Do Electricians Install?

The most common AFCIs that electricians install today are the circuit-breaker type. Circuit-breaker AFCIs are the same width as a standard circuit breaker so they will fit into a standard electrical panel. See Figure 6-13.

How to Connect AFCIs

When installing a 125-volt, single-pole AFCI circuit breaker, you first "snap" the breaker into the panel. Then the white lead from the AFCI circuit breaker is connected to the neutral bar in the panel.

FIGURE 6-13   A listed combination-type arc-fault circuit-interrupter (AFCI) circuit breaker. (*Courtesy of Schneider Electric*)

The black branch-circuit conductor is connected to the brass-colored load side terminal on the breaker. The white branch-circuit conductor is connected to the silver-colored load side terminal on the breaker. This is the same as for GFCI circuit breakers discussed earlier in this chapter.

Never share the neutral conductor of a multiwire branch circuit when using GFCIs and AFCIs unless they are listed for sharing a neutral. See Figure 6-11. Two-pole AFCIs and GFCIs are available. Always follow the instructions furnished with these devices.

May I Use a GFCI Receptacle on a Circuit Protected by an AFCI?

Yes! There should be no problem installing GFCI receptacles downstream from an AFCI circuit breaker in those locations where GFCI protection is required.

On branch circuits that are required by *210.12(A) or (B)* to be AFCI-protected, you might want to pick up a receptacle(s) that requires GFCI protection. This occurs a few times in the residence discussed in this book. The Front Bedroom branch circuit A16 picks up an outdoor receptacle on the front of the house. The Master Bedroom branch circuit A19 picks up an outdoor receptacle on the back of the house. The Living Room branch circuit B17 picks up an outdoor receptacle on the back of the house.

You could also install a dual-function AFCI/GFCI circuit breaker in the panel, which would provide arc-fault and ground-fault protection for an entire branch circuit with one device. It is a matter of choice and availability of AFCI, GFCI, and AFCI/GFCI dual-function breakers.

AFCIs in Existing Homes

The mandatory requirement in *210.12(A)* for AFCI protection is for new work only. It is not retroactive. But it's not a bad idea to consider installing AFCIs on existing installations. It is easy to do. Just replace existing circuit breaker(s) with an AFCI circuit breaker(s). We covered earlier in this chapter the new requirement in *210.12(B)* for adding AFCI protection for branch circuit extensions or modifications.

In older homes, the insulation on the original wiring is rated 60°C. Over the years, the insulation becomes brittle. For example, in and above a luminaire, the heat from the luminaire literally bakes and chars the insulation on the conductors. AFCI protection could "catch" an arcing problem in its early stages. Today, conductors used in house wiring are rated 90°C, so the deterioration of conductor insulation because of heat has become a thing of the past.

▶A requirement about AFCI protected receptacles has been added in 406.4(D)(4) of the 2011 *NEC* with an effective date of Jan. 1, 2014. It requires AFCI protection of receptacles that are replaced at a location where AFCI protection is required elsewhere in the *Code*. For example, a receptacle is replaced in a dwelling unit bedroom. AFCI protection is required for the branch circuit that supplies the receptacle by *210.12(A)*. As a result, the receptacle must have AFCI protection.

The AFCI protection can be provided in one of the following ways:

1. a listed outlet branch circuit type arc-fault circuit interrupter receptacle.

2. a receptacle protected by a listed outlet branch circuit type arc-fault circuit interrupter type receptacle.

3. a receptacle protected by a listed combination type arc-fault circuit interrupter type circuit breaker.◀

Testing AFCIs and GFCIs

AFCI/GFCI testers are used to test AFCI and GFCI receptacles and circuit breakers to make sure they are working properly. These devices simulate arc faults and ground faults. Figure 6-14(A) shows an arc-fault tester that also tests for proper polarity. Figure 6-14(B) shows a combination arc-fault/ground-fault tester that also tests for shared neutrals and verifies proper wiring connections.

OTHER SPECIAL PURPOSE RECEPTACLES

Tamper-Resistant Receptacles

A major step to reduce the number of injuries (burns), particularly to small children who always seem to find a way to insert "things" into the slots

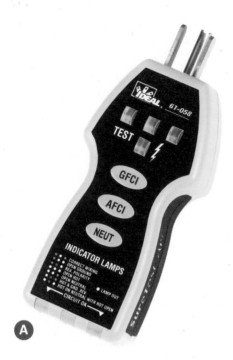

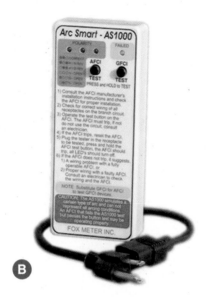

FIGURE 6-14   (A) An arc-fault tester that also tests for proper polarity. (B) A combination arc-fault/ground-fault tester that also tests for shared neutrals and verifies proper wiring connections. (*Photo (A) courtesy Ideal, photo (B) courtesy Fox Meter, Inc.*)

of a receptacle, was enacted for the 2008 *NEC* and has been revised for the 2011 edition of the *Code*. The requirements for tamper-resistant receptacles in dwellings are found in *406.12* and reads as follows:

▶In all areas specified in *210.52*, all nonlocking type 125-volt, 15- and 20-ampere receptacles shall be listed tamper-resistant receptacles.

- *Exception 1: Receptacles located more than 5 ½ ft (1.7 m) above the floor are not required to be of the tamper-resistant type.*

- *Exception 2: Receptacles that are part of a luminaire or appliance are likewise not required to be of the tamper-resistant type.*

- *Exception 3: A single receptacle for one appliance or a duplex receptacle for two appliances that are located within dedicated space for each appliance that in normal use is not easily moved from one place to another and that is cord-and-plug connected in accordance with 400.7(A)(6), (A)(7), or (A)(8) do not require tamper-resistant receptacles. (This exception recognizes that these receptacles would not be easily reached by children.)*

- *Exception 4: Nongrounding receptacles used for replacements as permitted in 406.4(D)(2)(a) are not required to be of the tamper-resistant type.*◀

Tamper-resistant receptacles are identified by the words "Tamper Resistant" or the letters "TR" where they will be visible after installation with a cover plate removed.

Recall that *210.52* lists the rooms, areas, and other locations inside and outside of a dwelling where receptacles are required to be installed, the spacing for the receptacles is given, and the number of receptacles required to be installed.

Tamper-resistant receptacles have a unique internal interlocking mechanism that energizes the blade contacts of the receptacle ONLY when a male attachment plug cap is inserted. If a paper clip or similar object is inserted into any one slot of the receptacle, no power is supplied to any of the blade contacts of the receptacle. Both slot shutters remain closed. Figure 6-15 is a typical tamper-resistant receptacle. Tamper-resistant receptacles have been required for many years in pediatric (children) areas of health care facilities, *517.18(C)*.

*Reprinted with permission from NFPA 70-2011.

FIGURE 6-15 Two styles of tamper-resistant receptacles. (*Photo courtesy of Leviton Manufacturing*)

Weather Resistant. All 15- and 20-ampere, 125- and 250-volt nonlocking type receptacles installed any damp or wet location are required to be listed weather-resistant type.

The construction of these receptacles makes them more durable if located where there is excessive moisture. These receptacles are required to be identified by the words "Weather Resistant" or the letters "WR" where they will be visible after installation with a cover plates secured as intended. See Figure 6-16.

Note that the definition of *damp location* and *wet locations* are found in NEC Article 100. It is quite obvious that an outdoor location subject to rain is a wet location. Locations below the carport roof and protected from weather are no doubt damp locations. To avoid dispute with inspectors, it make sense to use all weather-resistant receptacles outside of the dwelling.

Putting the Rules Together. A 15- or 20-ampere, 125-volt receptacle is installed outside the dwelling

FIGURE 6-16 Weather-resistant receptacle that is also tamper-resistant. (*Photo courtesy of Leviton Manufacturing*)

FIGURE 6-17   GFCI-type receptacle that is also weather resistant and tamper resistant. (*Photo courtesy of Leviton Manufacturing*)

in a wet location. This receptacle is required to have GFCI protection or be of the GFCI-type, to be weather-resistant, and tamper-resistant. You're in luck! Manufacturers of GFCI receptacles produce a version that complies with all of these requirements. See Figure 6-17.

REPLACING EXISTING RECEPTACLES

The *NEC* is very specific on the type of receptacle permitted to be used as a replacement for an existing receptacle. These *Code* rules are found in *406.4(D)*.

GFCI Protection. In *406.4(D)(3)* we find a retroactive rule requiring that when an existing receptacle is replaced in *any* location where the present *Code* requires GFCI protection (bathrooms, kitchens, outdoors, unfinished basements, garages, etc.), the replacement receptacle must be GFCI-protected. This could be a GFCI-type receptacle, or the branch circuit could be protected by a GFCI-type circuit breaker.

It is important to note that on existing wiring systems, such as "knob and tube," or older

nonmetallic-sheathed cable wiring that did not have an equipment grounding conductor, a GFCI receptacle will still properly function. It does not need an equipment grounding conductor. Thus, protection against lethal shock is still provided. See Figure 6-6.

Arc-Fault Circuit-Interrupters. ▶As detailed above, *406.4(D)(4)* requires AFCI protection where replacements are made at receptacle outlets that are required to be so protected elsewhere in this *Code*. Because AFCI receptacles are not available at this time, a delayed effective date for this requirement of January 1, 2014, has been included.◀

Tamper-Resistant Receptacles. ▶New section *406.4(D)(5)* requires that listed tamper-resistant receptacles be provided where replacements are made at receptacle outlets that are required to be tamper-resistant elsewhere in this *Code*.◀ *NEC 406.12* contains the requirements for tamper-resistant receptacles. See the discussion on tamper-resistant receptacles earlier in this chapter.

Weather-Resistant Receptacles. ▶New *NEC 406.4(D)(6)* requires that weather-resistant receptacles shall be provided where replacements are made at receptacle outlets that are required to be so protected elsewhere in the *Code*.◀ See the discussion on weather-resistant receptacles earlier in this chapter.

Replacing Existing Two-Wire Receptacles Where Grounding Means Does Exist

When replacing an existing receptacle [Figure 6-18(A)] where the wall box is properly grounded [Figure 6-18(E)] or where the branch-circuit wiring contains an equipment grounding conductor [Figure 6-18(D)], at least two easy choices are possible for the replacement receptacle:

1. The replacement receptacle must be of the grounding type (Figure 6-18[B]) unless …

2. The replacement receptacle is of the GFCI type [Figure 6-18(C)].

If the an equipment grounding conductor is not present in the enclosure for the receptacle, *250.130(C)* permits an equipment grounding

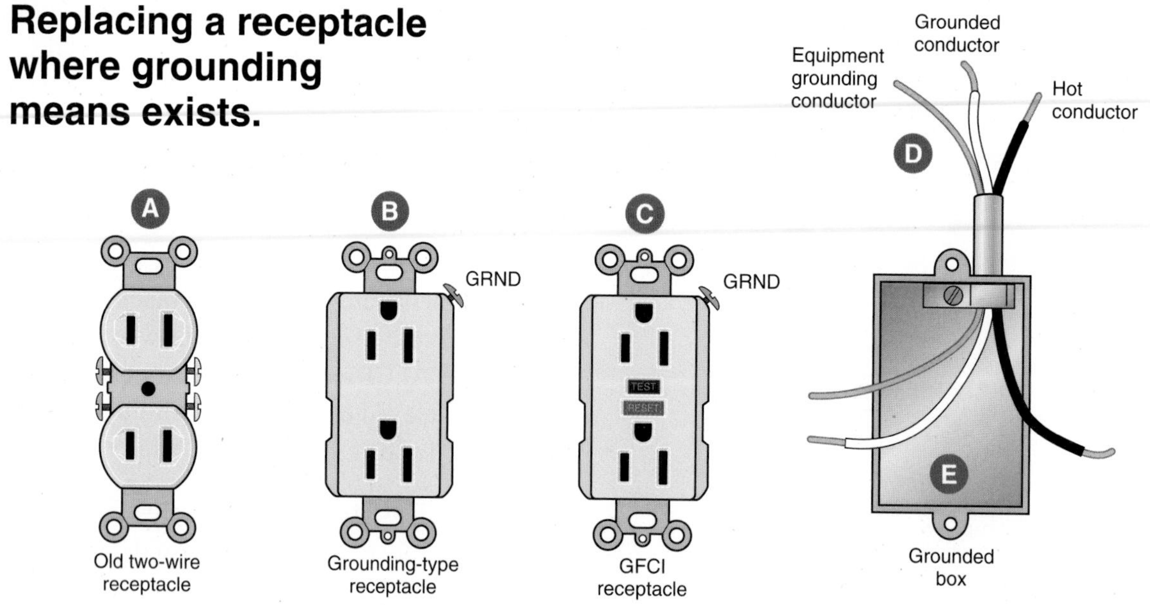

Replacing a receptacle where grounding means exists.

FIGURE 6-18 If a wall box (E) is properly grounded by any of the methods specified in *250.118*, or if the branch-circuit wiring contains an equipment grounding conductor (D), an existing receptacle of the type shown in (A) must be replaced with a grounding-type receptacle (B) or with a GFCI receptacle (C). See *406.4(D)(1)*. (*Delmar/Cengage Learning*)

conductor to be run from the enclosure to be grounded, or from the green hexagon grounding screw of a grounding-type receptacle, to any one of these four locations:

1. Any accessible point on the grounding electrode system

2. Any accessible point on the grounding electrode conductor

3. The equipment grounding terminal bar within the enclosure where the branch circuit for the receptacle originates

4. The grounded service conductor within the service-equipment enclosure

Note, however, that this permission is only for existing installations. See Figure 6-19.

Be sure that the equipment grounding conductor of the circuit is connected to the receptacle's green hexagon-shaped grounding terminal.

Do not connect the white grounded circuit conductor to the green hexagon-shaped grounding terminal of the receptacle.

Do not connect the white grounded circuit conductor to a metal box.

Replacing Existing Two-Wire Receptacles Where Grounding Means Does Not Exist

When replacing an existing 2-wire nongrounding-type receptacle [Figure 6-19(A)], where the box is not grounded [Figure 6-19(E)], or where an equipment grounding conductor has not been run with the circuit conductors [Figure 6-19(D)], four choices are possible for selecting the replacement receptacle:

1. The replacement receptacle may be a nongrounding type [Figure 6-19(A)].

2. The replacement receptacle may be a GFCI type [Figure 6-19(C)].

   - The GFCI replacement receptacle must be marked "No Equipment Ground," as in Figure 6-20.

   - The green hexagonal grounding terminal of the GFCI replacement receptacle does not have to be connected to any grounding means. It can be left "unconnected." The GFCI's trip mechanism will operate

Replacing a receptacle where grounding means does not exist.

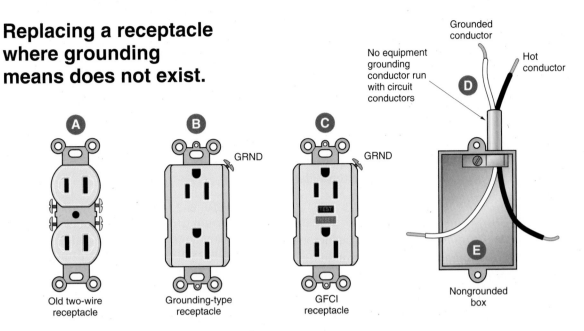

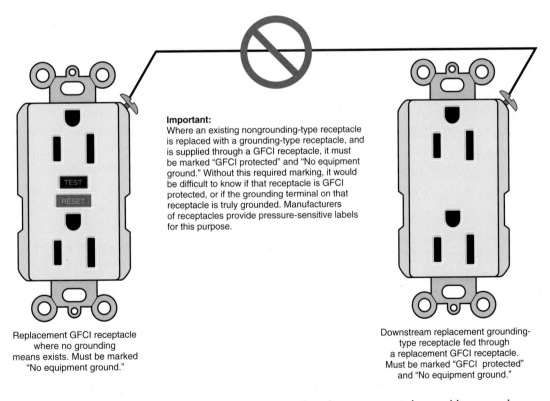

FIGURE 6-19 If a box (E) is *not* grounded, or if in an equipment grounding conductor has not been run with the circuit conductors (D), an existing nongrounding-type receptacle (A) may be replaced with a nongrounding-type receptacle (A), a GFCI-type receptacle (C), a grounding-type receptacle (B) if supplied through a GFCI receptacle (C), or a grounding-type receptacle (B) if a separate equipment grounding conductor is run from the receptacle to any accessible point on the grounding electrode system, *250.130(C)*. See Figure 6-21. (*Delmar/Cengage Learning*)

Important:
Where an existing nongrounding-type receptacle is replaced with a grounding-type receptacle, and is supplied through a GFCI receptacle, it must be marked "GFCI protected" and "No equipment ground." Without this required marking, it would be difficult to know if that receptacle is GFCI protected, or if the grounding terminal on that receptacle is truly grounded. Manufacturers of receptacles provide pressure-sensitive labels for this purpose.

Replacement GFCI receptacle where no grounding means exists. Must be marked "No equipment ground."

Downstream replacement grounding-type receptacle fed through a replacement GFCI receptacle. Must be marked "GFCI protected" and "No equipment ground."

FIGURE 6-20 Where a grounding means does not exist, there are certain marking requirements as presented in this diagram. *Do not* connect an equipment grounding conductor between these receptacles; see *406.3(D)(3)*. (*Delmar/Cengage Learning*)

properly when ground faults occur anywhere on the load side of the GFCI replacement receptacle. Ground-fault protection is still there. Refer to Figure 6-6.

- Do not connect an equipment grounding conductor from the green hexagonal grounding terminal of a replacement GFCI receptacle (see Figure 6-20) to any other downstream receptacles that are fed through the replacement GFCI receptacle.

    The reason this is not permitted is that if at a later date someone saw the conductor connected to the green hexagonal grounding terminal of the downstream receptacle, there would be an immediate assumption that the other end of that conductor had been properly connected to an acceptable grounding point of the electrical system. The fact is that the so-called equipment grounding conductor had been connected to the replacement GFCI receptacle's green hexagonal grounding terminal that was not grounded in the first place. We now have a false sense of security, a real shock hazard. The far better choice from a safety standpoint is to go to the time and expense of installing the equipment grounding conductor as permitted in *250.130(C)*.

3. The replacement receptacle may be a grounding type (Figure 6-19[B]) if it is supplied through a GFCI-type receptacle (Figure 6-19[C]).

- The replacement receptacle must be marked "GFCI Protected" and "No Equipment Ground," as in Figure 6-20.

- The green hexagonal equipment grounding terminal of the replacement grounding-type receptacle (Figure 6-19[B]) need not be connected to any grounding means. It may be left "unconnected." The upstream feed-through GFCI receptacle (Figure 6-19[C]) trip mechanism will work properly when ground faults occur anywhere on its load side. Ground-fault protection is still there. Refer to Figure 6-6.

- Do not connect an equipment grounding conductor from the green hexagonal grounding terminal of a replacement GFCI receptacle (Figure 6-20) to any other

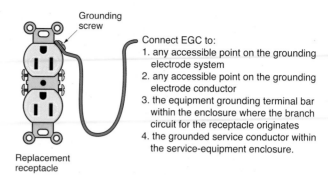

Grounding screw

Connect EGC to:
1. any accessible point on the grounding electrode system
2. any accessible point on the grounding electrode conductor
3. the equipment grounding terminal bar within the enclosure where the branch circuit for the receptacle originates
4. the grounded service conductor within the service-equipment enclosure.

Replacement receptacle

FIGURE 6–21 In old installations where an equipment grounding conductor does not exist in the box, one option is to install an equipment grounding conductor as permitted in *250.130(C)*. This is preferred over installing a GFCI receptacle without an equipment grounding conductor. (*Delmar/Cengage Learning*)

downstream receptacles that are fed through the replacement GFCI receptacle.

4. The replacement receptacle may be a grounding type (Figure 6-19[B]) if

- an equipment grounding conductor, sized per *Table 250.122*, is run from the replacement receptacle's green hexagonal grounding screw and properly connected to one of the four locations described earlier in this chapter. You are *not* permitted to make the equipment ground connection to the interior metal water piping anywhere beyond the first 5 ft (1.5 m). See *250.130(C)* and the *Informational Note*. Also read *250.50* for the explanation of a "grounding electrode system." This permission is only for replacing a receptacle in existing installations. See Figure 6-21.

PERSONNEL GROUND-FAULT PROTECTION FOR ALL TEMPORARY WIRING

Because of the nature of construction sites, there is a continual presence of shock hazard that can lead to serious personal injury or death through electrocution. Workers are standing in water, standing on damp or wet ground, or in contact with steel framing members. Electric cords and cables are lying on the ground, subject to severe mechanical abuse. All of these conditions spell danger.

▶All 125-volt, single-phase, 15-, 20-, and 30-ampere receptacle outlets that are not part of the permanent wiring of the building and that will be used by the workers on the construction site must be GFCI protected, *590.6(A)(1)*.◀

▶All 125-volt, single-phase, 15-, 20-, and 30-ampere receptacle outlets that are part of the actual permanent wiring of a building and are used by personnel for temporary power are also required to be GFCI protected *NEC 590.6A)(2)*.◀

▶*Receptacles on 15-kW or less Portable Generators.* All 125-volt and 125/250-volt, single-phase, 15-, 20-, and 30-ampere receptacle outlets that are a part of a 15-kW or smaller portable generator shall have listed ground-fault circuit-interrupter protection for personnel. All 15- and 20-ampere, 125- and 250-volt receptacles, including those that are part of a portable generator, used in a damp or wet location shall comply with *406.9(A)* and *(B)*. Listed cord sets or devices incorporating listed ground-fault circuit-interrupter protection for personnel identified for portable use shall be permitted for use with 15-kW or less portable generators manufactured or remanufactured prior to January 1, 2011.◀ It should be clear to all users of the *NEC* that the Code Panel responsible for this article is serious about ensuring that people who use electrical power on construction sites have GFCI protection.

For receptacles *other* than 125-volt, single-phase, 15-, 20-, and 30-ampere receptacles, there are two options for providing personnel protection. In *590.6(B)(1)*, GFCI protection is one of the choices. In *590.6(B)(2)*, a written *Assured Equipment Grounding Conductor Program* is the second choice. This program shall be continuously enforced on the construction site. A designated person must keep a written log, ensuring that all electrical equipment is properly installed and maintained according to the applicable requirements of *250.114, 250.138, 406.3(C)*, and *590.4(D)*. Because this option is difficult to enforce, many Authorities Having Jurisdiction (AHJs) require GFCI protection as stated in *590.6(B)(1)*.

Figure 6-22 is a listed manufactured power outlet for providing GFCI-protected temporary power on construction sites. It meets the requirements of *590.6(A)*.

FIGURE 6-22   A temporary power box used on construction sites. This particular type is connected with a 50-ampere, 120/240-volt power cord. It has six 20-ampere, GFCI-protected, 125-volt receptacles plus additional 30-ampere and 50-ampere Twist-Lock receptacles. (*Courtesy of Hubbell Lighting Outdoor & Industrial*)

Most electricians and other construction site workers carry their own listed portable ground-fault circuit-interrupter cord sets to be sure they are protected against shock hazard. These are shown in Figure 6-23 and also meet the requirements of *590.6(A)(2)* and *(A)(3)*.

Portable GFCI devices are easy to carry around. They have "open neutral" protection should the neutral conductor in the circuit supplying the GFCI open for whatever reason.

Portable GFCI devices are available with manual reset, which is advantageous should a power outage occur or if the GFCI is unplugged, so that equipment (drills, saws, etc.) will not start up again when the power is restored. This could cause injury to anyone using the equipment when the power is restored.

Portable devices are also available that reset automatically. These should be used on lighting, engine heaters, water (sump) pumps, and similar equipment where it would be advantageous to have the equipment start up as soon as power is restored.

Temporary Power for Construction Sites

In many parts of the country, it is common for the electrical contractor to furnish and install a

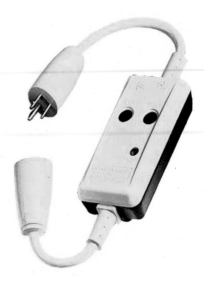

FIGURE 6-23 Cord-and-plug-connected "portable" GFCI devices are easy to carry around. They can be used anywhere when working with electrical tools and extension cords. These devices operate independently and are not dependent on whether the branch circuit is GFCI protected. These portable GFCI in-line cords are relatively inexpensive and should always be used when using portable electric tools. (*Courtesy of Hubbell Lighting Outdoor & Industrial*)

panel for all trades to use at a construction site of a new home. Figure 6-24 shows a typical pedestal that might be secured to a pole or set in the ground. It contains a meter base, a main breaker, and GFCI branch breakers for 120-volt receptacles.

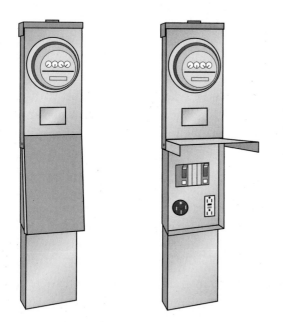

FIGURE 6-24 An example of a temporary construction site combination meter, breakers, and receptacles. The breakers and receptacles are under the lift-up cover. (*Delmar/Cengage Learning*)

IMMERSION DETECTION CIRCUIT INTERRUPTERS (IDCIs) AND APPLIANCE LEAKAGE CURRENT INTERRUPTERS (ALCIs)

Major culprits for electrical shock in the home are personal grooming appliances. The worst offender was found to be hand-held hair-drying appliances, most commonly used near a washbasin, or even worse, next to a water-filled bathtub with a person in it. Accidentally dropping or pulling the appliance into the tub can be deadly.

Because older homes probably do not have GFCI receptacles in bathrooms, *immersion detection circuit interrupters (IDCIs)* and *appliance leakage current interrupters (ALCIs)* were introduced to provide people with shock protection.

IDCI- and ALCI-protected appliances are easily recognized by the large attachment plug cap on the cord. IDCI and ALCI devices are an integral part of the attachment plug cap on hand-held hair-drying appliances such as electric hair dryers, stylers, and heated air combs. These devices might be manually resettable or nonresettable.

The operation of an IDCI or ALCI device depends on a "third" wire probe in the cord, which is connected to a sensor inside the appliance. This probe detects leakage current in the range of 4 to 6 milliamperes when conductive liquid (e.g., water) enters the appliance and makes contact with any live part inside the appliance and the internal "sensor." IDCI and ALCI devices cut off power to the appliance regardless of whether the appliance switch is in the "ON" or "OFF" position.

UL requires that the plug cap be marked: *Warning: To Reduce the Risk of Electric Shock, Do Not Remove, Modify, or Immerse This Plug.*

Never replace the special attachment plug cap on these appliances with a standard plug cap. That would make the appliance unsafe!

IDCIs and ALCIs are a requirement of *NEC 422.41* and UL Standard 859.

SURGE PROTECTIVE DEVICES (SPD)

The *Code* requirements for these devices that are permanently installed are found in *Article 285* in the *NEC*. Always install SPD devices that are listed under the requirements of UL Standard 1449. Carefully read and follow the instructions furnished with the SPD.

In today's homes, we find many electronic appliances (television, stereo, personal computer, fax equipment, word processor, VCR, VDR, CD player, digital stereo equipment, microwave oven), all of which contain many sensitive, delicate electronic components (printed circuit boards, chips, microprocessors, transistors, etc.).

Voltage transients, called *surges* or *spikes*, can stress, degrade, and/or destroy these components. They can cause loss of memory in the equipment or "lock up" the microprocessor.

The increased complexity of electronic integrated circuits makes this equipment an easy target for "dirty" power that can and will affect the performance of the equipment.

Voltage transients cause abnormal current to flow through the sensitive electronic components. This energy is measured in joules. A *joule* is the unit of energy when 1 ampere passes through a 1-ohm resistance for 1 second (like wattage, only on a much smaller scale).

Line surges can be line to neutral, line to ground, and line to line.

Transients

Transients are generally grouped into two categories:

- **Ring wave:** These transients originate within the building and can be caused by copiers, computers, printers, HVAC cycling, spark igniters on furnaces, water heaters, dryers, gas ranges, ovens, motors, and other inductive loads.

- **Impulse:** These transients originate outside of the building and are caused by utility company switching, lightning, and so on.

To minimize the damaging results of these transient line surges, service equipment, panelboards, load centers, feeders, branch circuits, and individual receptacle outlets can be protected with a SPD, as shown in Figure 6-25.

A SPD contains one or more metal-oxide varistors (MOV) that clamp the transients by absorbing the major portion of the energy (joules) created by the surge, allowing only a small, safe amount of energy to enter the actual connected load.

The MOV clamps the transient in times of less than 1 nanosecond, which is one billionth of a second, and keeps the voltage spike passed through to the connected load to a maximum range of 400 to 500 peak volts.

Typical SPD devices for homes are available as an integral part of receptacle outlets that mount in the same wall boxes as regular receptacles. They look the same as a normal receptacle and may have an audible alarm that sounds when an MOV has failed. The SPD device may also have a visual indication, such as a light-emitting diode (LED) that glows continuously until an MOV fails. Then the LED starts to flash on and off, as in Figure 6-26.

SPD devices are available in plug-in strips (Figure 6-27) and as part of desktop computer hardware power direction stands commonly placed under a monitor for plugging in a monitor, printer, scanner, and other peripheral equipment. These minimize the accumulation of the multitude of cords under and behind the equipment.

A SPD on a branch circuit will provide surge suppression for all of the receptacles on the same

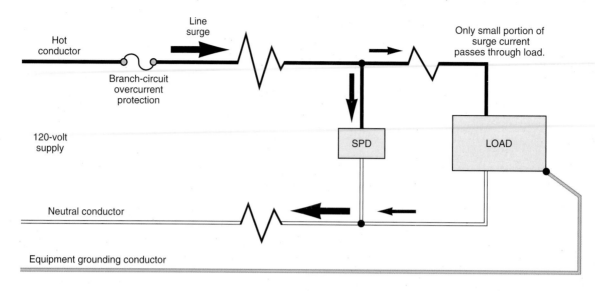

FIGURE 6-25 A surge protective device absorbs and bypasses (shunts) surge currents around the load. The MOV dissipates the surge in the form of heat. (*Delmar/Cengage Learning*)

FIGURE 6-26 Surge protective device in the form of a receptacle. (*Courtesy of Hubbell Lighting Outdoor & Industrial*)

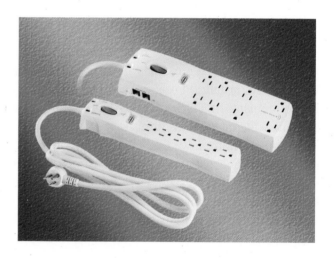

FIGURE 6-27 Surge protective device included the plug strip. (*Courtesy of Hubbell Lighting Outdoor & Industrial*)

circuit. These surges are classified as Category A low-level surges.

"Whole-house" surge protectors are available that offer surge protection for the entire house. See Figure 6-28. These are generally available at electrical supply houses and are installed at the main electrical panel of a home.

When a whole-house surge protector is installed, it is still a good idea to install surge protectors at the plug-in locations to more closely protect against low-level surges at the computer or other delicate electronic equipment location.

Noise

Electromagnetic "noise," not to be confused with audible noise, is external interference that can be generated in varying degrees by all types

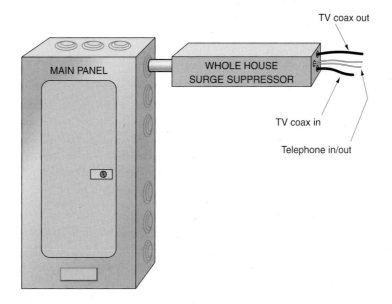

FIGURE 6-28   A whole-house transient voltage surge suppressor installed next to the main service panel. *Article 285* in the *NEC* requires that these devices be listed and have their short-circuit rating marked on the device. Be sure to follow the manufacturer's installation and connection instructions. (*Delmar/Cengage Learning*)

of electrical and electronic equipment. You might recognize "noise" as snow on a TV screen or as humming/buzzing/static on a radio or telephone. Some typical sources of noise are fluorescent lamps and ballasts, motors running or being switched "ON" and "OFF," loose electrical connections, or improper grounding. Even equipment running in the area such as X-ray equipment can be the culprit.

Although low-level noise is not physically damaging to electronic equipment, computers can lose memory, perform wrong calculations, have program malfunction, or lose Internet connection.

Noise comes from electromagnetic interference (EMI) and radio frequency interference (RFI).

EMI is usually caused by ground currents of very low values from such things as motors, utility switching loads, and lightning, and is transmitted through metal conduits.

RFI is noise, like the buzzing heard on a car radio when driving under a high-voltage transmission line. This interference "radiates" through the air from the source. Noise can be radiated or conducted from one residence to another. In commercial buildings, this radiated noise can be picked up by the grounding system of the building.

Consumer and commercial products today must meet Federal Communication Commission requirements for emission of electromagnetic and electronic noise so as not to interfere with communications equipment. Note: A cell phone has very limited amount of transmitting power.

In commercial work, isolated ground receptacles are installed in computer rooms to minimize "noise."

In residential work, isolated ground receptacles are rarely, if ever, used.

Various types of electronic "noise" eliminators are available from electronic stores that filter out and virtually eliminate undesirable interference.

Here are some related UL Standards:

- UL 498 Attachment Plugs and Receptacles
- UL 943 Ground-Fault Circuit Interrupters
- UL 1283 Electromagnetic Interference Filters
- UL 1363 Relocatable Power Taps
- UL 1449 Transient Voltage Surge Suppressors
- UL 1644 Immersion Detection Circuit Interrupters
- UL 1699 Arc-Fault Circuit Interrupters

*Reprinted with permission from NFPA 70-2011.

REVIEW

1. Explain the operation of a ground-fault circuit interrupter. Why are GFCI devices used? Where are GFCI receptacles required? _____

_____

_____

_____

_____

2. Residential GFCI devices are set to trip a ground-fault current above _____ milliamperes.

3. Where must GFCI receptacles be installed in residential garages? _____

_____

_____

_____

_____

4. The *Code* requires GFCI protection for certain receptacles in the kitchen. Explain where these are required. _____

_____

_____

5. Is it a *Code* requirement to install GFCI receptacles in a fully carpeted, finished recreation room in the basement? _____

6. A homeowner calls in an electrical contractor to install a separate circuit in the basement (unfinished) for a freezer. Is a GFCI receptacle required? _____

7. GFCI protection is available as (a) branch-circuit breaker GFCI, (b) feeder circuit breaker GFCI, (c) individual GFCI receptacles, (d) feed-through GFCI receptacles. In your opinion, for residential use, what type would you install? _____

8. Extremely long circuit runs connected to a GFCI branch-circuit breaker might result in

   a. nuisance tripping of the GFCI.

   b. loss of protection.

   c. the need to reduce the load on the circuit.

   Circle the correct answer.

9. If a person comes in contact with the "hot" and grounded conductors of a 2-wire branch circuit that is protected by a GFCI, will the GFCI trip "OFF"? Why?

_____

_____

_____

_____

10. What might happen if the line and load connections of a feed-through GFCI receptacle were reversed? _____

_____

_____

_____

_____

_____

11. May a GFCI receptacle be installed as a replacement in an old installation where the 2-wire circuit has no equipment grounding conductor? _____

12. What two types of receptacles may be used to replace a defective receptacle in an older home that is wired with knob-and-tube wiring where no equipment grounding means exists in the box? _____

_____

13. You are asked to replace a receptacle. On checking the wiring, you find that the wiring method is conduit and that the wall box is properly grounded. The receptacle is of the older style 2-wire type that does not have a grounding terminal. You remove the old receptacle and replace it with (circle the letter of the correct answer)

a. the same type of receptacle as the type being removed.

b. a receptacle that is of the grounding type.

c. GFCI receptacle.

14. What color are the terminals of a standard grounding-type receptacle? _____

_____

_____

15. What special shape and color are the grounding terminals of receptacle outlets and other devices? _____

_____

16. Construction sites can be dangerous because of the manner in which extension cords, portable electrical tools, and other electrical equipment are used and abused. To reduce this hazard, *590.6(A)* of the *NEC* requires that all 125-volt, single-phase, 15-, 20-, and 30-ampere receptacle outlets used for temporary wiring during construction must be _____ protected.

17. In your own words, explain why the *Code* does not require certain receptacle outlets in kitchens, garages, and basements to be GFCI protected. _____

_____

_____

_____

_____

_____

_____

_____

18. Circuit A1 supplies GFCI receptacles ① and ②. Circuit A2 supplies GFCI receptacles ③ and ④. Receptacles ① and ③ are feed-through type. Using colored pencils or marking pens, complete all connections.

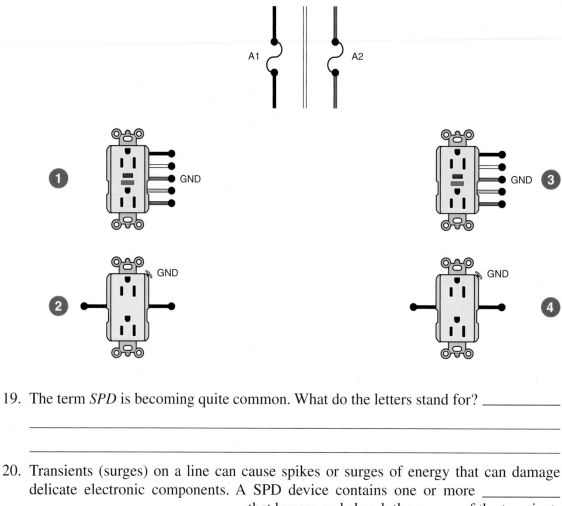

19. The term *SPD* is becoming quite common. What do the letters stand for? _____

_____

_____

20. Transients (surges) on a line can cause spikes or surges of energy that can damage delicate electronic components. A SPD device contains one or more _____ _____ that bypass and absorb the energy of the transient.

21. Undesirable "noise" on a circuit can cause computers to lock up, lose their memory, and/or cause erratic performance of the computer. This "noise" does not damage the equipment. The two types of this "noise" are EMI and RFI. What do these letters mean? _____

_____

_____

_____

22. Can SPD receptacles be installed in standard device boxes? _____

23. Some line transients are not damaging to electronic equipment but can cause the equipment to operate improperly. The effects of these transients can be minimized by installing _____.

24. Briefly explain the operation of an immersion detection circuit interrupter (IDCI).

_____

_____

_____

_____

_____

_____

25. What range of leakage current must trip an IDCI? _____

_____

_____

26. What amount of current flowing through a male human being will cause muscle contractions that will keep him from letting go of the live wire? _____ milliamperes

27. Other than for a few exceptions, the NEC requires that all systems having a voltage of over _____ volts shall be grounded.

28. In an old house, an existing nongrounding non-GFCI-type receptacle in a bathroom needs to be replaced. According to _406.4(D)(3)_, this replacement receptacle shall be (circle the letter of the correct answer):

    a. a nongrounding-type receptacle.

    b. a GFCI-type receptacle or be GFCI protected in some other way, such as changing the breaker that supplies the receptacle to a GFCI type.

29. Do all receptacle outlets in kitchens have to be GFCI protected? Explain. _____

_____

_____

_____

30. _NEC 210.12(A)_ requires AFCI protection for the branch-circuit wiring that supplies all electrical outlets in specific rooms in dwellings. The branch circuits serving what areas, locations, or rooms in dwellings are not required to have AFCI protection?

_____

_____

_____

31. When installing GFCI and AFCI circuit breakers in a panel, it is permissible to connect them to a multiwire branch circuit. (True) (False). Circle the correct answer.

Luminaires, Ballasts, and Lamps

OBJECTIVES

After studying this chapter, you should be able to

- understand luminaire terminology, such as Type IC and Type Non-IC.

- understand the *NEC* requirements for installing and connecting surface and recessed luminaires.

- realize that thermal insulation may have to be kept away from recessed luminaires.

- understand thermal protection requirements for recessed luminaires.

- know how to use "fixture whips."

- understand energy-saving ballasts and lamps.

- understand what a Class P ballast is.

WHAT IS A LUMINAIRE?

NEC defines a *luminaire* as *a complete lighting unit consisting of a light source such as lamp or lamps, together with the parts designed to position the light source and connect it to the power supply.** *Luminaire* is the international term for "lighting fixture" and is used throughout the *NEC*.

TYPES OF LUMINAIRES

There are literally thousands of different types of luminaries from which to choose to satisfy certain needs, wants, desires, space requirements, and, last but not least, price considerations.

Note in Table 7-1 that whether the luminaire is incandescent or fluorescent, the basic categories are surface mounted, recessed mounted, and suspended ceiling mounted.

The *Code* Requirements

Article 410 sets forth the requirements for installing luminaries. The electrician must "meet Code" with regard to mounting, supporting, grounding, live-parts exposure, insulation clearances, supply conductor types, maximum lamp wattages, and so forth.

Probably the two biggest contributing factors to fires caused by luminaries are installing lamp wattages that exceed that for which the luminaire has been designed, and burying recessed luminaries under thermal insulation when the luminaire has not been designed for such an installation.

---

*Reprinted with permission from NFPA 70–2011.

TABLE 7-1

Mountings for basic categories of luminaires.

Fluorescent	Incandescent
• Surface	• Surface
• Recessed	• Recessed
• Suspended Ceiling	• Suspended Ceiling

Nationally Recognized Testing Laboratories (NRTL) tests, lists, and labels luminaries that are in conformance with the applicable UL safety standards. Always install luminaries that bear the label from a qualified NRTL.

In addition to the *NEC*, the UL *Electrical Construction Materials Directory* (*Green Book*) and the UL *Guide Information for Electrical Equipment* (*White Book*), and manufacturers' catalogs and literature are excellent sources of information about luminaries.

Read the Label

NEC 110.3(B) states that *Listed or labeled equipment shall be installed and used in accordance with any instructions included in the listing or labeling.**

It is important to carefully read the label and any instructions furnished with a luminaire. Most *Code* requirements can be met by simply following this information.

Here are a few examples of label and instruction information:

- *Maximum lamp wattage*
- *Type of lamp*
- *For supply connections, use wire rated for at least _____°C*
- *Type-IC*
- *Type Non-IC*
- *Suitable for wet locations*
- *Thermally protected*

Installing and Connecting Luminaires

The circuit conductors in a wall or ceiling box where luminaries are to be installed are usually

- white—the "identified" grounded conductor, and
- black—the ungrounded "hot" conductor. A "hot" switch leg might also be red or another color, but never white or green.

Most surface-mounted luminaries will have a black and a white conductor in the canopy, making it easy to match these conductors to the circuit conductors in the box—white to white, black to black.

Chain-suspended luminaires usually come with a flexible, flat parallel conductor cord that weaves through the links of the chain. In a cord, it's a little more difficult to make a distinction between the "hot" conductor and the "identified" conductor.

To ensure proper polarity when making up the cord connections, generally connect as follows:

- The conductor with round insulation connects to the black "hot" circuit conductor.

- The conductor with grooved or raised insulation is the "identified" conductor that connects to the white circuit conductor.

- In some instances, the "identified" conductor will be tinned so it will have a silver color.

- The bare equipment grounding conductor (EGC) from the luminaire connects to the green hexagon-shaped screw in a metal electrical box or on the luminaire's mounting bar. Nonmetallic boxes will have a bare EGC from the nonmetallic-sheathed cable(s) to which the luminaire's bare EGC is connected.

Surface-Mounted Luminaires

These are easy to install. It is simply a matter of following the manufacturers' instructions furnished with the luminaires. The luminaire is attached to a ceiling outlet box or wall outlet box using luminaire studs, hickeys, bar straps, or luminaire extensions. Do not exceed the maximum lamp wattage marked on the luminaire.

Special attention must be given when installing surface-mounted luminaires on low-density cellulose fiberboard, Figure 7-1. Because of the potential fire hazard, *410.136(B)* states that fluorescent luminaires that are surface mounted on this material must be spaced at least 1½ in. (38 mm) from the fiberboard surface unless the fixture is marked "Suitable for Surface Mounting on Low-Density Cellulose Fiberboard."

The *Informational Note* to *410.136(B)* explains combustible low-density cellulose fiberboard as sheets, panels, and tiles that have a density of

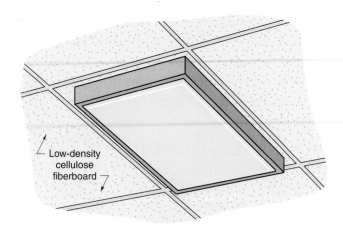

FIGURE 7-1 If they are to be mounted on *low-density cellulose fiberboard*, surface-mounted fluorescent luminaires must be marked "Suitable for Surface Mounting on Low-Density Cellulose Fiberboard." (*Delmar/Cengage Learning*)

20 lb/ft³ (320 kg/m³) or less that are formed of bonded plant fiber material. It does not include fiberboard that has a density of over 20 lb/ft³ (320 kg/m³) or material that has been integrally treated with fire-retarding chemicals to meet specific standards. Solid or laminated wood does not come under the definition of *combustible low-density cellulose fiberboard*.

To obtain fire-resistance ratings, most acoustical ceiling panels today are made from mineral wool fibers or fiberglass.

Recessed Luminaires

Recessed luminaires as shown in Figure 7-2 have an inherent heat problem. They must be suitable for the application and must be installed properly.

It is absolutely essential for the electrician to know early in the roughing-in stages of wiring a house what types of luminaires are to be installed. This is particularly true for recessed-type luminaires. To ensure that such factors as location, proper and adequate framing, and possible obstructions have been taken into consideration, the electrician must work closely with the general building contractor, carpenter, plumber, heating contractor, the thermal insulation installer, and the other building trades.

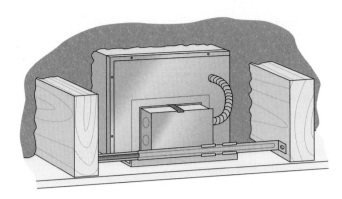

FIGURE 7-2   Typical recessed luminaire.
(*Delmar/Cengage Learning*)

When working with recessed luminaires, you will come across unique terms such as

- *Type Non-IC:* See Figure 7-3(A). This type of luminaire is marked "Type Non-IC" and is for installations in noninsulated ceilings. Where installed in an insulated ceiling, thermal insulation must be kept at least 3 in. (75 mm) away from any part of the luminaire.

- *Type IC:* See Figure 7-3(B). This type of luminaire is marked "Type IC" and is permitted to be buried in thermal insulation.

- *Inherently Protected:* This type of luminaire is designed so that the outside surfaces of the luminaire do not exceed temperatures greater than 194°F (90°C)—even if buried in thermal insulation, if mislamped (a lamp type not specified on the product), or if overlamped (lamp wattage exceeds the maximum wattage rating marked on the product). It is marked "Inherently Protected."

When a Type Non-IC recessed luminaire gets overheated because of overlamping, mislamping, or being too close to thermal insulation, the thermal protector (Figure 7-4) trips off. When the luminaire cools down, it comes on again. This on–off cycling will repeat until the problem is corrected. These luminaires are marked "Blinking Light of

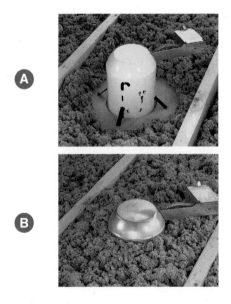

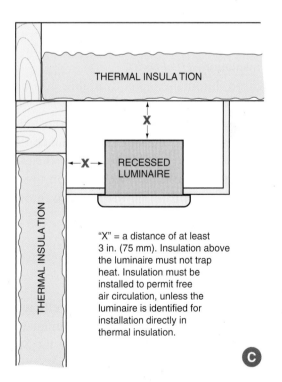

"X" = a distance of at least 3 in. (75 mm). Insulation above the luminaire must not trap heat. Insulation must be installed to permit free air circulation, unless the luminaire is identified for installation directly in thermal insulation.

FIGURE 7-3   (A) is a Type Non-IC recessed luminaire that requires clearance between the luminaire and the thermal insulation. (B) is a Type IC recessed luminaire that may be completely covered with thermal insulation. (C) shows the required clearances from thermal insulation for a Type Non-IC luminaire. See *410.66*. (*Photos courtesy of Cooper Lighting [A, B]*)

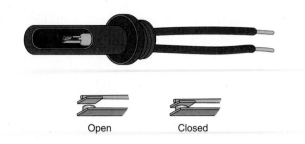

FIGURE 7-4 One type of recessed luminaire thermal protector. (*Delmar/Cengage Learning*)

This Thermally Protected Luminaire May Indicate Overheating."

In the trade, recessed luminaire housings are commonly referred to as "cans."

Installing recessed luminaire housings requires more attention than simple surface-mounted luminaires. The *Code* requirements for the installation of recessed luminaires are found in *410.110* through *410.122*. UL Standard 1571 covers recessed incandescent luminaires.

Figure 7-3(C) is a sketch that summarizes the minimum clearances for a Type Non-IC recessed luminaire that is not listed for direct burial in thermal insulation, *410.116(B)*.

Recessed luminaires are available for both new work and remodel work. Remodel work recessed luminaires can be installed from below an existing ceiling by cutting a hole in the ceiling, bringing power to the luminaire, making the electrical connections in the integral junction box on the luminaire, then installing the housing through the hole.

Energy-Efficient Housings

Many states have residential energy-efficiency requirements, particularly for recessed luminaires. The most widely followed regulations are the State of California Title 24 requirements, and the State of Washington Restricted Air Flow Requirements.

Basically, these recessed "cans" have double-walled gasketed airtight housings that prevent heated or cooled air to escape into attics and other unconditioned spaces in the house. They are listed

for use in insulated ceilings and in direct contact with insulation.

Although energy-efficiency requirements generally are not applicable to new one- and two-family homes, they might have been adopted by your state or community. Check with your local electrical inspector and/or building official to see whether such laws are applicable in your area for one- and two-family dwellings.

Thermal Protection

To protect against the hazards of overheating, UL and the *NEC* in *410.115(C)* require that recessed luminaires be equipped with an integral thermal protector, as in Figure 7-4. These devices will cycle on and off repeatedly until the heat problem has been resolved. There are two exceptions to this requirement. Thermal protection is not required for recessed incandescent luminaires designed for installation directly in a concrete pour, or for recessed incandescent luminaires that are constructed so as to not exceed "temperature performance characteristics" equivalent to thermally protected luminaires. These luminaires are so identified.

What the Junction Box (Wiring Compartment) on the Recessed Luminaire Housing Is For

Most recessed luminaires come equipped with a junction box (wiring compartment) that is an integral part of the luminaire housing. This is clearly shown in Figures 7-2 and 7-5. For most residential-type recessed luminaire installations, nonmetallic-sheathed cable or armored cable can be run directly into this junction box. These cables are required to have conductors rated 90°C. Check the marking on the luminaire for supply conductor temperature requirements.

Figure 7-6 shows the clearances normally needed between recessed luminaires and combustible framing members. This figure also summarizes the *NEC* requirements for connecting the supply conductors to a recessed luminaire. Note in one case

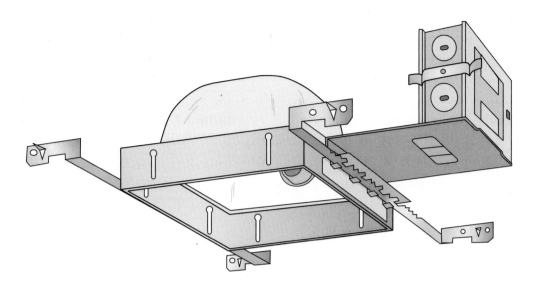

FIGURE 7-5 Roughing-in box of a recessed luminaire with mounting brackets and junction box (wiring compartment). (*Delmar/Cengage Learning*)

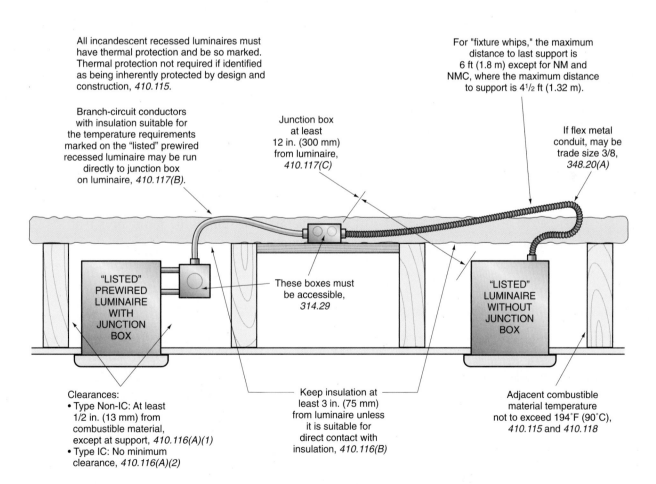

All incandescent recessed luminaires must have thermal protection and be so marked. Thermal protection not required if identified as being inherently protected by design and construction, *410.115*.

Branch-circuit conductors with insulation suitable for the temperature requirements marked on the "listed" prewired recessed luminaire may be run directly to junction box on luminaire, *410.117(B)*.

Junction box at least 12 in. (300 mm) from luminaire, *410.117(C)*

For "fixture whips," the maximum distance to last support is 6 ft (1.8 m) except for NM and NMC, where the maximum distance to support is 4¹/₂ ft (1.32 m).

If flex metal conduit, may be trade size 3/8, *348.20(A)*

"LISTED" PREWIRED LUMINAIRE WITH JUNCTION BOX

These boxes must be accessible, *314.29*

"LISTED" LUMINAIRE WITHOUT JUNCTION BOX

Clearances:
• Type Non-IC: At least 1/2 in. (13 mm) from combustible material, except at support, *410.116(A)(1)*
• Type IC: No minimum clearance, *410.116(A)(2)*

Keep insulation at least 3 in. (75 mm) from luminaire unless it is suitable for direct contact with insulation, *410.116(B)*

Adjacent combustible material temperature not to exceed 194°F (90°C), *410.115* and *410.118*

FIGURE 7-6 Requirements for installing recessed luminaires. (*Delmar/Cengage Learning*)

that the branch-circuit wiring is brought directly into the integral junction box on the luminaire. In the other case, a "fixture whip" is run from an adjacent junction box to the luminaire. In some instances, the luminaire comes with a flexible metal "fixture whip," and in other situations the "fixture whip" is supplied by the electrician. "Fixture whips" are discussed later.

In *410.21*, we find that *Luminaires shall be of such construction or installed so that the conductors in outlet boxes shall not be subjected to temperatures greater than that for which the conductors are rated.**

NEC *410.21* further states that *Branch-circuit wiring, other than 2-wire or multiwire branch circuits supplying power to luminaires connected together, shall not be passed through an outlet box that is an integral part of a luminaire unless the luminaire is identified for through-wiring.**

In *410.64*, we find that *Luminaires shall not be used as a raceway for circuit conductors unless listed and marked for use as a raceway.**

Some recessed luminaires are marked "identified for through-wiring," in which case branch-circuit conductors in addition to the conductors that supply the luminaire are permitted to be run through the outlet box or wiring compartment on the luminaire. A typical label will look like Figure 7-7.

The manufacturer of the luminaire, following UL requirements, will determine the

---

*Reprinted with permission from NFPA 70–2011.

Maximum of ___ ___ AWG through branch-circuit conductors suitable for at least ___°C (___°F) permitted in a junction box (___ in ___ out).

FIGURE 7-7 A typical label found on a recessed luminaire specifying the maximum number and size of conductors permitted to be run through the wiring compartment on the luminaire. (*Delmar/Cengage Learning*)

maximum number, size, and temperature rating for the conductors entering and leaving the outlet box or wiring compartment, and will so mark the label.

Only those luminaires that have been tested for the extra heat generated by the additional branch-circuit conductors will bear the label marking shown in Figure 7-7.

These fixtures have been tested for the added heat generated by the additional branch-circuit conductors.

Figure 7-8 shows three recessed luminaires, each with an integral outlet box. If these luminaires are marked "Identified for Through-Wiring," then it would be permitted to run conductors through the outlet boxes *in addition* to the conductors that supply the luminaires. If the "Identified for Through-Wiring" marking is not found on the luminaire, then it is a *Code* violation to run any conductors through the box other than the conductors that supply the luminaires that are "daisy chained."

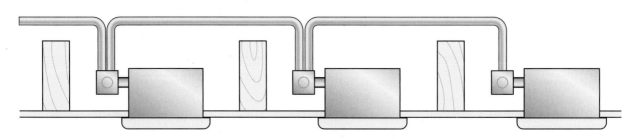

FIGURE 7-8 The only conductors permitted to run into or through the junction boxes on the recessed luminaires are those that supply the luminaires. Conductors other than those that supply the luminaires are permitted to run through the outlet boxes if the luminaire is marked "Identified for Through-Wiring." See *410.21* and *410.64*. (*Delmar/Cengage Learning*)

Routing the Branch Circuit

Some electricians prefer to run the branch-circuit wiring to the junction box on the recessed luminaire, then drop the switch loop to the switch box location. Others prefer to run the branch circuit to the switch location, then to the junction box on the luminaire. Either way is acceptable.

A new requirement was added to *404.2(C)*. It requires ▶*If switches control lighting loads supplied by a grounded general purpose branch circuit, a grounded circuit conductor shall be provided at the switch location.*◀ Two exceptions follow. This rule is extensively discussed in Chapter 5 of this text. As will be seen after reviewing the material, it will be simpler and save on the number of conductors required if the circuit wiring is routed through the switches rather than through the luminaire.

Recessed Luminaire Trims

There seems to be an endless choice of trims for recessed luminaires: eyeball, open, baffled, cone, pinhole, wall wash, and so on. Be very careful when choosing a trim for a recessed luminaire.

Recessed incandescent luminaires are listed by the NRTL with specific trims. Luminaire/trim combinations are marked on the label. Do not use trims that are not listed for use with the particular recessed luminaire. Installing a trim not listed for use with a specific recessed luminaire can result in overheating and possible on–off cycling of the luminaire's thermal protective device. Mismatching luminaires and trims is a violation of *110.3(B)* of the *NEC*, which states that *Listed or labeled equipment shall be installed or used in accordance with any instructions included in the listing or labeling.**

Sloped Ceilings

Be very careful when selecting recessed luminaires. Conventional recessed luminaires are listed for installation in flat ceilings, not in sloped ceilings! Recessed luminaires intended for installation in a sloped ceiling are submitted

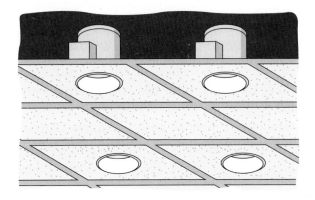

FIGURE 7-9 Suspended ceiling luminaires.
(*Delmar/Cengage Learning*)

to the NRTL specifically for testing in a sloped ceiling. Look for the marking "Sloped Ceiling." Instructions furnished with recessed luminaires provide the necessary information, such as being suitable for sloped ceilings having a 2/12, 6/12, or 12/12 pitch. Check the manufacturer's catalog and the NRTL listing.

Suspended Ceiling Lay-In Luminaires

Figure 7-9 illustrates luminaires installed in a typical suspended ceiling.

The recreation room of this residence has a "dropped" suspended acoustical paneled ceiling. Luminaires installed in a "dropped" ceiling must bear a label stating "Suspended Ceiling Luminaire." These types of luminaires are not classified as recessed luminaires because in most cases there is a great deal of open space above and around them.

Support of Suspended Ceiling Luminaires

In *410.36(C)*, we find that when framing members of a suspended ceiling grid are used to support luminaires, all framing members must be securely fastened together and to the building structure. It is the responsibility of the installer of the ceiling grid to do this.

*Reprinted with permission from NFPA 70–2011.

All lay-in suspended ceiling luminaires must be securely fastened to the ceiling grid members with bolts, screws, or rivets. Listed clips supplied by the manufacturer of the luminaire are also permitted to be used. Clips that meet this criteria are shown in Figure 17-1.

The logic behind all of the "securely fastening" requirements is that in the event of a major problem such as an earthquake or fire, the luminaires will not fall down and injure someone.

Connecting Suspended Ceiling Luminaires

The most common way to connect suspended ceiling lay-in luminaires is to complete all of the wiring above the ceiling using conventional wiring methods.

Then, from accessible outlet boxes strategically placed above the ceiling near the intended location of the lay-in luminaires, a flexible connection not more than 6 ft (1.8 m) long is made between the outlet boxes and the luminaires. See Figure 7-10. Lay-in ceiling tiles provides access to these outlet boxes.

In the electrical trade, these flexible connections are referred to as "fixture whips."

The *NEC* does not have a definition of a "fixture whip."

Don't confuse a "fixture whip" with a "tap." Most fixture whips in residential installations have conductors the same size as the branch-circuit conductors, so the "whip" is really an extension of the branch circuit, and by the *NEC* definition is not really a "tap."

A "tap" is defined in *240.2* as a conductor that has overcurrent protection ahead of its point of supply that exceeds the maximum overcurrent protection for that conductor. This is quite common in commercial installations where the branch-circuit overcurrent protection is rated 20 amperes and the tap conductors are rated 15 amperes. Tap conductors must be in a suitable raceway or Type AC or MC cable of at least 18 in. (450 mm) but not more than 6 ft (1.8 m) in length.

A fixture whip might be

- armored cable (AC), *320.30(D)(3)*.

- metal-clad cable (MC), *330.30(D)(2)*.

- nonmetallic-sheathed cable (NM and NMC), *334.30(B)(2)*.

- flexible metal conduit (FMC): Okay to use trade size ⅜, *348.30(A), Exceptions No. 3 and No. 4*. If trade size ⅜, length is not to exceed 6 ft (1.8 m), *348.20(A)(2)*.

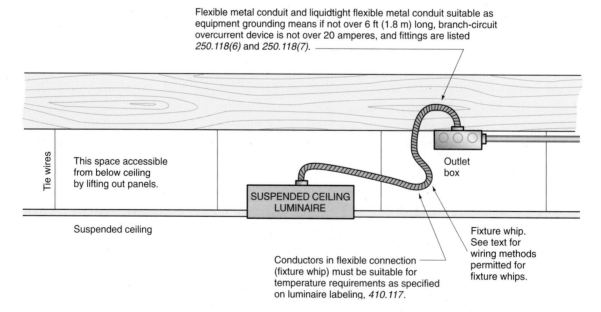

Flexible metal conduit and liquidtight flexible metal conduit suitable as equipment grounding means if not over 6 ft (1.8 m) long, branch-circuit overcurrent device is not over 20 amperes, and fittings are listed *250.118(6)* and *250.118(7)*.

Tie wires

This space accessible from below ceiling by lifting out panels.

SUSPENDED CEILING LUMINAIRE

Suspended ceiling

Outlet box

Fixture whip. See text for wiring methods permitted for fixture whips.

Conductors in flexible connection (fixture whip) must be suitable for temperature requirements as specified on luminaire labeling, *410.117*.

FIGURE 7-10 A suspended ceiling luminaire connected with a flexible fixture whip between an outlet box and the luminaire. (*Delmar/Cengage Learning*)

- liquidtight flexible metallic conduit (LFMC): Okay to use trade size ⅜, *350.30(A), Exceptions 3 and 4*. If trade size ⅜, length is not to exceed 6 ft (1.8 m), *350.20(A)*.

- liquidtight flexible nonmetallic conduit (LFNC): Okay to use trade size ⅜, *356.30(2)*. If trade size ⅜, length is not to exceed 6 ft (1.8 m), *356.20(A)*. Where grounding is required, a separate equipment grounding conductor shall be installed, *356.60*. Install according to *Article 250*. Size per *Table 250.122*.

- electrical nonmetallic-tubing (ENT), 362.30(A). Trade size smaller than ½ is not permitted, 362.20(A). Where grounding is required, a separate equipment grounding conductor shall be installed, 362.60. Install according to Article 250. Size per *Table 250.122*.

Note: For the above wiring methods, the maximum distance to the last point of support is 6 ft (1.8 m), except for NM and NMC, where the maximum distance to the last point of support is 4½ feet (1.32 m).

Grounding Luminaires

Generally, all luminaires must be grounded, *410.42*. Assuming that the wiring is a metallic wiring method or nonmetallic-sheathed cable with an equipment grounding conductor, you have an acceptable equipment grounding connection at the lighting outlet. Grounding of a surface-mounted luminaire is easily accomplished by securing the luminaire to the properly grounded metal outlet box with the hardware provided with the luminaire. Metal outlet boxes have a tapped No. 10-32 hole in which to insert a green hexagonal grounding screw. The bare copper equipment grounding conductor in the nonmetallic-sheathed cable is usually terminated under this screw. Grounding and bonding is discussed in Chapter 5. If the outlet box is nonmetallic, the small bare equipment grounding conductor from the luminaire is connected to the equipment grounding conductor in the outlet box. For suspended ceiling luminaires, grounding of the luminaire is accomplished by using metallic fixtures whips or nonmetallic-sheathed cable with ground between the outlet box and the luminaire. Always follow the manufacturers' instructions for making the electrical connections.

What about replacing a luminaire in an older home? If there is no equipment grounding means in the outlet box, then a luminaire that is made of insulating material and has no exposed conductive parts must be used, *410.44 Exception No. 1*. *Exception No. 2 to 410.44(B)* allows a metallic replacement luminaire to be installed and connected to a system that does not have an equipment grounding conductor. In this case, a separate equipment grounding conductor must be installed in conformance with *250.130(C)*. This requirement is discussed in Chapter 6 in the topic of replacing receptacles, so it is not repeated here. *Exception No. 3 to 410.44(B)* allows a metallic replacement luminaire to be installed and connected to a branch circuit that does not have an equipment grounding conductor if it is protected by GFCI device. Acceptable types of equipment grounding conductors are listed in *250.118*.

In Chapter 17, Figures 17-1 and 17-2 have additional information regarding recessed luminaire installations such as those in the recreation room of this residence.

Luminaires in Closets

There are special requirements for installing luminaires in closets. This is covered in Chapter 8.

Luminaires in Bathrooms

Because of the electrical shock hazards associated with electricity and water, there are special restrictions for luminaires installed in bathrooms. This is covered in Chapter 10.

FLUORESCENT BALLASTS AND LAMPS, INCANDESCENT LAMPS

The following is a brief introduction to a complex subject. For much more information, check out lamp and ballast manufacturers' Web sites listed in the Appendix of this text.

From an electrician's point of view, lamps and light bulbs mean the same thing. Use any one of the words, and everyone will know what you mean.

Always read the label on a luminaire for the type and maximum wattage of the lamp to be installed in that luminaire.

Fluorescent Ballasts

An incandescent lamp contains a filament that has a specific "hot" resistance value when operating at rated voltage. When energized, the lamp will provide the light output for which the lamp was designed.

A fluorescent lamp, on the other hand, cannot be connected directly to a circuit. It needs a ballast.

Fluorescent lamps do not have a filament running from end to end. Instead, they have a filament at each end. The filaments are connected to a ballast, as in Figure 7-11. The ballast is needed to (1) control power, (2) control voltage to heat the filaments, (3) control voltage across the lamp to start the arc within the tube, and (4) limit the current flowing through the lamp. The arc is needed to ionize the gas and vaporize the droplets of mercury inside the lamp.

Without a ballast, the lamp would probably not start. But if it did start, it would "run away" with itself. The current flowing through the gases within the lamp would rise rapidly, destroying the lamp in a very short time.

Ballast Types

Preheat. Preheat ballasts are connected in a simple series circuit, Figure 7-11. They are easily identified because they have a "starter." One type of starter is automatic and looks like a small roll of Lifesavers with two "buttons" on one end.

Another type of starter is a manual "ON–OFF" switch that has a momentary "make" position just beyond the "ON" position. When you push the switch on and hold it there for a few seconds, the lamp filaments glow. When the switch is released, the start contacts open, an arc is initiated within the lamp, and the lamp lights up.

Preheat lamps have two pins on each end.

Preheat lamps and ballasts are not used for dimming applications.

Rapid Start. Probably the most common type used today. Rapid start ballasts/lamps do not require a starter. The lamps start in less than 1 second. For reliable starting, ballast manufacturers recommend that there be a grounded metal surface within ½ in. (12.7 mm) of the lamp and running the full length of the lamp, that the ballast be grounded, and that the supply circuit originates from a grounded system. T5 rapid start lamps do not require a grounded surface for reliable starting.

Rapid start lamps have two pins on each end.

Rapid start lamps can be dimmed using a special dimming ballast. See Chapter 13.

Instant Start. Instant start lamps do not require a starter. These ballasts provide a high-voltage "kick" to

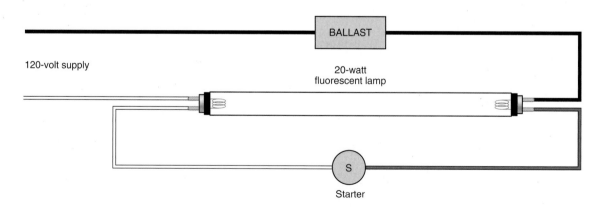

FIGURE 7-11 Simple series circuit. (*Delmar/Cengage Learning*)

start the lamp instantly. They require special fluorescent lamps that do not require preheating of the lamp filaments. Because instant start fluorescent lamps are started by brute force, they have a shorter life (as much as 40% less) than rapid start lamps when older style magnetic ballasts are used. With electronic ballasts, satisfactory lamp life can be expected.

Instant start lamps have one pin on each end.

Instant start ballasts/lamps cannot be used for dimming applications.

Dimming Ballasts.   Special dimming ballasts and dimmers are needed for controlling the light output of fluorescent lamps. Rapid start lamps are used. Incandescent lamp dimmers cannot be used to control fluorescent lamps. An exception to this is that dimmers marked "Incandescent Only" can be used to dim compact fluorescent lamps. Dimming ballasts are discussed in Chapter 13.

Mismatching of Fluorescent Lamps and Ballasts

Be sure to use the proper lamp for a given ballast. Mismatching a lamp and ballast may result in poor starting and poor performance, as well as shortened lamp and/or ballast life. The manufacturer's ballast and/or lamp warranty may be null and void.

T8 lamps are designed to be used interchangeably on magnetic or electronic rapid-start ballasts or electronic instant-start ballasts. Lamp life is reduced slightly when used with an instant-start ballast.

Voltage

Operating a ballast at an over-voltage condition will cause the it to run hot and shorten its life.

Operating a ballast at an under-voltage situation can result in premature lamp failure and unreliable starting.

Most ballasts today will operate satisfactorily within a range of +5% to −7% of their rated voltage. The higher quality CBM certified ballasts will operate satisfactorily within a range of ±10%.

Sound Rating

Most ballasts will hum, some more than others.

Ballasts are sound rated and are marked with letters "A" through "F." "A" is the quietest, and "F" is the noisiest. Look for an "A" or "B" sound rating for residential applications.

Magnetic ballasts (core and coil) hum when the metal laminations vibrate because of the alternating current reversals. This hum can be magnified by the luminaire itself, and/or the surface the luminaire is mounted on.

Electronic ballasts have little, if any, hum.

CAUTION: Do not insert spacers, washers, or shims between a ballast and the luminaire to make the ballast more quiet. This will cause the ballast to run much hotter and could result in shortened ballast life and possible fire hazard. Instead, replace the noisy ballast with a quiet, sound-rated one. Sometimes checking and tightening the many nuts, bolts, and screws of the luminaire will solve the problem.

Energy Saving by Control

We have all heard the words "Shut off the lights when you leave the room." Sometimes this works—sometimes not.

This text is used in all 50 states in this country and in Canada. It is not the intent of this text to go into the detail of specific state and/or local energy requirements. Just be aware that new developments are happening relative to energy conservation, and you need to pay attention. As an example, the State of California recently put into place Title 24, Part 6. The intent is to reduce power consumption in new homes. The States of Washington and Wisconsin have stringent energy conservation laws. As time passes, other states will follow.

The Energy Policy Act of 2005 stipulates in Title XII the electricity that types of lamps are permitted based on lumens per watt (efficacy). This law in Section 1252 also calls for utilities to provide "time-based" metering for residential, commercial, and industrial customers. Customers then can vary their electrical demands based on the rates for different times of usage. You can learn more about this by visiting the government's Web site: http://energycommerce .house.gov/.

As you install residential lighting, you need to be familiar with the terms "efficacy" and "lumens per watt." Simply stated, efficacy is measured in lumens per watt.

This law simply requires that lamps in permanently installed luminaires

- have high-efficacy rating (high lumens per watt), or
- be controlled by occupancy sensors, or
- have a combination of lamp efficacy, occupancy sensor control, and/or dimmers.

Cord-connected floor or table lamps are not governed by present current energy conservation laws.

Most fluorescent lamps with energy-saving ballasts, compact fluorescent lamps (CFL), and halogen lamps meet this high-efficacy requirement. Conventional incandescent lamps do not. See Table 7-2 to view the comparison of lumens per watt for different types of lamps.

The following chart shows what is required to meet the lamp requirement.

Lamp Wattage	Lamp Efficacy
Less than 15 watts	40 lumens per watt
15–40 watts	50 lumens per watt
Over 40 watts	60 lumens per watt

TABLE 7-2

Comparison of various lamps' characteristics. These characteristics are typical, and will vary by manufacturer. Always verify application with lamp and luminaire manufacturer's label, literature, and installation instructions.

Type of Lamp	Lumen per Watt	Dimming	Color and Application	Life (approx. hours)	Typical Shapes
Incandescent	14–18	Yes	Warm and natural. Great for general lighting. Brand names have various trade names for their lamps, such as Reveal, Soft White, etc.	500, 750, 1000, 1500, 3000 hours. Depends on type of lamp. Lamp life typically is based on operating the lamp an average of 3 hours of operation per start.	Standard, spots, floods, decorative, flame, tubes, globes, PAR (similar to standard spots and floods but stronger). Use rough service bulbs where there is vibration, like garage door openers and ceiling fans. Base types: candelabra, intermediate, medium, mogul.
Halogen	16–22 Are more efficient than conventional incandescent lamps. More lumen output per watt.	Yes	Brilliant white. Excellent for accent and task lighting. Are filled with halogen gas and floods and have an inner lamp, allowing the filament to run hotter (whiter).	2000–4000 hours. Lamp life typically is based on operating the lamp an average of 3 hours of operation per start.	PAR spots and floods, flame, crystal, mini-reflector spots. Base types: candelabra, intermediate, and medium. Can replace most incandescent lamps.
Fluorescent	T12 82 T8 92 T5 104	Yes, but only 40-watt rapid-start lamps using special dimming ballast. See Chapter 13.	Warm and deluxe warm white, cool and deluxe cool white, plus many other shades of white. Great for general lighting, like the Recreation Room in this residence. The higher the K rating, the cooler (whiter) is the color rendition.	6000 to 24,000 hours. Average life with lamps turned off and restarted once every 12 operating hours.	Straight, U-tube, circular. Single pin and double pin.

TABLE 7-2

(continued)

Type of Lamp	Lumen per Watt	Dimming	Color and Application	Life (approx. hours)	Typical Shapes
Compact Fluorescent	45/60 Because of their high lumens per watt, CFLs can save $20 to $50 over their lifetime when compared to equivalent incandescent lamps. Produce more lumens per watts (e.g., 27-watt CFL produces 1800 lumens vs. 1750 for a 100-watt incandescent.) Use about 80% less energy, and generate 90% less heat than equivalent incandescent lamp.	Yes, if so marked. CFLs have integral ballasts.	Soft warm white.	10,000–15,000 hours. Can last 9 to 13 times longer than incandescent. Lamp life typically is based on operating the lamp an average of 3 hours of operation per start. Read instructions carefully for restrictions such as *Not for use with dimmers, electronic timers, occupancy sensors, photocells, or lighted switches, Do not use in recessed or totally enclosed luminaires.*	Twisted (spiral) tubes, folded tubes, globe. R20, R30, R40, PAR38, "A" shape (like a standard incandescent). The number of tube loops are referred to as twin, double twin, triple twin, quad twin. Base types: medium. Can replace most incandescent lamps.
Light-Emitting Diodes (LED)	50+ An LED lamp contains a cluster (array) of many individual LEDs to produce this lumen output.	Yes or No. Check manufacturer's instructions and warnings.	White for LED lighting.	60,000 to 100,000 hours.	Base type: medium. Can replace an incandescent lamp.

The Bottom Line

In a nutshell, this amounts to the mandatory use of high-efficiency luminaires (air-tight), high-efficiency lamps, and various levels of control such as occupancy sensors and dimmers.

Today, occupancy sensor switches are available that install the same way as conventional toggle switches. They provide manual "ON–OFF" as well as automatic shut-off when no motion is detected after a preset time. As soon as motion is detected, they turn the lamps on. See Figure 7-12.

To learn more about the California energy conservation law, check out http://www.energy.ca.gov/title24 and http://www.haloltg.com. Browse for the summary of the California Title 24 law. Other lighting manufacturers' Web sites should also be checked out.

Energy-Saving Ballasts

The market for magnetic (core and coil) ballasts is shrinking!

The National Appliance Energy Conservation Amendment of 1988, Public Law 100-357 prohibited manufacturers from producing ballasts having a power factor of less than 90%. Ballasts that meet or exceed the federal standards for energy savings are marked with a letter "E" in a circle. Dimming

FIGURE 7-12   An occupancy (motion) sensor.
(*Photo courtesy of Leviton*)

When installing fluorescent luminaires, check the label on the ballast that shows the actual volt-amperes that the ballast and lamp will draw in combination. *Do not* attempt to use lamp wattage only when making load calculations because this could lead to an over-loaded branch circuit. For example, a high-efficiency ballast might draw a total of 42 volt-amperes, whereas an old-style magnetic ballast might draw 102 volt-amperes. Refer to Table 7-3 for other comparisons.

The higher the power factor rating of a ballast, the more energy efficient. Look for a power factor rating in the mid to high 90s.

Energy-Saving Lamps

The Energy Policy Act of 1992 enacted restrictions on lamps. In October 1995, the common 4-ft, 40-watt T12 linear medium bipin fluorescent lamp was eliminated. This was replaced by an energy-efficient 34-watt T12 lamp, a direct replacement for the discontinued lamp.

The Energy Policy Act was amended drastically in 2005, particularly in the electricity section entitled Title XII. Visit this Web site: http://www.gpo.gov/fdsys/pkg/PLAW-109publ58/pdf/PLAW-109publ58.pdf

Some lamps may be designated F40T12/ES, but the lamp draws 34 watts instead of 40 watts. The "ES" stands for "energy saving." ES is a generic designation. Manufacturers may use other designations such as "SS" for SuperSaver, "EW" for Econ-o-Watt, "WM" for Watt-Miser, and others.

The older, high wattage incandescent R30, R40, and PAR38 lamps were also discontinued and replaced with lower wattage lamps. See the sections on "Lamp Shapes" and "Lamp Diameters" in this chapter.

ballasts and ballasts designed specifically for residential use were exempted.

Today's electronic ballasts are much lighter in weight and considerably more energy efficient than older style magnetic ballasts (core and coil). Energy-saving ballasts might cost more initially, but the payback is in the energy consumption saving over time.

Old-style fluorescent ballasts get very warm and might consume 14 to 16 watts, whereas an electronic ballast might consume 8 to 10 watts. Combined with energy-saving fluorescent lamps that use 32 or 34 watts instead of 40 watts, energy savings are considerable. You are buying light, not heat.

TABLE 7-3

Various line currents, volt-amperes, wattages, and overall power factor for various single-lamp fluorescent ballasts.

Ballast	Line Current	Line Voltage	Line Volt-Amperes	Lamp Wattage	Line Power Factor
No. 1	0.35	120	42	40	0.95(95%)
No. 2	0.45	120	54	40	0.74(74%)
No. 3	0.55	120	66	40	0.61(61%)
No. 4	0.85	120	102	40	0.39(39%)
No. 5	0.22	120/277	26	30	0.99(99%)

These values were taken from actual ballast manufacturers' data. For two-lamp ballasts, double the values in this table. Note in line 5 how high the efficiency is of the latest type of electronic, high performance ballast for T8 fluorescent lamps.

FIGURE 7-13 An electronic Class P ballast, thermally protected as required in *410.130(E)*. Electronic ballasts provide improved performance in fluorescent lighting installations. Electronic ballast lighting systems are 25% to 40% more energy efficient than conventional magnetic (core & coil) ballast fluorescent systems. (*Courtesy of OSRAM Sylvania*)

T12 lamps are still found in 4-ft shop lights and square luminaires that use U-tube lamps. Most newer square luminaires have U-tube T8 lamps. In new commercial installations, the T8 lamp has taken over from the T12 lamp.

Energy-saving fluorescent lamps use up to 80% less energy than incandescent lamps of similar brightness. Fluorescent lamps can last 13 times or more longer than incandescent lamps.

It has been estimated that the total electric bill savings across the country will exceed $250 billion over the next 15 years. Table 7-3 compares the power factor of various types of ballasts. Note the difference in line current for the different types of ballasts.

Class P Ballasts

In *410.130(E)*, we find that all fluorescent ballasts installed indoors (except simple reactance-type ballasts), both for new and replacement installations, must have thermal protection built into the ballast by the manufacturer of the ballast, as shown in Figure 7-13. Ballasts provided with built-in thermal protection are listed by UL as Class P ballasts. Under normal conditions, the Class P ballast has a case temperature not exceeding 194°F (90°C). The thermal protector must open within 2 hours when the case temperature reaches 230°F (110°C).

Some Class P ballasts also have a nonresetting fuse integral with the capacitor to protect against capacitor leakage and violent rupture. The Class P ballast's internal thermal protector will disconnect the ballast from the circuit in the event of excessive temperature. Excessive temperatures can be caused by abnormal voltage and improper installation, such as being covered with insulation.

The reason for thermal protection is to reduce the hazards of possible fire due to an overheated ballast when the ballast becomes shorted, grounded, covered with insulation, lacking in air circulation, and so on. Ballast failure has been a common cause of electrical fires.

See Figure 7-14 for some important requirements for fluorescent and incandescent luminaires.

Cold Temperature and Fluorescent Lamps

Conventional fluorescent lamps will operate satisfactorily in ambient temperatures of 60°F (16°C) or more.

In cold temperatures, standard fluorescent lamps will start poorly or possibly not even start. Special ballasts and cold-weather fluorescent lamps are needed. Refer to the manufacturer's literature for instructions relating to cold temperature.

Lamp Designations for Rapid Start and Preheat Fluorescent Lamps

Example: F40T12/WWX/RS

"F"—fluorescent

"40"—wattage

"T"—tubular shape

"12"—diameter of lamp in eighths of an inch

"WWX"—color (warm white deluxe)

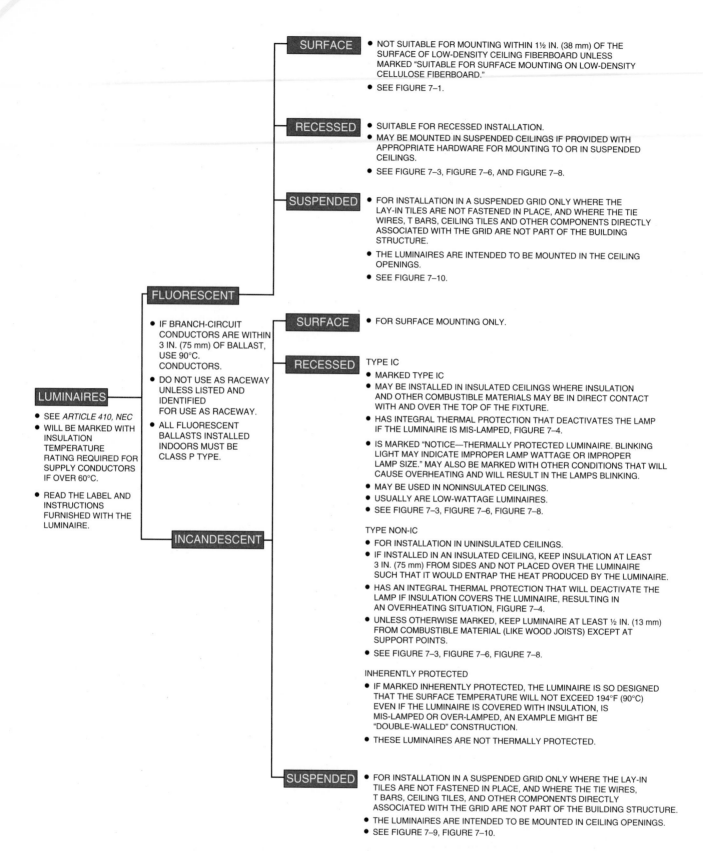

FIGURE 7-14 Some of the most important UL and *NEC* requirements for fluorescent and incandescent luminaires. Always refer to the UL Standards, the *NEC*, and the label and/or instructions furnished with the luminaire. (*Delmar/Cengage Learning*)

"RS"—rapid start

"FC"—shape (circular)

"FB" or "FU"—U-shaped, bent

Lamp Shapes

"A"—standard shape, general use

"P" or "PS"—pear shaped (150 watts and larger)

"C"—cone shape, like a night light or Christmas tree bulb

"F"—flame shape, decorative

"G"—globular

"PAR"—parabolic shape like a bowl, concentrates light

"R"—reflector type; might have reflective material near the base or at the bottom of the lamp; concentrates light

Burn Base Down.   Some lamps will be designated as "Burn Base Down." Be sure to follow these words of warning.

Lamp Diameters

There is a simple way to determine the diameter of a lamp at its widest measurement. The industry uses an "eighths of an inch" rule. For example, the diameter of a T5 fluorescent lamp is $5 \div 8 = 0.625$ in. The diameter of a T8 fluorescent lamp is $8 \div 8 = 1$ in. The diameter of a T12 lamp is $12 \div 8 = 1\frac{1}{2}$ in.

Incandescent lamps follow the same system. The diameter of an A21 lamp is $21 \div 8 = 2.625$ in. The diameter of an R30 lamp is $30 \div 8 = 3\frac{3}{4}$ in.

Watts versus Volt-Amperes

It is very easy to overload a branch circuit that supplies fluorescent lighting. The most common mistake of making load calculations for fluorescent lighting loads is to use lamp wattage instead of the volt-amperes and total current draw as marked on the ballast's label.

The culprits are old-style, low-power factor, low-efficiency ballasts!

Let's make a comparison of possibilities for the recreation room lighting where there are six recessed fluorescent luminaires, each containing two 2-lamp ballasts. The nameplate on each ballast indicates a line current rating of 0.70 ampere at 120 volts, which is 84 volt-amperes. The lamps are marked 40-watts.

The total current in amperes is

$$12 \times 0.70 = 8.4 \text{ amperes}$$

The total power in volt-amperes is

$$8.4 \times 120 = 1008 \text{ volt-amperes}$$

The total lamp wattage is

$$40 \times 24 = 960 \text{ watts}$$

If we had calculated the load for these luminaires based upon wattage, the result would have been

$$\frac{W}{E} = \frac{960}{120} = 8 \text{ amperes}$$

This result is not much different from the 8.4 amperes using data from the ballast nameplate.

Had we used low-cost, low-efficiency ballasts like No. 4 shown in Table 7-3, the result would be entirely different. The ballasts in Table 7-3 are single-lamp ballasts. We double the current values for a good approximation of a similar 2-lamp ballast. Using the low-power factor ballast, we find that the total current in amperes is

$$12 \times 0.85 \times 2 = 20.4 \text{ amperes}$$

The total power in volt-amperes is

$$20.4 \times 120 = 2,448 \text{ volt-amperes}$$
$$\text{or}$$
$$12 \times 102 \times 2 = 2,448 \text{ volt-amperes}$$

The total lamp wattage is

$$24 \times 40 = 960 \text{ watts}$$

As previously shown, if we use lamp wattage to calculate the current draw:

$$I = \frac{W}{E} = \frac{960}{120} = 8 \text{ amperes}$$

But we really have a current draw of 20.4 amperes, which would overload the 15-ampere branch circuit B12. In fact, a load of 20.4 amperes is too much for a 20-ampere branch circuit. Two 15-ampere branch circuits would have been needed to hook up the recreation room recessed fluorescent luminaires. Remember, *do not* load any circuit to more than 80% of the branch circuit's rating. That is 16 amperes for a 20-ampere branch circuit and 12 amperes for a 15-ampere branch circuit. And that is the maximum.

It is apparent that the *installed* cost using cheap luminaires is greater and more complicated than if high-quality luminaires using energy-efficient ballasts and lamps had been used. However, after the initial installation, the energy savings add up significantly.

Low-power factor ballasts mean higher light bills! High-efficiency, high-power factor ballasts should always be used. More on the wiring of the lighting in the recreation room is covered in Chapter 17.

Other Considerations

- Because heat trapped by insulation around and on top of a luminaire can shorten the life of a ballast, always follow the manufacturer's installation requirements and the requirements found in *Article 410*. There is a "nonscientific" rule of thumb referred to as the "half-life rule." It means that for every 21°F (10°C) above electrical equipments (motors, conductors, transformers, etc.) recommended maximum operating temperature, the expected life of that equipment is cut roughly in half. Raising the temperature another 21°F (10°C) will again cut the expected life in half. A 21°F (10°C) temperature rise (**Note:** This is a temperature difference, not an absolute temperature measurement), above the ballast's rated temperature of 90°C can reduce the ballast's life by one-half. This "half-life rule" is also true for conductors, motors, transformers, and other electrical equipment.

- Fluorescent lamps that are intensely blackened on both ends should be replaced. Operating a two-lamp ballast with only one lamp working will cause the ballast to run hot and will shorten the life of the ballast. Severe blackening of one end of the lamp can also ruin the ballast. A flickering lamp should be replaced.

- Poor starting of a fluorescent lamp can be caused by poor contact in the lampholder, poor

grounding, excessive moisture on the outside of the tube, or cold temperature (approximately 50°F [10°C]), as well as dirt, dust, and grime on the lamp.

- Most ballasts will operate satisfactorily within a range of +5% to −7% of their rated voltage. The higher quality CBM-certified ballasts will operate satisfactorily within a range of ±10%.

- Most T12 fluorescent lamps sold today are of the energy-efficient type. Generally, energy-efficient fluorescent lamps should not be used with old-style magnetic ballasts. To do so could result in poor starting, reduced lamp life, flickering, or spiraling. Home centers carry the most common T12 fluorescent lamps. Electrical supply houses usually carry different types of fluorescent lamps that can be matched with different types of ballasts. Check the markings on the ballast and the lamp carton, as well as the instructional literature available at the point-of-sale of the ballast or lamp, and/or in the manufacturer's catalogs.

- Lamp life generally is rated in "X" number of hours of operation based on 3 hours per start. Frequent switching results in shortened expected lamp life. Inversely, leaving the lamps on for long periods of time extends the expected lamp life. Vibration, rough handling, cleaning, and so on, shortens lamp life.

- Be careful of the type of conductors you use to connect fluorescent luminaires. Branch circuit conductors within 3 in. (75 mm) of a ballast must have an insulation temperature rating of not less than 90°C, *410.68*. Type THHN conductors, the conductors in nonmetallic-sheathed cable and in Type ACTHH armored cable are rated 90°C.

Voltage Limitations

The maximum voltage allowed for residential lighting is 120 volts between conductors, per *210.6(A)*.

In or on a home, lighting equipment that operates with an open-circuit voltage over 1000 volts

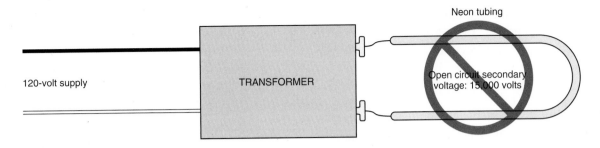

FIGURE 7-15   It is *not* permitted to install lighting in or on residences where the open-circuit voltage is over 1000 volts. This is a violation of *410.140(B)*. (*Delmar/Cengage Learning*)

is not permitted, *410.140(B)*. This pretty much eliminates most neon lighting systems for decorative lighting purposes, as shown in Figure 7-15.

Incandescent Lamp Life at Different Voltages

Operating an incandescent lamp at other than rated voltage will result in longer—or shorter—lamp life. The following formula predicts the approximate expected lamp life at different voltages. For example, assume that a 120-volt incandescent lamp has a published lamp life of 1000 hours. The calculations show the expected lamp life of this lamp when operated at 130 volts, and the expected lamp life when operated at 110 volts.

- At 130 volts:

$$\text{Expected Life} = \left(\frac{\text{Rated Volts}}{\text{Actual Volts}}\right)^{13} \times \text{Rated Life}$$

$$= \left(\frac{120}{130}\right)^{13} \times 1000$$

$$= (0.923)^{13} \times 1000$$

$$= 0.353 \times 1000$$

$$= 353 \text{ hours}$$

- At 110 volts:

$$\text{Expected Life} = \left(\frac{\text{Rated Volts}}{\text{Actual Volts}}\right)^{13} \times \text{Rated Life}$$

$$= \left(\frac{120}{110}\right)^{13} \times 1000$$

$$= (1.091)^{13} \times 1000$$

$$= 3.103 \times 1000$$

$$= 3103 \text{ hours}$$

Incandescent Lamp Lumen Output at Different Voltages

Operating an incandescent lamp at a voltage lower than the lamp's voltage rating will result in longer lamp life. A strong case might be made to install 130-volt lamps, particularly where they are hard to reach, such as flood lights located high up. However, the lumen output is reduced. For example, calculate the approximate lumen output of a 100-watt, 130-volt incandescent lamp that has an initial lumen output of 1750 lumens at rated voltage. The lamp is to be operated at 120 volts.

$$\text{Expected Lumens} = \left(\frac{\text{Actual Volts}}{\text{Rated Volts}}\right)^{3.4} \times \text{Rated Lumens}$$

$$= \left(\frac{120}{130}\right)^{3.4} \times 1750$$

$$= (0.923)^{3.4} \times 1750$$

$$= 0.762 \times 1750$$

$$= 1334 \text{ Lumens}$$

Formulae such as these are found in lamp manufacturer's catalogs. These formulae are useful in determining the effect of applied voltage on lamp wattage, line current, lumen output, lumens per watt, and lamp life. A calculator that has a y^x power function key is needed to solve these equations.

Exponents.   Confused? If your calculator does not have a y^x function key, take the easy route. Just multiply the number again and again for the value of the exponents (power). Cumbersome… but it works. For example, the number 2^3 can be read "2 to the third power" or "2 raised to the third power."

$$2 \times 2 \times 2 = 8$$

Another example, the number 8^5 can be read "eight to the fifth power" or "eight raised to the fifth power." We simply multiply that number 8 by itself five times.

$$8 \times 8 \times 8 \times 8 \times 8 = 32,768$$

Some simple exponents can be read in a certain way for example a^2 is usually read as "a squared" and a^3 as "a cubed."

Figure 7-16 is a graph showing the affect of different voltages on lamp life and lumen output.

Keep It Simple!

The formulae for lamp life and lumen output are complicated, but here is a quick summary:

- An incandescent lamp operating at a voltage *less* than its rated voltage will last longer but will not burn as bright as it should.

- An incandescent lamp operating at a voltage *greater* than its rated voltage will not last as long but will burn brighter.

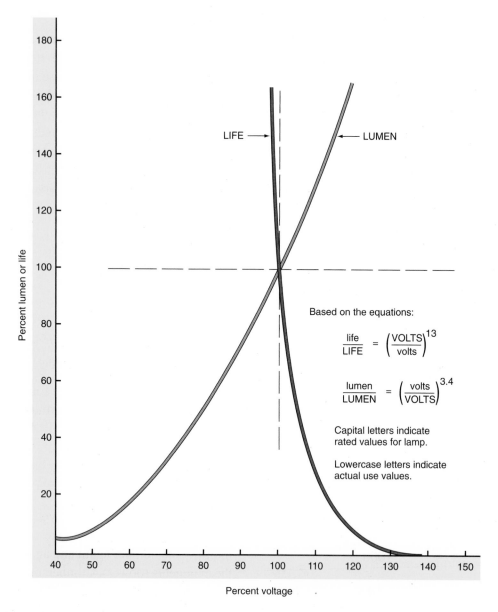

FIGURE 7-16 Typical operating characteristics of an incandescent lamp. (*Delmar/Cengage Learning*)

LAMP EFFICACY

A lumen is a measurement of visible light output from a lamp.

One lumen on 1 ft² of surface produces 1 *foot-candle*.

Another term you will hear in lighting is *efficacy*. Efficacy is the ratio of light output from a lamp to the electric power it consumes and is measured in *lumens per watt* (lm/w). In other words, efficacy is a measurement of input to output. Examples:

- Lamp #1: 26-watt compact fluorescent lamp (CFL) produces 1700 lumens at rated voltage, which equates to 65 lm/w.

- Lamp #2: 100-watt incandescent lamp produces 1200 lumens at rated voltage, which equates to 12 lm/w.

- Lamp #3: 100-watt lamp incandescent produces 1750 lumens at rated voltage, which equates to 17.5 lm/w.

The more lumens per watt, the greater the efficacy of the lamp. See Table 7-2.

It is obvious that typical incandescent lamps are very inefficient compared to compact fluorescent lamps. There is a dramatic trend toward the use of CFL lamps: more light for less energy, and less energy means lower electric bills!

LAMP COLOR TEMPERATURE

Lamp color temperature is rated in Kelvin degrees, and the term is used to describe the "whiteness" of the lamp light. In incandescent lamps, color temperature is related to the physical temperature of the filament. In fluorescent lamps where no hot filament is involved, color temperature is related to the light as though the fluorescent discharge is operating at a given color temperature. The lower the Kelvin degrees, the "warmer" the color tone. Conversely, the higher the Kelvin degrees, the "cooler" the color tone.

Incandescent lamps provide pleasant color tones, bringing out the warm red flesh tones similar to those of natural light. This is particularly true for the "soft" and "natural" white lamps.

Tungsten filament halogen lamps have a gas filling and an inner coating that reflects heat. This keeps the filament hot with less electricity. Their light output is "whiter." They are more expensive than the standard incandescent lamp.

Fluorescent lamps are available in a wide range of "coolness" to "warmth." Warm fluorescent lamps bring out the red tones. Cool fluorescent lamps tend to give a person's skin a pale appearance.

Fluorescent lamps might be marked daylight D (very cool), cool white CW (cool), white W (moderate), warm white WW (warm). These categories break down further into a deluxe X series (i.e., deluxe warm white—deluxe cool white), specification SP series, and specification deluxe SPX series.

Typical color temperature ratings for lamps are 2800K (incandescent), 3000K (halogen), 4100K (cool white fluorescent), and 5000K (fluorescent that simulates daylight). Note that a halogen lamp is "whiter" than a typical incandescent lamp.

Catalogs from lamp manufacturers provide detailed information about lamp characteristics.

Fluorescent lamps and ballasts are a moving target. In recent years, there have been dramatic improvements in both lamps and electronic ballast efficiency. First, the now-antiquated T12 fluorescent lamps (40 watts) were replaced by energy-saving T8 fluorescent lamps. These original T8 lamps are becoming a thing of the past. The latest T8 high-efficiency, energy-saving (25 watts vs. 32 watts) lamps have an expected 50% longer life than the original T8 lamps.

The newer T8 lamps use approximately 40% less energy than the older T12 lamps. At $0.06 per kWh, one manufacturer claims a savings of $27.00 per lamp over the life (30,000 hours) of the lamp. At $0.10 cents per kWh, the savings is said to be $45.00 per lamp over the life of the lamp. Using the newer T8 lamps on new installations and as replacements for existing installations makes the payback time pretty attractive. One electronic ballast can operate up to four lamps, whereas the older style magnetic ballast could operate only two lamps. For a three- or four-lamp luminaire, one ballast instead of two results in quite a saving. Some electronic ballasts can operate six lamps.

Hard to believe! You now can have reduced power consumption and increased light output using electronic ballasts.

Today's high-efficiency ballasts are available with efficiencies of from 98% to 99%.

The only way you can stay on top of these rapid improvements is to check out the Web sites of the various lamp and ballast manufacturers.

Today's magnetic and electronic ballasts handle most of the fluorescent lamp types sold, including standard and energy-saving preheat, rapid start, slimline, high output, and very high output. Again, check the label on the ballast.

LED Lighting

Light-emitting diode (LED) is pronounced "ell-eee-dee."

Reduce the electric bill! Save energy! Reduce energy consumption! Reduce the air-conditioning load!

It has been said that the incandescent bulb is from the dinosaur age, having been around since Thomas Edison applied for a patent on May 4, 1880.

Coming on strong is a new concept for lighting that uses light-emitting diodes (LEDs) as its source of light. It is called "solid-state lighting." LED lighting has very low power consumption. Electricians had better get ready for this new type of light source in luminaires. Although not quite yet here for general lighting of a room, that will come as the light output of LEDs increases.

LEDs are solid-state devices that have been around since 1962. When connected to a dc source, the electrons in the LED smash together, creating light. Think of an LED as a tiny light bulb, but with no filament. Figure 7-17(A) is a typical LED that luminaire manufacturers can cluster in their luminaires to obtain the amount of light output they are looking for. Figure 7-17(B) is a high light output LED that lighting manufacturers can use in their luminaires. Figure 7-18 shows two LED-powered luminaires.

Today, LEDs are all around us. They are commonly recognized by the tiny white, red, yellow, green, purple, orange, and blue lights found in the digital displays in TVs, radios, DVD and CD players, remotes, computers, printers, fax machines, telephones, answering machines, Christmas light strings, night lights, "locator" switches, traffic lights, digital clocks, meters, testers, tail lights on automobiles, strobe lights, occupation sensors, and other electronic devices, equipment, and appliances.

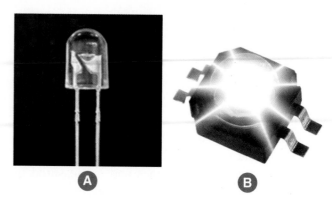

FIGURE 7-17 Some types of LEDs that luminaire manufacturers can use in their product. (A) is a typical single light-emitting diode (LED). (B) is a high light output LED. (*Courtesy of Philips Lumileds*)

FIGURE 7-18 LED-powered luminaires. (*Courtesy of Hubbell Lighting Outdoor & Industrial*)

LEDs for lighting are a rather recent concept. Because the lumens per watt in LEDs are on the increase, it is now making sense to use LEDs in luminaires.

Individual LEDs are rather small. Putting a cluster (called an array) of LEDs together (i.e., 5, 20, 30, 60, 120) produces a lot of light. The result is an

LED bulb, usable in a luminaire the same as a typical medium Edison-base lampholder incandescent bulb. Figure 7-18 shows three different styles of LED lamps.

LED Luminaires and the *NEC*

The *2008 NEC* recognized LED lighting for the first time in *Article 410*, specifically *410.16(A) (3)*, *410.16(C)(1)*, and *410.16(C)(3)*. Other *NEC* references include *410.24*, *410.68*, *410.74*, *410.116*, *410.136*, and *410.137*. LED lighting is beginning to appear in exit, recessed, surface, and under-cabinet luminaires; desk lamps; wall luminaires; down lights; stop-and-go lights; to name a few. LEDs in flashlights have been around for quite a while.

Lamp manufacturers have come out with many different types of lamps for accent, task, conventional shape, flood, spots, and so on. They are available with the standard Edison base and candelabra base to replace existing incandescent lamps. The light is white, but with different LEDs other colors are available. See Figure 7-19 for an assortment of LED lamps.

FIGURE 7-19   An assortment of light bulbs (lamps) powered by a number of individual light-emitting diodes (LEDs). These lamps screw into a standard medium Edison-base lampholder. (*Delmar/Cengage Learning*)

Today's LEDs last 60,000 to 100,000 hours. They produce virtually no heat, have no filament to burn out, can withstand vibration and rough usage, contain no mercury, operate better when cool, lose life and lumen output at extremely high temperatures, and can operate at temperatures as low as $-40°F$ ($-40°C$).

The lumen output of LEDs slowly declines over time. The decline varies. The industry seems to be settling on a lumen output rating of 70% after 50,000 hours of use.

LEDs use a tiny amount of electricity. Virtually all of the power consumption converts to light, whereas for an incandescent bulb, 5% of the power consumption converts to light and the rest converts to heat. The purpose of lighting is to have illumination—not heat. Heat is a waste.

Some LEDs may be dimmed, whereas others may not. Check with the manufacturer of the lamp (bulb) to verify the dimming or no dimming capability.

LEDs start instantly with no flickering. LEDs put out directional light as compared to the conventional incandescent lamp that shoots light out in almost all directions.

A study was recently made to compare an LED's predicted life of 60,000 hours (that's almost 7 years of continuous burning, or 21 years at 8 hours per day of usage) to a standard 60-watt incandescent lamp that has a rated life of 1000 hours. Over the 60,000 hour life:

- 60 standard incandescent lamps would be used compared to one LED bulb.

- the standard incandescent lamps would use 3600 kilowatt-hours, resulting in the cost of electricity at $360. The LED lamp would use 120 kilowatt-hours, resulting in a cost of electricity at $12.

- the total cost (lamps and cost of electricity) for the incandescent lamps was $400 compared to the cost of operating the LED lamp, which was $47.

Today, the lumen output of LED lamps is similar to CFLs, producing approximately 50 lm/w. This is expected to improve to 150 lm/w by 2010. Compare the LED lumen output to a typical incandescent lamp that has a lumen output of 14-18 lm/w. That's a significant increase in lumens per watt!

Things Are Changing Fast

Recently announced is a line of LED lamps that are direct replacements for the conventional 40-watt fluorescent lamp. Nothing has to be done other than replace the existing fluorescent lamp with an LED lamp. They work on both magnetic and electronic ballasts. When compared to a conventional 40-watt fluorescent lamp, the LED replacement lamp typically has a 10-year life as opposed to a 2- or 3-year life—their power consumption is 20% less and their lumen output is comparable or slightly greater—and they can operate at 32°F (0°C). These LED lamps are currently available in warm white, cool white, daylight, neutral, and bright white.

LED lighting is accelerating at a rapid pace. The *NEC* Code-Making Panels will be seeing more and more proposals for changes in the *NEC* relating to LED-type luminaires. Keep an eye out for these changes. As with all electrical equipment, carefully read the label on the luminaire to be sure your installation "meets Code."

For more information about LEDs and LED lighting, check out the Web site of the LED industry: http://www.ledsmagazine.com.

Outdoor Lighting

Before installing outdoor lighting, check with the local electrical inspector and/or building official to find out whether there are any restrictions regarding outdoor lighting.

A virtually unlimited array of outdoor luminaires is available that provide uplighting, downlighting, diffused lighting, moonlighting, shadow and texture lighting, accent lighting, silhouette lighting, and bounce lighting. After the luminaires are selected, the type and color of the lamp is then selected. Some luminaires have specific light "cutoff" data that are useful in determining whether the emitting light will spill over onto the neighbor's property. The method of control must also be considered. Switch control, timer control, dusk-to-dawn control, and motion sensors are ways of turning outdoor lighting on and off.

In recent years, more and more complaints are coming from neighbors claiming that they are being bothered by glare, brightness, and light spillover from their neighbor's outdoor luminaires. Security lighting, yard flood lights, driveway lighting, "moonlighting" in trees, and shrub lighting are examples of sources of light that might cross over the property line and be a "nuisance" to the next door neighbor.

Outdoor "lightscape" lighting considered by a homeowner to be aesthetically wonderful might be offensive to the neighbor. Nuisance lighting is also referred to as *light pollution, trespassing, intrusion, glare, spillover,* and *brightness.* Some quiet residential neighborhoods are beginning to look like commercial areas, used car lots, and airport runways because no restrictions govern outdoor lighting methods. This is not a safety issue and is not addressed in the *NEC*. However, the issue might be found in local building codes.

Many communities are being forced to legislate strict outdoor lighting laws, specifying various restrictions on the location, type, size, wattage, and/or footcandles for outdoor luminaires. Checking building codes in your area might reveal requirements such as these:

Light Source: The source of light (the lamp) must not be seen directly.

Glare: Glare, whether direct or reflected, such as from floodlights, and as differentiated from general illuminations, shall not be visible at any property line.

Exterior Lighting: Any lights used for exterior illuminations shall direct light away from adjoining properties.

The International Dark-Sky Association has a lot of information regarding light pollution issues on its Web site: http://www.darksky.org.

Replacing a Luminaire in an Older Home: Will There Be Problems?

Very possible!

Older homes were wired with conductors that were insulated with rubber (Type R) or thermoplastic (Type T). Both of these older style insulations were rated for maximum 60°C. The heat generated by the lamps over the years literally baked the conductor insulation to a hard, brittle material particularly is the luminaires were over-lamped.

If you aren't careful, pulling on or moving these conductors can easily break off the conductor insulation, which will result in a hazardous short-circuit or ground-fault situation.

Handle with care! Being very careful in handling the conductors while making up the splices and mounting the new luminaire—possibly sliding some readily available tubing-like electrical insulation—or taping over the conductor insulation with plastic electrical tape might be all that is needed.

Many modern luminaires are marked by the manufacturer with a requirement for wiring the luminaire with either 75°C or 90°C supply conductors. You may need to do some rewiring using the latest in nonmetallic-sheathed cables Type NM-B, Type NMC-B, or Type NMS-B. If the conductors are in a raceway, replacing the conductors with Type THHN might be a good idea. All of these conductors have a maximum temperature rating of 90°C.

Another common correction method is shown in Figure 7-20. Install a junction box approximately 24 inches from the location of the new luminaire. Connect a wiring method having conductors with an insulation system rated for the marked supply temperature from the junction box to the luminaire. It is

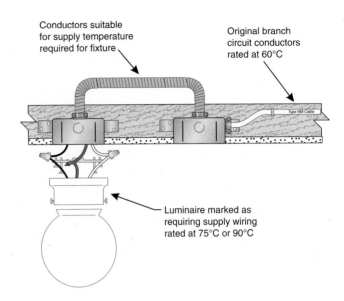

FIGURE 7-20 Installing luminaire requiring higher temperature supply conductors on a lower-rated older wiring system. (*Delmar/Cengage Learning*)

best if the junction box is located in the attic or other accessible location. Otherwise, install a blank canopy over the junction box if it is exposed on the ceiling.

More on this subject is found in Chapter 4 in the nonmetallic-sheathed cable section.

REVIEW

1. Is it permissible to install a recessed luminaire directly against wood ceiling joists when the label on the luminaire does not indicate that the luminaire is suitable for insulation to be in direct contact with the luminaire? This is a Type Non-IC fixture.

   _____

2. If a recessed luminaire without an integral junction box is installed, what extra wiring must be provided? _____

   _____

   _____

   _____

   _____

   _____

   _____

   _____

   _____

3. Thermal insulation shall not be installed within _____ in. (mm) of the top or _____ in. (mm) of the side of a recessed luminaire unless the luminaire is identified for use in direct contact with thermal insulation. This is a Type Non-IC fixture.

4. Recessed luminaires are available for installation in direct contact with thermal insulation. These luminaires bear the UL mark "Type _____."

5. Unless specifically designed, all recessed incandescent luminaires must be provided with factory-installed _____.

6. Plans require the installation of surface-mounted fluorescent luminaires on the ceiling of a recreation room that is finished with low-density ceiling fiberboard. What sort of mark would you look for on the label of the luminaire? _____

   _____

   _____

   _____

7. A recessed luminaire bears no marking indicating that it is "Identified for Through-Wiring." Is it permitted to run branch-circuit conductors *other* than the conductors that supply the luminaire through the integral junction box on the luminaire? _____

   _____

8. Fluorescent ballasts for all indoor applications must be _____ type. These ballasts contain internal _____ protection to protect against overheating.

9. Additional backup protection for ballasts can be provided by connecting a(an) _____ with the proper size fuse as recommended by the ballast manufacturer.

10. You are called upon to install a number of luminaires in a suspended ceiling. The ceiling will be dropped approximately 8 in. (200 mm) from the ceiling joists. Briefly explain how you might go about wiring these luminaires. _____

    _____

    _____

    _____

    _____

    _____

    _____

    _____

    _____

11. The *Code* places a maximum open-circuit voltage on lighting equipment in or on homes. This maximum voltage is (600) (750) (1000). (Circle the correct answer.) Where in the *NEC* is this voltage maximum referenced?

    _____

    _____

12. The letter "E" in a circle on a ballast nameplate indicates that the ballast _____
_____
_____

13. A 120-volt lamp fluorescent ballast for two 40-watt lamps is marked 85 volt-amperes. What is the power factor of the ballast? _____
_____
_____

14. Can an incandescent lamp dimmer be used to control a fluorescent lamp load?
_____

15. A good "rule-of-thumb" to estimate the expected life of a motor, a ballast, or other electrical equipment is that for every _____ °C above rated temperature, the expected life will be cut in _____ .

16. A post light has a 120-volt, 60-watt lamp installed. The lamp has an expected lamp life of 1000 hours. The homeowner installed a dimmer ahead of the post light and leaves the dimmer set so the output voltage is 100 volts. The lamp burns slightly dimmer when operated at 100 volts, but this is not a problem. What is the expected lamp life when operated at 100 volts? _____
_____
_____
_____

17. Does your community have exterior outdoor lighting restrictions?
If yes, what are they? _____
_____
_____

18. Define the following terms.
    a. A *tap conductor* is _____
    _____

    b. A *fixture whip* is _____
    _____

19. Circle the correct answer for the following statements:
    a. Always match a fluorescent lamp's wattage and type designation to the type of ballast in the luminaire. (True) (False)
    b. It really makes no difference what type of fluorescent lamp is used, just so the wattage is the same as marked on the ballast. (True) (False)

20. Circle the correct answer for the following statements.
    a. When selecting a trim for a recessed incandescent luminaire, select any trim that physically fits and can be attached to the luminaire. (True) (False)
    b. When selecting a trim for a recessed incandescent luminaire, select a trim that the manufacturer indicates may be used with that luminaire. (True) (False)

21. Have you installed LED lighting? Add comments on your experience involving LED lighting. _____
_____

Lighting Branch Circuit for the Front Bedroom

OBJECTIVES

After studying this chapter, you should be able to

- understand the meaning of general, accent, task, and security lighting.

- estimate loads for the outlets connected to a branch circuit.

- determine how many receptacles to connect to a branch circuit.

- determine how many branch circuits are needed.

- draw a cable layout and wiring diagram for a branch circuit.

- properly size outlet boxes based on the number of conductors and devices.

- understand the *NEC* requirements for luminaires in clothes closets.

In the first seven chapters of this text, you studied the *National Electrical Code (NEC)* requirements and the basics of house wiring. You will now apply all that you have learned to an actual house—room by room—circuit by circuit. You will use the text, the plans, the specifications that are in the back of this book, and the *NEC* to "bring it all together."

The home used in this text is a typical home—nothing prestigious or elegant. Other than the size of the service conductors and equipment, more branch circuits, and more gadgets, the *NEC* requirements are the same.

This might be a good time for you to refresh your memory about the arc-fault circuit-interrupter (AFCI) requirements for bedrooms. All 120-volt, single-phase, 15- and 20-ampere outlets in bedrooms require AFCI protection, covered in Chapter 6.

RESIDENTIAL LIGHTING

Residential lighting is a personal thing. The homeowner, builder, and electrical contractor must meet to decide on what types of luminaires are to be installed in the residence. Many variables (cost, personal preference, construction obstacles, etc.) must be taken into consideration.

Residential lighting can be segmented into four groups.

General Lighting

General lighting often is referred to as ambient lighting.

General lighting provides overall illumination for a given area, such as the front hall, the bedroom hall, the workshop, or the garage lighting in this residence. General lighting can be very basic, which means just getting the job done by providing adequate lighting for the area involved. Or, it can take the form of decorative lighting, such as a chandelier over a dining room table or other decorative luminaires that can be attached to ceiling paddle fans.

Lighting in hallways and stairways must not only be thought of as attractive general lighting, but also as safety lighting. These areas must have good lighting.

Accent Lighting

Accent lighting provides focus and attention to an object or area in the home. Examples are the recessed "spots" over the fireplace. Another example is the track lighting in the Living Room of this residence, which can accent a picture, photo, painting, or sculpture that might be hung on the wall. To be effective, accent lighting should be at least five times that of the surrounding general lighting.

Task Lighting

This is sometimes referred to as "activity" lighting. Task lighting provides proper lighting where "tasks" or "activities" are performed. The fluorescent luminaires above the workbench, the kitchen range hood, and the recessed luminaire over the kitchen sink are all examples of "task" lighting. To avoid too much contrast, task lighting should not exceed three times that of the surrounding general lighting.

Wall Washing. The term *wall washing* is used when luminaires are installed in a row, parallel to the wall that is to be "washed" with light. For example, if you wish to "wash" a 12-ft (3.7-m) wall, as illustrated in Figure 8-1, and you would like the luminaires to be 2 ft (600 mm) from the wall:

$$\frac{12}{2} = 6 \text{ liminaires recommended}$$

Security Lighting

Security lighting generally includes outdoor lighting, such as post lamps, wall luminaires, walkway lighting, and all other lighting that serves the purpose of providing lighting for security and safety reasons. Security lighting in many instances is provided by the normal types of luminaires found in the typical residence. In the residence discussed in this text, the outdoor bracket luminaires in front of the garage and next to the entry doors, as well as the post light, might be considered to be security lighting even though they add to the beauty of the residence. Security lighting can be controlled manually with regular wall switches, timers, light sensors, or motion detectors.

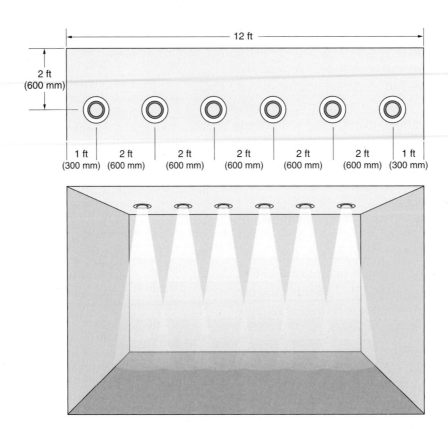

FIGURE 8-1 Illustration of how recessed luminaires can be installed in the ceiling close to a wall to provide beautiful "wall washing." (*Delmar/Cengage Learning*)

American Lighting Association

The American Lighting Association offers excellent residential lighting recommendations both online and hardcopy brochures. If you intend to become competent in residential lighting, a visit to its Web site at http://www.americanlightingassoc.com is a must!

This organization is comprised of many manufacturers of luminaires and electrical distributors that maintain extensive lighting showrooms across the country. A visit to one of these showrooms offers the prospective buyer an opportunity to select from thousands of luminaires. Most of the better-quality lighting showrooms are staffed with certified lighting consultants (CLCs), lighting specialists (LSs), and/or lighting associates (LAs) who are highly qualified to make recommendations to homeowners so that they will get the most value for their money.

Find copies of lighting manufacturers' catalogs. Visit their Web sites. They offer many lighting technique suggestions.

The lighting provided throughout this residence conforms to the residential lighting recommendations of the American Lighting Association. Certainly many other variations are possible. For the purposes of studying an entire wiring installation for this residence, all of the load calculations, wiring diagrams, and so on, have been accomplished using the luminaire selection as indicated on the plans and in the specifications.

LAYING OUT GENERAL-PURPOSE LIGHTING AND RECEPTACLE CIRCUITS

In most residential installations, circuit layout is usually done by the electrician, always keeping in mind the *NEC* requirements. In larger, more costly residences, the circuit design might be specified by the architect and included on the plans. Many circuit arrangements are possible.

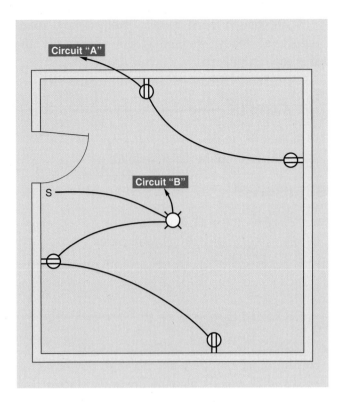

FIGURE 8-2 Wiring layout showing one room that is fed by two different circuits. If one of the circuits goes out, the other circuit will still provide electricity to the room.
(*Delmar/Cengage Learning*)

Here are a few thoughts:

• Supply general-purpose receptacle and lighting outlets with more than one circuit. Mix it up. If one circuit has a problem, the second circuit continues to supply power to the other outlets in that room, as shown in Figure 8-2.

• Keep general-purpose receptacles on one circuit and lighting outlets on another branch circuit.

• Economize on wiring materials by connecting receptacle outlets back to back, as shown in Figure 8-3.

• Run the branch circuit first to the switch then to the controlled lighting outlet. This keeps the outlet box at the lighting outlet or the wiring compartment on recessed cans less crowded. Having the grounded circuit conductor present at the switch location makes it easy to install switching devices that require a connection to the grounded circuit conductor. ▶Doing so satisfies the rule in *404.2(C)*.◀

• Install a receptacle below the room wall switch—an area never blocked by furniture.

• Think about where furniture might be placed. Try to locate receptacles to the side of these so as to be easily accessible after the furniture is in place. Don't let the 6-ft (1.8-m) receptacle requirement hinder your thinking.

• Although not a *Code* issue, it is poor practice to include outlets on different floors on the same circuit. Some local building codes limit this type of installation to lights at the head and foot of a stairway.

Wiring bedrooms became a little more complicated when AFCIs appeared on the scene, as

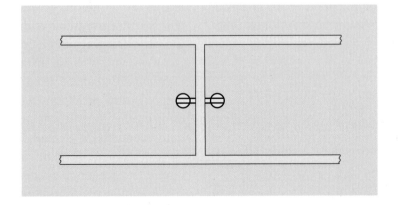

FIGURE 8-3 Receptacle outlets connected back to back. This can reduce the cost of the installation because of the short distance between the outlets. Do not do this in fire-rated walls unless special installation procedures are followed as illustrated in Figures 2-29, 2-30, and 2-31.
(*Delmar/Cengage Learning*)

required by *210.12*. The dilemma is putting all bedroom receptacles and lighting outlets on one AFCI-protected branch circuit, or connecting the receptacle and lighting outlets on different branch circuits, which means that more than one AFCI circuit breaker will be needed.

The number of receptacle and lighting outlets connected to one branch circuit is discussed a little later.

Cable Runs

A student studying the great number of wiring diagrams throughout this text will find many different ways to run the cables and make up the circuit connections. Sizing of boxes that will conform to the *Code* for the number of wires and devices is covered in great detail in this text. When a circuit is run "through" a recessed luminaire like the type shown in Figure 7-8, that recessed luminaire must be identified for "through-wiring." This subject is discussed in detail in Chapter 7. When in doubt as to whether the luminaire is suitable for running wires to and beyond it to another luminaire or part of the circuit, it is recommended that the circuitry be designed so as to end up at the recessed luminaire with only the two wires that will connect to that luminaire.

 ESTIMATING LOADS FOR OUTLETS

Although this text is not intended to be a basic electrical theory text, it should be mentioned that, when calculating loads,

$$\text{Volts} \times \text{Amperes} = \text{Volt-amperes}$$

Yet, many times we say that

$$\text{Volts} \times \text{Amperes} = \text{Watts}$$

What we really mean to say is that

$$\text{Volts} \times \text{Amperes} \times \text{Power factor} = \text{Watts}$$

In a pure resistive load such as a simple light bulb, a toaster, an iron, or a resistance electric heating element, the power factor is 100%. Then

$$\text{Volts} \times \text{Amperes} \times 1 = \text{Watts}$$

With transformers, motors, ballasts, and other "inductive" loads, wattage is not necessarily the same as volt-amperes.

 EXAMPLE

Calculate the wattage and volt-amperes of a 120-volt, 10-ampere resistive load.

Solutions:

a. $120 \times 10 \times 1 = 1200$ watts
b. $120 \times 10 = 1200$ volt-amperes

EXAMPLE

A 120-volt fluorescent ballast has an input current of 0.34 ampere and an input power rating of 22 watts. Calculate the power factor of the ballast.

Solution:

$$\text{Power Facter} = \frac{\text{Watts}}{\text{Volt-amperes}}$$

$$= \frac{22}{120 \times 0.34} = 0.54 \ (54\% \ \text{PF})$$

Therefore, to be sure that adequate ampacity is provided for in branch-circuit wiring, feeder sizing, and service-entrance calculations, the *Code* requires that we use the term *volt-amperes*. This allows us to ignore power factor and address the *true* current draw that will enable us to determine the correct ratings of electrical equipment.

However, in some instances the terms *watts* and *volt-amperes* can be used interchangeably without creating any problems. For instance, *220.55* recognizes that for electric ranges and other cooking equipment, the kVA and kW ratings shall be considered to be equivalent for the purpose of branch-circuit and feeder calculations.

A similar permission of equivalency of kilovolt-amperes (kVA) and kilowatts (kW) is given in *220.54* for electric clothes dryers.

The load calculation examples in *Annex D* of the *NEC* show kilowatt values in the list of connected loads, but the actual calculations are done using volt-ampere values. Throughout this text, *wattage* and *volt-amperes* are used when calculating and/or estimating loads.

For the residence in the plans, it is shown that six lighting circuits meet the minimum standards set by the *Code*. However, to provide sufficient capacity, 13 lighting circuits are to be installed in this residence.

How Many Outlets Are Permitted on One General-Purpose Lighting Branch Circuit?

This is a tough question, but it can be answered reasonably well using a few simple "rules of thumb" discussed next.

The *NEC* does not specify the maximum number of receptacle outlets or lighting outlets that may be connected to one 120-volt lighting or small-appliance branch circuit in a residence. It may not seem logical that 10 or 20 receptacle outlets and lighting outlets could be connected to one branch circuit and not be in violation of the *Code*. On our side is the fact that there is much load diversity in residential occupancies. We have no idea what will be plugged into the receptacles. We do know that rarely, if ever, will all receptacle outlets and lighting outlets be used at the same time. Having many "convenience" receptacles is safer because more receptacle outlets virtually eliminate or minimize the use of extension cords, one of the leading causes of electrical fires.

Recessed and surface-mounted incandescent luminaires are marked with their maximum lamp wattage. Fluorescent luminaires are marked with lamp wattage, and ballasts are marked in volt-amperes. In new construction, other than for recessed luminaires, we probably do not know what type of surface or hanging luminaires will be selected. Most luminaire manufacturers' catalogs provide excellent recommendations for lamp wattages for residential applications.

There are many items such as nightlights (1 to 7½ watts), smoke detectors (5 watts), clocks (5 watts), carbon monoxide detector (60 milliamperes), and clothes closet lights (60 to 75 watts) that use only a small portion of the branch circuit's capability. It

comes down to approximating, estimating, experience, and common sense! It is not an exact science.

Circuit Loading "Rules of Thumb"

Let's look at a few favorite "rules of thumb" that electricians use for determining how many outlets they connect to one general-purpose branch circuit.

Outlets per Circuit Method.   This is the most common method used by electricians across the country. It is very simple. No calculations are needed. Just count the outlets! Connect 10 outlets on a circuit where you anticipate fairly heavy wattage lighting loads. Connect 15 outlets per circuit for circuits that have a lot of low-wattage lighting loads. Somewhere in-between might be the right number. This method is really estimating a load of 1 to 1½ amperes per outlet.

An example of this might be a bedroom with six or seven receptacles, one or two closet lights, and possibly a ceiling paddle fan with a light kit. This would probably be one 15-ampere lighting branch circuit with not much leeway for additional outlets.

Common sense tells us there can be more outlets when the circuit consists of low-wattage loads and fewer outlets when the circuit consists of higher wattage loads. At best, it's a "guesstimate."

How Many Circuits Do You Need Using the Outlets per Circuit Method?   If we were to wire a new house that has a count of 80 general-purpose lighting and receptacle outlets, how many 15-ampere lighting branch circuits do we need?

Where you anticipate somewhat high-wattage loads, figure on connecting 10 outlets per circuit.

$$\frac{80}{10} = 8 \text{ (a minimum of eight 15-ampere general-purpose lighting circuits)}$$

Where you anticipate low-wattage loads, figure on connecting 15 outlets per circuit.

$$\frac{80}{15} = 5.3 \text{ (a minimum of six 15-ampere general-purpose lighting circuits)}$$

Certainly more circuits than the results of the preceding calculations would be better.

How Many Circuits Do You Need Using per Square Foot Method? This method is more complicated because it involves calculations. In Chapter 3, we discussed how typical 15-ampere lighting branch circuits in homes could be figured at one 15-ampere branch circuit for every 600 ft² (55.8 m²). Although not often used in residential wiring, if you choose to install 20-ampere lighting branch circuits, figure one 20-ampere branch circuit for every 800 ft² (74.4 m²). This results in a good approximation for the minimum number of lighting branch circuits needed and is based on 3 volt-amperes per square foot (33 volt-amperes per square meter of floor area).

A 2400 ft² house would have a minimum of

$$\frac{2400}{600} = \text{four 15-ampere general-purpose lighting branch circuit}$$

With so many electronic and electrical appliances in homes today, this method might prove to be inadequate. For this size house, you would probably want to install more than four 15-ampere general-purpose lighting branch circuits. It is better to be safe than sorry.

The 80% Rule. This method is also complicated because it involves calculations. Add up the wattages and/or the amperes of the loads you anticipate will be connected to a given branch circuit. Then limit the load to not more than 80% of the branch circuit's rating. Although this is a *Code* requirement for continuous loads for commercial and industrial applications as found in *NEC 210.19(A)*, *210.20(A)*, and *215.2(A)*, *215.3*, *230.42(A)*, it makes good sense to follow this guideline for most residential loads.

A 15-ampere, 120-volt lighting branch circuit would be calculated as

$$15 \times 0.80 = 12 \text{ amperes maximum connected load}$$

or

$$12 \text{ amperes} \times 120 \text{ volts} = 1440 \text{ watts (volt-amperes) maximum connected load}$$

A 20-ampere, 120-volt lighting branch circuit would be calculated as

$$20 \times 0.80 = 16 \text{ amperes (volt-amperes) maximum connected load}$$

or

$$16 \text{ amperes} \times 120 \text{ volts} = 1920 \text{ watts (volt-amperes) maximum connected load}$$

Wrap Up. In this residence, estimated loads for general-purpose outlets are figured at 120 volt-amperes (1 ampere). Receptacles in the garage and workshop are figured at 180 volt-amperes (1½ amperes) because these outlets will probably supply portable tools.

Obviously the 10 to 15 outlets (1 to 1½ amperes per outlet) would not apply to the required small-appliance circuits in the Kitchen, the Laundry area, Workshop, or similar areas that supply a considerable number of plug-in tools and appliances. These circuits are separate issues and are discussed later on in this text.

Divide Loads Evenly

It is mandatory that loads be divided evenly among the various circuits, *210.11(B)*. The obvious reason is to avoid overload conditions on some circuits when other circuits might be lightly loaded. This is common sense. At the Main Service, don't connect the branch circuits and feeders to result in, for instance, 120 amperes on phase A and 40 amperes on phase B. Look at the probable and/or calculated loads to attain as close a balance as possible, such as 80 amperes on each A and B phase.

SYMBOLS

The symbols used on the cable layouts in this text are the same as those found on the actual electrical plans for the residence. Refer to Figures 2-5, 2-6, 2-7, 2-8, and 2-9.

Pictorial illustrations are used on all wiring diagrams in this text to make it easy for the reader to complete the wiring diagrams, as in Figure 8-4.

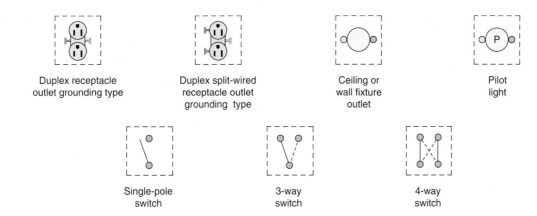

Duplex receptacle outlet grounding type

Duplex split-wired receptacle outlet grounding type

Ceiling or wall fixture outlet

Pilot light

Single-pole switch

3-way switch

4-way switch

FIGURE 8-4   Pictorial illustrations used on wiring diagrams in this text. (*Delmar/Cengage Learning*)

DRAWING A CABLE LAYOUT AND WIRING DIAGRAM

Typical house plans show the electrical symbols on the construction floor plans. Rarely are separate electrical plans provided such as those included in the back of this text. Most residential plans do not provide much detail other than showing the location of receptacles, lighting, and appliances.

For the most part, the electrician gets on the job—studies the plans then figures out how to make it work, according to *Code*, and how to do it in a cost-effective way. The electrician decides how to lay out the branch circuits, the number of outlets on a circuit, how to run the cables, and from where the branch circuit home runs will be taken.

Skilled electricians, after many years of experience, will generally not prepare detailed wiring diagrams because they have ability to "think" the connections through mentally. They can visualize the size and type of boxes and cables needed.

Most electricians sketch a Cable Layout, bearing in mind that they are limited to working with 2-wire, 3-wire, and sometimes 4-wire cables. With the requirement in *404.2(C)* for having a neutral at every switch location, how the circuit is run will determine the number of conductors required in each cable. We discussed this extensively in Chapter 5.

The following steps will guide you through the simple steps for preparing a cable layout and wiring diagram. In later chapters, you will be asked to draw cable layouts and wiring diagrams. This is an exercise in how to make a color-coded cable layout and wiring diagram. Certainly there are other ways to lay out the cable runs for this particular "typical" room.

Note that we are feeding the lighting portion of the branch circuit from the switch rather than from the lighting outlet. This allows us to comply with the rules in *404.2(C)* without using 4-conductor cable for a portion of the control circuit. That section requires a neutral conductor to be present in each switch box for possible connection of wiring devices such as timers and occupancy sensors.

DRAWING THE WIRING DIAGRAM OF A LIGHTING CIRCUIT

1. Refer to plans and make a cable layout of all lighting and receptacle outlets, as in Figure 8-5.

2. Draw a wiring diagram showing the traveler conductors for all 3-way switches, if any, as in Figure 8-6.

3. Draw a line between each switch and the outlet or outlets it controls. For 3-way switches, do this for one switch only. This is the "switch leg."

4. Draw a line from the grounded terminal in the lighting panel to each current-consuming outlet. This line may pass through switch boxes but must not be connected to any switches. This is the white grounded circuit conductor, often called the "neutral" conductor.

Note: An exception to step 4 may be made for double-pole switches. For these switches, all

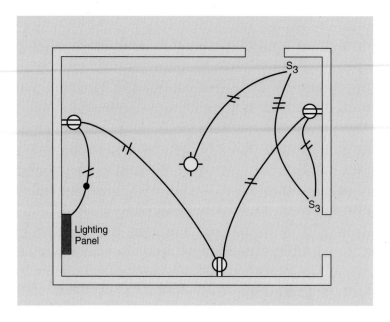

FIGURE 8-5 Typical cable layout. (*Delmar/Cengage Learning*)

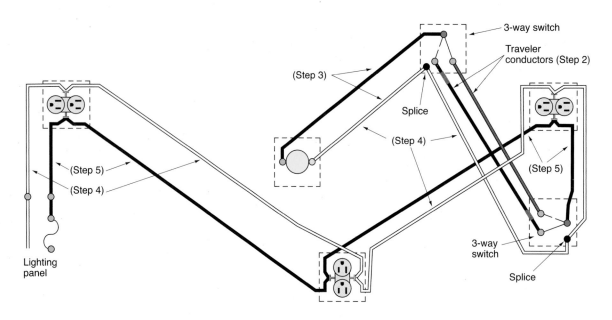

FIGURE 8-6 Wiring diagram of the circuit shown in Figure 8-5. (*Delmar/Cengage Learning*)

conductors of the circuit are opened simultaneously. They are rarely used in residential wiring.

5. Draw a line from the ungrounded "hot" terminal in the lighting panel to each switch and to each unswitched outlet. For 3-way switches, do for one switch only. This is the "feed." The "switch leg" (switch loop, switch return) is covered in step 3.

6. Show splices as small dots where the various wires are to be connected together. In the wiring diagram, the terminal of a switch or outlet may be used for the junction point of wires. In actual wiring practice, however, the *Code* does not permit more than one wire to be connected to a terminal unless the terminal is a type identified for use with

more than one conductor. The standard screw-type terminal is not acceptable for more than one wire, *110.14*.

7. The final step in preparing the wiring diagram is to mark the color of the conductors, as in Figure 8-6. Note that the colors selected—black (B), white (W), and red (R)—are the colors of 2- and 3-conductor cables (refer to Chapter 5). It is suggested that the reader use colored pencils or markers for different conductors when drawing the wiring diagram.

Use a fine line for the white conductor.

General comment: All cable layouts in this text are but one way of laying out the circuits. Certainly other layouts are possible. Some circuits were designed to give the student practice in hooking up similar circuits differently.

LIGHTING BRANCH CIRCUIT A16 FOR THE FRONT BEDROOM

The Front Bedroom is the beginning of our journey to study the wiring of a typical dwelling. Circuit A16 is a 15-ampere branch circuit. This branch circuit involves two key issues: (1) the requirement for GFCI protection dealt with in *210.8(A)(3)* for the receptacle located outside the front of the house, and (2) the requirement for AFCI protection dealt with in *210.12(A)*. In accordance with *210.12(A)*, arc-fault circuit protection for this circuit is provided by a 15-ampere AFCI circuit breaker in Panel A. GFCIs and AFCIs are discussed in detail in Chapter 6. As we look at this circuit, we find five split-wired or half-switched receptacle outlets. These will provide the general lighting through the use of table or other lamps that will be plugged into the switched receptacles, as shown in Figure 8-7.

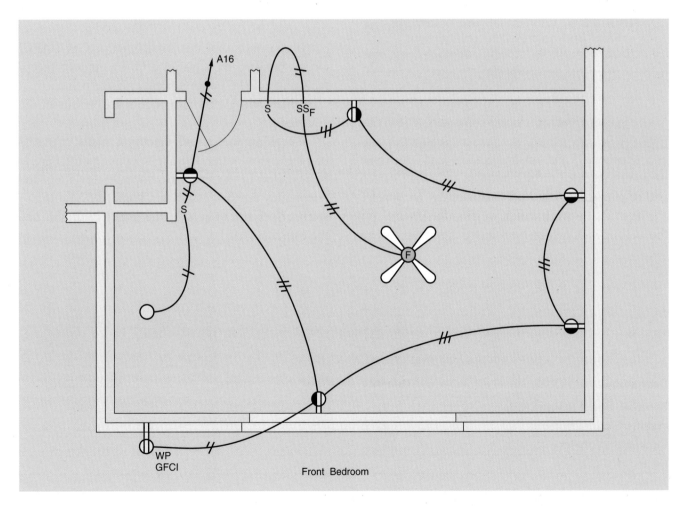

FIGURE 8-7 Cable layout for Front Bedroom circuit, A16. (*Delmar/Cengage Learning*)

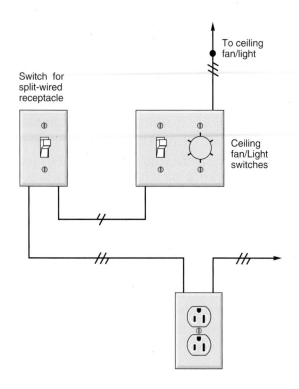

Switch for split-wired receptacle

To ceiling fan/light

Ceiling fan/Light switches

FIGURE 8-8 Conceptual view of how the Front Bedroom switch arrangement is to be accomplished. (*Delmar/Cengage Learning*)

Televisions, radios, clocks, computers, CD/DVD/ DVR/VCR/VHS players and burners, scanners, printers, telephones, and other electrical and electronic items not intended to be controlled by a wall switch will be plugged into the "hot continuously" receptacle.

Next to the wall switch we find the control for the ceiling fan, which also has a luminaire as an integral part of the fan, as shown in Figure 8-8. A single-pole switch controls the "ON–OFF" and speed of the fan motor.

The clothes closet light is controlled by a single-pole switch outside and to the right of the clothes closet door.

The outdoor weatherproof receptacle connected to this branch circuit meets the requirements of *210.52(E)*, covered in detail in Chapter 3. This receptacle is required to have GFCI protection, *210.8(A)(3)*, discussed in detail in Chapter 6.

For this particular installation, there are two ways to accomplish GFCI protection for this receptacle:

1. Install a GFCI receptacle.

2. Install a dual-function GFCI/AFCI circuit breaker in the main panel. Be careful when selecting the circuit breaker to be sure the nameplate indicates that it will provide both GFCI and AFCI protection.

If you cannot locate a dual-function GFCI/AFCI circuit breaker, then install a GFCI outdoor receptacle and an AFCI circuit breaker in the panel.

GFCI and AFCI protection are discussed in Chapter 6.

Checking the actual electrical plans, we find one television outlet and one telephone outlet are to be installed in the Front Bedroom. Televisions and telephones are discussed in Chapter 25. Table 8-1 summarizes the outlets in the Front Bedroom and the estimated load.

TABLE 8-1

Front Bedroom: outlet count and estimated load (Circuit A16).

Description	Quantity	Watts	Volt-Amperes
Receptacles @ 120 watts each	5	600	600
Weatherproof receptacle	1	120	120
Clothes closet fixture	1	75	75
Ceiling fan/light	1		
Three 50-W lamps		150	150
Fan motor (0.75 A @ 120 V)		80	90
TOTALS	8	1025	1035

DETERMINING THE SIZE OF OUTLET BOXES, DEVICE BOXES, JUNCTION BOXES, AND CONDUIT BODIES

Box fill is discussed in detail in Chapter 2. Here is more on the subject. One of the most common problems found in residential wiring are wires jammed into electrical boxes. Factors that have to be considered are the size and number of conductors, the number of wiring devices, cable clamps, luminaire studs, hickeys, and splices. *NEC 314.16* provides the rules for sizing boxes.

Experience has shown that *minimum* does not always make for easy working with the splices, conductors, wiring devices, and carefully putting them all into the box without crowding. Do not skimp on box sizing. The small additional cost of larger boxes is worth much more than the grief and added time it will take to trim out the box using *minimum* size boxes. You will be much happier with boxes that have more than the *minimum* size. When it comes to box sizing, think big—not small.

The following example shows how to calculate the *minimum* size box for a particular installation, as in Figure 8-9.

1. Add the 14 AWG circuit
    conductors:
    2 + 3 + 3 =   8
2. Add equipment grounding
    wires (count one only)   1
3. Add two conductors for
    the receptacle   _2_
    Total   11 Conductors

Note that the "pigtails" connected to the receptacle in Figure 8-9 need not be counted when determining the correct box size. *NEC 314.16(B)(1)* states, *A conductor, no part of which leaves the box, shall not be counted.*

Once the total number of conductors plus the volume count required for wiring devices, luminaire studs, hickeys, and clamps is known, refer to *Table 314.16(A)* and *(B)* and Figure 2-20 to select a box that has sufficient space. Also refer to Table 2-1 for a summary of box fill requirements.

The "pigtails" connected to the receptacle need not be counted when determining the correct box size. The *Code* in *314.16(B)(1)* states, *a conductor, no part of which leaves the box, shall not be counted.*

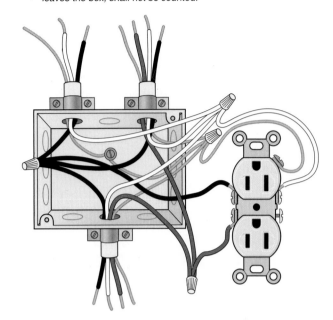

FIGURE 8-9   Determining size of box according to number of conductors. (*Delmar/Cengage Learning*)

For example, a 4 × 2⅛ in. square box with a suitable plaster ring can be used.

The volume of the box plus the space provided by plaster rings, extension rings, and raised covers may be used to determine the total available volume. In addition, it is desirable to install boxes with external cable clamps. Remember that if the box contains one or more devices such as cable clamps, fixture studs, or hickeys, the number of conductors permitted in the box shall be one less than shown in *Table 314.16(A)* for *each type* of device contained in the box.

GROUNDING OF WALL BOXES

The specifications for the residence and the *NEC* require that all metal boxes be grounded. Equipment grounding using nonmetallic-sheathed cable is accomplished by properly connecting the bare equipment grounding conductor in the cable to the metal box. An equipment grounding conductor is used *only* to ground the metal box and must *never* be used as a current-carrying conductor.

When using nonmetallic boxes, the equipment grounding conductor in nonmetallic-sheathed cable is connected to the green hexagon-shaped equipment grounding screw found on all switches and receptacles.

Improper connecting of the bare equipment grounding conductor has resulted in electrocutions!

Methods of connecting equipment grounding conductors are shown in many of the illustrations in Chapters 4 and 5 and in Figure 8-9.

Grounding-type receptacles must be installed on 15- and 20-ampere branch circuits, *406.3 (A)*.

POSITIONING OF SPLIT-WIRED RECEPTACLES

The receptacle outlets shown in the Front Bedroom are called split-wired or half-switched receptacles. The top portion of such a receptacle is "hot" at all times, and the bottom portion is controlled by the wall switch, as in Figure 8-10. It is recommended that the electrician wire the bottom half of the duplex receptacle as the switched section. As a result, when the attachment plug cap of a lamp is inserted into the bottom switched portion of the receptacle, the cord does not hang in front of the unswitched section.

When split-wired receptacles are horizontally mounted, which is common when 4-in. square boxes are used, because the plaster ring is easily fastened to the box in either a vertical or horizontal position, locate the switched portion to the right.

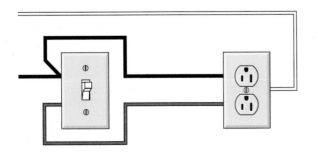

FIGURE 8-10   Split-wired or half-switched wiring for receptacles. The top receptacle is "hot" at all times. The bottom receptacle is controlled by the switch. (*Delmar/Cengage Learning*)

POSITIONING OF RECEPTACLES NEAR ELECTRIC BASEBOARD HEATING

Electric heating for a home can be accomplished with an electric furnace, electric baseboard heating units, heat pump, or resistance heating cables embedded in plastered ceilings or sandwiched between two layers of drywall sheets on the ceiling.

The important thing to remember at this point is that some baseboard electric heaters may not be permitted to be installed below a receptacle outlet, *210.52 Informational Note*.

According to *210.52*, receptacles that are part of an electric baseboard heating unit may be counted as one of the required number of receptacles in a given space, but only if the receptacles are not connected to the electric baseboard heating branch circuit.

See Chapter 23 for a detailed discussion of installing electric baseboard heating units and their relative positions below receptacle outlets.

The Front Bedroom does have a ceiling fan/light on the ceiling. Ceiling fans are discussed in Chapter 9.

LUMINAIRES IN CLOTHES CLOSETS

The *NEC* defines a *clothes closet* as *A nonhabitable room or space intended primarily for storage of garments and apparel.* Many closets in homes are obviously "clothes closets." Others might be called storage closets, linen closets, broom closets, game closets, and laundry closets. These closets become a judgment call as to how to apply the *NEC* rules.

The *NEC* does not require luminaires in clothes closets, *210.70*. However, some local codes and some building codes do. When luminaires are installed in clothes closets, they must be of the correct type and must be installed properly.

Clothing, boxes, and other material normally stored in clothes closets are a potential fire hazard. These items may ignite on contact with the hot surface of exposed lamps. Incandescent lamps have a

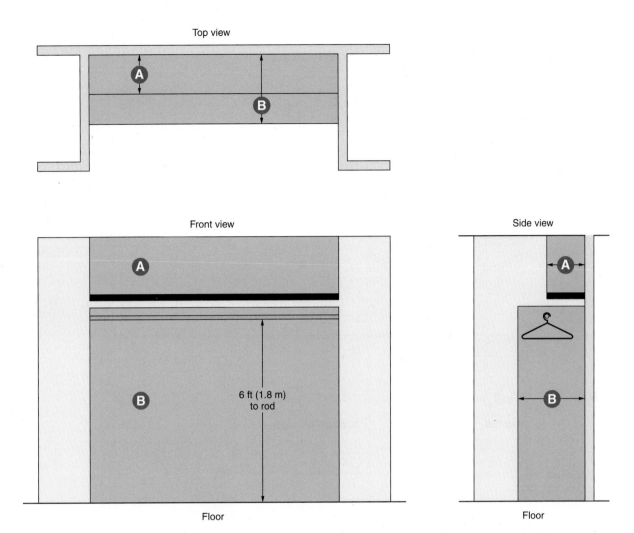

FIGURE 8-11   Typical clothes closet with one shelf and one rod. The shaded area defines "clothes closet storage space." Dimension A is width of shelf or 12 in. (300 mm) from wall, whichever is greater. Dimension B is below rod, 24 in. (600 mm) from wall; see *410.16*. (*Delmar/Cengage Learning*)

hotter surface temperature than fluorescent lamps. In *410.16*, there are very specific rules for the location and types of luminaires permitted to be installed in clothes closets. Figures 8-11, 8-12, 8-13, 8-14, and 8-15 illustrate the requirements of *410.16*.

In this residence, the closets in the bedrooms and front hall are 24 in. (600 mm) deep. The storage closet in the Recreation Room is 30 in. (750 mm) deep. The shelving is approximately 12 in. (300 mm) wide. Because of the required clearances between luminaires and the storage space, locating luminaires in closets can be difficult. In fact, some clothes closets are so small that luminaires of any kind are not permitted because of clearance requirements.

Figure 8-14 shows the possibilities: recessed incandescents, recessed fluorescents, surface-mounted incandescents, and surface-mounted fluorescents. These can be on the ceiling, or on the wall space above the clothes closet door. Whichever type is chosen, be sure that the minimum clearances are provided as required by *410.16(C)*.

Figure 8-16 shows typical bedroom-type luminaires.

Figure 8-17 shows unique luminaires referred to as *sconces*. They hug the wall and provide interesting uplighting. They are generally used to complement other lighting. They can be located just about anywhere—hallways, next to a fireplace, foyers, and similar places.

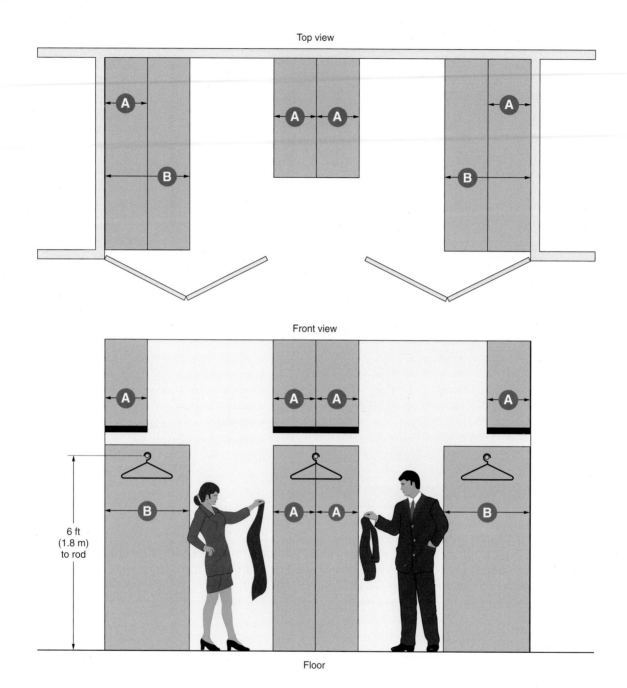

FIGURE 8-12 Large walk-in clothes closet where there is access to the center rod from both sides. The shaded area defines "clothes closet storage space." Dimension A is width of shelf or 12 in. (300 mm), whichever is greater. Dimension B is 24 in. (600 mm) from wall; see *410.16*.
(*Delmar/Cengage Learning*)

Luminaires in clothes closets

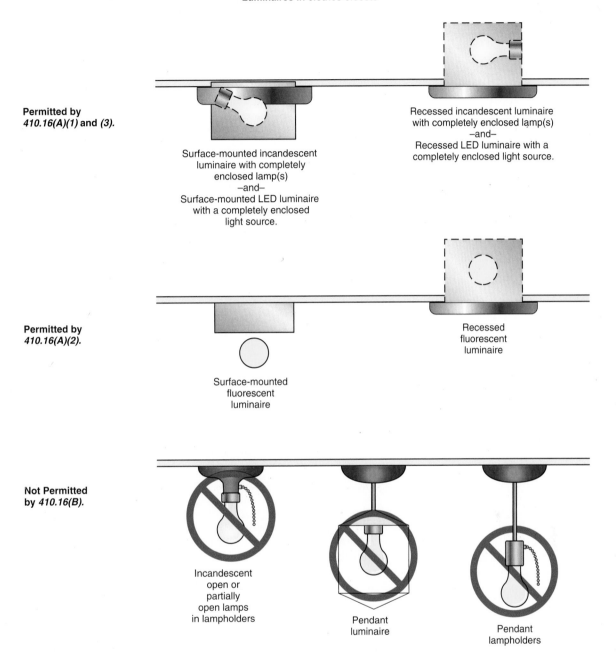

Permitted by 410.16(A)(1) and (3).

Surface-mounted incandescent luminaire with completely enclosed lamp(s)
–and–
Surface-mounted LED luminaire with a completely enclosed light source.

Recessed incandescent luminaire with completely enclosed lamp(s)
–and–
Recessed LED luminaire with a completely enclosed light source.

Permitted by 410.16(A)(2).

Surface-mounted fluorescent luminaire

Recessed fluorescent luminaire

Not Permitted by 410.16(B).

Incandescent open or partially open lamps in lampholders

Pendant luminaire

Pendant lampholders

FIGURE 8-13   Illustrations of the types of luminaires permitted in clothes closets. The *Code* does not permit bare incandescent lamps, lampholders, pendant luminaires, or pendant lampholders to be installed in clothes closets. (*Delmar/Cengage Learning*)

Location of luminaires in clothes closets

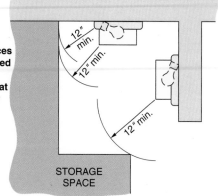

410.16(C)(1):
Minimum clearances for surface-mounted incandescent or LED luminaires that have a completely enclosed light source.

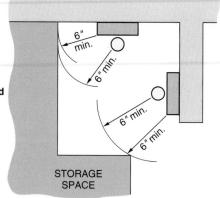

410.16(C)(2):
Minimum clearances for surface-mounted fluorescent luminaires.

410.16(C)(3) & (4):
Minimum clearances for recessed fluorescent luminaires and incandescent or LED luminaires that have a completely enclosed light source.

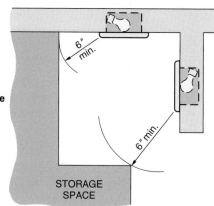

Note:
6 in. = 150 mm
12 in. = 300 mm

FIGURE 8-14 The above illustrations show the minimum clearances required between luminaires and the storage space in clothes closets. See Figure 8-11 and Figure 8-12 for a definition of storage space. *NEC 410.16(C)(3)* does permit surface-mounted fluorescent or LED luminaires to be installed within the storage space, but only if they are identified for this use. (*Delmar/Cengage Learning*)

FIGURE 8-15 Recessed incandescent closet luminaire with pull-chain switch that may be used where a separate wall switch is not installed. (*Courtesy of Progress Lighting*)

FIGURE 8-16   Typical bedroom-type luminaires. (*Courtesy of Progress Lighting*)

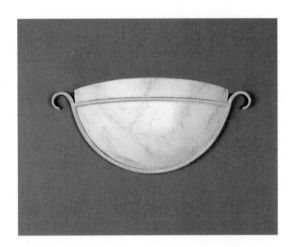

FIGURE 8-17   Sconces can provide attractive uplighting on a wall. Install approximately 66 in. (1.68 m) to the center from the floor. (*Courtesy of Progress Lighting*)

REVIEW

Note: Refer to the *Code* or the plans where necessary.

1. Can the outlets in a circuit be arranged in different groupings to obtain the same result? Why? _____

_____

_____

_____

_____

_____

2. Is it good practice to have outlets on different floors on the same circuit? Why?

_____

_____

_____

_____

_____

3. A good rule to follow is to never load a circuit to more than _____ % of the branch-circuit rating.

4. The *NEC* does not limit the number of lighting and receptacle outlets permitted on one branch circuit for residential installations. A guideline that is often used by electricians for house wiring is to allow _____ to _____ amperes per outlet.

   For a 15-ampere circuit, this results in _____ to _____ outlets on the branch circuit.

5. For residential wiring, not less than one 15-ampere lighting branch circuit should be provided for every _____ ft² of floor area. If 20-ampere lighting branch circuits are used, provide one 20-ampere lighting branch circuit for every _____ ft² of floor area.

6. For this residence, what are the estimated wattages used in determining the loading of branch circuit A16?

   Receptacles _____ watts (volt-amperes)

   Clothes closet fixture _____ watts (volt-amperes)

7. a. What is the ampere rating of circuit A16? _____

   b. What is the conductor size of circuit A16? _____

8. The *NEC* requires what type of unique protection for all 120-volt, 15- and 20-ampere branch-circuit outlets in residential homes? In which section of the *NEC* is this requirement found? _____

_____

_____

9. How many and what type of receptacles are connected to this circuit? _____

_____

_____

_____

_____

10. What main factors influence the choice of wall boxes? _____

_____

_____

_____

_____

11. How is a wall box grounded? _____

_____

_____

_____

12. What is a split-wired receptacle? _____

_____

_____

_____

13. Is the switched portion of an outlet mounted toward the top or the bottom? Why?

_____

_____

_____

_____

14. The following questions pertain to luminaires in clothes closets.

   a. Does the *Code* permit bare incandescent lamps to be installed in porcelain keyless or porcelain pull-chain lampholders? _____

   b. Does the *Code* permit bare fluorescent lamp luminaires to be installed? _____

   _____

   c. Does the *Code* permit pendant lampholders to be installed? _____

   _____

   d. What is the minimum clearance from the storage area to surface-mounted incandescent luminaires? _____

   e. What is the minimum clearance from the storage area to surface-mounted fluorescent luminaires? _____

   f. What is the minimum distance between recessed incandescent or recessed fluorescent luminaires and the storage area? _____

   g. Define the "storage area."

   _____

   _____

   _____

   _____

   _____

   _____

h. If a clothes hanging rod is installed where there is access from both sides, such as might be found in a large walk-in clothes closet, define the storage area under that rod.

_____

_____

_____

_____

15. How many switches are in the bedroom circuit, what type are they, and what do they control? _____

_____

16. The following is a layout of the lighting circuit for the Front Bedroom. Using the cable layout shown in Figure 8-7, make a complete wiring diagram of this circuit. Use colored pencils or pens.

17. When planning circuits, what common practice is followed regarding the division of loads? _____

_____

18. The *Code* uses the terms *watts, volt-amperes, kW,* and *kVA*. Explain their significance in calculating loads. _____

_____

_____

_____

_____

19. How many 14 AWG conductors are permitted in a device box that measures 3 in. $\times$ 2 in. $\times$ 2¾ in.? _____

20. A 4 in. $\times$ 1½ in. octagon box has one cable clamp and one luminaire stud. How many 14 AWG conductors are permitted? _____

_____

_____

_____

_____

CHAPTER

9

Lighting Branch Circuit for the Master Bedroom

OBJECTIVES

After studying this chapter, you should be able to

- draw the wiring diagram of the cable layout for the Master Bedroom.

- understand that AFCIs are required for bedrooms.

- study *Code* requirements for the installation of ceiling suspended (paddle) fans.

- estimate the probable connected load for a room based on the number of luminaires and outlets included in the circuit supplying the room.

- gain more practice in determining box sizing based on the number of conductors, devices, and clamps in the box.

The discussion in Chapter 3 and Chapter 8 about grouping outlets, estimating loads, selecting wall box sizes, and drawing wiring diagrams can also be applied to the circuit for the Master Bedroom.

The residence panel schedules show that the Master Bedroom is supplied by Circuit A19. Circuit A19 is a 15-ampere branch circuit. In accordance with *210.12(A)*, overcurrent protection for this circuit is provided by a 15-ampere AFCI circuit breaker in Panel A.

In Chapter 8, we discussed how GFCI and AFCI protection could be achieved in the Front Bedroom. The same principle applies to the Master Bedroom.

AFCI and GFCI requirements are discussed in detail in Chapter 6.

Because Panel A is located in the basement below the wall switches next to the sliding doors, the home run for Circuit A19 is brought into the outdoor weatherproof receptacle. This results in six conductors in the outdoor receptacle box. The home run could have been brought into the corner receptacle outlet in the bedroom. Again, it is a matter of studying the circuit to determine the best choice for conservation of cable or conduit in these runs and to economically select the correct size of wall boxes.

LIGHTING BRANCH CIRCUIT A19 FOR THE MASTER BEDROOM

Figure 9-1, Table 9-1, and the electrical plans show that Circuit A19 has four split-wired receptacle outlets in this bedroom, one outdoor weatherproof GFCI receptacle outlet, two closet luminaires, each on a separate switch, plus a ceiling fan/luminaire, one telephone outlet, and one television outlet.

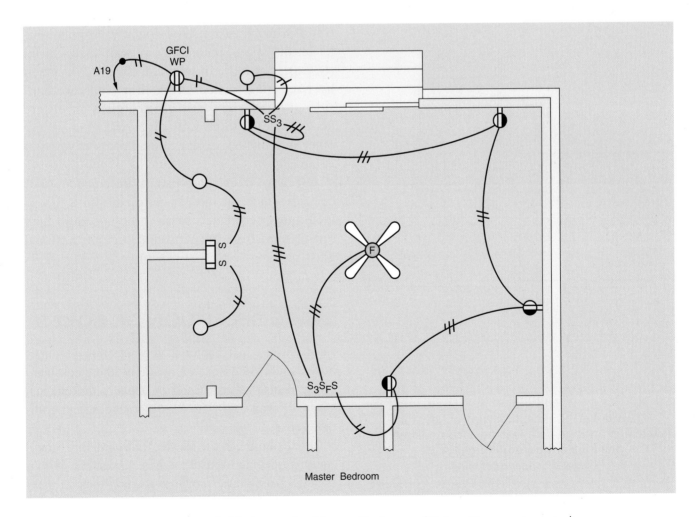

Master Bedroom

FIGURE 9-1 Cable layout for Master Bedroom. (*Delmar/Cengage Learning*)

TABLE 9-1

Master Bedroom outlet count and estimated load (Circuit A19).

Description	Quantity	Watts	Volt-Amperes
Receptacles @ 120 watts each	4	480	480
Weatherproof receptacle	1	120	120
Outdoor bracket luminaire	1	150	150
Clothes closet luminaires One 75-W lamp each	2	150	150
Ceiling fan/light	1		
Three 50-W lamps		150	150
Fan motor (0.75 @ 120 V)		80	90
TOTALS	9	1130	1140

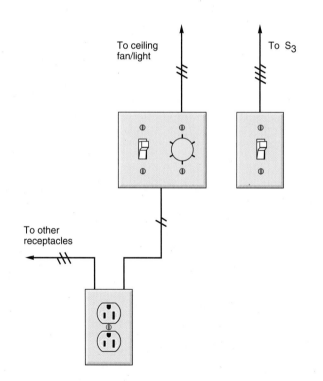

FIGURE 9-2 Conceptual view of how the switching arrangement is to be accomplished in the Master Bedroom Circuit A19.
(*Delmar/Cengage Learning*)

In addition, an outdoor bracket luminaire is located adjacent to the sliding door and is controlled by a single-pole switch just inside the sliding door.

The split-wired receptacle outlets are controlled by two 3-way switches. One is located next to the sliding door. As in the Front Bedroom, Living Room, and Study/Bedroom, the use of split-wired receptacles offers the advantage of having switch control of one of the receptacles at a given outlet, while the other receptacle remains "live" at all times. See Figures 3-4 and 12-14 for the definition of a receptacle.

Next to the bedroom door are a 3-way switch and the ceiling fan/light control, which is installed in a separate 2-gang box, as shown in Figure 9-2.

SLIDING GLASS DOORS AND FIXED GLASS PANELS

Receptacle outlets must be installed so that no point along the floor line in any wall space is more than 6 ft (1.8 m), measured horizontally, from an outlet in that space, including any wall space 24 in. (600 mm) or more in width. Wall space occupied by fixed glass panels in exterior walls are considered to be wall space. Sliding glass panels are not considered to be wall space, the same as any other interior or exterior door. This is covered in *210.52(A)(2)*.

Because fixed glass panels are considered to be wall space, in the Master Bedroom a receptacle must be located so as to be not more than 6 ft (1.8 m) from the left-hand edge of the fixed glass panel.

SELECTION OF BOXES

As discussed in Chapter 2, the selection of outlet boxes and switch boxes is made by the electrician. These decisions are based on *Code* requirements, space allowances, good wiring practices, and common sense.

For example, in the Master Bedroom, the electrician may decide to install a 4-in. square box with a 2-gang raised plaster cover at the box location next to the sliding door. Or two sectional switch boxes ganged together may be installed at that location, as shown in Figure 9-3 and Table 9-2. Two-gang

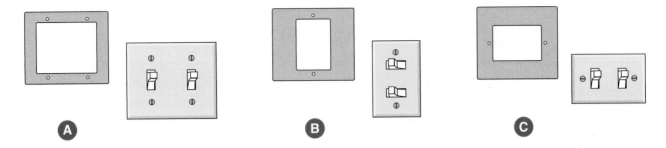

FIGURE 9-3   (A) shows two switches and a wall plate that attach to a 2-gang, 4 in.-square raised plaster cover or to two device boxes that have been ganged together. (B) and (C) show two interchangeable-type switches and wall plate that attach to a 1-gang, 4 in.-square raised plaster cover or to a single-device (switch) box. These types of wiring devices may be mounted vertically or horizontally.
(*Delmar/Cengage Learning*)

TABLE 9-2

Height and width of standard wall plates.

No. of Gangs	Height	Width
1	4½ in. (114.3 mm)	2¾ in. (69.85 mm)
2	4½ in. (114.3 mm)	4⁹⁄₁₆ in. (115.9 mm)
3	4½ in. (114.3 mm)	6³⁄₈ in. (161.9 mm)
4	4½ in. (114.3 mm)	8³⁄₁₆ in. (207.9 mm)
5	4½ in. (114.3 mm)	10 in. (254 mm)
6	4½ in. (114.3 mm)	11¹³⁄₁₆ in. (300 mm)

nonmetallic boxes are readily available and are particularly suitable for branch circuit wiring using nonmetallic cable.

The type of box to be installed depends on the number of conductors entering the box. The suggested cable layout, Figure 9-1, shows that four cables enter this box for a total of 14 conductors (ten 14 AWG circuit conductors and four equipment grounding conductors).

The requirements for calculating box fill are found in *314.16*. These requirements are summarized in Figure 2-36. Also refer to *Table 314.16(A), 314.16(B)*, and Figure 2-20.

In this example, we have 10 circuit conductors, four equipment grounding conductors (counted as one conductor), four cable clamps (counted as one conductor), one single-pole switch (counted as two conductors), and one 3-way switch (counted as two conductors), for a total of 16 conductors.

1. Count the circuit conductors
   2 + 2 + 3 + 3 =   10
2. Add one for one or more
   equipment grounding conductors   1
3. Add two for each switch (2 + 2)   4
4. Add one for one or more
   cable clamps   1
                                             TOTAL   16

Now look at the Quik-Chek Box Selection Guide (Figure 2-20), and select a box or a combination of gangable device boxes that are permitted to hold 16 conductors.

Possibilities are:

1. Gang two 3 in. × 2 in. × 3½ in. device boxes together (9 × 2 = eighteen 14 AWG conductors allowed).

2. Install a 4 in. × 4 in. × 2⅛ in. square box (fifteen 14 AWG conductors allowed). Note that the raised plaster cover, if marked with its cubic-inch volume, can increase the maximum number of conductors permitted for the combined box and raised cover. See Figure 2-37.

CEILING-SUSPENDED (PADDLE) FANS

For appearance and added comfort, ceiling-suspended (paddle) fans are very popular. Figure 9-4 is a typical ceiling-suspended (paddle) fan. Ceiling-suspended (paddle) fans rotate slowly in the range of approximately 60 r/min to 250 r/min for residential fans. Without forced air movement, warm air rises to

FIGURE 9-4 Typical home-type ceiling fan. (*Courtesy of Broan-NuTone, LLC*)

the ceiling and cool air drops to the floor. Air movement results in increased evaporation from our skin, and we feel cooler. Fans that hug the ceiling are not quite as efficient in moving air as those that are suspended a few inches or more from the ceiling.

Most ceiling-suspended (paddle) fans have a reversing switch, a "HIGH/MED/LOW/OFF" speed control pull chain switch for the fan, and an "ON–OFF" pull chain switch for the light accessory kit. Upscale models feature remote "ON–OFF" infinite speed and light control, eliminating the need for "hard wiring" of wall switches. These remote controls are similar to those used with television, VCR, stereo, and cable channel selectors. Wiring for wall switches is discussed later in this chapter.

What Direction Should a Ceiling-Suspended (Paddle) Fan Rotate?

Major manufacturers of ceiling-suspended (paddle) fans recommend directing the air downward (counterclockwise) in the summer to benefit from the "wind chill" effect (see Figure 9-5[A]), and upward (clockwise) in the winter (see Figure 9-5[B]). Some people like air blowing on them; others do not. Upward or downward, the key is to get the stratified air moving. Try different speeds and direction, and do whatever feels best. Read the manufacturer's suggestions.

SUGGESTED FAN SIZES	
For Rooms Up To:	**Fan Size**
144 ft²	42 in.
225 ft²	44–48 in.
400 ft²	52–54 in.
500 ft²	56 in.

Supporting Ceiling-Suspended (Paddle) Fans

This is a very important issue. There have been many reports of injuries (contusions, concussions, lacerations, fractures, head trauma, etc.) caused by the falling of a ceiling-suspended (paddle) fan. There is also the possibility of starting an electrical fire when the wires in the outlet box and fan canopy "short-circuit" as the fan falls from the ceiling. These problems are usually traced back to an improper installation.

The safe supporting of a ceiling-suspended (paddle) fan involves three issues:

1. the actual weight of the fan
2. the twisting and turning motion when the fan is started
3. vibration caused by unbalanced fan blades

Here's the scoop for installing ceiling-suspended (paddle) fans. These requirements are found in *NEC 314.27(C)* and *422.18*, and in the UL *White Book*.

The outlet box or outlet box system

- must be listed and marked as "Acceptable for fan support." Always look for this on the product.

- is permitted to be metallic or nonmetallic.

- must not support more than 70 lb. (32 kg).

- if designed to support more than 35 lb. (16 kg), but not more than 70 lb. (32 kg), will be marked "Acceptable for fan support up to 70 lb. (32 kg)."

- has a weight limitation that includes the fan and light kit.

- must be installed according to the manufacturer's instructions.

- ▶*Where spare, separately switched, ungrounded conductors are provided to a ceiling mounted outlet box, in a location acceptable for a ceiling-suspended (paddle) fan in single or multi-family dwellings, the outlet box or outlet box system shall be listed for sole support of a ceiling-suspended (paddle) fan.** ◀

What Is an Outlet Box System?

This includes such things as the outlet box, hanger, clamps, and other mounting hardware.

*Reprinted with permission from NFPA 70-2011.

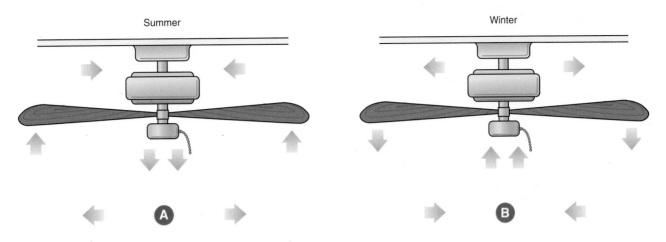

FIGURE 9-5   Suggested rotation of a ceiling-suspended (paddle) fan for winter and summer use. (*Delmar/Cengage Learning*)

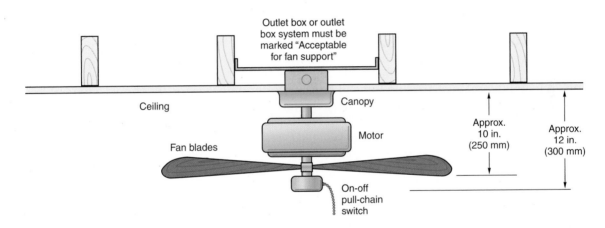

FIGURE 9-6   Typical mounting of a ceiling-suspended (paddle) fan. (*Delmar/Cengage Learning*)

New Work. Install ceiling boxes marked "Acceptable for fan support" in all locations where it is likely that a ceiling-suspended (paddle) fan might be installed. This generally includes all habitable rooms.

Ceiling boxes installed close to a wall, in hallways, in closets, and similar nonhabitable rooms, and ceiling boxes installed for smoke alarms, fire alarms, and security devices do not require fan support boxes.

Some electricians are tired of trying to outguess what *is likely* and what the homeowner might do at some later date. They install suitable fan support outlet boxes for all ceiling outlets other than those that are clearly intended for smoke alarms, fire alarms, and security devices.

Old Work. When replacing an existing luminaire with a ceiling-suspended (paddle) fan, make sure that the existing outlet box is listed and marked as suitable for fan support. In older homes, it is unlikely that you will find this marking. You are taking a big chance if you secure a ceiling fan to an outlet box that is not listed. It might fall on someone's head! You may have to support the fan from the building structure. Carefully read and follow the manufacturer's installation instructions.

See Figure 9-6 for a typical mounting of a ceiling-suspended (paddle) fan.

You always have the choice of hanging a ceiling-suspended (paddle) fan independent of the outlet box, as illustrated in Figure 9-7(E).

Figure 9-7 shows one type of box/hanger assembly that can be used for new work or for existing installations.

Figure 9-8 shows an outlet box that is listed for both luminaire and fan support. This box has No. 8-32 holes for luminaire mounting and No. 10-32 holes for fan support.

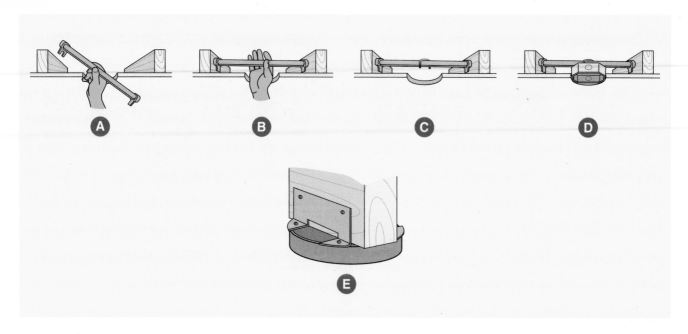

FIGURE 9-7 (A), (B), (C), and (D) show how a listed ceiling fan hanger and box assembly is installed through a properly sized and carefully cut hole in the ceiling. This is done for existing installations. Similar hanger/box assemblies are used for new work. The hanger adjusts for 16-in. (400-mm) and 24-in. (600 mm) joist spacing but can be cut shorter if necessary. (E) shows a type of box listed and identified for the purpose where the fan is supported from the joist, independent of the box, as required by *422.18* and *314.27(C)* for fans that weigh more than 35 lb. (16 kg). (*Courtesy of Legrand/Pass & Seymour*)

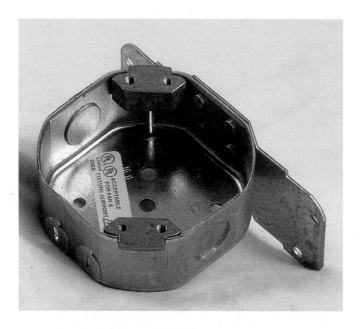

FIGURE 9-8 A special outlet box for the support of a luminaire or a ceiling-suspended (paddle) fan. There are no ears on the box. Instead, special strong mounting brackets are provided. Note that the mounting holes on the brackets are of two sizes: one set of holes is tapped for No. 8-32 screws (for a luminaire), and the other set of holes is tapped for No. 10-32 screws (for a ceiling-suspended [paddle] fan). (*Courtesy of Legrand/Pass & Seymour*)

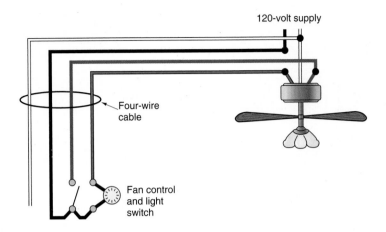

FIGURE 9-9   Fan/light combination with the supply at the fan/light unit. (*Delmar/Cengage Learning*)

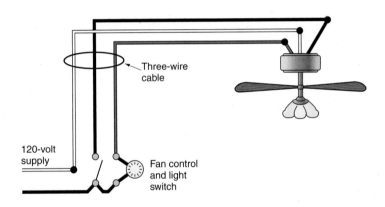

FIGURE 9-10   Fan/light combination with the supply at the switch. (*Delmar/Cengage Learning*)

Figure 9-9 shows the wiring for a fan/light combination with the supply at the fan/light unit. Figure 9-10 shows the fan/light combination with the supply at the switch. Figure 9-7 shows two types of ceiling fan hanger/box assemblies. Many other types are available.

Figure 9-11 illustrates a typical combination light switch and three-speed fan control. The electrician is required to install a deep 4-in. box with a 2-gang raised plaster ring for this light/fan control.

Figure 9-12 illustrates a "slider" switch for speed control of a fan, and a "dual control" that provides 4-speed fan control and two-level "hi/lo" dimming for the incandescent lamps. Figure 9-13 shows two other styles of fan/light controls. Dual controls are needed where the ceiling fan also has

FIGURE 9-11   Combination light switch and three-speed fan switch. (*Delmar/Cengage Learning*)

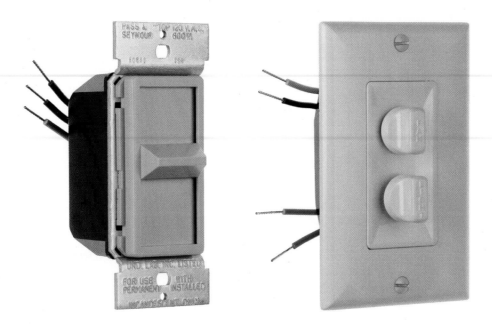

FIGURE 9-12 The "slider" switch is for speed control of a fan. The dual-control switch provides 4-speed fan control and a two-level "hi/lo" dimming for an incandescent lamp. (*Courtesy of Legrand/Pass & Seymour*)

FIGURE 9-13 Two types of fan/light controls. One has a toggle switch for turning a ceiling fan off and on, plus a "slider" that provides 3-speed control of the fan. The other has a toggle switch for turning a light off and on, plus two "sliders," one for "ON–OFF" and 3-speed control of a ceiling fan, the other for full range dimming of lighting. (*Courtesy of Lutron*)

a luminaire. These controls fit into a deep single-gang device box or, preferably, into a 4-in. square box with single-gang plaster cover. Be careful of "box fill." Make sure the wall box is large enough. Dimmer and speed control devices are quite large when compared to regular wall switches. Do not jam the control into the wall box.

A typical home-type ceiling fan motor draws 50 to 100 watts (50–100 volt-amperes). This is approximately 0.4 to slightly over 0.8 ampere.

Ceiling fan/light combinations increase the load requirements somewhat. Some ceiling fan/light units have one lamp socket, whereas others have four or five lamp sockets.

For the ceiling fan/light unit in the Master Bedroom, 240 volt-amperes were included in the load calculations. The current draw is

$$I = \frac{VA}{V} = \frac{240}{120} = 2 \text{ amperes}$$

For most fans, switching on and off may be accomplished by the pull-switch, an integral part of the fan ("HIGH/MED/LOW/OFF"), or with a solid-state switch having an infinite number of speeds. Some of the more expensive models offer a remote control similar to that used with television, VCRs, stereos, and cable channel selectors.

Most ceiling fan motors have an integral reversing switch to blow air downward or upward.

Do not use a standard incandescent lamp dimmer to control fan motors, fluorescent ballasts, transformers, low-voltage lighting systems (they use transformers to obtain the desired low voltage), motor-operated appliances, or other "inductive" loads. Serious overheating and damage to the motor can result. Fan speed controllers have internal circuitry that is engineered specifically for use on inductive load circuits. Fan speed controllers virtually eliminate any fan motor "hum." Read the label and the instructions furnished with a dimmer to be sure it is suitable for the application.

CAUTION! **Most fan manufacturers state that when using a wall speed control, the pull-chain speed control on the fan should be set at its highest speed setting. To use other pull-chain speed positions in combination with a wall speed control might damage the motor or the wall speed control. Several manufacturers of speed controls say that they do not test them at other than the high-speed setting on the fan. Others state in their instructions that their solid-state speed controls:**

- **are to be used only with fans marked "Suitable for use with solid-state fan speed controls only."**

- **are not to be used in circuits protected by GFCIs.**

- **are not to be used to control lighting—just as a lighting dimmer control might be marked not to be used as a fan speed control.**

NEC 110.3(B) requires that *Listed or labeled equipment shall be installed and used in accordance with any instructions included in the listing or labeling.**

You can readily see the importance of carefully reading and following the manufacturer's installation instructions.

*Reprinted with permission from NFPA 70-2011.

REVIEW

Note: Refer to the *Code* or the plans where necessary.

1. What circuit supplies the Master Bedroom? _____

_____

2. What unique type of electrical protection is required for all 120-volt, single-phase, 15- and 20-ampere branch circuits in certain habitable rooms and areas in new dwellings? Where in the *NEC* is this requirement found? _____

_____

3. What type of receptacles are provided in this bedroom? How many receptacles are there?

   _____

4. How many ceiling outlets are included in this circuit? _____

   _____

5. What wattage for the clothes closet luminaires was used for calculating their contribution to the circuit? _____

6. What is the current draw for the ceiling fan/light? _____

7. What is the estimated load in volt-amperes for the circuit supplying the Master Bedroom?

   _____

8. Which is installed first, the switch and outlet boxes or the cable runs? _____

   _____

9. How many conductors enter the ceiling fan/light wall box? _____

10. What type and size of box may be used for the ceiling fan/light wall box? _____

    _____

    _____

11. What type of covers are used with 4-in. square outlet boxes? _____

    _____

    _____

12. a. Does the circuit for the Master Bedroom have a grounded circuit conductor?

      _____

    b. Does it have an equipment grounding conductor? _____

    c. Explain the difference between a "grounded" conductor and an "equipment grounding" conductor. _____

      _____

      _____

13. Approximately how many feet (meters) of 2-wire cable and 3-wire cable are needed to complete the circuit supplying the Master Bedroom? Two-wire cable: _____ ft (_____ m). Three-wire cable: _____ ft (_____ m).

14. If the cable is laid in notches in the corner studs, what protection for the cable must be provided? _____

    _____

15. How high are the receptacles mounted above the finished floor in this bedroom?

    _____

16. Approximately how far from the bedroom door is the first receptacle mounted? (See the plans.) _____

    _____

17. What is the distance from the finished floor to the center of the wall switches in this bedroom? _____

18. The Master Bedroom features a sliding glass door. For the purpose of providing the proper receptacle outlets, answer the following statements (circle the correct answer).

   • Sliding glass panels are considered to be wall space.   (True) (False)

   • Fixed panels of glass doors are considered to be wall space.   (True) (False)

19. What type of receptacle will be installed outdoors, just outside of the Master Bedroom?_____

   _____

20. Show your calculations of how to select a proper wall box for the clothes closet luminaire switch. Keep in mind that the available space between the wood casings is small. See Figure 9-1. _____

   _____

   _____

   _____

   _____

21. When an outlet box is to be installed to support a ceiling fan, how must it be marked?

   _____

22. The following is a layout of the lighting circuit for the Master Bedroom. Using the cable layout shown in Figure 9-1, make a complete wiring diagram of this circuit. Use colored pencils or pens.

Master bedroom

Fan/light

Fan Light

23. Connect the 4-wire cable, fan, light, fan speed control, and light switch. Refer to *200.7* to review the permitted use of the white conductor in the cable.

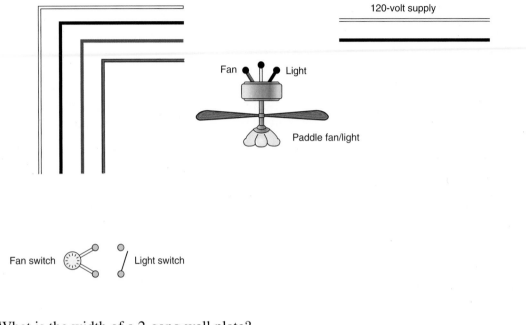

24. What is the width of a 2-gang wall plate? _____

25. May a standard electronic dimmer be used to control the speed of a fan motor?

    _____

26. Outlet boxes that pass the tests at Underwriters Laboratories for the support of ceiling-suspended (paddle) fans are listed and marked "Acceptable for Fan Support." When roughing-in the electrical wiring for a new home, this type of outlet box must be installed wherever there is a likelihood of installing a ceiling-suspended (paddle) fan. Name those locations in homes that you consider likely to support a ceiling-suspended (paddle) fan. _____

    _____

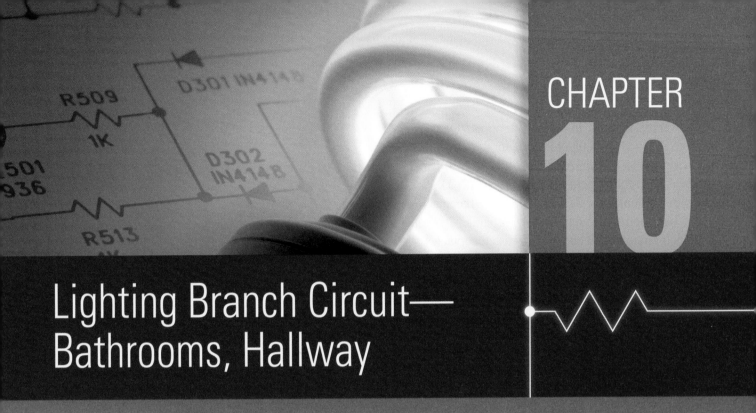

CHAPTER 10

Lighting Branch Circuit—Bathrooms, Hallway

OBJECTIVES

After studying this chapter, you should be able to

- list equipment grounding requirements for bathroom installations.
- draw a wiring diagram for the bathroom and hallway.
- understand *Code* requirements for receptacles installed in bathrooms.
- understand *Code* requirements for receptacle outlets in hallways.
- discuss fundamentals of proper lighting for bathrooms.

Circuit A14 is a 15-ampere branch circuit that supplies the hallway lighting, the hallway receptacle, and the vanity lighting in both bathrooms. Overcurrent protection and arc-fault protection for this circuit is provided by a 15-ampere AFCI circuit breaker in Panel A. The AFCI requirement is found in *210.12(A)*.

As required by *210.11(C)(3)*, the receptacles in the bathrooms are supplied by separate 20-ampere branch circuits A22 and A23. These receptacles shall be GFCI protected, either by installing GFCI receptacles, or by installing GFCI circuit breakers in the panel. The GFCI requirement is found in *210.8(A)(1)*.

Table 10-1 summarizes outlets and estimated load for the bathrooms and bedroom hall.

Note that each bathroom shows a ceiling heater/light/fan that is connected to separate circuit A12 (Special Purpose ▲ J) and separate circuit A11 (Special Purpose ▲ K). These are discussed in detail in Chapter 22.

A hydromassage bathtub is located in the bathroom serving the master bedroom. It is connected to a separate circuit A9 (Special Purpose ▲ A) and is also discussed in Chapter 22.

The attic exhaust fan in the hall is supplied by a separate circuit A10 (Special Purpose ▲ L), also covered in Chapter 22.

LIGHTING BRANCH CIRCUIT A14 FOR THE HALLWAY AND BATHROOMS

Figure 10-1 and the electrical plans for this area of the home show that each bathroom has a luminaire above the vanity mirror. Some typical luminaires are shown in Figure 10-2. Of course, the homeowner might decide to purchase a medicine cabinet complete with a self-contained luminaire. See Figure 10-3, which illustrates how to rough-in the wiring for each type. These luminaires are controlled by single-pole switches at the doors.

You must remember that a bathroom (powder room) should have proper lighting for shaving, combing hair, grooming, and so on. Mirror lighting can accomplish this, because a mirror will reflect what it "sees." If the face is poorly lit, with shadows on the face, that is precisely what will be reflected in the mirror.

A luminaire directly overhead will light the top of one's head but will cause shadows on the face.

TABLE 10-1

Bathrooms and bedroom hall: outlet count and estimated load (Circuit A14).

Description	Quantity	Watts	Volt-Amperes
Receptacles @ 120 W each	1	120	120
Vanity luminaires... @ 200 W each	2	400	400
Hall luminaire	1	100	100
TOTALS	4	620	620

The receptacles in the master bathroom and hall bathroom are not included in this table as they are connected to separate 20-ampere branch circuits A22 and A23.

Mirror lighting and/or adequate lighting above and forward of the standing position at the vanity can provide excellent lighting in the bathroom. Figures 10-4, 10-5, and 10-6 show pictorial as well as section views of typical soffit lighting above a bathroom vanity.

Bathroom Receptacles

Figure 10-7 shows the definition of a bathroom according to the *NEC*. This definition was revised and expanded in the 2011 *NEC* to read, ▶ *An area including a basin with one or more of the following: a toilet, a urinal, a tub, a shower, a bidet, or similar plumbing fixtures.** ◀ Note that the definition describes an *area*, not a *room*. The basin is the common denominator. As shown in Figure 10-7(D), the two rooms make up an area with the basin located in a space that is adjacent to the room with the tub and toilet.

As required by *210.11(C)(3)*, the receptacles in the bathrooms are supplied by separate 20-ampere branch circuits A22 and A23. These receptacles shall be GFCI protected, either by installing GFCI receptacles, or by installing GFCI circuit breakers in the panel. The GFCI requirement is found in *210.8(A)(1)*.

Although the *NEC* permits all bathroom receptacles to be on a single 20-ampere branch circuit so long as it doesn't supply other loads, we have chosen to run a separate 20-ampere branch circuit A22 to the receptacle in the Master Bedroom bathroom, a separate 20-ampere branch circuit A23 to the receptacle in the hall bathroom, and another separate

*Reprinted with permission from NFPA 70-2011.

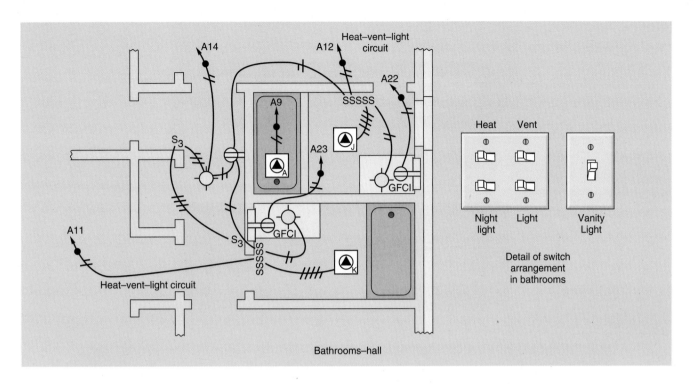

FIGURE 10-1   Cable layout for the master bathroom, hall bathroom, and hall. Layout includes lighting Circuit A14, two circuits (A11 and A12) for the heat/vent/lights, and two circuits (A22 and A23) for the receptacles in the bathrooms. The special-purpose outlets for the hydromassage tub, attic exhaust fan, and smoke detector are covered elsewhere in this text.
(*Delmar/Cengage Learning*)

20-ampere branch circuit B21 to the receptacle in the Powder Room located near the Laundry. These separate circuits are included in the general lighting load calculations, therefore, no additional load need be added, *220.14(J)*.

Receptacles in Bathtub and Shower Spaces

Because of the obvious hazards associated with water and electricity, receptacles are *not* permitted to be installed in bathtub and shower spaces, *406.9(C)*.

⏦ HANGING LUMINAIRES IN BATHROOMS

NEC 410.10(D) does not permit cord-connected luminaires; chain-, cable-, or cord-suspended luminaires; lighting track; pendants; or ceiling-suspended (paddle) fans to be located within a zone measured

3 ft (900 mm) horizontally and 8 ft (2.5 m) vertically from the top of a bathtub rim or top of a shower stall threshold. Figure 10-8 shows a top view of the 3-ft (900-mm) restriction. Figure 10-9 shows a side view of both permitted and not permitted installations.

Recessed or surface-mounted luminaires and recessed exhaust fans may be located within the restricted zone.

Luminaires Near Bathtub or Shower

Install luminaires listed for *wet* locations if subject to shower spray.

Install luminaires listed for *damp* or *wet* locations if located within the actual outside dimension of the bathtub or shower up to a height of 8 ft (2.5 m) vertically measured from the rim of the bathtub or shower threshold. The label on the luminaire will show the listing information.

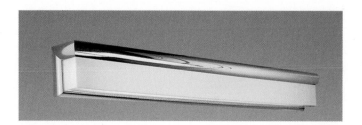

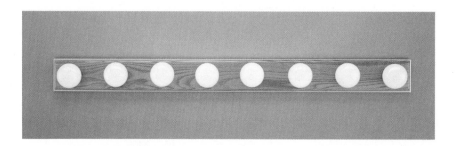

FIGURE 10-2 Typical vanity (bathroom) luminaires of the side bracket and strip types.
(*Courtesy of Progress Lighting*)

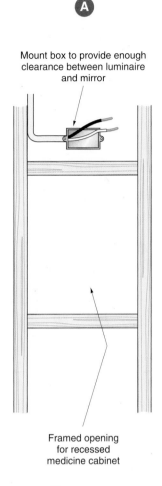

Mount box to provide enough clearance between luminaire and mirror

Framed opening for recessed medicine cabinet

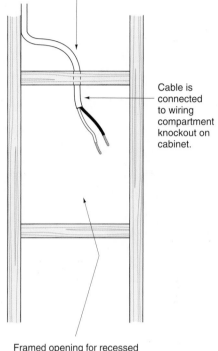

Be sure to bring cable or conduit in at proper location. It is best to have the actual medicine cabinet or installation instructions to determine exact location of cable knockout.

Cable is connected to wiring compartment knockout on cabinet.

Framed opening for recessed medicine cabinet that comes complete with a luminaire

FIGURE 10-3   Two ways to rough-in wiring for lighting above or on a vanity mirror or a medicine cabinet. (*Delmar/Cengage Learning*)

HALLWAY LIGHTING

The hallway lighting is provided by one ceiling luminaire that is controlled with two 3-way switches located at either end of the hall. The home run to Main Panel A has been brought into this ceiling outlet box.

RECEPTACLE OUTLETS IN HALLWAYS

One receptacle outlet has been provided in the hallway as required in *210.52(H)*, which states that *for hallways of 3 m (10 ft) or more in length, at least one receptacle outlet shall be required.**

For the purpose of determining the length of a hallway, the measurement is taken down the center-

line of the hall, turning corners if necessary, but not passing through a doorway, Figure 10-10.

EQUIPMENT GROUNDING

With few exceptions, fixed and fastened-in-place electrical equipment with exposed non-current-carrying metal parts that could become energized must be grounded. This includes such things as appliances, furnaces, air conditioners, heat pumps, heat exchangers, fans, luminaires, electric heaters, and medicine cabinets that have lighting. Equipment that must be grounded is itemized in *250.110, 250.112, 250.114* (cord-and-plug connected), and *Article 410, Part V* (luminaires).

Acceptable equipment grounding conductors are listed in *250.118*, which was previously discussed.

---

*Reprinted with permission from NFPA 70-2011.

Wrong

Right

FIGURE 10-4 Positioning of bathroom luminaires. Note the wrong way and the right way to achieve proper lighting. (*Delmar/Cengage Learning*)

CAUTION: When using nonmetallic-sheathed cable or armored cable, *never* use the bare equipment grounding conductor as a grounded circuit conductor (neutral), and *never* use the grounded circuit conductor as an equipment grounding conductor. These two conductors come together only at the main service panel, *never* in branch-circuit wiring. Refer to *250.142.*

Double Insulation in Lieu of Grounding

Appliances that are cord-and-plug connected with a 3-wire cord are grounded by the equipment grounding conductor in the cord. Instead of grounding, many small electrical appliances have a 2-wire cord, and are "double insulated." Examples of these are some portable electrical tools, razors, tooth brushes, radios, TVs, VCRs, CD players and burners, stereo components, and similar appliances.

Double-insulated appliances provide two levels of protection against electric shock. Accepted levels of protection can be the extra insulation on a conductor, added insulating material, the plastic nonmetallic enclosure of the appliance itself, or even a specified air-gap at an exposed terminal. With double-insulated products, it takes two failures to create a shock hazard to the user. These appliances are tested and clearly marked that they are double insulated in accordance to UL Standard 1097 *Double-Insulated Systems for Use in Electrical Equipment.*

Immersion Detection Circuit Interrupters

Another way to protect people from electrical shock is found on grooming appliances that are equipped with a special attachment plug cap that provides the protection should the appliance be immersed in water.

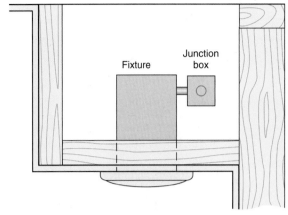

End cutaway of soffit showing recessed incandescent luminaires in typical soffit above bathroom vanity. Two or three luminaires generally installed to provide proper lighting.

Typical incandescent recessed soffit lighting over bathroom vanity

FIGURE 10-5   Incandescent soffit lighting. Refer to Chapter 7 for minimum-clearance *Code* requirements for installation of recessed luminaires. (*Delmar/Cengage Learning*)

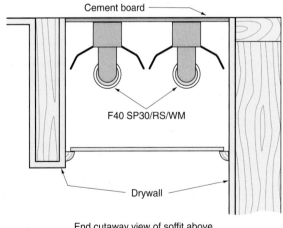

End cutaway view of soffit above bathroom vanity showing recessed fluorescent luminaires concealed above translucent acrylic lens

Typical fluorescent recessed soffit lighting over bathroom vanity. Note additional incandescent "side-of-mirror" lighting.

FIGURE 10-6   Combination fluorescent and incandescent bathroom lighting. (*Delmar/Cengage Learning*)

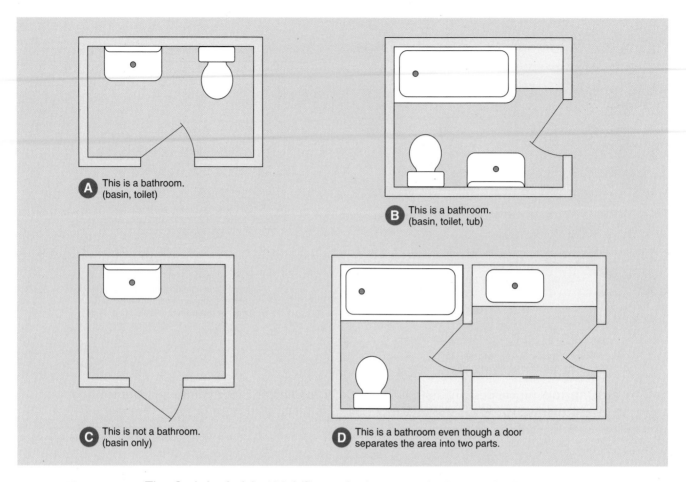

FIGURE 10-7 The *Code* in *Article 100* defines a bathroom as ▶ *An area including a basin with one or more of the following: a toilet, a urinal, a tub, a shower, a bidet, or similar plumbing fixtures.** ◀ (*Delmar/Cengage Learning*)

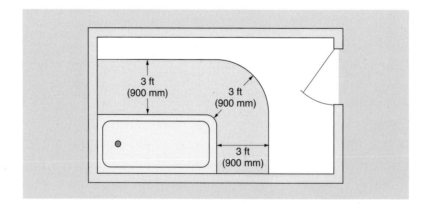

FIGURE 10-8 In a bathtub or shower area, no parts of cord-connected or cord-suspended luminaires or ceiling-suspended (paddle) fans are permitted in the shaded area. See text for details. Also see Figure 10-9. (*Delmar/Cengage Learning*)

*Reprinted with permission from NFPA 70-2011.

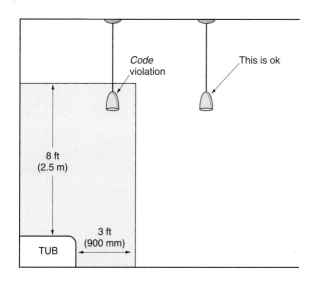

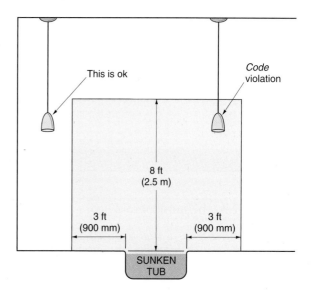

FIGURE 10-9 In a bathtub or shower area, no part of cord-connected or cord-suspended luminaires or ceiling-suspended (paddle) fans are permitted in the shaded area. See text for details. Also see Figure 10-8. (*Delmar/Cengage Learning*)

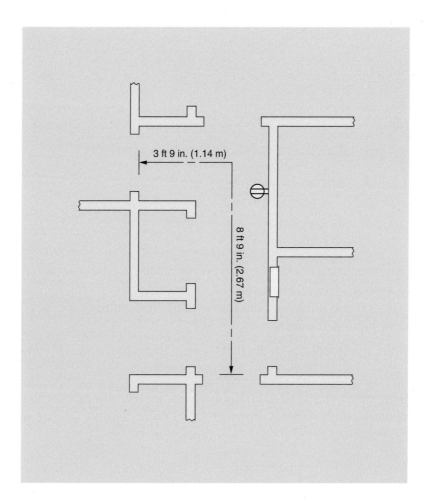

FIGURE 10-10 The centerline measurement of the bedroom hallway in this residence is 12 ft 6 in. (3.81 m), which requires at least one wall receptacle outlet, *210.52(H)*. This receptacle may be installed anywhere in the hall. The *Code* does not specify a location. (*Delmar/Cengage Learning*)

Switches in Wet Locations

Do *not* install switches in wet locations, such as in bathtubs or showers, unless they are part of a listed tub or shower assembly, in which case the manufacturer has taken all of the proper precautions and submitted the assembly to a recognized testing laboratory to undergo exhaustive testing to establish the safety of the equipment; see *404.4*.

REVIEW

1. List the number and types of switches and receptacles used in Circuit A14.

2. There is a 3-way switch in the bedroom hallway leading into the living room. Show your calculation of how to determine the box size for this switch. The box contains cable clamps.

3. What wattage was used for each vanity luminaire to calculate the estimated load on Circuit A14? _____

4. What is the current draw for the answer given in question 3?

5. Exposed non-current-carrying metallic parts of electrical equipment must be grounded if installed within _____ ft (_____ m) vertically or _____ ft (_____ m) horizontally of bathtubs, plumbing fixtures, pipes, or other grounded metal work or grounded surfaces.

6. What color are the faceplates in the bathrooms? Refer to the specifications.

   _____

7. Most appliances of the type commonly used in bathrooms, such as hair dryers, electric shavers, and curling irons, have 2-wire cords. These appliances are _____ insulated or _____ protected.

8. a. The *NEC* in Section _____ requires that all receptacles in bathrooms be _____ protected.

   b. The *NEC* in Section _____ requires that all receptacles in bathrooms be connected to one or more separate 20-ampere branch circuits that serve no other outlets.

c. The *NEC* in Section _____ permits the additional required 20-ampere branch circuit for bathroom receptacles to supply more than one bathroom.

d. The *NEC* in Section _____ permits other electrical equipment to be connected to the additional required 20-ampere branch circuit for bathroom receptacles, but only if the branch circuit supplies a single bathroom and the other equipment is located in that same bathroom.

e. The *NEC* in Section _____ prohibits mounting receptacles in bathrooms face-up in the countertops and work surfaces near basins.

9. Hanging luminaires must be kept at least _____ ft (_____ m) from the edge of the tub as measured horizontally. In bathrooms with high ceilings, where the hanging luminaire is installed directly over the tub, it must be kept at least _____ ft (_____ m) above the edge of the tub.

10. The following is a layout of a lighting circuit for the bathroom and hallway. Using the cable layout shown in Figure 10-1, make a complete wiring diagram of this circuit. Use colored pencils or pens.

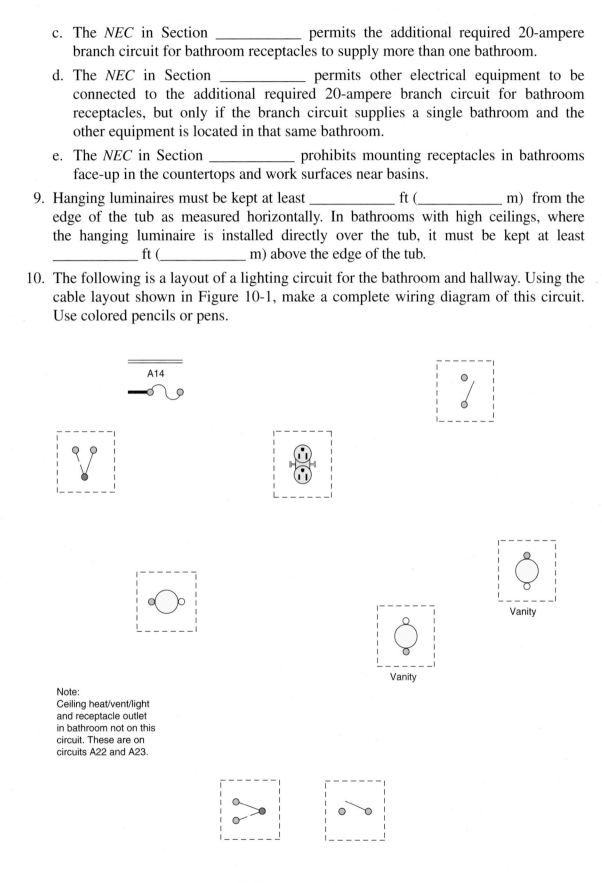

A14

Vanity

Vanity

Note:
Ceiling heat/vent/light
and receptacle outlet
in bathroom not on this
circuit. These are on
circuits A22 and A23.

Bathrooms – Hallway

11. Circle the correct answer as to whether a receptacle outlet is required in the following hallways.

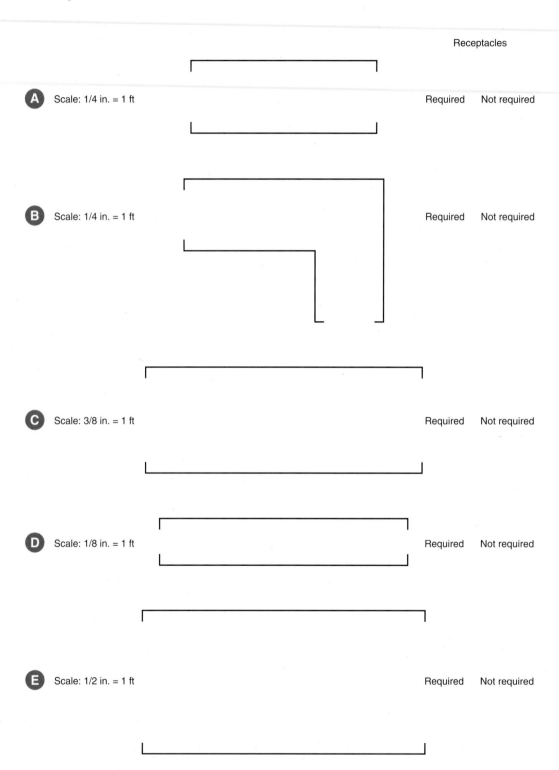

Receptacles

A Scale: 1/4 in. = 1 ft   Required   Not required

B Scale: 1/4 in. = 1 ft   Required   Not required

C Scale: 3/8 in. = 1 ft   Required   Not required

D Scale: 1/8 in. = 1 ft   Required   Not required

E Scale: 1/2 in. = 1 ft   Required   Not required

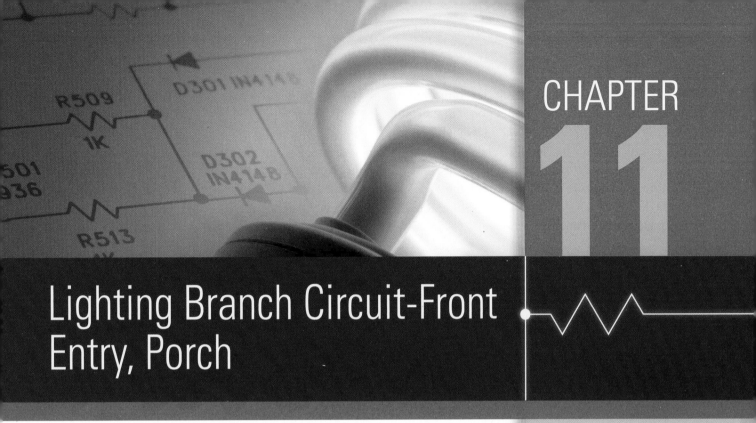

Lighting Branch Circuit-Front Entry, Porch

OBJECTIVES

After studying this chapter, you should be able to

- understand how to install a switch in a doorjamb for automatic "ON–OFF" when the door is opened or closed.

- discuss types of luminaires recommended for porches and entries.

- complete the wiring diagram for the entry-porch circuit.

- discuss the advantages of switching outdoor receptacles from indoors.

- define wet and damp locations.

- understand box fill for sectional ganged device boxes.

The front entry area and porch is connected to Circuit A15. ▶Overcurrent protection and arc-fault protection for this circuit is provided by a 15-ampere AFCI circuit breaker in Panel A. The AFCI requirement is found in *210.12(A)*.◀ GFCI protection for the outdoor receptacle could be provided by a GFCI receptacle, or by an AFCI/GFCI dual-function circuit breaker in the Panel A.

The home run enters the ceiling box in the front entry. From this box, the circuit spreads out, feeding the closet light, the porch bracket luminaire, the two bracket luminaires on the front of the garage, one receptacle outlet in the entry, and one outdoor weatherproof GFCI receptacle on the porch, as shown in Figure 11-1. Table 11-1 summarizes the outlets and estimated load for the entry and porch.

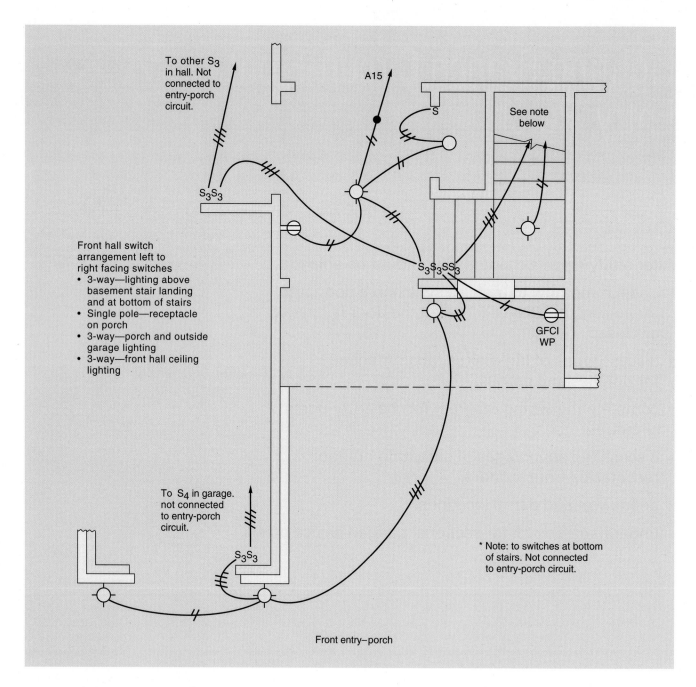

FIGURE 11-1 Cable layout of the front entry-porch circuit A15. Note that cables from some of the other circuits are shown so that you can get a better idea of exactly how many wires will be found at the various locations. (*Delmar/Cengage Learning*)

TABLE 11-1

Entry and porch: outlet count and estimated load (Circuit A15).

Description	Quantity	Watts	Volt-Amperes
Receptacles @ 120 W	1	120	120
Weatherproof receptacles @ 120 W	1	120	120
Outdoor porch bracket luminaire	1	100	100
Outdoor garage bracket luminaires @ 100 W each	2	200	200
Ceiling luminaire	1	150	150
Clothes closet recessed luminaire	1	75	75
TOTALS	7	765	765

Note that the receptacle on the porch is controlled by a single-pole switch just inside the front door. This allows the homeowner to plug in such things as strings of ornamental holiday lighting or decorative lighting and have convenient switch control of that receptacle from inside the house, a nice feature.

Although it adds cost, consider an outdoor weatherproof receptacle on both sides of the front entry door. This makes for easy plugging in of outdoor decorations. Some electricians install a receptacle in the soffit. Switched or nonswitched? The decision is up to you.

Typical ceiling luminaires commonly installed in front entryways, where it is desirable to make a good first impression on guests, are shown in Figure 11-2.

Typical outdoor wall-bracket-style porch and entrance luminaires are shown in Figure 11-3.

A location exposed to the weather is a wet location. A partially protected area such as an open porch, or under a roof or canopy, is a damp location. For more precise definitions of dry, wet, and damp locations, see *Article 100* of the *NEC*.

Luminaires installed in wet or damp locations must be installed so that water cannot enter or accumulate in wiring compartments, lampholders, or other electrical parts. Luminaires installed in

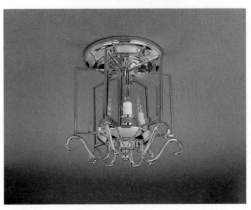

FIGURE 11-2   Hall luminaires: ceiling mount and chain mount. (*Courtesy Progress Lighting*)

FIGURE 11-3 Outdoor porch band entrance luminaires: wall bracket styles. (*Courtesy of Progress Lighting*)

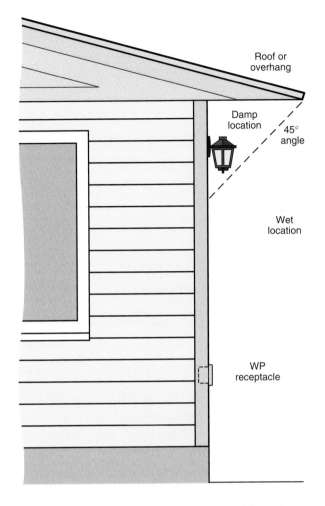

FIGURE 11-4 Outdoor areas considered
to be either a *damp* or *wet* location.
(Delmar/Cengage Learning)

wet locations shall be marked, "Suitable for Wet
Locations." Luminaires installed in damp locations
shall be marked, "Suitable for Wet Locations" or
"Suitable for Damp Locations." This information is
found in *410.10(A)*.

Some inspectors consider the area within a 45°
angle from a roof line or overhang to be a damp
location, and outside the 45° angle to be a wet location. See Figure 11-4.

⌁⌁⌁ CIRCUIT A15

Circuit A15 is a rather simple circuit that has two sets
of 3-way switches: one set controlling the front entry
ceiling luminaire, the other set controlling the porch
bracket luminaire and the two bracket luminaires on

the front of the garage. Also, one single-pole switch
controls the weatherproof porch receptacle outlet,
and another single-pole switch controls the clothes
closet light. In the Review questions, you will be
asked to follow the suggested cable layout for making up all of the circuit connections.

Probably the most difficult part of this circuit
is planning and making up the connections at the
4-gang switch location just inside the front door.
Refer to Figure 11-1 for details. Figure 11-5 shows
a 4-gang nonmetallic box of the type that can be
used for the front door switch location. Note that
the 3-way switch for the luminaire over the stairway
leading to the basement is part of the Recreation
Room lighting branch circuit B12.

There are a lot of conductors, wiring devices,
and cable clamps in this box. Let's figure out the
proper wall size wall box.

1. Add the circuit conductors
   $4 + 3 + 3 + 2 + 4 =$ 16

2. Add for equipment
   grounding wires (count 1 only)   1

3. Add eight for the four switches   8

4. Add for cable clamps
   (count 1 only)   <u>1</u>

   Total   26

To select a box that has adequate cubic-inch capacity for this example, refer to *Table 314.16(A)* in the
NEC, and Figure 2-20 and Table 2-1 in this text. One
possibility would be to use four 3 in. × 2 in. × 3½ in.
device boxes ganged together, providing 4 × 9 = 36
conductor count based on 14 AWG conductors.

When using nonmetallic boxes, refer to *Table
314.16(B)* and base the volume allowance on 2 in.³ for
each 14 AWG conductor and for the volume allowance
for the wiring devices, cable clamps, and equipment
grounding conductors. In the previous example, we are
looking for a cubic-inch volume of 26 × 2 = 52 in.³

Foyers. A new requirement was added to the
2011 *NEC* in *210.52(I)*. ▶This section requires
receptacles to be installed in compliance with the
placement rules in *210.52(A)(2)(1)* for foyers that
have an area greater than 60 ft². ◀ *NEC 210.52(A)
(2)(1)* requires that a receptacle be provided for wall
spaces 2 ft (600 mm) or more in width. This results
in one receptacle to be installed for the foyer.

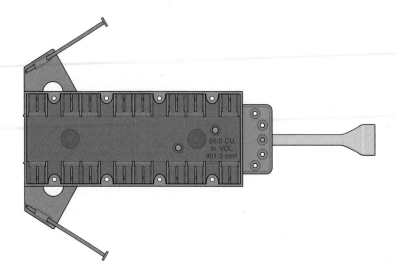

FIGURE 11-5 A 4-gang, nonmetallic box that has a marked volume of 55 in.³ (901.3 cm³). (*Delmar/Cengage Learning*)

DOORJAMB SWITCH

An interesting possibility presents itself for the front-entry clothes closet. Although the plans show that the clothes closet luminaire is turned on and off by a standard single-pole switch to the left as you face the closet, a doorjamb switch could have been installed.

A doorjamb switch comes as an assembly—the switch, the special box, and the cover.

The instructions furnished with the doorjamb switch specify the maximum number and size of conductors the box can accommodate. These instructions also show the roughing-in dimensions of the box. The box furnished with a doorjamb switch is generally suitable only for the switch loop conductors. Because the wiring space is very limited, do not plan on using this box for splices or through wiring other than those necessary to make up the connections to the switch.

A doorjamb switch is usually mounted about 6 ft (1.8 m) above the floor, on the inside of the 2 × 4 framing on the hinge side of the closet door. Run the switch loop cable to this point, and let it hang out until the carpenter cuts the proper size opening into the doorjamb for the box. After the finished woodwork is completed, the switch and cover plate can be installed. Figure 11-6 illustrates a door switch.

The plunger on the switch is pushed inward when the edge of the door pushes on it as the door is closed, shutting off the light. The plunger can be adjusted in or out to make the switch work properly.

Figure 11-6 has been revised to comply with ▶*NEC 404.2(C)*, which requires a grounded circuit conductor (neutral) at the box to the door switch.◀ Because space in the special box is very limited, it is suggested that the electrical inspector be questioned to determine whether he or she will permit the switch loop to be installed with a 2-wire cable. The neutral conductor simply takes up space in the box and will never have a purpose or function.

Here's another option. Install a low-voltage system. A simple, yet effective system can be created with a 24-volt transformer, a 120-volt lighting relay (with a 24-volt coil) and a 24-volt plunger-type switch. By using a Class 2 transformer, a box is not required for the switch. This can be accomplished quite simply by mounting a 4-square deep metal box on the surface of the wall in the corner of the closet approximately 6 ft (1.8 m) above the floor. Pre-wire the 24-volt doorbell cable to the switch location and the power for the light through the location for the metal box. A cable or EMT is pre-installed to the outlet box for the closet luminaire. At the time the trim is installed, connect the transformer and relay along with the switch and you're finished.

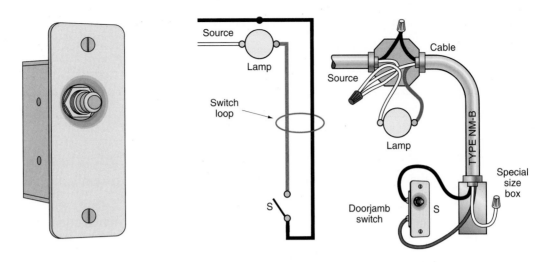

FIGURE 11-6 Doorjamb switch. Feed at light. Switch loop run from ceiling box to doorjamb box. (*Delmar/Cengage Learning*)

REVIEW

1. a. How many circuit wires enter the entry ceiling box? _____

   b. How many equipment grounding conductors enter the entry ceiling box?_____

2. Assuming that an outlet box for the entry ceiling has a fixture stud but has external cable clamps, what size box could be installed? Show calculations for both metal and nonmetallic boxes. _____

   _____

   _____

   _____

   _____

   _____

3. How many receptacle outlets and lighting outlets are supplied by Circuit A15? _____

   _____

4. Outdoor luminaires directly exposed to the weather must be marked as

   a. suitable for damp locations.

   b. suitable for dry locations only.

   c. suitable for wet locations.

   Circle the correct answer.

5. Make a material list of all types of switches and receptacles connected to Circuit A15.

   _____

   _____

   _____

   _____

6. From left to right, facing the switches, what do the switches next to the front door control?

_____

_____

_____

_____

_____

7. Who is to select the entry ceiling luminaire? _____

8. When installing the wiring for a doorjamb switch, the best choice is (circle the correct answer)

   a. run the supply conductors to the doorjamb box, then run the switched conductors to the luminaire.

   b. run the supply conductors to the outlet box where the luminaire is to be installed, then run the switch loop conductors to the doorjamb box.

9. The following layout is for Lighting Circuit A15, the entry, porch, and front garage lights. Using the cable layout shown in Figure 11-1, make a complete wiring diagram of this circuit. Use colored pencils to indicate the color of the conductors' insulation.

Entry–Porch

A15

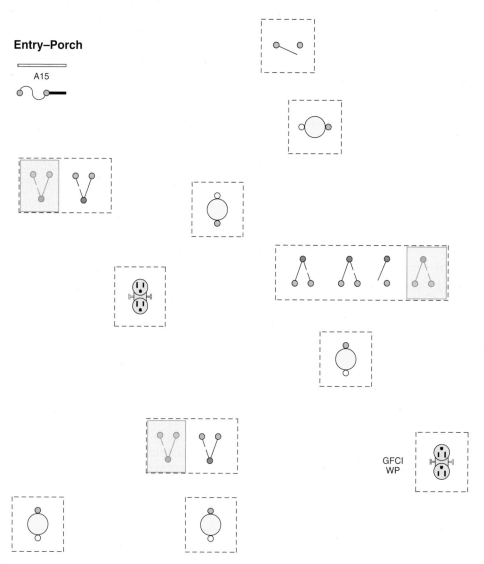

Lighting Branch Circuit and Small-Appliance Circuits for the Kitchen

OBJECTIVES

After studying this chapter, you should be able to

- understand lighting for a typical kitchen and dining room.

- understand the installation and operation of kitchen exhaust fans.

- know the *NEC* requirements for small-appliance branch circuits in kitchens.

- know the *NEC* requirements for GFCI protection for all receptacles that serve countertops in kitchens.

- decide whether an individual branch circuit should be installed for a refrigerator.

- decide where it might be desirable to install multiwire branch circuits and split-wired receptacles in areas where a high concentration of plug-in appliances might be used.

KITCHEN

The *NEC* defines a *kitchen* as *An area with a sink and permanent provisions for food preparation and cooking.** Although the *NEC* does not define *permanent*, it no doubt means *not temporary*. The concept is the permanent provisions are in place even though the appliance may not be permanent. For example, a dwelling is to have an electric range in the kitchen. The receptacle outlet for the range is permanent. However, the range can be placed in the dedicated space and the "pigtail" be plugged into the receptacle. The range can be moved and replaced with another appliance, yet the *provisions* for the supply to the range are permanent.

*Reprinted with permission from NFPA 70-2011.

LIGHTING CIRCUIT B7

The lighting circuit supplying the Kitchen originates at Panel B. Checking *210.12(A)*, the kitchen lighting branch circuit is not required to be AFCI protected. However, depending on how you lay out the branch circuits, there is no problem if you do provide AFCI protection for the kitchen lighting. The circuit number is B7. The cable layout for this circuit is shown in Figure 12-1. The home run is very short, leading from Panel B in the corner of the recreation room to the switch box to the right of the kitchen sink. From here, the circuit continues upward to the outlet box above one of the square surface-mounted luminaires on the ceiling, then across the ceiling to the outlet

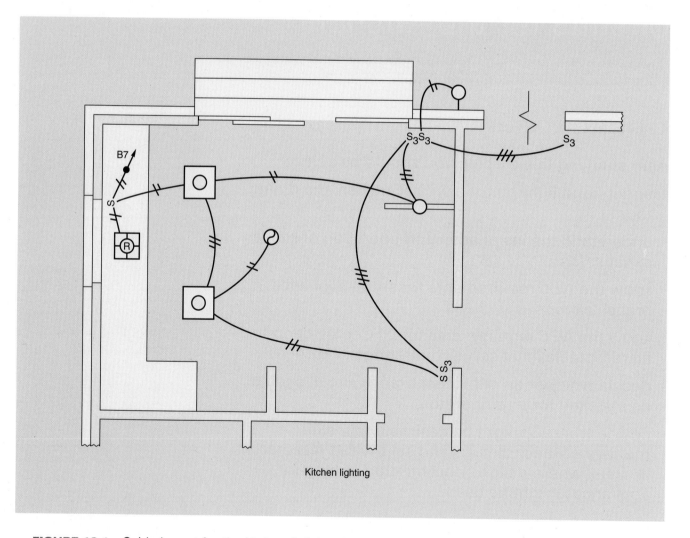

Kitchen lighting

FIGURE 12-1 Cable layout for the kitchen lighting. This is Circuit B7. The appliance circuit receptacles are connected to 20-ampere circuits B13, B15, and B16 and are not shown on this lighting circuit layout.
(Delmar/Cengage Learning)

TABLE 12-1

Kitchen: outlet count and estimated load (Circuit B7).

Description	Quantity	Watts	Volt-Amperes
Ceiling luminaires (eight 20-W FL lamps)	2	160	180
Recessed luminaire over sink	1	100	100
Lighting track	1	100	100
Outdoor rear bracket luminaire	1	100	100
Range hood fan/light Two 60-W lamps	1	120	120
Fan motor (1⅓ A @ 120 V)	1	160	180
TOTALS	7	740	780

FIGURE 12-2 Typical ceiling luminaire of the type installed in the Kitchen of this residence. This type of luminaire might have four 20-watt fluorescent lamps, two 40-watt U-shaped fluorescent lamps, or four 60-watt incandescent lamps. (*Courtesy of Progress Lighting*)

box above the eating area where a lighting track will be installed. From this outlet box, the circuit picks up the outdoor bracket luminaire. The range hood exhaust fan/light is fed from the outlet box above the other square luminaire on the ceiling.

As you study the cable layouts in this text, be assured that there are other ways to run the conductors. The thought process should be to sketch a few different cable-run possibilities. Then consider how many conductors enter the boxes, the ease or difficulty in making up the connections, and problems you might encounter because of obstructions (pipes and ducts) in the walls or ceilings. When installing recessed luminaires, be sure they are listed for "through wiring" if you intend to use the wiring compartment on the luminaire to continue on with the branch-circuit wiring. These luminaires are marked with the maximum number and size of conductors permitted in the wiring compartment.

Table 12-1 shows the outlets and estimated loads for the kitchen lighting Circuit B7.

KITCHEN LIGHTING

Kitchen lighting can become quite a challenge. Prior to installing the wiring, the electrician really has to consider the wishes of the homeowner. As stated

earlier, lighting is a personal choice. Arrange lighting to avoid shadows in work areas.

The general lighting for the Kitchen in this residence is provided by two square surface-mounted luminaires, as shown in Figure 12-2. These luminaires might have four straight or two U-bent fluorescent lamps. These luminaires are controlled by a single-pole switch located next to the doorway leading to the living room.

Another consideration for kitchen lighting could be a series of recessed luminaires installed in the ceiling approximately 18 in. (450 mm) from the edge of the cabinets, spaced about 3 ft (0.9144 m) to 4 ft (1.2192 m) apart. This would offer pleasant lighting for the kitchen area, the countertops, the oven, the pantry, and the refrigerator.

Still another possibility would be a track lighting system on the ceiling approximately 18 in. (450 mm) from the edge of the cabinets in front of the kitchen cabinets above the sink wall (West), and on the ceiling in front of the refrigerator and pantry wall (South). A number of track lighting heads can be installed and adjusted to provide good lighting without shadows on work surfaces.

Above the breakfast area table is a short section of track lighting controlled by two 3-way switches: one located adjacent to the sliding door,

FIGURE 12-3 Fixtures that can be hung either singly or on a track over dinette tables. A lighting track is shown in Figure 12-1 over the breakfast eating area. (*Courtesy of Progress Lighting*)

the other at the doorway leading to the Living Room. Figure 12-3 illustrates just a few of the many types of luminaires that might be secured to the track. A one-light stem-hung pendant luminaire or a few recessed luminaires could also be considered for this eating area.

Lighting for the electric range is in the range hood, with switching an integral part of the range hood.

Lighting over the sink is provided by the recessed luminaire above the sink, with switching to the right of the sink.

Figure 12-4 shows a variety of chain-suspended luminaires of the type commonly used in dining rooms.

Undercabinet Lighting

In some instances, particularly in homes with little outdoor exposure, the architect might specify strip fluorescent or LED luminaires under the kitchen cabinets. Several methods may be used to install under-cabinet luminaires. They can be fastened to the wall just under the upper cabinets (Figure 12-5[A]), installed in a recess that is part of the upper cabinets so that they are hidden from view (Figure 12-5[B]), or fastened under and to the front of the upper cabinets (Figure 12-5[C]). All three possibilities require close coordination with the cabinet installer to be sure that the wiring is brought out of the wall at the proper location to connect to the undercabinet luminaires.

Flat fluorescent and LED luminaires are available that mount on the underside of upper cabinets. Some of these luminaires are direct connected—others are cord-and-plug connected. These are pretty

much out of sight, but provide excellent lighting for countertop work surfaces.

Luminaire manufacturers also offer shallow undercabinet track lighting systems in both 120 volts and 12 volts. These systems include track, track lighting heads for small halogen lamps, end fittings, and transformer with cord and plug. Some have two-level, high-low switching to vary the light intensity.

Visiting a lighting showroom, an electrical supply house, a home center, or a hardware store is an exciting experience. The choice of luminaires is almost endless.

Lamp Types

Good color rendition in the kitchen area is achieved by using incandescent lamps, or by using warm white deluxe (WWX), warm white (WW), or deluxe cool white (CWX) 3000 K lamps wherever fluorescent lamps are used. See Chapter 7 for details regarding energy-saving ballasts and lamps.

FAN OUTLET

Ducted fans exhaust air to the outside. They remove cooking odors, steam, and smoke. Ductless fans remove cooking odors but do not remove steam or smoke.

In this residence, the range hood exhaust fan above the electric range is connected to lighting branch circuit B7. The speed control, light, and light switch are integral parts of the fan. The electric supply is brought down from the ceiling to the wiring compartment on the fan. For exhaust fans mounted under cabinets

FIGURE 12-4   Types of luminaires used over dining room tables. (*Courtesy of Progress Lighting*)

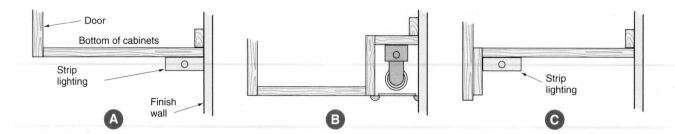

FIGURE 12-5   Methods of installing undercabinet fluorescent strip lighting. (*Delmar/Cengage Learning*)

along a wall, the electrical supply can be brought into the top or into the back of the fan housing wiring compartment. In either case, a sufficient length of cable must be left during the roughing-in stage to be able to make the final connections in the wiring compartment. Figure 12-6 is a typical range hood exhaust fan.

NEC 422.16(B)(4) permits a range hood to be cord-and-plug connected. The receptacle would be installed inside the cabinetry above the range. To do this, all of the following conditions must be met:

- The flexible cord must be identified as suitable for use on a range hood in accordance to the manufacturer's installation instructions.
- The flexible cord must have a grounding-type attachment plug cap.
- The flexible cord shall not be less than 18 in. (450 mm) and not over 36 in. (900 mm).

- The receptacle shall be located to avoid physical damage to the flexible cord.
- The receptacle shall be accessible.
- The receptacle is supplied by an individual branch circuit.

The logic behind the individual branch-circuit requirement is that a range hood installed today might later be replaced with a combination microwave/range hood appliance that is generally cord-and-plug connected.

The *NEC* defines an *individual branch circuit* as *A branch circuit that supplies only one utilization equipment.* To meet this requirement, either a single receptacle or a duplex receptacle that supplies no other load would have to be installed for the range hood. *NEC 210.21(B)(1)* requires that a single receptacle supplied by an individual branch circuit

FIGURE 12-6   Typical range hood exhaust fan. Speed control and light are integral parts of the unit. (*Courtesy of Broan-NuTone*)

shall have an ampere rating not less than the rating of the branch circuit. The range hood branch circuit could be a 15- or 20-ampere branch circuit. If you run a 15-ampere branch circuit, the single receptacle could be a 15- or a 20-ampere receptacle. If you run a 20-ampere branch circuit, the single receptacle shall be rated 20 amperes.

Instead of trying to outguess what the current draw of a future microwave/range hood combination might be, install a 20-ampere individual branch circuit.

Fan Noise

The fan motor and air movement through an exhaust fan produce a certain amount of noise. Exhaust fan manufacturers specify a sound-level rating in their descriptive literature. This provides some idea of the noise level one might expect from the fan after it is installed. The sound-level rating unit used to define fan noise is a *sone* (rhymes with tone).

For simplicity, one sone is the noise level of an average refrigerator. The lower the sone rating, the quieter the fan. All manufacturers of exhaust fans provide this information in their descriptive literature.

CLOCK OUTLETS

Battery-operated wall and mantle clocks for the most part have replaced electric clocks. They can be put just about anywhere. Choices are virtually unlimited.

Clock hanger receptacles are still available. In a clock hanger receptacle, the receptacle is set back from the front edge of the faceplate, providing room for the typical small plug cap on the end of a cord. This type of receptacle can be hidden behind a painting or photograph that you want to light up.

SMALL-APPLIANCE BRANCH CIRCUITS FOR RECEPTACLES IN THE KITCHEN

The *Code* requirements for small-appliance circuits and the spacing of receptacles in kitchens are covered in *210.11(C)(1)*, *210.52(B)*, and *220.52(A)*. This is discussed in detail in Chapter 3. Important points follow:

- At least two 20-ampere small-appliance circuits shall be installed, as in Figure 12-7.

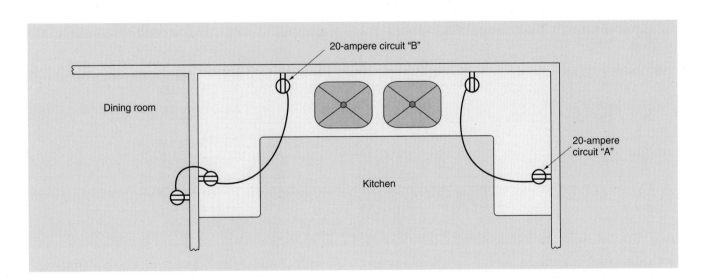

FIGURE 12-7   A receptacle outlet in a dining room may be connected to one of the two required small-appliance circuits in the Kitchen, *210.52(B)(1)*. Countertop receptacles must be supplied by at least two 20-ampere small-appliance circuits, *210.52(B)(3)*. Branch circuits for the receptacle and lighting outlets in dining rooms require listed combination AFCI devices per *210.12(A)*. If the Authority Having Jurisdiction (AHJ) considers a breakfast nook to be a "similar room or area" as stated in *210.12(A)*, then AFCI protection is required. (*Delmar/Cengage Learning*)

- Either (or both) of the two circuits required in the kitchen is permitted to supply receptacle outlets in other rooms, such as a dining room, breakfast room, or pantry, Figure 12-7.

- These small-appliance circuits shall be assigned a load of 1500 volt-amperes each when calculating feeders and service-entrance requirements.

- The receptacle for a refrigerator in a kitchen is permitted to be supplied by an individual 15- or 20-ampere branch circuit that is suitable for the load.

- A clock outlet may be connected to a small appliance branch circuit when this outlet is to supply and support the clock only (see Figure 3-3).

- Countertop receptacles must be supplied by at least two 20-ampere small-appliance circuits, as in Figure 12-7.

- All 125-volt, single-phase, 15- and 20-ampere receptacles installed in kitchens to serve the countertops shall be GFCI protected, *210.8(A)(6)*, Figure 12-8. A person can receive a deadly electrical shock by accidentally touching a "hot" wire

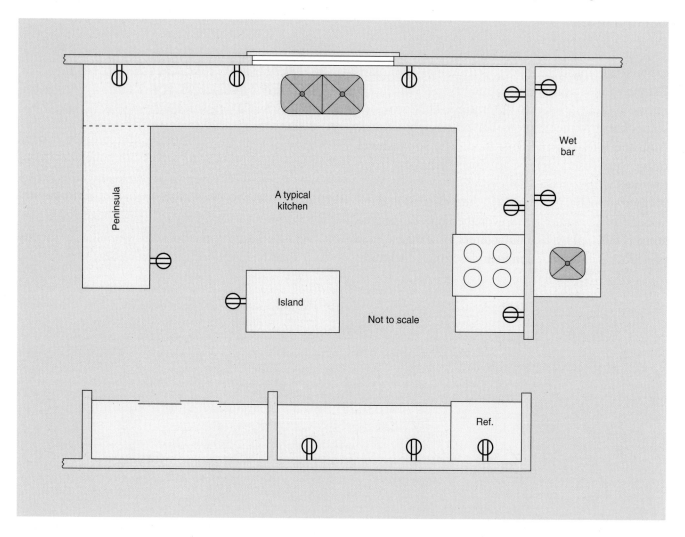

FIGURE 12-8 In kitchens, *all* 125-volt, single-phase, 15- and 20-ampere receptacles that serve countertop surfaces (walls, islands, or peninsulas) must be GFCI protected. The key words are "that serve countertop surfaces." GFCI is *not* required for the refrigerator receptacle, a receptacle under the sink for plugging in a food waste disposer, a receptacle installed inside cabinets above the countertops for plugging in a microwave oven attached to the underside of the cabinets, or a receptacle behind a range. ▶ *All 125-volt, single-phase, 15- and 20-ampere receptacles within 6 ft (1.8 m) of the outer edge of a sink shall be GFCI protected.* ◀ See *210.8(A)(7)*. (*Delmar/Cengage Learning*)

*Reprinted with permission from NFPA 70-2011.

or defective appliance and a grounded surface at the same time. A grounded surface could be a stainless steel sink, a faucet, frames of a range, an oven, a cook-top, a refrigerator, a microwave oven, a countertop food processor, a luminaire, a metal water pipe, or metal hot/cold air registers.

- A receptacle may be installed for plugging in a gas appliance to serve the ignition and/or clock devices on the appliance. See Figure 3-3.

- The 20-ampere small-appliance circuits are provided to serve plug-in portable appliances. They are not permitted by the *Code* to serve lighting or appliances such as dishwashers, garbage disposals, or exhaust fans.

- A receptacle installed below the sink for easy plug-in connection of a food waste disposer is *not* to be connected to the required small-appliance circuits, *210.52(B)(2)*, and it is *not* required to be GFCI protected, *210.8(A)(6)*.

Refer to Chapter 3 for spacing requirements for receptacles located in kitchens.

The 20-ampere small-appliance circuits prevent circuit overloading in the kitchen because of the heavy concentration and use of electrical appliances. For example, one type of cord-connected microwave oven is rated 1500 watts at 120 volts. The current required by this appliance alone is

$$I = \frac{W}{E} = \frac{1500}{120} = 12.5 \text{ amperes}$$

In *90.1(B)* we find a statement that *Code* requirements are considered necessary for safety. According to the *Code*, compliance with these rules will not necessarily result in an efficient, convenient, or adequate installation for good service or future expansion of electrical use.

Figure 12-9 shows four 20-ampere branch circuits feeding the receptacles in the kitchen area. These circuits are B13, B15, B16, and B22.

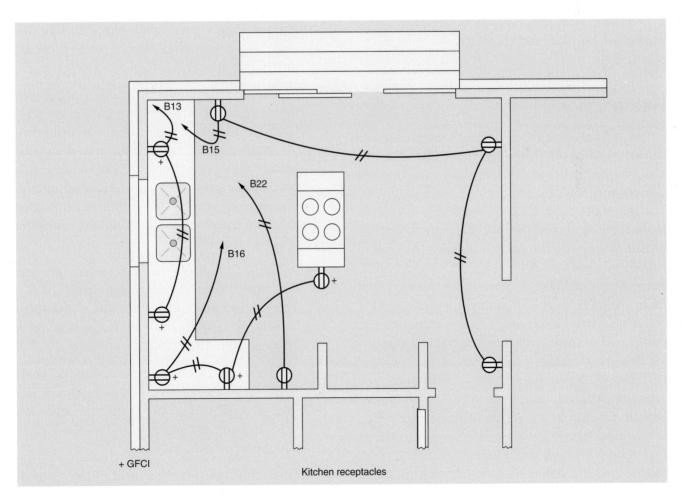

+ GFCI

Kitchen receptacles

FIGURE 12-9 Cable layout for kitchen receptacles. (*Delmar/Cengage Learning*)

Figure 12-9 shows one way to arrange the 20-ampere branch circuits in the kitchen area. Obviously, there are other ways to arrange these branch circuits. The intent is to arrange the circuits so as to provide the best availability of electrical power for appliances that will be used in the heavily concentrated work areas in the Kitchen.

All receptacles serving countertops in kitchens must be GFCI protected, *210.8(A)(6)*. See Chapter 6.

GFCI receptacles or GFCI circuit breakers can provide this protection.

In Figure 12-9, Circuits B13 and B16 are similar in that the circuit is run first to the line side terminals of a feedthrough GFCI receptacle. From the load-side terminals of the feedthrough GFCI receptacle, conductors are then run to the other grounding-type receptacle(s) on the circuit. GFCI circuit breakers could have been installed on these circuits, in which case the receptacles would have been grounding-type receptacles. The choice is yours!

Circuits B15 and B22 do not require GFCI protection because the receptacles do not serve countertops.

Refrigerator Receptacle

Many nuisance power outages in homes are caused by using too many high wattage small appliances in the kitchen at the same time—and then the refrigerator "kicks in."

The *NEC* does not help us in this respect. It *permits* the receptacle for the refrigerator to be connected to any one of the two required 20-ampere small-appliance branch circuits, *210.52(B)(1)*. Some home-type combination refrigerator/freezers with many features draw 11 amperes or more. That doesn't leave much room on a 20-ampere branch circuit for other small cord-and-plug-connected appliances in the kitchen.

The *NEC* also *permits* this receptacle to be connected to an individual 15- or 20-ampere branch circuit, *210.52(B)(1), Exception 2*. This is a much better way to do it. Many instructions furnished with these larger home-type refrigerator/ freezers require that an individual branch circuit be

*Reprinted with permission from NFPA 70-2011.

provided. In this residence, Circuit B22 is an individual (dedicated) 20-ampere branch circuit that supplies only the refrigerator receptacle. This is a single receptacle rated 20 amperes.

Following are a few more *NEC* requirements:

110.3(B) states, *Listed or labeled equipment shall be installed and used in accordance with any instructions included in the listing or labeling.**

This receptacle located behind the refrigerator does not require GFCI protection since it does not serve a countertop, *210.8(A)(6)*.

Branch Circuit, Individual: The *NEC* defines an *individual branch circuit* as *A branch circuit that supplies only one utilization equipment.*

- When an individual branch circuit is run for the refrigerator, some inspectors require that a single receptacle be installed to "meet Code." A single receptacle ensures that the receptacle will not supply other appliances, although this is highly unlikely since the receptacle is behind the refrigerator.

- Providing an individual branch circuit for the refrigerator diverts the refrigerator load from the receptacle branch circuits that serve the countertop areas. Because this individual branch circuit does not serve countertops, it is not considered a small-appliance branch circuit. The same logic would apply in other situations such as a receptacle located behind a gas range. It's up to the electrical inspector.

- A single receptacle must be rated not less than the branch-circuit rating, *210.21(B)(1)*. That's a 15-ampere receptacle on a 15-ampere branch circuit—a 20-ampere receptacle on a 20-ampere branch circuit.

- Cord-and-plug-connected appliances shall not exceed 80% of the branch-circuit rating, *210.23(A)(1)*. That's 12 amperes on a 15-ampere circuit, and 16 amperes on a 20-ampere circuit.

- This individual branch circuit is not considered to be additional load for service-entrance calculations, *220.14(J)*. The refrigerator load would have been there no matter what circuit it was connected to. The advantage of doing this is that it diverts the refrigerator load from the small-appliance branch circuits in the kitchen.

Sliding Glass Doors

See Chapters 3 and 9 for discussions on how to space receptacle outlets on walls where sliding glass doors are installed.

Small-Appliance Branch Circuit Load Calculations

The 20-ampere small-appliance branch circuits are figured at 1500 volt-amperes for each 2-wire circuit when calculating feeders and services. The separate 20-ampere branch circuit for the refrigerator is excluded from this requirement, *220.14(J)* and *220.53(A), Exception*. See Chapter 29 for complete coverage of load calculations.

SPLIT-WIRED RECEPTACLES AND MULTIWIRE CIRCUITS

Split-wired receptacles may be installed wherever desired or where a heavy concentration of plug-in load is anticipated, in which case each receptacle is connected to a separate circuit, as in Figure 12-10 and 12-11.

A popular use of split-wired receptacles is to control one receptacle with a switch and leave one receptacle "hot" at all times, as is done in the bedrooms and Living Room of this residence. However, this use is not common for kitchen receptacles.

Split-wired GFCI receptacles are not available. Refer to Chapter 6, where GFCIs are discussed in great detail.

Therefore, the use of split-wired receptacles is rather difficult to incorporate where GFCI receptacles are required. A multiwire branch

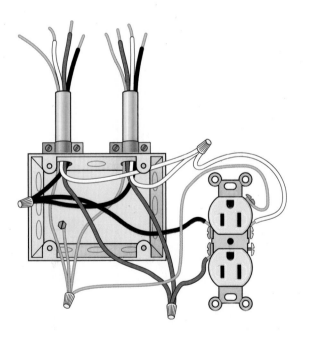

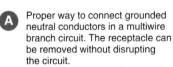

A Proper way to connect grounded neutral conductors in a multiwire branch circuit. The receptacle can be removed without disrupting the circuit.

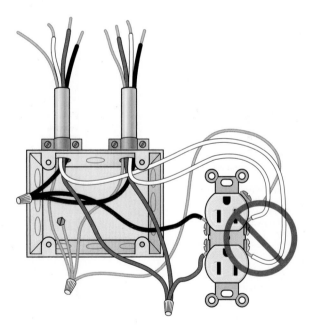

B Improper way to connect grounded neutral conductors in a multiwire branch circuit. Not permitted. Removing the receptacle will disrupt the circuit because the neutral bar on the receptacle is part of the circuit.

FIGURE 12-10   Connecting the grounded neutral conductors in a 3-wire (multiwire) branch circuit, *NEC 300.13(B)*. (*Delmar/Cengage Learning*)

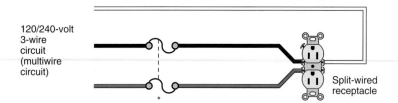

120/240-volt 3-wire circuit (multiwire circuit)

Split-wired receptacle

* NEC 210.4(B) and 210.7(B) require that a means must be provided to simultaneously disconnect both ungrounded conductors at the panelboard where the branch circuit originates. This could be a 2-pole switch with fuses, a 2-pole circuit breaker, or two single-pole circuit breakers with a listed handle tie.

FIGURE 12-11 Split-wired receptacle connected to multiwire circuit. (*Delmar/Cengage Learning*)

circuit could be run to a box where proper connections are made, with one circuit feeding one feedthrough GFCI receptacle and its downstream receptacles.

This would be used when it becomes more economical to run one 3-wire circuit rather than installing two 2-wire circuits. Again, it is a matter of knowing what to do, and when it makes sense to do it, as in Figure 12-12.

When multiwire circuits are installed, do not use the terminals of a receptacle or other wiring device to serve as the splice for the grounded neutral conductor. The *NEC* in *300.13(B)* states, *In multiwire branch circuits, the continuity of a grounded conductor shall not be dependent upon device connections . . .* Any splicing of the grounded conductor in a multiwire circuit must be made independently of the receptacle or lampholder, or other

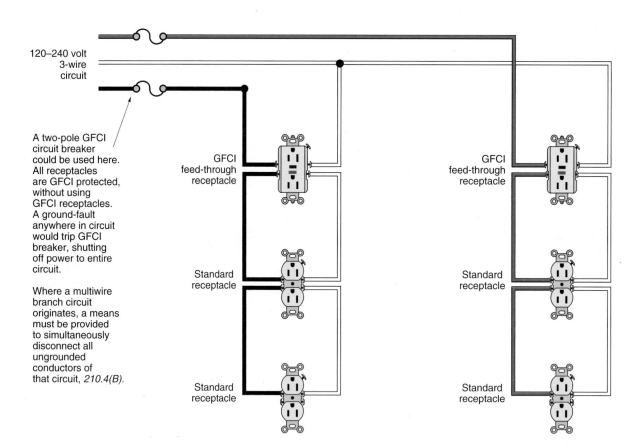

120–240 volt 3-wire circuit

A two-pole GFCI circuit breaker could be used here. All receptacles are GFCI protected, without using GFCI receptacles. A ground-fault anywhere in circuit would trip GFCI breaker, shutting off power to entire circuit.

Where a multiwire branch circuit originates, a means must be provided to simultaneously disconnect all ungrounded conductors of that circuit, 210.4(B).

GFCI feed-through receptacle

Standard receptacle

Standard receptacle

GFCI feed-through receptacle

Standard receptacle

Standard receptacle

FIGURE 12-12 This diagram shows how a multiwire circuit can be used to carry two circuits to a box, then split the 3-wire circuit into two 2-wire circuits. This circuitry requires simultaneous disconnect of the ungrounded conductors as shown in Figure 12-11. (*Delmar/Cengage Learning*)

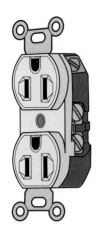 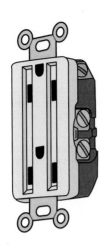

FIGURE 12-13 Grounding-type duplex receptacles, 15A-125V rating. Most duplex receptacles can be changed into "split-wired" receptacles by breaking off the small metal tab on the terminals. The figure clearly shows this feature. (*Delmar/Cengage Learning*)

wiring device. This requirement is illustrated in Figure 12-10. Note that the screw terminals of the receptacle *must* not be used to splice the neutral conductors. The hazards of an open neutral are discussed in Chapter 17. Figure 12-13 shows two types of grounding receptacles commonly used for installations of this type.

When multiwire branch circuits are used, a means must be provided to disconnect simultaneously (at the same time) all of the "hot" conductors at the panelboard where the branch circuit originates. Figure 12-11 shows a 120/240-volt, 3-wire multiwire branch circuit supplying a split-wired receptacle. Note that 240 volts are present on the wiring device, connected in such a way that 120 volts are connected to each receptacle on the wiring device. Multiwire branch circuits are also used to connect appliances such as electric clothes dryers and electric ranges. See *210.4(A)*, *210.4(B)*, and *210.7(B)*.

CAUTION: Care must be taken when connecting a GFCI to a multiwire branch circuit. As discussed in detail in Chapter 6, some of these devices are listed for use on multiwire branch circuits where the neutral is shared; others are not listed for this application.

Important: "Handle ties" are discussed in detail in Chapter 28. Two very important safety issues must be considered. Handle ties between two single-pole circuit breakers

- *do* provide simultaneous disconnection of both circuit breakers when they are manually turned off.

- *do not* provide simultaneous tripping (common tripping) of both circuit breakers in the event of an overload, short-circuit, or ground fault on only one of the circuits of a multiwire branch circuit supplied by the two "handle-tied" circuit breakers.

RECEPTACLES AND OUTLETS

Article 100 defines a receptacle and a receptacle outlet, as illustrated in Figures 3-4, 12-14, and 12-15.

Receptacles are selected for circuits according to the following guidelines. A single receptacle connected to a circuit must have a rating not less than the rating of the circuit, *210.21(B)*. In a residence, typical examples for this requirement for single receptacles are the clothes dryer outlet (30 amperes), the range outlet (50 amperes), or the freezer outlet (15 amperes).

Circuits rated 15 amperes supplying two or more receptacles shall not contain receptacles rated over 15 amperes. For circuits rated 20 amperes supplying two or more receptacles, the receptacles connected to the circuit may be rated 15 or 20 amperes. See *Table 210.21(B)(3)*.

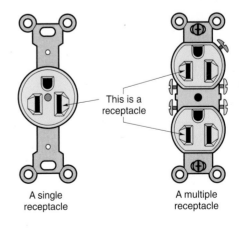

This is a receptacle

A single receptacle

A multiple receptacle

FIGURE 12-14 Receptacles: See *NEC* definitions, *Article 100*. (*Delmar/Cengage Learning*)

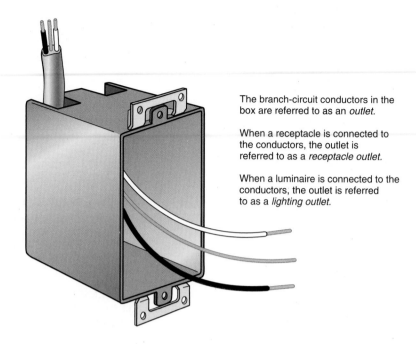

The branch-circuit conductors in the box are referred to as an *outlet*.

When a receptacle is connected to the conductors, the outlet is referred to as a *receptacle outlet*.

When a luminaire is connected to the conductors, the outlet is referred to as a *lighting outlet*.

FIGURE 12-15 An *outlet* is defined in the *NEC* as *A point on the wiring system at which current is taken to supply utilization equipment.* (*Delmar/Cengage Learning*)

REVIEW

Note: Refer to the *Code* or plans where necessary.

1. If everything on Circuit B7 were turned on, what would be the total current draw?

_____

2. From what panel does the kitchen lighting circuit originate? What size conductors are used? _____

_____

3. How many luminaires are connected to the kitchen lighting circuit?

_____

_____

4. What color fluorescent lamps are recommended for residential installations?

_____

_____

5. a. What is the minimum number of 20-ampere small-appliance circuits required for a kitchen according to the *Code*? _____

   b. How many are there in this kitchen? _____

6. How many receptacle outlets are provided in the kitchen? _____

7. What is meant by the term *two-circuit (split-wired) receptacles?* _____

_____

_____

_____

_____

8. Duplex receptacles connected to the 20-ampere small-appliance branch circuits in kitchens and dining rooms (circle the correct response):

   a. may be rated 15 amperes.   (True) (False)

   b. must be rated 15 amperes.   (True) (False)

   c. may be rated 20 amperes.   (True) (False)

   d. must be rated 20 amperes.   (True) (False)

9. In kitchens, a receptacle must be installed at each counter space _____ in. or wider.

10. A fundamental rule regarding the grounding of metal boxes, luminaires, and so on, is that they must be grounded when "in reach of _____."

11. How many circuit conductors enter the box

   a. where the range hood will be installed? _____

   b. in the ceiling box over which the track will be installed? _____

   c. at the switch location to the right of the sliding door? _____

12. How much space is there between the countertop and upper cabinets? _____

13. Where is the speed control for the range hood fan located? _____

14. Who is to furnish the range hood? _____

15. List the appliances in the kitchen that must be electrically connected. _____

_____

_____

_____

_____

_____

_____

16. Complete the wiring diagram, connecting feedthrough GFCI 2 to also protect receptacle 1, both to be supplied by Circuit A1. Connect feedthrough GFCI 3 to also protect

receptacle 4, both to be supplied by Circuit A2. Use colored pencils or markers to show proper color. Assume that the wiring method is EMT, where more freedom in the choice of insulation colors is possible.

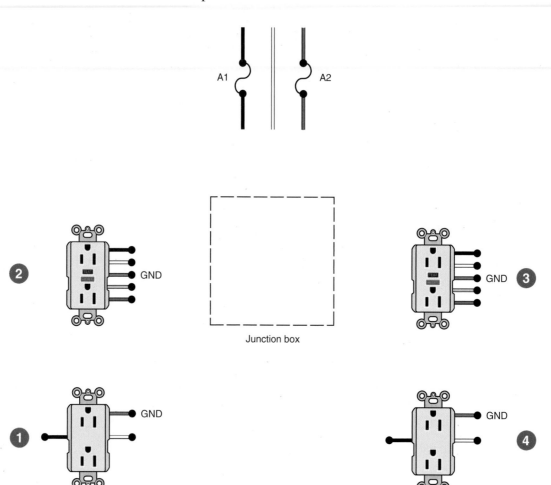

Junction box

17. Each 20-ampere small-appliance branch-circuit load demand shall be determined at _____. Choose one.

   a. 2400 volt-amperes   b. 1500 volt-amperes   c. 1920 volt-amperes

18. a. Does the *NEC* permit the receptacle for the refrigerator to be supplied by one of the required 20-ampere small-appliance branch circuits? Yes _____
   No _____ *Section* _____.

   b. Does the *NEC* permit the receptacle for the refrigerator to be supplied by an individual branch circuit? Yes _____ No _____
   Section _____.

   c. What is the practice in your area?

   Run an individual branch circuit? _____

   Connect it to one of the 20-ampere small-appliance branch circuits? _____

19. a. The *Code* requires a minimum of two small-appliance circuits in a kitchen. Is it permitted to connect a receptacle in a dining room to one of the kitchen small-appliance circuits? _____

What *Code* section applies to this situation? _____

b. May outdoor weatherproof receptacles be connected to a 20-ampere small-appliance circuit? _____

c. Receptacles located above the countertops in the kitchen must be supplied by at least _____ 20-ampere small-appliance circuits.

20. According to *210.52*, no point along the floor line shall be more than _____ ft (_____ m) from a receptacle outlet. A receptacle must be installed in any wall space _____ ft (_____ m) wide or greater.

21. The *Code* states that in multiwire circuits, the screw terminals of a receptacle must *not* be used to splice the neutral conductors. Why?

_____

_____

_____

22. Electric fans produce a certain amount of noise. It is possible to compare the noise levels of different fans prior to installation by comparing their _____ ratings.

23. Is it permitted to connect the white grounded circuit conductor to the equipment grounding terminal of a receptacle? _____

24. What is unique about 120-volt appliances that are 2-wire cord-and-plug connected instead of being 3-wire cord-and-plug connected? _____

_____

25. a. All 125-volt, single-phase 15- and 20-ampere receptacles installed in a kitchen that serve countertop surfaces must be GFCI protected. (True) (False) Circle the correct answer.

b. All 125-volt, single-phase 15- and 20-ampere receptacles installed within 6 ft (1.8 m) of the outside edge of a wet bar, laundry, utility, or similar sink must be GFCI protected. Circle the correct answer.   (True) (False)

26. The kitchen features a sliding glass door. For the purpose of providing the proper receptacle outlets, answer the following statements *true* or *false*.

a. Sliding glass panels are considered to be wall space. _____

b. Fixed panels of glass doors are considered to be wall space. _____

27. The following is a layout for the lighting circuit for the Kitchen. Complete the wiring diagram using colored pens or pencils to show the conductors' insulation color.

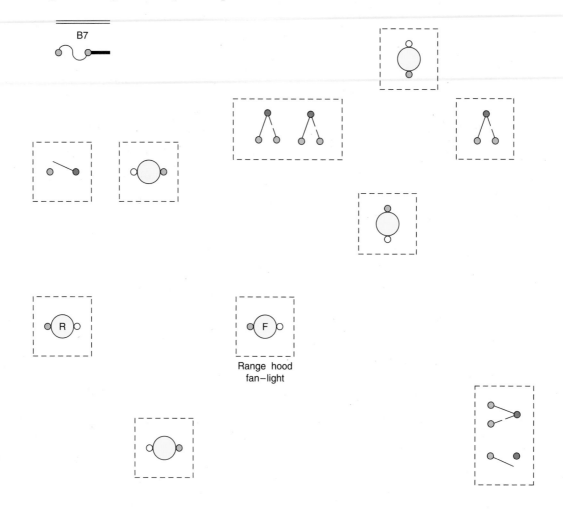

Kitchen

28. a. In your area, does the electrical inspector require that an individual branch circuit be installed for the refrigerator? (Yes) (No) Circle the correct answer.

   b. In your area, does the electrical inspector require that a single receptacle be installed for the refrigerator? (Yes) (No) Circle the correct answer.

Lighting Branch Circuit for the Living Room

OBJECTIVES

After studying this chapter, you should be able to

- lay out the wiring for a typical living room.
- understand the *NEC* requirements for track lighting.
- understand the basics for dimming incandescent and fluorescent lamps.

LIGHTING CIRCUIT B17 OVERVIEW

The feed for the living room lighting branch circuit is connected to Circuit B17, a 15-ampere circuit. Overcurrent protection and arc-fault protection for this circuit are provided by a 15-ampere AFCI circuit breaker in Panel B. The AFCI requirement is found in *210.12(A)*. GFCI protection required by *210.8(A)(3)* for the outdoor receptacle could be provided by a GFCI receptacle or by an AFCI/GFCI dual-function circuit breaker in Panel B.

The home run is brought from Panel B to the weatherproof receptacle outside of the Living Room next to the sliding door. The circuit is then run to the split-wired receptacle just inside the sliding door, almost back to back with the outdoor receptacle, Figure 13-1.

A 3-wire cable is then carried around the Living Room, feeding in and out of the eight receptacles. This 3-wire cable carries the black and white circuit conductors plus the red wire, which is the switched conductor. Because these receptacles will provide most of the general lighting for the Living Room

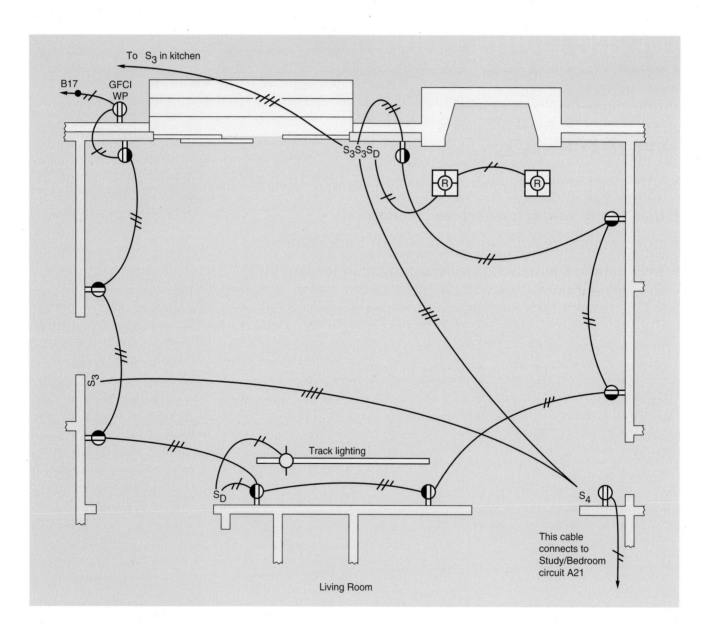

FIGURE 13-1 Cable layout for the Living Room. Note that the receptacle on the short wall where the 4-way switch is located is fed from the Study/Bedroom circuit A21. (*Delmar/Cengage Learning*)

through the use of floor lamps and table lamps, it is advantageous to have some of these lamps controlled by the three switches—two 3-way switches and one 4-way switch.

The receptacle below the 4-way switch in the corner of the Living Room is required because the *Code* requires that a receptacle be installed in any wall space 24 in. (600 mm) or more in width, *210.52(A)(2)(1)*. It is highly unlikely that a split-wired receptacle connected for switch control, as are the other receptacles in the Living Room, will ever be needed. Plus, the short distance to the Study/Bedroom receptacle makes it economically sensible to make up the connections as shown on the cable layout.

Accent lighting above the fireplace is in the form of two recessed luminaires controlled by a single-pole dimmer switch. See Chapter 7 for installation data and *Code* requirements for recessed luminaires.

Table 13-1 summarizes the outlets and estimated load for the living room circuit.

The spacing requirements for receptacle outlets and location requirements for lighting outlets are covered in Chapter 3, as is the discussion on the spacing of receptacle outlets on exterior walls where sliding glass doors are involved, *210.52(A)(2)(2)*.

You might want to consider running more than one branch circuit for the receptacles in the Living Room. You might also consider using 12 AWG for the "home run" or even for the entire branch circuit. This may be to provide for a concentration of stereo surround sound, TV, CD burners and players, video recorders and players, DVD recorders and players, personal computers, printers, fax machines, copy machines, adding machines, paper shredders, all sorts of other audio/video equipment, and desk lamps. It's a judgment call. Remember, the *Code* is a minimum!

TRACK LIGHTING (ARTICLE 410, PART XV)

The plans show that track lighting is mounted on the ceiling of the Living Room on the wall opposite the fireplace. The lighting track is controlled by a single-pole dimmer switch.

Track lighting provides accent lighting for a fireplace, a painting on the wall, or some item (sculpture or collection) that the homeowner wishes to focus attention on, or it may be used to light work areas such as counters, game tables, or tables in such areas as the kitchen eating area.

Differing from recessed luminaires and individual ceiling luminaires that occupy a very definite space, track lighting offers flexibility because the actual lampholders can be moved and relocated on the track as desired.

Lampholders from track lighting selected from hundreds of styles are inserted into the extruded aluminum or PVC track at any point (the circuit conductors are in the track), and the plug-in connector on the lampholder completes the connection. In addition to being fastened to the outlet box, the track is generally fastened to the ceiling with toggle bolts or screws. Various track light installations are shown in Figures 13-2, 13-3, and 13-4.

Note on the plans for the residence that the living room ceiling has wood beams. This presents a challenge when installing track lighting that is longer than the space between the wood beams.

There are a number of things the electrician can do:

1. Install pendant kit assemblies that will allow the track to hang below the beam, as in Figure 13-5.

TABLE 13-1

Living Room outlet count and estimated load (Circuit B17).

Description	Quantity	Watts	Volt-Amperes
Receptacles @ 120 W Note: Receptacle under S₄ is connected to Study/Bedroom circuit.	8	960	960
Weatherproof receptacle	1	120	120
Track light five lamps @ 40 W each	1	200	200
Fireplace recessed luminaires @ 75 W each	2	150	150
TOTALS	12	1430	1430

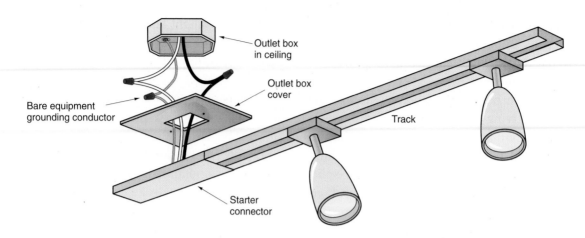

FIGURE 13-2 End-feed track light. (*Delmar/Cengage Learning*)

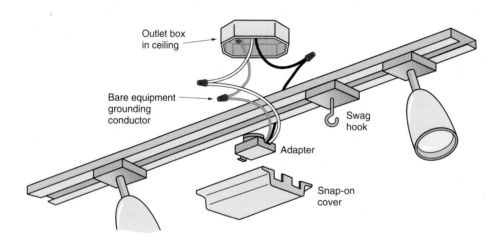

FIGURE 13-3 Center-feed track light. (*Delmar/Cengage Learning*)

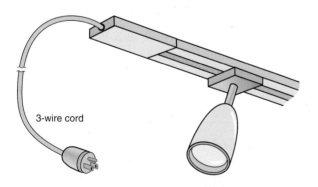

FIGURE 13-4 Plug-in track light.
(*Delmar/Cengage Learning*)

2. Install conduit conductor fittings on the ends of sections of the track, then drill a hole in the beam through which ½-in. EMT can be installed and connected to the conduit connector fittings, as in Figure 13-6.

3. Run the branch-circuit wiring concealed above the ceiling to outlet boxes located at the points where the track is to be fed by the branch-circuit wiring, as in Figure 13-7.

Where to Mount Lighting Track

Lighting track manufacturers' catalogs and Web sites provide excellent recommendations for achieving the desired lighting results.

Lighting Track and the *Code*

Lighting track is listed under UL Standard 1574.

The *Code* rules for lighting track are found in *410.151* through *410.155*.

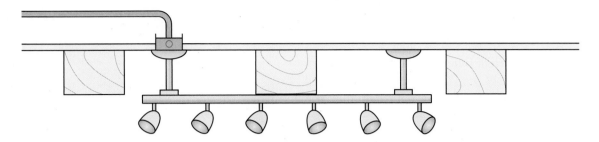

FIGURE 13-5 Track lighting mounted on the bottom of a wood beam. The supply wires to the track assembly are fed through the pendant. (*Delmar/Cengage Learning*)

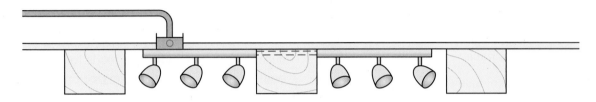

FIGURE 13-6 Track lighting mounted on the ceiling between the wood beams. The two track assembly sections are connected together by means of a conduit that has been installed through a hole drilled in the beam. (*Delmar/Cengage Learning*)

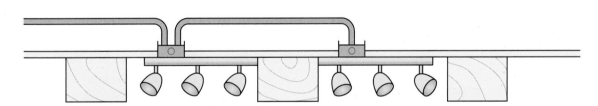

FIGURE 13-7 Track lighting mounted on the ceiling between the wood beams. Each of the two track assembly sections are connected to outlet boxes in the ceiling in the same manner that a regular luminaire is hung. (*Delmar/Cengage Learning*)

Lighting track in residential installations

- shall be listed by an NRTL.
- shall be installed according to the manufacturers' instructions, per *110.3(B)*.
- shall not be installed
  - where subject to physical damage.
  - in damp or wet locations.
  - concealed.
  - through walls or partitions.
  - less than 5 ft (1.5 m) above finished floor unless protected against physical damage, or if the track is of the low-voltage type operating at less than 30 volts rms open-circuit voltage. See *Article 411* for low-voltage system requirements.
  - in the zone 3 ft (900 mm) horizontally and 8 ft (2.5 m) vertically from the top of a bathtub rim or shower stall threshold. This is discussed in Chapter 10.

- shall be permanently installed and permanently connected to a branch circuit.

- shall be supported twice if the track is not more than 4 ft (1.2 m), and shall have one additional support for each additional 4 ft (1.2 m) of track.

- shall be grounded. Grounding of equipment is covered in previous chapters.

- shall not be cut to length in the field unless permitted by the manufacturer. Instructions will specify where and how to make the cut, and how to close off the cut with a proper end cap.

- shall have lampholders designed specifically for the particular track.

- shall not have so many lampholders that the load would exceed the rating of the track.

Figure 13-8 shows cross-sectional views of a typical lighting track. Figure 13-9 shows typical lighting track lampholders.

The track lighting in the Living Room of this residence is controlled by a single-pole dimmer switch.

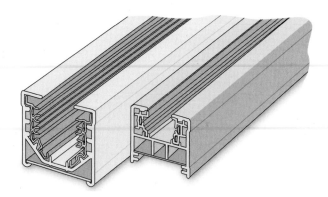

FIGURE 13-8 Tracks in cross-sectional views. (*Delmar/Cengage Learning*)

Load Calculations for Track Lighting

For track lighting installed in homes, it is not necessary to add additional loads for the purpose of branch circuit, feeder, or service-entrance calculations. Track lighting is just another form of general lighting. Refer to *220.43(B)*.

In a commercial building, a loading factor of 150 VA for each 2 ft (600 mm) of lighting track must be added to the calculations for branch circuit, feeder, and service-entrance calculations, *220.43(B)*.

For further information regarding track lighting, refer to *Part XV* of *Article 410*.

FIGURE 13-9 Typical lampholders that attach to track lighting systems. (*Courtesy of Progress Lighting*)

Portable Lighting Tracks

Portable lighting tracks, per UL Standard 153 shall

- not be longer than 8 ft (2.4 m).
- have a cord not less than 5 ft (1.5 m) permanently attached to the track by the manufacturer.
- have overcurrent protection (fuse or circuit breaker) built into the plug cap. Overcurrent protection in the plug cap is not required if the conductors in the cord are 12 AWG, rated 20 amperes.
- not have its length altered in the field. To do so would void the UL listing.

DIMMER CONTROLS FOR HOMES

Dimmers are used in homes to lower the level of light. There are two basic types: electronic and autotransformer.

Electronic Dimmers

Figure 13-10 shows two types (slide and rotate) of electronic dimmers. These are very popular.

They fit in a standard switch (device) box. They can replace standard single-pole or 3-way toggle switches. See Figure 13-11 for the typical hookup of a single-pole and a 3-way dimmer.

The most common type of electronic dimmer for residential use is rated 600 watts, 125 VAC, 60 hertz. They are also available in 1000-watt ratings. For larger wattage requirements, consult manufacturers' literatures and Web sites. The connected load must not exceed the marked wattage rating of the dimmer.

Electronic dimmers offer many features such as LED nightlight, soft start when changing from "Off" to the desired setting, and return to the previous setting after a power outage.

Dimmer wiring devices and fan speed controls are physically larger than typical wiring devices (receptacles and switches). Think ahead to make sure you install a large enough wall box to accommodate the dimmer, conductors, and splices that are in the box. To jam the dimmer into the box may cause trouble, such as short circuits and ground faults.

Dimmers that install in flush device boxes are intended to control only permanently installed incandescent lighting unless listed for other applications, *404.14(E)*. The instructions will state something

FIGURE 13-10 Some typical electronic dimmers. Both perform the same function of varying the voltage to the connected incandescent lamp load: (A) rotating knob; (B) slider knob. (*Courtesy of Legrand/Pass & Seymour*)

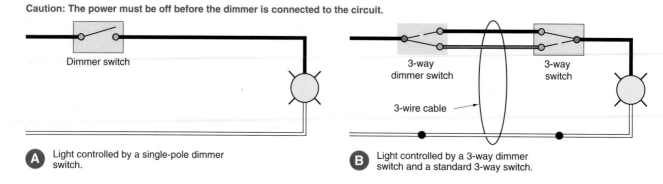

Caution: The power must be off before the dimmer is connected to the circuit.

A Light controlled by a single-pole dimmer switch.

B Light controlled by a 3-way dimmer switch and a standard 3-way switch.

FIGURE 13-11 Use of single-pole and 3-way electronic dimmer controls in circuits. (*Delmar/Cengage Learning*)

like "Do not use to control receptacle outlets." The reasons for this follow:

1. Serious overloading could result, because it may not be possible to determine or limit what other loads might be plugged in the receptacle.

2. Dimmers reduce voltage. Reduced voltage supplying a television, radio, stereo components, personal computer, vacuum cleaner, and so on, can result in costly damage to the appliance.

Electronic dimmers for use *only* to control incandescent lamps are marked with a *wattage* rating. Serious overheating and damage to the transformer and to the electronic dimmer will result if an "incandescent only" dimmer is used to control inductive loads.

For example, a dimmer marked 600 watts @ 120 volts is capable of safely carrying 5 amperes.

$$\text{Amperes} = \frac{\text{Watts}}{\text{Volts}} = \frac{600}{120} = 5 \text{ amperes}$$

Dimmers that are suitable for control of "inductive" loads, such as lighting systems that incorporate a transformer (as in low-voltage outdoor lighting and low-voltage track lighting), will be marked in *volt-amperes*. The instructions will clearly indicate that the dimmer can be used on inductive loads.

See Chapter 7 for a discussion of watts versus volt-amperes.

Do *not* hook up a dimmer with the circuit "hot." Not only is it dangerous because of the possibility of receiving an electrical shock, but the electronic

dimmer can be destroyed when the splices are being made as a result of a number of "make and breaks" before the actual final connection is completed. The momentary inrush of current each time there is a "make and break" can cause the dimmer's internal electronic circuitry to heat up beyond its thermal capability. As one lead is being connected, the other lead could come in contact with "ground." Unless the dimmer has some sort of built-in short-circuit protection, the dimmer is destroyed. The manufacturer's warranty will probably be null and void if the dimmer is worked "hot." It is well worth the time to take a few minutes to de-energize the circuit.

Electronic dimmers use triacs to "chop" the ac sine wave, so some incandescent lamps might "hum" because the lamp filament vibrates. This hum can usually be eliminated or greatly reduced by turning the dimmer setting to a different brightness level; changing the lamp, installing a "rough service lamp"; or, as a last resort, changing the dimmer.

Autotransformer Dimmers

Figure 13-12 shows an autotransformer (one winding) type of dimmer. These are physically larger than the electronic dimmer and require a special wall box. They are not ordinarily used for residential applications unless the load to be controlled is very large. They are generally used in commercial applications for the control of large loads. The lamp intensity is controlled by varying the voltage to the lamp load. When the knob is rotated, a brush contact moves over a bared portion of the transformer winding. As

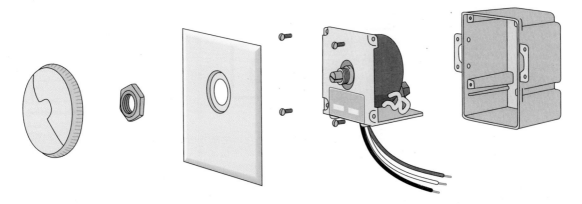

FIGURE 13-12 Dimmer control, autotransformer types. (*Delmar/Cengage Learning*)

with electronic dimmers, they will not work on direct current, but that is not a problem because all homes are supplied with alternating current. The maximum wattage to be controlled is marked on the nameplate of the dimmer. Overcurrent protection is built into the dimmer control to prevent burnout due to an overload.

Figure 13-13 illustrates the connections for an auto-transformer dimmer that controls incandescent lamps.

Dimming Fluorescent Lamps

To dim incandescent lamps, the voltage to the lamp filament is varied (raised or lowered) by the dimmer control. The lower the voltage, the less intense is the light. The dimmer is simply connected "in series" with the incandescent lamp load.

Fluorescent lamps are a different story. The connection is not a simple "series" circuit. To dim fluorescent lamps, special dimming ballasts and special dimmer switches are required. The dimmer control allows the dimming ballast to maintain a voltage to the cathodes that will maintain the cathodes' proper operating temperature, and also allows the dimming ballast to vary the current flowing in the arc. This in turn varies the intensity of light coming from the fluorescent lamp.

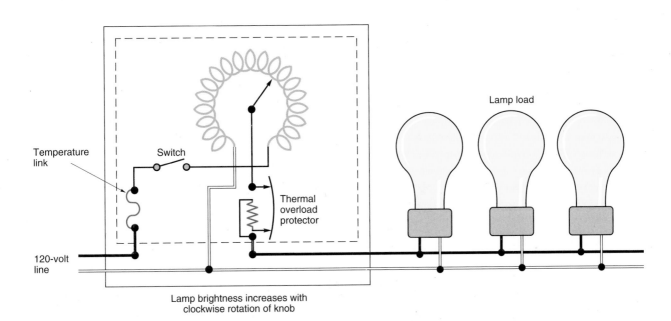

FIGURE 13-13 Dimmer control (autotransformer) wiring diagram for incandescent lamps. (*Delmar/Cengage Learning*)

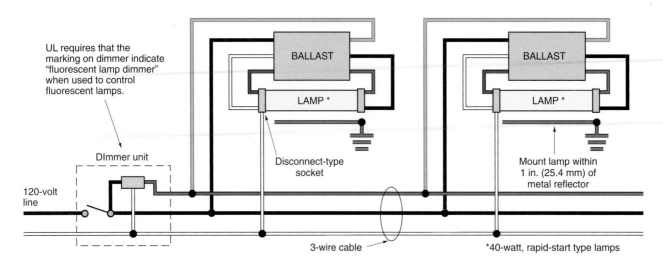

FIGURE 13-14 Dimmer control wiring diagrams for fluorescent lamps. (*Delmar/Cengage Learning*)

There is an exception to the need to use special fluorescent dimming ballasts and special dimmer controls. Some compact fluorescent lamps are available that have an integrated circuit built right into the lamp, allowing the lamp to be controlled by a standard incandescent wall dimmer. These dimmable compact fluorescent lamps can be used to replace standard incandescent lamps of the medium screw shell base type. No additional wiring is necessary. They are designed for use with dimmers, photocells, occupancy sensors, and electronic timers that are marked "Incandescent Only."

Figure 13-14 illustrates the connections for an autotransformer-type dimmer for the control of fluorescent lamps using a special dimming ballast. The fluorescent lamps are F40T12 rapid-start lamps. With the newer electronic dimming ballasts, 32-watt, rapid-start T8 lamps are generally used.

Figure 13-15 shows an electronic dimmer for the control of fluorescent lamps using a special dimming ballast.

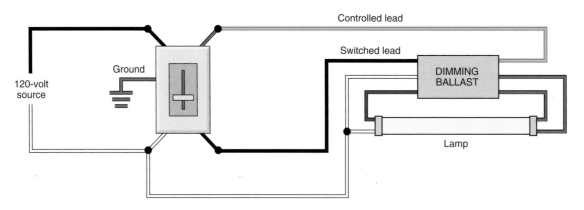

FIGURE 13-15 This wiring diagram shows the connections for an electronic dimmer designed for use with a special dimming ballast for rapid-start lamps. Always check the dimmer and ballast manufacturers' wiring diagrams, as the color coding of the leads and the electrical connections may be different than shown in this diagram. The dimmer shown is a "slider" type, and offers wide-range (20% to 100%) adjustment of light output for the fluorescent lamp. (*Delmar/Cengage Learning*)

Fluorescent dimming ballasts and controls are available in many types such as: simple "ON–OFF," 100% to 1% variable light output, 100% to 10% variable light output, and two-level 100% or 50% light output. Wireless remote control is also an option.

Here are some hints to ensure proper dimming performance of fluorescent lamps:

- Make sure that the lamp reflector and the ballast case are solidly grounded.
- Do not mix ballasts or lamps from different manufacturers on the same dimmer.
- Do not mix single- and two-lamp ballasts.
- "Age" the lamps for approximately 100 hours before dimming. This allows the gas cathodes in the lamps to stabilize. Without "aging" (sometimes referred to as "seasoning"), it is possible to have unstable dimming, lamp striation (stripes), and/or unbalanced dimming characteristics between lamps.

- Because incandescent lamps and fluorescent lamps have different characteristics, they cannot be controlled simultaneously with a single dimmer control.
- Do not control electronic ballasts and magnetic ballasts simultaneously with a single dimmer control.

Always read and follow the manufacturer's instructions furnished with dimmers and ballasts.

Without special dimming fluorescent ballasts and controls, a simple way to have two light levels is to use 4-lamp fluorescent luminaires, such as those installed in the Recreation Room. Then, control the ballast that supplies the center two lamps with one wall switch, and control the ballast that supplies the outer two lamps with a second wall switch. This is sometimes referred to as an "inboard/outboard" connection. Lamps and ballasts are discussed in Chapter 7.

REVIEW

Note: Refer to the *Code* or the plans where necessary.

1. To what circuit is the living room connected? _____

2. How many receptacles are connected to the living room circuit? _____

3. a. How many wires enter the switch box at the 4-way switch location? _____

   b. What type and size of box may be installed at this location? _____

   _____

   _____

   _____

4. How many wires must be run between an incandescent lamp and its dimmer control?

   _____

5. Complete the wiring diagram for the dimmer and lamp.

120-volt
supply

6. Is it possible to dim standard fluorescent ballasts? _____

7. a. How many wires must be run between a dimming-type fluorescent ballast and the dimmer control? _____

   b. Is a switch needed in addition to the dimmer control? _____

   _____

   _____

   _____

   _____

8. Explain why fluorescent lamps having the same wattage draw different current values.

   _____

   _____

   _____

   _____

   _____

9. What is the total current consumption of the track lighting and recessed luminaires above the fireplace? Show your calculations.

10. How many television outlets are provided in the Living Room? _____

11. Where is the telephone outlet located in the Living Room? _____

    _____

12. Electronic dimmers of the type sold for residential use (shall) (shall not) be used to control fluorescent luminaires. These dimmers (are) (are not) intended for speed control of small motors. (Circle the correct answers.)

13. When calculating a dwelling lighting load on the "volt-ampere per ft²" basis, the receptacle load used for floor lamps, table lamps, and so forth,

    a. must be added to the general lighting load at 1½ amperes per receptacle.

    b. must be added to the general lighting load at 1500 volt-amperes for every 10 receptacles.

    c. is already included in the calculations as part of the volt-amperes per ft².
       (Circle the letter of the correct answer.)

14. A layout of the outlets, switches, dimmers, track lighting, and recessed luminaires is shown in the following diagram. Using the cable layout of Figure 13-1, make a complete wiring diagram of this circuit. Use colored pencils or marking pens to indicate conductors.

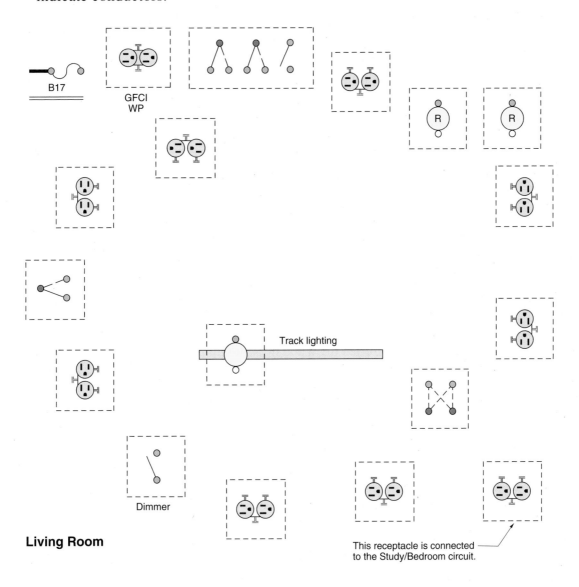

15. Prepare a list of television outlets, telephone outlets, and wiring devices shown on the plans and cable layout for the living room area. Include the number of each type present.

_____

_____

_____

_____

_____

_____

16. The Living Room features a sliding glass door. For the purpose of providing the proper receptacle outlets, answer the following statements *true* or *false*.

   a. Sliding glass panels are considered wall space. _____

   b. Fixed panels of glass doors are considered wall space. _____

17. a. May lengths of track lighting be added when the track is permanently connected?

   _____

   b. May lengths of track lighting be added when the track is cord connected?

   _____

18. Must track lighting always be fed (connected) at one end of the track? _____

   _____

19. May a standard electronic dimmer be used to control low-voltage lighting that incorporates a transformer, or for controlling the speed of a ceiling fan motor?

   Yes _____ No_____

20. Does the *Code* permit cutting lengths of portable track lighting to change their length?

   Yes_____ No_____

21. Why should the power be turned off when hooking up a dimmer? _____

   _____

   _____

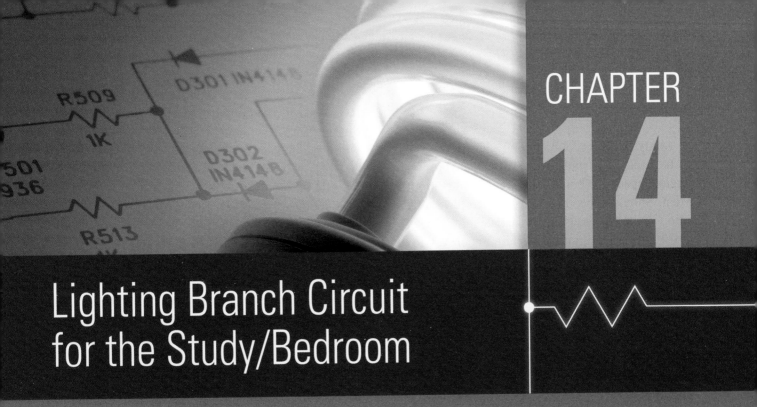

Lighting Branch Circuit for the Study/Bedroom

CHAPTER 14

OBJECTIVES

After studying this chapter, you should be able to

- discuss valance lighting.
- make all connections in the Study/Bedroom for the receptacles, switches, fan, and lighting.
- discuss surge suppressors.

339

CIRCUIT A21 OVERVIEW

The Study/Bedroom circuit is so named because it can be used as a study for the present, providing excellent space for home office, personal computer, or den. Should it become necessary to have a third bedroom, the change is easily made.

Circuit A21 originates at Main Panel A. Circuit A21 is a 15-ampere branch circuit. Overcurrent protection and arc-fault protection for this circuit are provided by a 15-ampere AFCI circuit breaker in Panel A as required by *210.12(B)*. AFCIs are discussed in detail in Chapter 6.

This circuit feeds the Study/Bedroom at the receptacle in the Living Room just outside of the Study. From this point, the circuit feeds the split-wired receptacle just inside the door of the Study. A 3-wire cable runs around the room to feed the other four split-wired receptacles. Figure 14-1 shows the cable layout for Circuit A21.

The black and white conductors carry the "live" circuit, and the red conductor is the "switch return" for the control of one portion of the split-wired receptacles.

There are four switched split-wired receptacles in the Study/Bedroom, controlled by two 3-way switches. A fifth receptacle on the short wall leading to the bedroom hallway is not switched. The ready

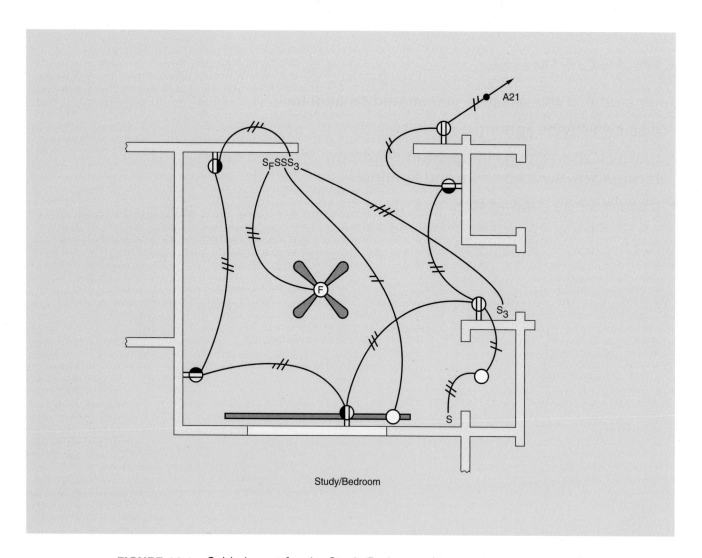

FIGURE 14-1 Cable layout for the Study/Bedroom. (*Delmar/Cengage Learning*)

TABLE 14-1

Study/Bedroom outlet count and estimated load (Circuit A21).

Description	Quantity	Watts	Volt-Amperes
Receptacles @ 120 W each	6	720	720
Closet luminaire	1	75	75
Paddle fan/light	1		
Three 50-W lamps		150	150
Fan motor		80	90
(0.75 A 120 V)			
Valance lighting	1		
Two 40-W fluorescent lamps		80	90
TOTALS	9	1105	1125

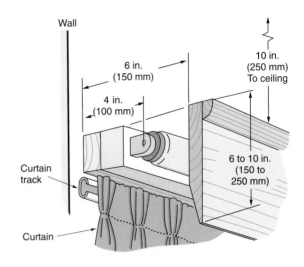

FIGURE 14-2 Fluorescent lighting behind the valance. (*Delmar/Cengage Learning*)

access to this receptacle makes it ideal for plugging in vacuum cleaners or similar appliances.

A ceiling fan/light is mounted on the ceiling. This installation is covered in Chapter 9. Table 14-1 summarizes the outlets and estimated load for the Study/Bedroom circuit.

As we all know, when someone decides to use any room, such as the Study/Bedroom, as a home office, a lot of electronic equipment will have to be plugged in. The current draw is very low, but the list for electronic equipment never ends! You will never have enough wall receptacles. Plug-in strips—probably of the surge protection type—are almost always needed.

You could consider:

- running more than one branch circuit for the receptacles and lighting in the Study/Bedroom.

- using 12 AWG for the home run or even for the entire branch circuit(s).

- that all 120-volt, single-phase, 15- and 20-ampere branch circuits that supply outlets in bedrooms require AFCI protection.

- wiring in quadplex receptacles instead of duplex receptacles.

- roughing in more wall receptacles than the *NEC* minimum.

- using only plug-in strips that are UL listed.

As previously mentioned, the arrival of AFCI requirements might require new thinking about how to do the wiring for bedrooms.

VALANCE LIGHTING

Figure 14-2 illustrates an interesting indirect fluorescent lighting treatment above the windows and behind the valance board.

If the fluorescent valance lighting is to have a dimmer control, then special dimming ballasts would be required. See the discussion on fluorescent lamp dimming in Chapter 13.

The Study/Bedroom wiring is rather simple. Most of the concepts are covered in previous chapters in this text.

Figure 14-3 suggests one way to arrange the wiring for the ceiling fan/light control, the valance lighting control, and the receptacle outlet control.

SURGE SUPPRESSORS

Transient voltage surge suppressors, covered in Chapter 6, minimize the possibility of voltage surges damaging sensitive electronic equipment.

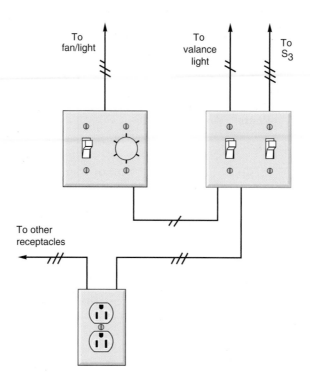

FIGURE 14-3 Conceptual view of how the Study/Bedroom switch arrangement is to be accomplished (Circuit A21). (*Delmar/Cengage Learning*)

REVIEW

Note: Refer to the *Code* or plans where necessary.

1. Based on the total estimated load calculations, what is the current draw on the Study/ Bedroom circuit? _____

2. a. The Study/Bedroom is connected to circuit _____ .

   b. The conductor size for this circuit is _____ .

3. a. What unique type of electrical protection is required for all 120-volt, single-phase, 15- and 20-ampere branch circuits for most habitable rooms and areas in new homes?

   _____

   b. What section of the *Code* covers this kind of protection? _____

4. Why is it necessary to install a receptacle in the wall space leading to the bedroom hallway? _____

   _____

   _____

   _____

   _____

5. Show the calculations needed to select a properly sized box for the receptacle outlet mentioned in Question 4. Nonmetallic-sheathed cable is the wiring method.

_____

_____

_____

_____

_____

6. Show the calculations necessary to select a properly sized box for the receptacle outlet mentioned in Question 4. The wiring method is EMT, usually referred to as "thin-wall conduit."

_____

_____

_____

_____

_____

7. Prepare a list of _all_ wiring devices used in Circuit A21.

_____

_____

_____

_____

_____

_____

_____

_____

_____

_____

_____

8. Using the cable layout shown in Figure 14-1, make a complete wiring diagram of Circuit A21. Use colored pencils or pens.

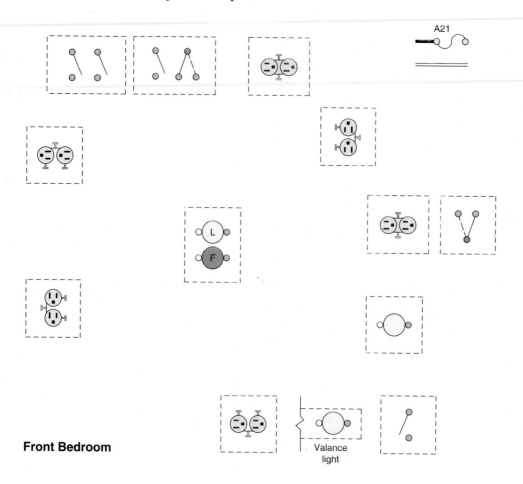

Front Bedroom

Valance light

CHAPTER 15

Dryer Outlet, Lighting and Receptacle Circuits for the Laundry, Powder Room, Rear Entry Hall, and Attic

OBJECTIVES

After studying this chapter, you should be able to

- understand the *Code* requirements for bathrooms receptacles.

- understand the *Code* requirements for making load calculations and electrical connections for electric dryers.

- understand the *Code* requirements for receptacle outlets in laundry areas.

- understand the principles of exhaust fans.

- understand the *Code* requirements for attic wiring.

TABLE 15-1

Laundry-Rear Entrance-Powder Room-Attic (Circuit B10).

Description	Quantity	Watts	Volt-Amperes
Rear Entry hall receptacle	1	120	120
Powder Room vanity luminaire	1	200	200
Rear Entrance hall ceiling luminaires @ 100 W each	2	200	200
Laundry Room fluorescent luminaire Four 32-W lamps	1	128	144
Attic luminaires @ 75 W each	4	300	300
Laundry exhaust fan	1	80	90
Powder Room exhaust fan	1	80	90
TOTALS	11	1108	1144

LIGHTING CIRCUIT B10

The estimated loads for the various receptacles and luminaires in the Laundry, Rear Entry, Powder Room, and Attic are shown in Table 15-1.

The lighting circuit is discussed a little later on in this chapter.

RECEPTACLE CIRCUIT B21

Receptacles in a residential bathroom require special treatment. This is discussed in detail in Chapter 3. Now would be a good time for you to review Chapter 3.

Circuit B21 is a separate 20-ampere branch circuit that supplies the receptacle in the Powder Room.

A bathroom is defined in *Article 100* as ▶*An area that has a basin, plus one or more of the following: a toilet, a urinal, a tub, a shower, a bidet, or similar plumbing fixtures.**◀ Therefore, by definition, the *NEC* considers the powder room to be a bathroom. As a result, special requirements apply to the receptacle outlet in the room.

*Reprinted with permission from NFPA 70-2011.

CLOTHES DRYER CIRCUIT ▲D

The electric clothes dryer requires a separate 120/240-volt, single-phase, 3-wire branch circuit. In this residence, the electric clothes dryer is indicated on the Electrical Plans in the laundry by the symbol ▲D. It is supplied by Circuit B(1–3) located in Panel B. Figure 15-1 shows the internal components and wiring of a typical electric clothes dryer.

Dryer Connection Methods

Electric clothes dryers can be connected in several ways.

One method is to run a 3-wire armored cable directly to a junction box on the dryer provided by the manufacturer, Figure 15-2. This flexible connection allows the dryer to be moved for servicing without disconnecting the wiring. The armor of the Type AC cable serves as the equipment grounding conductor. If Type AC cable is used for the branch circuit wiring, the disconnecting means must comply with *NEC 422.32* as covered later in this chapter.

Another method is to run a conduit (EMT) to a point just behind the dryer. A combination coupling makes the transition from the EMT to a length of flexible metal conduit 2 ft (600 mm) to 3 ft (900 mm) long, as in Figure 15-3. This flexible connection also allows the dryer to be moved for servicing without disconnecting the wiring. A separate equipment grounding conductor must be installed in the EMT to meet the grounding requirements found in *250.118(5)*. If the wiring method described in this paragraph is used for the branch circuit wiring, the disconnecting means must comply with *NEC 422.32* as covered later in this chapter.

Probably the most common method is to install a 4-wire, 30-ampere receptacle on the wall behind the dryer. A 4-wire, 30-ampere cord ("pigtail") is then connected to the dryer. This arrangement is referred to as a "cord-and-plug connection" and is shown in Figure 15-4. In conformance to *210.23*, the connected load shall not exceed the branch-circuit rating.

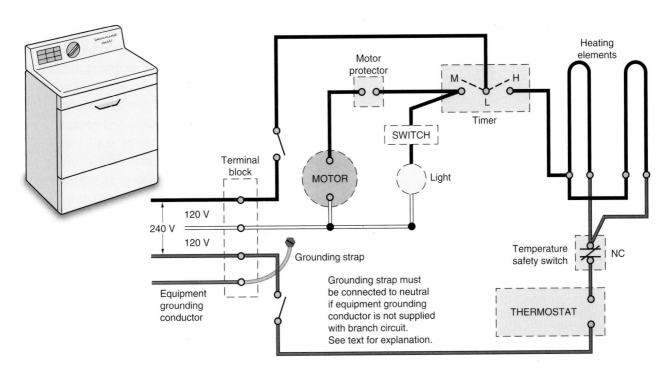

FIGURE 15-1 Clothes dryer: typical wiring and components. (*Delmar/Cengage Learning*)

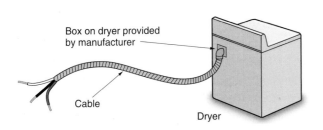

FIGURE 15-2 Dryer connected by armored cable. (*Delmar/Cengage Learning*)

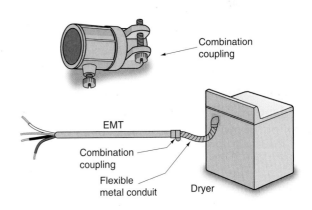

FIGURE 15-3 Dryer connected by EMT and flexible metal conduit. (*Delmar/Cengage Learning*)

Appliances

All new homes have appliances—some that operate entirely on electricity and some that require electricity as well as another energy source, such as a gas furnace.

Article 422 in the *NEC* is where to look for the majority of *Code* requirements for appliances. In *Article 422*, you will find quite a few references to other parts of the *NEC*, such as *Article 430* where motors are involved, and *Article 440* where hermetic-motor compressors (air conditioners and heat pumps) are involved.

As you continue studying this text, specific *Code* rules for appliances are discussed at the time it becomes necessary to mention the rule.

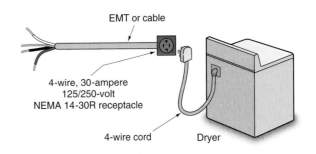

FIGURE 15-4 Dryer connection using a cord set. (*Delmar/Cengage Learning*)

Disconnecting Means

All electrical appliances are required to have a disconnecting means, *422.30*.

For permanently connected motor-driven appliances, *422.32* states that a disconnect switch or a circuit breaker in a panelboard may serve as the disconnecting means, provided the switch or circuit breaker is within sight of the appliance, or is capable of being locked in the "Off" position. Breaker manufacturers produce "lock-off" devices that fit over the circuit breaker handle. Disconnect switches have provisions for locking the switch "Off."

The "unit switch" on an appliance is acceptable as a disconnect for an appliance, provided the unit switch has a marked "Off" position and it disconnects all the conductors in the appliance; *422.34*. This does not normally apply to controllers of the type typically provided in dishwashers as they do not shut off all the power to the appliance. If the appliance does have a qualifying unit switch, the service disconnect serving as the "other" disconnect does not have to be within sight of the appliance, *422.32, Exception*.

Cord-and-Plug Connection. For cord-and-plug-connected appliances, *422.33(A)* accepts the cord-and-plug arrangement as the disconnecting means.

Figure 15-4 shows a 30-ampere, 4-wire receptacle configuration for an electric clothes dryer installation in conformance to the latest *NEC*. Figure 15-5 shows (A) a surface-mount 4-wire receptacle and (B) a flush mount 4-wire receptacle.

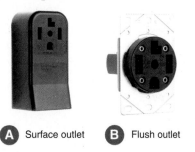

Ⓐ Surface outlet   Ⓑ Flush outlet

FIGURE 15-5 (A) shows a surface-mount 4-wire receptacle. (B) shows a flush-mount 4-wire receptacle. These are typically rated 30 amperes and are of the types required to supply electric dryers. Prior to the 1996 *NEC*, 3-wire receptacles and cords were permitted. This is no longer permitted. It is mandatory that a separate equipment grounding conductor be installed to ground the frames of electric clothes dryers and electric ranges. Also see Figure 20-2. (*Courtesy Leviton*)

For most residential electric clothes dryers, receptacles and cord sets are rated 30 amperes, 125/250 volts. Some large electric clothes dryers might require a 50-ampere, 125/250-volt receptacle and cord set.

Receptacles for dryers can be surface mounted or flush mounted. Flush-mounted receptacles are quite large and require a large wall box. A 4-in. square, 2⅛-in. deep outlet box with a suitable two-gang plaster ring generally works fine. Correct box sizing is determined by referring to *Table 314.16(A)* or *Table 314.16(B)*, or to Figure 2-20 and Table 2-1.

To avoid duplication of text, refer to Chapter 20 for a complete description of ratings and blade and slot configurations for 3-wire and 4-wire dryer and range receptacles and their associated cord sets.

Load Calculations

In this residence, the dryer is installed in the Laundry Room. The Schedule of Special Purpose Outlets in the specifications shows that the dryer is rated 5700 watts, 120/240 volts. The schematic wiring diagram in Figure 15-1 shows that the heating element is connected across 240 volts. The motor and lamp are connected to the 120-volt terminals.

It is impossible for the electrician to know the ampere rating of the motor, the heating elements, timers, lights, relays, and so on, and in what sequence they actually operate. This dilemma is answered in the *NEC* and UL listing requirements for all appliances.

The nameplate must show volts and amperes or volts and watts, *422.60(A)*. The nameplate must also show the minimum supply circuit conductor's ampacity and the maximum overcurrent protection, *422.62(B)(1)*.

A minimum load of 5000 watts (volt-amperes) or the nameplate rating of the dryer, whichever is larger, is used when calculating branch circuit, feeder, or service-entrance requirements, *220.54*.

The rating of an appliance branch circuit must not be less than the marked rating on the appliance, *422.10(A)*.

The electric clothes dryer in this residence has a nameplate rating of 5700 watts. This calculates out to be:

$$I = \frac{W}{E} = \frac{5700}{240} = 23.75 \text{ amperes}$$

Because wiring devices used for dryer circuits are rated at 30 amperes and to comply with *210.3*, we have a minimum 30-ampere branch-circuit rating requirement.

Conductor Sizing

The minimum branch-circuit rating for the electric clothes dryer was found to be 30 amperes, and that is the minimum rating of the overcurrent protective device—fuses or circuit breaker. We need to find a conductor that has an ampacity of 30 amperes that will serve as the branch circuit for the dryer. The conductors will be protected by the 30-ampere overcurrent device.

According to ▶*Table 310.15(B)(16),*◀ the allowable ampacity for a 10 AWG copper conductor is 30 amperes in the 60°C column. The maximum overcurrent protection for these conductors must not exceed 30 amperes, *240.4(D)*.

In this residence, Circuit B(1–3) is a 3-wire, 30-ampere, 240-volt branch circuit.

Nonmetallic-sheathed cable is run concealed in the walls from Panel B to a flush-mounted 4-in. square box behind the dryer location in the Laundry Room. The nonmetallic-sheathed cable contains three 10 AWG conductors plus a 10 AWG equipment grounding conductor.

Where EMT is used as the wiring method, *Table C1, Annex C* of the *NEC* shows that three 10 AWG THHN conductors require trade size ½ EMT. The EMT serves as the equipment ground per *250.118(4)*.

Overcurrent Protection for the Dryer

The motor of the dryer has integral thermal overload protection as required by *422.11(G)*. Note the motor protector in Figure 15-1. This protection is required by UL standards and prevents the motor from reaching dangerous temperatures as the result of an overload, bearing failure, or failure to start. Also note the temperature safety switch. This is a high-temperature cutoff that shuts off the heating element should the appliance's thermostat fail to operate. This high temperature cutoff is also a UL requirement.

Overcurrent Protection for Branch Circuit

NEC 422.62 requires the appliance to be marked with the minimum supply circuit conductor ampacity and the maximum size overcurrent device.

The rating of the branch circuit must not be less than the rating marked on the appliance, *422.10(A)*.

In *422.11(A)*, we find that the overcurrent protection for the branch-circuit conductors must be sized according to *240.4*, which refers us right back to *Article 422—Appliances*.

The maximum overcurrent protection for 10 AWG copper conductors is 30 amperes, *240.4(D)*.

Grounding Frames of Electric Dryers and Electric Ranges

For protection against electrical shock hazard, frames of electric clothes dryers and electric ranges must be grounded. Some key *Code* references are *250.134, 250.138, 250.140*, and *250.142(B)*.

There are now two methods for grounding the frames of electric dryers and electric ranges, namely *New Installations* and *Existing Installations*.

New Installations. One accepted grounding means is to connect the appliance with a metal raceway such as EMT, an equipment grounding conductor installed through flexible metal conduit (Greenfield), armored cable (BX), or other means listed in *250.118*. There are limitations on the use of flexible metal conduit as an equipment grounding conductor. See *250.118(5)* and *348.60*. This topic is covered in Chapter 4.

The separate equipment grounding conductor found in nonmetallic-sheathed cable is also an acceptable equipment grounding means. The equipment grounding conductor in nonmetallic-sheathed cable is sized according to *Table 250.122*. See Figure 15-6.

Service-entrance cable is also permitted as a branch circuit to an electric dryer or electric range. The *Code* rules governing service-entrance cable are covered in Chapter 4.

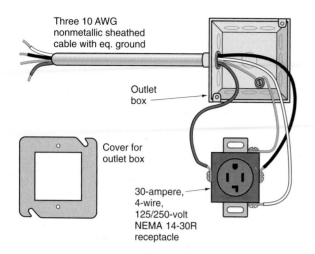

FIGURE 15-6 New installations of branch-circuit wiring for electric ranges, wall-mounted ovens, surface cooking units, and clothes dryers require that all equipment grounding be done according to *250.134, 250.138, 250.140,* and *250.142(B).* Receptacles and cord sets must be 4-wire (three circuit conductors *plus* equipment grounding conductor). (*Delmar/Cengage Learning*)

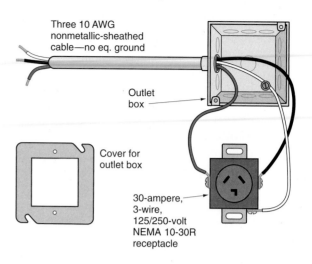

FIGURE 15-7 Prior to the 1996 *National Electrical Code,* it was permitted to ground the junction box to the neutral conductor only if the box was part of the circuit for electric ranges, wall-mounted ovens, surface cooking units, and clothes dryers. The receptacle was permitted to be a 3-wire type because the appliance grounding was accomplished by making a connection between the neutral conductor and the metal frame of the appliance. *This is no longer permitted.* (*Delmar/Cengage Learning*)

When electric clothes dryers and electric ranges are cord-and-plug connected, the receptacle and the cord set must be 4-wire, the fourth wire being the separate equipment grounding conductor.

The small copper bonding strap (link) that is furnished with an electric clothes dryer or electric range *must not* be connected between the neutral terminal and the metal frame of the appliance.

Existing Installations. Prior to the 1996 edition of the *NEC, 250.60* permitted the frames of electric ranges, wall-mounted ovens, surface cooking units, and clothes dryers to be grounded to the neutral conductor, as in Figure 15-7. By way of *Tentative Interim Amendment 53,* this special permission was put into effect in July of 1942, and was supposedly an effort to conserve raw materials during World War II. In effect, this special permission allowed the neutral conductor to serve a dual purpose: (1) the neutral conductor and (2) the equipment grounding conductor. This special permission remained in effect until the 1996 *NEC.*

Since then, grounding equipment such as ranges and dryers to a neutral conductor is not permitted!

Existing installations need not be changed, if they were in conformance to the *NEC* at the time the branch circuit was installed.

If an electric clothes dryer or electric range is to be installed in a residence where the dryer branch-circuit wiring had been installed according to pre-1996 *NEC* rules, the wiring *does not* have to be changed. Merely use the grounding strap furnished by the dryer manufacturer to make a connection between the neutral conductor and the frame of the dryer.

RECEPTACLE OUTLETS—LAUNDRY

At least one receptacle outlet shall be installed for the laundry equipment, *210.52(F).*

For this laundry receptacle, a dedicated 20-ampere branch circuit shall be provided to supply the laundry receptacle outlet or outlets. This circuit *shall have no other outlets, 210.11(C)(2).* In this residence, the clothes washer receptacle is supplied by 20-ampere Circuit B18. This branch circuit is permitted to supply the other receptacle outlets in the room intended for laundry equipment but is not permitted to supply the outdoor weatherproof receptacle.

▶*All 125-volt, single-phase, 15- and 20-ampere receptacles within 6 ft (1.8 m) of the outside edge of a sink shall be GFCI protected, 210.8(A)(7).**◀ There are no exceptions to this rule. Because all three receptacles in the Laundry Room are within 6 ft (1.8 m) of the outside edge of the laundry sink, they are all required to be GFCI protected. As repeated a number of times in this text, GFCI protection can be provided by either a GFCI circuit breaker or a GFCI receptacle. This is covered in detail in Chapter 6.

Figure 15-8 shows the cable layout for the Laundry Room.

Circuit B20 is an additional 20-ampere branch circuit for the laundry area. It supplies the other two receptacles in the Laundry Room that can be used

---

*Reprinted with permission from NFPA 70-2011.

---

for plugging in an iron, sewing machine, TV, radio, or other plug-in appliance.

This circuit also serves the weatherproof receptacle on the outside wall of the laundry area. This branch circuit is *in addition* to the required separate 20-ampere circuit (B18) for the laundry equipment.

Circuits B18 and B20 are included in the calculations for service-entrance and feeder conductor sizing at a calculated load of 1500 volt-amperes per circuit, *220.52(B).* This load is included with the general lighting load for the purpose of calculating the service-entrance and feeder conductor sizing and is subject to the demand factors that are applicable to these calculations. See the complete calculations in Chapter 29.

Depth of Box—Watch Out!

Determining the cubic volume of a box is not the end of the story. Outlet and device boxes have to be deep enough so that conductors will not be damaged when installing the wiring device or equipment

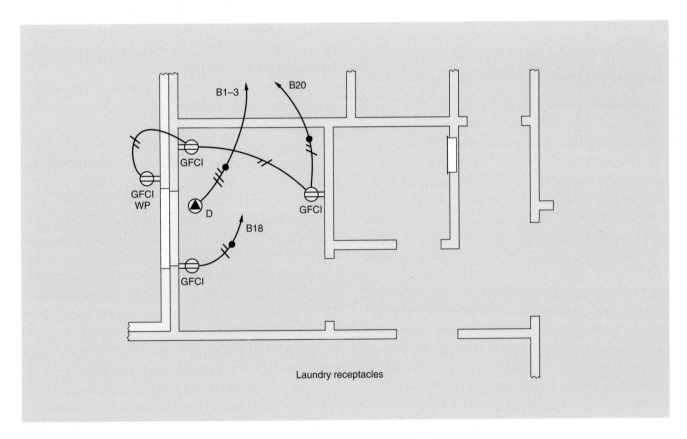

FIGURE 15-8 Cable layout for Laundry Room receptacles. (*Delmar/Cengage Learning*)

into the box. Merely calculating and providing the proper volume of a box is not always enough. The volume calculation might prove adequate, yet the size (depth) of the wiring device or equipment might be such that conductors behind it may possibly be damaged. See *NEC 314.24*.

Width of Box—Watch Out!

Large wiring devices, such as a 30-ampere, 3-pole, 4-wire dryer or range receptacle will not fit into a single gang box that is 2 in. (50.8 mm) wide. Dryer receptacles measure 2.10 in. (53.3 mm) in width. Likewise, a 50-ampere, 3-pole, 4-wire receptacle measures 2.75 in. (69.9 mm) in width. Consider using a 4-in. (101.6-mm) square box with a 2-gang plaster device ring for flush mounting or a 2-gang raised cover for surface mounting. Or, use a 2-gang device box. The center-to-center mounting holes of both the 30-ampere and the 50-ampere receptacles are 1.81 in. (46.0 mm) apart, exactly matching the center-to-center holes of a 2-gang plaster ring, raised cover, or gang device box. See *NEC 314.16(B)(4)*.

COMBINATION WASHER/DRYERS

Combination washer/dryers take up about half the floor space of a traditional washer and dryer pair. Some models have the washer/dryer units combined into one appliance, using the same drum for both wash and dry cycles. Others stack the dryer above the washer. Electric washer/dryer combinations generally require a 30-ampere branch circuit similar to that required for a typical electric clothes dryer. The receptacle would be a 30-ampere, 4-wire, 125/250-volt, NEMA 14-30R, Figure 15-6.

Gas washer/dryer combinations draw 10 or 12 amperes and plug into a standard grounding-type receptacle. If the receptacle is located within 6 ft (1.8 m) of the outside edge of a sink, then it must be GFCI protected, *210.8(A)(7)*. This receptacle and the branch circuit for the laundry appliances must conform to *210.52(F)* and *210.11(C)(2)*.

For installations that require a 30-ampere branch circuit and receptacle for a combination washer/dryer, a 120-volt receptacle and separate 20-ampere branch circuit must still be provided in conformance to *210.52(F)* and *210.11(C)(2)*.

LIGHTING CIRCUIT

The general lighting circuit for the Laundry, Powder Room, and hall area is supplied by Circuit B10. This circuit must be AFCI protected as required by *210.12(A)*.

Note on the cable layout, Figure 15-9, that in order to help balance loads evenly, the attic lights are also connected to Circuit B10.

The types of vanity and ceiling luminaires and wall receptacle outlets, as well as the circuitry and switching arrangements, are similar to types discussed in other chapters of this text and will not be repeated.

The receptacle in the Powder Room is supplied by Circuit B21, a separate 20-ampere branch circuit as required by *210.52(D)*, discussed in Chapter 10.

Exhaust Fans

Ceiling exhaust fans are installed in the Laundry and in the Powder Room. These exhaust fans are connected to Circuit B10.

The exhaust fan in the Laundry will remove excess moisture resulting from the use of the clothes washer, dryer, and ironing.

Exhaust fans may be installed in walls or ceilings, as in Figure 15-10. The wall-mounted fan can be adjusted to fit the thickness of the wall. If a ceiling-mounted fan is used, sheet-metal duct must be installed between the fan unit and the outside of the house. The fan unit terminates in a metal hood or grille on the exterior of the house. The fan has a shutter that opens as the fan starts up and closes as the fan stops. The fan may have an integral pull-chain switch for starting and stopping, or it may be used with a separate wall switch. In either case, single-speed or multispeed control is available. The fan in use has a very small power demand, 90 volt-amperes.

To provide better humidity control, both ceiling-mounted and wall-mounted fans may be controlled with a humidistat. This device starts the fan when the humidity reaches a certain value. When the humidity drops to a preset level, the humidistat turns the fan off.

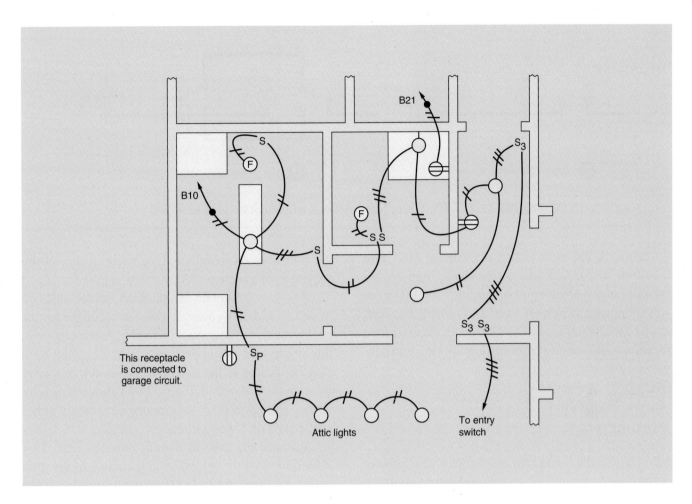

This receptacle is connected to garage circuit.

Attic lights

To entry switch

FIGURE 15-9 Cable layout for Laundry, Rear Entrance, Powder Room, and Attic (*Circuit B10*). (*Delmar/Cengage Learning*)

WALL TYPE

CEILING TYPE

FIGURE 15-10 Exhaust fans. (*Courtesy of Broan-NuTone*)

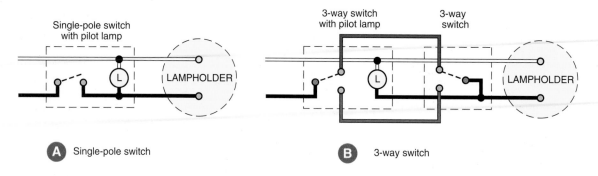

A Single-pole switch

B 3-way switch

FIGURE 15-11 Pilot lamp connections. (*Delmar/Cengage Learning*)

Some of the more expensive exhaust fans have a built-in sensor that will automatically turn on the exhaust fan at a predetermined humidity level. The fan will automatically turn off when the humidity has been lowered to a preset level. This eliminates the need for a separate wall-mounted humidity control (humidistat).

ATTIC LIGHTING AND PILOT LIGHT SWITCHES

NEC 210.70(A)(3) requires that at least one lighting outlet must be installed in an attic and that this lighting outlet contain a switch such as a pull-chain lampholder, or be controlled by a switch located near the entry to the attic. This section of the *Code* addresses attics that are used for storage and attics that contain equipment that might need servicing, such as air-conditioning equipment. *NEC 210.70(A)(3)* requires that the lighting outlet(s) be installed at or near this equipment.

A 125-volt, single-phase, 15- or 20-ampere receptacle outlet shall be installed in an accessible location within 25 ft (7.5 m) of air-conditioning or heating equipment in attics, in crawl spaces, or on the roof, *210.63*. Connecting this receptacle outlet to the load side of the equipment's disconnecting means is not permitted because this would mean that if you turned off the equipment, the power to the receptacle would also be off and thus would be useless for servicing the equipment.

The residence discussed in this text does not have air-conditioning, heating, or refrigeration equipment in the attic or on the roof.

The porcelain lampholders are available with a receptacle outlet for convenience in plugging in an extension cord. However, most electrical inspectors (authority having jurisdiction) would not accept the porcelain lampholder's receptacle in lieu of the required receptacle outlet as stated in *210.63*.

The four porcelain lampholders in the attic are turned on and off by a single-pole switch on the garage wall close to the attic storable ladder. Associated with this single-pole switch is a pilot light. The pilot light may be located in the handle of the switch or it may be separately mounted. Figure 15-11 shows how pilot lamps are connected in circuits containing either single-pole or 3-way switches.

Because the attic in this residence is served by a folding storable ladder, switch control in the attic is not required. Where a permanent stairway of six or more risers is installed, *210.70(A)(2)(c)* requires switch control at both levels.

If a neon pilot lamp in the handle (toggle) of a switch does not have a separate grounded conductor connection, then it will glow only when the switch is in the "Off" position, as the neon lamp will then be in series with the lamp load, Figure 15-12.

The voltage across the load lamp is virtually zero, so it does not burn, and the voltage across the neon lamp is 120 volts, allowing it to glow. When the switch is turned on, the neon lamp is bypassed (shunted), causing it to turn off and the lamp load to turn on.

Use this type of switch when it is desirable to have a switch glow in the dark to make it easy to locate.

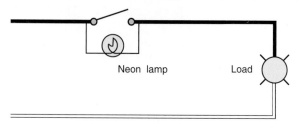

FIGURE 15-12 Example of neon pilot lamp in handle of toggle switch. (*Delmar/Cengage Learning*)

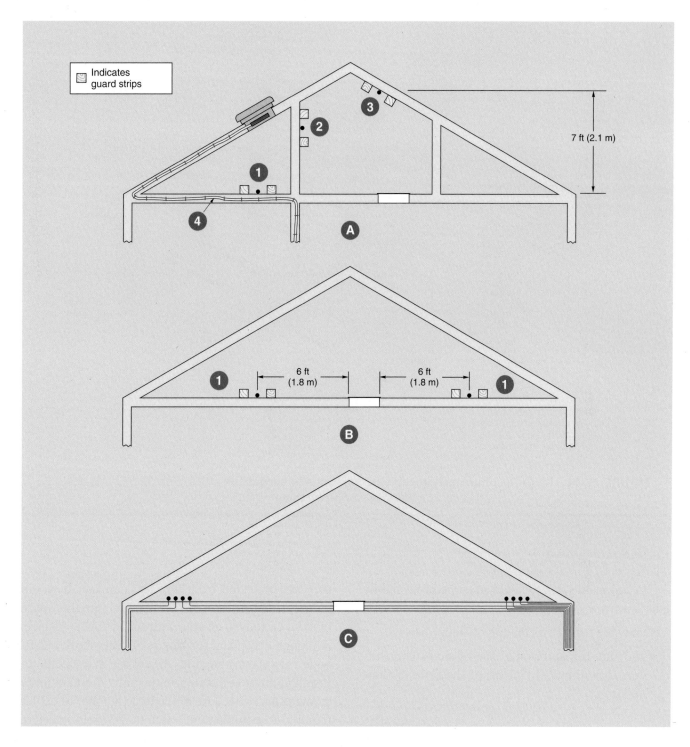

FIGURE 15-13 Protection of nonmetallic-sheathed cable or armored cable in an attic. Refer to *334.23* and *320.23*. (*Delmar/Cengage Learning*)

Installation of Cable in Attics

When nonmetallic-sheathed cable is installed in accessible attics, the installation must conform to the requirements of *334.23*. This section refers the reader directly to *320.23*, which describes how the cable is to be protected. See Figures 15-13 and 15-14.

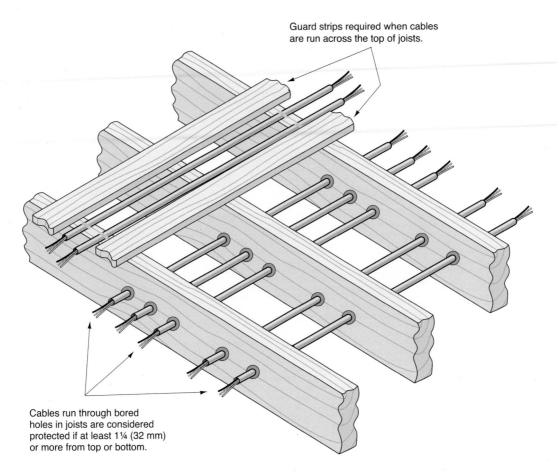

Guard strips required when cables
are run across the top of joists.

Cables run through bored
holes in joists are considered
protected if at least 1¼ (32 mm)
or more from top or bottom.

FIGURE 15-14 Methods of protecting cable installations in accessible attics. See *320.23* and *334.23*.
(*Delmar/Cengage Learning*)

In accessible attics, cables must be protected by guard strips if

- they are run across the top of floor joists ①.
- they are run across the face of studs ② or rafters ③ within 7 ft (2.1 m) of the floor or floor joists.

Guard strips are *not* required if the cable is run along the sides of rafters, studs, or floor joists ④.

In attics not accessible by permanent stairs or ladders, as in Figure 15-13(B), guard strips are required only within 6 ft (1.8 m) of the nearest edge of the scuttle hole or entrance.

Figure 15-13(C) illustrates a cable installation that most electrical inspectors consider to be safe. Because the cables are installed close to the point where the ceiling joists and the roof rafters meet,

they are protected from physical damage. It would be very difficult for a person to crawl into this space or store cartons in an area with such a low clearance. Although the plans for this residence show a 2-ft-wide catwalk in the attic, the owner may decide to install additional flooring in the attic to obtain more storage space. Because of the large number of cables required to complete the circuits, it would interfere with flooring to install guard strips wherever the cables run across the tops of the joists, as in Figure 15-14. However, the cables can be run through holes bored in the joists and along the sides of the joists and rafters. In this way, the cables do not interfere with the flooring.

When running cables parallel to framing members or furring strips, be careful to maintain at least 1¼ in. (32 mm) between the cable and the edge of the framing member. This is a *Code* requirement referenced in *300.4(D)*, to minimize the possibility of driving nails into the cable. All of *300.4* is devoted to the subject of *Protection Against Physical Damage.*

See Chapter 4 for more detailed discussion and illustrations relating to the physical protection of cables.

REVIEW

Note: Refer to the *Code* or plans where necessary.

1. List the switches, receptacles, and other wiring devices that are connected to Circuit B10._____

   _____

   _____

   _____

2. a. Fill in the blank spaces. At least 120-volt receptacle must be installed in a bathroom within _____ in. of the basin, and it must be on the wall _____ to the basin. The receptacle must not be installed _____ on the countertop. The receptacle must be _____ protected.

   b. Circle the correct answers. A separate (15) (20)-ampere branch circuit must supply the receptacle(s) in bathrooms and (shall) (shall not) serve other loads. If a separate 20-ampere branch circuit serves a single bathroom, then that circuit (is permitted) (is not permitted) to supply other loads in that bathroom.

3. What special type of switch is controlling the attic lights? _____

   _____

   _____

4. When installing cables in an accessible attic along the top of the floor joists _____ must be installed to protect the cables.

5. The total estimated volt-amperes for Circuit B10 has been calculated to be 1300 volt-amperes. How many amperes is this at 120 volts? _____

6. If an attic is accessible through a scuttle hole, guard strips are installed to protect cables run across the top of the joists only within (6 ft [1.8m]) (12 ft [3.7m]) of the scuttle hole. Circle the correct answer.

7. What section(s) of the *Code* refers (refer) to the receptacle required for the laundry equipment? State briefly the requirements.

   _____

   _____

   _____

   _____

   _____

   _____

   _____

8. What is the current draw of the exhaust fan in the Laundry?_____

9. The following is a layout of the lighting circuit for the Laundry, Powder Room, Rear Entry Hall, and Attic. Complete the wiring diagram using colored pens or pencils to indicate conductor insulation color. The GFCI receptacle in the Powder Room is not shown in the diagram because it is connected to a separate circuit, B21.

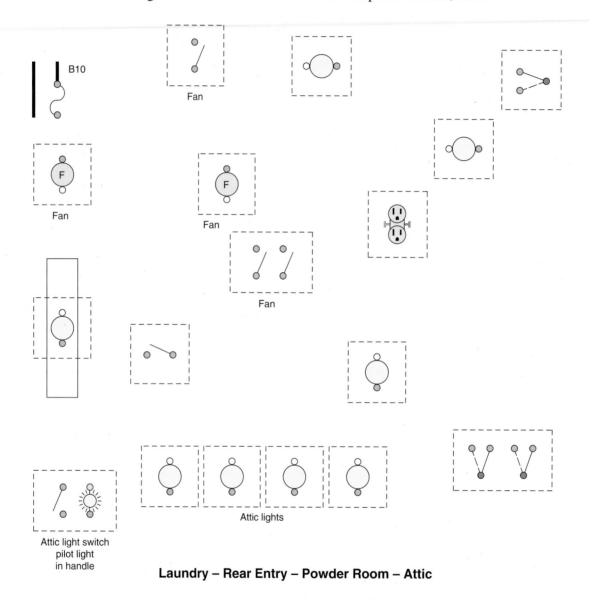

Laundry – Rear Entry – Powder Room – Attic

10. Laundry receptacle outlets are included in the residential load calculations at a value of _____ volt-amperes per circuit. Choose one: 1000, 1500, 2000 volt-amperes.

11. List the various methods of connecting an electric clothes dryer. _____

_____

12. a. What is the minimum power demand allowed by the *Code* for an electric dryer if no actual rating is available for the purpose of calculating feeder and service-entrance conductor sizing? _____

   b. What is the current draw? _____

13. What is the maximum permitted current rating of a portable appliance on a 30-ampere branch circuit? _____

14. What provides motor running overcurrent protection for the dryer?_____

   _____

15. a. Must the metal frame of an electric clothes dryer be grounded?_____

   b. May the metal frame of an electric clothes dryer be grounded to the neutral conductor?

   _____

16. An electric dryer is rated at 7.5 kW and 120/240 volts, 3-wire, single-phase. The terminals on the dryer and panelboard are marked 75°C.

   a. What is the wattage rating?_____

   b. What is the current rating?_____

   c. What minimum size type THHN copper conductors are required? _____

   _____

   _____

   d. If EMT is used, what minimum size is required? _____

   _____

   _____

   _____

   _____

   _____

17. When a metal junction box is installed as part of the cable wiring to a clothes dryer or electric range, may this box be grounded to the circuit neutral conductor?

18. A residential air-conditioning unit is installed in the attic.

   a. Is a lighting outlet required? _____

   b. What *Code* section applies? _____

   c. If a lighting outlet is required, how shall it be controlled? _____

   _____

   d. What *Code* section applies? _____

   e. Is a receptacle outlet required? _____

   f. What *Code* section applies? _____

19. The *Code* states that a means must be provided to disconnect an appliance. In the case of an electric clothes dryer, the disconnecting means can be

    a. a separate disconnect switch located within sight of the appliance.

    b. a separate disconnect switch located out of sight from the appliance. The switch is capable of being locked in the "OFF" position.

    c. a cord-and-plug-connected arrangement.

    d. a circuit breaker in a panel that is within sight of the dryer or capable of being locked in the "Off" position.

    Circle the correct statement(s).

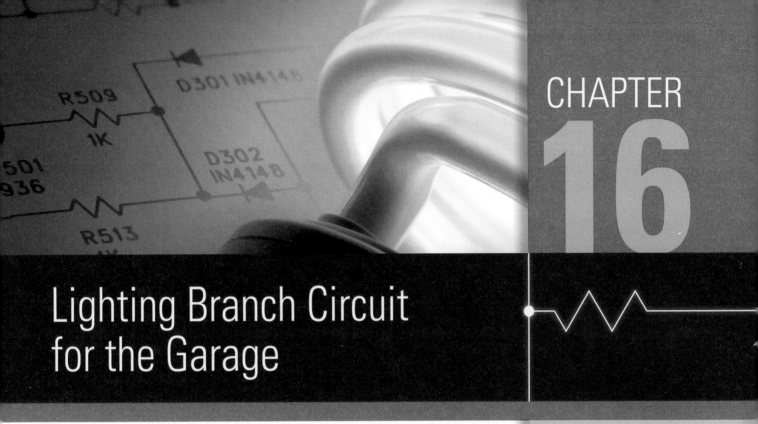

Lighting Branch Circuit for the Garage

OBJECTIVES

After studying this chapter, you should be able to

- understand the basics for lighting a residential garage.

- understand the *NEC* requirements for receptacle outlets in a residential garage.

- become familiar with simple landscape lighting.

- understand the *NEC* requirements for underground cable and conduit wiring.

- understand the operation and electrical connections for overhead door operators.

LIGHTING BRANCH CIRCUIT B14

Circuit B14 is a 15-ampere circuit that originates at Panel B in the recreation room and is brought into the wall receptacle box on the front wall of the garage. This branch circuit is not required to have AFCI protection, *210.12(A)*. The receptacles connected to this branch circuit are required to have GFCI protection, *210.8(2)*. Figure 16-1 shows the cable layout for Circuit B14. From this point, the circuit is then carried to the switch box

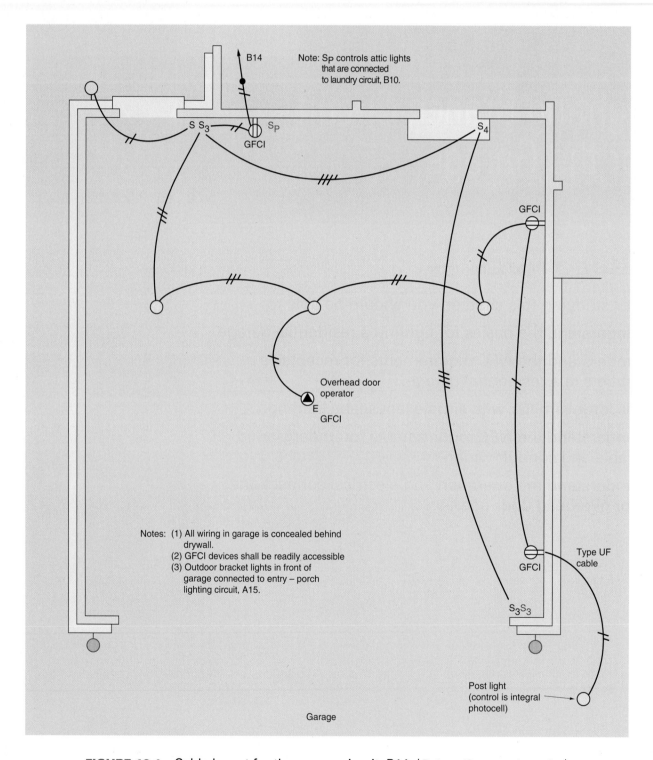

FIGURE 16-1 Cable layout for the garage circuit, B14. (*Delmar/Cengage Learning*)

adjacent to the side door of the garage. From there the circuit conductors and the switch leg are carried upward to the ceiling lampholders and then to the receptacle outlets on the right-hand wall of the garage.

The garage ceiling's porcelain lampholders are controlled from switches located at all three entrances to the garage.

The post light in the front of the house is fed from the receptacle located just inside the overhead door. The post light turns on and off by a photocell, an integral part of the post light.

Note that the luminaires on the outside next to the overhead garage door are controlled by two 3-way switches, one just inside the overhead door and the other in the front entry. These luminaires are not connected to the garage lighting circuit.

All of the wiring in the garage will be concealed. Building codes for one- and two-family dwellings require that the walls separating habitable areas and a garage have a fire-resistance rating. In this residence, the walls and ceilings have a fire-resistance rating of 1 hour. Fire resistance rating requirements are discussed in Chapter 2.

Table 16-1 summarizes the outlets and the estimated load for the garage circuit.

Want More Than Code Minimum?

The garage cable layout in Figure 16-1 supplied by Circuit B14 "Meets *Code*" but could certainly be termed "minimum." You might want to consider running an individual 20-ampere branch circuit for the overhead door operator (some localities require this) and a separate 20-ampere branch circuit for the other receptacles in the garage. These receptacles could very well be used for powering workbench power tools, a refrigerator/freezer, and similar uses in the garage.

UL requirements for residential overhead door operators are covered later in this chapter.

LIGHTING A TYPICAL RESIDENTIAL GARAGE

The recommended approach to provide adequate lighting in a residential garage is to install one 100-watt lamp on the ceiling above each side of an automobile, Figure 16-2. Figure 16-3 shows typical lampholders commonly mounted in garages.

TABLE 16-1

Garage outlet count and estimated load (Circuit B14).

Description	Quantity	Watts	Volt-Amperes
Receptacles @ 180 W each	3	540	540
Ceiling lampholders @ 100 W each	3	300	300
Outdoor garage bracket luminaire (rear)	1	100	100
Post light	1	100	100
Overhead door opener motor:	1		
4.8 A × 120 V			576
4.8 A × 120 V × 0.9 PF		518	
two 60-W lamps		120	120
TOTALS	9	1678	1736

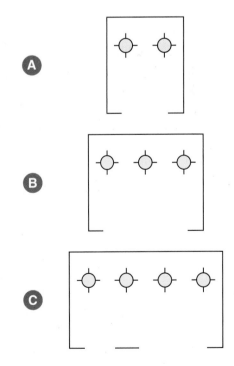

FIGURE 16-2 Positioning of lights in (A) a one-car garage, (B) a two-car garage, and (C) a three-car garage. (*Delmar/Cengage Learning*)

FIGURE 16-3 Typical porcelain or nonmetallic lampholders that would be installed in garages, attics, basements, crawl spaces, and similar locations. Note that one has a pull chain and one does not. (*Delmar/Cengage Learning*)

- For a one-car garage, a minimum of two lampholders is recommended.

- For a two-car garage, a minimum of three lampholders is recommended.

- For a three-car garage, a minimum of four lampholders is recommended.

Lampholders arranged in this manner eliminate shadows between automobiles. Shadows are a hazard because they hide objects that can cause a person to trip or fall. It is highly recommended that these ceiling lampholders be mounted toward the front end of the automobile as it is normally parked in the garage. This arrangement will provide better lighting where it is most needed when the owner is working under the hood. Do not place lights where they will be covered by the open overhead door.

Fluorescent luminaires are quite often used in the warmer parts (south and southwest) of the country. For example, instead of the three porcelain lampholders for conventional incandescent lamps indicated on the electrical plans on the ceiling of the garage, fluorescent strip lights or fluorescent "shop" lights could be installed.

Standard fluorescent lamps installed in a cold garage (northern parts of the country) would be difficult if not impossible to start. If they do start, the lamps will probably "flutter," and the light output will be significantly less than when operated at normal room temperatures. Special lamps and ballasts are required at temperatures below 50–60°F (10–16°C). It is generally cost-

effective to use incandescent lamps in an unheated garage. LED lamps work extremely well in cold temperatures and should be considered for this application.

RECEPTACLE OUTLETS IN A GARAGE

At least one receptacle must be installed in an attached residential garage, *210.52(G)*. The garage in this residence has three wall receptacles. If a detached garage has no electricity provided, then obviously no rules are applicable, but just as soon as a detached garage is wired, then all pertinent *Code* rules become effective and must be followed. These would include rules relating to grounding, lighting outlets, receptacle outlets, GFCI protection, and so on.

NEC 210.8(A)(2) requires that all 125-volt, single-phase, 15- or 20-ampere receptacles installed in garages shall have ground-fault circuit-interrupter protection. GFCIs are discussed in detail in Chapter 6.

To provide GFCI protection for the receptacles in the garage and for the underground wiring to the post light, there are two possibilities. One is to install a GFCI circuit breaker for Circuit B14. Another simple possibility is to install a GFCI feed-through receptacle at the point where Circuit B14 feeds the garage. See Figure 16-1.

▶Note that all GFCI devices are required to be readily accessible, *210.8.*◀ This rule is intended to allow these safety devices to be tested at least monthly, as required in manufacturers' instructions. GFCI devices are not permitted behind appliances such as refrigerators or freezers or on the ceiling near the garage door operator.

All receptacles and switches in the garage are to be mounted 46 in. (1.15 m) to center according to the notation on the First Floor Electrical Plan.

LANDSCAPE LIGHTING

One of the nice things about residential wiring is the availability of a virtually unlimited variety of sizes, types, and designs of outdoor luminaires for gardens, decks, patios, and for accent lighting. Luminaire showrooms at electrical distributors and the electrical departments of home centers offer a wide range of outdoor luminaires, both 120-volt and low-voltage types. Figure 16-4 and Figure 16-5 illustrate landscape lighting mounted on the ground. See also the following section on outdoor wiring.

UL Standard 1838 covers *Low-Voltage Landscape Lighting Systems* generally used outdoors in wet locations.

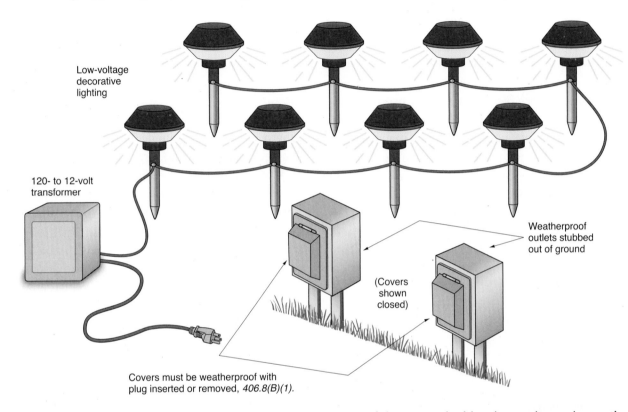

Low-voltage decorative lighting

120- to 12-volt transformer

Weatherproof outlets stubbed out of ground

(Covers shown closed)

Covers must be weatherproof with plug inserted or removed, *406.8(B)(1)*.

FIGURE 16-4   Weatherproof receptacle outlets stubbed out of the ground; either low-voltage decorative landscape lighting or 120-volt PAR luminaires may be plugged into these outlets. *(Delmar/Cengage Learning)*

FIGURE 16-5   Luminaires used for decorative purposes outdoors, under shrubs and trees, in gardens, and for lighting paths and driveways. *(Courtesy of Progress Lighting)*

UL Standard 2108 covers *Low-Voltage Lighting Systems* for use in dry locations.

Both of these UL standards require the manufacturer to include installation instructions with the product. It is extremely important to read these instructions to be sure the luminaires and associated wiring are installed and used in a safe manner.

Low-voltage lighting products will bear markings such as *Outdoor Use Only, Indoor Use Only,* or *Indoor/Outdoor Use.* If marked for use in damp or wet locations, the maximum voltage rating is 15 volts unless live parts are made inaccessible to contact during normal use. Low-voltage luminaires are intended for specific applications, such as surface mounting, suspended installations, or recessed installations, and are so marked. If of the recessed type, they must be installed according to the provisions of *410.116,* as discussed in Chapter 7.

More about Low-Voltage Lighting Systems

Low-voltage lighting systems are covered in *Article 411.* The basic requirements for these systems are that they shall:

- operate at 30 volts rms (42.4 volts peak) or less.

- have one or more secondary circuits, each limited to 25 amperes maximum.

- be supplied through an isolating transformer. A standard step-down transformer does not meet this criterion. Look carefully at the nameplate on the transformer.

- be listed for the purpose.

- not have the wiring concealed or extended through a building unless the wiring meets the requirements of *Chapter 3, NEC.* An example of this is running Type NM nonmetallic-sheathed cable for the low-voltage portion of the wiring that is concealed or extended through the structure of the building.

- not be installed within 10 ft (3.0 m) of pools, spas, or water fountains unless specifically permitted in *Article 680.*

- not have the secondary of the isolating transformer grounded.

- not be supplied by a branch circuit that exceeds 20 amperes.

As stated previously, low-voltage lighting systems operate at 30 volts or less and are covered in *Article 411.* These systems are supplied through an isolating (two-winding) transformer that is connected to a 120-volt, 20-ampere-or-less branch circuit.

Careful consideration must be given to the length of the low-voltage wiring. When the low-voltage wiring is long, larger conductors are necessary to compensate for voltage drop. If voltage drop is ignored, the lamps will not get enough voltage to burn properly; the lamps will burn dim. Voltage drop calculations are discussed in Chapter 4.

You must also be concerned with the current draw of a low-voltage lighting system. For the same wattage, the current draw of a low-voltage lamp will be much greater than that of a 120-volt lamp. If not taken into consideration, the higher current can cause the conductors to run at dangerous temperatures; that could cause a fire!

 EXAMPLE ———————————

A 480-watt load at 120 volts draws:

$$480 \div 120 = 4 \text{ amperes}$$

This is easily carried by a 14 AWG conductor.

A 480-watt load at 12 volts draws:

$$480 \div 12 = 40 \text{ amperes}$$

This requires an 8 AWG conductor, as determined by *Table 310.15(B)(16).*

You can readily see that for this 480-watt example, the conductors must be considerably larger for the 12-volt application to safely carry the current.

Prewired low-voltage lighting assemblies that bear the UL label have the correct size and length of conductors. Adding conductors in the field must be done according to the manufacturers' instructions.

Voltage drop calculations are covered in Chapter 4.

OUTDOOR WIRING

If the homeowner requests that 120-volt receptacles be installed away from the building structure, the electrician can provide weatherproof receptacle outlets as illustrated in Figure 16-4.

Outdoor receptacles must have GFCI protection. See Chapter 6 for the complete discussion of GFCI protection.

These types of boxes have trade size ½ female threaded openings or hubs in which conduit fittings secure the conduit "stub-ups" to the box. Any unused openings are closed with trade size ½ plugs that are screwed tightly into the threads. Refer to Figure 16-6.

All receptacles installed in a wet location must have an enclosure that is weatherproof, whether or not an attachment plug cap is inserted, *406.8(B)(1)*. This requires a self-closing cover that is deep enough to shelter the attachment plug cap of the cord. See Figure 16-4, Figure 16-6, and Figure 16-7.

What about the Receptacles?

Straight blade receptacles rated 15- and 20-amperes, 125- and 250-volts installed in wet locations shall be listed as weather-resistant type, *406.8(B)(1)*.

Where flush-mounted weatherproof receptacles are desired for appearance purposes, a recessed weatherproof receptacle, as illustrated in Figure 16-7, may be used. This type of installation takes up more depth than a standard receptacle and requires special mounting, as indicated in the "exploded" view.

Raceway supported enclosures that do not contain wiring devices or support luminaires:
- Shall be not larger than 100 in.³ (1650 cm³).
- Shall have threaded entries or identified hubs.
- Shall be supported by at least two conduits.
- Conduits shall be secured within 18 in. (450 mm) of the enclosure if all conduit entries are on the same side.
- Use RMC or IMC for support. Extend conduit at least 18 in. (450 mm) into ground or pour concrete around conduits at grade.

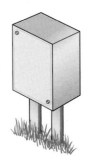

Raceway supported enclosures that do contain wiring devices or support luminaires:
- Shall be not larger than 100 in.³ (1650 cm³).
- Shall have threaded entries or identified hubs.
- Shall be supported by at least two conduits.
- Conduits shall be secured within 18 in. (450 mm) of the enclosure if all conduit entries are on the same side.
- Use RMC or IMC for support. Extend conduit at least 18 in. (450 mm) into ground or pour concrete around conduits at grade.

Not permitted to support an enclosure with one conduit. The enclosure could easily twist, resulting in conductor insulation damage, and a poor connection between the conduit and the enclosure.
Do not use EMT or PVC to support enclosures. PVC conduit permitted if enclosure is supported independently by a post, strut, or other suitable methods. EMT not suitable for underground.

RMC = Rigid metal conduit (*Article 344*)
IMC = Intermediate metal conduit (*Article 342*)
PVC = Rigid nonmetallic conduit (*Article 352*)
EMT = Electrical metallic tubing (*Article 358*)

FIGURE 16-6 Raceway-supported enclosures with and without devices, *314.23(E)* and *(F)*.
(*Delmar/Cengage Learning*)

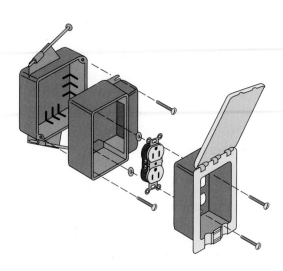

FIGURE 16-7 A recessed type of weatherproof receptacle that meets the requirements of *406.8(B)(1)* and *(B)(2)*, which require that the enclosure be weatherproof while in use if unattended, where the cord will be plugged in for long periods of time. The "exploded" view illustrates one method of "roughing-in" this type of weatherproof receptacle. Straight blade receptacles rated 15- and 20-amperes, 125- through 250-volts installed in wet locations shall be listed as weather-resistant type. (*Photos courtesy of TayMac Corporation*)

Careful thought must be given during the "roughing-in" stages to ensure that the proper depth "roughing-in" box is installed.

Wiring with Type UF Cable (Article 340)

The plans show that Type UF underground cable, Figure 16-8, is used to connect the post light. The current-carrying capacity (ampacity) of Type UF cable is found in *Table 310.16*, using the 60°C column. According to *Article 340*, Type UF cable:

- is marked underground feeder cable.

- is available in sizes from 14 AWG through 4/0 AWG (for copper conductors) and from 12 AWG through 4/0 AWG (for aluminum conductors).

- may be used in direct exposure to the sun if the cable is of a type listed for sunlight resistance or is listed and marked for sunlight resistance.

- may be used with nonmetallic-sheathed cable fittings.

- is flame-retardant.

- is moisture-, fungus-, and corrosion-resistant.

- may be buried directly in the earth.

- may be used for branch-circuit and feeder wiring.

- may be used in interior wiring for wet, dry, or corrosive installations.

- is installed by the same methods as nonmetallic-sheathed cable (*Article 334*).

FIGURE 16-8 Type UF underground cable. (*Courtesy of Southwire Company*)

- must not be used as service-entrance cable.

- must not be embedded in concrete, cement, or aggregate.

- must be buried in the same trench where single-conductor cables are installed.

- shall be installed according to *300.5*.

You will also find underground feeder cable with the marking UF-B. This indicates that the conductors are rated 90°C dry and 60°C wet. The jacket is rated 75°C. The conductor ampacity is that of a 60°C conductor. The conductors meet the requirements of THWN but are not marked along the entire length of the conductors.

Equipment fed by Type UF cable is grounded by properly connecting the bare equipment grounding conductor found in the UF cable to the equipment to be grounded, Figure 16-9.

For underground wiring, it is significant to note that whereas the *Code* does permit running up the side of a tree with conduit or cable (when protected from physical damage), as in Figure 16-10, the *Code* does *not* permit supporting conductors between trees, as in Figure 16-11. Figure 16-12 shows one type of luminaire that is permitted for the

FIGURE 16-10 This installation conforms to the *Code*. *NEC 225.26* does not permit spans of conductors to be run between live or dead trees, but it does not prohibit installing decorative lighting as shown in this figure. *NEC 410.36(G)* permits outdoor luminaires and associated equipment to be supported by trees. Recognized wiring methods to carry the conductors up the tree must be used. (*Delmar/Cengage Learning*)

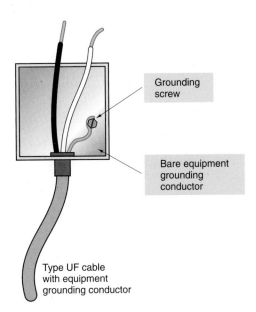

Grounding screw

Bare equipment grounding conductor

Type UF cable with equipment grounding conductor

FIGURE 16-9 Illustration showing how the bare equipment grounding conductor of a Type UF cable is used to ground the metal box. (*Delmar/Cengage Learning*)

installation shown in Figure 16-10. Splices are permitted to be made directly in the ground, but only if the connectors are listed by a Nationally Recognized Testing Laboratory for such use, *110.14(B)* and *300.5(E)*.

UNDERGROUND WIRING

Underground wiring is common in residential applications. Examples include wiring for decorative landscape lighting, post lamps, detached buildings

FIGURE 16-11 *Violation:* Permanent overhead conductor spans are not permitted to be supported by live or dead vegetation, such as trees. Temporary (not over 90 days) holiday decorative lighting branch-circuit conductors or cables are permitted to be supported by trees. See *225.26, 230.10, 590.3(B),* and *590.4(J).*
(*Delmar/Cengage Learning*)

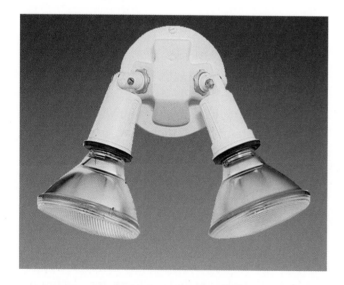

FIGURE 16-12 Outdoor luminaire with lamps.
(*Courtesy of Progress Lighting*)

such as a garage or tool shed, and low-voltage wiring for lawn sprinkling (irrigation) systems.

Grounding Equipment Supplied by Underground Wiring

For life safety, proper grounding is important! Figure 16-13 illustrates a lethal *Code* violation. The post lantern is grounded to the white grounded circuit conductor. This violation results in a parallel return path for the current—one path through the ground and the second path through the person. That person might be a child playing around the post. This could result in a possible electrocution!

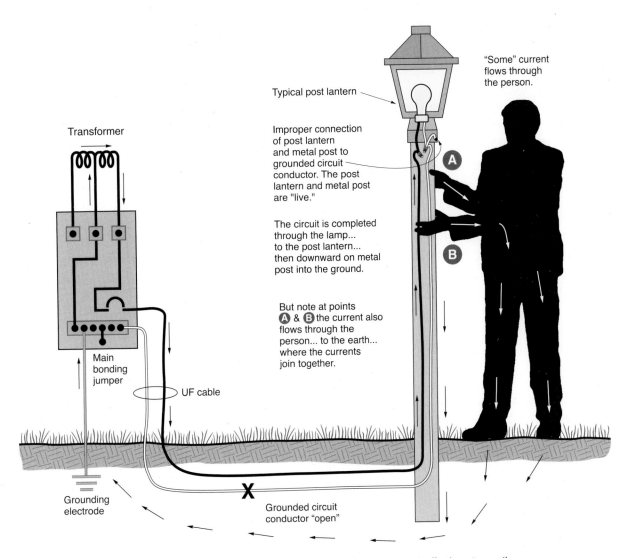

"Some" current flows through the person.

Typical post lantern

Improper connection of post lantern and metal post to grounded circuit conductor. The post lantern and metal post are "live."

The circuit is completed through the lamp... to the post lantern... then downward on metal post into the ground.

But note at points Ⓐ & Ⓑ the current also flows through the person... to the earth... where the currents join together.

Transformer

Main bonding jumper

UF cable

Grounding electrode

Grounded circuit conductor "open"

X

If the grounded circuit conductor "opens" for any reason, the earth provides an ineffective return path for the current. The current travels through the ground back to the grounding electrode... to the panel grounded neutral bus... through the service neutral conductor... through the transformer... back to the service panel... then through the "hot" conductor to post lantern and metal post.

FIGURE 16-13   VIOLATION! This diagram clearly illustrates the electric shock hazard associated with the *Code* violation practice of grounding metal objects to the grounded circuit conductor. In this illustration, the post lantern and metal post have been improperly grounded (connected) to the grounded circuit conductor at (A). The grounded circuit conductor is "open." One path for the return current is through the metal post into the ground. The second path is through the person. This is a violation of *250.24(A)(5)* and *250.142(B)*.
(*Delmar/Cengage Learning*)

NEC *250.4(A)(5)* is very clear about this scenario. It states that *The earth shall not be considered as an effective ground-fault current path.**

Proper grounding of the metal post is achieved using the equipment grounding conductor in the UF cable. Read the instructions that come with the post lantern.

NEC *250.24(A)(5)* and *250.142(B)* prohibit making any connections of equipment to the grounded circuit conductor anywhere on the load side of the service disconnect.

How Deep Should Underground Wiring Be?

Digging, aerating, rototilling, and similar yard work tasks can easily damage underground wiring.

*Reprinted with permission from NPFA 70-2011.

This is extremely hazardous from the electric shock standpoint. It can also result in costly repairs. The *NEC* addresses minimum burial depths of conduit and cable in great detail.

Table 300.5 shows the minimum depths required for the various types of wiring methods. Refer to Figures 16-14 and 16-15.

Across the top of *Table 300.5* are listed wiring methods such as direct burial cables or conductors, rigid metal conduit, intermediate metal conduit, rigid PVC conduit that is approved for direct burial, special considerations for residential branch circuits rated 120 volts or less having GFCI protection and overcurrent protection not over 20 amperes, and low-voltage landscape lighting supplied with Type UF cable.

Down the left-hand side of the table we find the location of the wiring methods that are listed across the top of the table.

The notes to *Table 300.5* are very important. *Note 4* allows us to select the shallower of two depths when the wiring methods in Columns 1, 2, or 3 are combined with Columns 4 or 5. For example, the basic rule for direct buried cable is that it be covered by a minimum of 24 in. (600 mm). However, this minimum depth is reduced to 12 in. (300 mm) for residential installations when the branch circuit is not over 120 volts, is GFCI protected, and the overcurrent protection is not over 20 amperes. This is for residential installations only.

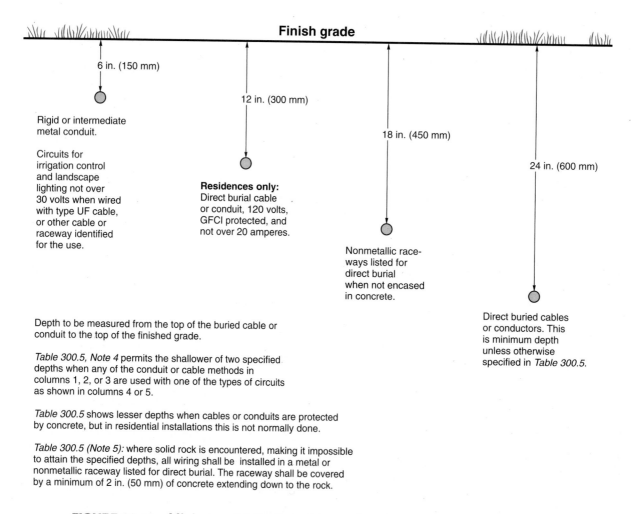

Finish grade

6 in. (150 mm)

Rigid or intermediate metal conduit.

Circuits for irrigation control and landscape lighting not over 30 volts when wired with type UF cable, or other cable or raceway identified for the use.

12 in. (300 mm)

Residences only: Direct burial cable or conduit, 120 volts, GFCI protected, and not over 20 amperes.

18 in. (450 mm)

Nonmetallic raceways listed for direct burial when not encased in concrete.

24 in. (600 mm)

Direct buried cables or conductors. This is minimum depth unless otherwise specified in *Table 300.5.*

Depth to be measured from the top of the buried cable or conduit to the top of the finished grade.

Table 300.5, Note 4 permits the shallower of two specified depths when any of the conduit or cable methods in columns 1, 2, or 3 are used with one of the types of circuits as shown in columns 4 or 5.

Table 300.5 shows lesser depths when cables or conduits are protected by concrete, but in residential installations this is not normally done.

Table 300.5 (Note 5): where solid rock is encountered, making it impossible to attain the specified depths, all wiring shall be installed in a metal or nonmetallic raceway listed for direct burial. The raceway shall be covered by a minimum of 2 in. (50 mm) of concrete extending down to the rock.

FIGURE 16-14 Minimum depths for cables and conduit installed underground. See *Table 300.5* for depths for other conditions. (*Delmar/Cengage Learning*)

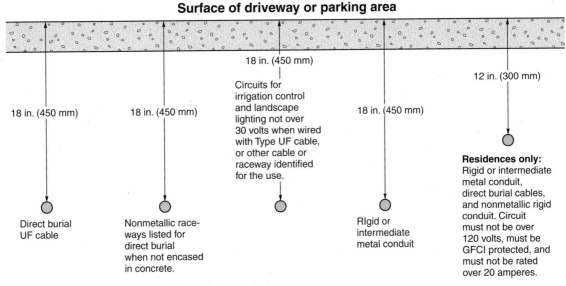

FIGURE 16-15   Depth requirements under 1- and two-family dwelling driveways and parking areas. These depths are permitted for one- and two-family residences. See *Table 300.5* for depths for other conditions. (*Delmar/Cengage Learning*)

The measurement for the depth requirement is from the top of the raceway or cable to the top of the finished grade, concrete, or other similar cover.

Do not back-fill a trench with rocks, debris, or similar coarse material, *300.5(F)*.

Protection from Damage

Underground direct buried conductors and cables must be protected against physical damage according to the rules set forth in *300.5(D)*.

Installation of Conduit Underground

By definition, an underground installation is a wet location. Cables and insulated conductors installed in underground raceways shall be listed for use in wet locations. Conductors suitable for use in wet locations will have the letter "W" in their type designation. For example RHW, TW, THW, THHW, THWN, and XHHW. See *300.5(B)* and *310.10(C)*.

All conduit installed underground must be protected against corrosion, *300.6(A)(3)*. The manufacturer of the conduit and UL Standards will furnish information as to whether the conduit is suitable for direct burial. Additional supplemental protection of metal conduit can be accomplished by "painting" the conduit with a nonmetallic coating. See Table 16-2.

When metal conduit is used as the wiring method, the metal boxes, post lantern, and so forth that are properly connected to the metal raceways are considered to be grounded because the conduit serves as the equipment ground, *250.118* (Figure 16-16). Note that many electrical inspection authorities require an equipment grounding conductor of the wire type be installed inside all metal conduits that are installed underground and in conduit installed in concrete on grade. This is to address the problem of conduit that may rust and fail due to corrosive conditions in the soil. See *Article 410, Part V,* for details on grounding requirements for luminaires.

When nonmetallic raceways are used, then a separate equipment grounding conductor, either

TABLE 16–2

Table 300.5 Minimum Cover Requirements, 0 to 600 Volts, Nominal, Burial in Millimeters (Inches)

Location of Wiring Method or Circuit	Type of Wiring Method or Circuit									
	Column 1 Direct Burial Cables or Conductors		Column 2 Rigid Metal Conduit or Intermediate Metal Conduit		Column 3 Nonmetallic Raceways Listed for Direct Burial Without Concrete Encasement or Other Approved Raceways		Column 4 Residential Branch Circuits Rated 120 Volts or Less with GFCI Protection and Maximum Overcurrent Protection of 20 Amperes		Column 5 Circuits for Control of Irrigation and Landscape Lighting Limited to Not More Than 30 Volts and Installed with Type UF or in Other Identified Cable or Raceway	
	mm	in.	mm	in.	mm	in.	mm	in.	mm	in.
All locations not specified below	600	24	150	6	450	18	300	12	150	6
In trench below 50-mm (2-in.) thick concrete or equivalent	450	18	150	6	300	12	150	6	150	6
Under a building	0 (in raceway or Type MC or Type MI cable identified for direct burial)	0	0	0	0	0	0 (in raceway or Type MC or Type MI cable identified for direct burial))	0	0 (in raceway or Type MC or Type MI cable identified for direct burial))	0
Under minimum of 102-mm (4-in.) thick concrete exterior slab with no vehicular traffic and the slab extending not less than 152 mm (6 in.) beyond the underground installation	450	18	100	4	100	4	150 (direct burial) 100 (in raceway)	6 4	150 (direct burial) 100 (in raceway)	6 4
Under streets, highways, roads, alleys, driveways, and parking lots	600	24	600	24	600	24	600	24	600	24
One- and two-family dwelling driveways and outdoor parking areas, and used only for dwelling-related purposes	450	18	450	18	450	18	300	12	450	18
In or under airport runways, including adjacent areas where trespassing prohibited	450	18	450	18	450	18	450	18	450	18

Notes:
1. Cover is defined as the shortest distance in millimeters (inches) measured between a point on the top surface of any direct-buried conductor, cable, conduit, or other raceway and the top surface of finished grade, concrete, or similar cover.
2. Raceways approved for burial only where concrete encased shall require concrete envelope not less than 50 mm (2 in.) thick.
3. Lesser depths shall be permitted where cables and conductors rise for terminations or splices or where access is otherwise required.
4. Where one of the wiring method types listed in Columns 1–3 is used for one of the circuit types in Columns 4 and 5, the shallowest depth of burial shall be permitted.
5. Where solid rock prevents compliance with the cover depths specified in this table, the wiring shall be installed in metal or nonmetallic raceway permitted for direct burial. The raceways shall be covered by a minimum of 50 mm (2 in.) of concrete extending down to rock.

Reprinted with permission from NFPA 70-2011.

IS SUPPLEMENTAL CORROSION PROTECTION REQUIRED?

	In Concrete above Grade?	In Concrete below Grade?	In Direct Contact with Soil?
Rigid Conduit[1]	No	No	No[2]
Intermediate Conduit[1]	No	No	No[2]
Electrical Metallic Tubing[1]	No	Yes	Yes[3]

[1]Severe corrosion can be expected where ferrous metal conduits come out of concrete and enter the soil. Some electrical inspectors and consulting engineers might specify the application of some sort of supplemental nonmetallic corrosion protection.

[2]Unless subject to severe corrosive effects. Different soils have different corrosive characteristics.

[3]In most instances, electrical metallic tubing is not permitted to be installed underground in direct contact with the soil because of corrosion problems.

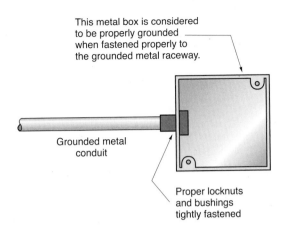

This metal box is considered to be properly grounded when fastened properly to the grounded metal raceway.

Grounded metal conduit

Proper locknuts and bushings tightly fastened

FIGURE 16-16 This illustration shows that a box is grounded when properly fastened to the grounded metal raceway. This is acceptable in *250.118*. More and more *Code*-enforcing authorities and consulting engineers are requiring that a separate equipment grounding conductor be installed in all raceways. This ensures effective grounding of the installation.
(*Delmar/Cengage Learning*)

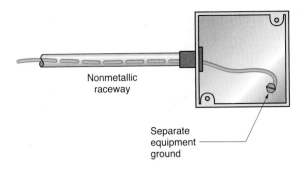

Nonmetallic raceway

Separate equipment ground

FIGURE 16-17 Grounding a metal box with a separate equipment grounding conductor.
(*Delmar/Cengage Learning*)

green or bare, must be installed to accomplish the adequate grounding of the equipment served. See Figure 16-17. This equipment grounding conductor must be sized according to *Table 250.122, NEC*.

The last sentence of *250.4(A)(5)* states that *The earth shall not be considered as an effective ground-fault current path.* See Figure 16-18.

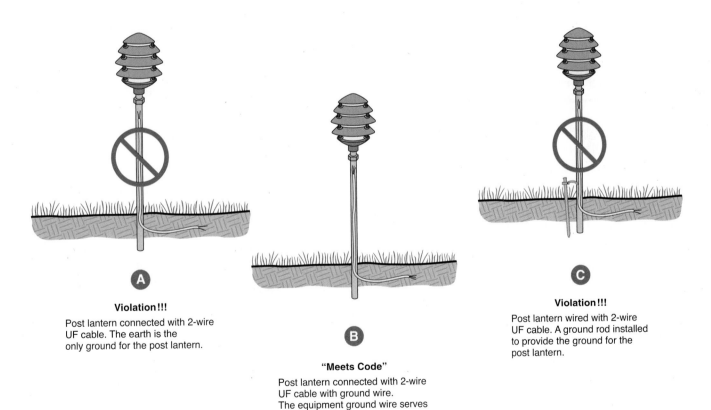

A

Violation!!!

Post lantern connected with 2-wire UF cable. The earth is the only ground for the post lantern.

B

"Meets Code"

Post lantern connected with 2-wire UF cable with ground wire. The equipment ground wire serves as the grounding means for the post lantern.

C

Violation!!!

Post lantern wired with 2-wire UF cable. A ground rod installed to provide the ground for the post lantern.

FIGURE 16-18 These drawings illustrate the intent of the last sentence of *250.4(A)(5)*, which states *The earth shall not be considered as an effective ground-fault current path.*
(*Delmar/Cengage Learning*)

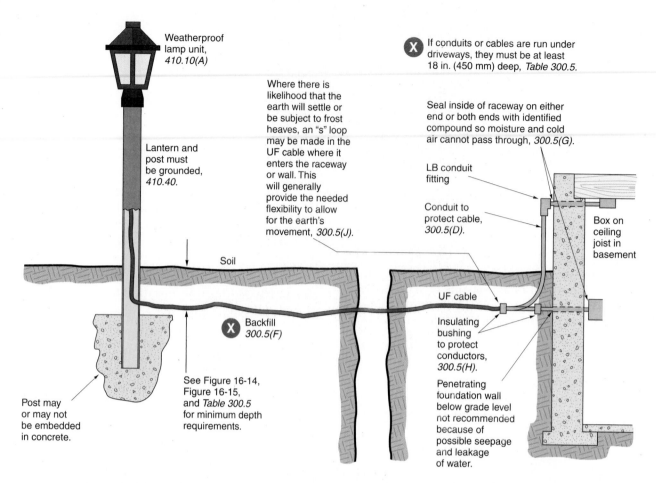

FIGURE 16-19 Methods of bringing cable and/or conduit through a concrete wall and/or upward from a concrete wall into the hollow space within a framed wall. (*Delmar/Cengage Learning*)

Figure 16-19 shows two methods of bringing the conduit into the basement: (1) It can be run below ground level and then can be brought through the basement wall, or (2) it can be run up the side of the building and through the basement wall at the ceiling joist level. When the conduit is run through the basement wall, the opening must be sealed to prevent moisture from seeping into the basement. The electrician must decide which of the two methods is more suitable for each installation.

According to the *Code*, any unprotected location exposed to the weather is considered a wet location.

Luminaires installed in wet locations shall be marked "Suitable for Wet Locations." These luminaires are constructed so that water cannot enter or accumulate in lampholders, wiring compartments, or other electrical parts, *410.10(A)*. Manufacturers' installation instructions may require the application of caulking to ensure protection against the entrance of water.

A damp location is defined as being protected from the weather but subject to moderate degrees of moisture. Partially protected areas such as roofed open porches or areas under canopies are damp locations. Luminaires in these locations must be marked "Suitable for Damp Locations." A luminaire that is "Suitable for Wet Locations" is also suitable for damp locations. A luminaire that is "Suitable for Damp Locations" is not suitable for wet locations.

Check the UL label on a luminaire to determine the suitability of the fixture for a wet or damp location.

The post in Figure 16-19 may or may not be embedded in concrete, depending on the consistency of the soil, the height of the post, and the size of the post lantern. Most electricians prefer to embed the base of the post in concrete to prevent rotting of wood posts and rusting of metal posts.

Figure 16-20 illustrates some typical post lights.

FIGURE 16-20   Post-type lanterns. (*Courtesy of Progress Lighting*)

OVERHEAD GARAGE DOOR OPERATOR ⓐE

The receptacle for the garage overhead door operator is connected to Circuit B14. As you are about to learn, the maximum current draw for a typical overhead door operator is 5 amperes. Although not required by the *NEC*, some electrical inspectors insist that a separate branch circuit be run. This is an issue you will have to check out with your local electrical inspector.

Garage door operators are listed under UL Standard 325. This is the *Standard for Door, Drapery, Gate, Louver, and Window Operators and Systems*. Residential overhead garage door operators are required to have an inherent "entrapment protection system" that reverses the door in 2 seconds maximum when sensing an obstruction on its way down.

Standard 325 for residential overhead garage door operators also requires that under standard test conditions:

- The *maximum* current draw of the operator shall not exceed 5 amperes, excluding lamps or other devices.*

- The *maximum* current of the motor in locked rotor condition plus the lamps and other devices* shall not exceed 15 amperes.

- The maximum length of the power supply cord is 6 ft (1.8 m).

- The cord must have an equipment grounding conductor and have a grounding-type attachment plug cap.

- The photoelectric sensor shall be mounted not higher than 6 in. (150 mm) above the garage floor.

- Always follow the manufacturer's installation instructions.

Typical residential operators are advertised as ⅓ and ½ horsepower at 120 volts. The larger the horsepower rating, the more "lift" power. The manufacturer is permitted to show horsepower on the carton but *not* on the nameplate. Because the horsepower rating is a mechanical rating, the manufacturers' advertised horsepower rating cannot simply be converted to full-load amperes by referring to *NEC Table 430.248* as we do for conventional electric motors. The ampere rating marked on the nameplate is what counts. This nameplate ampere marking is based on standard test conditions and is the sum of the motor

---

*Such as a capacitor, microprocessor, or photoelectric sensors.

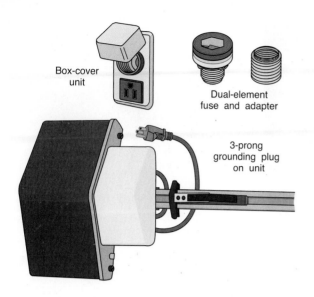

FIGURE 16-21 Overhead door operator with 3-wire cord connector. (*Delmar/Cengage Learning*)

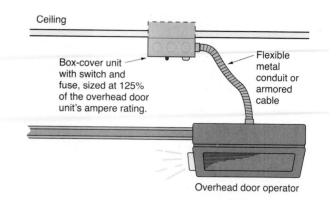

FIGURE 16-22 Permanent wiring for overhead door operator. (*Delmar/Cengage Learning*)

running current, the current draw of the light bulbs, and any other small current draw of additional devices.

The overhead door operator in this residence is marked 5.8 amperes, 120 volts, and is cord-and-plug connected as shown in Figure 16-21. The carton the unit came in was marked ⅓ horsepower. Overload protection for the motor is an integral part of the motor.

Where to Locate the Receptacle

For a typical residential overhead door operator, a receptacle is roughed-in on the ceiling approximately 3 ft (0.9144 m) greater than the height of the garage door header from the floor, in line with the center of the header. The approximate distance from the header to the receptacle would be:

$$7 \text{ ft} + 3 \text{ ft} = 10 \text{ ft}$$
$$(2.1336 \text{ m} + 0.9144 \text{ m} = 3.048 \text{ m})$$
$$8 \text{ ft} + 3 \text{ ft} = 11 \text{ ft}$$
$$(2.4384 \text{ m} + 0.9144 \text{ m} = 3.3528 \text{ m})$$
$$9 \text{ ft} + 3 \text{ ft} = 12 \text{ ft})$$
$$(2.7432 \text{ m} + 0.9144 \text{ m} = 3.6576 \text{ m})$$
$$10 \text{ ft} + 3 \text{ ft} = 13 \text{ ft}.$$
$$(3.048 \text{ m} + 0.9144 \text{ m} = 3.9624 \text{ m})$$

The cord supplied with the operator is long enough to reach this receptacle.

This receptacle is required to be GFCI protected. See *210.8(A)(2)*.

Although not commonly done, most residential overhead door operators can be "hard-wired" as shown in Figure 16-22.

Do Lamps in Overhead Door Opener Burn Out Often?

If you are experiencing too much lamp burnout in your garage door opener, the culprit is probably vibration. Garage door opener lamps are available. These are rough service bulbs. They have a more rugged filament than standard lamps and can withstand vibration for a longer period of time. They can also be used in portable work lights. Premature lamp burnout is often found where home laundry equipment is located above a luminaire, such as a washer and dryer located on the second floor.

Push Buttons

Push buttons are generally installed next to each door leading into the garage. Key-operated weatherproof switches or radio-transmitter type controls can also be installed outside next to the overhead door. The wiring between the overhead door operator and the push buttons is Class 2 wiring, which is similar to that used for chime and thermostat wiring. See Figure 16-23.

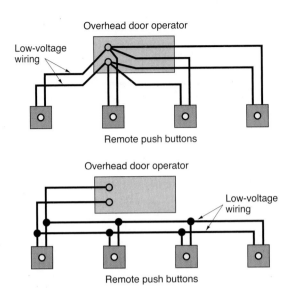

FIGURE 16-23 Push-button wiring for overhead door operator. (*Delmar/Cengage Learning*)

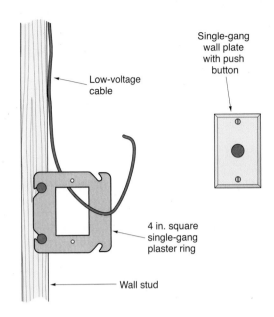

FIGURE 16-24 A sketch of how a 4-in. square cover can be used to provide an opening for low-voltage wiring. (*Delmar/Cengage Learning*)

Push buttons should be installed at least 5 ft (1.5 m) or higher above the garage floor so as to be out of reach of small children.

Quite often, electricians will install the low-voltage, Class 2 wiring during the "rough-in" stage so as to conceal this wiring between the push-button locations and the overhead door operator. A neat way to do this is to nail a single-gang, 4-in. square plaster ring to the stud at the desired push-button locations. When the house is "trimmed out," a single-gang faceplate with the push button is installed. See Figure 16-24.

Box-Cover Unit

The overhead door operator unit has overload protection built into it according to UL Standard 325. Additional short-circuit "backup" protection can be installed in the field. Here's how.

Instead of a standard receptacle on the ceiling near the operator, a box-cover unit can be installed. Box-cover units are available with *switch/fuse* or *receptacle/fuse* combinations, as in Figures 16-21, 16-22, 16-25, and 16-26. Box-cover units provide a convenient disconnecting means in addition to having a fuse holder in which to install a dual-element, time-delay Type S fuse.

Size the time-delay fuse in the range of *approximately* 125%–150% of the overhead door opener's marked ampere rating.

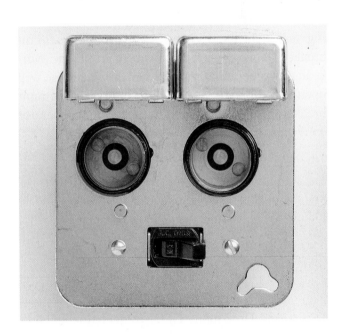

FIGURE 16-25 Box-cover unit that provides disconnecting means plus individual overcurrent protection. (*Courtesy of Bussmann, Inc.*)

For example, the overhead door operator in this residence is marked 5.8 amperes. Then,

$$5.8 \times 1.25 = 7.25 \text{ amperes}$$
$$5.8 \times 1.5 = 8.7 \text{ amperes}$$

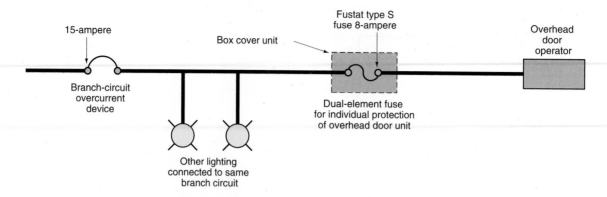

FIGURE 16-26 This diagram shows the benefits of installing a small-ampere-rated dual element time-delay fuse that is sized to provide backup overload protection for the overhead door operator. If an electrical problem occurs at the overhead door operator, only the 8-ampere fuse will open. The 15-ampere branch circuit is not affected. All other loads connected to the 15-ampere circuit remain energized. (*Delmar/Cengage Learning*)

Therefore, installing an 8-ampere, time-delay, dual-element Type S fuse in the box-cover unit would be appropriate.

The objective is to have a lower ampere rating fuse than the 15- or 20-ampere branch-circuit breaker. Should a short circuit occur within the overhead door operator, the lower ampere rated fuse will open faster than the 15- or 20-ampere branch-circuit breaker. This takes the overhead door operator off the branch circuit. The power to the rest of the branch circuit stays on.

NEC 210.8(A)(2) requires that all receptacles in a dwelling garage be GFCI protected. If you choose to install a receptacle switch box-cover unit on the ceiling for the overhead door operator(s), make sure that you lay out the garage circuitry so GFCI protection is provided for the receptacle/switch box-cover unit.

REVIEW

Note: Refer to the *Code* or plans where necessary.

1. What circuit supplies the garage? _____

   _____

2. What is the circuit rating? _____

3. How many GFCI-protected receptacle outlets are connected to the garage circuit?

   _____

4. a. How many cables enter the wall switch box located next to the side garage door?

      _____

      _____

   b. How many circuit conductors enter this box?

      _____

   c. How many equipment grounding conductors enter this box?

      _____

5. Show calculations on how to select a proper size box for Question 4. What kind of box would you use? _____

_____   _____

_____

6. a. How many lights are recommended for a one-car garage? _____

     for a two-car garage? _____ for a three-car garage? _____

  b. Where are these lights to be located? _____

_____

7. From how many points in the garage of this residence are the ceiling lights controlled?

_____

8. GFCI breakers or GFCI receptacles are relatively expensive. How would *you* arrange the wiring of the garage circuit to make as economical an installation as possible that complies with the *Code*? _____

_____

_____

_____

9. The total estimated volt-ampere load of the garage lighting branch circuit B14 draws how many amperes? Show your calculations. _____

_____

_____

10. How high from the floor are the switches and receptacles to be mounted? _____

_____

11. What type of cable feeds the post light? _____

12. All raceways, cables, and direct burial-type conductors require a "cover." Explain the term "cover." _____

_____

_____

_____

13. In the spaces provided, fill in the cover (depth) for the following residential underground installations. See *NEC Table 300.5.*

  a. Type UF. No other protection. _____ in.

  b. Type UF below driveway. _____ in.

  c. Rigid metal conduit under lawn. _____ in.

  d. Rigid metal conduit under driveway. _____ in.

  e. Electrical metallic tubing between house and detached garage. _____ in.

  f. UF cable under lawn. Circuit is 120 volts, 20 amperes, GFCI protected. _____ in.

g. Rigid PVC conduit approved for direct burial. No other protection. _____ in.

h. Rigid PVC conduit. Circuit is 120 volts, 20 amperes, GFCI protected. _____ in.

i. Rigid conduit passing over solid rock and covered by 2 in. (50 mm) of concrete extending down to rock. _____ in.

14. What section of the *Code* prohibits embedding Type UF cable in concrete? _____

_____

15. The following is a layout of the garage circuit. Using the suggested cable layout, make a complete wiring diagram of this circuit, using colored pens or pencils to indicate conductor insulation colors.

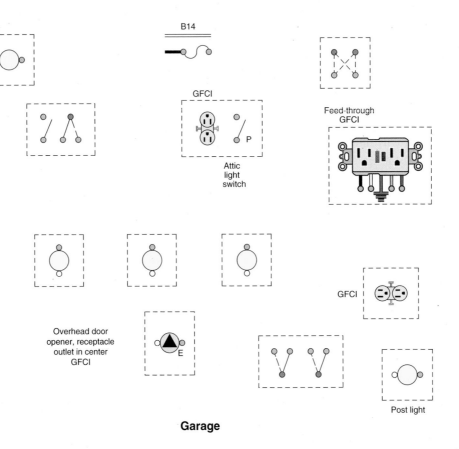

Garage

16. When a box-cover unit is installed for an overhead door operator, a time-delay dual-element fuse ampere rating would generally be sized at _____ % of the unit's nameplate current rating.

17. When outdoor receptacles are installed where a cord will be plugged in permanently, such as in the case of decorative lighting, the cover must provide weatherproof integrity when the cord is plugged in. True or false? _____ *NEC* _____ .

18. A homeowner purchased some single-conductor cable that was marked as suitable for direct burial in the ground to wire a post light. A separate equipment grounding conductor to the post light was not carried. Instead, the bottom of the metal post buried 2 ft (0.6 m) into the ground was used as a ground for the post light. The inspector "red tagged" (turned down) the installation. Who is right? The homeowner or the electrical inspector? Explain.

_____

_____

_____

_____

19. What concerns are there when installing a low-voltage lighting system?

_____

_____

_____

Recreation Room

OBJECTIVES

After studying this chapter, you should be able to

- understand 3-wire (multiwire) branch circuits.
- understand how to install lay-in luminaires.
- calculate watts loss and voltage drop in 2-wire and 3-wire circuits.
- understand the term *fixture whips*.
- understand the advantages of installing multiwire branch circuits.
- understand problems that can be encountered on multiwire branch circuits as a result of open neutrals.

RECREATION ROOM LIGHTING (B9, 11, 12)

The Recreation Room is well lighted through the use of six lay-in 2 ft × 4 ft fluorescent luminaires. These fixtures are exactly the same size as two 2 ft × 2 ft ceiling tiles. The luminaires rest on top of the ceiling tee bars, as in Figure 17-1.

Junction boxes are mounted above a dropped (suspended) ceiling usually within 12 to 24 in. (300 to 600 mm) of the intended luminaire location. A flexible conduit referred to as a *fixture whip*, 4 to 6 ft (1.2 to 1.8 m) long, is installed between the junction box and the luminaire. These fixture whips contain the correct type and size of conductor suitable for the temperature ratings and load required by the *Code* for recessed luminaires. See Chapter 7 for a complete explanation of *Code* regulations for the installation of recessed luminaires, as in Figure 17-2.

Important: To prevent the luminaire from inadvertently falling, *410.36(B)* of the *Code* requires that (1) suspended ceiling framing members that support recessed luminaires must be securely fastened to each other, and must be securely attached to the building structure at appropriate intervals, and (2) recessed luminaires must be securely fastened to the suspended ceiling framing members by bolts, screws, rivets, or special listed clips provided by the manufacturer of the luminaire for the purpose of attaching the luminaire to the framing member.

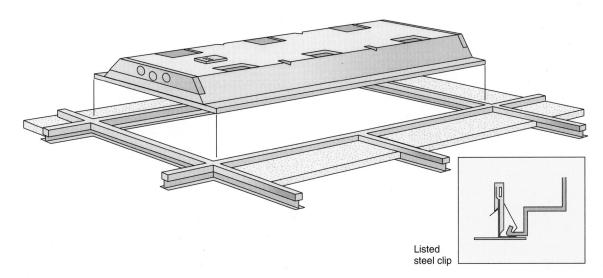

Listed steel clip

FIGURE 17-1 Typical lay-in fluorescent luminaire commonly used in conjunction with dropped suspended ceilings. See *410.36(B)*. (*Delmar/Cengage Learning*)

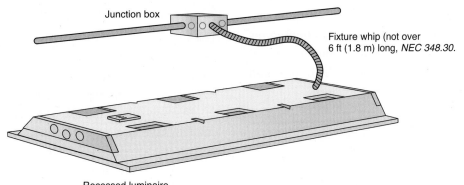

Junction box

Fixture whip (not over 6 ft (1.8 m) long, *NEC 348.30*.

Recessed luminaire

FIGURE 17-2 Typical suspended ceiling "lay-in" fluorescent luminaire showing flexible fixture whip connection. Conductors in whip must have temperature rating as required on label in luminaire; for example, "For Supply Use 90°C conductors." (*Delmar/Cengage Learning*)

Fluorescent luminaires of the type shown in Figure 17-1 might bear a label stating "Recessed Fluorescent Luminaire." The label might also state "Suitable for Use in Suspended Ceilings." Although they might look the same, there is a difference.

Recessed fluorescent luminaires are intended for installation in cavities in ceilings and walls and are to be wired according to *Article 410, Part XI*. These luminaires may also be installed in suspended ceilings if they have the necessary mounting hardware.

If marked "Suitable for Use in Suspended Ceilings," fluorescent luminaires are intended only for installation in suspended ceilings where the acoustical tiles, lay-in panels, and suspended grid are not part of the actual building structure.

Underwriters Laboratories (UL) Standard 1570 covers recessed and suspended ceiling luminaires in detail.

The six recessed "lay-in" fluorescent luminaires in the Recreation Room are 4-lamp luminaires with warm white energy-saving F32SPX30/RS lamps installed in them. The ballasts in these luminaires are high power factor, energy-saving ballasts. Five of the luminaires are controlled by a single-pole switch located at the bottom of the stairs. One luminaire is controlled by the 3-way switches that also control the stairwell luminaire hung from the ceiling above the stair landing midway up the stairs.

This was done so that the entire Recreation Room lighting would not be on continually whenever someone came down the stairs to go into the workshop. This is a practical energy-saving feature.

Circuit B12 feeds the fluorescent luminaires in the Recreation Room. See the cable layout in Figure 17-3. Table 17-1 summarizes the outlets and estimated load for the Recreation Room.

How to Connect Recessed and Lay-In Luminaires

Figures 17-1 and 17-2 show typical lay-in fluorescent luminaires. The installation and wiring connections for recessed luminaires and lay-in suspended-ceiling luminaires are covered in Chapter 7.

Recessed Luminaires between Ceiling Joists

Roughing-in recessed luminaires that will be installed between ceiling joists can be quite a challenge, particularly if you are trying to design spacing that makes sense. The conflict is between the 16 in. (400 mm) on center joists and the 24 in. (600 mm) square acoustic ceiling panels and their metal "T" bars and supports. Multiples of 16 in. (400 mm) and 24 in. (600 mm) don't agree except at 4 ft (1.2 m) intervals.

Dropped Ceiling. If there is a "dropped ceiling" that provides sufficient space between the finished ceiling and the bottom edge of the ceiling joists, you probably can space recessed incandescent luminaires easily to be in the exact centers of typical 2 ft × 2 ft (600 mm × 600 mm) acoustic panels.

In the case of recessed fluorescent luminaires, their 2 ft × 4 ft (600 mm × 1.2 m) size nicely coincides with the ceiling panels.

No Dropped Ceiling. If there is no drop, or if the drop is not as much as the height of the recessed luminaires, you have to give a lot of thought to laying out the lighting. You will find that the center-to-center spacing of recessed luminaires will have to be in increments of 4 ft (1.2 m). This spacing will find space between the ceiling joists that are 16 inches on center. With care, the recessed luminaires can be located to be exactly in the center of 2 ft × 2 ft (600 mm × 600 mm) acoustic ceiling panels.

RECEPTACLES AND WET BAR (B9–11)

The circuitry for the Recreation Room wall receptacles and the wet bar lighting area introduces a new type of circuit. This is termed a *multiwire branch circuit* or a *3-wire branch circuit*.

In many cases, the use of a multiwire branch circuit can save money in that one 3-wire branch circuit will do the job of two 2-wire branch circuits.

Also, if the loads are nearly balanced in a 3-wire branch circuit, the neutral conductor carries only the unbalanced current. This results in less voltage drop and watts loss for a 3-wire branch circuit as compared to similar loads connected to separate 2-wire branch circuits.

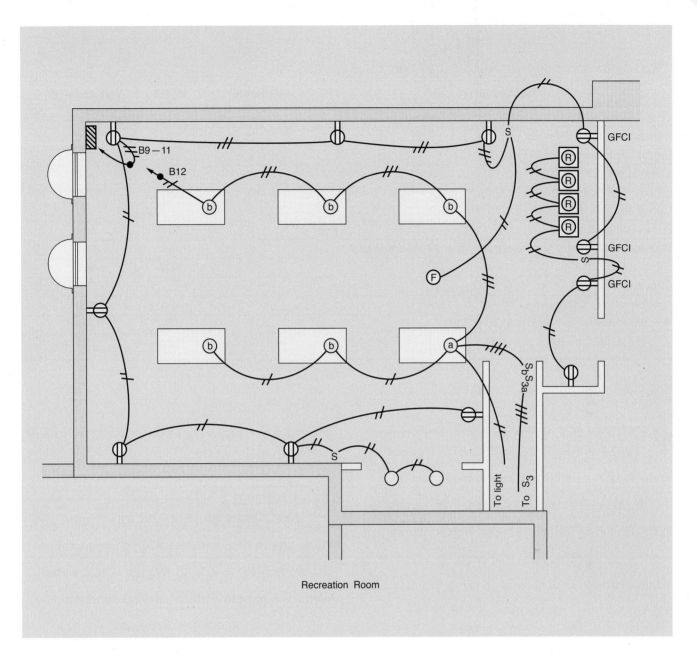

FIGURE 17-3 Cable layout for Recreation Room. The six fluorescent luminaires are connected to Circuit B12. The receptacles located around the room are connected to Circuit B11. The four recessed luminaires over the bar, the three receptacles located on the bar wall, and the receptacles located to the right as you come down the steps are connected to Circuit B9. Circuit B9 and Circuit B11 together make up a 3-wire circuit. Junction boxes are located above the dropped ceiling and to one side of each fluorescent luminaire. The electrical connections are made in these junction boxes. A flexible fixture whip connects between each junction box and each luminaire. See Figure 17-2. (*Delmar/Cengage Learning*)

Figures 17-4 and 17-5 illustrate the benefits of a multiwire branch circuit relative to watts loss and voltage drop. The example, for simplicity, is a purely resistive circuit, and loads are exactly equal.

The distance from the load to the source is 50 ft. The conductor size is 14 AWG solid, uncoated

copper. From *Table 8* in *Chapter 9* in the *NEC*, we find that the resistance of 1000 ft (300 m) of a 14 AWG uncoated copper conductor is 3.07 ohms (10.1 ohms/km). The resistance of 50 ft (15 m) is

$$\frac{3.07}{1000} \times 50 = 0.1535 \text{ ohms (round off to 0.154)}$$

TABLE 17-1

Recreation Room outlets and estimated load (three circuits).

Description	Quantity	Watts	Volt-Amperes
Circuit Number B11			
Receptacles @ 120 W each	7	840	840
Closet luminaires @ 75 W each	2	150	150
Exhaust fan (2.5 A × 120V)	1	270	300
TOTALS	10	1260	1290
Circuit Number B12			
Luminaire on stairway	1	100	100
Recessed fluorescent luminaires with four 32-W lamps each	6	768	806
TOTALS	7	868	906
Circuit Number B9			
Receptacles @ 120 W each	4	480	480
Wet bar recessed luminaires @ 75 W each	4	300	300
TOTALS	8	780	780

EXAMPLE 1

TWO 2-WIRE BRANCH CIRCUITS (Figure 17-4)

Watts loss in each current-carrying conductor is

$$\text{Watts} = I^2R = 10 \times 10 \times 0.154 = 15.4 \text{ watts}$$

Watts loss in all four current-carrying conductors is

$$15.4 \times 4 = 61.6 \text{ watts}$$

Voltage drop in each current-carrying conductor is

$$\begin{aligned} E_d &= IR \\ &= 10 \times 0.154 \\ &= 1.54 \text{ volts} \end{aligned}$$

E_d for both current-carrying conductors in each circuit:

$$2 \times 1.54 = 3.08 \text{ volts}$$

Voltage available at load:

$$120 - 3.08 = 116.92 \text{ volts}$$

Note in Figure 17-4 that the two 2-wire branch circuits result in four current-carrying conductors. If the wiring method is a raceway or cable, the ampacity of the conductors must be derated because there are more than three current-carrying conductors in the raceway. Adjustment and correction factors are discussed in Chapter 18 of this text. Refer to *NEC Table 310.15(B)(3)(c)*.

EXAMPLE 2

ONE 3-WIRE BRANCH CIRCUIT (Figure 17-5)

Watts loss in each current-carrying conductor is

$$\text{Watts} = I^2R = 10 \times 10 \times 0.154 = 15.4 \text{ watts}$$

Watts loss in both current-carrying conductors is

$$15.4 \times 2 = 30.8 \text{ watts}$$

Voltage drop in each current-carrying conductor is

$$\begin{aligned} E_d &= IR \\ &= 10 \times 0.154 \\ &= 1.54 \text{ volts} \end{aligned}$$

E_d for both current-carrying conductors is

$$2 \times 1.54 = 3.08 \text{ volts}$$

Voltage available at loads:

$$240 - 3.08 = 236.92 \text{ volts}$$

Voltage available at each load:

$$\frac{236.92}{2} = 118.46 \text{ volts}$$

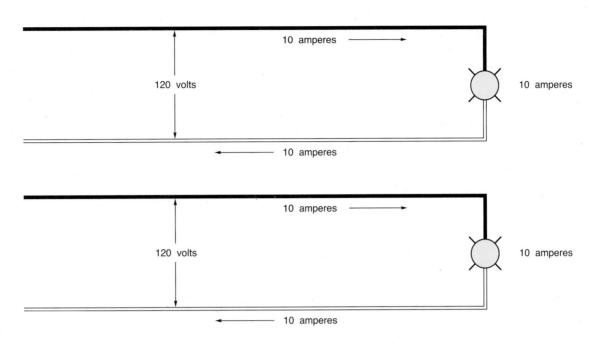

FIGURE 17-4   The 2-wire branch circuits. (*Delmar/Cengage Learning*)

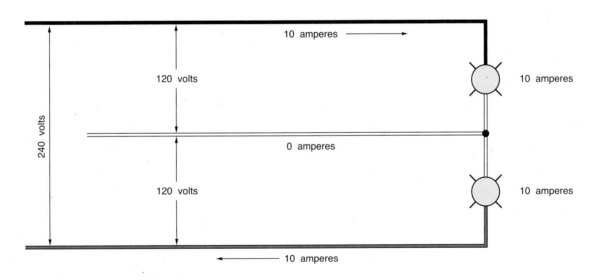

FIGURE 17-5   A 3-wire branch circuit. (*Delmar/Cengage Learning*)

Advantages of a 3-Wire Branch Circuit

1. Uses less wire for two circuits
2. There are three conductors instead of four. Might result in smaller raceway (if wiring in a raceway). Less need for correcting (derating) the conductor's ampacity due to having more than three current-carrying conductors in the raceway.
3. Less voltage drop in the conductors
4. Less watts loss in the conductors

These advantages are why there are 3-wire services and feeders, as shown in Panel B in this residence.

Disadvantages of a 3-Wire Branch Circuit

1. Possible burnout of equipment if the neutral conductor opens

2. Having to work with 240 volts present in outlet and device boxes

3. Possible unwanted power outage because *210.4(B)* requires all ungrounded conductors of a branch circuit to be disconnected simultaneously at the point the branch circuit originates

4. If you inadvertently open the neutral of a multiwire branch circuit and touch the open the conductors, you could be electrocuted. Remember the fundamental rule: line voltage appears across an open circuit.

5. Very important! When installing a 3-wire branch circuit where AFCIs and GFCIs are required, you must check the instructions furnished with these devices. Single-pole units are not designed to function on branch circuits where the neutral is shared. Double-pole units are designed to operate on shared-neutral branch circuits. This can be the key issue when choosing between 2-wire and 3-wire branch circuits.

The Recreation Room Wiring

At the beginning of this chapter, we pointed out that Circuit B12 supplies the ceiling lighting in the Recreation Room. Circuit B9 supplies the wet bar lighting and four receptacles. Circuit B11 powers the other receptacles in the Recreation Room, the ceiling exhaust fan, and the lighting in the storage closet.

Branch circuits B9, B11, and B12 present a challenge. The reason for this challenge is that these circuits are confronted with two distinct types of unique protection required by the *NEC*.

▶In accordance with *NEC 210.12(A)*, these branch circuits are required to have AFCI protection. AFCIs are discussed in Chapter 6.◀

▶In addition to the AFCI requirement, *NEC 210.8(A)* requires that receptacles *installed within 1.8 m (6 ft) of the outside edge of a sink be GFCI protected.* GFCIs are discussed in Chapter 6.◀

Depending on the availability of AFCI/GFCI products from different circuit breaker manufacturers, there are a couple of ways to "meet *Code*" for connecting branch circuits B9, B11, and B12. As with all electrical devices, they shall be listed by a Nationally Recognized Testing Laboratory (NRTL).

Option 1:

1. For B12—install a single-pole AFCI circuit breaker.

2. For B9 and B11—install a 2-pole, independent-trip, dual-function AFCI/GFCI circuit breaker. The receptacles in the wet bar area could be conventional receptacles because GFCI protection for these receptacles is provided by the dual-function circuit breaker.

Option 2:

1. For B12—install a dual-function AFCI circuit breaker in Panel B.

2. For B9 and B11—install a 2-pole, independent-trip dual-function AFCI circuit breaker in Panel B.

3. Install GFCI receptacles where the receptacles are within 6 ft (1.8 m) of the outside edge of the wet bar sink.

Other possibilities for wiring the Recreation Room are discussed in the Instructor's Guide. Now let's take a look at the actual installation as suggested in the cable layout.

In the Recreation Room, a 3-wire cable carrying Circuit B9 and Circuit B11 feeds from Panel B to the receptacle closest to the panel. This 3-wire cable continues to each receptacle wall box along the same wall. At the third receptacle along the wall, the multiwire circuit splits, where the wet bar Circuit B9 continues on to feed into the receptacle outlet to the left of the wet bar.

The white conductor of the 3-wire cable is common to both the receptacle circuit and the wet bar circuit. The black conductor feeds the receptacles. The red conductor is spliced straight through the boxes and feeds the wet bar area.

Care must be used when connecting a 3-wire circuit to the panel. The black and red conductors

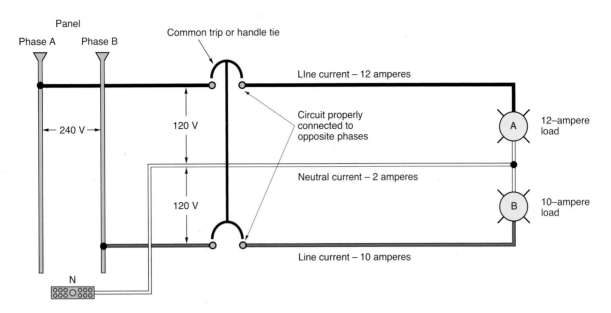

FIGURE 17-6 Correct wiring connections for a 3-wire (multiwire) branch circuit.
(*Delmar/Cengage Learning*)

must be connected to the opposite phases in the panel to prevent heavy overloading of the neutral grounded (white) conductor. In addition, the circuit breaker must be either of the 2-pole, common trip type or have an identified tie-bar for two single-pole breakers installed. See *210.4(B)*. This is a safety issue and is intended to prevent an electric shock. A shock hazard exists when one circuit breaker of a multiwire branch circuit (shared neutral circuit) is opened and the neutral connection is broken. The neutral can be carrying current for the second circuit. You can get a severe electric shock if your body completes the neutral circuit.

The neutral (grounded) conductor of the 3-wire cable carries the unbalanced current. This current is the difference between the current in the black wire and the current in the red wire. For example, if one load is 12 amperes and the other load is 10 amperes, the neutral current is the difference between these loads (2 amperes), Figure 17-6.

If the black and red conductors of the 3-wire cable are connected to the same phase in Panel A, Figure 17-7, the neutral conductor must carry the total current of both the red and black conductors rather than the unbalanced current. As a result, the neutral conductor will be overloaded. All single-

phase, 120/240-volt panels are clearly marked to help prevent an error in phase wiring. The electrician must check all panels for the proper wiring diagrams.

Figure 17-7 shows how an improperly connected 3-wire branch circuit results in an overloaded neutral conductor.

If an open neutral occurs on a 3-wire branch circuit, some of the electrical appliances in operation may experience voltages higher than the rated voltage at the instant the neutral opens.

For example, Figure 17-8 shows that for an open neutral condition, the voltage across load A decreases and the voltage across load B increases. If the load on each circuit changes, the voltage on each circuit also changes. According to Ohm's law, the voltage drop across any device in a series circuit is directly proportional to the resistance of that device. In other words, if load B has twice the resistance of load A, then load B will be subjected to twice the voltage of load A for an open neutral condition. To ensure the proper connection, care must be used when splicing the conductors.

An example of what can occur should the neutral conductor of a 3-wire multiwire branch circuit open is shown in Figure 17-9. Trace the flow of

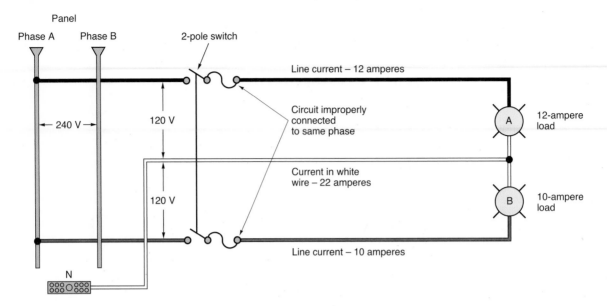

FIGURE 17-7 Improperly connected 3-wire (multiwire) branch circuit. (*Delmar/Cengage Learning*)

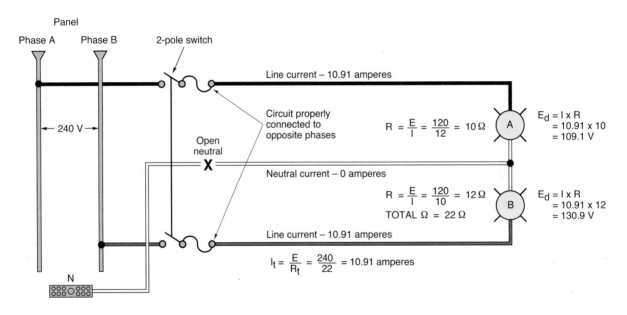

FIGURE 17-8 Example of an open neutral conductor. (*Delmar/Cengage Learning*)

current from phase A through the television set, then through the toaster, then back to phase B, thus completing the circuit. The following simple calculations show why the television set (or stereo or home computer) can be expected to burn up.

$$R_t = 8.45 + 80 = 88.45 \text{ ohms}$$

$$I = \frac{E}{R} = \frac{240}{88.45} = 2.71 \text{ amperes}$$

Voltage appearing across the toaster:

$$IR = 2.71 \times 8.45 = 22.9 \text{ volts}$$

Voltage appearing across the television:

$$IR = 2.71 \times 80 = 216.8 \text{ volts}$$

This example illustrates the problems that can arise with an open neutral conductor on a 3-wire, 120/240-volt multiwire branch circuit.

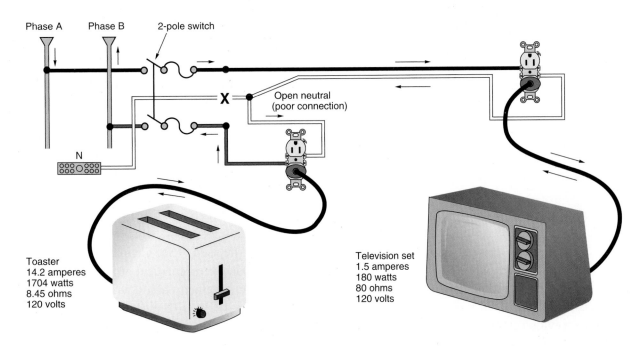

Phase A   Phase B   2-pole switch

X ● Open neutral (poor connection)

N

Toaster
14.2 amperes
1704 watts
8.45 ohms
120 volts

Television set
1.5 amperes
180 watts
80 ohms
120 volts

FIGURE 17-9 Problems that can occur with an open neutral on a 3-wire (multiwire) branch circuit. (*Delmar/Cengage Learning*)

The same problem can arise when the neutral of the utility company's incoming service-entrance conductors (underground or overhead) opens. The problem is minimized because the neutral point of the service is solidly grounded to the metal water piping system within the building. However, there are cases on record where poor service grounding has resulted in serious and expensive damage to appliances within the home because of an open neutral in the incoming service-entrance conductors.

For example, poor service grounding results when relying on a driven iron pipe (that rusts in a short period of time) as the only means of obtaining the service equipment ground. This was quite often done years ago when the electrical code was not as stringent as today's *NEC*. When working on older homes, check the service ground. It could very well be inadequate, unsafe, and in need of updating to the present *NEC*.

Keep in mind the purpose of the grounding electrode system is never to carry current for faulty circuits in the building or between the building and the utility transformer. In fact, the *NEC* states in *250.4(A)(5)* that the path through the earth is not an effective ground-fault return path.

Receptacles Near Sinks

▶All 125-volt, single-phase, 15- and 20-ampere receptacles within 6 ft (1.8 m) of the outside edge of a sink shall be GFCI protected. See *210.8(A)(7)*.◀

To meet this requirement for the receptacles located within 6 ft (1.8 m) of the wet bar sink, a feed-through GFCI receptacle could be installed at the first receptacle (upper right-hand in Figure 17-3). With this arrangement, that receptacle and everything beyond it would be GFCI protected. The other choice would be to install a GFCI receptacle at each location as required to "Meet *Code*." Three GFCI receptacles are required to do this. As with all circuit designs, it becomes a cost issue (time and material) as well as keeping the wiring as simple as possible.

Single-pole GFCI and AFCI devices will not work properly when sharing a neutral, such as in the multiwire branch circuit for Circuits B9 and B11. A single-pole GFCI breaker in Panel B for Circuit B9 is not an option. Installing a feed-through GFCI receptacle for the first receptacle above the wet bar countertop will provide the required GFCI protection for the receptacles as well as the rest of the circuit. The confusion of how to "Meet *Code*" regarding a room that involves both GFCI and AFCI protection requirements is solved by using a 2-pole, independent-trip, AFCI/GFCI dual-function circuit breaker in the panel for Circuits B9 and B11.

Receptacles shall not be installed face up on the countertop surfaces or work surfaces. See *406.4(E)*.

A single-pole switch to the right of the wet bar controls the four recessed luminaires above it. See Chapter 7 for details pertaining to recessed luminaires.

An exhaust fan of a type similar to the one installed in the laundry is installed in the ceiling of the Recreation Room to exhaust stale, stagnant, and smoky air from the room.

The basic switch, receptacle, and luminaire connections are repetitive of most wiring situations discussed in previous chapters.

The wall boxes for the receptacles are selected based on the measurements of the furring strips on the walls. Two-by-two furring strips will require the use of 4-in. square boxes with suitable raised covers. If the walls are furred with 2 × 4s, then possibly sectional device boxes could be used. Select a box that can contain the number of conductors, devices, and clamps to "Meet *Code*." After all, the installation must be safe.

Probably the biggest factor in deciding which wiring method (cable or EMT) will be used to wire the Recreation Room is "what size and type of wall furring will be used?" See Chapter 4 for the in-depth discussion on the mechanical protection required for nonmetallic sheathed cables where the cables will be less than 1¼ in. (32 mm) from the edge of the framing members.

REVIEW

Note: Refer to the *Code* or plans where necessary.

1. What is the total current draw when all six fluorescent luminaires are turned on?

   _____

   _____

2. The junction box that will be installed above the dropped ceiling near the fluorescent luminaires closest to the stairway will have _____ (number of) 14 AWG conductors.

3. Why is it important that the "hot" conductor in a 3-wire branch circuit be properly connected to opposite phases in a panel? _____

   _____

4. In the diagram, Load A is rated at 10 amperes, 120 volts. Load B is rated at 5 amperes, 120 volts.

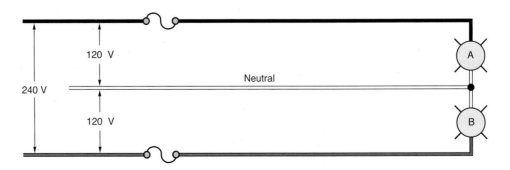

a. When connected to the 3-wire branch circuit as indicated, how much current will flow in the neutral conductor? _____

b. If the neutral conductor should open, to what voltage would each load be subjected, assuming both loads were operating at the time the neutral conductor opened? Show all calculations.

5. Calculate the watts loss and voltage drop in each conductor in the following circuit. Show calculations. Obtain resistance data from *Table 8, Chapter 9*, of the *NEC*.

50 ft 12 AWG THHN solid copper conductor

15 ampere

50 ft 12 AWG THHN solid copper conductor

6. Unless specifically designed, all recessed incandescent luminaires must be provided with factory-installed _____ protection (see Chapter 7).

7. Where in the *NEC* would you look for the installation requirements regarding flush and recessed lighting luminaires?

_____

8. What is the current draw of the recessed luminaires above the bar? _____

_____

_____

_____

9. a. VA load for Circuit B9 is _____volt-amperes.

b. VA load for Circuit B11 is _____ volt-amperes.

c. VA load for Circuit B12 is _____ volt-amperes.

10. Calculate the total current draw for Circuit B9, Circuit B11, and Circuit B12.

11. Complete the wiring diagram for the Recreation Room. Follow the suggested cable layout. Use colored pencils or pens to identify the various colors of the conductor insulation.

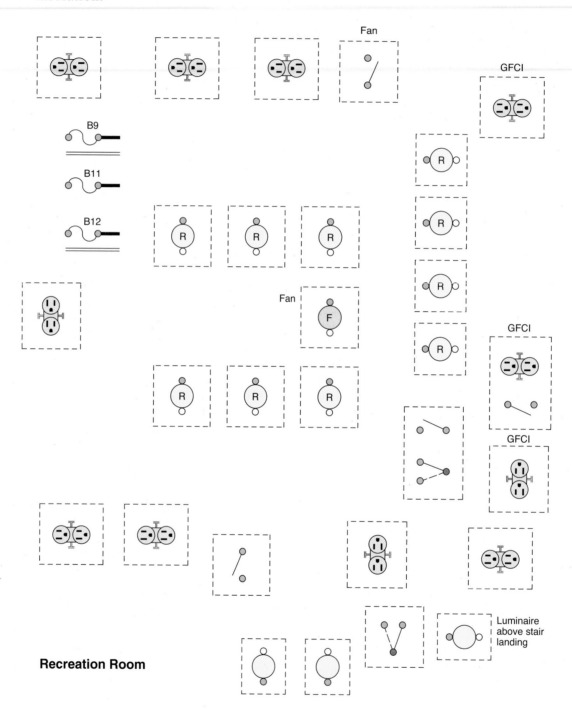

Recreation Room

12. a. May a fluorescent luminaire that is marked "Recessed Fluorescent Luminaire" be installed in a suspended ceiling? _____

   b. May a fluorescent luminaire that is marked "Suspended Ceiling Fluorescent Luminaire" be installed in a recessed cavity of a ceiling? _____

13. The *NEC* requires GFCI protection for receptacles located within 6 ft (1.8 m) of the outside edge of a wet bar sink. This requirement is found in *NEC* _____

   _____.

14. Lay-in fluorescent luminaires that rest on the metal framing members of a suspended ceiling cannot just lay on the framing members. *NEC 410.36(B)* requires that (1) framing members used to support recessed luminaires must be securely fastened to each other and to the building structure at appropriate intervals, and (2) the luminaires must _____

   _____.

15. List the rooms and/or areas in a newly constructed single-family residence where arc-fault circuit-interrupter (AFCI) protection is required by the *NEC*.

   _____

   _____

   _____

CHAPTER 6

Lighting Branch Circuit, Receptacle Circuits for Workshop

OBJECTIVES

After studying this chapter, you should be able to

- install cable and conduit wiring for a typical nonfinished basement.
- understand *NEC* location requirements for GFCI receptacles in basements.
- make conductor fill calculations for raceways.
- use derating and correction factors for determining conductor current-carrying capacity.
- understand maximum ratings of overcurrent protection for conductors.
- install multioutlet assemblies.

The workshop area is supplied by more than one circuit:

A13   Separate branch circuit for freezer
A17   Lighting
A18   Plug-in strip (multioutlet assembly)
A20   Two receptacles on window wall

AFCI protection is not required for the unfinished basement area, *NEC 210.12(A)*. GFCI protection is required for the receptacles located in an unfinished basement, *NEC 210.8(A)(5)*. As discussed in prior chapters, this can be accomplished by installing GFCI circuit breakers for Circuits A13, A18, and A20 in Panelboard A or by installing GFCI receptacles. However, keep in mind the GFCI devices are required to be installed in a readily accessible location. This means you have to be able to walk up to the device and check it for proper operation without using a ladder or moving objects such as a refrigerator or freezer. See *NEC 210.8*.

The wiring method in the workshop is electrical metallic tubing (EMT). Figure 18-1 shows the branch circuit layout for the workshop. An option if you prefer to use cable is to install a drop of EMT from the joist space to metal boxes that are secured to the wall. These boxes are often secured with masonry anchors. This is illustrated in Figure 18-2. This option is provided in *NEC 334.15(C)*. The installation is made by securing the box and EMT to the wall, installing a fitting or bushing at the top to protect the cable, and routing the cable to the box leaving an adequate length to connect to the wiring device. The cable is required to be secured to within 12 in. (300 mm) of the point where the cable enters the EMT. The Type NM cable is not required to be secured to the EMT or to the box.

A single-pole switch at the entry controls all five ceiling porcelain lampholders. However, note on the plans that three of these lampholders have pull-chains, which would allow the homeowner to turn these lampholders on and off as needed (Figure 18-3). They would not have to be on all of the time, which would save energy.

In addition to supplying the lighting, Circuit A17 also feeds the chime transformer and ceiling exhaust fan. The current draw of the chime transformer is extremely small. Refer to Table 18-1. Refer to Table 18-2 for estimated loads on Circuit A13, Circuit A18, and Circuit A20. Note that these tables are used only for estimating the loading on the branch circuits. Actual load calculations for dwellings for lighting and receptacle loads used for general purposes are performed on a volt-amperes per square foot area. Detailed explanations of the load calculation are covered in Chapter 29 of this text.

WORKBENCH LIGHTING

Two 2-lamp, 32-watt fluorescent luminaires are mounted above the workbench to reduce shadows over the work area. The electrical plans show that these luminaires are controlled by a single-pole wall switch. A junction box is mounted immediately above or adjacent to the fluorescent luminaire so that the connections can be made readily.

When armored cable or flexible metal conduit is used as the wiring method between the luminaires and the junction box or boxes on the ceiling, either wiring method provides the necessary equipment grounding for the fluorescent luminaires, provided the metal outlet box is properly grounded. See *250.118* for the many acceptable means to accomplish equipment grounding.

Many workshop fluorescent luminaires are furnished with a 3-wire (black, white, green) flexible cord suitable for the purpose that has the attachment plug cap factory-attached to the cord. These luminaires can be plugged into a grounding-type receptacle mounted on the ceiling immediately above the fluorescent luminaires. This provides proper grounding of the luminaire.

RECEPTACLE OUTLETS

NEC 210.8(A)(4) and *210.8(A)(5)* require that all 125-volt, single-phase, 15- or 20-ampere receptacles installed in unfinished basements—such as workshop and storage areas that are not considered habitable, and crawl spaces that are at or below ground level—must be GFCI protected.

See Chapter 6 for a full discussion of GFCIs.

NEC 210.52(G) makes it mandatory to install at least one receptacle outlet in each unfinished basement area. If the laundry area is in the basement,

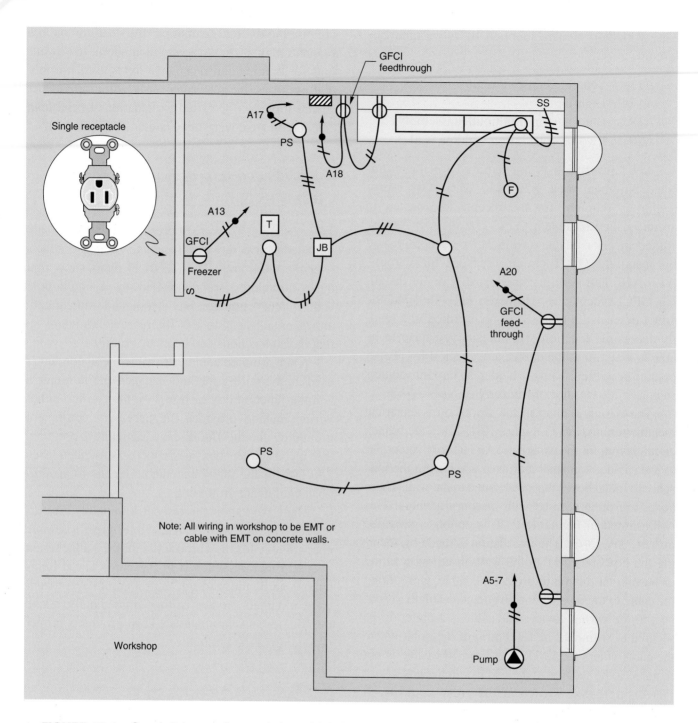

FIGURE 18-1 Conduit layout for workshop. Lighting is Circuit A17. Freezer single receptacle is Circuit A13. The two receptacles on the window wall are connected to Circuit A20. Plug-in strip receptacles and receptacle to the right of Main Panelboard are connected to Circuit A18. All receptacles are GFCI protected. A junction box mounted on the ceiling near the fluorescent luminaires above the workbench will provide a convenient place to make up the necessary electrical connections. (*Delmar/Cengage Learning*)

then a receptacle outlet must *also* be installed for the laundry equipment, *210.52(F)*. This additional receptacle must be within 6 ft (1.8 m) of the intended location of the washer, *210.50(C)*, and must be a 20-ampere circuit.

The circuit feeding the required laundry area receptacle(s) shall have no other outlets connected to it, *210.11(C)(2)*.

When a single receptacle is installed on an individual branch circuit, the receptacle rating must

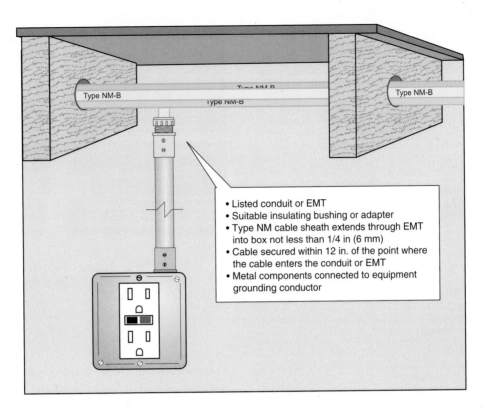

FIGURE 18-2 One method of installing Type NM cable on masonry walls in unfinished basement. (*Delmar/Cengage Learning*)

The callout box in the figure reads:

- Listed conduit or EMT
- Suitable insulating bushing or adapter
- Type NM cable sheath extends through EMT into box not less than 1/4 in (6 mm)
- Cable secured within 12 in. of the point where the cable enters the conduit or EMT
- Metal components connected to equipment grounding conductor

not be less than the rating of the branch circuit, *210.21(B)(1)*. So, if a single receptacle is installed on a 20-ampere branch circuit, as is shown for the freezer in Figure 18-2, the receptacle must have a rating not less than 20 amperes.

FIGURE 18-3 Porcelain and nonmetallic lampholders of both the keyless and pull-chain types. These lampholders are also available with a receptacle outlet, which must be GFCI protected when installed in locations where GFCI protection is required by the *Code*. (*Delmar/Cengage Learning*)

TABLE 18-1

Workshop: Outlet count and estimated load. Lighting load (Circuit A17).

Description	Quantity	Watts	Volt-Amperes
Ceiling lights @ 100 W each	5	500	500
Fluorescent luminaires Two 32-W lamps each	2	128	160
Chime transformer	1	8	10
Exhaust fan	1	80	90
TOTALS	9	716	760

CABLE INSTALLATION IN BASEMENTS

Chapter 4 covers the installation of nonmetallic-sheathed cable (Type NM) and armored cable (Type AC).

Some additional rules apply when nonmetallic-sheathed cable and armored cable are run

TABLE 18-2

Workshop: Outlet count and estimated load for three 20-A circuits A13, A18, A20.

Description	Quantity	Watts	Volt-Amperes
20-A Circuit, Number A18 Receptacles next to main panelboard	1	180	180
A six-receptacle plug-in multioutlet assembly at 1½ A per outlet (180 VA)	1	1080	1080
TOTALS	2	1260	1260
20-A Circuit, Number A20 Receptacles on window wall @ 180 W each	2	360	360
TOTALS	2	360	360
20-A Circuit, Number A13 Single receptacle for freezer	1	—	696
TOTALS	1	1620	2316

FIGURE 18-4 Types of electrical raceways. (*Courtesy of Allied Tube & Conduit Corp.*)

exposed, because of possible physical damage to the cables.

In unfinished basements, nonmetallic-sheathed cables made up of conductors smaller than two 6 AWG or three 8 AWG must be run through holes bored in joists, or on running boards, *334.15(C)*. In unfinished basements, exposed runs of armored cable must closely follow the building surface or running boards to which it is fastened, *320.15. NEC 320.15* also permits armored cable (Type AC) to be run on the undersides of joists in basements, attics, or crawl spaces, where supported on each joist and so located as not to be subject to physical damage. This exception is not permitted when using nonmetallic-sheathed cable. The local inspection authority usually interprets the meaning of "subject to physical damage."

Review Chapter 4 and Chapter 15 for additional discussion regarding protection of cables installed in exposed areas of basements and attics.

CONDUIT INSTALLATION IN BASEMENTS

Where metal raceways are used in house wiring, in most cases the raceways will be EMT, as shown in Figure 18-4. EMT is also referred to in the trade as "Thinwall," which is defined as a raceway and is not considered to be a conduit. Use EMT fittings marked "Raintight" or "Wet Location" where they will be exposed to weather, rain, and/or water. Use EMT fittings marked "Concrete Tight" where used in poured concrete. The rules for EMT are found in *Article 358*.

In most instances, rigid metal conduit (*Article 344*) will be used for a mast service; refer to Chapter 27. The following discussion only briefly touches on metal raceways, because most house wiring across the country is done with nonmetallic-sheathed cable, which is covered in detail in Chapter 4.

Local electrical codes may not permit cable wiring in unfinished basements other than where absolutely necessary. For example, a cable may be dropped into a basement from a switch at the head of the basement stairs. Transitions between cable and conduit also must be made. Usually this will be a junction box. The electrician must check the local codes before selecting conduit or cable for the installation to prevent costly wiring errors.

The electrician should never assume that the *NEC* is the recognized standard everywhere. Any city or state may pass electrical installation and licensing laws. In many cases these laws are more stringent than the *NEC*.

Outlet boxes may be fastened to masonry walls with lead or plastic anchors, shields, concrete nails, or power-actuated studs.

FIGURE 18-5 A one-hole conduit strap.
(*Delmar/Cengage Learning*)

Conduit and EMT may be fastened to masonry or wood surfaces using straps, such as the one-hole strap illustrated in Figure 18-5.

The conduits on the workshop walls and on the ceiling are exposed to view. The electrician must install the exposed wiring in a neat and skillful manner while complying with the following practices:

- The conduit runs must be straight.
- Bends and offsets must be true.
- Vertical runs down the surfaces of the walls must be plumb.

- To ensure that the conductor insulation will not be injured, be sure to ream the cut ends of rigid conduit and electrical metallic tubing to remove rough edges and burrs.

All conduits and boxes on the workshop ceiling are fastened to the underside of the wood joists, as in Figure 18-6. Raceways shall be installed as a complete system between pull points before pulling in the wires, as shown in Figure 18-7.

If the electrician can get to the residence before the basement concrete floor is poured, he might find it advantageous to install conduit runs in or under the concrete floor from the main panelboard to the freezer outlet, the window wall receptacles, and the water pump disconnect switch location. This can be a cost-saving (labor and material) benefit.

Now might be a good time for you to review the *NEC* rules in *Article 358* for electrical metallic tubing.

Metal raceways and armored cables (Type AC) are acceptable equipment grounding conductors, *250.118*. The runs would have to be continuous between other effectively grounded boxes, cabinets, and/or equipment. When a nonmetallic wiring method (NM, NMC, UF, etc.) is installed where

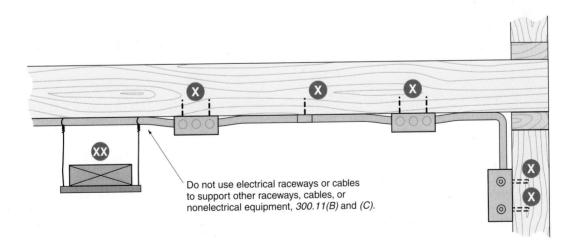

Do not use electrical raceways or cables to support other raceways, cables, or nonelectrical equipment, *300.11(B)* and *(C)*.

FIGURE 18-6 Raceways, boxes, fittings, and cabinets must be securely fastened in place (points X), *300.11(A)*. It is not permissible to hang other raceways, cables, or nonelectrical equipment from an electrical raceway or cable (point XX), *300.11(B)* and *(C)*. Refer to Figure 24-10 for an exception that allows low-voltage thermostat cables to be fastened to the conduit that feeds a furnace or other similar equipment. See Chapter 17 for a discussion of the use of suspended-ceiling support wire for supporting equipments. (*Delmar/Cengage Learning*)

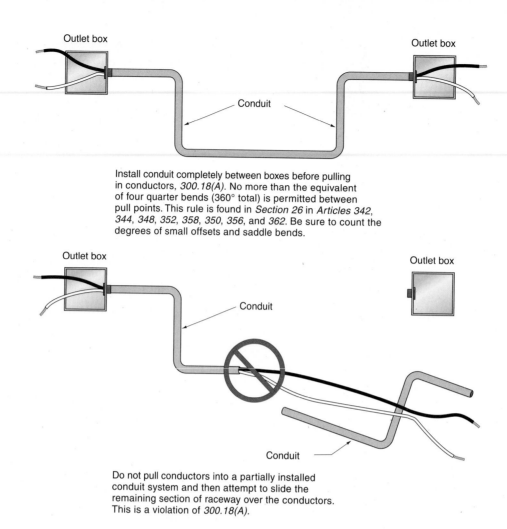

Install conduit completely between boxes before pulling in conductors, *300.18(A)*. No more than the equivalent of four quarter bends (360° total) is permitted between pull points. This rule is found in *Section 26* in *Articles 342, 344, 348, 352, 358, 350, 356,* and *362*. Be sure to count the degrees of small offsets and saddle bends.

Do not pull conductors into a partially installed conduit system and then attempt to slide the remaining section of raceway over the conductors. This is a violation of *300.18(A)*.

FIGURE 18-7 Do not pull wires into the raceway until the conduit system is completely installed between pull points. (*Delmar/Cengage Learning*)

short sections of metal raceway are provided for mechanical protection—or for appearance—these short sections of raceway need not be grounded, *250.86, Exception No. 2.*

Cable in Ceilings—EMT on Walls

In many parts of the country, the wiring method in unfinished basements is nonmetallic-sheathed cable in the ceiling and EMT on the walls for receptacles and switches. This protects and provides a neat appearance for the cable. Figures 18-2 and 4-20 clearly illustrates the requirements for this type of installation. See *NEC 334.15(C)* and *300.18(A), Exception.*

Outlet Boxes for Use in Exposed Installations

Figure 18-8 illustrates outlet boxes and plaster rings that are used for concealed wiring, such as for the recreation room and first floor wiring in the residence. For surface wiring such as the workshop, various types and sizes of outlet boxes may be used. For example, a handy box, Figure 18-9, could be used for the freezer receptacle. The remaining workshop boxes could be 4-in. square outlet boxes with raised covers, Figure 18-10. The box size and type is determined by the maximum number of conductors contained in the box, *Article 314.*

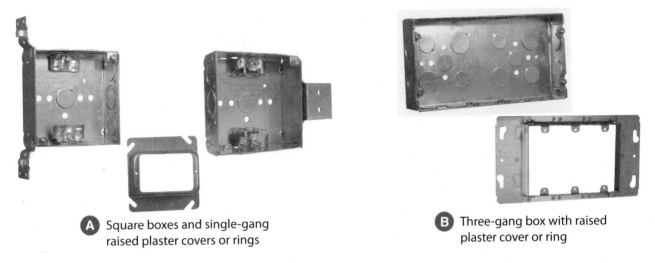

A Square boxes and single-gang raised plaster covers or rings

B Three-gang box with raised plaster cover or ring

FIGURE 18-8 Typical outlet boxes and raised covers. (*Courtesy of Appleton Electrical Company*)

FIGURE 18-9 Handy box and covers. (*Courtesy of Appleton Electric Company*)

FIGURE 18-10 Four-inch square boxes and raised covers. (*Courtesy of Appleton Electric Company*)

NEC 406.4(C) prohibits mounting and supporting a receptacle to a cover (flat or raised) by only one screw unless the device assembly or box cover is listed and identified for securing by a single screw.

The listing and identifying for the purpose is a decision that is made by an NRTL, such as Underwriters Laboratories, Inc. In the past, securing a receptacle with a single screw often resulted in a loose receptacle when used over a long period of time. Many short-circuits and ground-faults have been reported because of this.

For the typical raised cover, as illustrated in Figure 18-10, the duplex receptacle would be fastened to the raised cover by two small No. 6-32 bolts and nuts, usually furnished with the raised cover. The raised cover has two small "knockout" holes for these screws. The center No. 6-32 screw could also be used.

If a receptacle is installed on a surface-mounted box where there is direct metal-to-metal contact between the grounded metal box and the metal yoke of the receptacle, the receptacle is considered to be properly grounded without connecting an additional equipment grounding conductor to the equipment grounding terminal on the receptacle, *250.146.*

To make sure that there is direct metal-to-metal contact between the metal yoke of a receptacle and a surface-mounted metal box, *250.146(A)* accepts the automatic self-grounding device on the yoke of the receptacle or the removal of at least one of the insulating washers from one of the No. 6-32 screws on the receptacle. Automatic self-grounding devices on the yoke of a receptacle are clearly shown in Figures 5-3, 5-4, and 5-5.

Most commonly used outlet, device, and junction boxes have knockouts for trade size ½, ¾, and 1 connectors. The knockout size selected depends on the size of the conduits entering the box. The conduit size is determined by the maximum percentage of fill of the conduit's total cross-sectional area by the conductors pulled into the conduit.

Supporting equipment and/or raceways to the joists, walls, and so on, is accomplished with proper nails, screws, anchors, and straps. Be careful when trying to support other equipment from raceways as this is not always permitted. Refer to Figure 18-5.

Conduit Fill Calculations

For simplicity, the following conduit fill calculations show in. and in.² To show calculations in in., in.², mm, and mm² would be very confusing.

An electrician must be able to select the proper size raceways based on the number, size, and type of conductors to be installed in a particular type of raceway. To do this, the electrician must become familiar with the use of certain tables found in *Chapter 9* and *Annex C* of the *NEC.*

Percent Fill. The *NEC* has established the following "percent fill" values for conduit and tubing. These percentages relate to the area of conduit and tubing as shown in *Table 4* of *NEC Chapter 9.* A separate table is provided for each type of circular raceway.

- one conductor   53% fill
- two conductors   31% fill
- three or more conductors   40% fill
- conduit or nipples
  not over 24 in. (600 mm)   60% fill

Derating of conductor ampacity not required for the raceway that is not over 24 in. (600 mm) in length.

These percentages are taken from *Chapter 9, Table 1* and the accompanying notes to the table.

All Conductors the Same Size and Type. When all conductors installed in the raceway are the same size and type:

- Determine how many conductors of one size and type are to be installed in the raceway.
- Refer to *Tables C1* through *C12A* in the Annex of the *NEC* for the type of conduit, tubing, or raceway and for the specific size and type of conductors.
- Select the proper size raceway directly from the table.

These tables are not difficult to use. Here are a few examples of conductor fill for EMT. Look these up in the tables to confirm the accuracy of the examples.

- Not over five 10 AWG THHN conductors may be pulled into a trade size ½ EMT.
- Not over eight 12 AWG THW conductors may be pulled into a trade size ¾ EMT.
- Not over three 8 AWG THHN conductors may be pulled into a trade size ½ EMT.
- Not over six 8 AWG THW conductors may be pulled into a trade size 1 EMT.

Conductors of Different Sizes and/or Types.

When conductors are different sizes or have different insulation types:

- Refer to *Chapter 9*, *Table 5* of the *NEC*.
- Find the size and type of conductors used to determine the approximate cross-sectional area in in.² of the conductors.
- Refer to *Chapter 9*, *Table 8* for the in.² area of bare conductors.
- Total the in.² areas of all conductors.
- Refer to *Chapter 9*, *Table 4* for the type and size of raceway used.
- Determine the minimum size raceway based on the total in.² area of the conductors and the allowable in.² area fill for the particular type of raceway used.
- The sum must not exceed the allowable percentage fill of the cross-sectional in.² area of the raceway used.

For illustrative purposes, only the data for a few types of wires from *Table 5* are shown in Table 18-3. The dimensions and areas for some EMT from *Table 4* are shown in Table 18-4. You will need to refer to *Chapter 9* and *Annex C* of the *NEC* for the many other sizes and types of conductors and raceways.

EXAMPLE

If three 6 AWG THW conductors and two 8 AWG THW conductors are to be installed in one EMT, determine the proper size EMT.

Solution

1. Find the in.² area of the conductors in *NEC Chapter 9, Table 5*.

Three 6 AWG THW conductors	0.0726 in.²
	0.0726 in.²
	0.0726 in.²
Two 8 AWG THW conductors	0.0437 in.²
	0.0437 in.²
Total Area	0.3052 in.²

2. In *NEC Chapter 9, Table 4*, look at the first chart entitled *Article 358*. Now look in the last column marked "Over 2 Wires—40%." Note that a trade size ¾ EMT holds up to 0.213 in.² of conductor fill, and a trade size 1 EMT holds up to 0.346 in.² of conductor fill. Therefore, a trade size 1 EMT is the minimum size for the combination of conductors in the example.

TABLE 18-3

This sampling of typical conductors and their dimensions is taken from *Table 5, Chapter 9, NEC*. See *Table 5* for metric values.

Type	Size	Approx. Diam. In.	Approx. Area In.²
THHN/THWN	14	0.111	0.0097
THHN/THWN	12	0.130	0.0133
THHN/THWN	10	0.164	0.0211
THHN/THWN	8	0.216	0.0366
THHN/THWN	6	0.254	0.0507
THHN/THWN	3	0.352	0.0973
THW	6	0.304	0.0726
THW	8	0.236	0.0437

TABLE 18-4

Dimensional and percent fill data for EMT. This is one of the twelve charts found in *NEC Chapter 9, Table 4*, for various types of raceways. See *Table 4* for metric values.

ELECTRICAL METALLIC TUBING

Trade Size (in.)	Internal Diameter (in.)	Total Area 100% (in.²)	2 Wires 31% (in.²)	Over 2 Wires 40% (in.²)	1 Wire 53% (in.²)
½	0.622	0.304	0.094	0.122	0.161
¾	0.824	0.533	0.165	0.213	0.283
1	1.049	0.864	0.268	0.346	0.458
1¼	1.380	1.496	0.464	0.598	0.793
1½	1.610	2.036	0.631	0.814	1.079
2	2.067	3.356	1.040	1.342	1.778
2½	2.731	5.858	1.816	2.343	3.105
3	3.356	8.846	2.742	3.538	4.688
3½	3.834	11.545	3.579	4.618	6.119
4	4.334	14.753	4.573	5.901	7.819

Many electrical ordinances require that a separate equipment grounding conductor be installed in all raceways, whether rigid or flexible, whether metallic or nonmetallic. This requirement resulted from inspectors finding loose or missing set screws on set-screw type connectors and couplings. Corrosion of fittings also contributes to poor grounding. A separate equipment grounding conductor does take up space, and therefore must be counted when considering conduit fill.

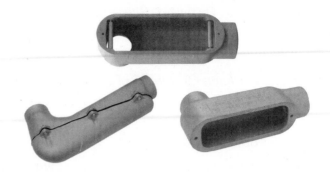

FIGURE 18-11 A variety of common conduit bodies and covers. (*Delmar/Cengage Learning*)

 EXAMPLE

A feeder to an electric furnace is fed with two 3 AWG THHN conductors plus one green insulated equipment grounding conductor. This feeder is installed in EMT. The feeder overcurrent protection is 100 amperes. What is the minimum size EMT to be installed?

Solution

According to *Table 250.122*, the minimum size equipment grounding conductor for a 100-ampere feeder is 8 AWG copper. Find the in.² area of the conductors in *Table 5, Chapter 9, NEC.*

Two 3 AWG THHN	0.0973 in.²
	0.0973 in.²
One 8 AWG THHN	0.0366 in.²
Total Area	0.2312 in.²

From *Table 4, Chapter 9, NEC:*

40% fill for trade size ¾ EMT = 0.213 in.²
40% fill for trade size 1 EMT = 0.346 in.²
Therefore, install trade size 1 EMT.

Remember, *300.18(A)* states that a conduit system must be completely installed before pulling in the conductors. Refer to Figure 18-7.

Conduit Bodies [*314.16(C)*]

Conduit bodies are used with conduit installations to provide an easy means to turn corners where there is not adequate space to install an elbow with a standard radius as well as to route the conduit or tubing around sharp corners and to terminate conduits. They are also used to provide access to conductors, to provide space for splicing (when permitted), and to provide a means for pulling conductors (Figure 18-11).

A conduit body:

- must have a cross-sectional area not less than twice the cross-sectional area of the largest conduit to which it is attached, as shown in Figure 18-12(A). This is a requirement only when the conduit body contains 6 AWG conductors or smaller.

- may contain the same maximum number of conductors permitted for the size raceway attached to the conduit body.

- must not contain splices, taps, or devices unless the conduit body is marked with its in.³ capacity, as in Figure 18-12(B), and the conduit body must be supported "in a rigid and secure manner."

- the conductor fill volume must be properly calculated according to *Table 314.16(B)*, which lists the free space that must be provided for each conductor within the conduit body. If the conduit body is to contain splices, taps, or devices, it must be supported rigidly and securely.

- must be sized according to *NEC 314.28* when it contains conductors 4 AWG or larger that are required to be insulated.

Conduit bodies are often referred to in the trade as Condulets and Unilets. The terms *Condulets* and *Unilets* are copyrighted trade names of Cooper

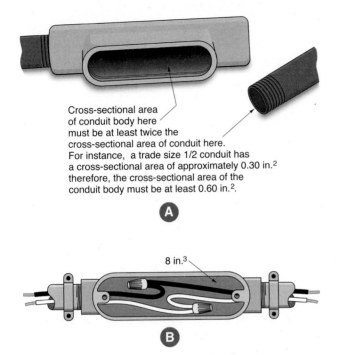

Cross-sectional area of conduit body here must be at least twice the cross-sectional area of conduit here. For instance, a trade size 1/2 conduit has a cross-sectional area of approximately 0.30 in.² therefore, the cross-sectional area of the conduit body must be at least 0.60 in.².

A

8 in.³

B

FIGURE 18-12 Requirements for conduit bodies.
(*Delmar/Cengage Learning*)

Crouse-Hinds and Appleton Electric, respectively, for their line of conduit bodies. Conduit bodies are excellent for turning corners and changing directions of a conduit run. They are available with threaded hubs for threaded conduit and threaded fittings, set screw hubs for direct insertion of EMT without needing an EMT connector, gasketed covers, built-in rollers to make it easy to pull wires, and in jumbo sizes to provide more room for pulling, splicing, and making taps. The chart that follows shows how typical conduit bodies are identified.

When Looking at the Front Opening, if the Direction is:	The Conduit Body is Called:
to the back	LB
Left	LL
Right	LR
straight through	C
straight through in both directions	X
straight through and to the left or right	T
straight through and to the back	TB
one conduit entry only	E

ADJUSTMENT AND CORRECTION (DERATING) FACTORS FOR MORE THAN THREE CURRENT-CARRYING CONDUCTORS IN CONDUIT OR CABLE

These requirements apply equally if more than three current-carrying conductors are installed in a raceway or multiconductor cables are installed for a continuous length longer than 24 in. (600 mm) without maintaining spacing. For more than three current-carrying wires in conduit or cable, refer to Table 18-5.

NEC 310.15(B)(3)(a) states that for the conductors listed in the tables, the maximum allowable ampacity must be reduced when more than three current-carrying conductors are installed in a raceway or multiconductor cables are installed for a continuous length longer than 24 in. (600 mm) without maintaining spacing. The reduction factors based on the number of conductors are noted in Table 18-5. When the length of the grouped or bundled conductors does not exceed 24 in. (600 mm), the reduction factor as shown in Table 18-5 does not apply.

According to *310.15(B)(5)(a)*, a neutral conductor of a multiwire branch circuit carrying only unbalanced currents shall not be counted in determining the number of current-carrying conductors, as shown in Table 18-5.

TABLE 18-5

Derating factors necessary to determine conductor ampacity when there are more than three current-carrying conductors in a raceway or cable. See *310.15(B)(3)(a)*.

Number of Current-Carrying Conductors	Percent of Values in *Table 310.15(B)(16)*. If High Ambient Temperatures Are Present, Also Apply Correction Factors Found in *Table 310.15(B)(2)(a)*.
4 through 6	80
7 through 9	70
10 through 20	50
21 through 30	45
31 through 40	40
41 and above	35

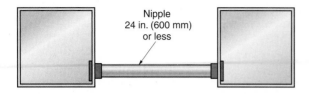

FIGURE 18-13 The derating factors in Table 18-5 are not required for short lengths of conduit that do not exceed 24 in. (600 mm), *310.15(B)(3)(a)(2)*. Conductor fill not to exceed 60% according to *Chapter 9, Table 1, Note 4*. (*Delmar/Cengage Learning*)

 EXAMPLE

What is the correct ampacity for size 6 AWG THHN/THWN when there are six current-carrying conductors in the conduit or cable and it is installed in a dry location? Note that we can begin the derating from the ampacity in the 90°C column of *Table 310.15(B)(16)* in a dry location because the THHN insulation has a 90°C rating.

$$75 \times 0.80 = 60 \text{ amperes}$$

CORRECTION FACTORS DUE TO HIGH TEMPERATURES

When conductors are installed in locations where the temperature is higher than 86°F (30°C), the correction factors noted in *Table 310.15(B)(2)(a)* must be applied.

For example, consider that four current-carrying 3 AWG THWN copper conductors are to be installed in one raceway or cable in an ambient temperature of approximately 90°F (32°C). "Ambient" temperature means "surrounding" temperature. This temperature is frequently exceeded in attics where the sun shines directly onto the roof.

The maximum ampacity for these conductors is determined as follows:

- The allowable ampacity of 3 AWG THWN copper conductors from *Table 310.15(B)(16)* = 100 amperes.

- Apply the correction factor for 90°F (32°C) found in *Table 310.15(B)(2)(a)*. The first column shows Celsius; the last column shows Fahrenheit.

$$100 \times 0.94 = 94 \text{ amperes}$$

- Apply the derating factor for four current-carrying conductors in one raceway.

$$94 \times 0.80 = 75.2 \text{ amperes}$$

- Therefore, 75.2 amperes is the new corrected ampacity for the example given.

Some AHJs in extremely hot areas of the country, such as the southwestern desert climates, will require that service-entrance conductors running up the side of a building or above a roof exposed to direct sunlight be "corrected" per the correction factors in *Table 310.15(B)(2)(a)*. Check this out before proceeding with an installation where this situation might be encountered.

OVERCURRENT PROTECTION FOR BRANCH CIRCUIT CONDUCTORS

The basic rule is that overcurrent protection for conductors shall not be more than the allowable ampacity of the conductor, *240.4*. Allowable ampacities for conductors are found in *Table 310.15(B)(16)*.

If the allowable ampacity of a conductor does not match a standard ampere rating of a fuse or circuit breaker, *240.4(B)* permits the use of the next higher rated fuse or circuit breaker, 800 amperes or less. For example, a conductor having an ampacity of 55 amperes could be protected by a 60-ampere overcurrent device. This permission does not apply if the conductors of the branch circuit supply a number of receptacles for cord-and-plug connection of portable loads, because it would be impossible to know just how much load might be plugged in.

Small-size branch-circuit conductors must be protected according to the values shown in Table 18-6. This table is based on *240.4(D)*.

An exception to this basic maximum size overcurrent protection rule is the protection of motor branch-circuit conductors. For motor branch circuits, the maximum size overcurrent protection values are found in *Table 430.52*.

TABLE 18-6

Maximum size overcurrent protection for small-size conductors, *240.4(D)*.

Conductor Size (Copper)	Maximum Ampere Rating of Overcurrent Device
14 AWG	15 amperes
12 AWG	20 amperes
10 AWG	30 amperes

If the allowable ampacity of a given conductor is *derated* and/or *corrected*, the overcurrent protection must be based on the new *derated* or *corrected* ampacity.

BASIC *CODE* CONSIDERATIONS FOR CONDUCTOR SIZING AND OVERCURRENT PROTECTION

The following is a quick review of *Code* requirements for conductor sizing and conductor overcurrent protection encountered in house wiring.

- *210.3:* The rating of a branch circuit is based on the rating of the overcurrent device.

- *210.19(A):* Branch-circuit conductors shall have an ampacity not less than the maximum load to be served.

- *210.20:* An overcurrent device shall not be less than 125% for continuous loads on branch circuits.

- *210.20(B):* This section refers us to *240.4*, where we find that overcurrent protection for conductors shall not exceed the ampacity of a conductor. See *240.4* for exceptions such as "next standard size permitted" if the OCD and the conductor ampacity do not match.

- *210.23:* In no case shall the load exceed the branch-circuit ampere rating.

- *215.2(A):* Feeder conductors shall have an ampacity not less than required to supply the load.

- *215.3:* An overcurrent device shall not be less than 125% for continuous loads on feeders. This section refers us to *Part 1* of *240.4*, where we find that overcurrent protection for feeders

shall not exceed the ampacity of a conductor. Refer to *240.4* for specific requirements and exceptions to this basic rule.

- *220.4:* The total load shall not exceed the rating of the branch circuit.

- *240.4:* This section contains many of the overcurrent protection requirements for conductors.

- *230.42:* Service-entrance conductors, before application of any adjustment or correction factors shall not be less than 125% of continuous loads for services.

- *424.3(B):* For fixed electric space-heating equipment such as an electric furnace, the branch circuit loads are considered to be continuous (the load is likely to continue for 3 hours or more); as a result, the branch-circuit conductors and the overcurrent protective devices shall not be less than 125% of the total load of the motor(s) and heater(s).

- *426.4:* Fixed outdoor electric deicing and snow-melting equipment, such as heating cables buried in the concrete of a driveway, are considered to be continuous loads. As such, the branch-circuit conductors and the overcurrent protective devices shall not be less than 125% of the total load of the heaters.

- *430.22:* Branch-circuit conductors for motors shall not be less than 125% of the motor's full-load current as shown in *NEC Table 430.248*. Motor branch circuit short-circuit and ground-fault protection is sized according to the percentages listed in *Table 430.52*.

- *Article 440, Part III:* Branch-circuit short-circuit and ground-fault protection for air-conditioning equipment is covered in Chapter 23 of this text.

- *Article 440, Part IV:* Conductor sizing for air-conditioning equipment is also covered in Chapter 23 of this text.

A few of the above *Code* sections make reference to *continuous loads*. The *Code* defines a *continuous load* as *where the maximum current is expected to continue for three hours or more.*

Loads in homes such as electric water heaters, electric furnace, and snow-melting heating cables buried in the concrete of a driveway are classified as continuous loads because they could be on for periods of 3 hours or more. The 125% factor for these loads must be applied to comply with *422.13, 424.3(B)* and *426.4*.

Many loads in commercial and industrial installations can be considered to be continuous loads. For example, an electric range in a restaurant might be used continuously 24 hours a day, 7 days a week. An electric range in a home would not be used in this manner.

For continuous loads, the overcurrent protective device and the conductors must not be less than 125% of the continuous load. For example, a continuous load is 40 amperes. The minimum overcurrent device rating and the minimum conductor ampacity is $40 \times 1.25 = 50$ amperes.

For combination continuous and noncontinuous loads, the overcurrent protective device and the conductors must not be less than 125% of the continuous load plus the noncontinuous load. For example, a feeder supplies a continuous load of 40 amperes and a noncontinuous load of 30 amperes. The minimum overcurrent device rating and the minimum conductor ampacity is $(40 \times 1.25) + 30 = 80$ amperes.

EXAMPLE OF DERATING, CORRECTING, ADJUSTING, OVERCURRENT PROTECTION, AND CONDUCTOR SIZING

High ambient temperatures and/or more than three current-carrying conductors in the same raceway or cable call for the application of adjustment or correction factors relative to a conductor's ability to safely carry current.

Suppose that eight THHN/THWN copper conductors are installed in one raceway or cable(s)—the conditions are dry and the temperature is expected to reach 100°F (38°C). All conductors are considered to be current-carrying. The expected noncontinuous load on each conductor is 18 amperes. Let's analyze this.

- Consider 14 AWG THHN/THWN copper conductors:
  - The allowable ampacity in the 90°C column from ▶*NEC Table 310.15(B)(16)*◀ is 25 amperes.

  - The maximum overcurrent protection from Table 18-6 in this text or from *NEC 240.4(D)* is 15 amperes—not adequate to carry the 18-ampere load.
  - To find the corrected ampacity when eight current-carrying conductors are in one raceway or cable(s), apply the 0.70 adjustment factor (penalty) from Table 18-5 in this text or from *NEC Table 310.15(B)(3)(a)*.

    $$25 \times 0.70 = 17.5 \text{ amperes}$$

  - To find the corrected ampacity for the high ambient temperature, apply the 0.91 correction factor (penalty) from ▶*NEC Table 310.15(B)(2)(a)*◀.

    $$17.5 \times 0.91 = 15.9 \text{ amperes}$$

  - You can do this all in one step:

    $$25 \times 0.70 \times 0.91 = 15.9 \text{ amperes}$$

  - Analysis: 14 AWG THHN/THWN copper conductors are *too small* for the 18-ampere load. The 15-ampere maximum overcurrent protection *will not carry* the 18-ampere load.

- Consider 12 AWG THHN/THWN copper conductors:
  - The ampacity from the 90°C column of ▶*NEC Table 310.15(B)(16)*◀ is 30 amperes.
  - The maximum overcurrent protection from Table 18-6 in this text or from *NEC 240.4(D)* is 20 amperes.
  - To find the corrected ampacity when eight current-carrying conductors are in one raceway or cable(s), apply the 0.70 adjustment factor (penalty) from Table 18-5 in this text or from ▶*NEC Table 310.15(B)(3)(A)*.◀
  - To find the corrected ampacity for the high ambient temperature, apply the 0.91 correction factor (penalty) from ▶*NEC Table 310.15(B)(3)*.◀

Let's do the calculation in one step:

$$30 \times 0.70 \times 0.91 = 19.1 \text{ amperes}$$

  - Analysis: 12 AWG THHN copper conductors are all right for the 18-ampere load. The 20-ampere maximum overcurrent protection *will* carry the 18-ampere load. *NEC*

240.4(B) allows the next higher rated overcurrent device to be used. The next standard rating above 19.1 amperes is 20 amperes.

Watch Out! The Overcurrent Device Depends on the Type of Load

The basic rule is that a conductor overcurrent protection shall not exceed its ampacity, *240.4*. There are exceptions.

Read *240.4(B)* closely! The permission to use the next higher standard size overcurrent device has tough restrictions. The key is whether the conductors are supplying a fixed load, or supplying receptacles for cord-and-plug-connected portable loads. Let's take a closer look at the above example.

In the previous calculations, 14 AWG THHN copper conductors are obviously too small.

The 12 AWG THHN copper conductors are adequate for the 18-ampere load, and the 20-ampere overcurrent device is proper but *only* if the conductors serve a fixed load.

If the conductors supply receptacles for cord-and-plug-connected portable loads, we have no control over what loads might be plugged in. In this situation, the conductor's final allowable ampacity of 19.1 is adequate for the anticipated load of 18 amperes but is not permitted to be protected by a 20-ampere overcurrent device.

We would need to consider using 10 AWG THHN copper conductors.

From ▶*Table 310.15(B)(16),*◀ using the 90° column, we find that the allowable ampacity of a 10 AWG THHN copper conductor is 40 amperes. We then apply the adjustment and correction factors. The new allowable ampacity becomes

$$40 \times 0.70 \times 0.91 = 25.5 \text{ amperes}$$

The conclusion here is that 10 AWG THHN conductors are more than adequate to serve the 18-ampere noncontinuous load and are properly protected by a 20-ampere fuse or circuit breaker.

As you can readily see, when derating, correcting, and adjusting the ampacity of a conductor, sizing the conductors and sizing the overcurrent protection become a little more complicated. On top of this, one must consider voltage drop where long runs are concerned.

MULTIOUTLET ASSEMBLY

A multioutlet assembly (Wiremold) has been installed above the workbench. For safety reasons, as well as for adequate wiring, it is recommended that any workshop receptacles be connected to a separate circuit. In the event of a malfunction in any power tool commonly used in the home workshop (saw, planer, lathe, or drill), only receptacle Circuit A18 is affected. The lighting in the workshop is not affected by a power outage on the receptacle circuit.

As shown in Figure 18-15, the GFCI feed-through receptacle to the left of the workbench provides GFCI protection for the entire plug-in strip, *210.8(A)(5)*. GFCI protection could also have been provided by installing a GFCI circuit breaker in the circuit breaker panelboard.

The installation of multioutlet assemblies must conform to the requirements of *Article 380* of the *NEC*.

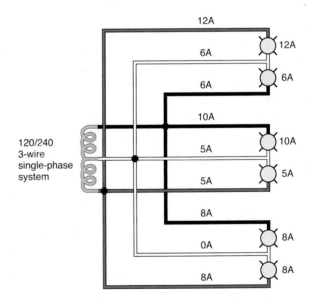

FIGURE 18-14 A neutral conductor carrying the unbalanced currents from the other conductors need not be counted when determining the ampacity of the conductor, according to *310.15(B)(5)(a)*. The *Code* would recognize this example as two current-carrying conductors. Thus, in the three multiwire circuits shown, the actual total is nine conductors, but only six current-carrying conductors are considered when the derating factor of Table 18-5 is applied.

(Delmar/Cengage Learning)

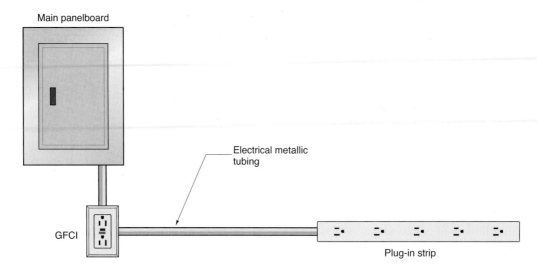

FIGURE 18-15 Detail of how the workbench receptacle plug-in strip is connected by feeding through a GFCI feed-through receptacle located below the main panelboard. This is Circuit A18. (*Delmar/Cengage Learning*)

Load Considerations for Multioutlet Assemblies

NEC Table 220.12 gives the general lighting loads in volt-amperes per ft² (0.093 m²) for types of occupancies. *NEC 220.14(J)* indicates that *all* general-use receptacle outlets in one, two, and multifamily dwellings are to be considered outlets for general lighting, and as such have been included in the volt-amperes per ft² calculations. Therefore, no additional load need be added.

Because the multioutlet assembly in Figure 18-15 has been provided in the workshop for the purpose of plugging in portable tools, a separate circuit, A18, is provided. To provide capacity above and beyond the minimum requirements of the NEC, this circuit has been shown in the service-entrance calculations at 1500 volt-amperes, similar to the load requirements for small-appliance circuits.

Many types of receptacles are available for various sizes of multioutlet assemblies. For example, duplex, single-circuit grounding, split-wired grounding, and duplex split-wired receptacles can be used with multioutlet assemblies.

Wiring a Multioutlet Assembly

A multioutlet assembly can be connected by bringing the supply into one end as in

Figure 18-16, or into the back of the assembly, as in Figure 18-17.

FIGURE 18-16 A typical metal multioutlet assembly. Multioutlet assemblies are permitted to be run through (not within) a dry partition when no outlets are located within the partition, and all exposed portions of the assembly can have the cover removed. Nonmetallic multioutlet assemblies are also available. (*Courtesy of The Wiremold Company*)

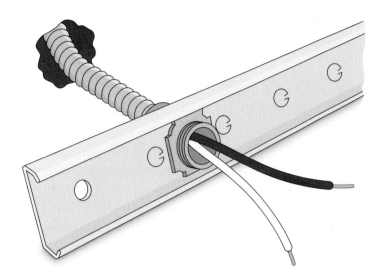

FIGURE 18-17   A multioutlet assembly back plate with the armored cable supply feeding into the back of the assembly. (*Delmar/Cengage Learning*)

Receptacles for multioutlet assemblies can be spaced and wired on the job, in which case the covers are cut to the desired length. Receptacles can also be factory prewired, with six receptacles in a 3-ft (900-mm) long strip, and 10 receptacles in a 5-ft (1.5-m) long strip.

For other strip lengths, standard spacing of receptacles for precut assemblies are 6 ft (1.8 m), 9 in. (225 mm), 12 in. (300 mm), or 18 in. (450 mm). These measurements are center to center. Prewired multioutlet assemblies are generally wired with 12 AWG THHN conductors and have either 15- or 20-ampere receptacles, depending on how they are specified and ordered.

Multioutlet assemblies are not difficult to install. There is a wide variety of fittings, such as connectors, couplings, ground clamps, blank and end fittings, elbows for turning corners (flat, inside, outside, twist), flush plate adapters, starter boxes, raceway adapters, mounting clips, and so on, to fit just about any application.

Figure 18-18 shows a cord-connected multi-outlet assembly of the type used in workshops or other locations where there is a need for the convenience of having a number of receptacles in close proximity. These prewired assemblies are available with six receptacles and a circuit breaker, 40 in. long, or 10 receptacles and a circuit breaker,

FIGURE 18-18   A cord-connected, prewired multioutlet assembly installed above a workbench. (*Courtesy of The Wiremold Company*)

52 in. long. These can also be used under kitchen cabinets. Figure 18-17 shows a multioutlet assembly back plate with the electrical supply feeding into the back of the assembly. Figure 18-19 shows

FIGURE 18-19 A multioutlet assembly installed above the backsplash of a kitchen countertop. (*Courtesy of The Wiremold Company*)

a multioutlet assembly above the backsplash of a kitchen countertop.

EMPTY CONDUITS

Although not truly part of the workshop wiring, *Note 8* to the first-floor electrical plans indicates that two empty trade size 1 EMT raceways are to be installed, running from the workshop to the attic. This installation is unique but certainly welcomed if at some later date additional wiring for telephone, security, computer, structured wiring, home automation cables, or other similar future needs might be required. The empty raceways will provide the necessary route from the basement to the attic, thus eliminating the need to fish through the wall partitions.

REVIEW

1. a. What circuit supplies the workshop lighting? _____

   b. What circuit supplies the plug-in strip over the workbench? _____

   c. What circuit supplies the freezer receptacle? _____

2. Approximately how much trade size ½ EMT is used for connecting Circuit A13, Circuit A17, Circuit A18, and Circuit A20? _____

3. a. How is EMT fastened to masonry? _____
   _____

   b. The cut ends of rigid and electrical metallic tubing must be _____ to prevent the insulation on conductors from being damaged.

4. What type of luminaires are installed on the workshop ceiling?
   _____

5. A check of the plans indicates that two empty trade size 1 EMT conduits are installed between the basement and the attic. In your opinion, is this is a good idea? Briefly explain your thoughts. _____
   _____
   _____

6. To what circuit are the smoke detectors connected?
   _____

7. Is it a *Code* requirement to connect smoke detectors to a GFCI-protected circuit?

_____

8. When a freezer is plugged into a receptacle, be sure that the receptacle is protected by a GFCI. Is this statement true or false? Explain.

_____

_____

9. A sump pump is plugged into a single receptacle outlet. This receptacle shall
   a. be GFCI protected.
   b. not required to be GFCI protected.

   Circle the correct answer.

10. When a 120-volt receptacle outlet is provided for the laundry equipment in an unfinished basement, what is the *minimum* number of additional receptacle outlets required in that basement? What *Code* reference? _____

   _____

11. Derating factors for conductor ampacities must always be taken into consideration when (a) _____ are contained in one raceway.

   Correction factors are applied when (b) _____

12. What current is flowing in the neutral conductor at
   a. _____ amperes?
   b. _____ amperes?
   c. _____ amperes?
   d. _____ amperes?

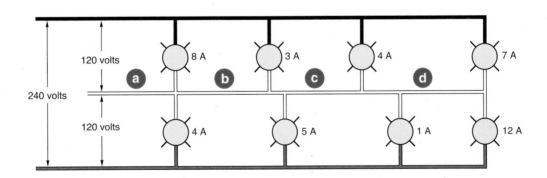

13. When a single receptacle is installed on an individual branch circuit, the receptacle must have a rating _____ than the rating of the branch circuit.

14. Calculate the total current draw of Circuit A17 if all luminaires and the exhaust fan were turned on.

15. A conduit body must have a cross-sectional area not less than (two) (three) (four) times the cross-sectional area of the (largest) (smallest) conduit to which it is attached.

   Circle the correct answers.

16. A conduit body may contain splices if marked with

    a. the UL logo.

    b. its in.³ area.

    c. the size of conduit entries.

    Circle the correct answer.

17. What size box would you use where Circuit A17 enters the junction box on the ceiling? _____

    _____

18. List the proper trade size of electrical metallic tubing for the following. Assume that THHN insulated conductors are used. See *Chapter 9, Table C1, NEC.*

    a. Three 14 AWG _____

    b. Four 14 AWG _____

    c. Five 14 AWG _____

    d. Six 14 AWG _____

    e. Three 12 AWG _____

    f. Four 12 AWG _____

    g. Five 12 AWG _____

    h. Six 12 AWG _____

    i. Six 10 AWG _____

    j. Four 8 AWG _____

19. According to the *Code*, what trade size EMT is required for each of the following combinations of conductors? Assume that these are new installations. Show all calculations. Refer to *Chapter 9, NEC, Table 4, Table 5,* and *Table 8.*

    a. Three 14 AWG THHN, four 12 AWG THHN

    _____

    b. Two 12 AWG THHN, three 8 AWG THHN

    _____

    c. Three 3/0 AWG THHN, two 8 AWG THHN

    _____

20. a. When more than three current-carrying conductors are installed in one raceway, their allowable ampacities must be reduced according to *NEC* _____.

    b. The following is a project. We need four 15-ampere branch circuits. We will use two 3-wire multiwire branch circuits and 14 AWG THHN conductors. We will install the conductors in EMT. The room temperature will not exceed 86°F (30°C). The connected load is nonmotor and is noncontinuous.

    1. What is the minimum trade size EMT? _____

    2. What is the ampacity of the conductors before derating? _____

    _____

    3. What is the ampacity of the conductors after derating? _____

    _____

4. What is the maximum overcurrent protection permitted for these conductors?

_____

c. Let's increase the conductor size to 12 AWG THHN for the same installation to see what happens.

1. What is the minimum trade size EMT? _____

2. What is the ampacity of the conductors before derating? _____

_____

3. What is the ampacity of the conductors after derating? _____

_____

4. What is the maximum overcurrent protection permitted for these conductors?

_____

d. It (is) (is not) necessary to count equipment grounding conductors, a neutral conductor that carries the unbalanced current from other conductors of the same circuit, and/or "travelers" when installing 3-way and 4-way switches for the purposes of derating the ampacity of these conductors. Circle the correct answer. Explain.

_____

_____

21. What is the minimum number of receptacles required by the *Code* for basements?

_____

22. For laundry equipment in basements, what sort of electrical circuitry is required by the *Code*? _____

_____

_____

23. How many receptacles are required to complete the multioutlet assembly above the workbench? _____

24. A Type NM cable carries a branch circuit to the outside of a building, where it terminates in a metal weatherproof junction box. From this box, a short length of EMT is installed upward to another metal weatherproof junction box on which is attached a cluster of outdoor floodlights. The first box is located approximately 24 in. (600 mm) above the ground. To "Meet *Code*," do the metal boxes and metal raceway have to be grounded? Yes _____ No_____ NEC _____ .

25. When raised covers are used for receptacles, does the *Code* permit fastening the receptacle to the cover with the one center No. 6-32 screw? Yes _____ No_____ NEC _____ .

26. The following is a layout of the lighting circuits for the workshop. Using the conduit layout in Figure 18-1, make a complete wiring diagram. Use colored pencils or pens to indicate conductors.

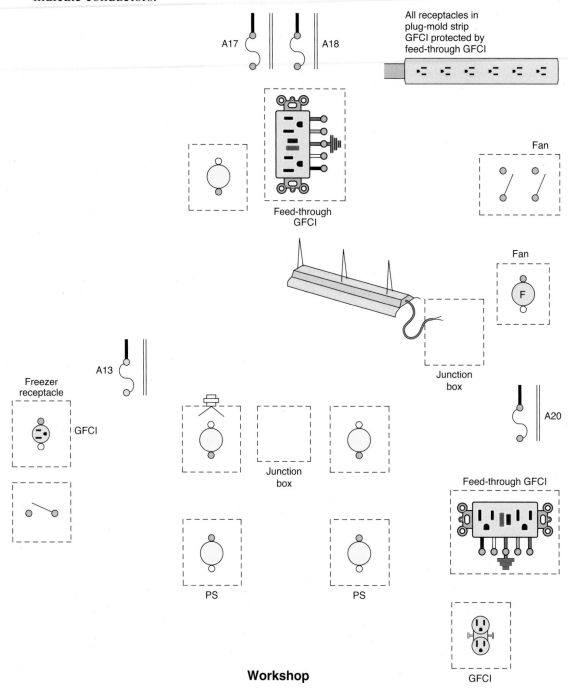

Workshop

27. *NEC 210.8(A)(5)* requires that GFCI protection be provided for all receptacles in unfinished basements. Are there any exceptions?

_____

28. Where loads are likely to be on continuously, the *calculated* load for branch circuits and feeders must be figured at (100%) (125%) (150%) of the continuous load, (plus) (minus) the noncontinuous load. This requirement generally (does) (does not) apply to residential wiring. Circle the correct answers.

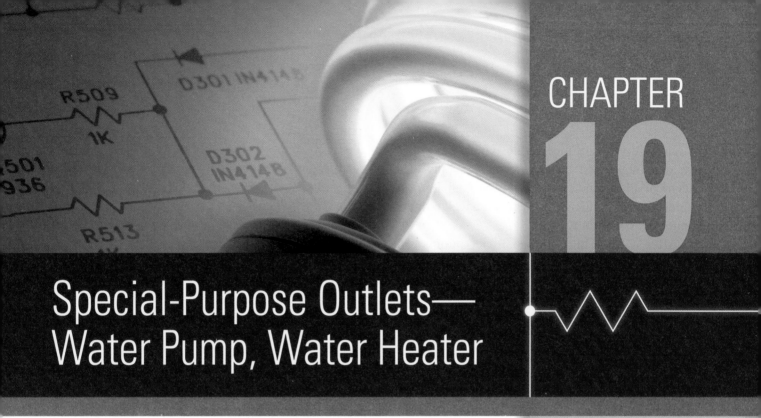

Special-Purpose Outlets— Water Pump, Water Heater

OBJECTIVES

After studying this chapter, you should be able to

- understand the operation of jet pumps, submersible pumps, and their components.

- understand the operation of electric water heaters and their components.

- be familiar with the *NEC* requirements for designing the branch circuit, including conductors, cables, raceway, motor branch-circuit short-circuit ground-fault and overload protection, disconnecting means, and grounding for water pumps and electric water heaters.

- figure out various electrical connections for "Time-of-Use" metering.

- calculate the effect of voltage variation on heating elements and motors.

- understand the hazards of possible scalding.

- discuss heat pump water heaters.

WATER PUMP CIRCUIT ▲B

In rural areas where there is no public water supply, dwellings need their own water supply. The Electrical Plans and Schedule of Special Purpose Outlets for this residence show that the well pump is connected to Circuit A 5-7.

After a brief introduction to jet pumps and submersible pumps, we will discuss the key elements that make up a typical motor branch circuit. For the most part, submersible pumps dominate the residential market.

The manufacturer's installation instructions must always be followed.

JET PUMPS

Figure 19-1 shows the major components of a typical deep-well jet pump.

The pump impeller wheel, ①, forces water down a drive pipe, ②, at a high velocity and pressure to a point just above the water level in the well casing. Just above the water level, the drive pipe curves sharply upward and enters a larger vertical suction pipe, ③. The drive pipe terminates in a small nozzle or "jet," ④. The water emerges from the jet with great force and flows upward though the suction pipe.

Water rises in the suction pipe, drawn up through the tailpipe, ⑤, by the action of the jet. The water rises to the pump inlet and passes through the impeller wheel of the pump. Some of the water is forced down through the drive pipe again. The remaining water passes through a check valve and enters the storage tank, ⑥.

The foot valve, ⑦, prevents water in the pumping equipment from draining back into the well when the pump is not operating. The strainer, ⑧, keeps debris out of the system. The pressure switch, ⑨, starts and stops the pump motor.

Figure 19-2 shows the wiring of a typical water pump. Always follow the wiring instructions furnished by the specific pump manufacturer.

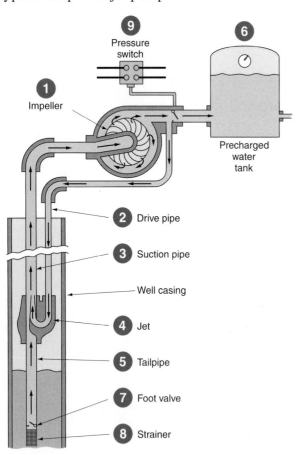

FIGURE 19-1 Components of a jet pump.
(*Delmar/Cengage Learning*)

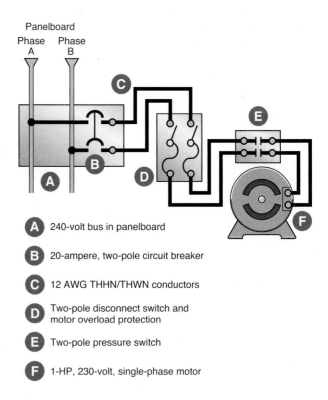

A 240-volt bus in panelboard

B 20-ampere, two-pole circuit breaker

C 12 AWG THHN/THWN conductors

D Two-pole disconnect switch and motor overload protection

E Two-pole pressure switch

F 1-HP, 230-volt, single-phase motor

FIGURE 19-2 The pump circuit.
(*Delmar/Cengage Learning*)

⎍⎍⎍ SUBMERSIBLE PUMPS

Figure 19-3 shows the major components of a typical submersible pump.

A submersible pump consists of a centrifugal pump, ①, driven by an electric motor, ②. The pump and the motor are contained in one housing, ③, submersed below the permanent water level, ④, within the well casing, ⑤. The pump housing is a cylinder 3–5 in. in diameter and 2–4 ft long. When running, the pump raises the water upward through the piping, ⑥, to the water tank, ⑦. Proper pressure is maintained in the system by a pressure switch, ⑧. The disconnect switch, ⑨, pressure switch, ⑧, limit switches, ⑩, and controller, ⑪, are installed in a logical and convenient location near the water tank.

Pumps commonly called "2-wire pumps" contain all required starting and protection components and connect directly to the pressure switch—with no other aboveground controller. Generally, there are no moving electrical parts within the submersible pump, such as the centrifugal starting switch found in a typical single-phase, split-phase induction motor.

Other pumps, referred to as "3-wire pumps," require an aboveground controller that contains any required components not in the motor, such as a starting relay, overload protection, starting and running capacitors, lightning arrester, and terminals for making the necessary electrical connections.

Most water pressure tanks have a precharged air chamber in an elastic bag (bladder) that separates the air from the water. This ensures that air is not absorbed by the water. Water pressure tanks compress air to maintain delivery pressure in a usable range without cycling the pump at too fast a rate. When in direct contact with the water, air is gradually absorbed into the water, and the cycling rate of the pump will become too rapid for good pump life unless the air is periodically replaced. With the elastic bag, the initial air charge is always maintained, so recharging is unnecessary.

Submersible pumps are covered by UL Standard 778.

Submersible Pump Cable

Power to the motor is supplied by a "drop" cable, ⑫, especially designed for use with submersible pumps. This cable is marked "submersible pump cable," and it is generally supplied with the pump. This cable can also be purchased separately. The cable is cut to the proper length to reach between the pump and its controller, as shown in Figure 19-3, or between the pump and the well cap, Figure 19-4. When needed, the cable may be spliced according to the manufacturer's specifications.

Submersible water pump cable is "tag-marked" for use within the well casing for wiring deep-well water pumps, where the cable is not subject to repetitive handling caused by frequent servicing of the pump units.

A submersible pump cable is not designed for direct burial in the ground unless it is marked "Type USE" or "Type UF." Should it be required to run the

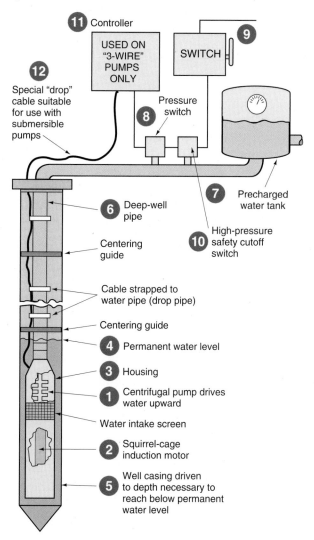

FIGURE 19-3   Submersible pump.
(*Delmar/Cengage Learning*)

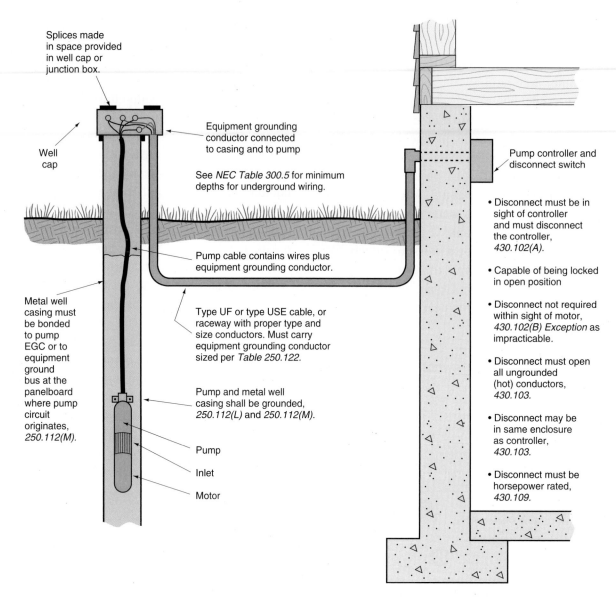

Splices made in space provided in well cap or junction box.

Well cap

Equipment grounding conductor connected to casing and to pump

See *NEC Table 300.5* for minimum depths for underground wiring.

Metal well casing must be bonded to pump EGC or to equipment ground bus at the panelboard where pump circuit originates, *250.112(M)*.

Pump cable contains wires plus equipment grounding conductor.

Type UF or type USE cable, or raceway with proper type and size conductors. Must carry equipment grounding conductor sized per *Table 250.122*.

Pump and metal well casing shall be grounded, *250.112(L)* and *250.112(M)*.

Pump

Inlet

Motor

Pump controller and disconnect switch

- Disconnect must be in sight of controller and must disconnect the controller, *430.102(A)*.

- Capable of being locked in open position

- Disconnect not required within sight of motor, *430.102(B) Exception* as impracticable.

- Disconnect must open all ungrounded (hot) conductors, *430.103*.

- Disconnect may be in same enclosure as controller, *430.103*.

- Disconnect must be horsepower rated, *430.109*.

FIGURE 19-4 Grounding and disconnecting means requirements for submersible water pumps. (*Delmar/Cengage Learning*)

pump's circuit underground for any distance, it is necessary to install Type UF or Type USE cable or a raceway with suitable conductors, and then make up the necessary splices in a listed weatherproof junction box, as in Figure 19-4. Chapter 16 covers underground wiring in great detail.

MOTOR CIRCUIT DESIGN

We will now take a look at the *NEC* requirements for all of the basics for a typical electrical motor branch circuit.

The pump circuit in this residence originates in Main Service Panelboard A, Circuit A5-7.

Current and Voltage

This information is found on the nameplate and instructions furnished with the pump.

The water pump in this residence is a 1-horsepower (hp), single-phase 115/230-volt pump. Dual-voltage-rated pumps can be supplied by either 115 volts or 230 volts. When connected for the higher voltage, the current draw will be half of that when connected for the lower voltage. For example, in *Table 430.248*, the full-load current rating of a 1-horsepower, 230-volt, single-phase motor is 8 amperes. The same motor connected to 115 volts will draw 16 amperes. Current is a big factor in determining conductor size, watts loss in the conductors, voltage drop, disconnect switch size, controller size, motor overload protection, and the size and type of branch-circuit overcurrent protection. Some pump controllers have a simple slide switch to change the connections from 115 volts to 230 volts. Pumps sized ½ horsepower are usually prewired for 115 volts. Pumps ¾ horsepower and larger are usually prewired for 230 volts.

For the purpose of our motor circuit design, we will connect this motor for 230 volts.

Table 430.248 for 115-volt motors covers the voltage range of 110–120 volts. For 230-volt motors, the voltage range is 220–240 volts. You will also find that some motors are marked 115/208–230 volts.

The NEMA Standard MG-1 *Information Guide for General Purpose Industrial AC Small and Medium Squirrel-Cage Induction Motor Standards* is a condensed version of the much larger NEMA Standard MG 1-2003 *Motors and Generators*. These standards show that most general-purpose NEMA-rated motors are designed to operate at ±10% of the motor's rated voltage.

Important: For determining the conductor size, switch size, and the branch-circuit short-circuit and ground-fault protection, use the current values from the tables in the *NEC*. See *430.6(A)(1)*.

For overload protection sizing, use the motor nameplate current value. See *430.6(A)(2)*. For the typical submersible pump motor, the running overload protection is almost always furnished by the manufacturer in the controller.

The nameplate rating of the pump motor is 8 amperes, 230 volts.

Conductor Size *(430.22)*

From *Table 430.248*, the full-load current rating of a 1-horsepower, 230-volt, single-phase motor is 8 amperes. To calculate the conductor size, use 125% of 8 amperes.

$$1.25 \times 8 = 10 \text{ amperes}$$

Use Type THHN/THWN conductors per the Specifications.

Checking ▶*Table 310.15(B)(16)*◀ of the *NEC*, we find that a 14 AWG Type THHN/THWN conductor has an ampacity in the 60°C column of 15 amperes and 20 amperes in the 75°C column, more than ample for the pump motor circuit. However, the Schedule of Special Purpose Outlets found in this text, which is part of the Specifications for the residence, indicates that 12 AWG THHN/THWN conductors are to be used for the pump circuit. Installing larger size conductors can minimize voltage drop and might be suitable at a later date should the need arise to install a larger-capacity pump.

The two 12 AWG THHN/THWN conductors are connected to a 2-pole, 20-ampere branch-circuit breaker in Panelboard A, Circuit A(5-7).

Conduit Size

Two 12 AWG Type THHN/THWN conductors require trade size ½ EMT, *NEC Annex C, Table Cl*. This EMT will run across the ceiling from Panelboard A to the disconnecting means located in the southeast corner of the workshop. The controller and other equipment relating to the water pump are located in this corner of the workshop. Type UF cable or PVC conduit is buried underground to the well casing location. The cable or conduit is protected at both ends from the burial depth given in *Table 300.5* to the point of entry.

Motor Branch-Circuit Short-Circuit and Ground-Fault Protection *(430.52, Table 430.52)*

Fuses and circuit breakers are the most commonly used forms of motor branch-circuit short-circuit and ground-fault protection. Should the motor windings or branch-circuit conductors short-circuit or go to ground, the motor's branch-circuit short-circuit and ground-fault overcurrent device, either fuses or circuit breakers, will open.

Motor branch-circuit short-circuit and ground-fault protection does not protect the motor against overload burnout! That protection is provided in the motor controller for the submersible pump.

The various choices for motor branch-circuit short-circuit and ground-fault protection are found in *430.52* and *Table 430.52*. In almost all instances, the values found in *Table 430.52* will work just fine. In the event that the sizing calculation results in an unavailable ampere rating fuse or circuit breaker, select the next standard ampere rating, but do not exceed the absolute maximum sizing percentages shown in *430.52*. This will be discussed in more detail after we do the calculations.

From *Table 430.248*, the full-load current rating of a 1-horsepower, 230-volt, single-phase motor is 8 amperes. Here are the possibilities:

- Non-Time-Delay Fuses (300% desired—maximum size: 400%)

    $8 \times 3 = 24$ amperes (use 25-ampere fuses)

- The disconnect switch would be a 30-ampere, 2-pole, 250-volt switch.

- Dual-Element, Time-Delay Fuses (175% desired—maximum size: 225%)

    $8 \times 1.75 = 14$ amperes (use 15-ampere fuses)

    The disconnect switch would be a 30-ampere, 2-pole, 250-volt switch.

- Inverse Time Breakers (250%)

    $8 \times 2.5 = 20$-ampere rating

    Maximum setting: 400% for FLA of 100 amperes or less.

- Instant Trip Breakers are never used in residential applications.

The motor branch-circuit overcurrent device must be capable of allowing the motor to start, *430.52(B)*. Where the values for branch-circuit protective devices (fuses or breakers) as determined by *Table 430.52* do not correspond to the standard sizes, ratings, or settings as listed in *240.6*, the next higher size, rating, or setting is permitted, *430.52(C)(1), Exception No. 1*.

If, after applying the values found in *Table 430.52* and after selecting the next higher standard size, rating, or setting of the branch-circuit overcurrent device, the motor still will not start, *430.52(C)(1), Exception No. 2* permits using an even larger size, rating, or setting. The maximum values (percentages) are shown in this exception.

A nuisance opening of a fuse, or tripping of a breaker, can easily be predicted by referring to time/current curves for the intended fuse or breaker. This topic is covered in Chapter 28.

Motor Overload Protection (*430.31* through *430.44*)

Usually the manufacturer provides the necessary motor overload protection in the controller, and no additional protection is required.

Overload protection is needed to keep the motor from dangerous overheating due to overloads or failure of the motor to start. *NEC 430.31* through *430.44* cover virtually every type and size of motor, starting characteristics, and duty (continuous or noncontinuous) in use today. The overload protection for a motor might be thermal overloads (sometimes called *heaters*) in the controller, electronic sensing overload devices, built-in (inherent) thermal protection, or time-delay fuses.

We find the sizing requirements for motor overload protection in *430.32(A)(1)* and *430.32(B)(1)*. Although there are exceptions, most of the common installations require motor overload protection not to exceed 125% of the motor's full-load current draw, as indicated on the nameplate. If you need to provide this running overload protection, the running load must be obtained from the manufacturer's literature or from the nameplate on the motor. We will assume the nameplate current to be 7 amperes. For the pump motor this would be:

$$1.25 \times 7 = 8.8 \text{ amperes}$$

Therefore, for backup motor overload protection, we could install 10-ampere, time-delay, dual-element fuses in the disconnect switch. If we are interested only in motor branch-circuit short-circuit and ground-fault protection, we refer to *Table 430.52* and apply the multiplier of 175 percent.

$$7 \times 1.75 = 12.3 \text{ amperes}$$

We are permitted to go to the next standard size, which are 15-ampere time-delay, dual-element fuses.

Instructions furnished with a particular motor or motor-operated appliance might indicate a maximum size overload protection and a maximum size branch-circuit protection. These values must be followed.

Disconnecting Means

The *NEC* requires that all motors be provided with a means to disconnect the motor from its electrical supply, *Article 430, Part IX*.

NEC 430.103 requires that a disconnecting means for a motor shall

- open all ungrounded conductors,
- be designed so all poles operate together (simultaneously), and
- be designed so that it cannot be closed automatically.

The disconnecting means for the water pump is a 30-ampere, two-pole, 250-volt switch mounted on the wall next to the pump controller. See Figures 19-3 and 19-4.

Follow these two simple rules for locating the disconnect switch.

Rule #1. *NEC 430.102(A):* An individual disconnect must be in sight of the controller and must disconnect the controller. The *NEC* definition of "In Sight" means that the controller must be visible and not more than 50 ft (15 m) from the disconnect.

Rule #2. *NEC 430.102(B):* The disconnect must be in sight of the motor and driven machinery. If the disconnect, as required in *430.102(A)*, is in sight of the controller, the motor, and driven machinery, then that disconnect meets the requirements of both *430.102(A)* and *430.102(B)*.

How to Use the Exception to the "In-Sight" Rule

As with most good rules, there are exceptions! The exception and the *Informational Notes* to *430.102(B)* tell us that the disconnect need not be in sight of the motor and driven machinery if the disconnect can be individually locked in the open "OFF" position. The locking provision must be of a permanent type installed on the switch or circuit breaker. Most disconnect switches have this "LOCK-OFF" feature. Listed circuit breaker "LOCK-OFF" devices that fit over the top of the circuit breaker handle also meet this requirement.

Anyone working on the pump needs to be assured that the power is off and stays off until he or she is ready to turn the power back on. This is particularly important in our installation because the motor is out of sight from the controller.

The disconnect switch for the submersible pump for this residence is located next to and in sight of the controller. It is not in sight of the motor and can't be, because the motor is located inside the well casing below the water level. Here's where the *Exception* comes into play. The *Exception* provides that disconnecting means is not required within sight of the motor if such a location is impracticable. The disconnecting means located on the line or supply side of the controller must be capable of being locked in the open position. See Figure 19-4.

GROUNDING

Figure 19-4 illustrates how to provide proper grounding of a submersible water pump and the well casing. Proper grounding of the well casing and submersible pump motor will minimize or eliminate stray voltage problems that could occur if the pump motor is not grounded.

Key *Code* sections that relate to grounding water-pumping equipment:

- *NEC 250.4(A)(5):* The last sentence states that *The earth shall not be considered as an effective ground-fault current path.*
- *NEC 250.86:* Metal enclosures for circuit conductors shall be grounded, other than short sections of metal enclosures that provide support or physical protection.
- *NEC 250.112(L):* Motor-operated water pumps, including submersible pumps, must be grounded.
- *NEC 250.112(M):* Bond metal well casing to the pump circuit's equipment grounding conductor.
- *NEC 250.134(B):* Equipment fastened in place or connected by permanent wiring must have the equipment grounding conductor run with the circuit conductors.
- *NEC 430.241:* Grounding of motors—general.
- *NEC 430.242:* Grounding of motors—stationary motors.
- *NEC 430.244:* Grounding of controllers.
- *NEC 430.245:* Acceptable methods of grounding.

Because so much nonmetallic (PVC) water piping is used today, some local electrical codes or interpretations require that a grounding electrode conductor be installed between the neutral bar of the main service disconnecting means and the metal well casing. The interpretation often related to the proximity of the well casing to the main building or structure. The *NEC* does not provide a distance beyond which a connection to the well casing is no longer required. Connecting to the metal well casing would satisfy the requirements of *250.104* (bonding), *250.50* (grounding electrode system), *250.52* (grounding electrode descriptions), and *250.53* (installation of grounding electrodes). If installed, the grounding electrode conductor would be sized in compliance with *Table 250.66* of the *Code*. At the metal casing, the grounding electrode conductor is attached to a lug termination, or by means of exothermic (Cadweld) welding.

WATER HEATER CIRCUIT Ⓐᴄ

Residential electric water heaters are listed under UL Standard 174, *Household Electric Storage Tank Water Heaters*.

Electrical contractors many times are also in the plumbing, heating, and appliance business. They need to know more than how to do the electrical hookup. The following text discusses electrical as well as other data about electric water heaters that will prove useful.

All homes require a supply of hot water. To meet this need, one or more automatic water heaters are generally installed as close as practical to the areas having the greatest need for hot water. Water piping carries the heated water from the water heater to the various plumbing fixtures and to appliances such as dishwashers and clothes washers.

For safety reasons, in addition to the regular temperature control thermostat that can be adjusted by the homeowner, the *NEC* and UL Standards require that electric water heaters be equipped with a high-temperature cutoff. This high-temperature control limits the maximum water temperature to 190°F (88°C). This is 20°F (11°C) below the 210°F (99°C) temperature that causes the "temperature/pressure"

relief valve to open. The high-temperature cutoff is factory preset and should never be tampered with or modified in the field. In conformance to *422.47* of the *NEC*, the high-temperature limit control must disconnect all ungrounded conductors.

In most residential electric water heaters, the high-temperature cutoff and upper thermostat are combined into one device, as in Figure 19-5.

Combination Pressure/Temperature Safety Relief Valves

Look at Figure 19-5. A pressure/temperature relief valve is an important safety device installed into an opening in the water heater tank within 6 in. (150 mm) of the top. The opening is provided for and clearly marked by the manufacturer. Many water heater pressure/temperature relief valves are factory installed, although some models still require installation of the valve by the installing contractor. Proper discharge piping must be installed from the pressure/temperature relief valve downward to within 6 in. (150 mm) of the floor, preferably near a floor drain. Terminating the discharge pipe to within 6 in. (150 mm) of the floor is important to protect people from injury from scalding water or steam discharge should the relief valve activate for any reason. This also protects electrical equipment in and around the water heater from becoming soaked should the relief valve operate.

Pressure/temperature relief valves are installed for two reasons. Their pressure/temperature sensing features are interrelated.

- *Pressure:* Water expands when heated. Failure of the normal temperature thermostat and the high-temperature-limiting device to operate could result in pressures that exceed the rated working pressure of the water heater tank and the plumbing system components. The relief valve, properly installed and properly rated to local codes, will open to relieve the excess pressure in the tank to the predetermined setting marked on the relief valve. The rating is stamped on the valve's identification plate.

- *Temperature:* Should runaway water heating conditions occur within the water heater, temperatures far exceeding the atmospheric boiling

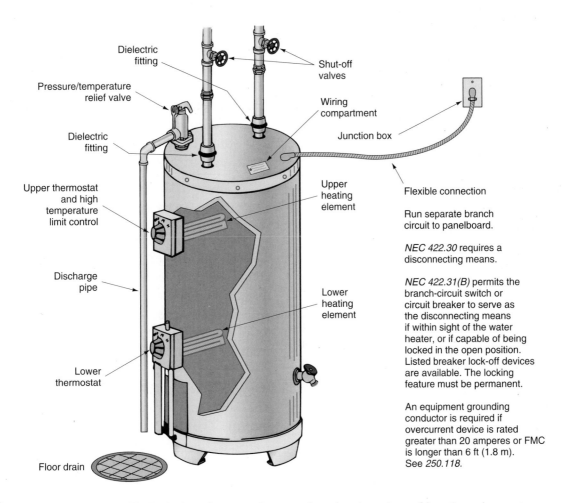

Dielectric fitting

Pressure/temperature relief valve

Dielectric fitting

Shut-off valves

Wiring compartment

Junction box

Flexible connection

Run separate branch circuit to panelboard.

NEC 422.30 requires a disconnecting means.

NEC 422.31(B) permits the branch-circuit switch or circuit breaker to serve as the disconnecting means if within sight of the water heater, or if capable of being locked in the open position. Listed breaker lock-off devices are available. The locking feature must be permanent.

An equipment grounding conductor is required if overcurrent device is rated greater than 20 amperes or FMC is longer than 6 ft (1.8 m). See *250.118.*

Upper thermostat and high temperature limit control

Upper heating element

Discharge pipe

Lower heating element

Lower thermostat

Floor drain

FIGURE 19-5   Typical electric water heater showing location of heating elements and thermostats and electrical connection. (*Delmar/Cengage Learning*)

point of water (212°F [100°C]) can take place. Under high pressures, water will stay liquid at extremely high temperatures. For a typical municipal water supply pressure of 45 psi, water will not boil until it reaches approximately 290°F (143°C). At 290°F (143°C), opening a faucet, a broken pipe, or a ruptured water heater tank would allow the pressure to instantly drop to normal atmospheric pressure. The superheated water would immediately flash into steam, having about 1600 times more volume than the liquid water occupied. This blast of steam can be catastrophic, and could result in severe burns or death.

Thus, the function of the temperature portion of the pressure/temperature relief valve is to open when a temperature of 210°F (99°C) occurs within the water

heater tank. Properly rated and maintained, the pressure/temperature relief valve will allow the water to flow from the water heater through the discharge pipe at a rate faster than the heating process, thereby keeping the water temperature below the boiling point. Should the cold water supply be shut off, the relief valve will allow the steam to escape without undue buildup of temperature and pressure in the tank.

Requirements for relief valves for hot water supply systems are found in the ANSI Standard Z21.22.

Sizes

Residential electric water heaters are available in many sizes, such as 6, 12, 15, 20, 30, 32, 40, 42, 50, 52, 66, 75, 80, 82, 100, and 120 gallons.

Corrosion of Tank

To reduce corrosion of the steel tank, most water heater tanks are glass lined. Glass lining is a combination of silica and other minerals (rasorite, rutile, zircon, cobalt, nickel oxide, and fluxes). This special mix is heated, melted, cooled, crushed, then sprayed onto the interior steel surfaces and baked on at about 1600°F (871°C). To further reduce corrosion, aluminum or magnesium anode rods are installed in special openings, or as part of the hot water outlet fitting, to allow for the flow of a protective current *from* the rod(s) *to* any exposed steel of the tank. This is commonly referred to as *cathodic protection*. Corrosion problems within the water heater tank are minimized as long as the aluminum or magnesium rod(s) remain in an active state. Anodes should be inspected at regular intervals to determine when replacement is necessary, usually when the rod(s) is reduced to about one-third of its original diameter, or when the rod's core wire is exposed. Further information can be obtained by contacting the manufacturer, and consulting the installation manual.

Some water heater tanks are fiberglass or plastic lined.

Heating Elements

The wattage ratings of electric water heaters can vary greatly, depending on the size of the heater in gallons, the speed of recovery desired, local electric utility regulations, and codes. Typical wattage ratings are 1500, 2000, 2500, 3000, 3800, 4500, and 5500 watts.

A resistance heating element might be dual rated. For example, the element might be marked 5500 watts at 250 volts, and 3800 watts at 208 volts.

Most residential-type water heaters are connected to 240 volts except for the smaller 2-, 4-, and 6-gallon *point of use* sizes generally rated 1500 watts at 120 volts. Commercial electric water heaters can be rated single-phase or 3-phase, 208, 240, 277, or 480 volts.

UL requires that the power (wattage) input must not exceed 105% of the water heater's nameplate rating. All testing is done with a supply voltage equal to the heating element's *rated* voltage. Most heating elements may burn out prematurely if operated at voltages 5% higher than for which they are rated.

To help reduce the premature burnout of heating elements, some manufacturers will supply 250-volt heating elements, yet will mark the nameplate 240 volts, with its corresponding wattage at 240 volts. This allows a safety factor if slightly higher than normal voltages are experienced.

Heating Element Construction

Heating elements generally contain a nickel-chrome (nichrome) resistance wire embedded in compacted powdered magnesium oxide to ensure that no grounds occur between the wire and the sheath. The magnesium oxide is an excellent insulator of electricity, yet it effectively conducts the heat from the nichrome wire to the sheath.

The sheath can be made of tin-coated copper, stainless steel, or an iron-nickel-chromium alloy. The latter alloy comes under the trade name of Incoloy. The stainless steel and Incoloy types can withstand higher operating temperatures than the tin-coated copper elements. They are used in premium water heaters because of their resistance to deterioration, lime scaling, and dry firing burnout.

High watt density elements generally have a U-shaped tube, and give off a great amount of heat per square inch of surface. Figure 19-6(A) illustrates a flange-mounted high-density heating element. Figure 19-6(B) illustrates a high-density heating element that screws into a threaded opening in the tank. High-density elements are very susceptible to lime-scale burnout.

Low watt density elements, as in Figure 19-6(C), generally have a double loop, like a U tube bent in half. They give off less heat per in.^{2}, have a longer life, and are less noisy than high-density elements.

An option offered by at least one manufacturer of electric water heaters is a dual-wattage heating element. A dual-wattage heating element actually contains two heating elements in the one jacket. One element is rated 3800 watts and one element is rated 1700 watts. Three leads are brought out to the terminal block. The 3800-watt heating element is connected initially, with the option of connecting the second 1700-watt heating element if the wiring to the water heater is capable of handling the higher

FIGURE 19-6   Heating elements. (*Delmar/Cengage Learning*)

total combined wattage of 5500 watts. To attain the higher wattage, a jumper (provided by the manufacturer) is connected between two terminals on the terminal block, Figure 19-7.

Any electric heating element coated with lime scale may be noisy during the heating cycle. Limed up heating elements should be cleaned or replaced, following the manufacturer's instructions.

Do not energize the electrical supply to a water heater unless you are sure it is full of water. Most water heater heating elements are designed to operate only when submersed in water.

Residential electric water heaters are available with one or two heating elements. Figure 19-5 shows two heating elements. Single-element water heaters will have the heating element located near the bottom of the tank. If the water heater has two heating elements, the upper element will be located about two-thirds of the way up the tank.

Speed of Recovery

Speed of recovery is the time required to bring the water temperature to satisfactory levels in a

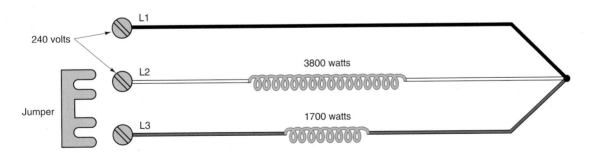

FIGURE 19-7   A dual-wattage heating element. The two wires of the incoming 240-volt circuit are connected to L1 and L2. These leads connect to the 3800-watt element. To connect the 1700-watt element, the jumper is connected between L2 and L3. With both heating elements connected, the total wattage is 5500 watts. (*Delmar/Cengage Learning*)

given length of time. The current accepted industry standard is based on a 90°F (32.22°C) rise. For example, referring to Table 19-1, we find that a 3500-watt element can raise the water temperature 90°F (32.22°C) of a little more than 16 gallons an hour. A 5500-watt element can raise water temperature 90°F (32.22°C) of about 25 gallons an hour. Speed of recovery is affected by the type and amount of insulation surrounding the tank, the supply voltage, and the temperature of the incoming cold water.

TABLE 19-1

Electric water heater recovery rate per gallon of water for various wattages.

HEATING ELEMENT WATTAGE	GALLONS PER HOUR RECOVERY FOR INDICATED TEMPERATURE RISE					Equivalent NET Btu Output
	60°F	70°F	80°F	90°F	100°F	
0750	5.2	4.4	3.9	3.5	3.1	02559
1000	6.9	5.9	5.2	4.6	4.1	03412
1250	8.6	7.4	6.5	5.7	5.2	04266
1500	10.3	8.9	7.8	6.9	6.2	05119
2000	13.8	11.8	10.3	9.2	8.3	06825
2250	15.5	13.3	11.6	10.3	9.3	07678
2500	17.2	14.8	12.9	11.5	10.3	08531
3000	20.7	17.7	15.5	13.8	12.4	10,238
3500	24.1	20.7	18.1	16.1	14.5	11,944
4000	27.6	23.6	20.7	18.4	16.6	13,650
4500	31.0	26.6	23.3	20.7	18.6	15,356
5000	34.5	29.6	25.9	23.0	20.7	17,063
5500	37.9	32.5	28.4	25.3	22.5	18,769
6000	41.4	35.5	31.0	27.6	24.6	20,475

Notes: If the incoming water temperature is 40°F (Northern states), use the 80° column to raise the water temperature to 120°F.

If the incoming water temperature is 60°F (Southern states), use the 60° column to raise the water temperature to 120°F.

For example, approximately how long would it initially take to raise the water temperature in a 42-gallon electric water heater to 120°F where the incoming water temperature is 40°F? The water heater has a 2500-watt heating element.

Answer: 42 ÷ 12.9 = 3.26 hours (approximately 3 hours and 16 minutes).

To keep this table simple, noncluttered, and easy to read, only Fahrenheit temperatures are shown.

Table 19-2 shows the approximate time it takes to run out of hot water using typical shower heads with various storage capacity residential electric water heaters. Current standards for new showerheads are typically 2.5 gallons per minute. When the flow restrictors are removed, the flow rate can increase to 6–18 gallons per minutes per showerhead. There are very low-flow-rate showerheads available in the range of 1–2 gallons per minute. All of this depends on the water pressure.

An easy way to determine a showerhead's flow rate is to turn on the water, hold a large bucket of known capacity under it, and time how long it takes to fill the bucket.

Scalding from Hot Water

The following are plumbing code issues but are discussed briefly because of their seriousness.

Temperature settings for water heaters are problematic!

Set the water heater thermostat high enough to kill bacteria on the dishes in the dishwasher and possibly get scalded in the shower. Or set the thermostat to a lower setting and not get scalded in the shower, but live with the possibility that bacteria will still be present on washed dishes.

The Consumer Product Safety Commission (CPSC) reports that 3800 injuries and 34 deaths occur each year due to scalding from excessively hot tap water. Unfortunately, the victims are the elderly and children under the age of 5. They recommend that the setting should not be higher than 120°F (49°C). Some state laws require that the thermostat be preset at no higher than 120°F (49°C). Residential water heaters are all preset by the manufacturer at 120°F (49°C).

Manufacturers post caution labels on their water heaters that read something like this:

SCALD HAZARD: WATER TEMPERATURE OVER 120°F (49°C) CAN CAUSE SEVERE BURNS INSTANTLY OR DEATH FROM SCALDS. SEE INSTRUCTION MANUAL BEFORE CHANGING TEMPERATURE SETTINGS.

To reduce the possibility of scalding, water heaters can be equipped with a thermostatic nonscald mixing

TABLE 19-2

This chart shows the approximate time in minutes it takes to run out of hot water when using various shower heads, different incoming water temperatures, and different capacity water heaters.

LENGTH OF CONTINUOUS SHOWER TIME—IN MINUTES—FOR TYPICAL SIZES OF RESIDENTIAL ELECTRIC WATER HEATERS. CHART BASED ON 70% DRAW EFFICIENCY. RECOVERY RATE IS NOT FIGURED IN BECAUSE IT IS NORMALLY NOT A FACTOR IN CONTINUOUS DRAW OF WATER SITUATIONS.

SHOWER-HEAD(S) RATE IN GALLONS PER MINUTE	HOT WATER FLOW RATE IN GALLONS PER MINUTE BASED ON 120°F. STORED HOT WATER TO PROVIDE 105°F. MIXED WATER OUT OF SHOWERHEAD. 40°F. Incoming water	HOT WATER FLOW RATE IN GALLONS PER MINUTE BASED ON 120°F. STORED HOT WATER TO PROVIDE 105°F. MIXED WATER OUT OF SHOWERHEAD. 60°F. Incoming water	100 Gallon		80 Gallon		66 Gallon		50 Gallon		40 Gallon		30 Gallon	
			40°F. Incoming water	60°F. Incoming water	40°F. Incoming water	60°F. Incoming water	40°F. Incoming water	60°F. Incoming water	40°F. Incoming water	60°F. Incoming water	40°F. Incoming water	60°F. Incoming water	40°F. Incoming water	60°F. Incoming water
02.0	1.6	1.5	43	47	34	37	28	31	22	23	17	19	13	14
02.5	2.0	1.9	34	38	28	30	23	25	17	19	14	15	10	11
03.0	2.4	2.3	29	31	23	25	19	21	14	16	11	12	09	09
04.0	3.3	3.0	22	23	17	19	14	15	11	12	09	09	06	07
05.0	4.0	3.8	17	19	14	15	11	12	09	09	07	07	05	06
06.0	4.9	4.5	14	16	11	12	09	10	07	08	06	06	04	05
08.0	6.5	6.0	11	12	09	09	07	08	05	06	04	05	03	04
12.0	9.8	9.0	07	08	06	06	05	05	04	04	03	03	02	02

Example #1: A 40-gallon electric water heater has a 40°F incoming water supply. How many minutes can a 3 GPM shower head be used to obtain 105°F hot water?

Answer: 11 minutes

Example #2: A residence has an 80-gallon electric water heater. The incoming water supply is approximately 60°F. How many minutes can two 2 GPM shower heads be used at the same time to obtain 105°F hot water? Hint: Two 2 GPM shower heads are the same as one 4 GPM shower head.

Answer: 19 minutes

valve or pressure-balancing valve that holds a selected water temperature to within one degree regardless of incoming water pressure changes. This prevents sudden unanticipated changes in water temperatures.

Table 19-3 shows time/temperature relationships relating to scalding.

What about Washing Dishes?

Older automatic dishwashers required incoming water temperature of 140°F (60°C) or greater to get the dishes clean. A water temperature of 120°F (49°C) is not hot enough to dissolve grease, activate power detergents, and kill bacteria. That is why commercial and restaurant dishwashers boost temperatures to as high as 180°F (82°C).

Newer residential dishwashers have their own built-in water heaters to boost the incoming water temperature in the dishwasher from 120°F (49°C) to 140°F–145°F (60°C–63°C). This heating element also provides the heat for the drying cycle. Today, a thermostat setting of 120°F (49°C) on the water heater should be satisfactory.

Thermostats/High-Temperature Limit Controls

All electric water heaters have a thermostat(s) and a high-temperature limit control(s).

TABLE 19-3

Time/temperature relationships related to scalding of an adult subjected to moving water. For children, the time to produce a serious burn is less than for adults. (*Courtesy of Shriners Burn Institute, Cincinnati Unit*)

Temperature	Time to Produce Serious Burn
120°F (48.89°C)	Approx. 9½ minutes
125°F (51.67°C)	Approx. 2 minutes
130°F (54.44°C)	Approx. 30 seconds
135°F (57.22°C)	Approx. 15 seconds
140°F (60.00°C)	Approx. 5 seconds
145°F (62.78°C)	Approx. 2½ seconds
150°F (65.76°C)	Approx. 1 8/10 seconds
155°F (68.33°C)	Approx. 1 second
160°F (71.11°C)	Approx. ½ second

FIGURE 19-8 Typical electric water heater controls. The 7-terminal control combines water temperature control plus high-temperature limit control. This is the type commonly used to control the upper heating element as well as provide a high-temperature-limiting feature. The 2-terminal control is the type used to control water temperature for the lower heating element. (*Courtesy of THERMO-DISC, Inc., subsidiary of Emerson Electric*)

Thermostats and high-temperature limit controls are available in many configurations. They are available as separate controls and as combination controls.

Figure 19-8 illustrates a combination thermostat/high-temperature limit control. This particular control is an interlocking type sometimes referred to as a snap-over type. It is used on 2-element water heaters and controls the upper heating element as well as providing the safety feature of limiting the water temperature to a factory-preset value. Another thermostat is used to control the lower heating element. When all of the water in the water heater is cold, the upper element heats the upper third of the tank. The lower thermostats are closed, calling for heat, but the interlocking characteristic of the upper thermostat/high limit will not allow the lower heating element to become energized. With this connection, both heating elements cannot be on at the same time. This is referred to as *nonsimultaneous* or *limited demand* operation.

When the water in the upper third of the tank reaches the temperature setting of the upper thermostat, the upper thermostat "snaps over," completing the circuit to the lower element. The lower element then begins to heat the lower two-thirds of the tank. Most of the time, it is the lower heating element that keeps the water at the desired temperature. Should a large amount of water be used in a short time, the upper element comes on. This provides fast recovery of the water in the upper portion of the tank, which is where the hot water is drawn from the tank.

Safety Note: Most residential water heater thermostats are of the single-pole type. They do not open both ungrounded conductors of the 240-volt supply. They open one conductor only, which is all that is needed to open the 240-volt circuit to the heating element. What this means is that a reading of 120 volts to ground will always be present at the terminals of the heating element on a 240-volt, single-phase system such as found in residential wiring. If the heating element becomes grounded to the metal sheath, the element can continue to heat even when the thermostat is in the "OFF" position because of the presence of the 120 volts.

The high-temperature limit control opens both ungrounded conductors.

Various Types of Electrical Connections

There are many variations for electric water heater hookups. Most electric utilities have their own unique time-of-use programs and specific electrical connection requirements for electric water heaters. It is absolutely essential that you contact your local electric utility for this information.

Because of the cumulative large power consumption of electric water heater loads, electric utilities across the country have been innovative in creating special lower rate structures (programs) for electric water heaters, air conditioners, and heat pumps. Electric utilities want to be able to shed some of this load during their peak periods. They will offer the homeowner choices of how the water heater, air conditioner, or heat pump is to be connected and, in some cases, will provide financial incentives (lower rates) to those who are willing to use these restricted "time-of-

day" connections. "Time-of-day," "time-of use," and "off-peak" are terms used by different utilities. If homeowners have a large-storage-capacity electric water heater, they will have an adequate supply of hot water during those periods of time the power to the water heater is off. If they cannot live with this, they have other options. Check with the local power company for details on their particular programs.

The following text discusses some of the methods in use around the country for connecting electric water heaters, and in some cases air-conditioning and heat pump equipment.

Utility Controlled

The world of electronics has opened up many new ways for an electric utility to meter and control residential electric water heater, air-conditioning, and heat pump loads. Electronic watt-hour meters can be read remotely, programmed remotely, and can detect errors remotely. The options are endless.

The utility can install watt-hour meters that combine a watt-hour meter and a set of switching contacts. The contacts are switched on and off by the utility during peak hours, using a power line carrier signal. Older style versions of this type of watt-hour meter incorporated a watt-hour meter, time clock, and switching contacts. The time clock would be set by the utility for the desired times of the On–Off cycle. Today, the metering and the programming for timing and control of the On–Off cycling is accomplished through electronic devices. Electronic meters contain an optical assembly that scans the rotating disc and a microprocessor that records the number of turns and speed of the disc. These electronic digital meters can record the energy consumption during normal hours and during peak hours, allowing the electric utility to bill the homeowner at different rates for different periods of time. One type of residential single-phase watt-hour meter can register four different "time-of-use" rates. This type of meter is shown in Figure 29-2.

One major utility programs their residential meters for "peak periods" from 9 AM to 10 PM, Monday through Friday—except for holidays.

Their "off-peak" periods are weekend days and all other hours. Programming of electronic meters can extend out for more than 5 years. The possibilities for programming electronic meters are endless. Older style mechanical meters used gears to record these data, and did not have the programming capabilities that electronic meters do as in Figures 19-9, 19-10, and 19-11.

The switching contacts in the meter are connected in *series* with the water heater, air conditioner, or heat pump load that is to be controlled for the special "time-of-day" rates, Figures 19-9, 19-10, and 19-11. The utility determines when they want to have the electric water heater on or off.

Figure 19-9 shows a separate combination meter/time clock for the electric water heater load. The utility sets the clock to be "OFF" during their peak hours of high power consumption. Thus the term *off-peak metering*. Off-peak metering is oftentimes called "time-of-use" or "time-of-day" metering. In exchange for lower electric rates, the homeowner runs the risk of being without hot water. Should homeowners run out of hot water during these off periods, they have no recourse but

to wait until the power to the water heater comes on again. For this type of installation, large-storage-capacity water heaters are needed so as to be able to have hot water during these periods.

A 240/120-volt service

B Main watt-hour meter

C Water heater watt-hour meter

D Service conductors to main panelboard

E Two-pole disconnect switch or circuit breaker

F Combination high temperature limit and upper interlocking thermostat

G Upper heating element

H Lower thermostat

I Lower heating element

Notations used in Figures 19-9, 19-10, 19-12, and 19-13.

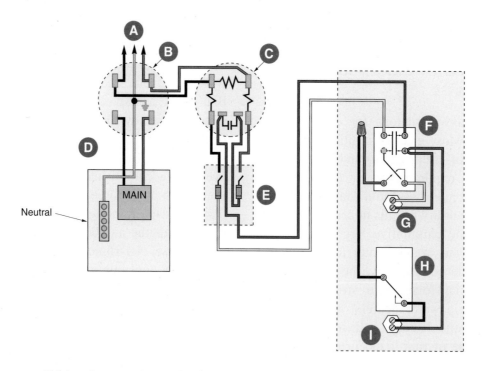

FIGURE 19-9 Wiring for a typical "off-peak" electric water heater circuit. A separate watt-hour meter/time clock controls the circuit to the water heater. The utility sets the time clock to turn the power off during certain peak periods of the day. This type of circuitry is called "time-of-day" programming. (*Delmar/Cengage Learning*)

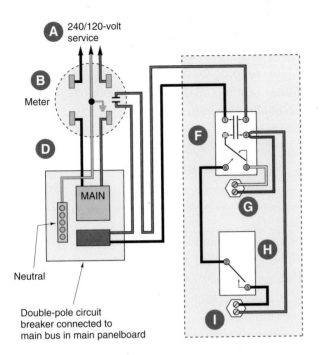

FIGURE 19-10 The meter contains a set of contacts that can be controlled by the utility. The water heater circuit is fed from the main panelboard, through the contacts, then to the water heater. (*Delmar/Cengage Learning*)

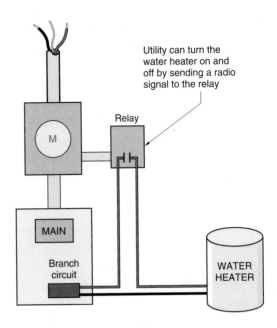

FIGURE 19-11 In this diagram, the utility sends a radio signal to the relay, which in turn can "shed" the load of a residential water heater, air conditioner, or heat pump. If the relay has two sets of contacts, four conductors would be required. Low-voltage relays are also used, in which case low-voltage conductors are run between the radio control unit and relay in the appliance. For the control of an air conditioner or heat pump, the radio control unit is connected "in series" with the room thermostat. This arrangement can be used with one watt-hour meter, using a sliding rate schedule with a "time-of-day" control. Contact the electric utility for technical data regarding how it wants the electrical connections to be made for its particular programs. (*Delmar/Cengage Learning*)

Figure 19-10 illustrates a type of meter that contains an integral set of contacts, controlled by the utility via signals sent over the power lines. These meters are available with either 30-ampere or 50-ampere contacts, the contact rating selected to switch the load to be controlled. The electric utility determines when to shed the load. The homeowner must be willing to live with the fact that the power to the water heater, air conditioner, or heat pump might be turned off during the peak periods.

In Figure 19-11, the utility mounts a radio-controlled relay to the side of the meter socket. One major utility using this method makes use of 10 different radio frequencies. When the utility needs to shed load, it can signal one or all of the relays to turn the water heater, air conditioner, or heat pump load off and on. This might only be a few times each year, as opposed to time-clock programs that operate daily. This method is a simple one. One watt-hour meter registers the total power consumption. The homeowner signs up for this program. The program offers a financial incentive, and the homeowner knows up front that the utility might occasionally turn off the load during crucial peak periods.

Customer Controlled

There are some rate schedules where customers can somewhat control their energy costs by doing their utmost to use electrical energy during off-peak hours. For example, they would not wash clothes, use the electric clothes dryer, take showers, or use the dishwasher during peak (premium) hours. The choice is up to homeowners. To accomplish this, watt-hour meters can be installed that have two sets of dials: one set showing the *total kilowatt hours*, and the second set showing *premium time kilowatt hours*. See Figure 29-2. With this type of meter, the utility will have the total kilowatt hours used as well as the premium time kilowatt hours used. With this information, the utility bills customers at one rate for the energy used during peak hours, and at another rate for the difference between total kilowatt hours and premium time kilowatt hours. The electric utility will determine and define *premium time*.

Some homeowners have installed time-clocks on their water-heater circuit to shut off during peak premium time hours to be sure they are not using energy during the periods with a high rate per kilowatt-hour.

Uncontrolled

The simplest and most common uncontrolled method is shown in Figure 19-12. There are no extra meters, switching contacts in the meter, extra disconnect switches, or controllable relays. The water heater circuit is connected to a 2-pole branch-circuit breaker in the main distribution panelboard. The electrical power to the water heater is on 24 hours a day. This is the connection for the water heater in the residence in this text. Instead of separate "low-rate" metering that requires an additional watt-hour meter, the utility offers a sliding scale of rate steps.

Kilowatt-hour power consumption of water heater is metered by same meter used for normal household loads. Only one meter is required. Power rates are usually on a sliding scale; that is, the more power in kilowatt-hours used, the lower rate per kilowatt-hour.

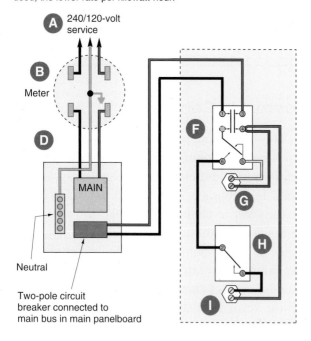

FIGURE 19-12 Water heater connected to 2-pole circuit in Main Panelboard. (*Delmar/Cengage Learning*)

EXAMPLE

Meter reading 06-01-20XX	00532	
Meter reading 05-01-20XX	00007	
Total kilowatt-hours used	00525	
Basic monthly energy charge		$11.24
1st 400 kWh × 0.10206		40.82
Next 125 kWh × 0.06690		8.36
Amount Due		$60.42

State, local, and applicable regulatory taxes would be added to the amount due.

Certain utilities will use the higher energy charge during the June, July, August, and September summer months.

Another uncontrolled method is to use two watt-hour meters, Figure 19-13. The water heater has power 24 hours per day. One meter will register the normal lighting load at the regular residential energy charge. The second meter will register the energy consumed by the water heater, air conditioner, or heat pump at some lower energy rate.

Are Water Heater Control Conductors Permitted in the Same Raceway as Service-Entrance Conductors?

Conductors other than service-entrance conductors are not permitted in the same raceway as the service-entrance conductors, *230.7*. However, *Exception No. 2* to this section does permit load management control conductors that have overcurrent protection to be installed in the service raceway.

ELECTRIC WATER HEATER BRANCH CIRCUITS

Branch-Circuit Rating

NEC 422.13 applies to all fixed storage electric water heaters having a capacity of 120 gallons (454.2 L) or less. This would include most typical residential electric water heaters. An electric water heater is a continuous load per *422.13*,

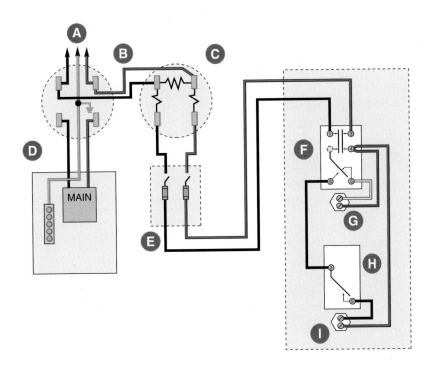

FIGURE 19-13   Twenty-four-hour water heater installation. The total energy demand for the water heater is fed through a separate watt-hour meter at a lower kWh rate than the regular meter. (*Delmar/Cengage Learning*)

which would require that the branch-circuit rating must not be less than 125% of the nameplate rating.

Disconnecting Means

For obvious safety reasons, *NEC 422.30* requires that an appliance have a means of disconnecting it from the power source. In most instances, this is a separate disconnect switch.

NEC 422.31(B) tells us that for permanently connected appliances rated greater than 300 volt-amperes or ⅛ horsepower, the disconnecting means is permitted to be the branch-circuit switch or circuit breaker for the appliance if the switch or breaker is within sight of the appliance *or* is capable of being locked in the "OFF" position.

The locking provision must be permanently installed on or at the switch or circuit breaker used as the disconnecting means. The locking provision must remain in place with or without the lock installed.

Workers have been seriously shocked when working on appliances that they thought were de-energized, because after turning off the power,

someone came along wondering why the switch was off and turned the power back on.

Listed lock-off devices are available from the various manufacturers of circuit breakers. Disconnect switches have brackets with holes through which a padlock is installed. OSHA lock-off requirements are discussed in Chapter 1.

NEC 422.35 requires that switches and circuit breakers used as the disconnecting means be of the indicating type, meaning they must clearly show that they are in the "ON" or "OFF" position.

Overcurrent Protection and Conductor Size

There are two separate issues that must be considered.

- The overcurrent protective device ampere rating must be calculated.

- The conductor size must be determined.

Do both calculations. Then compare the results of both calculations to make sure that both overcurrent device and conductors are *Code* compliant.

Branch-Circuit Overcurrent Protection

NEC 422.11(E) states that for single non-motor-operated electrical appliances, the overcurrent protective device shall not exceed the protective device rating marked on the appliance. If the appliance has no such marking, then the overcurrent device is to be sized as follows:

- If the appliance does not exceed 13.3 amperes—20 amperes
- If the appliance draws more than 13.3 amperes—150% of the appliance rating

For a single non-motor-operated appliance, if the 150% sizing does not result in a standard size overcurrent device rating as listed in *240.6(A)*, then it is permitted to go to the next standard size.

EXAMPLE

A water heater nameplate indicates 4500 watts, 240 volts. What is the maximum size fuse permitted by the Code?

Solution

$$I = \frac{W}{E} = \frac{4500}{240} = 18.8 \text{ amperes}$$

An electric water heater is a continuous load per *422.13*. The branch-circuit rating must not be less than 125% of the water heater's nameplate rating. The minimum branch-circuit rating is

$$18.75 \times 1.25 = 23.4 \text{ amperes}$$

This would require a minimum 25-ampere fuse or circuit breaker, which is the next standard size larger than the calculated 23.4 amperes. See *240.6(A)* for standard ampere ratings for fuses and circuit breakers.

The maximum overcurrent device for the water heater is

$$18.75 \times 1.5 = 28.1 \text{ amperes}$$

This would be a 30-ampere fuse or circuit breaker, which is the next standard size larger than the calculated 28.1 amperes. But there is no reason to use a 30-ampere branch-circuit fuse or circuit breaker because we have already determined that a 25-ampere fuse or circuit breaker is suitable for the 4500-watt load.

Conductor Size

The conductors supplying the 4500-watt water heater will be protected by the 25-ampere or breaker as determined above.

In *Table 310.15(B)(16)*, we find the allowable ampacity of conductors in the 60°C column as required by *110.14(C)*. In *240.4(D)*, we find the maximum overcurrent protection for small conductors. Putting this all together, we have the information shown in Table 19-4.

For the example, we selected a 25-ampere overcurrent device. A 30-ampere OCD would have been acceptable. Next, we need to find a conductor that is properly protected by a 25- or 30-ampere OCD. The conductor would have to be a minimum 10 AWG Type THHN, which has an allowable ampacity of 30 amperes, more than adequate for the minimum branch-circuit rating of 23.4 amperes. A 10 AWG conductor is properly protected by a maximum 30-ampere overcurrent device.

If a 12 AWG Type THHN had been selected, it would have been suitable for the load but would not have been properly protected by the 25-ampere OCD. The maximum OCD for a 12 AWG is 20 amperes unless the branch circuit is for a special application such as for motors.

TABLE 19-4

Table showing typical small conductors, their allowable ampacities, and the maximum rating overcurrent devices permitted for these conductors.

Conductor Size	Allowable Ampacity Ampere from 60°C Column of Table 310.15(B)(16)	Maximum Rating of Overcurrent Device
14 AWG	15 amperes	15 amperes
12 AWG	20 amperes	20 amperes
10 AWG	30 amperes	30 amperes

You might want to review Chapter 4 for a refresher as to why we use the 60°C column of *Table 310.15(B)(16)* for selecting the conductor size.

The Water Heater for This Residence

The water heater for the residence is connected to Circuit A6-8. This is a 30-ampere, 2-pole circuit breaker located in the main panelboard in the workshop. This circuit is a straight 240 volts and does not require a neutral conductor.

Checking the Schedule of Special-Purpose Outlets, we find that the electric water heater has two heating elements: a 4500-watt element upper and a 4500-watt lower element. Thus it is considered a quick-recovery unit. Because of the thermostats on the water heater, both elements cannot be energized at the same time. Therefore, the maximum load demand is

$$\text{Amperes} = \frac{\text{Watts}}{\text{Volts}} = \frac{4500}{240} = 18.75 \text{ or } 18.8 \text{ amperes}$$

For water heaters having two heating elements connected for nonsimultaneous operation, the nameplate on the water heater will be marked with the largest element's wattage at rated voltage. For water heaters having two heating elements connected for simultaneous operation, the nameplate on the water heater will be marked with the total wattage of both elements at rated voltage.

Conductor Size and Raceway Size for the Water Heater in This Residence

The conductor and overcurrent device is required to be not smaller than 125 percent of the current.

18.8 amperes × 1.25 = 23.5 amperes

This results in a 10 AWG copper conductor and a 30 ampere overcurrent device because the small-conductor rule in *240.4(D)(4)* generally requires a 12 AWG conductor to have overcurrent protection not greater than 20 amperes.

Table C1 in *Annex C* of the *NEC* indicates that trade size ½ EMT will be okay for two 10 THHN/

THWN conductors. For ease of installation, a short length of trade size ½ flexible metal conduit 18 to 24 in. (450 to 600 mm) long is attached to the EMT and is connected into the knockout provided for the purpose on the water heater. A 10 AWG equipment grounding conductor is required through the flexible metal conduit as required by the rules in *NEC 250.118(5)*.

Cord/Plug Connections Not Permitted for Water Heaters

NEC 400.7(A) and *422.16* list the uses where flexible cords are permitted. A key *Code* requirement, often violated, is that flexible cords shall be used only *where the fastening means and mechanical connections are specifically designed to permit ready removal for maintenance and repair, and the appliance is intended or identified for flexible cord connection.* Certainly the plumbing does not allow for the "ready removal" of the water heater. The conductors in flexible cords cannot handle the high temperatures encountered on the water heater terminals. Check the instruction manual for proper installation methods.

Flexible cords are not permitted to be

- used as a permanent wiring method.
- run through holes in walls, ceilings, etc.
- run through doorways, windows, or similar openings.
- attached to building surfaces.
- concealed above ceilings, in walls, or under floors.
- installed in raceways.
- used where subject to physical damage.

Equipment Grounding

The electric water heater is grounded through the trade size ½ EMT and an internal 10 AWG equipment grounding conductor installed through the flexible metal conduit (FMC) or by the equipment grounding conductor contained within the Type NM cable if used as the wiring method. *Code* references are *250.110, 250.118, 250.134,* and

348.60. The subject of equipment grounding is covered in Chapter 4.

EFFECT OF VOLTAGE VARIATION ON RESISTIVE HEATING ELEMENTS

The heating elements in the water heater in this residence are rated 240 volts. A resistive heating element will only produce rated wattage at rated voltage. It will operate at a lower wattage with a reduction in voltage. If connected to voltages above their rating, heating elements will have a very short life.

Ohm's law and the wattage formula show how the wattage and current depend on the applied voltage. Always use *rated* voltage to calculate the resistance, current draw, and wattage of a resistive heating element. Check manufacturers' specifications.

Nichrome wire, commonly used to make heating elements, has a "hot" resistance approximately 10% higher than its "cold" resistance.

CAUTION: When actually testing a heating element, never take an ohms reading on an energized circuit. Remove any other connected wires from the heating element terminal block on a water heater to prevent erratic ohms readings. Always check from the heating element leads to ground to confirm that there is no ground fault in the heating element.

Let's take a look at a 3000-watt, 240-volt heating element and calculate its resistance and current draw.

$$R = \frac{E^2}{W} = \frac{240 \times 240}{3000} = 19.2 \text{ ohms}$$

$$I = \frac{E}{R} = \frac{240}{19.2} = 12.5 \text{ amperes}$$

If a different voltage is substituted, the wattage and current values change accordingly.

At 220 volts:

$$W = \frac{E^2}{R} = \frac{240 \times 220}{19.2} = 2521 \text{ watts}$$

$$I = \frac{W}{E} = \frac{2521}{220} = 11.5 \text{ amperes}$$

Another way to calculate the effect of voltage variance is

$$\text{Correction Factor} = \frac{\text{Applied Voltage Squared}}{\text{Rated Voltage Squared}}$$

Using the previous example:

$$\text{Correction Factor} = \frac{220 \times 220}{240 \times 240} = \frac{48,400}{57,600} = 0.84$$

then $3000 \times 0.84 = 2521$ watts

In resistive circuits, current is directly proportional to voltage and can be simply calculated using a ratio and proportion formula. For example, if a 240-volt heating element draws 12.7 amperes at rated voltage, the current draw can be determined at any other applied voltage, say 208 volts.

$$\frac{208}{240} = \frac{X}{12.7}$$

$$\frac{208 \times 12.7}{240} = 11.0 \text{ amperes}$$

To calculate this example if the applied voltage is 120 volts:

$$\frac{120}{240} = \frac{X}{12.7}$$

$$\frac{120 \times 12.7}{240} = 6.4 \text{ amperes}$$

Also, in a resistive circuit, wattage varies as the square of the current. Therefore, when the voltage on a heating element is doubled, the current also doubles and the wattage increases four times. When the voltage is reduced to one-half, the current is halved and the wattage is reduced to one-fourth.

A 240-volt heating element connected to 208 volts will have approximately three-quarters of the wattage rating than at rated voltage.

EFFECT OF VOLTAGE VARIATION ON MOTORS

The above formulas work only for resistive circuits. Formulas for inductive circuits such as an electric motor are much more complex than this book is intended to cover. For typical electric motors, the following information will suffice (see Table 19-5).

As previously mentioned, NEMA rated electric motors are designed to operate at ±10% of their nameplate voltage.

TABLE 19-5

Table showing the approximate change in full-load current and starting-current for typical electric motors when operated at under voltage (90%) and over voltage (110%) conditions.

Voltage Variation	Full-Load Current	Starting Current
110%	7% decrease	10–12% increase
90%	11% increase	10–12% decrease

HEAT PUMP WATER HEATERS

With so much attention given to energy savings, heat pump water heaters have entered the scene.

Residential heat pump water heaters can save 50% to 60% over the energy consumption of resistance-type electric water heaters, depending on energy rates. The heat energy transferred to the water is three to four times greater than the electrical energy used to operate the compressor, fan, and pump.

Heat pump water heaters work in conjunction with the normal resistance-type electric water heater. Located indoors near the electric water heater and properly interconnected, the heat pump water heater becomes the primary source for hot water. The resistance electric heating elements in the electric water heater are the backup source should it be needed.

A heat pump water heater removes low-grade heat from the surrounding air, but instead of transferring the heat outdoors as does a regular air-conditioning unit, it transfers the heat to the water. The unit shuts off when the water reaches 130°F (54.44°C). Moisture is removed from the air at approximately 1 pint per hour, reducing the need to run a dehumidifier. The exhausted cool air from the heat pump water heater is about 10° to 15°F (−12.22°C to −9.44°C) below room temperature. The cool air can be used for limited cooling purposes, or it can be ducted outdoors.

A typical residential heat pump water heater rating is 240 volts, 60 Hz, single-phase, 500 watts.

Heat pump water heaters operate with a hermetic refrigerant motor compressor. The electric hookup is similar to a typical 240-volt, single-phase, residential air-conditioning unit. *Article 440* of the *NEC* applies. This is covered in Chapter 23.

Electric utilities and numerous Web sites are good sources to learn more about the economics of installing heat pump water heaters to save energy.

REVIEW

Note: Refer to the *Code* or the plans where necessary.

WATER PUMP CIRCUIT B

1. Does a jet pump have any electrical moving parts below the ground level? _____

2. Which is larger, the drive pipe or the suction pipe? _____

3. Where is the jet of the pump located? _____

4. What does the impeller wheel move? _____

5. Where does the water flow after leaving the impeller wheel?

   a. _____

   b. _____

6. What prevents water from draining back into the pump from the tank? _____

_____

7. What prevents water from draining back into the well from the equipment? _____

_____

8. What is compressed in the water storage tank? _____

9. Explain the difference between a "2-wire" submersible pump and a "3-wire" submersible pump. _____

_____

_____

_____

10. What is a common speed for jet pump motors? _____

11. Why is a 240-volt motor preferable to a 120-volt motor for use in this residence?

_____

_____

12. How many amperes does a 1-horsepower, 240-volt, single-phase motor draw? (See *Table 430.248.*) _____

13. What size are the conductors used for this circuit? _____

14. What is the branch-circuit protective device? _____

15. What provides the running overload protection for the pump motor? _____

_____

_____

16. What is the maximum ampere setting permitted for running overload protection of the 1-horsepower, 240-volt pump motor? _____

17. Submersible water pumps operate with the electrical motor and actual pump located (circle the correct answer)

    a. above permanent water level.

    b. below permanent water level.

    c. half above and half below permanent water level.

18. Because the controller contains the motor starting relay and the running and starting capacitors, the motor itself contains _____.

19. What type of pump moves the water upward inside of the deep-well pipe? _____

_____

20. Proper pressure of the submersible pump system is maintained by a _____

_____.

21. Fill in the data for a 16-ampere electric motor, single-phase, no *Code* letters.

    a. Branch-circuit protection, non-time-delay fuses:

    Normal size _____ A   Maximum size _____ A

    Switch size _____ A   Switch size _____ A

    b. Branch-circuit protection dual-element, time-delay fuses:

    Normal size _____ A   Maximum size _____ A

    Switch size _____ A   Switch size _____ A

   c. Branch-circuit protection instant-trip breaker:

      Normal setting _____ A

      Maximum setting _____ A

   d. Branch-circuit protection—inverse time breaker:

      Normal rating _____ A

      Maximum rating _____ A

   e. Branch-circuit conductor size, Type THHN _____

      Ampacity _____ A

   f. Motor overload protection using dual-element time-delay fuses:

      Maximum size _____ A

22. The *NEC* is very specific in its requirement that submersible electric water pump motors be grounded. Where is this specific requirement found in the *Code*?

    _____

    _____

23. Does the *NEC* allow submersible pump cable to be buried directly in the ground?

    _____

    _____

24. Must the disconnect switch for a submersible pump be located next to the well?

    _____

    _____

25. A metal well casing (shall) (shall not) be bonded to the pump's equipment grounding conductor or by grounding it with a separate equipment grounding conductor run all the way back to the same ground bus in the panelboard that supplies the pump circuit, *NEC*

    _____

WATER HEATER CIRCUIT

1. According to *422.47* the high-temperature limit control must disconnect _____ of the ungrounded conductors. The high-temperature control limits the maximum water temperature to _____ °F (_____ °C).

2. A major hazard involved with water heaters is that they operate under whatever pressure the serving water utility supplies. Water stays liquid at temperatures higher than the normal boiling point of 212°F (100°C). If the high-temperature limit control failed to operate for whatever reason, should a pipe burst or a faucet be opened, the pressure would instantly drop to normal atmospheric pressure, causing the superheated water to turn into _____ that could result in _____. To prevent this from happening, water heaters are equipped with pressure/temperature relief valves.

3. Magnesium rods are installed inside the water tank to reduce _____.

4. The heating elements in electric water heaters are generally classified into two categories. These are _____ density and _____ density elements.

5. An 80-gallon electric water heater is energized for the first time. Approximately how many hours would it take to raise the water temperature 80°F (26.67°C) (from 40°F to 120°F [4.44°C to 48.89°C])? The water heater has a 3000-watt heating element.

   _____

   _____

6. What term is used by utilities when the water heater power consumption is measured using different rates during different periods of the day?_____

7. Explain how the electric utility in your area meters residential electric water heater loads.

   _____

   _____

   _____

8. For residential water heaters, the Consumer Product Safety Commission suggests a maximum temperature setting of _____ °F (_____°C). Some states have laws stating a maximum temperature setting of _____ °F (_____ °C).

9. An 80-gallon electric water heater has 60°F (15.56°C) incoming water. How many minutes can a 3-gallon/minute showerhead be used to draw 105°F (40.56°C) water if the water heater thermostat is set at 120°F (48.89°C)? _____

10. Approximately how long would it take to produce serious burns to an adult with 140°F (60°C) water? _____

11. Two thermostats are generally used in an electric water heater.

    a. What is the location of each thermostat?_____

    _____

    b. What type of thermostat is used at each location? _____

    _____

12. a. How many heating elements are provided in the heater in the residence discussed in this text? _____

    b. Are these heating elements allowed to operate at the same time? _____

13. When does the lower heating element operate? _____

    _____

14. The *Code* states that water heaters having a capacity of 120 gallons (450 L) or less shall be considered _____ duty and, as such, the circuit must have a rating of not less than _____ percent of the rating of the water heater.

15. Why does the storage tank hold the heat so long? _____

16. The electric water heater in this residence is connected for "limited demand" so that only one heating element can be on at one time

    a. What size wire is used to connect the water heater? _____

    b. What size overcurrent device is used? _____

17. a. If both elements of the water heater in this residence are energized at the same time, how much current will they draw? (Assume the elements are rated at 240 volts.)

    b. What size and Type THHN wire is required for the load of both elements? Show calculations.

18. a. How much power in watts would the two elements in Question 17 use if connected to 220 volts? Show calculations.

    b. What is the current draw at 220 volts? Show calculations.

19. A condominium owner complains of not getting enough hot water. The serviceman checks the voltage at the water heater and finds the voltage to be 208 volts. The serviceman checks further and finds the electrical service on the building to be a 120/208-volt, three-phase, 4-wire system. The main electrical panelboard for each condominium unit is fed with a 3-wire supply. The nameplate on the electric water heater is marked 4500 watts, 240 volts. What is the wattage output of the heating element in this water heater when connected to a 208-volt supply? _____

20. For a single, nonmotor-operated electrical appliance rated greater than 13.3 amperes that has no marking that would indicate the size of branch-circuit overcurrent device, what percentage of the nameplate rating is used to determine the maximum size branch-circuit overcurrent device? _____

21. A 7000-watt resistance-type heating appliance is rated 240 volts. What is the maximum size fuse permitted to protect the branch-circuit supplying this appliance?

    _____

22. The *Code* requirements for standard resistance electric water heaters are found in *Article 422* of the *Code*. Because heat pump water heaters contain a hermetic refrigerant motor compressor, the *Code* requirements are found in *Article* _____ of the *Code*.

23. In your area, is the branch-circuit breaker or switch for the electric water heater branch circuit acceptable as the disconnecting means, or are you required to install a separate disconnect within sight of the water heater? Provide a brief explanation.

    _____

    _____

Special-Purpose Outlets for Ranges, Counter-Mounted Cooking Unit ⓐ$_G$, and Wall-Mounted Oven ⓐ$_F$

OBJECTIVES

After studying this chapter, you should be able to

- understand the *NEC* requirements for installing and connecting freestanding ranges, counter-mounted cooking units, wall-mounted ovens, microwave ovens, and light energy ovens.

- make load calculations to determine proper size conductors, overcurrent protection, disconnecting means, and how to achieve proper grounding.

- be familiar with the different configurations for 30- and 50-ampere NEMA receptacles and cords.

- be aware of the many types of temperature controls found on cooking equipment.

- understand significant differences of color coding and terminal identification on foreign-made appliances.

BASIC CIRCUIT REQUIREMENTS FOR ELECTRIC RANGES, COUNTER-MOUNTED COOKING UNITS, AND WALL-MOUNTED OVENS

Electric ranges are covered in *Article 422* of the *NEC* and by UL Standard 858, *Household Electric Ranges.*

UL Standard 1082 covers *Electric Household Cooking and Food Serving Appliances.*

Connection Methods

Most recognized wiring methods can be used to hook up appliances, such as nonmetallic-sheathed cable, armored cable, EMT, flexible metal or non-metallic conduit, or service-entrance cable. Check local electrical codes to see whether there are any restrictions on the use of any of the wiring methods listed above.

Here are some of the more common methods for connecting electric ranges, wall-mounted ovens, and counter-mounted cooktops.

Direct Connection. One method is to run the branch-circuit wiring directly to the wiring compartment on the appliance. When "roughing-in" the branch-circuit wiring, sufficient length must be provided so as to allow flexibility to make up the connections and place the appliance into position, and for servicing.

The time and material required to run an individual branch circuit to each of the major electrical appliances might very well be the most economical way to do the hookups, instead of using junction boxes, tap conductors, and extra splices, as discussed next. Don't forget, "time" includes the time it takes to calculate the proper size tap conductors and junction boxes, plus the actual installation time.

Junction Box. Another method is to run the branch-circuit wiring to a junction box (an outlet box) behind, under, or adjacent to the range, oven, or cooktop location. By definition, this junction box is accessible because, should the need arise, the junction box can be reached by removing the appliance. The "whip" on the appliance is then connected to the junction box where the splicing of the tap conductors in the whip and the branch-circuit conductors are made. According to UL Standard No. 858, the flexible metal whip must not be less than 3 ft (900 mm) and no longer than 6 ft (1.8 m). The conductors in the whip are rated for the high temperatures encountered in the appliance.

Depending on the location of the appliances, it might be possible to connect more than one cooking appliance to a single branch circuit from a single junction box or from two junction boxes. This is illustrated in Figure 20-1.

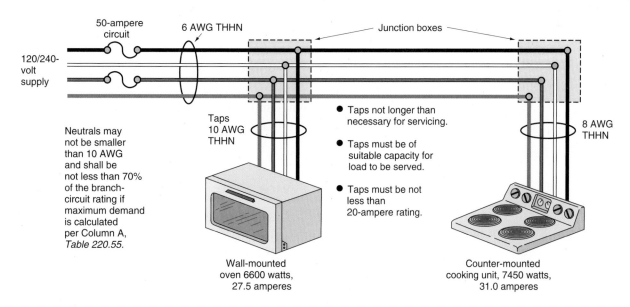

FIGURE 20-1 A counter-mounted cooktop unit and a wall-mounted oven connected to one branch circuit. Depending on the location of the appliances, it might be possible to make the installation with one junction box. (*Delmar/Cengage Learning*)

Cord-and-Plug Connection.   Still another very common method is to use a cord-and-plug connection, *422.33(A)*. This is particularly true for free-standing ranges but is uncommon for wall-mounted ovens or counter-mounted cooktops. A cord-and-plug connection also serves as the disconnecting means for household cooking equipment. The range receptacle is permitted to be at the rear base of an electric range. Removing the bottom drawer makes this connection accessible, *422.33(B)*. The receptacle must have an ampere rating not less than the rating of the appliance, *422.33(C)*. For household cooking equipment, the demand factors of *Table 220.55* are also applicable to the receptacle, *422.33(C)*, *Exception*. Range and dryer receptacles are similar but often have different ampere ratings; they are discussed later in this chapter.

Cord sets rated 30 amperes are referred to as *dryer cords*. Cord sets rated 40, 45, or 50 amperes are referred to as *range cords*.

Receptacles rated 30 amperes are referred to as *dryer receptacles*. Receptacles rated 50 amperes are referred to as *range receptacles*.

Nameplate on Appliance

NEC 422.60(A) requires that the nameplate must show the appliance's rating in volts and amperes or in volts and watts.

Load Calculations

In *210.19(A)(3)* and *Exceptions*, we find that

- the branch-circuit conductors must have an ampacity not less than the branch-circuit rating.
- the branch-circuit conductors' ampacity must not be less than the maximum load being served.
- the branch-circuit rating must not be less than 40 amperes for ranges of 8¾ kW or more.
- for ranges 8¾ kW or more, the neutral conductor of a 3-wire branch circuit may be 70% of the branch-circuit rating, but not smaller than 10 AWG.
- tap conductors connecting ranges, ovens, and cooking units to a 50-ampere branch circuit must be rated at least 20 amperes, must be adequate

for the load, and must not be any longer than necessary for servicing the appliance.

- *NEC 220.55* permits the use of the values found in *Table 220.55* and its footnotes for the load calculation.
- *Table 220.55* shows the kW-demand for different kW-rated household cooking appliances that are over 1¾ kW. Column C is used in most situations. Note that the maximum demand is considerably less than the actual appliance's nameplate rating. This is because rarely, if ever, will all of the heating elements be used at the same time. *Note 3* to the table tells us when we are permitted to use Columns A and B. This is discussed later in this chapter.
- *NEC 220.61* states that for calculating a feeder or service, the neutral conductor load can be figured at 70% of the load on the ungrounded conductors for ranges, ovens, and counter-mounted cooktops, using *Table 220.55* as the basis for the load calculation.

In *422.10(A)* we find that the branch-circuit rating shall not be less than the marked rating of the appliance. However, there is an exception for household cooking appliances. The last paragraph of this section states that *Branch circuits for household cooking appliances shall be permitted to be in accordance with Table 220.55*.

Separate Branch Circuits

When a separate branch circuit is run to a wall-mounted oven or to a counter-mounted cooking unit, the branch-circuit size is based on the appliance's full nameplate rating, *Table 220.55*, *Note 4*.

Wire Size

After making the load calculations to determine the appliance's calculated load in amperes, we then refer to *Table 310.15(B)(16)* to select a conductor that has an allowable ampacity equal to or greater than the calculated load. As mentioned many times in this text, *110.14(C)(1)* as well as the UL Standards require that we use the 60°C column regardless of the conductor's insulation rating.

An exception to this basic rule is when the equipment is marked for 75°C. This is discussed in detail in Chapter 4 in the section entitled "The Weakest Link of the Chain."

Overcurrent Protection

NEC 422.11(A) states that the branch-circuit overcurrent protection shall not exceed the rating marked on the appliance.

For appliances that have no marked maximum overcurrent protection, the branch-circuit overcurrent protection is sized at the ampacity of the branch-circuit conductors supplying the appliance. See *240.4(B)*, *(D)*, and *(E)*.

Disconnecting Means

All appliances require a means of disconnecting them from the power source, *422.30*.

NEC 422.33(A) accepts a cord-and-plug arrangement as the required disconnecting means.

A branch-circuit switch or circuit breaker is acceptable as the disconnecting means if it is within sight of the appliance or is capable of being locked in the "Off" position, *422.31(B)*. Breaker manufacturers supply "lock-off" devices that fit over the circuit breaker. Individual disconnect switches also have a lock-off provision. The lock-off provision must be permanently installed on the breaker or switch.

NEC 422.34 recognizes "unit switches" that have a marked "Off" position to be an acceptable disconnect for the appliance. Controllers such as clocks, timers, temperature controllers, and element controls do not qualify as "unit switches" as they do not disconnect all the ungrounded conductors in the appliance. A unit switch with a marked-off position disconnects all ungrounded conductors. This enables a person to work on a specific component of the appliance, knowing that the power is off to that particular component. Obviously, the line side terminals of the appliance are still "hot." Electric ranges, wall-mounted ovens, and counter-mounted cooktops are not normally equipped with unit switches, so another disconnecting means is required.

Using the branch-circuit switch or circuit breaker as the disconnecting means shuts off total power to the appliance and is much safer than relying on the unit switches. Be sure to use the required lock-off requirement.

GROUNDING FRAMES OF ELECTRIC RANGES, WALL-MOUNTED OVENS, AND COUNTER-MOUNTED COOKING UNITS

The frames of electric ranges, wall-mounted ovens, and counter-mounted cooking units must be grounded, *250.140*. Grounding is accomplished through the equipment grounding conductor in nonmetallic-sheathed cable, the metal armor of Type AC cable, or with metal raceways such as EMT and flexible metal conduit. See *250.118* and *250.134* for acceptable equipment grounding means.

The frames of electric ranges, wall-mounted ovens, counter-mounted cooking units, and clothes dryers *are not* permitted to be grounded to the neutral branch-circuit conductor, *250.140*.

Because of the ever-present confusion over using a separate equipment grounding conductor versus the old fashioned way of grounding the frame to the neutral conductor of the circuit, you had better review Chapter 15, where this subject is covered in detail. You might also want to review Chapter 4 where the grounding qualifications for various wiring methods are discussed.

Receptacles for Electric Ranges and Dryers

Because of their similarities, the following text applies to both range and dryer receptacles.

Residential electric ranges, electric dryers, wall-mounted ovens, and surface-mounted cooking units may be connected with cords designed specifically for this use. Depending on the appliance's wattage rating, the receptacles are usually rated 30 amperes or 50 amperes, 125/250 volts. Cord sets rated 40 amperes or 45 amperes contain 40- or 45-ampere conductors, yet the plug cap is rated 50 amperes. It becomes a cost issue as to whether to use a full 50-ampere-rated cord set in all cases or whether to match the cord ampere rating as closely as possible to that of the appliance being connected.

50-ampere
3-pole, 3-wire
125/250-volt
NEMA 10-50R
permitted
prior to 1996 *NEC*

50-ampere
4-pole, 4-wire
125/250-volt
NEMA 14-50R
required by 1996 *NEC*

30-ampere
3-pole, 3-wire
125/250-volt
NEMA 10-30R
permitted
prior to 1996 *NEC*

30-ampere
4-pole, 4-wire
125/250-volt
NEMA 14-30R
required by 1996 *NEC*

FIGURE 20-2 Illustrations of 30-ampere and 50-ampere receptacles used for cord-and-plug connection of electric ranges, ovens, counter-mounted cooking units, and electric clothes dryers. Four-wire receptacles and 4-wire cord sets are required for installations according to the *NEC*. Three-wire receptacles and 3-wire cord sets were permitted prior to the 1996 *NEC*. See *250.140*. (*Delmar/Cengage Learning*)

Range and dryer receptacles are available in both surface mount and flush mount.

Blade and Slot Configuration for Plug Caps and Receptacles

Figure 20-2 shows the different configurations for receptacles commonly used for electric range and dryer hookups. NEMA uses the letter "R" to indicate a receptacle, and the letter "P" for plug cap.

- **30-ampere, 3-wire:** The L-shaped slot on the receptacle (NEMA 10-30R) and the L-shaped blade on the plug cap (NEMA 10-30P) are for the white neutral conductor.

- **30-ampere, 4-wire:** The L-shaped slot on the receptacle (NEMA 14-30R) and the L-shaped blade on the plug cap (NEMA 14-30P) are for the white neutral conductor. The horseshoe-shaped slot on the receptacle and the round or horseshoe-shaped blade on the plug cap are for the equipment ground.

- **50-ampere, 3-wire:** The wide flat slot on the receptacle (NEMA 10-50R) and the matching wide blade on the plug cap (NEMA 10-50P) are for the white neutral conductor.

- **50-ampere, 4-wire:** The wide flat slot on the receptacle (NEMA 14-50R) and the matching wide blade on the plug cap (NEMA 14-50P) are for the white neutral conductor. The horseshoe-shaped slot on the receptacle and the round or horseshoe-shaped blade on the plug cap are for the equipment ground.

Cord Sets for Electric Ranges and Dryers

A 30-ampere cord set contains:

- **3-wire:** three 10 AWG conductors. The attachment plug cap (NEMA 10-30P) is rated 30 amperes.

- **4-wire:** four 10 AWG conductors. The attachment plug cap (NEMA 14-30P) is rated 30 amperes.

A 40-ampere cord set contains:

- **3-wire:** three 8 AWG conductors. The attachment plug cap (NEMA 10-50P) is rated 50 amperes.

- **4-wire:** three 8 AWG conductors and one 10 AWG conductor. The attachment plug cap (NEMA 14-50P) is rated 50 amperes.

A 45-ampere cord set contains:

- **3-wire:** two 6 AWG conductors and one 8 AWG conductor. The attachment plug cap (NEMA 10-50P) is rated 50 amperes.

- **4-wire:** three 6 AWG conductors and one 8 AWG conductor. The attachment plug cap (NEMA 14-50P) is rated 50 amperes.

A 50-ampere cord set contains:

- **3-wire:** two 6 AWG conductors and one 8 AWG conductor. The attachment plug cap (NEMA 10-50P) is rated 50 amperes.

- **4-wire:** three 6 AWG conductors and one 8 AWG conductor. The attachment plug cap (NEMA 14-50P) is rated 50 amperes.

Terminal Identification for Receptacles and Cords

Receptacle and cord terminals are marked as follows:

- "X" and "Y" for the ungrounded conductors.

- "W" for the white grounded conductor. The "W" terminals will generally be whitish or silver (tinned) in color, in accordance with *200.9* and *200.10* of the *NEC*.

- "G" for the equipment grounding conductor. This terminal is green colored, hexagon shaped, and is marked "G," "GR," "GRN," or "GRND" in accordance with *250.126* and *406.9(B)* of the *NEC*. The grounding blade on the plug cap must be longer than the other blades so that its connection is made before the ungrounded conductors make connection, in accordance with *406.9(D)* of the *NEC*.

WALL-MOUNTED OVEN CIRCUIT ⒶF

The wall-mounted oven in this residence is located to the right of the kitchen sink.

The branch-circuit wiring, using nonmetallic-sheathed cable, is run from Panelboard B to the oven location, coming into the cabinet space where the oven will be installed from behind or from the bottom. Earlier in this chapter we discussed the various methods for hooking up appliances. The choices are direct connection, junction box, or cord-and-plug connection. In most cases, a built-in wall-mounted oven will have a flexible metal conduit attached to it, which is run to a junction box where the electrical connections are completed.

Circuit Number. The circuit number is B (6-8). This information is found in the Schedule of Special Purpose Outlets contained in the Specifications at the rear of this text, and on the Directory for Panelboard B found in Chapter 27.

Nameplate Data. This appliance is rated 6.6 kW (6600 watts) @ 120/240 volts.

Load Calculations. (Use nameplate rating per *220.55, Note 4.*)

$$I = \frac{W}{E} = \frac{6600}{240} = 27.5 \text{ amperes}$$

Wire Size. The conductors in nonmetallic-sheathed cable are rated for 90°C. However, in conformance to *334.80*, the allowable ampacity is determined using the 60°C column of *Table 310.15(B)(16)*. Here we find that a 10 AWG conductor has an allowable ampacity of 30 amperes—more than adequate for the appliance's calculated load of 27.5 amperes. This will be a 10 AWG 3-conductor plus equipment grounding conductor nonmetallic-sheathed cable.

Overcurrent Protection. A 2-pole, 30-ampere circuit breaker located in Panelboard B.

Disconnecting Means. The disconnecting means is the 30-ampere, 2-pole circuit breaker in Panelboard B. If a cord-and-plug connection were made, this also would serve as a disconnecting means.

Grounding. The equipment grounding conductor contained in the nonmetallic-sheathed cable provides the means for grounding the frame of the oven. According to *Table 250.122*, the equipment grounding conductor for a circuit protected by a 30-ampere overcurrent device is a 10 AWG copper conductor. That is the size EGC contained in 10 AWG nonmetallic-sheathed cable.

COUNTER-MOUNTED COOKING UNIT CIRCUIT ⒶG

The cooktop in this residence is located in the island.

The branch-circuit wiring is run from Panelboard B to the island. The nonmetallic-sheathed cable is concealed above the recreation room ceiling, upward into the island cabinetry, ending in a junction box below the cooking unit. Earlier in this chapter, we discussed the various methods for hooking up appliances. The choices are direct connection, junction box, or cord-and-plug connection. In most cases, the cooktop will have a flexible metal conduit attached to it, which also runs to the junction box where the electrical connections are completed.

Note on the plans that a receptacle outlet is to be installed on the side of the island. This receptacle outlet is supplied by small-appliance branch circuit B16. This receptacle is required by *210.52(C)(2)*.

Circuit Number. The circuit number is B (2-4). This information is found in the Schedule of Special Purpose Outlets contained in the Specifications at the rear of this text, and on the Directory for Panelboard B found in Chapter 27.

Nameplate Data. This appliance is rated 7450 watts @ 120/240 volts.

Load Calculations. (Use nameplate rating per *Table 220.55, Note 4*.)

$$I = \frac{W}{E} = \frac{7,450}{240} = 31 \text{ amperes}$$

Wire Size. The conductors in nonmetallic-sheathed cable are rated for 90°C. However, according to *334.80*, the allowable ampacity is determined using the 60°C column of *Table 310.15(B)(16)*. Here we find that an 8 AWG conductor has an allowable ampacity of 40 amperes—more than adequate for the appliance's calculated load of 31 amperes. This will be an 8 AWG 3-conductor plus equipment grounding conductor nonmetallic-sheathed cable.

Overcurrent Protection. A 2-pole, 40-ampere circuit breaker located in Panelboard B.

Disconnecting Means. The disconnecting means is the 40-ampere, 2-pole circuit breaker in Panelboard B. A cord-and-plug connection would also be a disconnecting means.

Grounding. The equipment grounding conductor contained in the nonmetallic-sheathed cable provides the means for grounding the frame of the appliance. According to *Table 250.122*, the equipment grounding conductor for a circuit protected by a 40-ampere overcurrent device is a 10 AWG copper conductor. That is the size EGC contained in an 8 AWG nonmetallic-sheathed cable.

FREESTANDING RANGE

Load Calculations

The load calculations for a freestanding electric range are simple.

According to *Table 220.55*, the demand is permitted to be figured at 8 kW for an electric range that is rated 12 kW or less. For example, if an electric range is marked 11.4 kW, the maximum demand can be figured at 8 kW because, under normal situations, the surface heating elements, broiler element, baking element, and lights would not all be on at the same time.

Checking *Table 310.15(B)(16)*, we find that an 8 AWG 3-conductor plus equipment grounding conductor nonmetallic-sheathed cable with an ampacity of 40 amperes is more than adequate.

Larger Ranges

Load calculations for double-oven ranges are a little more complex than for typical one-oven electric ranges.

Larger double-oven freestanding ranges are probably rated more than 12 kW. Let us consider a range that has a rating of 14,050 watts. This is the same wattage as the previously discussed wall-mounted oven and counter-mounted cooking unit.

The total connected load of this freestanding electric range is

$$I = \frac{W}{E} = \frac{14,050}{240} = 58.5 \text{ amperes}$$

However, *Table 220.55* permits the calculated load to be considerably less than the connected load.

Table 220.55 and the footnotes provide the information we need to make the load calculation. The column C load of 8 kW is permitted to be used with 5% added for each kW or major fraction thereof by which the rating of the cooking appliances exceed 12 kW.

1. 14,050 − 12,000 watts = 2050 watts (2 kW)

2. According to *Note 1* of *Table 220.55*, a 5% increase for each kW and major fraction in

excess of 12 kW must be added to the load demand in Column C. 2 kW (the kW over 12 kW) × 5% per kW = 10%.

3. 8 kW (from Column C) × 0.10 = 0.8 kW

4. The calculated load is 8 kW + 0.8 kW = 8.8 kW.

5. In amperes, the calculated load is

$$I = \frac{W}{E} = \frac{8,800}{240} = 36.7 \text{ amperes}$$

Wire Size. The conductors in nonmetallic-sheathed cable are rated for 90°C. However, in conformance to *334.80*, the allowable ampacity is determined using the 60°C column of *Table 310.15(B)(16)*. Here we find that an 8 AWG conductor has an allowable ampacity of 40 amperes—more than adequate for the appliance's calculated load of 36.7 amperes. This will be an 8 AWG 3-conductor plus equipment grounding conductor nonmetallic-sheathed cable.

NEC 210.19(A)(3), Exception No. 2, states that for ranges of 8¾ kW or more, where the maximum demand has been calculated using Column C of *Table 220.55*, the neutral conductor may be reduced to not less than 70% of the branch-circuit rating but never smaller than 10 AWG.

$$40 \times 0.70 = 28 \text{ amperes}$$

Checking *Table 310.15(B)(16)*, we find that a 10 AWG conductor has an allowable ampacity of 30 amperes—more than adequate to carry the calculated neutral load. If the wiring method were a raceway method, this electric range could be served by two 8 AWG phase conductors and one 10 AWG neutral conductor.

Using the 70% multiplier is a moot point when installing nonmetallic-sheathed cable or armored cable, because the insulated phase conductors in these cables are all the same size.

Why the Reduced Neutral?

The heating elements are connected across the 240-volt source. Therefore, no neutral current from these elements is flowing in the 240/120-volt supply to the range.

Some loads in a range are 120-volt loads. These might be controls, timers, convection fans, rotisseries, and similar items. These loads do result in a current flow in the neutral conductor of the 240/120-volt supply to the range.

The permission to reduce the neutral conductor supplying an electric range to 70% is more than adequate to carry the above mentioned "line-to-neutral" connected loads within the range.

Some manufacturers supply a small 240/120-volt transformer inside the range that is connected across the two "hot" 240-volt conductors. The 120-volt secondary supplies the 120-volt loads within the range. With this arrangement, there is no neutral current flowing in the 240/120-volt supply to the range. In fact, the branch circuit to this type of electric range really only needs the two "hot" ungrounded conductors. Because you have no way of knowing what type of electric range will be installed, you must install a 4-wire branch circuit consisting of two "hot" ungrounded conductors, one grounded neutral conductor, and one equipment grounding conductor.

When using cable, reducing the neutral conductor is a moot point because all conductors in the cable are the same size, other than the equipment grounding conductor. When using a raceway wiring method, it is possible to take advantage of reducing the size of the neutral conductor.

Overcurrent Protection. A 2-pole, 40-ampere circuit breaker.

Disconnecting Means. The disconnecting means would be the 2-pole, 40-ampere circuit breaker. A cord-and-plug connection behind the base of a freestanding electric range would also serve as a disconnecting means, *422.33(B)*.

Grounding. The equipment grounding conductor contained in the nonmetallic-sheathed cable provides the means for grounding the frame of the appliance. According to *Table 250.122*, the equipment grounding conductor for a circuit protected by a 40-ampere overcurrent device is a 10 AWG copper conductor. That is the size EGC contained in an 8 AWG nonmetallic-sheathed cable.

CALCULATIONS WHEN MORE THAN ONE WALL-MOUNTED OVEN AND COUNTER-MOUNTED COOKING UNIT ARE SUPPLIED BY ONE BRANCH CIRCUIT

The following discusses how to make the load calculations when more than one wall-mounted oven and counter-mounted cooking unit that are supplied by one branch circuit.

Load Calculations

Note 4 to *Table 220.55* states that when a single branch circuit supplies a counter-mounted cooking unit and not more than two wall-mounted ovens, all located in the same room, the calculation is made by adding up the nameplate ratings of the individual appliances, then treating this total as if it were one electric range.

Figure 20-1 shows a wall-mounted oven and a counter-mounted cooking unit supplied by one 50-ampere branch circuit.

The total connected load of these two appliances is

$$I = \frac{W}{E} = \frac{14,050}{240} = 58.5 \text{ amperes}$$

However, because we are permitted to treat the two individual cooking appliances as one, we use the same calculations as we did for the freestanding range previously discussed.

The resulting conductor size, overcurrent protection, disconnecting means, and grounding are the same as for the equivalent freestanding range.

The cost of installing one 50-ampere branch circuit to supply both appliances might be higher than the cost of installing separate branch circuits to each appliance. There will be additional junction boxes, different size cables, cable connectors, conduit fittings, and splices. In addition, there are restrictions in the *NEC* on how to size the smaller tap conductors.

Factory-installed flexible metal conduit furnished and connected to these built-in cooking appliances will have the proper size and temperature-rated conductors for the particular appliance.

If you provide the flexible connections (whips) in the field, the minimum conductor sizing would be calculated as discussed above for each appliance. After the minimum conductor size is determined, we then must make sure that the conductors meet the "tap rule." This is important because the taps will have overcurrent protection rated greater than the ampacity of the tap conductors.

NEC 210.23(C) states that fixed cooking appliances that are fastened in place may be connected to 40- or 50-ampere branch circuits. If taps are to be made to a 40- or 50-ampere branch circuit, the taps must be able to carry the load, and in no case may the taps be less than 20 amperes, *210.19(A)(3)*.

For example, the 50-ampere circuit shown in Figure 20-1 has 10 AWG THHN tap conductors for the oven, and 8 AWG THHN tap conductors for the counter-mounted cooking unit. Each of these taps has an ampacity of greater than 20 amperes.

According to *Table 310.15(B)(16)*, a 6 AWG THHN conductor has an allowable ampacity of 55 amperes, using the 60°C column of the table.

USING A LOAD CENTER

The terms **load center** and **panelboard** are used interchangeably in the electrical trade. They are really one and the same thing. *Load center* is pretty much a manufacturers' term for panelboard. Panelboards might contain additional features that a typical residential load center might not have. Refer to Key Terms in the Appendix in this text for a more complete definition of "load center" and "panelboard."

In this residence, the built-in oven and range are very close to Panelboard B, in which case it is economical to install a separate branch circuit to each appliance. However, there are situations where the panelboard and the built-in cooking appliances are far apart.

The decision becomes one of installation cost (labor and material) as to whether to run separate branch circuits, or to run one larger branch circuit and use the tap rules.

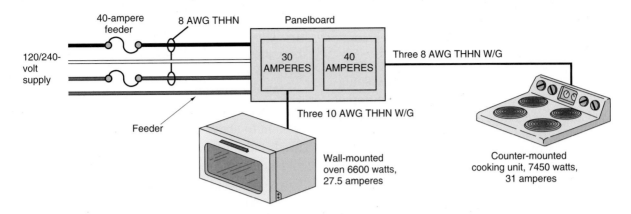

FIGURE 20-3   Appliances connected to a panelboard. (*Delmar/Cengage Learning*)

The tap rules can be ignored if a smaller panelboard is installed, as illustrated in Figure 20-3. There will be no taps—just branch circuits to each appliance sized for that particular appliance.

The calculations for the feeder from the main panelboard to another smaller panelboard are based on the two cooking appliances being treated as one. The conductors from the smaller panelboard are calculated for each appliance as previously discussed.

In homes with a basement, the panelboard could be installed below the kitchen where the appliances are installed.

CALCULATIONS WHEN MORE THAN ONE ELECTRIC RANGE, WALL-MOUNTED OVEN, OR COUNTER-MOUNTED COOKING UNIT IS SUPPLIED BY A FEEDER OR SERVICE

Although generally not an issue for wiring one- and two-family dwellings, the calculations for services and feeders of larger multifamily dwellings is more complex where a number of household cooking units are installed.

We need to refer to *Table 220.55.* This table has three columns. Column C is almost always used, with one exception. *Exception No. 3* shows that for household cooking units having a nameplate rating of more than 1¾ kW through 8¾ kW, Columns A and B are permitted to be used.

Column A is for household cooking units rated more than 1¾ kW but less than 3½ kW. Column B is for household cooking units rated 3½ kW through 8¾ kW.

When using Columns A and B, apply the demand factors for the number of household cooking units whose kW ratings fall within that column. Then add the results of each column together.

 EXAMPLE

Calculate the demand for a feeder that has four 3-kW wall-mounted ovens and four 6-kW counter-mounted cooktops connected to it.

Step 1: 4 ovens × 3 kW = 12 kW
12 kW × 66% = 7.92 kW
Step 2: 4 cooktops × 6 kW = 24 kW
24 kW × 50% = 12 kW
Step 3: 7.92 kW + 12 kW = 19.92 kW

Therefore, 19.92 kW is the demand load for the four ovens and four cooktops, even though the actual connected load is 36 kW. Diversity plays a large factor in arriving at the calculated demand load.

Review of Choices for Hooking Up Electric Cooking Appliances

- Run a separate branch circuit from the main panelboard to each appliance.
- Run one large branch circuit to a junction box(es), then make taps to serve each appliance. See Figure 20-1.

- Run one large feeder from the main panelboard to a panelboard located near the appliances, then run a separate branch circuit to each appliance. See Figure 20-3.

When one considers the complicated calculations, additional splices, taps, working with different size conductors, extra time and material, the best and simplest choice is probably to run a separate branch circuit from the main panelboard to each appliance.

MICROWAVE OVENS

Microwave ovens are available in countertop, over-the-range, and over-the-counter models and as part of a freestanding electric range.

The electrician's primary concern is to install the proper size and type of branch circuit and to provide proper equipment grounding for the appliance.

A freestanding electric range that includes a microwave oven is connected according to the methods previously discussed in this chapter.

Countertop microwave ovens are plugged into the most convenient receptacle. If the receptacles that serve the countertops have been wired according to the *NEC*, they would be supplied by at least two 20-ampere, small-appliance branch circuits, *210.11(C)(1)* and *210.52(B)(1)*.

Operating a microwave oven and another appliance at the same time might be too much load for a 20-ampere, small-appliance branch circuit. The branch-circuit breaker might trip, or the branch-circuit fuse might open. That is why microwave cooking appliance manufacturers suggest a dedicated 15- or 20-ampere branch circuit for the appliance.

Microwave ovens fastened to the underside of a cabinet above a countertop or range generally are cord-and-plug connected. They come with a rather short power supply cord. A receptacle outlet must be provided inside the upper cabinet. Instructions furnished with these microwave ovens describe where to position this receptacle. A receptacle installed inside the cabinet is *in addition* to the receptacles required to serve countertop surfaces, *210.52*. This receptacle must not be connected to any of the 20-ampere, small-appliance branch circuits that serve the countertop areas, *210.52(B)(2)*.

If a receptacle is installed inside the upper cabinet, the best choice is to connect the receptacle to an individual, dedicated, separate 120-volt, 20-ampere branch circuit for the microwave oven.

This receptacle is not required to be GFCI protected because it does not serve the countertop, *210.8(A)(6)*.

NEC 210.52(C)(5) states that receptacles that serve countertops shall not be located more than 20 in. (500 mm) above the countertop. A receptacle located inside the upper cabinets does not serve the countertop and therefore is not subject to *210.52(C)(5)*.

A cord-and-plug-connected appliance must not exceed 80% of the branch-circuit rating, *210.23(A)(1)*. A 20-ampere branch circuit may supply a maximum cord connected load of $20 \times 0.80 = 16$ amperes.

LIGHTWAVE ENERGY OVENS

A recent innovation is an oven that combines and cycles on and off lightwave energy from high-wattage halogen lamps and microwave energy. Cooking time is about 4 to 8 times faster than a conventional oven. These ovens are mounted to the underside of an upper cabinet or in a wall cabinet. Some require a dedicated 3-wire, 30-ampere, 120/240-volt circuit. A 4-wire, 30-ampere cord and plug is furnished with these ovens. A 30-ampere, 4-wire, NEMA 14-30R receptacle must be installed in the cabinet above the oven. Lower wattage units require a dedicated conventional 20-ampere, 120-volt branch circuit.

UL Standard 923 covers microwave and lightwave energy cooking appliances.

SURFACE HEATING ELEMENTS

Figures 20-4 and 20-5 illustrate typical electric range surface units.

Electric heating elements generate a large amount of radiant heat. Heat is measured in British Thermal Units (BTU). A BTU is defined as the amount of heat required to raise the temperature of one pound of water by 1 degree Fahrenheit. A 1000-watt heating element generates approximately 3412 BTU of heat per hour.

Heating elements in surface-type cooking units typically consist of spiral-wound nichrome

FIGURE 20-4 Typical 240-volt electric range surface heating element used with infinite heat controls. (*Delmar/Cengage Learning*)

FIGURE 20-5 Typical electric range surface unit of the type used with 7-position controls. (*Delmar/Cengage Learning*)

resistance wire carefully imbedded in magnesium oxide—a white, chalklike powder. The wire/magnesium oxide is then encased in a nickel-steel alloy sheath, flattened under very high pressure, and then formed into coils. In the world of electric ranges, we use the term *surface heating elements*. In the world of gas ranges, the term *burners* is used.

Surface heating elements with one-coil (one heating element) are used with infinite heat controls. A one-coil element is shown in Figure 20-6. Two-coil (two heating elements) surface unit heating elements are used with fixed-heat 3-, 5-, and 7-position controls. A 2-coil element is shown in Figure 20-7. Three-coil (three heating elements) surface unit heating elements are used with 3-position switches.

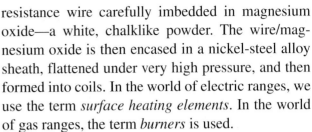

TEMPERATURE CONTROLS

Different types of controls are used to adjust the heating elements on electric ranges.

Infinite-Position Controls

Most electric ranges are equipped with infinite-position temperature controls. These controls contain electrical contacts that open and close constantly, "pulsing" the power supply to the heating element, thus maintaining the desired temperature of the surface heating unit.

Figure 20-6 shows the inner workings of this type of control.

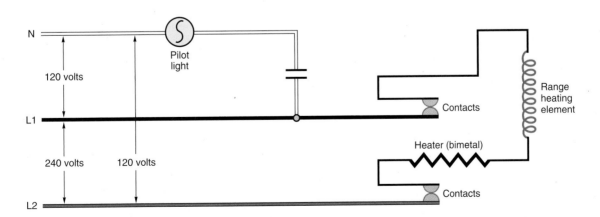

FIGURE 20-6 Typical internal wiring of an infinite heat surface element control. (*Delmar/Cengage Learning*)

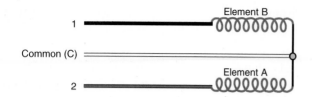

FIGURE 20-7 Typical surface unit wiring. (*Delmar/Cengage Learning*)

Fixed-Heat Controls

Still in use are rotary controls that have specific "indents" that lock the switch contacts into place as the knob is rotated to the various heat positions. Similar switching arrangements are found in push-button-type controls. These switches vary the connections to the heating elements individually, in series, in parallel, and in series/parallel to attain the various heat levels (high-low-medium, medium-high, etc.). These switches are commonly available in three, five, and seven positions.

Coil Size Selector Controls

Coil size selector controls are used with 3-coil heating elements. In position one, the center heating element comes on. In position two, the center and middle heating elements come on. In the third position, the center, middle, and outer heating elements come on. The fourth position is "Off."

Automatic Sensor Controls

This type of surface heating element has a sensor in the center of the unit. A special 12-volt control circuit is provided through a small transformer. When the control knob is set at a specific temperature, a "responder" pulses the heating element off and on to maintain the set temperature.

Caution When Selecting Conductor Size and Type

Recall from previous coverage in this chapter and in Chapter 4 that, according to UL Standards and *NEC 110.14(C)*, when conductor sizes 14 AWG through 1 AWG are installed, their allowable ampacity (current-carrying capacity) is found in the 60°C column of *Table 310.15(B)(16)*. If you are using nonmetallic-sheathed cable, the ampacity shall be in accordance with the 60°C conductor temperature rating, *334.80*.

That's why it is so important to read the appliance manufacturer's installation data carefully for recommended branch-circuit rating, conductor size, and conductor type.

Connecting Foreign-Made Appliances

Be careful when connecting foreign-made appliances. It is quite possible that the color coding of the terminals and conductors is different from what you are accustomed to. Throughout this text, we have discussed color coding as white for the grounded circuit conductor, black and red for the ungrounded "hot" conductor, and green or bare for the equipment grounding conductor. In this country, we use the term *ground*. In Europe, "ground" is referred to as "earth." Instructions might indicate something like this:

Green, green/yellow, or bare = earth

    Blue   = neutral

    Brown   = live

The equipment grounding terminal might be marked with the letter "E." Just be aware that there could be differences in the identification of the terminals and conductors.

REVIEW

Note: Refer to the *Code* or the plans where necessary.

⟞⟋\/\⊣ COUNTER-MOUNTED COOKING UNIT CIRCUIT ⒶG

1. a. What circuit supplies the counter-mounted cooking unit in this residence?

    _____

    b. What is the rating of this circuit? _____

2. What three methods may be used to connect counter-mounted cooking units?

    _____

    _____

3. Is it permissible to use standard 60°C insulated conductors to connect all counter-mounted cooking units? Why? _____

    _____

4. What is the maximum operating temperature (in degrees Celsius) for a

    a. Type TW conductor? _____

    b. Type THW conductor? _____

    c. Type THHN conductor? _____

5. The *NEC* permits the following wiring methods to supply household built-in ovens and ranges. True or false? (Circle the correct answer.)

    a. Nonmetallic-sheathed cable   (True)   (False)

    b. Armored cable   (True)   (False)

    c. Flexible metal conduit   (True)   (False)

6. For electric ranges that have a calculated demand of 8¾ kW or more, the neutral conductor can be reduced to _____ % of the branch-circuit rating but shall not be smaller than _____ AWG.

7. One kilowatt equals _____ BTU per hour.

8. Older style electric ranges had heat control knobs that had "indents" that could be felt when rotating the knob to adjust temperature to three, five, or seven different heat positions. Most new electric ranges have control knobs that do not have "indents." This type of heat control is known as an _____ heat control.

9. When a separate circuit supplies a counter-mounted cooking unit, what *Code* reference tells us that it is "against *Code*" to apply the demand factors of *Table 220.55* and requires us to calculate the load based on the appliance's actual nameplate rating?

    _____

⎓⎓⎓ WALL-MOUNTED OVEN CIRCUIT Ⓐ_F AND FREESTANDING RANGE

1. To what circuit is the wall-mounted oven connected? _____

2. An oven is rated at 7.5 kW. This is equal to

   a. _____ watts.

   b. _____ amperes at 240 volts.

3. a. What section of the *Code* governs the grounding of a wall-mounted oven? _____

   _____

   b. By what methods may wall-mounted ovens be grounded? _____

   _____

   _____

4. What is the type and ampere rating of the overcurrent device protecting the wall-mounted oven in this residence? *Hint:* Refer to the Schedule of Special Purpose Outlets in the Specifications. _____

   _____

5. Approximately how many ft (m) of cable are required to connect the oven in the residence? _____

   _____

6. When connecting a wall-mounted oven and a counter-mounted cooking unit to one feeder, how long are the taps to the individual appliances? _____

   _____

7. The branch-circuit load for a single wall-mounted self-cleaning oven or counter-mounted cooking unit shall be the _____ rating of the appliance.

8. A 6-kW counter-mounted cooking unit and a 4-kW wall-mounted oven are to be installed in a residence. Calculate the maximum demand according to *Column C, Table 220.55*. Show all calculations. Both appliances will be connected to the same branch circuit.

9. The size of the neutral conductor supplying an electric range may be based on _____ % of the branch-circuit rating.

10. A freestanding electric range is rated 11.8 kW, 120/240 volts. The wiring method is nonmetallic-sheathed cable. Use the 60°C column of *Table 310.15(B)(16)* in conformance to *110.14(C)*. Answer the following questions:

   a. According to Column C, *Table 220.55*, what is the maximum demand? _____ kW

   b. What is the minimum size for the ungrounded copper conductors? _____ AWG

   c. What is the minimum size neutral copper conductor? _____ AWG

   d. What is the correct rating for the branch-circuit overcurrent protection? _____ amperes

11. A double-oven electric range is rated at 18 kW, 120/240 volts. Calculate the maximum demand according to *Table 220.55*. Show all calculations.

12. For the range discussed in Question 11:

    a. What size 60°C ungrounded conductors are required? _____

    b. What size 60°C neutral conductor is required? _____

    c. What size 75°C ungrounded conductors are required? The terminal block and lugs on the range, the branch-circuit breaker, and the panelboard are marked as suitable for 75°C wire. _____

    d. What size 75°C neutral conductor is required? Terminations are the same as in part (c). _____

13. For ranges of 8¾ kW or higher rating, the minimum branch-circuit rating is _____ amperes.

14. A nonmetallic-sheathed cable is used to connect a wall-mounted oven. The insulated conductors are 10 AWG. What is the size of the equipment grounding conductor in this cable? _____

15. Match the following statements with the correct letter. Some letters may be used more than once.

    _____ A 30-ampere, 3-wire receptacle   a. for the equipment grounding conductor

    _____ A 50-ampere, 4-wire receptacle   b. for the white neutral conductor

    _____ The L-shaped blade on a 30-ampere plug cap   c. for the "hot" ungrounded conductors

    _____ The horseshoe-shaped slot on a 50-ampere receptacle   d. NEMA 10-30R

                      e. NEMA 14-50R

    _____ A 50-ampere, 4-wire plug cap   f. NEMA 14-50P

    _____ Terminals "X" and "Y"

    _____ Terminal "W"

    _____ Terminal "G"

16. A receptacle is mounted inside of a kitchen cabinet for the purpose of plugging in a microwave oven. Circle the correct answer.

    a. The receptacle may be connected to the same circuit that supplies the other receptacles that serve the countertop areas.

    b. The receptacle shall not be connected to the same circuit that supplies the other receptacles that serve the countertop areas.

Special-Purpose Outlets— Food Waste Disposer ▲H, Dishwasher ▲I

OBJECTIVES

After studying this chapter, you should be able to

- install circuits for a typical food waste disposer and dishwasher.

- know the meaning of "continuous feed" and "batch feed" food waste disposers.

- understand direct connections (hard wired) and cord-and-plug connections.

- understand the meaning of branch-circuit protection and overload protection.

- be aware of the requirements for providing a means to disconnect appliances.

- understand the various acceptable methods for grounding appliances.

- understand the significance of water temperature for dishwashers.

This chapter discusses the circuit requirements for food waste disposers (also called kitchen waste disposer, garbage disposer, garbage disposal, waste disposer, In-Sink-Erator®, Disposall®) and dishwashers of the type normally installed in homes. *Code* requirements for appliances are found in *Article 422* of the *NEC*.

These 120-volt major appliances are *not* permitted to be connected to the 20-ampere small-appliance branch circuits that serve the receptacles above the countertops in the kitchen.

FOOD WASTE DISPOSER ⌁ₕ

Types of electrical connections are shown in Figures 21-1 and 21-2. The food waste disposer outlet is shown on the plans by the symbol ⌁ₕ. It is rated 120 volts, 7.2 amperes, and it is connected to Circuit B19, a 20-ampere, 120-volt branch circuit that originates in Panelboard B.

Controls

Continuous feed food waste disposers are usually controlled by a single-pole wall switch conveniently located above the countertop, as shown in Figure 21-2. The branch-circuit wiring is run first to the switch box. A second cable is then run from the switch box to the junction box on the disposer.

"Batch-feed" food waste disposers are equipped with an integral switch that starts and stops the disposer when the user twists the drain cover into place after filling the disposer with waste food. A wall switch is not required. See Figure 21-3.

Although rarely used in residential installations, a plumber could install a flow switch in the cold water line under the sink to ensure that the food waste disposer is not run without water, Figure 21-4. Generally, water should be left running for at least 15 seconds after shutting off the disposer to flush the drain clear of food waste particles.

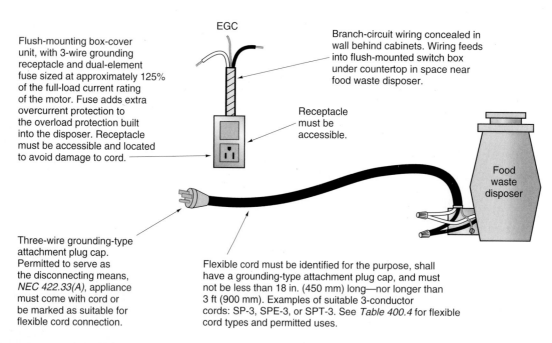

EGC

Flush-mounting box-cover unit, with 3-wire grounding receptacle and dual-element fuse sized at approximately 125% of the full-load current rating of the motor. Fuse adds extra overcurrent protection to the overload protection built into the disposer. Receptacle must be accessible and located to avoid damage to cord.

Branch-circuit wiring concealed in wall behind cabinets. Wiring feeds into flush-mounted switch box under countertop in space near food waste disposer.

Receptacle must be accessible.

Food waste disposer

Three-wire grounding-type attachment plug cap. Permitted to serve as the disconnecting means, *NEC 422.33(A)*, appliance must come with cord or be marked as suitable for flexible cord connection.

Flexible cord must be identified for the purpose, shall have a grounding-type attachment plug cap, and must not be less than 18 in. (450 mm) long—nor longer than 3 ft (900 mm). Examples of suitable 3-conductor cords: SP-3, SPE-3, or SPT-3. See *Table 400.4* for flexible cord types and permitted uses.

FIGURE 21-1   A cord-and-plug-connected food waste disposer, *422.16(A)* and *(B)(1)*. The same rules apply to built-in dishwashers and trash compactors, except the cord length must be 3 to 4 ft (900 mm to 1.2 m). The receptacle/fuse box-cover unit or a conventional receptacle could be used. A receptacle under the kitchen sink is not required to be GFCI protected because it does not serve the countertop, *210.8(A)(6)*. (*Delmar/Cengage Learning*)

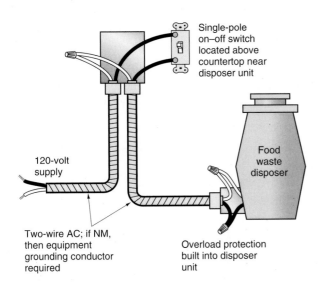

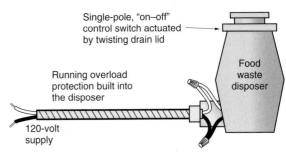

FIGURE 21-2   Wiring for a food waste disposer operated by a separate switch located above the countertop near the sink. Many local codes require this wall switch for all installations to be sure that there is a safe means of easily disconnecting the appliance. (*Delmar/Cengage Learning*)

FIGURE 21-3   Wiring for a "batch-feed" food waste disposer with an integral "ON–OFF" switch. (*Delmar/Cengage Learning*)

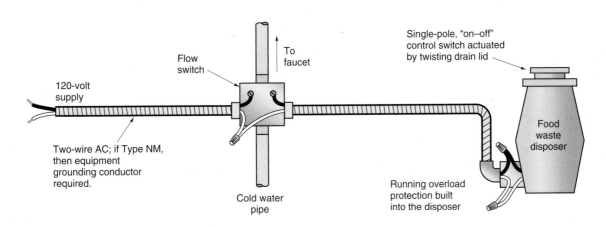

FIGURE 21-4   "Batch-feed" disposer with a flow switch in the cold water line. (*Delmar/Cengage Learning*)

DISHWASHER ▲_I

The dishwasher outlet is shown on the plans by the symbol ▲_I. The dishwasher is connected to Circuit B5, a 20-ampere, 120-volt branch circuit that originates in Panelboard B. A typical dishwasher contains a motor; an electric heating element; and, in some models, a small fan to assist in circulating the hot air during the drying cycle. For example, one manufacturer's dishwasher total connected load as marked on the nameplate is 9.2 amperes. This includes the motor, heating element, controls, relays, and an 875-watt heating element.

In most residential dishwashers, the motor and the booster heater operate at the same time. The heating element boosts the incoming hot water from a typical water heater setting of 120°F (49°C) to approximately 140°F (60°C) or higher so as to better clean, dissolve detergents, and somewhat sanitize the dishes. During the dry cycle, only the heating element is on, plus a circulating fan if so equipped.

CODE RULES COMMON TO THE FOOD WASTE DISPOSER AND THE DISHWASHER

Branch-Circuit Rating

The manufacturer will specify the recommended rating of the branch circuit. Although it is possible to connect both appliances to one branch circuit, much depends on the current draw of the appliances. The preferred way to connect these appliances is to supply each appliance with a separate 20-ampere branch circuit. This provides a little spare capacity if the homeowner should decide to add a booster water heater at the sink location later on.

Do not connect the food waste disposer or dishwasher to any of the required 20-ampere small-appliance branch circuits in the kitchen. This is a violation of *210.52(B)(2)*. Small-appliance circuits are intended to serve countertops for cord-connected *portable* appliances only. See Chapter 12.

Conductors

Per the Specifications in the back of this text, the branch-circuit conductors are 12 AWG. The wiring method might be nonmetallic-sheathed cable, armored cable, or EMT, depending on the wiring method required by the code in your locality.

Direct Connection

See Figures 21-2, 21-3, and 21-4. This connection is sometimes referred to as being "hard wired." These appliances have a junction box (wiring compartment) for the electrical hookup. The branch-circuit supply (cable or flex conduit) is run directly to this box. During the rough-in stage of new wiring, the electrician must make sure that the supply cable or flex conduit is brought in at the proper location. This might be under the sink, brought up from underneath, or brought in behind the appliance. A

sufficient length of wiring is necessary to be able to make the electrical connections and to position the appliance into place. The manufacturer's installation instructions provide all of the necessary information as to where to bring in the power supply.

Cord-and-Plug Connection

See Figure 21-1. Cord-and-plug connection is permitted by *422.16(B)(2)*.

If 3-wire cord-and-plug connection is desired for easy disconnecting and servicing, the appliance manufacturers make available a power supply cord kit for this purpose. The appliances would have to be listed as being intended or identified for flexible cord connection, *NEC 400.7* and *422.16(A)* and *(B)*.

UL lists these appliances for connection with and without a cord-and-plug assembly.

Cord lengths are restricted by the *NEC* and UL Standards:

- Food waster disposers: no shorter than 18 in. (450 mm) and no longer than 3 ft (900 mm)
- Dishwashers: no shorter than 3 ft (0.9 m) and no longer than 4 ft (1.2 m)

A duplex grounding-type receptacle is usually installed in an accessible location under the sink for plugging in the dishwasher and food waste disposer. This receptacle does not have to be GFCI protected because it does not serve the countertop, *210.8(A)(6)*.

Remember that *210.23(A)(2)* limits cord-connected equipment to 80% of the branch-circuit ampere rating.

Overload Protection

Overload protection is an integral part of the appliance. Overload protection prevents the motor from burning out if it stalls or overheats for any reason. This protection is required in the UL Standard.

Branch-Circuit Protection

Branch-circuit short-circuit and ground-fault protection is provided by the branch-circuit fuse or circuit breaker, sized according to the manufacturers' recommendations. In our case, each appliance is fed by a separate 20-ampere branch circuit.

Disconnecting Means

All appliances must be provided with some means of disconnecting the appliance, *Article 422, Part III.* The branch-circuit breaker can serve as the disconnecting means if it is within sight of the appliance or is capable of being locked in the "OFF" position. Rarely, if ever, are these appliances within sight of the breaker panel. Lock-off provisions must be permanently installed on the breaker or switch, *422.31(B).* Some local codes actually require that these appliances be cord-and-plug connected so the appliances can easily be disconnected for servicing by pulling out the plug. This requires that a receptacle be installed—usually under the sink in an accessible location.

Figure 21-1 shows a cord-and-plug connection that serves as the disconnecting means, *422.33(A).*

Figure 21-2 shows a wall switch that serves as the disconnecting means.

If you do install a receptacle(s) below the sink or behind the appliance for cord-and-plug connection of a dishwasher or food waste disposer, the receptacle(s) installed do not have to be GFCI protected because they are not there to serve countertops.

Grounding

- *Article 250, Part VI,* covers equipment grounding and equipment grounding conductors. In *Part VI,* we find:
  - *250.114*: Grounding requirements for cord-and-plug-connected appliances.
  - *250.118*: Listing of all the acceptable methods of grounding equipment, such as the equipment grounding conductor in nonmetallic-sheathed cable, armored cable, or flexible metal conduit.
  - *250.119*: Requirements for the equipment grounding conductor to be bare or have green insulation, or have green insulation with one or more yellow stripes.
  - *250.122*: The minimum size for equipment grounding conductors.
  - *250.126*: Requirement for the equipment grounding conductor terminal to be bare or green and hexagon shaped.

- *Article 250, Part VII* covers methods of equipment grounding. In *Part VI,* we find:
  - *250.134*: Discussion of grounding of equipment fastened in place, or "hard-wired"; referral back to *250.118.*
  - *250.138*: Acceptance of the equipment grounding conductor in flexible cords for cord-and-plug-connected appliances.
  - *250.142(B)*: Grounding appliances to the grounded circuit conductor is not permitted.

Supplemental Overcurrent Protection and Disconnecting Means

Instead of installing a duplex grounding-type receptacle under the sink, electricians will sometimes install a box-cover "switch/fuse unit" under the sink of a type similar to that shown in Figure 16-25. For a cord-and-plug connection, the box-cover unit would be a "receptacle/fuse" type, as shown in Figure 21-1. This receptacle/fuse unit does not have to be GFCI protected because it does not serve the countertop. Box-cover units provide both supplemental overload and short-circuit protection. They also serve as a convenient accessible disconnecting means, as required by *422.30,* making it easy for the technician to be assured that the circuit is off when working on the dishwasher. Time-delay fuses in this box-cover unit would be sized at approximately 125% of the nameplate rating of the dishwasher.

Read the Installation Instructions

Always read the installation instructions, the owner's manual, and the nameplate for specific electrical requirements.

PORTABLE DISHWASHERS

In addition to built-in dishwashers, portable models are available. Most portable models are convertible to become built-in units. Portable dishwashers have one hose that connects to the water faucet and a water drainage hose that hangs in the sink. The dishwasher probably will be plugged into the receptacle nearest the sink.

Portable dishwashers are supplied with a 3-wire cord and plug. If the 3-wire plug cap is plugged into a properly connected 3-wire grounding-type receptacle, the dishwasher is adequately grounded.

In older homes that do not have grounding-type receptacles, the use of a portable cord-and-plug-connected dishwasher will require the installation of a grounding-type receptacle. Assuming the appliance will be used in the kitchen, the receptacle will also have to be GFCI protected, *210.8(A)(6)*. This topic is covered in Chapter 6.

Cord-and-plug-connected portable dishwashers are generally rated 10 amperes or less at 120 volts.

WATER TEMPERATURE

Electric water heaters are shipped with the thermostat set as low as possible. When the water heater is installed, the thermostat should be reset to recommended settings such as 120°F (49°C). In fact, because of the problem of people being scalded, many municipalities and states have implemented code requirements that water heater thermostats be set at not over 120°F (49°C). Although this temperature is acceptable for baths and showers, 120°F (49°C) is not hot enough to properly clean dishes and glasses. Some manufacturers of dishwashers provide a feature that turns on the heating element for the final rinse cycle. This raises the water temperature from 120°F (49°C) to the desired 140°F (60°C)–145°F (63°C).

See Chapter 19 for a discussion on water heaters and the dangers of scalding with hot water of varying temperatures.

REVIEW

Note: Refer to the *Code* or the plans where necessary.

FOOD WASTE DISPOSER CIRCUIT ⒶH

1. How many amperes does the food waste disposer draw? _____

2. a. To what circuit is the food waste disposer connected?

   _____

   b. What size wire is used to connect the food waste disposer?

   _____

3. Means must be provided to disconnect the food waste disposer. The homeowner need not be involved in electrical connections when servicing the disposer if the disconnecting means is _____

   _____

4. How is motor overload protection provided in most food waste disposer units?

   _____

5. When running overcurrent protection is not provided by the manufacturer or if additional backup overcurrent protection is wanted, dual-element, time-delay fuses may be installed in a separate box-cover unit. These fuses are sized at not over _____ % of the full-load rating of the motor.

6. Why are flow switches sometimes installed on food waste disposers? _____

_____

7. Where in the *NEC* would you look for basic *Code* rules relating to grounding appliances?

_____

8. Do the plans show a wall switch for controlling the food waste disposer?

_____

9. A separate circuit supplies the food waste disposer in this residence. Approximately how many ft (m) of cable will be required to connect the disposer?

_____

10. If a receptacle is installed underneath the sink for the purpose of plugging in a cord-connected food waste disposer, must the receptacle be GFCI protected?

_____

─⋀⋀─ DISHWASHER CIRCUIT Ⓐ I

1. a. To what circuit is the dishwasher in this residence connected? _____

    b. What size wire is used to connect the dishwasher? _____

2. The dishwasher in this residence is rated (circle the correct answer)

    a. 5 A.

    b. 9.2 A.

    c. 15 A.

3. The heating element is rated at (circle the correct answer)

    a. 875 watts.

    b. 1000 watts.

    c. 1250 watts.

4. How many amperes at 120 volts do the following heating elements draw?

    a. 750 watts _____

    b. 1000 watts _____

    c. 1250 watts _____

5. How is the dishwasher in this residence grounded? _____

_____

_____

_____

_____

6. What type of cord is used on most portable dishwashers?

_____

7. How is a portable dishwasher grounded? _____

_____

_____

8. Who is to furnish the dishwasher?

_____

_____

9. What article of the *Code* specifically addresses electrical appliances?

_____

Special-Purpose Outlets for the Bathroom Ceiling Heat/Vent/Lights ▲K ▲J, the Attic Fan ▲L, and the Hydromassage Tub ▲A

OBJECTIVES

After studying this chapter, you should be able to

- explain the operation and control of heat/vent/lights and make electrical connections in conformance to the *NEC*.

- understand the operation and control of attic exhaust fans and make electrical connections in conformance to the *NEC*.

- understand humidity problems, solutions, and humidistats.

- understand *NEC* requirements for hydromassage bathtub branch circuits.

BATHROOM CEILING HEATER CIRCUITS ▲K ▲J

Both bathrooms contain a combination heater, light, and exhaust fan installed in the ceiling. The heat/vent/light is shown in Figure 22-1. The symbols ▲K and ▲J represent the outlets for these units.

Each heat/vent/light contains a heating element similar to the surface burners of electric ranges but usually has fins attached that are similar in purpose to the fins on a radiator for a car or truck. The appliances also have a single-shaft motor with a blower wheel, a lamp with a diffusing lens, and a means of discharging air to the outside of the dwelling.

The heat/vent/lights specified for this residence are rated 1500 watts at 120 volts. The unit in the Master Bedroom bathroom ▲J is connected to Circuit A12, a 20-ampere, 120-volt circuit. In the Front Bedroom bathroom, the unit is connected to Circuit A11, also a 20-ampere, 120-volt circuit.

The current rating of the 1500-watt unit is

$$I = \frac{W}{E} = \frac{1500}{120} = 12.5 \text{ amperes}$$

Article 424 of the *NEC* covers fixed electric space heating. *NEC 424.3(B)* considers fixed electric space heating to be a continuous load. Simply stated, this means that the 125% multiplier must be applied for the branch-circuit conductors, *210.19(A)*, and the branch-circuit overcurrent device, *210.20(A)*.

Therefore,

12.5 amperes × 1.25 = 15.625 amperes.

We have chosen to supply each of the bathroom ceiling heaters with a 20-ampere, 120-volt branch circuit. The 12 AWG conductors and the 20-ampere circuit breakers "Meet *Code*."

Wiring

At first, the connections look rather complicated. Let's walk through it!

In Figure 22-2, we see the wiring diagram for the heat/vent/light. The 120-volt line is run from Panel A to the wall box. A raceway (flexible or electrical metallic tubing) is run between the wall box where the switches are located to the heat/vent/light unit. In this raceway, five wires are needed—one common white grounded circuit conductor, and one conductor each for the heating element, fan motor, 100-watt lamp, and the 7-watt night light. The four ungrounded switch leg wires may be any color except white, gray, or green. Color coding of conductors is discussed in detail in Chapter 5. Because the circuit rating is 20 amperes, 12 AWG conductors are installed. See the Schedule of Special-Purpose Outlets contained in the Specifications for this text.

Nontrained electricians often violate the code by running one 3-wire and one 2-wire cable between the wall box and heat/vent/light. *NEC 300.3(B)*

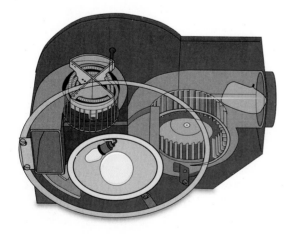

FIGURE 22-1 Heat/vent/light. (*Photo courtesy of Broan-NuTone*)

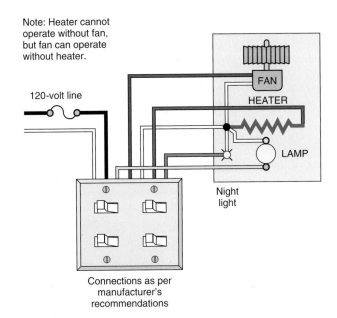

Note: Heater cannot operate without fan, but fan can operate without heater.

120-volt line

Night light

Connections as per manufacturer's recommendations

FIGURE 22-2 Wiring for heat/vent/light.
(*Delmar/Cengage Learning*)

states that *all conductors of the same circuit and, where used, the neutral and all equipment grounding conductors shall be contained within the same raceway, cable tray, trench, cable, or cord.**

The switch location requires a 2-gang opening, with the wall box being large enough to accommodate seven 12 AWG conductors, four switches, and cable clamps. For this installation, a 4-in. square, 2⅛ in. deep box with a 2-gang raised plaster ring is used. This provides ample room for the conductors and switches. Box sizing is covered in Chapter 2 and Chapter 8.

Operation of the Heat/Vent/Light

The heat/vent/light is controlled by four switches, as shown in Figure 22-2:

one switch for the heater
one switch for the light
one switch for the exhaust fan
one switch for the night light

When the heater switch is turned on, the heating element begins to give off heat. The heat activates a bimetallic coil attached to a damper section in the housing of the unit. The heat-sensitive coil expands

*Reprinted with permission from NFPA 70-2011.

until the damper closes the discharge opening of the exhaust fan. Air is taken in through the outer grille of the unit. The air is blown downward over the heating element. The blower wheel circulates the heated air back into the room area.

If the heater is turned off so that only the exhaust fan is on, the air is pulled into the unit and is exhausted to the outside of the house by the blower wheel.

Exhaust fans and luminaires are available with integral automatic moisture sensors and motion detectors. Step into a dark bathroom and the light goes on. Step into and run the shower, and the exhaust fan turns on at a predetermined moisture level.

Grounding the Heat/Vent/Light

The heat/vent/light must be properly grounded!

Proper grounding is accomplished by using any of the accepted equipment grounding methods found in *250.118*. For residential wiring, the equipment grounding conductor in nonmetallic-sheathed cable, the armor and bonding wire in armored cable, or flexible metal conduit provides the necessary equipment ground. As stated many times before, NEVER ground an appliance to the white grounded circuit conductor, *250.142(B)*.

ATTIC EXHAUST FAN CIRCUIT ⓐL

Read the Installation Instructions!

Before beginning any installation of electrical equipment, always refer to manufacturers' literature for specific installation instructions.

The attic exhaust fan in this residence has a 2-speed, ¼-horsepower direct-drive motor. "Direct-drive" means that the fan blade is attached directly to the shaft of the motor without the use of pulleys and belts.

The exhaust fan is mounted in the hall ceiling between the Master Bedroom and the Front Bedroom, as shown in Figure 22-3(A). Exhaust fans can also be installed in a gable of a house, as shown in Figure 22-3(B), or in the roof.

When the fan is not running, the louvers on the fan and house gable remain in the "CLOSED" position.

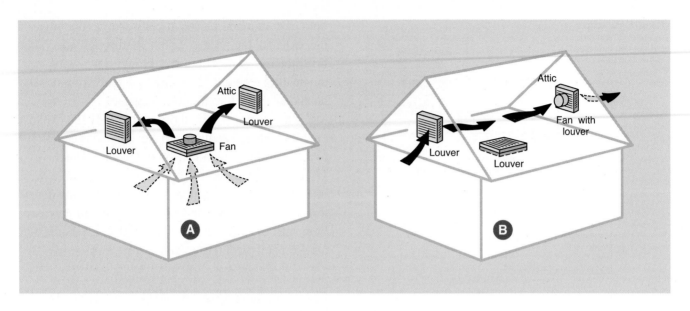

FIGURE 22-3 Exhaust fan installation in an attic. (*Delmar/Cengage Learning*)

When running, the louvers on the fan and house gable "OPEN," removing hot, stagnant, humid, or smoky air from the house by drawing in fresh air through open windows and doors, in turn exhausting the air through the louvers.

An exhaust fan can lower the indoor temperature of the house as much as 10°F to 20°F (–6°C to –11°C). On hot days, the air in an attic can reach superheated temperatures of 150°F (66°C) or more. The heat from the attic will radiate and conduct through the ceiling into the living areas. This results in increased loading of the air-conditioning system. These problems are minimized by properly vented exhaust fans. Humidity issues are discussed later on.

Bathroom exhaust fans, vents for clothes dryers, and similar high-humidity sources are *never* to be vented directly into the attic or other dead spaces. They are to be vented to the outside.

Sizing a Home-Type Exhaust Fan

Exhaust fan ventilating capability is rated in cubic feet of air per minute (CFM).

To determine the minimum amount of CFM needed, multiply the square footage of the house by 3. For example, an 1800-ft^2 house would need an exhaust fan rated $1800 \times 3 = 5400$ CFM.

To determine the minimum amount of attic exhaust louver square footage area, divide the CFM by 750. In this example, $5400 \div 750 = 7.2$ ft^2.

Never exhaust air into spaces within walls, ceilings, attics, crawl spaces, or garages. The humidity may damage the structure and insulation.

Figure 22-4 shows a typical ceiling-mounted whole-house ventilator exhaust fan.

Branch Circuit

The exhaust fan in this residence is rated 115 volts nominal, 5.8 amperes (696 VA). It is connected to Circuit A10, a 120-volt, 15-ampere circuit breaker in the main panel. This branch circuit supplies the attic exhaust fan only. It supplies no other loads.

Conductors (*430.22*)

The *NEC* requires that the minimum size conductor is to be not less than 125% of the motor's full-load current rating.

Therefore,

$$5.8 \times 1.25 = 7.25 \text{ amperes}$$

Table 310.106(A) shows a minimum size conductor for general wiring to be 14 AWG copper.

FIGURE 22-4   Ceiling exhaust fan as viewed from the attic. (*Courtesy of Broan-NuTone*)

Table 310.15(B)(16) shows that a 14 AWG THHN copper has an allowable ampacity of 15 amperes in the 60°C column.

In conformance to these requirements, the Schedule of Special-Purpose Outlets tells us that the branch-circuit conductors are 14 AWG conductors. The branch-circuit cable is run to a 4-in. square deep outlet box installed by the electrician near the exhaust fan. A switch/fuse box-cover unit is to be mounted on this outlet box. From this box, a flexible connection like armored cable, flexible metal conduit, or nonmetallic-sheathed cable is made to the factory-installed junction box on the frame of the fan. Depending on the size of junction box on the fan assembly, it might be possible to run the branch-circuit cable directly into that junction box.

Overload Protection (*430.31 through 430.44*)

The exhaust fan motor is provided with integral (built into the motor) overload protection. Overload protection prevents the motor from burning out if it stalls or overheats for any reason. This protection is required in the UL Standard.

NEC 430.32(C) states that for automatically started motors, the overload protection must not exceed 125% of the full-load running current of the motor. *Table 430.248* shows that a ¼-horsepower motor has a full-load current draw of 5.8 amperes. Thus, the overload device rating is 5.8 × 1.25 = 7.25 amperes. The motor's built-in overload protection meets this requirement.

Motor Branch-Circuit Short-Circuit and Ground-Fault Protection (*430.52, Table 430.52*)

Motor branch-circuit short-circuit and ground-fault protection is covered in Chapter 19. Here is a brief review.

For a circuit breaker, *Table 430.52* indicates a maximum rating of 250% of the motor's full-load current rating. Therefore, $5.8 \times 2.5 = 14.5$ amperes. The next standard rating is permitted by *430.52(C)(1)*. This is a 15-ampere circuit breaker according to *240.6(A)*.

For a dual-element, time-delay fuse, *Table 430.52* indicates a maximum rating of 175% of the motor's full-load current rating. Therefore, $5.8 \times 1.75 = 11+$. The next standard rating is permitted by *430.52(C)(1)*. This is a 15-ampere, dual-element, time-delay fuse according to *240.6(A)*.

There are exceptions to the above rules that permit even higher rated fuses or circuit breakers under certain conditions as stated in *430.52(C)*.

Disconnecting Means (*Article 430, Part IX*)

A switch/fuse box-cover unit is mounted on the 4-in. square outlet box. This meets the requirement that the disconnecting means shall be in sight of the motor, *430.102(B)*. Anyone working on the exhaust fan can easily turn off the power. The 15-ampere circuit breaker in the main panel also serves as a disconnecting means since it can be locked in the "Off" position.

Grounding (*Article 250, Part VI*)

The exhaust fan must be properly grounded!

Proper grounding is accomplished by using any of the accepted equipment grounding methods found in *250.118*. For residential wiring, the equipment grounding conductor in nonmetallic-sheathed cable, the armor and bonding wire in armored cable, or flexible metal conduit provides the necessary equipment ground. DO NOT ground the exhaust fan metal housing to the white grounded circuit conductor (neutral), *250.142(B)*.

Supplemental Overcurrent Protection

Integral overload protection in the motor is the first line of defense against motor burnout and potential fire. A dual-element, time-delay fuse installed in the switch/fuse box-cover unit adjacent to the exhaust fan per the specifications provides

backup overload protection as the second line of defense against motor burnout. Dual-element, time-delay fuses are available in many ampere ratings, such as 7½, 8, 9, 10, and 12 amperes.

Should the motor fail to start, short-circuit, or go to ground, the fuse in the box-cover switch/fuse unit will open without tripping the branch-circuit breaker back in Panel A. This makes it easy for servicing the unit.

Refer to Table 28-1 in this text for much more information relating to fuses and circuit breakers.

Fan Control

Typical residential fan controls are illustrated in Figures 22-5 and 22-6. These may be mounted adjacent to or above the other wall-mounted lighting switches.

Figure 22-7 shows many additional options available for controlling exhaust fans.

A. A simple "On–Off" switch.

B. A speed control switch that allows multiple and/or an infinite number of speeds. (The exhaust fan in this residence is controlled by this type of control.) See also Figure 22-5.

C. A timer switch (S_T) that allows the user to select how long the exhaust fan is to run, up to 12 hours. See also Figure 22-6.

FIGURE 22-5 A typical 2-speed fan control. Also available in 3-speed, 4-speed, and totally variable speed electronic types. (*Courtesy of Broan-NuTone*)

A   **B**

FIGURE 22-6 (A) A manual timer fan control. Some timer controls have a continuous "On" position. (*Courtesy of Broan-NuTone*) (B) An electronic timer fan control. (*Courtesy of Leviton*)

Figure 22-6(A) shows a spring-wound timer and Figure 22-6(B) an electronic timer. The electronic timer requires a connection to the branch circuit neutral.

D. A humidity control (H) switch that senses moisture buildups. See "Humidity Control" later in this chapter for further details on this type of switch.

E. Exhaust fans mounted into end gables or roofs of residences are available with an adjustable temperature control on the frame of the fan. This thermostat generally has a start range of 70°F to 130°F (21°C to 54°C), and will automatically stop at a temperature 10°F (−12°C) below the start setting, thus providing totally automatic "On–Off" control of the exhaust fan.

F. A high temperature automatic heat sensor that will shut off the fan motor when the

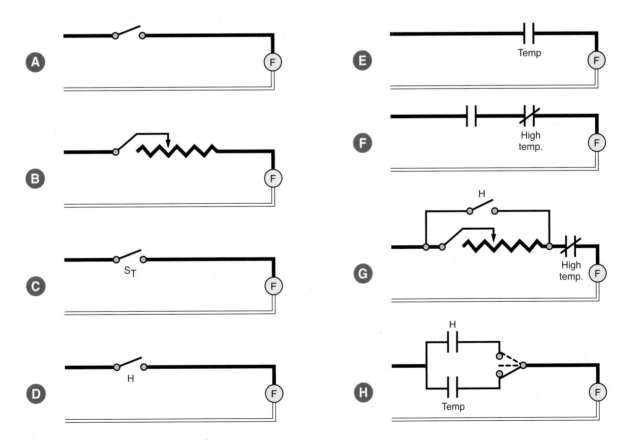

FIGURE 22-7 Different options for the control of exhaust fans. (*Delmar/Cengage Learning*)

temperature reaches 200°F (93°C). This is a safety feature so that the fan will not spread a fire. Connect in series with other control switches.

G. A combination of controls. This circuit shows an exhaust fan controlled by an infinite-speed switch, a humidity control (H), and a high temperature heat sensor. Note that the speed control and the humidity control are connected in parallel so that either can start the fan. The high temperature heat sensor is in series so that it will shut off the power to the fan when it senses 200°F (93°C), even if the speed control or humidity control is in the "On" position.

H. This circuitry combines a humidistat (D) and a thermostat (E) in such a way that the homeowner can select between controlling humidity and temperature in the attic. A selector switch is installed in a readily accessible place for ease of use by the homeowner. The switch has a "Center off" position. The humidity and temperature sensors are mounted in the attic.

Any of the above variations for controlling an exhaust fan could also include a pilot light to give clear visual indication that the fan is running. This pilot light could be mounted in the same switch box as the speed control or timer control, or it could be mounted next to these switches.

〰️🗻 HUMIDITY

Humidity can be a big problem in homes.

Well-insulated homes can experience problems with excess humidity due to the "tightness" of the house. Thermal insulation and vapor barriers that keep moisture out of the insulation must be properly installed. Insulation must be kept dry or its efficiency decreases. Whole house vinyl wraps (Tyvek®) stop airflow through wall cavities, as well as holding out wind-driven rain from getting behind wood, vinyl, brick, and stucco siding.

High humidity is uncomfortable, but a bigger problem is that high humidity promotes the growth of mold and the deterioration of fabrics and floor coverings. Framing members, wall panels, plaster, or drywall may also deteriorate because of the humidity.

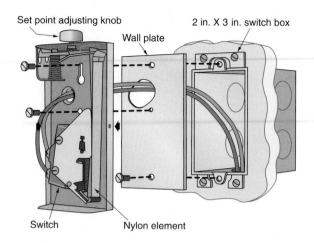

FIGURE 22-8   Details of a humidity control used with an exhaust fan. (*Delmar/Cengage Learning*)

Instead of manual "On–Off" control of an exhaust fan, a desired humidity level can be maintained automatically with a device called a humidistat. Humidistats have a humidity-sensitive element such as human hair (moisture causes hair to contract, which is why hair frizzes when exposed to high humidity), nylon strips as in Figure 22-8 (nylon expands and contracts with a change in humidity), or electronic components (that change resistance as the humidity changes) for the more sophisticated applications. When the relative humidity reaches a certain set level, the humidity-sensing component reacts, in turn activating the actual switching mechanism such as a small microswitch, which turns the motor on and off. A bimetallic element cannot be used because it reacts to temperature changes only.

Adjustable settings are from 5% to 95% relative humidity. A comfort level of about 50% relative humidity is considered acceptable.

Humidistats having proper voltage rating (120 or 240 volts) and ampere rating are used to automatically switch the motor directly. See Figure 22-9(A). They are available in appearance similar to typical speed and timer controls with a knob and faceplate to match the other switch and receptacle faceplates in a home. The wiring is as simple as wiring in a switch leg for switching lighting outlets.

Where the 120-volt power wiring becomes very lengthy because of the distance between the

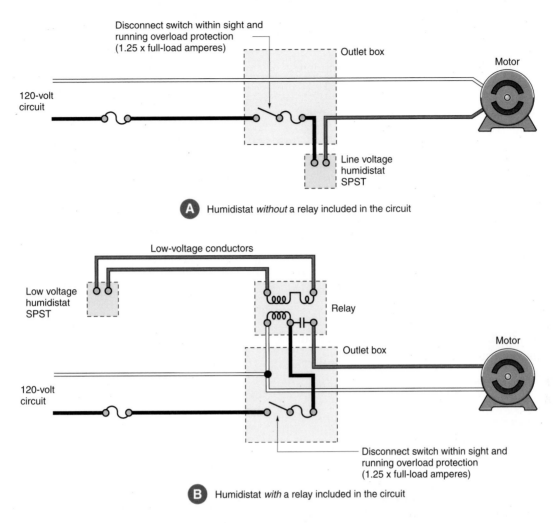

A Humidistat *without* a relay included in the circuit

B Humidistat *with* a relay included in the circuit

FIGURE 22-9   Wiring for the humidistat control. (*Delmar/Cengage Learning*)

controller and the fan, or for large motors having a high current draw, a relay can be used. See Figure 22-9(B). This relay is mounted near the whole house exhaust fan. The branch-circuit wiring is run from the Main Panel to a 4-in. square, 2⅛-in. deep outlet box conveniently located near the fan. A box-cover switch/fuse unit on this outlet box serves as the disconnecting means within sight of the motor, as required by *422.30* and *430.102*, and has a dual-element, time-delay fuse sized at approximately 125% of the motor's full-load current rating for motor overload backup protection. Typical home-type whole house exhaust fans do not require relays.

Dehumidifiers used in basements and other damp locations and humidifiers on furnaces use a humidistat to turn the power to the motor on and off.

APPLIANCE DISCONNECTING MEANS

To clean, adjust, maintain, or repair an appliance, it must be disconnected to prevent personal injury. *NEC 422.30* through *422.35* outline the basic methods for disconnecting fixed, portable, and stationary electrical appliances. Note that each appliance in the dwelling conforms to one or more of the disconnecting methods listed. The more important *Code* rules for appliance disconnects are as follows:

Each appliance must have a disconnecting means.

- The disconnecting means may be a separate disconnect switch if the appliance is permanently connected.

- The disconnecting means may be the branch-circuit switch or circuit breaker if the appliance is permanently connected and not over ⅛ horsepower or 300 volt-amperes.

- The disconnecting means may be an attachment plug cap if the appliance is cord connected.

- The disconnecting means may be the unit switch on the appliance only if other means for disconnection are also provided. In a single-family residence, the service disconnect serves as the other means.

- The disconnect must have a positive "On–Off" position.

- The disconnect must be capable of being locked in the "Off" position or be within sight of a motor-driven appliance where the motor is more than ⅛ horsepower. The lock-off provision must be permanent.

Read *422.30* through *422.35* for the details of appliance disconnect requirements.

HYDROMASSAGE BATHTUB CIRCUIT ⒶA

The Master Bedroom is equipped with a hydromassage bathtub. A hydromassage bathtub is sometimes referred to as a whirlpool bath.

NEC 680.2 defines a *hydromassage bathtub* as *a permanently installed bathtub equipped with a recirculating piping system, pump and associated equipment. It is designed so that it can accept, circulate, and discharge water upon each use.**

In other words, fill—use—drain.

The significant difference between a hydromassage bathtub and a regular bathtub is the recirculating piping system and the electric pump that circulates the water. Both types of tubs are drained completely after each use.

Spas and hot tubs are intended to be filled, then used. They are *not* drained after each use, because they have a filtering and heating system.

Electrical Connections

The *NEC* requirements for hydromassage tubs are found in Chapter 30 of this text.

*Reprinted with permission from NFPA 70-2011.

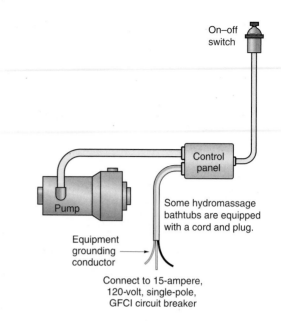

FIGURE 22-10 Typical wiring diagram of a hydromassage tub showing the pump motor, control panel, and electrical supply leads. (*Delmar/Cengage Learning*)

The hydromassage bathtub in this residence is supplied by a dedicated branch circuit A9, a 15-ampere, 120-volt circuit. If you are not sure what the actual current draw is, run a 20-ampere branch circuit.

The Schedule of Special-Purpose Outlets indicates that the hydromassage bathtub in this residence has a ½-horsepower motor that draws 10 amperes.

The hydromassage bathtub electrical control is prewired by the manufacturer. All the electrician needs to do is run the dedicated 15-ampere, 120-volt, GFCI-protected circuit to the end of the tub where the pump and control are located.

Some manufacturers supply a short flexible conduit that contains one black, one white, and one green equipment grounding conductor, as in Figure 22-10. The electrician installs a junction box near the pump. A dedicated branch circuit is run to this box, where the branch-circuit wiring is connected to the hydromassage conductors. This junction box must be accessible. The pump and power panel may also need servicing. Access may be from underneath or the end, whichever is convenient for the installation, as in Figure 22-11.

Make sure that the equipment grounding conductor of the circuit is properly connected to the green equipment grounding conductor of the hydromassage tub. *Never connect the green equipment*

FIGURE 22-11   The basic roughing-in of a hydromassage bathtub is similar to that of a regular bathtub. The electrician runs a separate 15-ampere, 120-volt GFCI-protected circuit to the area where the pump and control are located. Check the manufacturer's specifications for these data. An access panel from the end or from below is necessary to service the wiring, the pump, and the control panel. (*Delmar/Cengage Learning*)

grounding conductor to the grounded (white) branch-circuit conductor.

Some hydromassage tubs come equipped with a cord and plug, in which case the electrician installs a receptacle within 24 in. (600 mm) of the pump. Usually the branch circuit is protected by a GFCI circuit breaker, although a GFCI receptacle could be used. ▶If the hydromassage bathtub is cord- and plug-connected with the supply receptacle accessible only through a service access opening, the receptacle is required to be installed so its face is within direct view and not more than 1 ft (300 mm) of the opening, *680.73.*◀

Read and follow the manufacturer's installation instructions.

Figure 22-12 is a photograph of a typical hydromassage tub.

FIGURE 22-12   Top view of a typical hydromassage bathtub, sometimes referred to as a whirlpool. (*Courtesy of Trajet*)

REVIEW

Note: Refer to the *Code* or the plans where necessary.

BATHROOM CEILING HEATER CIRCUITS ⓐκ ⓐⱼ

1. What is the wattage rating of the heat/vent/light? _____

2. To what circuits are the heat/vent/lights connected? _____

3. Why is a 4-in. square, $2\frac{1}{8}$-in. deep box with a 2-gang raised plaster ring or similar deep box used for the switch assembly for the heat/vent/light? _____

   _____

   _____

   _____

   _____

4. a. How many wires are required to connect the wall switches and the heat/vent/light? _____

   b. What size wires are used? _____

5. Can the heating element be energized when the fan is not operating? _____

   _____

6. Can the fan be turned on without the heating element? _____

7. What device can be used to provide automatic control of the heating element and the fan of the heat/vent/light? _____

8. Where does the air enter the heat/vent/light? _____

9. Where does the air leave this unit? _____

10. Who is to furnish the heat/vent/light? _____

    _____

    _____

11. For a ceiling heater rated 1200 watts at 120 volts, what is the current draw? _____

    _____

12. An electrician wired up a heat/vent/light as described in this chapter. The wiring diagram showed that the electrician needed to run five conductors between the switches and the heat/vent/light. The electrician used one 12/3 nonmetallic-sheathed cable and one 12/2 nonmetallic-sheathed cable. Both cables had equipment grounding conductors. During the rough-in inspection, the electrical inspector cited "Noncompliance with *NEC 300.3(B)*." What was the problem?_____

    _____

    _____

    _____

ATTIC EXHAUST FAN CIRCUIT ▲L

1. What is the purpose of the attic exhaust fan? _____
   _____
   _____

2. At what voltage does the fan operate? _____

3. What is the horsepower rating of the fan motor? _____

4. Is the fan direct- or belt-driven? _____

5. How is the fan controlled? _____

6. What is the rating of the dual-element, time-delay fuse in the switch/fuse box-cover
   unit? _____

7. What is the rating of the running overcurrent protection if the motor is rated at
   10 amperes? _____

8. What is the basic difference between a thermostat and a humidistat? _____
   _____
   _____

9. What size conductors are to be used for this circuit? _____

10. How many ft (m) of cable are required to complete the wiring for the attic exhaust fan
    circuit? _____

11. May the metal frame of the fan be grounded to the grounded circuit conductor? _____
    _____

12. Which section of the *Code* prohibits grounding equipment to a grounded circuit
    conductor? _____

HYDROMASSAGE BATHTUB CIRCUIT ▲A

1. Which circuit supplies the hydromassage bathtub? _____

2. Which of the following statements "Meets *Code*"? (Circle the correct answer.)

   a. Hydromassage tubs and all of their electrical components shall be GFCI protected.

   b. Hydromassage tubs and all of their electrical components shall not be GFCI
      protected.

3. Which conductor size feeds the hydromassage tub in this residence?
   _____

4. What is the fundamental difference between a hydromassage bathtub and a spa? _____
   _____
   _____

5. Which sections of the *Code* reference hydromassage bathtubs? _____
   _____

6. Must the metal parts of the pump and power panel of the hydromassage tub be grounded? _____

_____

_____

_____

7. Is it permissible to connect the hydromassage tub's green equipment grounding conductor to the branch circuit's grounded (white) conductor?

_____

8. All 120-volt, single-phase receptacles within (6 ft [1.8 m]) (10 ft [3.0 m]) (15 ft [4.5 m]) of a hydromassage tub must be GFCI protected. (Circle the correct answer.)

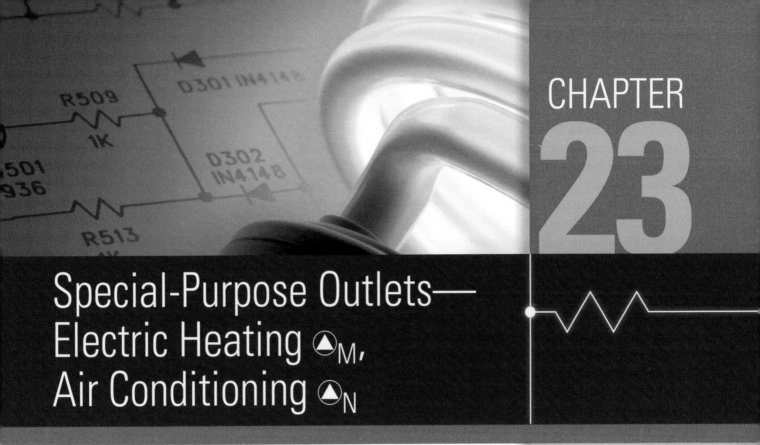

Special-Purpose Outlets—
Electric Heating ▲M,
Air Conditioning ▲N

OBJECTIVES

After studying this chapter, you should be able to

- understand the *NEC* requirements for embedded resistance heating cable, electric furnaces, electric baseboards, and heat pumps.

- understand how the heating output of resistive heating elements is affected by voltage variation.

- understand the data found on the nameplate of HVAC equipment, and determine electrical installation requirements in conformance to the *NEC*.

- realize that a 120-volt receptacle must be installed near HVAC equipment.

- understand energy rating terminology.

There are many types of electric heat available for heating homes (e.g., heating cable, unit heaters, boilers, electric furnaces, duct heaters, baseboard heaters, heat pumps, and radiant heating panels). This chapter discusses many of these types. The residence discussed in this text is heated by an electric furnace located in the workshop.

Detailed *Code* requirements for fixed electric space-heating equipment are found in *Article 424*.

This text cannot cover in detail the methods used to calculate heat loss and the wattage required to provide a comfortable level of heat in the building. For this residence, the total estimated wattage is 13,000 watts. Depending on the location of the residence (in the Northeast, Midwest, or South, for example), the heating load will vary.

Electric heating has gained wide acceptance when compared with other types of heating systems. It has a number of advantages. Electric heating is flexible when baseboard heating is used, because each room can have its own thermostat. Thus, one room can be kept cool while an adjoining room is warm. This type of zone control for an electric, gas-, or oil-fired central heating system is more complex and more expensive.

Electric heating is safer than heating with fuels. The system does not require storage space, tanks, or chimneys. Electric heating is quiet. Electric heat does not add or remove anything from the air. As a result, electric heat is cleaner. This type of heating is considered to be healthier than fuel heating systems that remove oxygen from the air. The only moving part of an electric baseboard heating system is the thermostat. This means that there is a minimum of maintenance.

As with all heating systems, adequate insulation must be provided. Proper insulation can keep electric bills to a minimum. Insulation also helps to keep the residence cool during the hot summer months. The cost of extra insulation is offset through the years by the decreased burden on the heating and air-conditioning equipment.

RESISTANCE HEATING CABLES

Resistance heating cables can be embedded in the plaster or between two layers of drywall. Installation requirements are found in *424.34* through *424.45*

of the *NEC*. Because premises wiring (nonmetallic-sheathed cable, etc.) will be located above these ceilings and will be subjected to the heat created by the heating cables, the *Code* requires the following:

- Keep the wiring not less than 2 in. (50 mm) above the heated ceiling, and *also* reduce the ampacity of the wiring to the 122°F (50°C) correction factors found in *Table 310.15(B)(2)(a)*, or

- Keep the wiring above the thermal insulation that is at least 2 in. (50 mm) thick—in which case the correction factors do not have to be applied.

For ease in identification, *424.35* requires that the leads on heating cables be color coded as follows:

- 120 volts   yellow
- 208 volts   blue
- 240 volts   red
- 277 volts   brown
- 480 volts   orange

ELECTRIC FURNACES

In an electric furnace, the source of heat is the resistance heating elements. The blower motor assembly, filter section, condensate coil, refrigerant line connections, fan motor speed control, high temperature limit controls, relays, and similar components are much the same as those found in gas furnaces.

As with all appliances, many of the *NEC* requirements have already been met by virtue of the product being listed by a recognized testing laboratory. Your job is to make sure the premises wiring is installed according to the *NEC*. The recognized testing laboratory takes care of the safety requirements of the appliance.

Summary of Important *NEC* Rules

Many of the following *NEC* requirements are emphasized in Figure 23-1.

- *NEC 220.51*: When making a calculation for sizing a service or feeder, include 100% of the rating of the total connected fixed space-heating load. This is shown in Chapter 29.

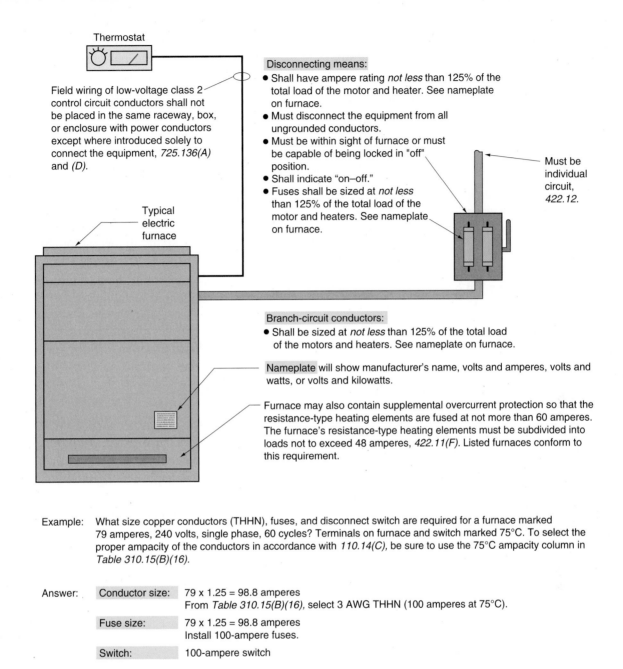

Thermostat

Field wiring of low-voltage class 2 control circuit conductors shall not be placed in the same raceway, box, or enclosure with power conductors except where introduced solely to connect the equipment, *725.136(A)* and *(D)*.

Typical electric furnace

Disconnecting means:
- Shall have ampere rating *not less* than 125% of the total load of the motor and heater. See nameplate on furnace.
- Must disconnect the equipment from all ungrounded conductors.
- Must be within sight of furnace or must be capable of being locked in "off" position.
- Shall indicate "on–off."
- Fuses shall be sized at *not less* than 125% of the total load of the motor and heaters. See nameplate on furnace.

Must be individual circuit, *422.12*.

Branch-circuit conductors:
- Shall be sized at *not less* than 125% of the total load of the motors and heaters. See nameplate on furnace.

Nameplate will show manufacturer's name, volts and amperes, volts and watts, or volts and kilowatts.

Furnace may also contain supplemental overcurrent protection so that the resistance-type heating elements are fused at not more than 60 amperes. The furnace's resistance-type heating elements must be subdivided into loads not to exceed 48 amperes, *422.11(F)*. Listed furnaces conform to this requirement.

Example: What size copper conductors (THHN), fuses, and disconnect switch are required for a furnace marked 79 amperes, 240 volts, single phase, 60 cycles? Terminals on furnace and switch marked 75°C. To select the proper ampacity of the conductors in accordance with *110.14(C)*, be sure to use the 75°C ampacity column in *Table 310.15(B)(16)*.

Answer: | Conductor size: | 79 x 1.25 = 98.8 amperes |
From *Table 310.15(B)(16)*, select 3 AWG THHN (100 amperes at 75°C). |
| Fuse size: | 79 x 1.25 = 98.8 amperes |
Install 100-ampere fuses. |
| Switch: | 100-ampere switch |

FIGURE 23-1   An electric furnace is considered to be a *continuous load*, subject to sizing the conductors, disconnecting means, and overcurrent devices as 125% of the load, *NEC 424.3(B)*. Central electric heating equipment must be supplied by an individual branch-circuit, *NEC 422.12*. (*Delmar/Cengage Learning*)

- *Article 250, Part VII*: Contains most of the requirements for equipment and appliances grounding. Grounding is covered in Chapters 4, 15, 20, 21, and 22.

- *NEC 250.118*: Lists the acceptable methods for equipment and appliance grounding. This is generally the equipment grounding conductor in nonmetallic-sheathed cable, the metal armor and bonding strip in armored cable, or a metal raceway such as EMT or FMC. Severe limitations are imposed on FMC in *250.118(5)*.

- *NEC 422.10(A)*: The branch-circuit rating for an appliance must not be less than the marked rating.

- *NEC 422.11*: The branch-circuit overcurrent protection must be sized according to the value indicated on the nameplate. The electrician need do nothing more than size the conductors and overcurrent protection according to the nameplate data and instructions furnished with the furnace. Overcurrent protection for internal components is provided by the manufacturer.

- *NEC 422.12*: Central heating equipment must be supplied by an individual branch circuit. *Associated* equipment such as humidifiers and electrostatic air cleaners is permitted to be connected to the same branch circuit. The logic behind this "individual circuit" requirement is that if the heating system were to be connected to some other branch circuit, such as a lighting branch circuit, a fault on that circuit would shut off the power to the heating system. We find this same requirement in *8.6.4* of *NFPA 54*, the *National Fuel Gas Code*.

- *NEC 422.30*: A disconnecting means must be provided.

- *NEC 422.31(B)*: The disconnecting means must be within sight of the furnace. For electric furnaces, the disconnect might be mounted on the side of the furnace, on an adjacent wall, or the circuit breaker in the panelboard, if the panelboard is within sight of the furnace. For gas furnaces, the disconnect could simply be a snap switch, switch/fuse mounted in a handy box on the side of the furnace, or the circuit breaker in the panelboard, if the panelboard is in sight of the furnace. See *430.102(A)* and *(B)*.

- *NEC 422.32*: This is another reminder that the disconnect for motor-driven appliances having a motor greater than ⅛ horsepower must be within sight of the appliance. See *430.102(A)* and *(B)*.

- *NEC 422.62(B)(1)*: The nameplate shall specify the minimum size supply circuit conductor's ampacity and the maximum branch-circuit over current protection.

- *NEC 424.3(B)*: Fixed electric space heating is considered to be a continuous load. The branch-circuit conductors and the rating or setting of overcurrent protective devices supplying fixed electric space-heating equipment consisting of resistance elements with or without a motor shall not be less than 125% of the total load of the motors and the heaters. If applying the 125% rule results in a nonstandard rating of

overcurrent protective device, then *240.4(B)* permits the use of the next higher rating. See Figure 23-1 for the 125% calculations.

- *NEC 430.60(B)*: The nameplate must show volts and watts or volts and amperes.

- *NEC 430.102(A)*: The disconnect shall be within sight of the controller and shall disconnect the controller. With a residential furnace, the controls are an integral part of the furnace.

- *NEC 430.102(B)*: The disconnect shall be located within sight of the motor and driven machinery. There is an exception to the "in sight" requirement, but the exception does not pertain to residential installations.

Figure 23-1 illustrates the key *NEC* requirements for hooking up an electric furnace.

Supply Voltage

Because proper supply voltage is critical to ensure full output of the furnace's heating elements, voltage must be maintained at not less than 98% of the furnace's rated voltage. The effect of voltage drop is discussed in Chapters 4 and 19. Mathematically, for every 1% drop in voltage, there will be a 2% drop in wattage output of a resistance heating element. Manufacturers' installation instructions usually have tables that show the wattage or kilowatt output at different voltages. Table 23-1 is a typical table.

A simple way to calculate the effect of voltage variance on a resistive heating element is:

$$\text{Correction Factor} = \frac{\text{Applied Voltage Squared}}{\text{Rated Voltage Squared}}$$

TABLE 23-1

Kilowatt rating.

@ 240 volts	@ 208 volts
12.0	9.0
15.0	11.3
20.0	15.0
25.0	18.8
30.0	22.5

EXAMPLE

The rating of the heating elements in an electric furnace is 30 kW at 240 volts. What is the kW output if the supply voltage is 208 volts?

Step 1: Correction Factor $= \dfrac{208 \times 208}{240 \times 240}$

$$= \dfrac{43,264}{57,600} = 0.75$$

Step 2: $30\ kW \times 0.75 = 22.5\ kW$

Therefore, the actual voltage at the supply terminals of an electric furnace or electric baseboard heating unit will determine the true wattage output of the heating elements.

Cord-and-Plug Connection Not Permitted

Flexible cords are not a recognized wiring method.

NEC 400.7(A)(8) and *422.16* describe the permitted uses for flexible cords. A key requirement, oftentimes violated, is that flexible cords shall be used only *where the fastening means and mechanical connections are specifically designed to permit ready removal for maintenance and repair, and the appliance is intended or identified for flexible cord connection.** Certainly the gas piping and the size

*Reprinted with permission from NFPA 70–2011.

of a gas or electric furnace do not allow for "ready removal" of the furnace. Furthermore, the conductors in flexible cords cannot withstand the high temperature encountered on the terminals of a furnace.

ANSI Standard Z21.47 does not allow cord-and-plug connection for gas furnaces.

CONTROL OF ELECTRIC BASEBOARD HEATING UNITS

Line-voltage and low-voltage thermostats can be used to control electric baseboard heating units, as shown in Figure 23-2. When the current rating or ampere rating of the load exceeds the rating of a line-voltage thermostat, then it is necessary to utilize a relay in conjunction with the low-voltage thermostat. This is illustrated in Figure 23-3. Color-coding requirements are explained later in this chapter, and are covered in detail in Chapter 5.

Most line-voltage thermostats are ampere rated for noninductive resistive loads. You may need to convert watts to amperes to select the right size thermostat, for example, if you are controlling a 2500-watt, 120-volt resistive heating element or a 5000-watt, 240-volt resistive heating element:

$$I = \frac{W}{E} = \frac{2500}{120} = 20.8\ \text{amperes}$$

—or—

$$I = \frac{W}{E} = \frac{5000}{240} = 20.8\ \text{amperes}$$

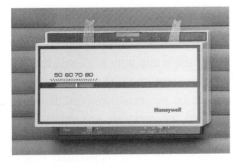

FIGURE 23-2 Thermostats for electric heating systems. A thermostat or other switching device that is marked with an "Off" position must open all ungrounded conductors of the circuit when turned to the "Off" position. Typical mounting height is 52 in. (1.3 m) to center. (*Courtesy of Honeywell International Inc.*)

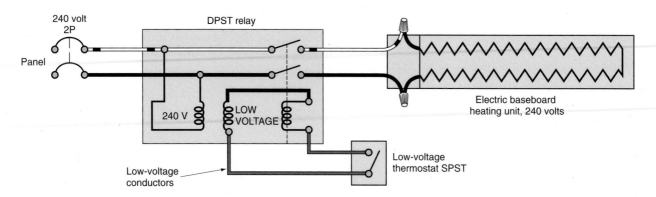

FIGURE 23-3 Wiring for an electric baseboard heating unit having a low-voltage thermostat and relay. *NEC 200.7* requires that a white wire in cable systems used as an ungrounded conductor must be permanently reidentified at points where connections are made or where it is visible and accessible. This becomes necessary when nonmetallic-sheathed cable or armored cable is the wiring method. Reidentification can be accomplished with black paint, black tape, or heat-shrink tubing. Color coding and reidentification of conductors is discussed in Chapter 5 and later in this chapter. (*Delmar/Cengage Learning*)

The total connected load for this thermostat must not exceed 20.8 amperes.

Disconnection of Baseboard Heaters. Like other fixed electric space-heating equipment, baseboard heaters are required to be provided with disconnecting means. The line-voltage thermostat is permitted serve as both the controller and the disconnecting means provided it meets all the following conditions:

1. The thermostat has a marked "Off" position.

2. It directly opens all ungrounded conductors when manually placed in the "Off" position.

3. It is designed so the circuit cannot be energized automatically after the device has been manually placed in the "Off" position.

4. It is located within sight of the heaters it controls and disconnects.

As can be seen, a 2-pole thermostat is required to serve as a disconnecting means for baseboard heaters. A single-pole thermostat does not have a marked "off" position and will not satisfy these requirements to serve as a disconnecting means. See *424.20*.

If 2-pole thermostats are not used to satisfy the disconnecting means requirements, the installation must comply with *424.19(B)*. This permits the circuit breaker to serve as the disconnecting means

if it's within sight from the heater or is capable of being locked in the open position. If the circuit breaker is located out of sight from the baseboard heaters, it must be provided with a means of being locked open.

MARKING THE CONDUCTORS OF CABLES

Two-wire cable contains one white and one black conductor. Because 240-volt circuits are permitted to supply electric baseboard heaters, it would appear that the use of 2-wire cable is in violation of the *Code*.

According to *200.7(C)*, 2-wire cable is permitted for the 240-volt heaters provided the white conductor is permanently reidentified by paint, colored tape, or other effective means. This is necessary because it might be assumed that an unmarked white conductor is a grounded conductor having no voltage to ground. Actually, the white wire is connected to a "hot" phase in the panel and has 120 volts to ground on it. A person could be subjected to a lethal shock by touching this conductor and the grounded baseboard heater (or any other grounded object) at the same time. The white conductor must be permanently reidentified at the electric heater terminals and at the panels where these cables originate. Figures 23-3, 23-4, and 23-5 show this reidentification. Review the section on "Color Coding" in Chapter 5.

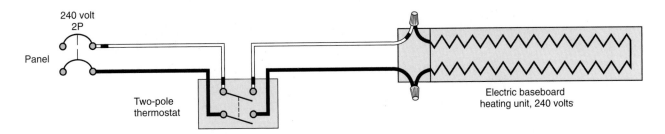

FIGURE 23-4 Wiring for a single electric baseboard heating unit. The thermostat is a line-voltage thermostat. See Figure 23-3 caption for color-coding requirements. (*Delmar/Cengage Learning*)

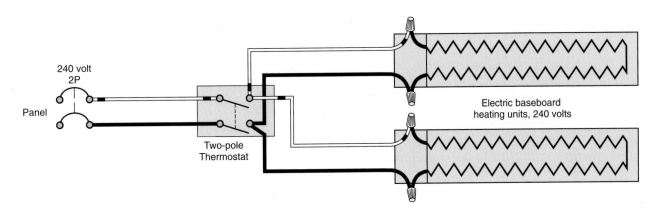

FIGURE 23-5 Wiring for two electric baseboard heating units. Note the circuit connections at the thermostat location. The thermostat is a line-voltage thermostat. See Figure 23-3 caption for color-coding requirements. (*Delmar/Cengage Learning*)

CIRCUIT REQUIREMENTS FOR ELECTRIC BASEBOARD HEATING UNITS

Figure 23-3 shows an electric baseboard heater controlled by a low-voltage thermostat. The low-voltage contacts of the relay are connected to the thermostat. The line-voltage contacts of the relay are used to switch the actual heater load. Relays are used whenever the load exceeds the ampere or wattage limitations of a line-voltage thermostat. Color coding for conductors is explained in Chapter 5 and is also discussed later in this chapter.

Figure 23-6 shows typical electric baseboard heating units. The heater to the right has an integral thermostat on it. Electric baseboard heating units are available in both 240-volt and 120-volt ratings and in a variety of wattage ratings.

The wiring for an individual baseboard heating unit or group of units is shown in Figures 23-4 and 23-5. A 2-wire cable (armored cable or non-metallic-sheathed cable with ground) would be run from a 240-volt, 2-pole circuit in the main panel to the outlet box or switch box installed at the thermostat location. A second 2-wire cable runs from the thermostat to the junction box on the heater unit. The proper connections are made in this junction box. Most heating unit manufacturers provide knockouts at the rear and on the bottom of the junction box. The supply conductors can be run through these knockouts. Most baseboard units also have a channel or wiring space running the full length of the unit, usually at the bottom. When two or more heating units are joined together, the conductors are run in this wiring channel. Most manufacturers indicate the type of wire required for these units because of conductor temperature limitations.

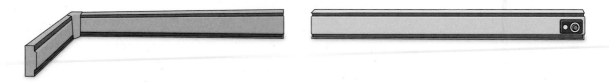

FIGURE 23-6   Electric baseboard heating systems. (*Delmar/Cengage Learning*)

Supply conductors generally must be rated 90°C. These would be Type THHN conductors, or nonmetallic-sheathed cable or armored cable that has 90°C conductors. The nameplate and/or instructions furnished with the electric baseboard heating unit provides this information.

Most wall and baseboard heating units are available with built-in thermostats, as shown in Figure 23-6. These thermostats are also available as an accessory that can be added in the field. The branch-circuit supply cable for such a unit runs from the main panelboard to the junction box on the unit.

Some heaters have 120-volt, 15- or 20-ampere receptacle outlets. UL states that when receptacle sections are included with the other components of baseboard heating systems, they must be supplied separately using conventional wiring methods. See Figure 23-7.

A variety of fittings such as internal and external elbows (for turning corners) and blank sections are available from the manufacturer.

Manufacturers of electric baseboard heaters list "blank" sections ranging in length from 2 to 10 ft (600 mm to 3.0 m). These blank sections give the installer the flexibility needed to spread out the heater sections, and to install these blanks where wall receptacle outlets are encountered, as in Figure 23-8.

LOCATION OF ELECTRIC BASEBOARD HEATERS IN RELATION TO RECEPTACLE OUTLETS

Listed electric baseboard heaters shall not be installed below wall receptacle outlets unless the instructions furnished with the baseboard heaters indicate that they may be installed below receptacle outlets. See the *Informational Note* to *210.52*. The reasoning for generally not allowing electric baseboard heaters to be installed below receptacle outlets is the possible fire and shock hazard resulting when a cord hangs over and touches the heated electric baseboard unit. The insulation on the cord can melt from the heat. See Figures 23-9 and 23-10.

The UL Guide Card KLDR makes the following statement: "Electrical cords, drapes, and other furnishings should be kept away from baseboard

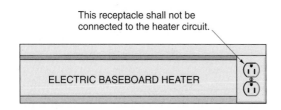

FIGURE 23-7   Factory-installed receptacle outlets or receptacle outlet assemblies provided by the manufacturer for use with its electric baseboard heaters may be counted as the required receptacle outlet for the space occupied by a permanently installed heater. See the second paragraph of *210.52*. (*Delmar/Cengage Learning*)

| HEATER | BLANK | HEATER |

FIGURE 23-8   Use of blank baseboard heating sections. (*Delmar/Cengage Learning*)

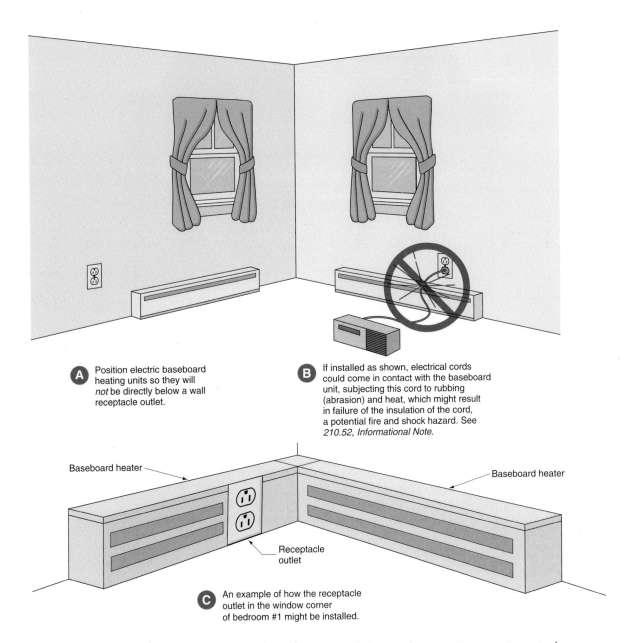

A Position electric baseboard heating units so they will *not* be directly below a wall receptacle outlet.

B If installed as shown, electrical cords could come in contact with the baseboard unit, subjecting this cord to rubbing (abrasion) and heat, which might result in failure of the insulation of the cord, a potential fire and shock hazard. See *210.52, Informational Note.*

Baseboard heater

Baseboard heater

Receptacle outlet

C An example of how the receptacle outlet in the window corner of bedroom #1 might be installed.

FIGURE 23-9 Position of electric baseboard heaters. (*Delmar/Cengage Learning*)

ELECTRIC BASEBOARD HEATER

FIGURE 23-10 *Code violation.* Unless the instructions furnished with the electric baseboard heater specifically state that the unit is listed for installation below a receptacle outlet, then this installation is a violation of *110.3(B).* In this type of installation, cords attached to the receptacle outlet would hang over the heater, creating a fire hazard and possible shock hazard should the insulation on the cord melt. See *Informational Notes* to *210.52.* (*Delmar/Cengage Learning*)

FIGURE 23-11 A toe-kick heater designed to be located near the bottom of a floor-mounted cabinet. (*Courtesy of Cadet*)

heaters. To reduce the likehood of cords contacting the heater, the heater is not to be located beneath electrical receptacles."

Electricians must pay close attention to the location of receptacle outlets and the electric baseboard heaters. They must study the plans and specifications carefully. They may have to fine-tune the location of the receptacle outlets and the electric baseboard heaters, and/or install blank spacer sections, keeping in mind the *Code* requirements for spacing receptacle outlets and just how much (wattage and length) baseboard heating is required (Figure 23-7 and Figure 23-10).

The *Code* does permit factory-installed or factory-furnished receptacle outlet assemblies as part of permanently installed electric baseboard heaters. These receptacle outlets shall not be connected to the heater circuit (see second paragraph of *210.52*, Figure 23-7 and Figure 23-9).

WALL-MOUNTED HEATERS

Electric space heaters are available in many sizes or heating capacities and designed for a variety of locations. For example, heaters in bathrooms can be installed in the wall or in the toe-kick space under the cabinet. Figure 23-11 shows a heater designed for installation in the toe-kick space near the floor. These heaters are designed for bathrooms or kitchens or other areas having floor-mounted cabinets.

FIGURE 23-12 Wall-mounted space heaters are designed for heating from small to large rooms or areas. (*Courtesy of Cadet*)

Figure 23-12 shows a wall heater designed for larger rooms or areas. The heater shown in Figure 23-13 is particularly well suited for bathrooms or nurseries as it is provided with a timer as well as a thermostat.

FIGURE 23-13   A wall-mounted heater well suited for bathrooms and nurseries as it has both a thermostat and timer. (*Courtesy of Cadet*)

These heaters typically have a can or enclosure that is installed at the rough-in stage of the project. The branch circuit is routed to the unit either through the thermostat location or directly to the rough-in can.

Wall-mounted heaters should be located carefully to avoid conflict with furniture layout. At times, this can best be accomplished if the heater is installed near the entry door to the room.

Wall-mounted heaters are available with or without a built-in thermostat and in a variety of heating capacities to match the heat loss of the room or area. Smaller capacity heaters are available in 120-volt models. Heaters designed to operate on 240 volts are available in ratings from 750 to 4000 watt capacities. Heater manufacturers provide on-line calculators to assist in proper sizing of heaters.

HEAT PUMPS

A heat pump is dual-purpose equipment that functions as a heating unit and an air conditioner operating in reverse. A "reversing valve" inside the unit changes the direction of the flow of the system's refrigerant.

A "split" heat pump system consists of an outdoor unit with a hermetic motor-compressor, a fan, and a coil. The indoor unit consists of a refrigerant coil and usually makeup resistance heating inside of the air handler. The movement of air inside the home is accomplished by the blower fan on the air handler. The "split system" is the most common type for residential use.

When all of the components are housed in one outdoor unit, the system is referred to as a "packaged system."

The hermetic motor-compressor is the heart of the system. The refrigerant circulates between the indoor and outdoor units, through the coils, tubing, and compressor, absorbing and releasing heat as it travels through the system.

In the winter, the heat pump (the outdoor coil) extracts heat from the outdoor air and distributes it through the warm air ducts in the home. In extremely cold climates, a heat pump is supplemented by electric heating elements in the air handler.

In the summer, the process reverses. The heat pump (indoor coil) absorbs heat and condenses humidity from the indoor air. The heat is transferred to the outside by the refrigerant, and the humidity (condensation) is removed by piping it to a suitable drain.

Heat pump water heaters are discussed in Chapter 19.

Because a heat pump contains a hermetic motor-compressor, the *Code* requirements found in *Article 440* apply. These are discussed later in this chapter.

GROUNDING

Grounding of electrical equipment has already been discussed and will not be repeated here. Grounding is covered in *Article 250* of the *NEC*. To repeat a caution, *never* ground electrical equipment to the grounded (white) circuit conductor of the circuit.

CIRCUIT REQUIREMENTS FOR ROOM AIR CONDITIONERS

For homes that do not have central air conditioning, window or through-the-wall air conditioners may be installed. These types of room air conditioners are available in both 120-volt and 240-volt ratings. Because room air conditioners are plug-and-cord connected, the receptacle outlet and the circuit capacity must be selected and installed according to applicable *Code* regulations. The *Code* rules for air conditioning are found in *Article 440*. The *Code* requirements for room air-conditioning units are found in *440.60* through *440.65* of the *NEC*.

The basic *Code* rules for installing these cord-and-plug-connected units and their receptacle outlets are as follows:

- The air conditioners must be grounded.

- The air-conditioner rating may not exceed 40 amperes at 250 volts, single phase.

- The rating of the branch-circuit overcurrent device must not exceed the branch-circuit conductor rating or the receptacle rating, whichever is less.

- The air-conditioner load shall not exceed 80% of the branch-circuit ampacity if no other loads are served.

- The air-conditioner load shall not exceed 50% of the branch-circuit ampacity if other loads are served.

- The attachment plug cap may serve as the disconnecting means.

- The maximum cord length is 10 ft (3.0 m) for 120-volt units, and 6 ft (1.8 m) for 208- or 240-volt units. This is also a UL requirement.

- There is an ever-present potential arcing-fault hazard related to the power supply cords on room air conditioners that could cause a fire or create an electrical shock hazard. In addition to requiring that the air-conditioner unit be properly grounded, *NEC 440.65* requires that factory-installed leakage current detector interrupter (LCDI) or arc-fault circuit interrupter (AFCI) devices be provided on single-phase cord-and-plug-connected room air conditioners. These devices are permitted to be an integral part of the attachment plug cap, or may be located in the power supply cord within 12 in. (300 mm) of the attachment plug cap.

LCDIs are *Leakage Current Detection and Interruption Protection* devices. They can sense leakage current flowing between or from the cord conductors and interrupt the circuit at a predetermined level of leakage current.

AFCIs are discussed in Chapter 6.

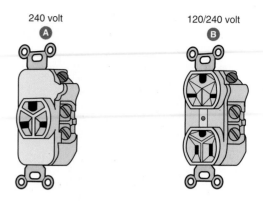

FIGURE 23-14 Types of receptacles.
(*Delmar/Cengage Learning*)

Receptacles for Room Air Conditioners

Figure 23-14(A) is a straight 240-volt receptacle. Figure 23-14(B) is a combination receptacle. The lower portion is for 120-volt use, and the upper portion is for 240-volt use. Note the different slot configurations that meet the noninterchangeable requirements of *406.3(F)*.

CENTRAL HEATING AND AIR CONDITIONING

The residence discussed throughout this text has central electric heating and air conditioning consisting of an electric furnace and a central air-conditioning unit.

The wiring for central heating and cooling systems is shown in Figure 23-15. Basic circuit requirements are shown in Figure 23-16. Note that one branch circuit runs to the electric furnace and another branch circuit runs to the air conditioner or heat pump outside the dwelling. Low-voltage wiring is used between the inside and outside units to provide control of the systems. The low-voltage Class 2 circuit wiring shall not be run in the same raceway as the power conductors, *725.136(A)*.

Refer to Chapter 24 for more information about low-voltage circuitry.

"Time-of-Use" Lower Energy Rates

Many utilities offer lower energy rates for air-conditioner and/or heat pump loads through special

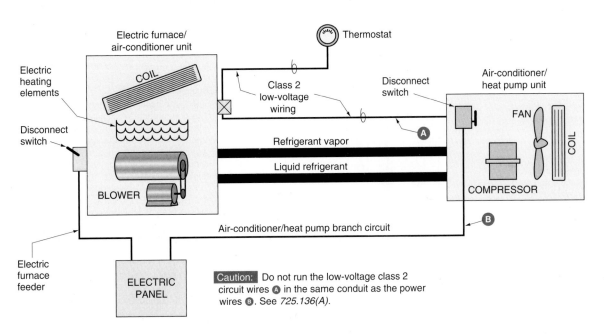

FIGURE 23-15 Connection diagram showing typical electric furnace and air-conditioner/heat pump installation. (*Delmar/Cengage Learning*)

programs, such as "time-of-use" usage and sliding scale energy rates. This is covered in Chapter 19.

UNDERSTANDING THE DATA FOUND ON AN HVAC NAMEPLATE

The letters HVAC stand for heating, ventilating, and air-conditioning.

Article 440 applies when HVAC equipment employs a hermetic refrigerant motor-compressor(s). *Article 440* is supplemental to the other articles of the *NEC* and is needed because of the unique characteristics of hermetic refrigerant motor-compressors. The most common examples are air conditioners, heat pumps, and refrigeration equipment.

Hermetic refrigerant motor-compressors

- combine the motor and compressor into one unit.

- have no external shaft.

- operate in the refrigerant.

- do not have horsepower and full-load current (often referred to as full-load amperes or FLA) ratings like standard electric motors. This is because as the compressor builds up pressure, the current increases, causing the windings to

get hot. At the same time, the refrigerant gets colder, passes over the windings, and cools them. Because of this, a hermetically sealed motor can be "worked" much harder than a conventional electric motor of the same size.

Special terminology is needed to understand how to properly make the electrical installation. These special terms are found in *440.2* and in UL Standard 1995. An air-conditioning nameplate from Shiver Manufacturing Company helps explain these special terms.

- **Rated-Load Current (RLC):** Rated-load current is established by the hermetic motor-compressor manufacturer under actual operation at rated refrigerant pressure and temperature, rated voltage, and rated frequency. RLC is marked on the nameplate. In most instances, the marked RLC is at least equal to 64.1% of the hermetic refrigerant motor-compressor's maximum continuous current (MCC). In our example, the RLC is 17.8 amperes. See *440.2*.

- **Maximum Continuous Current (MCC):** This is the maximum current that the hermetic refrigerant motor-compressor can draw before damage occurs. Manufacturers know the damage capabilities of their equipment and generally provide the necessary internal

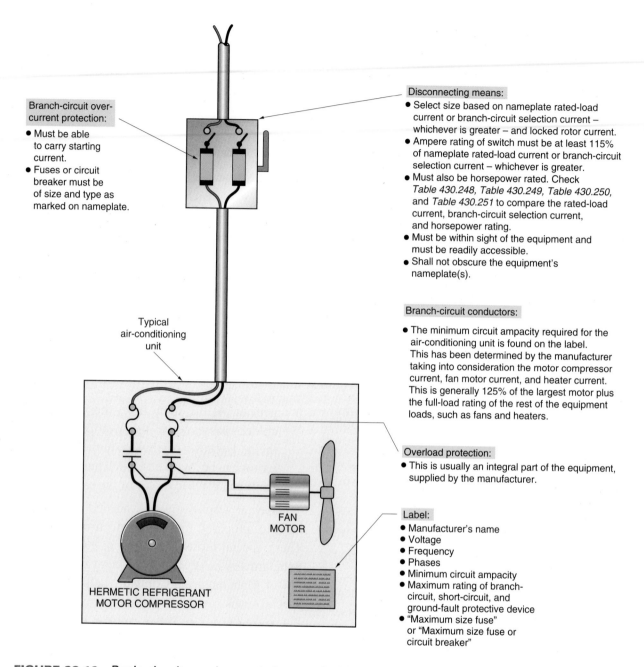

Branch-circuit over-current protection:

- Must be able to carry starting current.
- Fuses or circuit breaker must be of size and type as marked on nameplate.

Disconnecting means:

- Select size based on nameplate rated-load current or branch-circuit selection current – whichever is greater – and locked rotor current.
- Ampere rating of switch must be at least 115% of nameplate rated-load current or branch-circuit selection current – whichever is greater.
- Must also be horsepower rated. Check *Table 430.248*, *Table 430.249*, *Table 430.250*, and *Table 430.251* to compare the rated-load current, branch-circuit selection current, and horsepower rating.
- Must be within sight of the equipment and must be readily accessible.
- Shall not obscure the equipment's nameplate(s).

Branch-circuit conductors:

- The minimum circuit ampacity required for the air-conditioning unit is found on the label. This has been determined by the manufacturer taking into consideration the motor compressor current, fan motor current, and heater current. This is generally 125% of the largest motor plus the full-load rating of the rest of the equipment loads, such as fans and heaters.

Overload protection:

- This is usually an integral part of the equipment, supplied by the manufacturer.

Typical air-conditioning unit

FAN MOTOR

HERMETIC REFRIGERANT MOTOR COMPRESSOR

Label:
- Manufacturer's name
- Voltage
- Frequency
- Phases
- Minimum circuit ampacity
- Maximum rating of branch-circuit, short-circuit, and ground-fault protective device
- "Maximum size fuse" or "Maximum size fuse or circuit breaker"

FIGURE 23-16 Basic circuit requirements for a typical residential-type air conditioner or heat pump. Reading the label is important in that the manufacturer has determined the minimum circuit ampacity and the maximum size and type of fuse or circuit breakers. (*Delmar/Cengage Learning*)

overcurrent protection that keeps the hermetic motor-compressor from burning out. This value is 156% of the marked RLC or branch-circuit selection current (BCSC). The overload protective system for the hermetic refrigerant motor-compressor must operate for currents in excess of the 156% value. Do not look for the MCC value on the nameplate of the end-use equipment. The MCC value might be marked on the

motor-compressor. It is nice but not necessary for the electrician to know the MCC value of the motor-compressor. See *440.52*.

- **Branch-Circuit Selection Current (BCSC):** The manufacturer of the HVAC end-use equipment might design better cooling and better heat dissipation into the equipment, in which case the hermetic refrigerant motor-compressor might be capable of being continuously "worked"

TABLE 23-2

Shiver Manufacturing Company.

MODEL NO. XYZ – ELECTRICAL RATINGS

	VAC	HZ	PH	RLC	LRC	FLA
Compressor	230	60	1	17.8	107	—
Outdoor fan motor ¼ HP	230	60	1	—	—	1.5
Branch-circuit selection current				19.9 amperes		
Minimum circuit ampacity				26.4 amperes		
Maximum fuse or HACR type breaker				45.0 amperes		
Operating voltage range:				197 min. 253 max.		

harder than other equipment not so designed. "Working" the hermetic refrigerant motor-compressor harder will result in a higher current draw. This is safe insofar as the motor-compressor is concerned, but the conductors, disconnect switch, and branch-circuit overcurrent devices must also be capable of safely carrying this higher current draw. This higher value of current is marked on the nameplate as "BCSC."

Because the BCSC ampere value is always equal to or greater than the RLC ampere value, this BCSC value must be used instead of the RLC value when selecting conductors, disconnects, overcurrent devices, and other associated electrical equipment. Table 23-2 shows that the BCSC is 19.9 amperes. See *440.2* and *440.4(C)*.

- **Minimum Circuit Ampacity (MCA):** MCA is the minimum circuit ampacity requirement to determine the conductor size and switch rating. MCA data are always marked on the nameplate of the end-use equipment and are determined by the manufacturer of the end-use equipment as follows:

$$MCA = (RLC \text{ or } BCSC \times 1.25) + \text{other loads}$$

Other loads would include condensing fans, electric heaters, coils, and so forth, that operate concurrently (at the same time).

In our example, the marked MCA is 26.4 amperes, calculated by the manufacturer of the end-use equipment as follows:

$$(19.9 \times 1.25) + 1.5 = 26.375 \text{ amperes}$$
$$\text{(round off to 26.4)}$$

Referring to *Table 310.15(B)(16)*, using the 60°C column, 10 AWG copper conductors would be adequate for this air-conditioning unit. See *Article 440, Part IV*.

A mistake that electricians often make is to multiply the MCA by 1.25 again. Certainly there is no hazard in doing this, but it might result in an unnecessary higher installation cost. For long runs, larger conductors would keep voltage drop to a minimum.

Outdoor installations of air-conditioning equipment might be in very hot parts of the country. You might want to review Chapter 18 for the *NEC* correction factor requirements where high ambient temperatures are encountered.

- **Maximum Overcurrent Protection (MOP):** The MOP value is the maximum size overcurrent protection, fuse, or circuit breaker. The MOP value is marked on the nameplate and is determined by the manufacturer of the end-use equipment as follows:

$$(RLC \text{ or } BCSC) \times 2.25 + \text{other loads}$$

Other loads include condensing fans, electric heaters, coils, and so on, that operate at the same time. In our example, the marked MOP is 45 amperes, determined by the manufacturer as follows:

$$(19.9 \times 2.25) + 1.5 = 46.3 \text{ amperes}$$

Rounding down to the next lower standard size, the MOP is 45 amperes. See *Article 440, Part III*.

- **Type of Overcurrent Protection:** Check the nameplate carefully and do what it says! HACR (Heating, Air Conditioning, and Refrigeration) equipment nameplate might indicate "Maximum Size Fuse," "Maximum Size Fuse or Circuit Breaker," or "Maximum Size Fuse or HACR Circuit Breaker."

In the past, circuit breakers were subjected to specific tests unique to HVAC equipment and were marked with the letters "HACR." Today, no additional tests are made. All currently listed circuit breakers are now suitable for HVAC application. These circuit breakers and HVAC equipment may or may not show the letters "HACR." Because of existing inventories, it will take many years for the marking "HACR" to disappear from the scene. In the meantime, read and follow the information found on the nameplate of the equipment.

See Figures 23-17, 23-18, 23-19, and 23-20.

- **Disconnecting Means to Be in Sight:** *NEC 440.14* of the *Code* requires that the disconnecting means must be within sight of the unit, and it must be *readily accessible*. See Figures 23-18, 23-19, and 23-20. Mounting the disconnect on the side of the house instead of on the air conditioner itself allows for easy replacement of

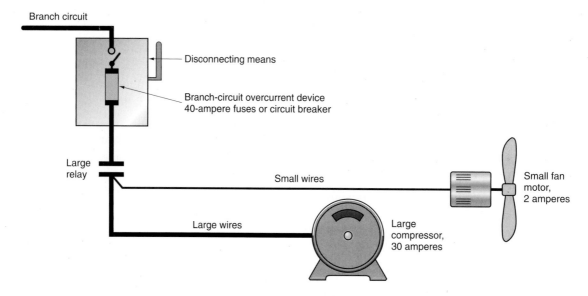

FIGURE 23-17 In a typical air-conditioner unit, the branch-circuit overcurrent device must protect the large components (large wires, large relay, hermetic motor-compressor) as well as the small components (small wires, fan motor, crankcase heater) under short-circuit and/or ground-fault situations. It is extremely important to install the proper size and type of overcurrent device. Read the nameplate on the equipment and the instructions furnished with the equipment. (*Delmar/Cengage Learning*)

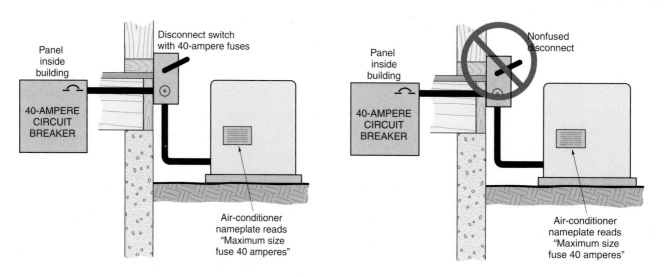

FIGURE 23-18 This installation "Meets *Code*." The disconnect is within sight of the unit. The disconnect contains fuses as specified on the air-conditioner nameplate. Refer to *NEC 110.3(B)*. (*Delmar/Cengage Learning*)

FIGURE 23-19 This installation does not "Meet *Code*" because the overcurrent device inside the building is not of the type specified on the air-conditioner nameplate. Refer to *NEC 110.3(B)*. (*Delmar/Cengage Learning*)

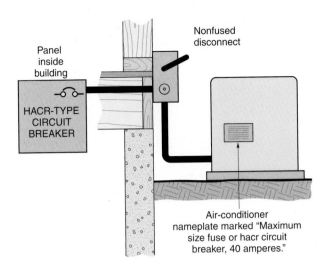

Panel inside building

HACR-TYPE CIRCUIT BREAKER

Nonfused disconnect

Air-conditioner nameplate marked "Maximum size fuse or hacr circuit breaker, 40 amperes."

FIGURE 23-20 In the past, air-conditioning equipment and circuit breakers were marked with the letters "HACR." As time passes and current inventory is used up, this marking will disappear from the nameplates of equipment and on the circuit-breaker label. In the meantime, follow the manufacturer's installation instructions and the data found on the nameplate. Refer to *NEC 110.3(B)*. (*Delmar/Cengage Learning*)

the air-conditioning unit, should that become necessary. Mounting the disconnect behind the air conditioner or heat pump would not be considered "readily accessible" by most electrical inspectors. Generally, there is just not enough space to work safely on the disconnect. Squeezing behind or leaning across the top of the air conditioner to gain access to the disconnect is certainly not a safe practice. Mounting the disconnect on the outside wall of the house, to one side of the AC unit, is the accepted practice. See *Article 440, Part II*, and *110.26*.

NEC 440.14 states that if you mount the disconnect on the equipment, do not obscure the equipment's nameplate(s). It is hard to believe that some installers mount the disconnect switch on top of the equipment's nameplate, making it impossible to read the nameplate.

- **Disconnecting Means Rating:** Figure 23-21 shows a fusible and a nonfusible "pull-out" disconnect commonly used for residential air-conditioning and/or heat pump installations. The horsepower rating of the disconnecting means must be at least equal to the sum of all of the individual loads within the end-use equipment,

at rated load conditions, and at locked rotor conditions, *440.12(B)(1)*.

The ampere rating of the disconnecting means must be at least 115% of the sum of all of the individual loads within the end-use equipment, at rated load conditions, *440.12(B)(2)*. Installing a disconnect switch that equals or exceeds the MOP value will in almost all cases be the correct choice. The MOP in our example is 40 amperes. Thus, a 60-ampere disconnect switch is the correct size. In cases where the MOP value is close to the ampere rating of the disconnect switch (i.e., 30, 60, 100, 200, etc.), *locked-rotor current* values must be considered.

- **Locked-Rotor Current (LRC):** This is the maximum current draw when the motor is in a "LOCKED" position. When the rotor is locked, it is not turning. The disconnect switch and controller (if used) must be capable of safely interrupting locked-rotor current. See *440.12* and *440.41*. In our example, the nameplate indicates that the hermetic refrigerant motor-compressor has an LRC of 107 amperes. The fan motor has an FLA of 1.5 amperes and would have an LRC approximately 6 times higher. According to *440.12(B)(1)(b)*, all locked-rotor currents and other loads are added together when combined loads are involved. Therefore, the total locked-rotor current that the disconnect switch in our example would be called on to interrupt is

$$(1.5 \times 6) + 107 = 116 \text{ locked-rotor amperes}$$

Checking *Table 430.251(A)* for the conversion of locked-rotor current to horsepower, we find that the locked-rotor current for a 230-volt, single-phase, 3-horsepower motor is 102 amperes, and the locked-rotor current for a 230-volt, single-phase, 5-horsepower motor is 168 amperes. Selecting a disconnect switch on the basis of the locked-rotor current in our example, the air-conditioning unit is considered to be a 5-horsepower unit, because 116 falls between 102 and 168. Manufacturers' technical literature indicates that a 30-ampere, 240-volt heavy-duty, single-phase disconnect switch has a 3-horsepower rating, and a 60-ampere, 240-volt heavy-duty, single-phase disconnect switch has a 10-horsepower rating. Thus, a 60-ampere disconnect is selected. Note

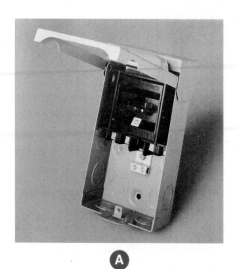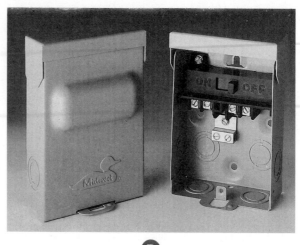

FIGURE 23-21 (A) is a pull-out fusible disconnect. Fuses are inserted into fuse clips on the pull-out device. Insert the pull-out device for "On," and remove the pull-out device for "Off." (B) is a nonfusible disconnect, available in 30- and 60-ampere ratings at 240 volts. Both have padlock provisions to prevent unauthorized tampering. Air-conditioner disconnects are available that have a GFCI receptacle as part of the disconnect enclosure. The GFCI receptacle must be wired to a separate 20-ampere, 120-volt branch circuit. (*Courtesy of Midwest Electric Products*)

that this is the same size disconnect selected previously on the basis of MOP.

ENERGY RATINGS

Energy ratings indicate the efficiency of heating and cooling equipment. Basically, these ratings are a comparison of output (heating or cooling) to input (electricity, gas, or oil).

Energy Efficiency Rating (EER): Cooling efficiency rating for room air conditioners. The ratio of the rated cooling capacity in Btu per hour divided by the amount of electrical power used in kilowatt-hours. The higher the EER number, the greater the efficiency.

Seasonal Energy Efficiency Rating (SEER): Cooling efficiency rating for central air conditioners and heat pumps. SEER is determined by the total cooling of an air conditioner or heat pump in Btu during its normal usage period for cooling divided by the total electrical energy input in kilowatt-hours during the same period. The higher the SEER number, the greater the efficiency.

Annual Fuel Utilization Efficiency (AFUE): Tells you how efficiently a furnace converts

fuel (gas or oil) to heat. For example, an AFUE of 85% means that 85% of the fuel is used to heat your home, and the other 15% goes up the chimney. The higher the efficiency, the lower the operating cost. Old furnaces might have an AFUE rating as low as 60%. Mid-efficiency ratings are approximately 80%. High-efficiency ratings are 90% or higher. Maximum furnace efficiency available is approximately 96.6%. The higher the AFUE number, the greater the efficiency.

Heating Seasonal Performance Factor (HSPF): Heating efficiency of a heat pump. HSPF is determined by the total heating of a heat pump in Btu during its normal usage period for heating divided by the total electrical energy input in kilowatt-hours during the same period. The higher the HSPF rating, the greater the efficiency.

NONCOINCIDENT LOADS

Loads such as heating and air conditioning are not likely to operate at the same time. The *NEC* recognizes this diversity in *220.60*. Therefore, when

calculating a feeder that supplies both types of loads, or when sizing service equipment, only the larger of the two loads need be considered. This is discussed in Chapter 29 where service-entrance calculations are presented. Of course, the branch circuit supplying the heating load is sized for that particular load, and the branch circuit supplying the air-conditioning load is sized for that particular load.

RECEPTACLE NEEDED FOR SERVICING HVAC EQUIPMENT

For servicing HVAC equipment, *NEC 210.63* requires that a 125-volt, single-phase, 15- or 20-ampere-rated receptacle be installed

- at an accessible location,
- on the same level as the HVAC equipment,
- within 25 ft (7.5 m) of the equipment, and
- the receptacle shall not be connected to the load side of the equipment disconnecting means.

The outdoor receptacles required by *210.52(E)* (one in front and one in back) *might* or *might not* meet the requirement of *210.63*. If the HVAC equipment is located on a roof, in an attic, or a similar location, a receptacle must be installed in that location so as to be on the same level as the equipment.

A receptacle is not required if the equipment served is an evaporative cooler in one- and two-family dwellings. This type of equipment requires very little servicing.

Some air-conditioner disconnects have a GFCI receptacle as an integral part of the disconnect enclosure, Figure 23-22. The GFCI receptacle must be supplied by a separate 15- or 20-ampere, 120-volt branch circuit. Installing this type of air-conditioner disconnect eliminates the need for installing a receptacle as part of the premises wiring. But this could also become a nightmare trying to figure out how to wire it. When wiring with this type of disconnect, both the AC branch circuit *and* the 20-ampere 120-volt branch circuit for the GFCI receptacle must somehow be run to the disconnect. This might require more time and material than wiring the required receptacle for the AC

FIGURE 23-22 Photo of an AC disconnect that has an integral GFCI receptacle. (*Courtesy of Midwest Electric Products, Inc.*)

as required by *210.63* in the customary way. The choice is yours.

Adjustment factors might have to be applied when more than three current-carrying conductors are installed in the same raceway or cable, *310.15(B)(2)*. All of this depends on the wiring method: is it NM cable or is it a raceway?

GAS EXPLOSION HAZARD

Although it might seem a bit out of place in an electrical book to talk about gas explosions, we must talk about it.

Often overlooked is a requirement in 2.7.2(c) of NFPA 54, National Fuel Gas Code, that gas meters be located at least 3 ft (900 mm) from sources of ignition. An electric meter or a disconnect switch are possible sources of ignition. Some utility regulations require a minimum of 3 ft (900 mm) clearance between electric metering equipment and gas meters and gas-regulating equipment. It is better to be safe than sorry! Check this issue out with the local electrical inspector and/or the local electric utility before installing the air-conditioner or heat-pump disconnect on the outside of the house. Also consider the location of the service watt-hour meter.

REVIEW

Note: Refer to the *Code* or the plans where necessary.

∿ ELECTRIC HEAT

1. a. What is the allowance in watts made for electric heat in this residence? _____

   b. What is the value in amperes of this load? _____

   _____

2. What are some of the advantages of electric heating? _____

   _____

   _____

   _____

3. List the different types of electric heating system installations. _____

   _____

   _____

   _____

4. There are two basic voltage classifications for thermostats. What are they? _____

   _____

5. What device is required when the total connected load exceeds the maximum rating of a thermostat? _____

6. The electric heat in this residence is provided by what type of equipment? _____

   _____

7. At what voltage does the electric furnace operate? _____

8. A certain type of control connects electric heating units to a 120-volt supply or a 240-volt supply, depending on the amount of the temperature drop in a room. These controls are supplied from a 120/240-volt, 3-wire, single-phase source. Assuming that this type of device controls a 240-volt, 2000-watt heating unit, what is the wattage produced when the control supplies 120 volts to the heating unit? Show all calculations.

   _____

   _____

   _____

9. What advantages does a 240-volt heating unit have over a 120-volt heating unit?

   _____

   _____

   _____

10. The white wire of a cable may be used to connect to a "hot" ungrounded circuit conductor only if _____

   _____

11. Receptacle outlets furnished as part of a permanently installed electric baseboard heater, when not connected to the heater's branch circuit, (may) (may not) be counted as the required receptacle outlet for the space occupied by the baseboard heater. Electric baseboard heaters (shall) (shall not) be installed beneath wall receptacle outlets unless the instructions furnished with the heaters indicate that it is acceptable to install the heater below a receptacle outlet. Circle the correct answers.

12. The branch circuit supplying a fixed electric space heater must be sized to at least _____ percent of the heater's rating according to *NEC* _____.

13. Calculate the current draw of the following electric furnaces. The furnaces are all rated 240 volts.

    a. 7.5 kW _____ amperes   b. 15 kW _____ amperes   c. 20 kW _____ amperes

14. For "ballpark" calculations, the wattage output of a 240-volt electric furnace connected to a 208-volt supply will be approximately 75% of the wattage output had the furnace been connected to a 240-volt supply. In Question 13, calculate the wattage output of (a), (b), and (c) if connected to 208 volts. _____

    _____

    _____

    _____

15. A central electric furnace heating system is installed in a home. The circuit supplying this furnace

    a. may be connected to a circuit that supplies other loads.

    b. shall be connected to a circuit that supplies other loads.

    c. shall be connected to a separate circuit.

16. What section of the *Code* provides the correct answer to Question 15? _____

17. Electric heating cable embedded in plaster, or "sandwiched" between layers of dry wall, creates heat. Therefore, any nonmetallic-sheathed cable run above these ceilings and buried in insulation must have its current-carrying capacity derated according to the (40°C) (50°C) (60°C) correction factors found at the bottom of *Table 310.15(B)(16)* in the *NEC*. Circle the correct answer.

AIR CONDITIONING

1. a. When calculating air-conditioner load requirements and electric heating load requirements, is it necessary to add the two loads together to determine the combined load on the system? _____

    b. Explain the answer to part (a). _____

    _____

    _____

2. The total load of an air conditioner shall not exceed what percentage of a separate branch circuit? (circle the correct answer)

    a. 75%   b. 80%   c. 125%

3. The total load of an air conditioner shall not exceed what percentage of a branch circuit that also supplies lighting? (circle the correct answer)

   a. 50%   b. 75%   c. 80%

4. a. Must an air conditioner installed in a window opening be grounded if a person on the ground outside the building can touch the air conditioner? _____

   b. What *Code* section governs the answer to part (a)? _____

5. A 120-volt air conditioner draws 13 amperes. What size is the circuit to which the air conditioner will be connected? _____

6. What is the *Code* requirement for receptacles connected to circuits of different voltages and installed in one building? _____

7. When a central air-conditioning unit is installed and the label states "Maximum Size Fuse 50 Amperes," is it permissible to connect the unit to a 50-ampere circuit breaker?_____

8. When the nameplate on an air-conditioning unit states "Maximum Size Fuse or HACR Circuit Breaker," what type of circuit breaker must be used? _____

   _____

9. What section of the *NEC* prohibits the installation of Class 2 control circuit conductors in the same raceway as the power conductors? _____

   _____

10. Match the following terms with the statement that most closely defines the term. Enter the letters in the blank space.

    RLC _____   a. The maximum ampere rating of the overcurrent device

    MCA _____   b. The current value to be used instead of the rated-load ampere

    BCSC _____   c. The value used to determine the minimum ampere rating for the branch circuit

    MCC _____   d. A current value for the hermetic motor-compressor that was determined by operating it at rated temperature, voltage, and frequency

    MOP _____   e. The maximum current that the compressor can draw continuously without damage

11. The disconnect for an air conditioner or heat pump must be installed _____ of the unit.

Gas and Oil Central Heating Systems

OBJECTIVES

After studying this chapter, you should be able to

- understand the basics of typical home warm air and hot water (hydronic) heating systems.

- understand and apply the *NEC* requirements for branch-circuit wiring for central heating systems.

- define all of the major components of typical heating systems.

- understand and apply the *NEC* requirements for Class 2 control circuit wiring.

In Chapter 23, we discussed various types of electric heating systems. In this chapter, we will take a look at gas- and oil-fired systems.

FORCED–WARM AIR FURNACES

Gas- and oil-fired heating systems provide the heat source for forced-warm air furnaces and hot water systems. Gas-fired, forced-warm air furnaces are the most common. Forced-warm air systems move hot air from the furnace through hot air ducts that in turn connect to hot air registers strategically located throughout the home. The room air is then pulled into cold air registers, returning to the furnace through cold air return ducts, where the air is again filtered, heated, and forced out through the hot air ducts.

Air-conditioning, humidity control, and fresh air intake are but a few of the things that can be added to a typical forced-warm air heating system.

HOT WATER SYSTEMS

Hot water systems move hot water through pipes to radiators (i.e., baseboards) and, in some instances, through tubing embedded in the ceiling or in the floor or on the underside of the floor to heat the floors from below. The water is returned to the boiler through a return pipe. Hot water systems are very adaptable to zone control by using a combination of one or more circulating pumps and/or valves for the different zones. Hot water heating systems are referred to as *hydronic* (wet) systems.

PRINCIPLE OF OPERATION

Most residential heating systems operate as follows. A room thermostat is connected to the proper electrical terminals on the furnace or boiler. Inside the equipment, the various controls and valves are interconnected in a manner that will provide safe and adequate operation of the furnace.

Most residential heating systems are "packaged units" in which the burner, hot surface igniter or hot spark igniter, safety controls, valves, fan controls, sensors, high-temperature-limit switches, fan blower motor, blower fan speed control, draft inducer motor, primary and secondary heat exchanger, and so on, are preassembled, prewired parts of the furnace or boiler. Many of these items are included in an integrated printed circuit board module, which is the "brain" of the system. This module provides proper sequencing and safety features.

Older models have standing gas pilot lights and thermocouples.

Similar electronic circuitry is used in oil burners where the ignition, the oil burner motor, blower fan motor, high-temperature limit controls, hot water circulating pump in the case of hot water systems, and the other devices previously listed are all interconnected to provide a packaged oil burner unit.

Evaporator coils for air-conditioning purposes might be an integral part of the unit, or they might be mounted in the ductwork above the furnace.

Wiring a Residential Central Heating System

Gas- and oil-fired heating systems are usually "packaged units," which means that all of the internal wiring of the components has been done by the manufacturer of the unit.

In Chapter 23, we discussed the major *NEC* requirements for wiring an electric furnace.

For the electrician, the "field wiring" for a typical residential furnace or boiler consists of the following:

1. Installing and connecting the low-voltage wiring between the furnace and the thermostat. This might be a cable consisting of two, three, four, or five small 18 AWG or 20 AWG conductors. Small knockouts are provided on the furnace through which the low-voltage Class 2 wires are brought into the furnace's wiring compartment (see Figure 24-1). Class 2 wiring must not be run through the same raceway or cable as power wiring, *725.136(A)*.

2. Installing and connecting the branch-circuit power supply to the furnace. Knockouts are provided on the furnace through which the line voltage power supply is brought into the wiring

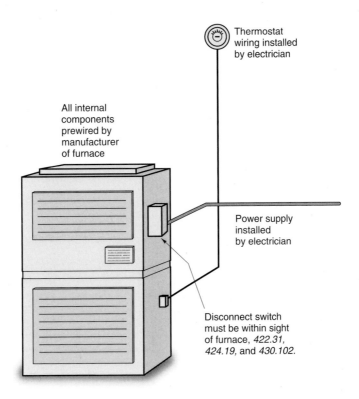

All internal
components
prewired by
manufacturer
of furnace

Thermostat
wiring installed
by electrician

Power supply
installed
by electrician

Disconnect switch
must be within sight
of furnace, *422.31,*
424.19, and *430.102.*

FIGURE 24-1 Diagram showing thermostat wiring, power supply wiring, and disconnect switch.
ANSI Standard Z21.47 does not allow cord-and-plug connections for a gas furnace.
(*Delmar/Cengage Learning*)

compartment of the furnace. Depending on local codes, the line voltage power supply wiring might be in EMT, armored cable, or other accepted wiring methods (Figure 24-1).

Disconnecting Means

The disconnecting means must be within sight of the furnace, *422.31, 422.32, 424.19,* and *430.102.* The *NEC* defines "within sight" as being visible and not more than 50 ft (15 m) from the equipment the disconnect controls.

Some furnaces come with a disconnect switch. If not, a toggle switch can be mounted on a box on the side of the furnace. This is simple for gas-fired furnaces. Electric furnaces require a larger ampere-rated disconnect. Some electric furnaces come complete with circuit breakers that provide the overcurrent protection for the heating elements as required by UL Standard, and to serve as the disconnecting means required by the *NEC.*

The disconnecting means shall have an ampere rating not less than 125% of the total load of the motors and the heaters. If the disconnecting means (switch or circuit breaker) is required to have a lock-off position, such as when the disconnecting means is not within sight of the equipment, then that disconnect provision for lock-off must remain in place with or without a lock installed.

Individual Branch Circuit Required

NEC 422.12 requires that central heating equipment be supplied by an individual branch circuit. By definition in the *NEC*, an individual branch circuit is *A branch circuit that supplies only one utilization equipment.**

The size of the branch circuit depends on the requirements of the particular furnace. A typical gas-forced warm air furnace might require only a 120-volt, 15-ampere branch circuit, whereas an electric furnace requires a 240-volt branch circuit and much larger conductors. Instructions furnished with a given furnace will indicate the branch-circuit requirements.

---

*Reprinted with permission from NFPA 70-2011.

Some typical wiring diagrams are shown in Figures 24-2, 24-3, 24-4(A), and 24-4(B). These wiring diagrams show the individual electrical components of a typical residential heating system. In older systems, many of these components were wired by the electrician. In newer systems, most components other than the thermostat are prewired as part of a packaged furnace or boiler. The manufacturer's literature includes detailed wiring diagrams for a given model.

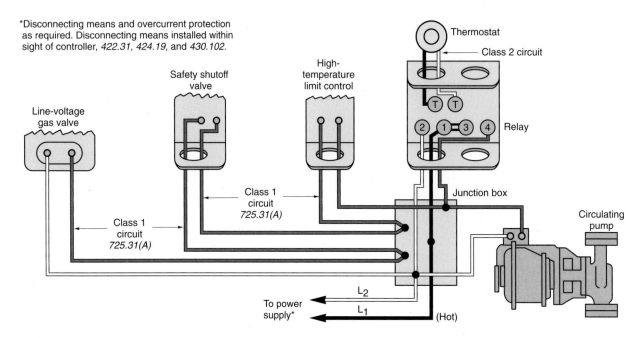

FIGURE 24-2 Typical wiring diagram for a gas burner, forced hot water system. Always consult the manufacturer's wiring diagram relating to a particular model heating unit. (*Delmar/Cengage Learning*)

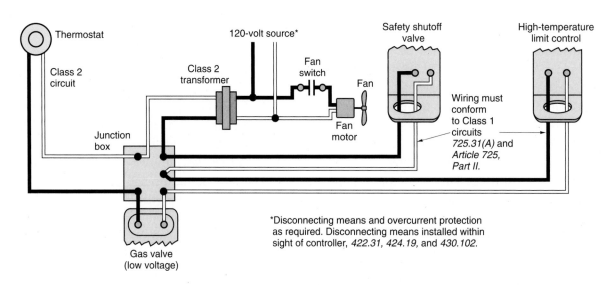

FIGURE 24-3 Typical wiring diagram for a gas burner, forced-warm air system. Always consult the manufacturer's wiring diagrams relating to a particular model heating unit. (*Delmar/Cengage Learning*)

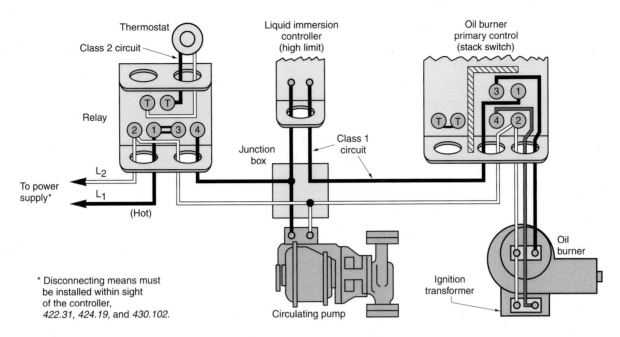

FIGURE 24-4(A)   Typical wiring diagram of an oil burner heating installation. Always consult the manufacturer's wiring diagram relating to a particular model heating unit.
(*Delmar/Cengage Learning*)

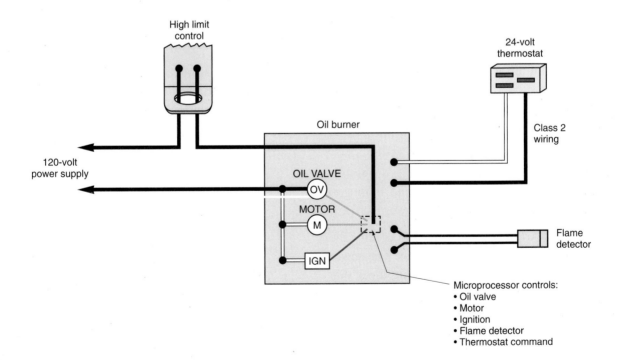

FIGURE 24-4(B)   An oil burner featuring an electronic microprocessor that controls all facets of the burner operation. Most of the components are an integral part of the unit, thus keeping field wiring to a minimum. All manufacturers provide detailed wiring diagrams for their products. Always consult the manufacturer's wiring diagrams relating to a particular model heating unit.
(*Delmar/Cengage Learning*)

⏤〜〜⏤ MAJOR COMPONENTS

Gas- and oil-fired warm air systems and hot water boilers contain many individual components. Here is a brief description of these components.

Aquastat: An aquastat is a direct immersion water temperature thermostat that regulates boiler or tank temperature in hydronic heating systems, ensuring that the circulating water maintains proper, satisfactory temperature.

Cad Cell: A cad cell is a cadmium sulfide sensor for sensing flame. Cad cells have pretty much replaced the older style stack-mounted bimetal switches.

Circulating Pump: A circulating pump circulates hot water in a hydronic central heating system.

Control Circuit: A control circuit is any electric circuit that controls any other circuit through a relay or an equivalent device; also referred to as a remote-control circuit.

Combustion Chamber: The combustion chamber surrounds the flame and radiates heat back into the flame to aid in combustion.

Combustion Head: A combustion head creates a specific pattern of air at the end of the air tube. The air is directed in such a way as to force oxygen into the oil spray so the oil can burn. A combustion head might also be referred to as the turbulator, fire ring, retention ring, or end cone.

Draft Regulator: A draft regulator is a counter-weighted swinging door that opens and closes to help maintain a constant level of draft over the fire.

Fan Control: A fan control is used to control the blower fan on forced-warm air systems. On newer furnaces, the fan control is an electronic timer that starts the fan motor a given time in seconds after the main burner has come on. This timing ensures that the blower will blow warm air. Older type adjustable temperature fan controls are still around.

Fan Motor: The fan motor is the electric motor that forces warm air through the heating ducts. In typical residential heating systems, the motor is provided with integral overload protection. The motor may be single, multiple, or variable speed.

Flue: A flue is a channel in a chimney or a pipe for conveying flame and smoke to the outer air.

Heat Exchanger: A heat exchanger transfers the heat energy from the combustion gases to the air in the furnace or to the water in a boiler.

High-Temperature Limit Control: A high-temperature limit control is a safety device that limits the temperature in the plenum of the furnace to some predetermined safe value as determined by the manufacturer. When this predetermined temperature is reached, the high-temperature limit control shuts off the power to the burner. In newer furnaces, the high-temperature limit control is part of an integrated circuit board and is communicated to by a sensor strategically located in the plenum. Older furnaces might have individual high-temperature limit controls and a separate fan control, or they might have combination high-temperature limit/fan controls.

Hot Surface Igniter: In furnaces that do not have a standing constant pilot light, the hot surface igniter heats to a cherry red/orange color. When it reaches the proper temperature, the main gas valve is allowed to open. If the hot surface igniter does not come on, the main gas valve will not open. Most energy-efficient furnaces use this concept for ignition.

Hydronic System: The hydronic system is a system of heating or cooling that involves transfer of heat by circulating fluid such as water in a closed system of pipes.

Ignition:

- **Constant:** The igniter is designed to stay on continuously.

- **Intermittent Duty:** Defined by UL 296 as "ignition by an energy source that is continuously maintained throughout the time the burner is firing." In other words, the igniter is on the entire time the burner is firing.

- **Interrupted Duty:** Defined by UL 296 as "an ignition system that is energized each time the main burner is to be fired and de-energized at the end of a timed trial for ignition period or after the main flame is proven to be established." In other words, the igniter comes on to light the flame; then, after the flame is established, the igniter is turned off and the main flame keeps burning.

Induced Draft Blower: When the thermostat calls for heat, the induced draft blower starts first to expel any gases remaining in the combustion chamber from a previous burning cycle. It continues to run, pulling hot combustion gases through the heat exchangers, then vents the gases to the outdoors.

Integrated Control: An integrated control is a printed circuit board containing many electronic components. When the thermostat calls for heat, the integrated circuit board takes over to manage the sequence of events that allow the burner to operate safely.

Liquid Immersion Control: This is also referred to as an aquastat. A liquid immersion control controls high water and circulating water temperature in a hot water system. A high-temperature limit control shuts off the burner when a predetermined dangerous high temperature is reached. When acting as a circulating water temperature control, this control makes sure that the water being circulated through the piping system is at the desired temperature—not cold.

Low-Water Control: Also referred to as a low-water cutoff, this device senses low-water levels in a boiler. When a low-water situation occurs, the low-water control shuts off the electrical power to the burner.

Main Gas Valve: The main gas valve is the valve that allows the gas to flow to the main burner of a gas-fired furnace. It may also function as a pressure regulator, a safety shut-off, and as the pilot and main gas valve. Should there be a "flame-out," if the pilot light fails to operate, or in the event of a power failure, this valve will shut off the flow of gas to the main burner.

Nozzle: A nozzle produces the desired spray pattern for the particular appliance in which the burner is used.

Oil Burner: The function of an oil burner is to break fuel oil into small droplets, mix the droplets with air, and ignite the resulting spray to form a flame.

Primary Control: A primary control controls the oil burner motor, ignition, and oil valve in response to commands from a thermostat. If the oil fails to ignite, the controller shuts down the oil burner.

Pump and Zone Controls: Pump and zone controls regulate the flow of water or steam in boiler systems to specific "zones" in the building.

Safety Controls: Safety controls such as pressure relief valves, high-temperature limit controls, low-water cutoffs, and burner primary controls protect against appliance malfunction.

Solenoid: A solenoid is an electrically operated device, which, when the valve coil is energized, a magnetic field is developed, causing a spring-loaded steel valve piston to overcome the resistance of the spring and immediately pull the piston into the stem. The valve is now in the open position. When de-energized, the solenoid coil magnetic field instantly dissipates, and the spring-loaded valve piston snaps closed, stopping oil flow or gas flow to the nozzle. The flame is extinguished, allowing fuel to flow. Usually a spring returns the valve back to the closed position.

Spark Ignition: In spark ignition systems, the spark comes on while pilot gas flows to the pilot orifice. Once the pilot flame is "proven" through an electronic sensing circuit, the main gas valve opens. Once ignition takes place, the sensor will monitor and prove existence of a main flame. If for some reason the main flame goes off, the furnace shuts down. Energy-efficient furnaces may use this concept for ignition.

Switching Relay: A switching relay is used between the low-voltage wiring and line-voltage wiring. When a low-voltage thermostat calls for heat, the relay "pulls in," closing the line-voltage contacts on the relay, which turns on the main power to the furnace or boiler. Usually relays have a built-in transformer that provides the low-voltage Class 2 control circuit.

Thermocouple: A thermocouple is used in systems having a standing pilot to hold open a gas valve pilot solenoid magnet. A thermocouple consists of two dissimilar metals connected together to form a circuit. The metals might be iron and copper, copper and iron constantan, copper-nickel alloy and chrome-iron alloy, platinum and platinum-rhodium alloy, or chromel and alumel. The types of metals used depend on the temperatures involved. When one of the junctions is heated, an electrical current flows (Figure 24-5). To operate, there must be a temperature difference between the metal junctions. In a gas burner, the source of heat for the thermocouple is the pilot light. The cold junction of the thermocouple remains open and is connected to the pilot safety shut-off gas valve circuit. A single

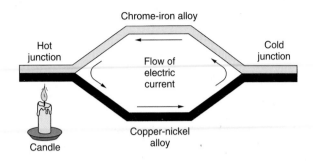

FIGURE 24-5 Principle of a thermocouple.
(*Delmar/Cengage Learning*)

thermocouple develops a voltage of about 25 to 30 dc millivolts. Because of this extremely low voltage, circuit resistance is kept to a very low value.

Thermopile: More than one thermocouple connected in series is a thermopile. See Figure 24-6. The power output of a thermopile is greater than that of a single thermocouple—typically either 250 or 750 dc millivolts. For example, 10 thermocouples (25 dc millivolts each) connected in series result in a 250 dc millivolt thermopile. Twenty-six thermocouples connected in series result in a 750 dc millivolt thermopile.

Thermostat: A thermostat turns the heating/cooling system on and off. Programmable thermostats might have day/night setback settings, a clock, a digital thermometer, humidity control, ventilation, and filtration control. Thermostats are usually mounted approximately 52 in. (1.3 m) above the finished floor. Do not mount thermostats where influenced by drafts, air currents from hot or cold air registers, near fireplaces, near concealed hot or cold water pipes

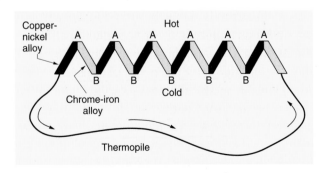

FIGURE 24-6 Principle of a thermopile.
(*Delmar/Cengage Learning*)

FIGURE 24-7 A digital electronic programmable thermostat that provides proper cycling of the heating system, resulting in comfort as well as energy savings. This type of thermostat, depending on the model, can be programmed to control temperature, time of operation, humidity, ventilation, filtration, circulation of air, and zone control. Installation and operational instructions are furnished with the thermostat.
(*Courtesy of Honeywell International Inc.*)

or ducts, in the direct rays of the sun, or on outside walls. See Figure 24-7. Some thermostats contain a small vial of mercury that "tips," so the mercury closes the circuit by bridging the gap between the electrical contacts inside the vial. These thermostats must be kept perfectly level to ensure accuracy.

WARNING: Do not throw mercury thermostats into the trash. The mercury is considered a hazardous waste material. They must be returned to a waste management organization for proper disposal.

Transformer: The transformer converts (transforms) the branch-circuit voltage to low voltage. For residential applications these are Class 2 transformers, 120 volts to 24 volts. The low-voltage wiring on the secondary of this transformer is Class 2 wiring.

Water Circulating Pump: In hydronic (wet) systems, a circulating pump circulates hot water through the piping system. Multizone systems have more than one pump, each controlled by a thermostat for a particular zone. In multizone systems, a high-temperature control (aquastat) might be set to maintain the boiler water temperature to a specific temperature. Some multizone systems have one circulating pump and use electrically operated valves that open and close on command from the thermostat(s) located in a particular zone(s).

CLASS 2 CIRCUITS

Remote-control, signaling, and power-limited circuits fall into three categories—Class 1, Class 2, and Class 3.

Class 1, Class 2, and 3 circuits offer alternative wiring methods regarding voltage, power limitations, wire size, derating factors, overcurrent protection, physical protection, insulation, and materials that differentiate these types of circuits from conventional wiring methods, as covered in *Chapter 1* through *Chapter 4* in the *NEC*. Figure 24-8 defines remote control, signaling, and power-limited terminology.

Low-voltage wiring in homes is generally Class 2 wiring. Class 2 wiring is very tolerant. We will limit our discussion to Class 2 wiring. Class 1 and Class 3 circuits are more restrictive than Class 2 circuits and are more commonly found in commercial and industrial applications. You can learn about Class 1 and Class 3 circuits in *Article 725* in the *NEC*.

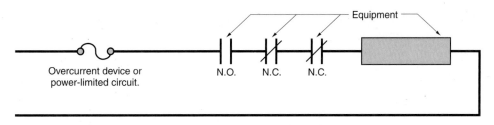

(A) A remote-control, signaling, or power-limited circuit is that portion of the wiring system between the load side of the overcurrent device or the power-limited supply and the connected equipment, *725.2*.

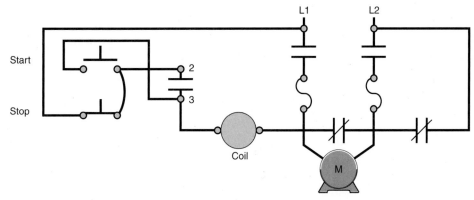

(B) A remote-control circuit is any electrical circuit that controls any other circuit through a relay or equivalent device, *Article 100, Definitions*.

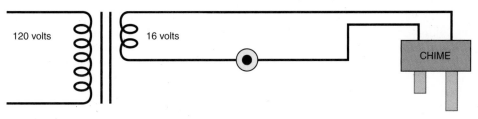

(C) A signal circuit is any electrical circuit that energizes signaling equipment, *Article 100, Definitions*.

FIGURE 24-8 The above diagrams explain a remote-control circuit and a signaling circuit.
(*Delmar/Cengage Learning*)

Wiring for Class 2 Circuits

Electricians describe wire and cable used for low-voltage wiring *bell wire* or *thermostat wire*. What you are really looking for are listed Class 2 conductors and cables, referred to as CL2 or CL2X cables. Examples of Class 2 wiring are the low-voltage conductors and cables used for heating and air conditioning, thermostats, security systems, remote control and signal wiring (chimes), intercom wiring, and other low-voltage applications.

Class 2 conductors are generally made of copper, have a thermoplastic insulation, and have a voltage rating of not less than 150 volts as required by *725.129*. Because the current required for Class 2 circuits is rather small, 18 AWG and 20 AWG conductors are most commonly used. Conductors as small as 24 AWG also are available; however, small conductors and long runs can result in voltage drop problems.

Multiconductor Class 2 cables consist of two to as many as 12 single conductors. These cables are available with or without a protective PVC outer jacket. The type with the outer jacket is preferred because there is less chance of damage to individual conductors, and it gives a neat appearance.

The conductors within these cables are color coded to make circuit identification easy. Table 24-1 is a table showing the color coding for the conductors in Class 2 cables.

Here is a summary of the *NEC* requirements for installing of Class 2 wiring:

- Class 2 circuits are "power-limited" and are considered to be safe from fire hazard and electric shock hazard.

- Class 2 wiring does not have to be in a raceway. If a failure in the wiring should occur, such as a short circuit or an open circuit, there would not be a direct fire or shock hazard.

- Class 2 wiring is not permitted in the same raceway, cable, compartment, outlet box, or fitting with light and power conductors, *725.136(A)*. Even if the Class 2 conductors have the same 600-volt insulation as the power conductors, this is not permitted. See Figures 24-9 and 24-10. Manufacturers of heating and cooling equipment provide separate openings for bringing in the Class 2 wiring and separate openings for bringing in the power supply. *NEC 725.136(D)* addresses this issue.

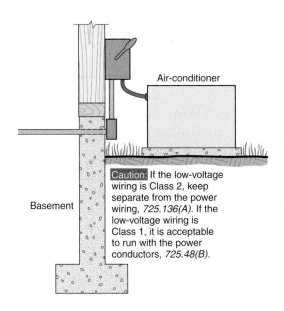

FIGURE 24-9   Most residential air-conditioning units, furnaces, and heat pumps have their low-voltage circuitry classified as Class 2. *NEC 725.136(A)* prohibits installing low-voltage Class 2 conductors in the same raceway as the power conductors, even if the Class 2 conductors have 600-volt insulation. Therefore, the conduit running out of the basement wall in the above diagram would not be permitted to contain both the 240-volt power conductors and the 24-volt low-voltage Class 2 conductors. Always read the instructions furnished with these types of appliances to be sure your wiring is in compliance with these instructions, per *110.3(B)* and the *NEC*. Review Chapter 4 for grounding methods.
(Delmar/Cengage Learning)

TABLE 24-1

Table showing the color coding of Class 2 conductors and cables.

Number of Conductors	Color	Number of Conductors	Color
1	Red	7	Orange
2	White	8	Black
3	Green	9	Pink
4	Blue	10	Gray
5	Yellow	11	Tan
6	Brown	12	Purple

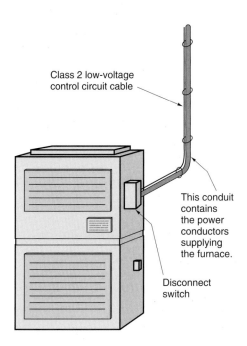

Class 2 low-voltage control circuit cable

This conduit contains the power conductors supplying the furnace.

Disconnect switch

FIGURE 24-10 The *Code* in *300.11(B)(2)* and *725.143* allows Class 2 control circuit conductors (cables) to be supported by the raceway that contains the power conductors supplying electrical equipment, such as the furnace shown. (*Delmar/Cengage Learning*)

- Keep Class 2 conductors at least 2 in. (50 mm) away from light or power wiring. This really pertains to old, open-knob-and-tube wiring. Today, however, where light and power wiring is installed in Type NMC, Type AC, or in conduit, there is no problem, *725.136(I)*.

- Class 2 wiring shall not enter the same enclosure unless there is a barrier in the enclosure that separates the Class 2 wiring from the light and power conductors, *725.136(B)*.

- Class 2 circuits are inherently current limiting and do not require separate overcurrent protection. A Class 2 transformer is an example of this.

- Transformers that are intended to supply Class 2 circuits are listed and marked "Class 2 Transformer." These transformers have built-in overcurrent protection to prevent overheating should a short circuit occur somewhere in the secondary winding or in the secondary circuit wiring. These transformers may be connected to branch circuits

having overcurrent protection not more than 20 amperes, *725.127*.

- Class 2 cables shall have a voltage rating of not less than 150 volts, *725.179(G)*.

Securing Class 2 Cables

- Class 2 cables must be installed in a neat and workmanlike manner and shall be adequately supported by the building structure in such a manner that they will not be damaged by normal building use, *725.24*.

- Class 2 cables shall be attached to structural components by straps, staples, hangers, or similar fittings designed and installed so as not to damage the cable, *725.24*.

- Class 2 cables may be secured directly to surfaces with insulated staples or may be installed in raceways.

- The insulation and jacket of Class 2 cables and conductors are rather thin. Be careful during installation not to pierce or crush the wires. Secure these cables and conductors with care, using the proper type of staples. Staple "guns" that use rounded staples are available. They straddle the cable nicely instead of flattening it, which might result in shorted-out wires.

- Do not support Class 2 cables from other raceways, cables, or nonelectric equipment. Taping, strapping, tie wrapping, hanging, or securing Class 2 cables to electrical raceways, piping, or ducts is generally prohibited, *725.143*.

- *NEC 725.143* refers back to *300.11(B)(2)*, which allows Class 2 conductors to be supported by the raceway that contains the power supply conductors for the same equipment that the Class 2 conductors are connected to. See Figure 24-10.

- Do not run Class 2 conductors or cables through the same holes in studs and joists, holes that contain nonmetallic-sheathed cable, armored cable, raceways, conduits, or other pipes. There is too much chance of physical damage to the Class 2 cables.

Class 2 Power Source Requirements

Class 2 power source voltage and current limitations, and overcurrent protection requirements, are found in *Chapter 9, Table 11(A)* and *Table 11(B)* of the *NEC*. You should refer to the nameplate data on Class 2 transformers, power supplies, and other equipment to verify that they are listed by a Nationally Recognized Testing Laboratory (NRTL) for a particular class.

REVIEW

1. The residence in this text is heated with

   a. gas.

   b. electricity.

   c. oil.

   Circle the correct answer.

2. The *NEC* in_____, _____, _____, and _____ requires that a disconnecting means be _____ _____ of the furnace.

3. The *NEC* in _____ requires that a central heating system be supplied by a _____branch circuit.

4. In a home, the low-voltage wiring between the thermostat and furnace is

   a. Class 1.

   b. Class 2.

   c. Class 3.

   Circle the correct answer.

5. Class 2 circuits are _____ -limited and are considered to be _____ from _____ hazard and from _____.

6. The nameplate on a transformer will indicate whether or not it is a Class 2 transformer.

   (True)   (False)   Circle the correct answer.

7. If Class 2 conductors are insulated for 150 volts, and the power conductors are insulated for 600 volts, does the *Code* permit pulling these conductors through the same raceway?

   (Yes)   (No)   Circle the correct answer. Give the *NEC* section number. _____

8. If you use 600-volt rated conductors for Class 2 wiring, does the *Code* permit pulling these conductors through the same raceway as the power conductors? (Yes)   (No) Circle the correct answer. Give the *NEC* section number. _____

9. The *NEC* is very strict about securing anything to electrical raceways. Name the *Code* sections that prohibit this practice. *NEC* _____

   _____

10. Does the *Code* permit attaching the low-voltage thermostat cable to the EMT that brings power to a furnace? (Yes) (No) Circle the correct answer. Give the *NEC* section numbers. _____

11. Explain what a thermocouple is and how it operates. _____

_____

_____

_____

_____

CHAPTER 25

Television, Telephone, and Low-Voltage Signal Systems

OBJECTIVES

After studying this chapter, you should be able to

- install residential telephone and television wiring, antennas, and CATV cables, in conformance to *NEC* requirements.

- describe the basic operation of satellite antennas.

- install typical low-voltage wiring for chimes.

⟨ᴧ⟩ INSTALLING THE WIRING FOR HOME TELEVISION

This chapter discusses the basics of installing outlets, cables, receivers, antennas, amplifiers, multiset couplers, and some of the *NEC* rules for home television. Television is a highly technical and complex field. To ensure a trouble-free installation of a home system, a competent television technician should do the work. These individuals receive training and certification through the Electronic Technicians Association in much the same way that electricians receive their training through apprenticeship and journeyman training programs.

Some cities and states require that, when more than three television outlets are to be installed in a residence, the television technician making the installation be licensed and certified. Just as electricians can experience problems because of poor connections, terminations, and splices, poor reception problems can arise as a result of poor terminations. In most cases, poor crimping is the culprit. Electrical shock hazards are present when components are improperly grounded and bonded.

According to the plans for this residence, television outlets are installed in the following rooms:

Front Bedroom	2
Kitchen	1
Laundry	1
Living Room	3
Master Bedroom	2
Recreation Room	3
Study/Bedroom	2
Workshop	1
Total	15

Because TV is a low-voltage system, metallic or nonmetallic standard single-gang device boxes, 4-in. square outlet boxes with a plaster ring, plaster rings only, or special mounting brackets can be installed during the rough-in stage at each likely location of a television set (more on this later). See Figure 25-1. For new construction, shielded coaxial cables are installed concealed in the walls. For remodel work, cables can be concealed in the walls by fishing the cables through the walls and installing mounting brackets that are inserted and snapped into place through a hole cut into the wall.

Shielded 75-ohm RG-6 coaxial cable is most often used to hook up television sets to minimize interference and keep the color signal strong. There are different kinds of shielded coaxial cable. Double-shielded cable that has a 100% foil shield covered with a 40% or greater woven braid is recommended. This coaxial cable has a PVC outer jacket. The older style, flat 300-ohm twin-lead cable, can still be found on existing installations, but the reception might be poor. To improve reception, 300-ohm cables can be replaced with shielded 75-ohm coaxial cables.

For this residence, 15 television outlets are to be installed. A shielded 75-ohm RG-6 coaxial cable will be run from each outlet location back to one central point, as in Figure 25-2.

This central point could be where the incoming Community Antenna Television (CATV) cable comes in, where the cable from an antenna comes in, or where the TV output of a satellite receiver is carried. For rooftop or similar antenna mounting, the cable is run down the outside of the house, then into the basement, garage, or other convenient location where the proper connections are made. CATV companies generally run their incoming coaxial

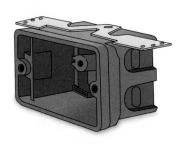

FIGURE 25-1 Nonmetallic boxes and a nonmetallic plaster ring. (*Delmar/Cengage Learning*)

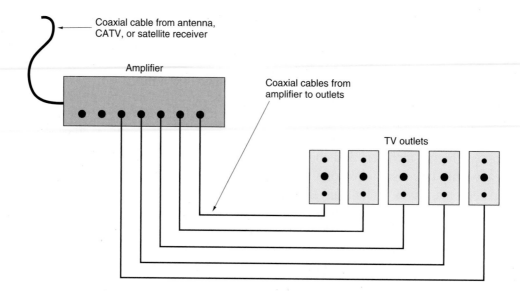

FIGURE 25-2 A television master amplifier distribution system may be needed where many TV outlets are to be installed. This will minimize the signal loss. In simple installations, a multiset coupler can be used, as shown in Figure 25-3. (*Delmar/Cengage Learning*)

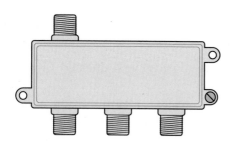

FIGURE 25-3 A 3-way CATV cable splitter: one in, three out. Two-way and 4-way splitters are also commonly available. (*Delmar/Cengage Learning*)

cables underground to some convenient point just inside of the house. Here, the technician hooks up all of the coaxial cables coming from the TV outlets to an amplifier that boosts the signal and improves reception, Figure 25-2, or to a multiset splitter, Figure 25-3. An amplifier may not be necessary if *only* those coaxial cables that will be used are hooked up.

Cable television does not require an antenna on the house because generally the cable company has the proper antenna to receive signals from satellites. The cable company then distributes these signals throughout the community they have contractually agreed to serve. These contracts usually require the cable company to run their coaxial cable to a point just inside the house. Inside the house, they will complete the installation by furnishing cable boxes, controls, and the necessary wiring. In geographical areas where cable television is not available, antennas, as shown in Figure 25-4 and Figure 25-5, are needed.

Faceplates for the TV outlets are available in styles to match most types of electrical faceplates in the home.

Hazards of Mixing Different Voltages

For safety reasons, voice/data/video (VDV) wiring must be separated from the 120-volt wiring. If both 120-volt and low-voltage circuits are run into the same box, then a permanent barrier must be provided in the box to separate the two systems. A better choice is to keep the two systems totally separate using two wall boxes.

Another popular choice is where VDV cables are run to a location right next to a 120-volt receptacle. Mounting brackets that are commercially available can be installed around the electrical device box during the rough-in stage. This lets you trim out a 120-volt receptacle and VDV

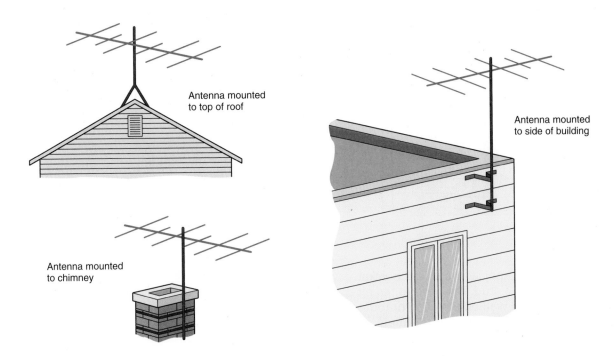

FIGURE 25-4 Older style antennas. Although still available, they have given way to the more popular digital satellite "dish," as shown in Figure 25-5(A). (*Delmar/Cengage Learning*)

FIGURE 25-5 (A) A digital satellite system 18-in. (450 mm) antenna, a receiver, and a remote control. (B) A large satellite antenna securely mounted to a post that is anchored in the ground. These large antennas are rarely used today. (*Photo A courtesy RCA, photo B Delmar/Cengage Learning*)

jacks with one faceplate, as shown in Figure 25-6. These mounting brackets keep the center-to-center measurements of the device mounting holes for the electrical box and the VDV wiring precisely in alignment. With this method,

a wall box is not provided for the VDV wiring. The VDV cables merely come out of the drywall next to the electrical box, and the faceplate takes care of trimming out the receptacle and the VDV wiring jacks.

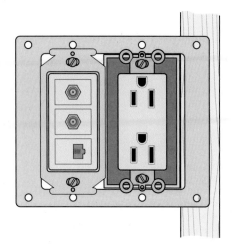

FIGURE 25-6 A special mounting bracket fastened to the wood stud. The bracket fits nicely over the electrical wall box to accommodate a 120-volt receptacle, two coaxial outlets, and one telephone outlet. A single two-gang faceplate is used for this installation. Many other combinations are possible. (*Delmar/Cengage Learning*)

When wiring a new house, it certainly makes sense to run at least one coaxial cable and one Category 5 cable to a single wall box wherever TV, telephone, and/or computers might be used. This could be a single box located close to a 120-volt receptacle outlet, or it could be a mounting bracket as shown in Figure 25-6. Run a separate coaxial cable and a separate Category 5 cable from each location to a common point in the house. You will use a lot of cable, but this will enable you to interconnect the cables as required, such as for a local area network (LAN) to serve more than one computer located in different rooms, or other home automation systems. This is the beginning of "structured wiring," as discussed in Chapter 31.

For adding VDV wiring to an existing home, many types of nonmetallic brackets are available that snap into a hole cut in the drywall and lock in place.

More *NEC* rules are discussed later in this chapter.

Chapter 31 discusses structured wiring using Category 5 unshielded twisted pair (UTP) cable and coaxial shielded cable.

Although shielded coaxial cable minimizes interference, it is good practice to keep the coaxial cables on one side of the stud space and keep the light and power cables on the other side of the stud space.

Intersystem Bonding Termination

An intersystem bonding termination is required to be installed at or near the service equipment and at or near the building disconnecting means if remote buildings are supplied with electric power, Figure 25-7. The intersystem bonding termination is intended to provide a convenient location where systems such as telephone, television, and antenna systems can be bonded together. This ensures that all the electrical systems that supply the building or structure are bonded (connected together). This helps reduce dangerous flashover should overvoltages

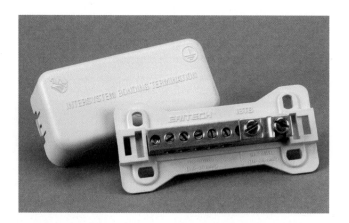

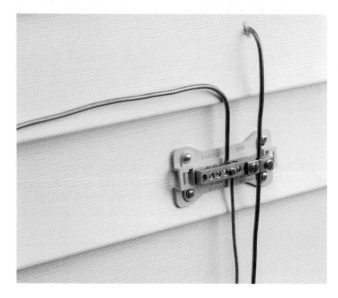

FIGURE 25-7 Intersystem bonding termination equipment. (*Courtesy of Erico*)

be imposed on the electrical systems. These overvoltages originate from lightning events or from problems related to the electric utility system.

The intersystem bonding termination equipment is required to

1. ►be accessible for connection and inspection.

2. consist of a set of terminals with the capacity for connection of not less than three intersystem bonding conductors.

3. not interfere with opening the enclosure for a service, building, or structure disconnecting means, or metering equipment.

4. at the service equipment, be securely mounted and electrically connected to an enclosure for the service equipment, to the meter enclosure, or to an exposed nonflexible metallic service raceway, or be mounted at one of these enclosures and be connected to the enclosure or to the grounding electrode conductor with a minimum 6 AWG copper conductor.

5. at the disconnecting means for a building or structure, be securely mounted and electrically connected to the metallic enclosure for the building or structure disconnecting means, or be mounted at the disconnecting means and be connected to the metallic enclosure or to the grounding electrode conductor with a minimum 6 AWG copper conductor.

6. be listed as grounding and bonding equipment. ◄

Code Rules for Cable Television (CATV) (Article 820)

The letters CATV stand for "Community Antenna Television," often simply referred to as "Cable TV."

CATV systems are installed both overhead and underground in a community. Then, to supply an individual customer, the CATV company runs coaxial cables through the wall of a residence at some convenient point. Up to this point of entry (Article 820, Part II) and inside the building (Article 820, Part V), the cable company must conform to the requirements of Article 820, plus local codes if applicable.

Coaxial cable generally used for one- and two-family dwellings is Type CATV and CATVX. The cable shall be listed. See 820.154, 820.179(C), and 820.179(D). Other types of acceptable cables are shown in Table 820.154(b).

Here are some of the key rules to follow when making a coaxial cable installation:

1. The outer conductive shield of the coaxial cables must be grounded as close to the point of entry as possible, 820.93.

2. Coaxial cables shall not be run in the same conduits or box with electric light and power conductors, 820.133(A)(1)(c).

3. Do not support coaxial cables from raceways that contain electrical light and power conductors, 820.133(B).

4. Keep the coaxial cable at least 2 in. (50 mm) from light and power conductors unless the conductors are in a raceway, nonmetallic-sheathed cable, armored cable, or UF cable, 820.133(A)(2). This clearance requirement really pertains to old knob-and-tube wiring. The 2-in. (50-mm) clearance is not required if the coaxial cable is in a raceway.

5. Where underground coaxial cables are run, they must be separated by at least 12 in. (300 mm) from underground light and power conductors, unless the underground conductors are in a raceway, Type UF cable, or Type USE cable, 820.47(B). The 12-in. (300-mm) clearance is not required if the coaxial cable has a metal cable armor.

6. The bonding and grounding conductor (Article 820, Part III and Part IV):

   a. outer conductive shield shall be grounded as close as possible to the coaxial cable entrance or attachment to the building, 820.93(A).

   b. shall not be smaller than a 14 AWG copper or other corrosion-resistant conductive material. It need not be larger than 6 AWG copper. The bonding or grounding electrode conductor is permitted to be insulated, covered or bare. It shall have a current-carrying capacity not less than that of the coaxial cable's outer metal shield. See *NEC 810.100(A)(1), (A)(2) and (A)(3)*.

c. length shall be as short as possible, but not longer than 20 ft (6.0 m). If impossible to keep the grounding conductor to this maximum permitted length, then the *Exception to 820.100(A)(4)* permits installing a separate grounding electrode. These two electrodes shall be bonded together with a bonding conductor not smaller than 6 AWG copper.

d. may be solid or stranded, *820.100(A)(2)*.

e. shall be run in a line as straight as practical, *820.100(A)(5)*.

f. is required to be protected against physical damage. If run in a metal raceway, the metal raceway shall be bonded to the bonding or grounding electrode conductor at both ends, *820.100(A)(6)*.

g. shall be connected to the nearest accessible location on one of the following, *820.100(B)*:

- bonded to the intersystem bonding termination if one exists, *820.100(B)(1)*.

- If the building or structure has no intersystem bonding means such as a terminal bar, the bonding conductor or grounding electrode conductor is to be connected to one of the following:

  1. the building grounding electrode system.

  2. the grounded interior metal water piping pipe, within 5 ft (1.5 m) of where the pipe enters the building. Refer to *250.52*.

  3. the power service accessible means external to the enclosure.

  4. the metallic power service raceway.

  5. the service equipment enclosure.

  6. the grounding electrode conductor or its metal enclosure.

  7. the grounding electrode conductor or grounding electrode of a disconnecting means that is grounded to an electrode according to *250.32*. This pertains to a second building on the same property served by a feeder or branch circuit from the main electrical service in the first building.

h. If *none* of the options in part (g) is available, then ground to any of the electrodes, per *250.52*, such as metal underground water pipes, the metal frame of the building, a concrete encased electrode, or a ground ring. Watch out for the maximum length permitted for the bonding or grounding electrode conductor, as mentioned above.

When installing CATV wiring and equipment, do it in a neat and workmanlike manner, not subject to physical damage when run on building surfaces. Secure cables with listed hardware, including straps, cable ties, and so on, *820.24*. The requirements found in *300.4(D)* (Cables and Raceways Parallel to Framing Members and Furring Strips) and *300.11* (Securing and Supporting) also apply to CATV cables.

CAUTION: DO NOT simply drive a separate ground rod to ground the metal shielding tape on the coax cable. Difference of potential between the coax cable shield and the electrical system ground during lightning strikes could result in a shock hazard as well as damage to the electronic equipment. If for any reason one grounding electrode is installed for the CATV shield grounding and another grounding electrode for the electrical system, bonding these two electrodes together with no smaller than a 6 AWG copper conductor will minimize the possibility that a difference of potential voltage might exist between the two electrodes.

SATELLITE ANTENNAS

A satellite antenna is often referred to as a "dish." See Figure 25-5. A satellite dish has a parabolic shape that concentrates and reflects the signal beamed down from one of the many stationary satellites orbiting 22,245 miles (35,800 kilometers) above the equator. Stationary means that the satellite is traveling at the same speed that the Earth rotates. The orbit in which satellites travel is called the "Clarke Belt," as shown in Figure 25-8.

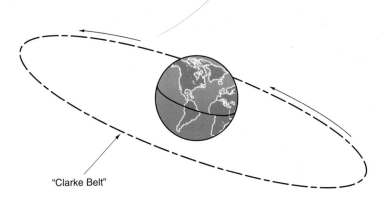

"Clarke Belt"

FIGURE 25-8 All satellites travel around the Earth in the same orbit called the "Clarke Belt." They appear to be stationary in space because they are rotating at the same speed at which the Earth rotates. This is called "geosynchronous orbit (geostationary orbit)." The satellite receives the uplink signal from Earth, amplifies the signal, and transmits it back to Earth. The downlink signal is picked up by the satellite antenna. (*Delmar/Cengage Learning*)

The latest satellite technology is the digital satellite system (DSS). A digital system allows the use of small antennas, approximately 18 in. (450 mm) in diameter, which can easily be mounted on a roof, a chimney, the side of the house, a pipe, or a pedestal, using the proper mounting hardware.

Although rarely used today for residential TV reception, large dish satellite antennas of the type shown in Figure 25-5(B) are still in use. Because of their huge size and weight, proper installation is important.

The basic operation of a satellite system is for a transmitter on Earth to beam an "uplink" signal to a satellite in space. Electronic devices on the satellite reamplify and convert this signal to a "downlink" signal, then retransmit the downlink signal back to earth, as shown in Figure 25-9.

Regardless of the type of antenna used, the line of sight from the antenna to the satellite must not be obstructed by trees, buildings, utility poles, or other structures. Instructions furnished with an antenna

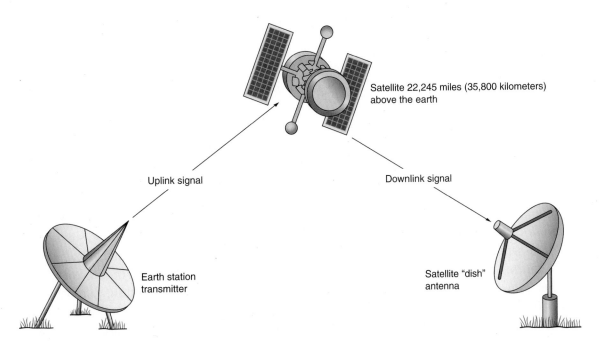

Satellite 22,245 miles (35,800 kilometers) above the earth

Uplink signal

Downlink signal

Earth station transmitter

Satellite "dish" antenna

FIGURE 25-9 The Earth station transmitter beams the signal up to the satellite, where the signal is amplified, converted, and beamed back to Earth. (*Delmar/Cengage Learning*)

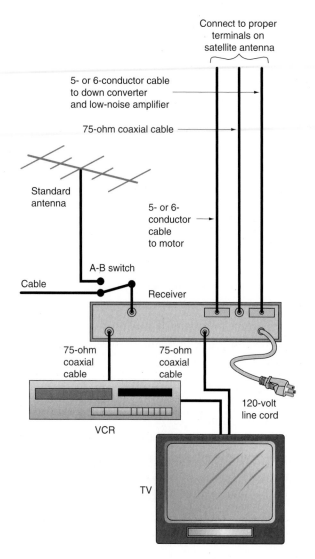

FIGURE 25-10 Typical satellite antenna/cable/ standard antenna wiring. Connection to a standard antenna cable, CATV cable, or to a satellite cable enables the homeowner to watch one channel while recording another channel. The installation manual for the receiver usually has a number of different hookup diagrams. In this diagram, the A-B switch allows a choice of connecting to a standard antenna or to the CATV cable. This hookup can be desirable should the CATV or satellite reception fail.
(*Delmar/Cengage Learning*)

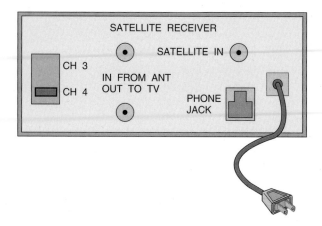

FIGURE 25-11 The terminals on the back of a digital satellite receiver. The connections are similar to those in Figure 25-10. Note the difference in that the digital satellite receiver has a modular telephone plug, a feature that uses a toll-free number to update the access card inside the receiver to ensure continuous program service. The telephone can also handle program billing.
(*Delmar/Cengage Learning*)

antenna. The manufacturer's installation instructions are always to be followed.

Figure 25-12 illustrates one method of installing a large antenna post in the ground in accordance to a manufacturer's instruction.

CODE RULES FOR THE INSTALLATION OF ANTENNAS AND LEAD-IN WIRES (ARTICLE 810)

Home television and AM and FM radios generally come complete with built-in antennas. For those locations in outlying areas on the fringe or out of reach of strong signals, it is quite common to install a separate antenna system or a satellite antenna system.

You will want to read *NEC 250.94* to learn more about the requirements for intersystem bonding.

Indoor or outdoor antennas may be used with televisions. The front of an outdoor antenna is aimed at the television transmitting station. When there is more than one transmitting station and they are located in different directions, a rotor is installed. A rotor turns the antenna on its mast so that it can face in the direction of each transmitter. The rotor is controlled from inside the building. The rotor

provide the necessary data for direction and up angle based on ZIP codes.

Before installing an antenna, check with your local inspection department. Local codes might have restrictions and requirements in addition to the *NEC*.

Figures 25-10 and 25-11 show typical connections between a television set, receiver, VCR, and

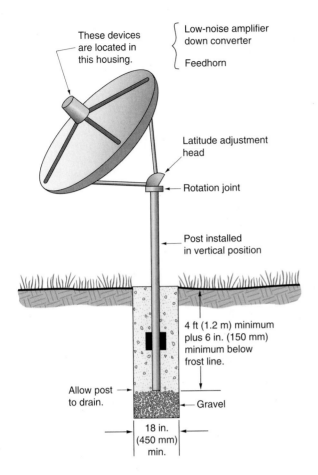

These devices are located in this housing.

Low-noise amplifier down converter

Feedhorn

Latitude adjustment head

Rotation joint

Post installed in vertical position

4 ft (1.2 m) minimum plus 6 in. (150 mm) minimum below frost line.

Allow post to drain.

Gravel

18 in. (450 mm) min.

FIGURE 25-12 Satellite antenna solidly installed in the ground according to the manufacturer's instructions. The popularity of these large antennas has given way to the small digital satellite system antenna, as shown in Figure 25-5. (*Delmar/Cengage Learning*)

controller's cord is plugged into a regular 120-volt receptacle to obtain power. A 3-wire or 4-conductor cable is usually installed between the rotor motor and the control unit. The wiring for a rotor may be installed during the roughing-in stage of construction, running the rotor's cable into a device wall box, allowing 5 to 6 ft (1.5 to 1.8 m) of extra cable, then installing a regular single-gang switchplate when finishing.

Article 810 covers radio and television equipment. Although instructions are supplied with antennas, the following key points of the *Code* regarding the installation of antennas and lead-in wires should be followed:

1. Antennas and lead-in conductors shall be securely supported, *810.12.*

2. Antennas and lead-in conductors shall not be attached to the electric service mast, *810.12.*

3. Antennas and lead-in conductors shall not be attached to any poles that carry light and power wires over 250 volts between conductors, *810.12.*

4. Lead-in conductors shall be securely attached to the antenna, *810.12.*

5. Antennas and lead-in conductors shall be kept away from all light and power conductors to avoid accidental contact with the light and power conductors, *810.13.*

6. Outdoor antennas and lead-in conductors shall not cross over light and power conductors, *810.13.*

7. Outdoor antennas and lead-in conductors shall be kept at least 24 in. (600 mm) away from open light and power conductors, *810.13.*

8. Where practicable, antenna conductors shall not be run under open light and power conductors, *810.13.*

9. On the outside of a building:

   a. Position and fasten lead-in conductors so they cannot swing closer than 24 in. (600 mm) to light and power conductors having *not* over 250 volts between conductors; 10 ft (3.0 m) if *over* 250 volts between conductors, *810.18(A).*

   b. Keep lead-in conductors at least 6 ft (1.8 m) away from a lightning rod system, *810.18(A),* or bonded together according to *250.60.*

   c. Underground lead-in radio and television conductors and cables shall be separated by at least 12 in. (300 mm) from underground light and power conductors.

   Note: The clearances in *a, b,* and *c* are not required if the light and power conductors or the lead-in conductors are in a metal raceway or metal cable armor.

10. On the inside of a building:

   a. Keep the antenna and lead-in conductors at least 2 in. (50 mm) from other open wiring (as in old houses) unless the other wiring is in a metal raceway or cable, *810.18(B).*

b. Keep lead-in conductors out of electric boxes unless there is an effective, permanently installed barrier to separate the light and power wires from the lead-in wire, *810.18(C)*.

11. Grounding:

a. All metal masts and metal structures that support antennas shall be grounded, *810.21*. See Figure 25-12.

b. The grounding conductor must be copper, aluminum, copper-clad steel, bronze, or a similar corrosion-resistant material, *810.21(A)*.

c. The bonding or grounding electrode conductor need not be insulated. It must be securely fastened in place, may be attached directly to a surface without the need for insulating supports, shall be protected from physical damage or be large enough to compensate for lack of protection, and shall be run in as straight a line as is practicable, *810.21(B), (C), (D), and (E)*.

d. In buildings with an intersystem bonding termination, the bonding conductor shall be connected to the intersystem bonding termination. See Figure 25-13.

e. If the building or structure does not have an intersystem bonding means, the bonding conductor or grounding electrode conductor is required to be connected to the nearest accessible location on one of the following:

- the building or structure grounding electrode system. Refer to *250.50* for more details.

- the grounded interior metal water pipe, within 5 ft (1.52 m) of where the water pipe enters the building. Refer to *250.53* and *250.68(C)* for more details.

- the metallic power service raceway.

- the service equipment enclosure.

- the grounding electrode conductor or the grounding electrode conductor metal enclosure of the power service.

f. If neither *d* nor *e* is available, then connect a grounding electrode conductor to any one of the grounding electrodes, per *250.52*, such as metal underground water pipe, metal frame of building, concrete-encased electrode, or ground ring, *810.21(F)(3)*.

g. The grounding conductor may be run inside or outside of the building, *810.21(G)*.

h. The grounding conductor shall not be smaller than 10 AWG copper or 8 AWG aluminum, *810.21(H)*.

i. Protect the grounding conductor from physical damage. If run in a metal raceway, the metal raceway shall be bonded to the grounding conductor at both ends, *800.21(D)*.

CAUTION: DO NOT simply drive a separate ground rod to ground the metal mast, structure, or antenna. Difference of potential between the metal mast, structure, or antenna and the electrical system ground during a lightning strike could result in a shock hazard as well as damage to the electronic equipment. If any reason permitted by the *Code* results in one grounding electrode for the antenna and another grounding electrode for the electrical system, bond the two electrodes together with a bonding jumper not smaller than 6 AWG copper or equivalent, *810.21(J)*.

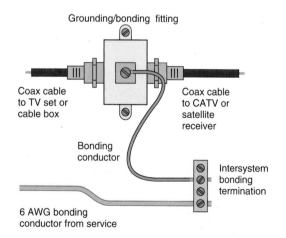

Grounding/bonding fitting

Coax cable to TV set or cable box

Coax cable to CATV or satellite receiver

Bonding conductor

Intersystem bonding termination

6 AWG bonding conductor from service

FIGURE 25-13 Typical connection for the required grounding of the metal shield on coaxial cable to the intersystem bonding termination.
(*Delmar/Cengage Learning*)

The objective of grounding and bonding to the same grounding electrode as the main service ground is to reduce the possibility of having a difference of voltage potential between the two systems.

TELEPHONE WIRING (ARTICLE 800)

Since the deregulation of telephone companies, residential "do-it-yourself" telephone wiring has become quite common. The following is an overview of residential telephone wiring.

The *NEC* in *800.156* requires that for new dwelling unit construction, *a minimum of one communications outlet shall be installed within the dwelling and cabled to the service provider demarcation point.* Prior to the *2008 NEC*, telephone outlets were not required but if installed, *Article 800* spelled out the installation requirements.

The telephone company will install the service line to a residence and terminate at a protector device, as in Figure 25-14. The protector protects the system from hazardous voltages. The protector may be mounted either outside or inside the home. Different telephone companies have different rules. Always check with the phone company before starting your installation.

The point where the telephone company ends and the homeowner's interior wiring begins is called the *demarcation point*, as shown in Figure 25-14. The preferred demarcation point is outside of the home, generally near the electric meter where proper grounding and bonding can be done. A *network interface* might be of the type depicted in Figure 25-14. This is a combination unit that provides the compartment for the utility with controlled access and an owner's compartment. The utility makes a connection from the grounding/bonding terminal in the network interface unit to the intersystem bonding termination that is installed near the service equipment. The utility provides the dial tone to the customer's compartment. The customer can make connections of individual or multipair cables to the terminals provided.

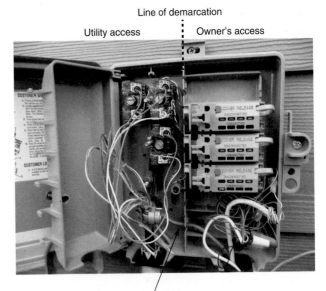

Line of demarcation

Utility access   Owner's access

Bonding conductor to intersystem bonding termination

FIGURE 25-14 Typical residential telephone network interface unit (NIU) installation. The telephone company installs the NIU and connects the underground cable to the protector in the NIU. The utility installs the bonding conductor to the intersystem bonding terminal bar that is installed near the service equipment. The customer installs the cable to the telephone outlet(s) and makes the connections in the customer section of the NIU.
(*Delmar/Cengage Learning*)

In the past, interior telephone wiring in homes was installed by the telephone company. Today, the responsibility for roughing-in the interior wiring is usually left up to the electrician. Telephone wiring is *not* to be run in the same raceway with electrical wiring. In addition, it cannot be installed in the same outlet and junction boxes as electrical wiring. Whatever the case, the rules found in *Article 800* apply. Refer to Figures 25-14, 25-15, and 25-16.

For a typical residence, the electrician will rough in wall boxes wherever a telephone outlet is wanted. The boxes can be of the single-gang types illustrated in Figures 25-1, 2-11, 2-17, 2-19, and 2-32. Concealed telephone wiring in new homes is generally accomplished using multipair cables that are run from each telephone outlet to the customer compartment of the network interface unit. The cable can also take the form of a "loop system," as

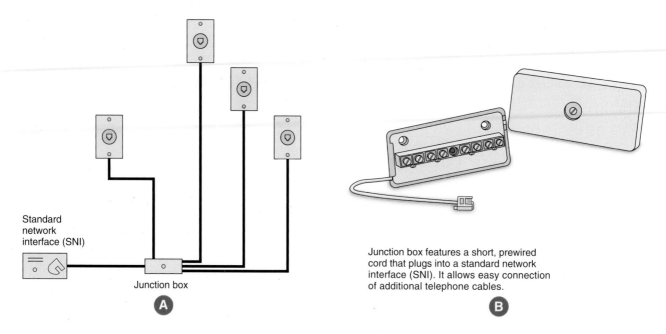

Standard network interface (SNI)

Junction box

A

Junction box features a short, prewired cord that plugs into a standard network interface (SNI). It allows easy connection of additional telephone cables.

B

FIGURE 25-15 (A) shows individual telephone cables run to each telephone outlet from the common connection point. This is sometimes referred to as a star-wired system. (B) shows a junction box for easy connection of multiple telephone cables. (*Delmar/Cengage Learning*)

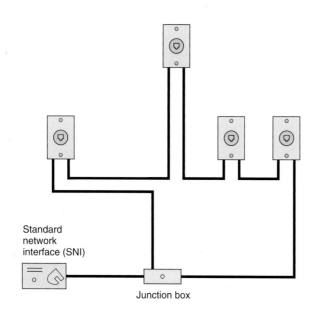

Standard network interface (SNI)

Junction box

FIGURE 25-16 A complete "loop system" (daisy-chain) of the telephone cable. If something happens to one section of the cable, the circuit can be fed from the other direction. (*Delmar/Cengage Learning*)

in Figure 25-16. A 3-pair, 6-conductor twisted cable is shown in Figure 25-17. There are some localities that require that these cables be run in a metal raceway, such as EMT.

The outer jacket is usually of a thermoplastic material. This outer jacket most often is a neutral color, such as white, light gray, or beige, that blends in with decorator colors in the home. This is particularly important in existing homes where the telephone cable may have to be exposed. Cables, mounting boxes, junction boxes, terminal blocks, jacks, adaptors, faceplates, cords, hardware, plugs, and so on, are all available through electrical distributors, builders' supply outlets, telephone stores, electronic stores, hardware stores, and similar wholesale and retail outlets. The selection of types of the preceding components is endless.

Telephone Conductors

The color coding of telephone cables is shown in Figure 25-17. Figure 25-18 shows some of the many types of telephone cords available.

Cables are available with various numbers of conductors. Here are a few:

2 pair . . . . 4 conductors
3 pair . . . . 6 conductors
4 pair . . . . 8 conductors
6 pair . . . 12 conductors
12 pair . . . 24 conductors

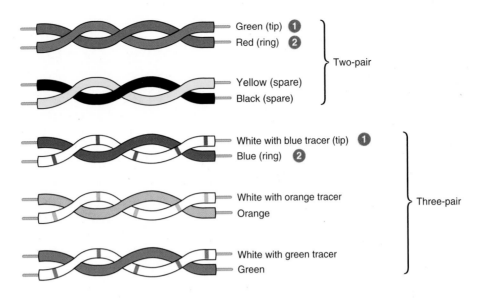

Green (tip) ❶
Red (ring) ❷
} Two-pair

Yellow (spare)
Black (spare)

White with blue tracer (tip) ❶
Blue (ring) ❷

White with orange tracer
Orange
} Three-pair

White with green tracer
Green

❶ *Tip* is the conductor that is connected to the telephone company's "positive" terminal. It is similar to the neutral conductor of a residential wiring circuit.

❷ *Ring* is the conductor that is connected to the telephone company's "negative" terminal. It is similar to the "hot" ungrounded conductor of a residential wiring circuit.

FIGURE 25-17 Color coding for a 2-pair and a 3-pair, 6-conductor twisted telephone cable. The color coding for 2-pair telephone cable stands alone. For three or more pair cables, the color coding becomes WHITE/BLUE, WHITE/ORANGE, WHITE/GREEN, WHITE/BROWN, WHITE/SLATE, RED/BLUE, RED/ORANGE, RED/GREEN, RED/BROWN, RED/SLATE, BLACK/BLUE, BLACK/ORANGE, BLACK/GREEN, BLACK/BROWN, BLACK/SLATE, YELLOW/BLUE, YELLOW/ORANGE, YELLOW/GREEN, YELLOW/BROWN, YELLOW/SLATE, VIOLET/BLUE, VIOLET/ORANGE, VIOLET/GREEN, VIOLET/BROWN, VIOLET/SLATE. In multipair cables, the additional pairs can be used for more telephones, fax machines, security reporting, speakerphones, dialers, background music, and so on. (*Delmar/Cengage Learning*)

Telephone circuits require a separate pair of conductors from the telephone all the way back to the phone company's central switching center. To keep to a minimum interference that could come from other electrical equipment, such as electric motors and fluorescent fixtures, each pair of telephone wires (two wires) is twisted.

Installing Communication Wiring

When installing communication wiring and equipment, do it in a neat and workmanlike manner, not subject to physical damage when run on building surfaces. Secure cables with listed hardware, including straps, cable ties, and so on, *800.24*. The requirements found in *300.4(D)* (Cables and Raceways Parallel to Framing Members and Furring Strips) and *300.11* (Securing and Supporting) also apply to communication cables.

Cross-Talk Problems

A common telephone interference problem in homes is "cross-talk." This is apparent if you hear someone faintly talking in the background when you are using the telephone. Another problem is losing an online connection to your computer. The signals traveling in one pair of conductors are picked up by the adjacent pair of conductors in the cord or cable. Quite often, these problems can be traced to older style flat (untwisted) 2-line telephone lines. Older style cables could carry audio signals quite well, but they are unsatisfactory for data transmission. These problems are virtually eliminated by using cables and cords that have conductors twisted at proper intervals, such as Category 5 cables. In Figure 25-17, each pair of conductors is twisted, then all of the pairs are twisted again. Generally, you should avoid flat 2-line cords.

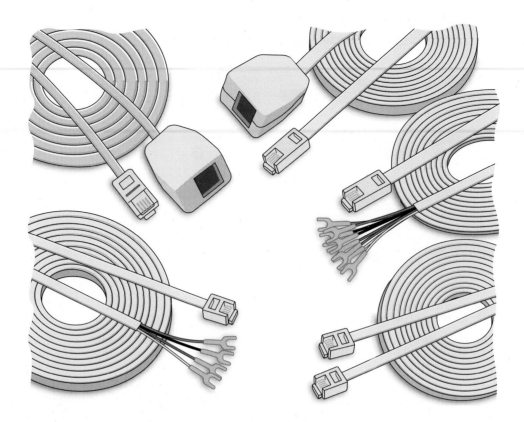

FIGURE 25-18 Some of the many types of telephone cable available are illustrated. Note that for ease of installation, the modular plugs and terminals have been attached by the manufacturer of the cables. These cords are available in round and flat configurations, depending upon the number of conductors in the cable. (*Delmar/Cengage Learning*)

The recommended circuit lengths are as follows:

24 AWG gauge—not over 200 ft (60 m)
22 AWG gauge—not over 250 ft (75 m)

Ringer Equivalence Number (REN)

Telephones, fax machines, answering machines, ringers, flashers, and so on, that have a sounder (ringer) are assigned an REN number. Generally, the maximum number of RENs on any single telephone line is 5. This maximum number of RENs ensures that all telephones will ring. When too many telephones are connected, the ring signal can become unreliable, resulting in devices failing to respond.

Older style electromechanical ringers were considered to be 1 REN. Electronic ringers have a minimal REN value. Mixing electromechanical and electronic ringers could present a problem as the electromechanical ringers tend to hog the current. It all depends on your telephone line. Try it and see what happens.

Pay close attention to the total RENs connected in your home. REN values are found on the label of the device.

 EXAMPLE ——————————

Four electronic telephones:	4 × 0.3 = 1.2 RENs
One fax machine:	1.5 RENs
One desk top telephone/ Answering machine:	0.7 REN
One desk top telephone:	0.4 RENs
Total	3.8 RENs

Call your telephone company if you think you have a problem because of too many telephones connected to one telephone line.

Plug-in or Permanent Connection?

Most residential telephones plug into a jack of the type illustrated in Figures 25-19 and 25-20. It might be a good idea to have at least one permanently connected telephone, such as a wall phone in the kitchen. If all of the phones were of the plug-in type, it is possible (but highly unlikely) that all of them could be unplugged. As a result, there would be no audible signal.

FIGURE 25-19 Telephone modular jack.
(*Delmar/Cengage Learning*)

FIGURE 25-20 Three styles of wall plates for modular telephone jacks: rectangular stainless steel, weatherproof for outdoor use, and circular. (*Delmar/Cengage Learning*)

The electrical plans show that nine telephone outlets are provided as follows:

Front Bedroom	1
Kitchen	1
Laundry	1
Living Room	1
Master Bedroom	1
Rear Outdoor Patio	1
Recreation Room	1
Study/Bedroom	1
Workshop	1

Wireless Telephones

Wireless telephones have become all the rage! A master telephone unit is plugged into the telephone outlet somewhere in the house. The unit usually requires a power-supply transformer that plugs into a 120-volt receptacle outlet to power the telephone master unit. The master unit acts as a transmitter/receiver for the other remote telephones located at desirable locations. The master unit often provides services for the entire system, including answering machine, telephone number list, and intercom. Each of the remote telephones sits on a charging cradle and communicates to the wired telephone network by high-frequency radio signal.

Though very flexible, these telephones rely on power from the electric utility for operation. You should always have a telephone set that will function without reliance on 120-volt power for use during power outages, storms, and emergencies.

Installation of Telephone Cables (Article 800, Part V)

Telephone cables

1. are generally Type CM or Type CMX (for one- and two-family dwellings only), listed for telephone installations as being resistant to the spread of fire. See *800.154, 800.179(D)*, and *800.179(E)*. Also refer to *Table 800.154(b)* and *Table 800.179* for other acceptable types.

2. shall be separated by at least 2 in. (50 mm) from light and power conductors unless the light and power conductors are in a raceway, or in nonmetallic-sheathed cable, Type AC cable, or Type UF cable, *800.133(A)(2)*.

3. shall not be placed in any conduit or boxes with electric light and power conductors unless the conductors are separated by a partition, *800.133(A)(1)(c)*.

4. do not support telephone cables from raceways that contain light and power conductors, *800.133(B)*.

5. should not share the same bored holes as electrical wiring, plumbing, gas pipes, and so on.

6. may be terminated in either metallic or non-metallic boxes or plaster rings.

7. should be secured using rounded or depth-stop plastic staples. Do not use metal staples.

8. should not be installed using nail guns. Do not crush the cable.

9. should be kept away from hot water pipes, hot air ducts, and other heat sources that might harm the insulation.

10. should not be run in the same stud space as electrical branch-circuit wiring and should be kept at least 12 in. (300 mm) from power wiring where the cables are run parallel to the power wiring. These cautions were written before twisted pairs and category-rated cables entered the scene. With the advent of *properly* installed twisted and shielded cables, these "old wives' tale" recommendations from the past are probably not an issue.

Follow the specifications and instructions for the cable you are installing.

Raceway

In some communities, electricians prefer to install regular device boxes at a telephone outlet location, then "stub" a trade size ½ EMT to the basement or attic from this box. Then, at a later date, the telephone cable can be fished through the raceway.

Grounding (Article 800, Part IV)

The telephone company will provide the proper grounding of their incoming cable sheath and primary protector, generally using an insulated conductor not smaller than 14 AWG copper or other corrosion resistant material, and not longer than 20 ft (6.0 m).

For one- and two-family dwellings, where it might be impractical to keep the grounding conductor 20 ft (6.0 m) or less, an additional communications grounding electrode must be installed. This additional ground rod must be bonded to the power grounding electrode system with a bonding conductor not smaller than a 6 AWG copper. In just about all instances, all grounding and bonding of telephone equipment is done by the telephone company personnel. Many times, they clamp the primary protector to the grounded metal service raceway conduit to establish "ground." See *250.94*.

Safety

Open-circuit voltage between conductors of an idle pair of telephone conductors is approximately 48 volts dc. The superimposed ringing voltage can reach 90 volts ac. Therefore, always work carefully with insulated tools and stay clear of bare terminals and grounded surfaces. Disconnect the interior telephone wiring if work must be done on the circuit, or take the phone off the hook, in which case the dc voltage level will drop to approximately 7 to 9 volts dc, and there should be no ac ringing voltage delivered.

Wiring for Computers and Internet Access

Wiring for telephones, high-speed Internet access, computers, television, printers, modems, security systems, intercoms, and similar home automation equipment could be thought of as one big project. Planning ahead is critical. Deciding on where this equipment is likely to be located will help you to design an entire voice/data/video system.

You will want to install Category 5 or Category 5e cables between each telephone outlet and a central distribution point so as to have the flexibility to connect the voice/data/video equipment in any configuration.

In addition to the previously mentioned equipment, there very well might be a scanner, an answering machine, desk lamps, floor lamps, a clock, an adding machine, a calculator, a television, a radio, a printer, CD/DVD/DVR/VCR/VHS players and burners, ZIP drives, and others. They draw little current, but all need to be plugged into a 120-volt receptacle! Managing and plugging in that many power cords is a major problem. Give consideration to installing two or three duplex or quadplex 120-volt receptacles at these locations. Multioutlet plug-in strips with surge protection will most likely be needed.

Chapter 31 contains much information about home automation and structured wiring.

Bringing Technology into Your Home

In most areas, the telephone or CATV company provides the entire "package" for homes. This might include digital local and long distance telephone service, Internet access, cable modems for personal computers, cable television, digital subscriber lines (DSL) (through existing copper telephone lines), and whatever other great new things may come next.

SIGNAL SYSTEM (CHIMES)

A signaling circuit is described in the *NEC* as *any electric circuit that energizes signaling equipment.* Signaling equipment includes such devices as chimes, doorbells, buzzers, code-calling systems, and signal lights.

Door Chimes (Symbol CH)

Present-day dwellings use chimes rather than bells or buzzers to announce that someone is at a door. A musical tone is sounded rather than a harsh ringing or buzzing sound. Chimes are available in single-note, 2-note, 8-note (4-tube), and repeater tone styles. In a repeater tone chime, both notes sound as long as the push button is depressed. In an 8-note chime, contacts on a motor-driven cam are arranged in sequence to sound the notes of a simple melody when the chime button is pushed. This type of chime is usually installed in dwellings having three entrances. The chime can be connected so that the 8-note melody sounds for the front door, two notes sound for the side door, and a single note sounds for the rear door. Chimes are also available with clocks and lights.

Electronic chimes may relay their chime tones through the various speakers of an intercom system. When any chime is installed, the manufacturer's instructions must be followed.

The plans show that two chimes are installed in the residence. Two-note chimes are used. Each chime has two solenoids and two iron plungers. When one solenoid is energized, the iron plunger is drawn into the opening of the solenoid. A plastic peg in the end of the plunger strikes one chime tone bar. When the solenoid is de-energized, spring action returns the plunger, where it comes to rest against a soft felt pad so that it does not strike the other chime tone bar. Thus, a single chime tone sounds. As the second solenoid is energized, one chime tone bar is struck. When the second solenoid is de-energized, the plunger returns and strikes the second tone bar. A two-tone signal is produced. The plunger then comes to rest between the two tone bars. Generally, two notes indicate front door signaling, and one note indicates rear or side door signaling.

Figure 25-21 shows various push-button styles used for chimes. Many other styles are available. Figure 25-22 shows several typical residential-type wall-mounted chimes. The symbols used to indicate push buttons and audible signals on the plans are shown in Figure 25-23.

FIGURE 25-21 Push buttons for door chimes. (*Courtesy of Broan-NuTone, Inc.*)

FIGURE 25-22 Typical residential door chimes. (*Courtesy of Broan-NuTone, Inc.*)

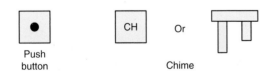

FIGURE 25-23 Symbols for chimes and push buttons. (*Delmar/Cengage Learning*)

Figure 25-24 shows how to provide proper backing for chimes. This is done during the rough-in stages of the electrical installation.

Chimes for the Hearing Impaired

Chimes for the hearing impaired are available with accessory devices that can turn on a dedicated lamp at the same time the chime is sounded. Figure 25-25 shows the devices needed to accomplish this. Figure 25-26 shows the wiring details.

Chime Transformers

Chime transformers are covered under UL Standard 1585. Because of their low-voltage rating

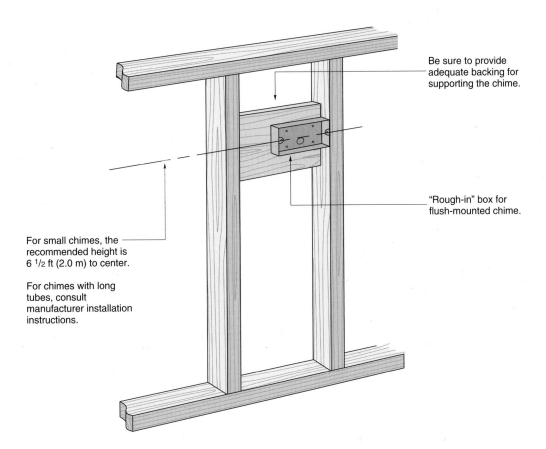

Be sure to provide
adequate backing for
supporting the chime.

"Rough-in" box for
flush-mounted chime.

For small chimes, the
recommended height is
6 ¹/₂ ft (2.0 m) to center.

For chimes with long
tubes, consult
manufacturer installation
instructions.

FIGURE 25-24   Roughing-in for a flush-mounted chime. (*Delmar/Cengage Learning*)

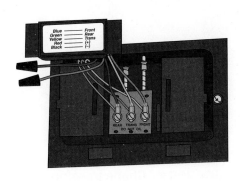

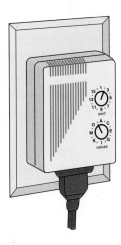

FIGURE 25-25   Accessory devices can be attached to new or existing chimes to provide visual signals when the chime is sounded. The first illustration shows the module that is connected and mounted inside the chime. The second illustration shows a transmitter that is plugged into a nearby receptacle. The third illustration shows a receiver into which a lamp is plugged.
(*Delmar/Cengage Learning*)

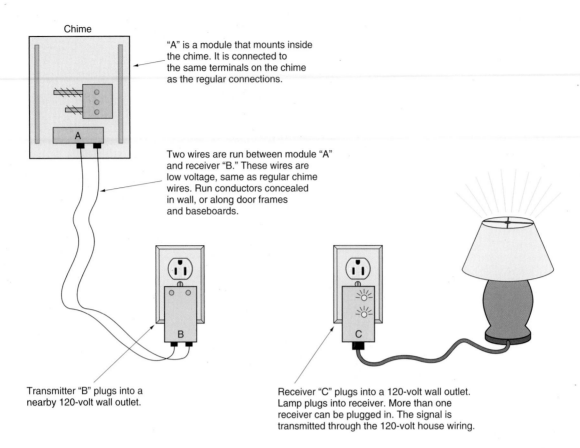

Chime

"A" is a module that mounts inside the chime. It is connected to the same terminals on the chime as the regular connections.

A

Two wires are run between module "A" and receiver "B." These wires are low voltage, same as regular chime wires. Run conductors concealed in wall, or along door frames and baseboards.

B

C

Transmitter "B" plugs into a nearby 120-volt wall outlet.

Receiver "C" plugs into a 120-volt wall outlet. Lamp plugs into receiver. More than one receiver can be plugged in. The signal is transmitted through the 120-volt house wiring.

FIGURE 25-26 Installation diagram for auxiliary devices for use with chimes that can help the hearing impaired. When the chime sounds, the signal is transmitted to the receiver, turning on the lamp that is plugged into the receiver. The "On" time is adjustable at "C." The lamp should be a "dedicated" lamp that lights up only when the chime sounds. (*Delmar/Cengage Learning*)

and power limitation, they are listed as Class 2 transformers. The low-voltage wiring on the secondary side of the transformer is Class 2 wiring. Class 2 circuits are discussed in Chapter 24 and in *Article 725*.

Figure 25-27 shows a chime transformer. The top view shows the 120-volt leads. Note the small set screw for ease of mounting into a conduit knockout in an outlet box. The bottom view shows the low-voltage terminals.

Chime transformers used in dwellings are generally rated 16 volts. They have built-in thermal overload protection. Should a short circuit occur in the low-voltage wiring, the overload device opens and closes repeatedly until the short circuit is cleared.

The UL standard requires that all metal parts of a transformer be properly grounded. To accomplish this, some chime transformers have a bare copper equipment grounding conductor in addition to the black and white supply conductors. Internal to the

FIGURE 25-27 Chime transformer. (*Courtesy of Broan-NuTone, Inc.*)

transformer, this equipment grounding conductor is connected to the metal parts (laminations) of the transformer. This bare equipment grounding conductor must be connected to the equipment grounding conductor or grounding screw in the outlet box.

Do not connect this equipment grounding conductor to the branch-circuit grounded (white) conductor.

Some chime transformers are marked "Install in Metal Box Only."

Additional Chimes

To extend a chime system to cover a larger area, a second or third chime may be added. In the residence, two chimes are used: one chime is mounted in the front hall and a second (extension) chime is mounted in the recreation room. The extension chime is wired in parallel to the first chime. The wires are run from one chime terminal board to the terminal board of the other chime. The terminals are connected as follows: transformer to transformer, front to front, and rear to rear, Figure 25-28.

When more than one chime is installed, it will probably be necessary to install a transformer with a higher volt-ampere (wattage) rating. Do not install a transformer with a higher voltage rating. Check the manufacturers' instructions furnished with the chime transformer.

Do not add another transformer to an existing transformer to solve the problem of multiple chimes not responding properly. The *NEC* in *725.121(B)* prohibits connecting transformers in parallel unless specifically listed for interconnection. Residential chime transformers are not listed for interconnection.

If a buzzer (or bell) and a chime are connected to a single transformer and are used at the same time, the transformer will put out a fluctuating voltage. This condition does not allow either the buzzer or the chime to operate properly. The use of a transformer with a larger rating may solve this problem.

The wattage consumption of chimes varies with the manufacturer. Typical ratings are shown in Table 25-1.

Transformers with ratings of 5, 10, 15, 20, and 30 watts (VA) are available. For a multiple chime installation, wattage ratings for the individual chimes are added. The total value is the minimum transformer rating needed to do the job properly.

Low-Voltage Wiring

The low-voltage wiring for a chime(s) is classified as Class 2 wiring, as discussed in Chapter 24.

Wiring the Chime

The circuit shown in Figure 25-28 is recommended for this chime installation because it provides a "hot" low-voltage circuit at the front hall location. However, it is not the only way in which these chimes may be connected.

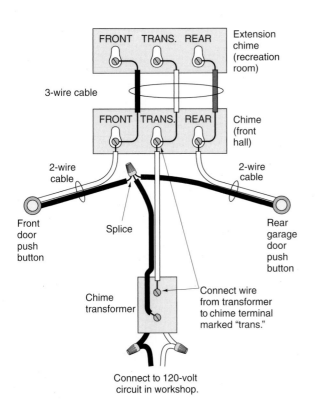

FIGURE 25-28 Circuit for chime installation. (*Delmar/Cengage Learning*)

TABLE 25-1	
Typical chimes with their power consumption.	
Type of Chime	**Power Consumption**
Standard 2-note	10 watts
Repeating chime	10 watts
Internally lighted, two lamps	10 watts
Internally lighted, four lamps	15 watts
Combination chime and clock	15 watts
Motor-driven chime	15 watts
Electronic chime	15 watts
Programmable musical	15 watts

Figure 25-28 shows that a 2-wire cable runs from the transformer in the utility room to the front hall chime. A 2-wire cable then runs from the chime to both the front and rear door push buttons. A 3-wire cable also runs between the front hall chime and the recreation room chime. Because of the "hot" low-voltage circuit at the front hall location, a chime with a built-in clock can be used. A 4-conductor cable may be run to the extension chime so that the owner may install a clock-chime at this location also.

The plans show that the chime transformer is mounted into the knockout on one of the ceiling boxes in the utility room. The line-voltage connections are easy to make here. Some electricians prefer to mount the chime transformer on the top or side of the distribution panelboards. The electrician decides where to mount the transformer after considering the factors of convenience, economy, and good wiring practice.

Don't Conceal Boxes!

It may seem obvious, but the issue of accessibility bears repeating. Electrical equipment junction boxes and conduit bodies must be accessible, *314.29*. It is permitted to have certain electrical equipment such as a chime transformer, junction boxes, and conduit bodies above a dropped lay-in ceiling because these are accessible by dropping out a ceiling panel. Never install electrical equipment, junction boxes, or conduit boxes above a permanently closed-in ceiling or within a wall. There may come a time when the electrical equipment and junction boxes need to be accessed. This includes chime transformers.

REVIEW

Note: Refer to the *Code* or the plans where necessary.

TELEVISION CIRCUIT

1. How many television outlets are installed in this residence? _____

2. Which type of television cable is commonly used and recommended? _____

   _____

3. What determines the design of the faceplates used? _____

   _____

4. What must be provided when installing a television outlet and receptacle outlet in one wall box? _____

   _____

   _____

   _____

   _____

5. From a cost standpoint, which system is more economical to install: a master amplifier distribution system or a multiset coupler? Explain the basic differences between these two systems. _____

   _____

   _____

   _____

6. How many wires are in the cable used between a rotor and its controller?

7. Digital satellite systems use an antenna that is approximately (18 in. [450 mm]) (36 in. [900 mm]) (72 in. [1.8 mm]) in diameter. Circle the correct answer.

8. List the requirements for cable television inside the house. _____

_____

_____

_____

9. Which article of the *Code* references the requirements for cable community television installation? _____

10. It is generally understood that grounding and bonding together all metal parts of an electrical system and the metal shield of the cable television cable to the same grounding reference point in a residence will keep both systems at the same voltage level should a surge, such as lightning, occur. Therefore, if the incoming cable that has been installed by the CATV cable company installer has the metal shield grounded to a driven ground rod, does this installation conform to the *NEC*? _____

_____

11. All television satellites rotate above the Earth in (the same orbit) (different orbits). Circle the correct answer.

12. Television satellites are set in orbit (10,000) (18,000) (22,245) miles above the Earth, which results in their rotating around the Earth at (precisely the same) (different) rotational speed as the Earth rotates. This is done so that the satellite "dish" can be focused on a specific satellite (once) (one time each month) (whenever the television set is used). Circle the correct answers.

13. Which section of the *Code* prohibits supporting coaxial cables from raceways that contain light or power conductors? *NEC* _____

14. When hooking up one CATV cable and another cable from an outdoor antenna to a receiver that has only one antenna input terminal, a(n) _____ switch is usually installed.

⌐∿⌐ TELEPHONE SYSTEM

1. How many locations are provided for telephones in the residence? _____

2. At what height are the telephone outlets in this residence mounted? Give measurement to center.

_____

_____

_____

_____

3. Sketch the symbol for a telephone outlet.

4. Is the telephone system regulated by the *NEC*? _____

5. a. Who is to furnish the outlet boxes required at each telephone outlet? _____

   b. Who is to furnish the faceplates? _____

6. Who is to furnish the telephones? _____

7. Who does the actual installation of the telephone equipment? _____

8. How are the telephone cables concealed in this residence? _____

   _____

   _____

   _____

9. The point where the telephone company's cable ends and the interior telephone wiring meets is called the _____ point. The device installed at this point is called a(n) _____.

10. What are the colors contained in a four-conductor telephone cable assembly and what are they used for? _____

    _____

    _____

    _____

    _____

    _____

    _____

11. Itemize the *Code* rules for the installation of telephone cables in a residence. _____

    _____

    _____

    _____

12. If finger contact were made between the red conductor and green conductor at the instant a "ring" occurs, what shock voltage would be felt? _____

13. What section of the *Code* prohibits supporting telephone wires from raceways that contain light or power conductors? *NEC* _____

14. The term *cross-talk* is used to define the hearing of the faint sound of voices in the background when you are using the telephone. Cross-talk can be reduced significantly by running (flat cables) (twisted pair cables) to the telephones. Circle the correct answer.

15. Which section of the *Code* prohibits telephone cables from being installed in the same box or enclosure or from being pulled into the same raceway as light and power circuits? _____

⊸〰⊢ SIGNAL SYSTEM

1. What is a signal circuit? _____

   _____

2. What style of chime is used in this residence? _____

3. a. How many solenoids are contained in a 2-tone chime? _____

   b. What closes the circuit to the solenoid of a chime? _____

4. Explain briefly how two notes are sounded by depressing one push button (when two solenoids are provided). _____

   _____

   _____

   _____

   _____

   _____

   _____

   _____

5. a. Sketch the symbol for a push button. _____

   b. Sketch the symbol for a chime. _____

6. a. At what voltage do residence chimes generally operate? _____

   b. How is this voltage obtained? _____

7. What is the maximum volt-ampere rating of transformers supplying Class 2 systems?

   _____

8. What two types of chime transformers for Class 2 systems are listed by UL?

   _____

   _____

9. Is the extension chime connected with the front hall chime in series or in parallel?

   _____

10. How many bell wires terminate at

    a. the transformer? _____

    b. the front hall chime? _____

    c. the extension chime? _____

    d. each push button? _____

11. a. What change in equipment may be necessary when more than one chime is connected to sound at the same time on one circuit? _____

    _____

    _____

    _____

    _____

    b. Why? _____

    _____

    _____

    _____

    _____

12. What type of insulation is usually found on low-voltage wires? _____

13. What size wire is installed for signal systems of the type in this residence? _____

14. a. How many wires are run between the front hall chime and the extension chime in the recreation room? _____

   b. How many wires are required to provide a "hot" low-voltage circuit at the extension chime? _____

15. Why is it recommended that the low-voltage secondary of the transformer be run to the front hall chime location and separate 2-wire cables be installed to each push button?

   _____

   _____

16. a. Is it permissible to install low-voltage Class 2 systems in the same raceway or enclosure with light and proper wiring? _____

   b. Which *Code* section covers this? _____

17. a. Where is the transformer in the residence mounted? _____

   b. To which circuit is the transformer connected? _____

18. a. How many feet of 2-conductor bell wire cable are required? _____

   b. How many feet of 3-conductor bell wire cable are required? _____

19. How many insulated staples are needed for the bell wire if it is stapled every 24 in. (600 mm)? _____

20. Should low-voltage bell wire be pulled through the same holes in studs and joists that also contain nonmetallic-sheathed cable? _____

21. What section of the *Code* prohibits supporting fire alarm conductors from raceways that contain light or power conductors? *NEC* _____ .

22. Complete this typical chime wiring diagram using colored pencils. The 2-wire, low-voltage cables contain one red and one white conductor. Use yellow to represent the white conductors.

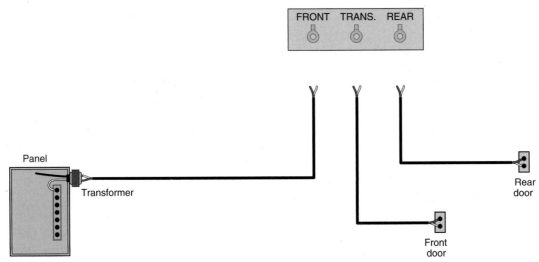

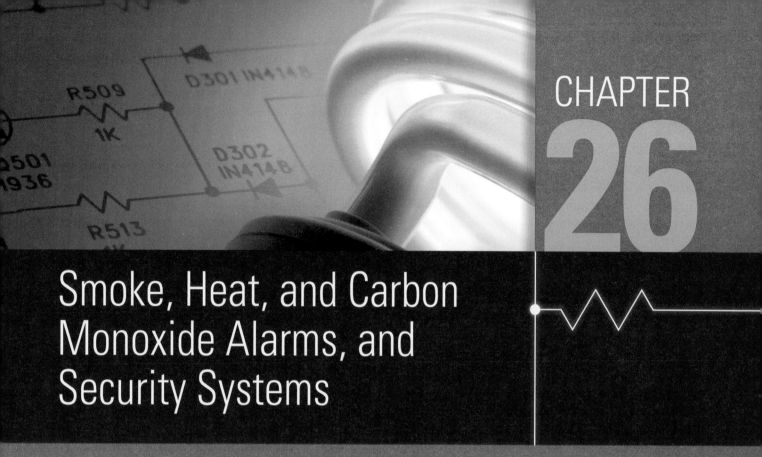

Smoke, Heat, and Carbon Monoxide Alarms, and Security Systems

OBJECTIVES

After studying this chapter, you should be able to

- understand the basics of the *National Fire Alarm Code NFPA 72 (2010)*, the *Standard for the Installation of Carbon Monoxide (CO) Detection and Warning Equipment NFPA 720 (2009)*, and the *National Electrical Code NFPA 70 (2011)*.

- understand the basics of smoke, heat, and carbon monoxide alarms.

- understand the location requirements for the installation of smoke, heat, and carbon monoxide alarms for *minimum* acceptable levels of protection.

- understand the location requirements for the installation of smoke and heat alarms that *exceed* the minimum acceptable levels of protection.

- discuss general requirements for the installation of security systems.

- be aware of important UL Standards covering fire warning equipment.

NATIONAL FIRE ALARM CODE (*NFPA 72*)

NFPA 72 is the *National Fire Alarm Code.*

The importance of installing smoke and fire alarms in homes is supported by results from exhaustive investigations of home fires indicating that measurable quantities of smoke come before detectable quantities of heat. In other words, smoke generally comes before fire. Because this is true, smoke alarms are considered the primary means of protection against fire in homes.

Following are some interesting facts about home fires.

Fires in homes were the third leading cause of deaths in homes—more deaths from asphyxiation than from burns.

Half of fires in homes occurred between 10 PM and 8 AM. That's sleep time.

Chapter 29 of the *National Fire Alarm Code* specifically covers *Single- and Multiple-Station Alarms* and *Household Fire Alarm Systems*. It contains the minimum requirements for the proper selection, installation, operation, and maintenance of fire warning equipment that will provide *reasonable* fire safety. *Chapter 29* also covers small residential board and care occupancies, which are defined as *A building or portion thereof that is used for lodging and boarding of four or more residents, not related by blood or marriage to the owners or operators, for the purpose of providing personal care services.*

In *NFPA 72, Section 29.1.2*, we find that the primary purpose of fire warning equipment is *to provide a reliable means to notify the occupants of the presence of a threatening fire and the need to escape to a place of safety before such escape might be impeded by untenable conditions in the normal path egress. NFPA 72* is not intended to provide protection to property, *29.1.5.*

This chapter deals with the types of smoke and heat alarms typically installed in one- and two-family dwellings. These smoke and heat alarms are not classified as fire alarm systems.

This chapter does not involve the more complex household fire alarm systems, but rather the devices and a fire alarm control unit(s) that receive, monitor, and process signals from detection devices. These types of fire alarm systems communicate via telephone or wireless with a central station that in turn calls the fire department. Fire alarm systems are required to meet stricter rules regarding the sound level and time of audible signals. There are also tougher rules regarding sleeping area requirements. Fire alarm systems require rechargeable batteries from the ac power source supplying the control unit(s). Typical home-type smoke and heat alarms are permitted to have replaceable batteries.

Chapter 29 of the *National Fire Alarm Code* tells us where to install smoke and heat detectors and alarms in homes.

Article 760 in the *National Electrical Code* (*NEC*) tells us how to install the wiring for fire alarm systems.

Section 29.3.2 of the *National Fire Alarm Code* states that fire warning equipment shall be installed according to the listing and manufacturer's instructions.

There is nothing in the *NEC* that deals with carbon monoxide. To learn about carbon monoxide detectors and alarms in homes, we refer to *NFPA 720*, the *Standard for the Installation of Carbon Monoxide (CO) Detection and Warning Equipment.*

Because carbon monoxide is said to be the number one killer, many communities have passed legislation adopting all or parts of *NFPA 720. NFPA 720* requires carbon monoxide detectors and alarms for all dwellings. Homeowners and landlords had better pay close attention to this! How often have you heard on the news or read in the newspaper that the carbon monoxide detectors did not work! The batteries were either dead or had been removed.

Equally important is that there is nothing in the *National Fire Alarm Code* that will prevent injury or death if proper escape routes have not been planned, *29.4.2.*

Fire warning devices commonly used in a residence are heat detectors, Figure 26-1, and combination smoke alarms/detectors, Figure 26-2.

The more elaborate systems connect to a central monitoring customer service center through a telephone line. These systems offer instant contact with police or fire departments, a panic button for emergency medical problems, low-temperature detection, flood (high-water) detection, perimeter protection, interior motion detection, and other features.

FIGURE 26-1 Heat detector.
(*Delmar/Cengage Learning*)

FIGURE 26-2 Combination smoke alarm/detector.
(*Delmar/Cengage Learning*)

All detections are transmitted first to the company's customer service center, where personnel on duty monitor the system 24 hours a day, 7 days a week. They will verify that the signal is valid. After verification, they will contact the police department, the fire department, and individuals whose names appear on a previously agreed-on list that the homeowner prepared and submitted to the company's customer service center.

NFPA Standards are not law until adopted by a city, state, or other governmental body. Some NFPA Standards are adopted totally, while only portions of others are adopted. Installers must be aware of all requirements in the locality in which they are doing work. Most communities require a special permit for the installation of fire warning equipment and security systems in homes, and require registration of the system with the police and/or fire departments. In most instances, fire protection requirements are found in the building codes of a community and are not necessarily spelled out in the electrical code.

Are Smoke and Heat Alarms Required?

Pay particular attention to the words found in *NFPA 72 29.1.2*, which states, ▶*Smoke and heat alarms shall be installed in all occupancies where required by governing laws, codes, or standards.*◀ Similar wording is found in *29.5.1.1*. This puts the responsibility on the local building department. It might seem confusing to have so many codes. Without question, you must become familiar with your local applicable laws, codes, or standards.

Most communities adopt building codes published by the International Code Council (ICC). These codes contain requirements for the installation of smoke alarms. Your local building department officials can explain what is required for new work and remodel work (alterations, repairs, and additions) where a permit is required. Whereas new work requires interconnected, hard-wired, battery backup smoke alarms, it might be acceptable on remodel jobs to install battery-powered-only smoke alarms for existing areas where interior walls or ceiling finishes are not removed to expose structural framing members. To install, interconnect, and hard-wire smoke alarms in these existing areas could result in damage to walls and ceilings, requiring patching, repainting, or other repairs—an uncalled for and tremendous expense. If there is access from an attic, crawl space, or basement, the inspector might require interconnected, hard-wired, battery backup smoke alarms if you can do the job without having to remove interior finishes. Check this out with your local inspector!

DEFINITIONS

Here are some very important definitions from *NFPA 72* and *NFPA 720*:

An **alarm** is *A warning of danger.**

A **heat alarm** is *A single- or multiple station alarm responsive to heat.**

A **smoke alarm** is *A single- or multiple station alarm responsive to smoke.**

A **detector** is *A device suitable for connection to a circuit that has a sensor that responds to a physical stimulus such as heat or smoke.**

*Reprinted with permission from NFPA 70-2011, 72-2010 or NFPA 720-2009.

A **heat detector** is *A fire detector that detects either abnormally high temperature or rate of temperature rise, or both.* *

A **smoke detector** is *A device that detects visible or invisible particles of combustion.* *

A **household fire alarm system** is *A system of devices that uses a fire alarm control unit to produce an alarm signal in the household for the purpose of notifying the occupants of the presence of a fire so that they will evacuate the premises.* *

Shall *Indicates a mandatory requirement.*

A **combination detector** is *A device that either responds to more than one of the fire phenomenon or employs more than one operating principle to sense one of these phenomenon. Typical examples are a combination of a heat detector with a smoke detector or a combination rate-of-rise and fixed-temperature heat detector. This device has listings for each sensing method employed.* *

A **single station alarm** is *A detector comprising an assembly that incorporates a sensor, control components, and an alarm notification appliance in one unit operated from a power source either located in the unit or obtained at the point of installation.* *

A **multiple station alarm** is *A single station alarm capable of being interconnected to one or more additional alarms so that the actuation of one causes the appropriate alarm signal to operate in all interconnected alarms.* *

A **dwelling unit** is *A single unit, providing complete and independent living facilities for one or more persons, including permanent provisions for living, sleeping, cooking, and sanitation (NEC).* *

A **living area** is *Any normally occupiable space in a residential occupancy, other than sleeping rooms or rooms that are intended for combination sleeping/living, bathrooms, toilet compartments, kitchens, closets, halls, storage or utility spaces, and similar areas.* *

A **separate sleeping area** is *The area of a dwelling unit where the bedrooms or sleeping rooms are located.* *

In this chapter, rather than going back and forth between the words "alarm" and "detector," we refer to fire warning devices as alarms, knowing full well

that before an alarm can sound, the cause must be detected. Most home-type fire warning devices combine the detector and alarm in one device. The location requirements for smoke and heat alarms are pretty much the same.

SMOKE, HEAT, AND CARBON MONOXIDE ALARMS

This chapter covers the basic requirements for protection in homes against the hazards of fire, heat, smoke, and carbon monoxide. Fire alarm systems for commercial installations are much more complicated than the typical household system.

Fire is the third leading cause of accidental death. Home fires account for the biggest share of these fatalities, most of which occur at night during sleeping hours. Rapidly developing high-heat fires and slow smoldering fires are the culprits. Both produce smoke and deadly gases.

Smoke, heat, and carbon monoxide alarms are installed in a residence to give the occupants *early warning* of the presence of fire or toxic fumes. Fires produce smoke and toxic gases that can overcome the occupants while they sleep. Most fatalities result from the inhalation of smoke and toxic gases, rather than from burns. Heavy smoke reduces visibility.

In nearly all home fires, detectable smoke precedes detectable levels of heat. People sleeping are less likely to smell smoke than people who are awake. The smell of smoke probably will not awaken a sleeping person. Therefore, smoke alarms are considered to be the primary devices for protecting lives.

Heat alarms DO NOT take the place of smoke alarms. Heat alarms are installed *in addition* to smoke detectors.

Another lifesaving device is the carbon monoxide (CO) alarm, which senses dangerous carbon monoxide emitted from a malfunctioning furnace or other source. *NFPA 720* is the *Standard for the Installation of Carbon Monoxide (CO) Detection and Warning Equipment.*

Home-type smoke, heat, and carbon monoxide devices as a rule have both the detector and the alarm in one device. There also are alarms that

---

*Reprinted with permission from NFPA 70-2011, 72-2010 or NFPA 720-2009.

combine more than one type of sensor into one unit, such as smoke/heat, smoke/carbon monoxide, or carbon monoxide/explosive gases.

In larger, more complex commercial installations, detectors and alarms are generally separate devices that, when triggered, send a signal to a central control panel, which in turn sounds the general alarm.

As with all electrical installations, always install fire warning equipment that has been listed by a nationally recognized testing laboratory (NRTL).

DETECTOR TYPES

AC/Battery: Required in new construction. These conform to the *NFPA 72 Section 29.6* requirement that alarms have two power supplies—regular household 120-volt ac for the primary source of power and 9-volt battery for the secondary source. The alarm "chirps" when the battery gets low. See Figure 26-3.

Interconnected: Required in new construction. Should any smoke, heat, or carbon monoxide

FIGURE 26-3 An ac-dc smoke alarm that operates on 120 volts ac as the primary source of power and a 9-volt battery as the secondary source of power. In the event of a power outage, the alarm continues to operate on the battery. (*Courtesy of BRK Electronics*)

alarm trigger, that alarm will sound and simultaneously send a signal to set off all other interconnected alarms. See *Section 29.5.2*. Check the manufacturer's instructions to determine how many units may be interconnected. Never connect more units than the number specified by the manufacturer. Never "mix" different manufacturer's devices. Testing laboratories do not perform tests using more than one manufacturer's devices in the test. The alarm "chirps" when the battery gets low. These units usually have a small 3-wire power connector—the wires connect to the field wiring and the other end has a connector that plugs into the back of the unit. See Figure 26-4. The pigtail may have a red or orange-colored "interconnect" wire. The field wiring between alarms is 3-wire nonmetallic-sheathed cable that contains a red insulated conductor that is used as the "interconnect" wire.

Plug-in: Not permitted in new construction. These alarms plug into a wall outlet that is not controlled by a wall switch. They operate on 120 volts as the primary source of power and on a 9-volt battery as the secondary source. The alarm "chirps" when the battery gets low. If of the plug-in type, they must have some sort of restraining means so as to reduce the possibility of being unplugged.

Battery-Operated Only: Not permitted in new construction. These alarms operate on a 9-volt battery only. The alarm "chirps" when the battery gets low. These alarms are commonly used in existing homes.

For the Hearing Impaired: These alarms have a bright strobe light that provides a visual alarm for the hearing impaired. This complies with the requirements found in the Americans with Disabilities Act (ADA).

WIRELESS SYSTEMS (LOW-POWER RADIO)

Wireless technology has come a long way in recent years. Here is an overview of what wireless means as related to smoke and fire alarms.

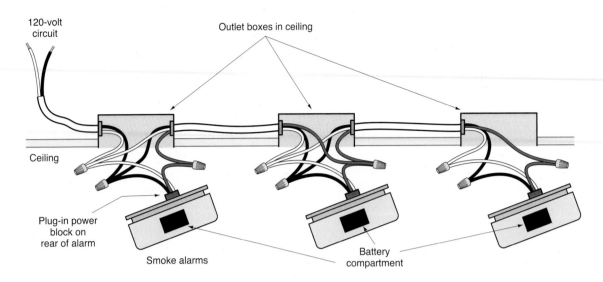

FIGURE 26-4 Interconnected smoke or heat alarms. The 2-wire branch circuit is run to the first alarm, then 3-wire nonmetallic-sheathed cable interconnects the other alarms. If one alarm in the series is triggered, all other alarms will sound off. See *NFPA 72, 29.5.2.1.1.* (*Delmar/Cengage Learning*)

When one alarm sounds off, all other alarms in the house are required to sound off at the same time. Stated another way, the alarms must be interconnected so as to "communicate" with one another. In new installations, this is generally done with hard-wiring, as indicated in Figure 26-4.

Hard-wiring smoke alarm systems—or adding more smoke alarms to an existing installation and trying to fish wires through walls and ceilings—can be messy, time-consuming, and costly.

Today, wireless systems that are listed in conformance to UL Standard 217, the standard for single- and multiple-station smoke alarms, are available. Both 120-volt ac- and battery-powered devices are used. Wireless systems might be the answer to installing a smoke alarm system economically and should be considered.

120-volt ac Powered: For existing 120-volt ac-powered systems, one interconnected smoke alarm is replaced with a 120-volt ac-powered smoke alarm that is designed specifically to send a wireless signal to other wireless smoke alarms of the same manufacturer. This triggers all of the other wireless smoke alarms to sound off. Additional smoke alarms can be installed anywhere in the house. That's all there's to it!

Battery Powered: For existing homes that do not have a smoke alarm system, the world of wireless technology can be quite advantageous. Battery-powered wireless smoke alarms can be installed. When one smoke alarm senses smoke, it triggers and sends a signal (wireless) to all other wireless battery-powered smoke alarms in the home. They are "interconnected" via wireless technology. No hard-wiring is required.

The maximum number of smoke alarms that can be interconnected is specified in the instructions furnished by the manufacturer, based on the testing and listing of the devices.

IMPORTANT: Local codes may or may not accept wireless smoke alarm systems. Check this out with your local building officials.

IMPORTANT: Whether the system is hard-wired or wireless, never mix different manufacturers' components.

IMPORTANT: Check the system regularly in accordance with manufacturer's instructions.

IMPORTANT: Be sure to install equipment "listed for the purpose," *Section 29.3.1.*

See *NFPA 72, Section 6.17* for a complete list of requirements for wireless systems. Also see *Section 29.7.7.*

TYPES OF SMOKE ALARMS

Two common types of smoke alarms are the **photoelectric type** (sometimes called photoelectronic) and the **ionization type**. They usually contain an indication light to show that the unit is functioning properly. They also may have a test button that simulates smoke, so that when the button is pushed, the detector's smoke-detecting ability as well as its circuitry and alarm are tested.

Some alarms are tested with a magnet. Others are tested with a listed spray from an aerosol can.

Smoke alarms generally do not sense heat, flame, or gas. However, some smoke alarms can be set off by acetylene and propane gas.

Some alarms have a "hush button" that can temporarily silence a nuisance alarm for about 15 minutes.

Photoelectric Type

The photoelectric type of smoke alarm has a light sensor that measures the amount of light in a chamber. When smoke is present, an alarm sounds, indicating a reduction in light due to the obstruction of the smoke. This type of sensor detects smoke from burning materials that produce large quantities of smoke, such as furniture, mattresses, and rags. The photoelectric type of alarm is less effective for gasoline and alcohol fires, which do not produce heavy smoke. This type of alarm can become more sensitive to smoke as it gets older.

Photoelectric sensors are a good choice for areas subject to steam, such as in or outside bathrooms, utility rooms, and kitchens.

There are two types of photoelectric devices: light obscuration and light scattering.

Ionization Type

The ionization type of alarm contains a low-level radioactive source (less than used in luminescent watch and clock dials), which supplies particles that ionize the air in the detector's smoke chamber. Plates in this chamber are oppositely charged. Because the air is ionized, an extremely small amount of current (millionths of an ampere) flows between the plates.

Smoke entering the chamber impedes the movement of the ions, reducing the current flow, which triggers the alarm. This type often sets off a nuisance alarm because cooking routinely gives off small "invisible" smoke particles. This type of alarm can become more sensitive to smoke as it get older.

The ionization type of alarm is effective for detecting small amounts of smoke, as in gasoline and alcohol fires, which are fast flaming with little or no smoke.

Some smoke alarms are available with both ionization and photoelectric sensors combined. Other smoke alarms are available with smoke and carbon monoxide detection capabilities or smoke and heat detection capabilities in one unit.

TYPES OF HEAT ALARMS

Heat alarms respond to heat—not smoke!

Three types of heat alarms are described as follows.

- Fixed temperature heat detectors sense a specific fixed temperature, such as 135°F (57°C) or 200°F (93°C). They shall have a temperature rating of at least 25°F (14°C) above the normal temperature expected, but not to exceed 50°F (28°C) above the expected temperature. Fixed temperature detectors are sometimes combined with smoke alarms and carbon monoxide alarms.

- Rate-of-rise heat detectors sense rapid changes in temperature (12°F to 15°F per minute) such as those caused by flash fires.

- Fixed/rate-of-rise temperature detectors are available as a combination unit.

- Fixed/rate-of-rise/smoke combination detectors are available in one unit.

INSTALLATION REQUIREMENTS

The following information includes specific recommendations for installing smoke alarms and heat alarms in homes. Remember that smoke alarms are the primary fire warning devices, and heat detectors are installed in addition to the required

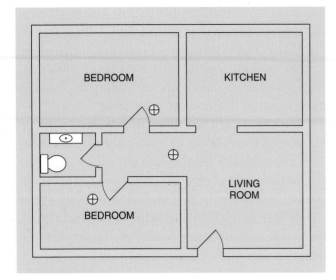

FIGURE 26-5 Smoke alarms are required in all sleeping areas, outside all sleeping areas, and on every level. Acceptable locations are on the ceiling or on the wall space above the door. See Figures 26-7, 26-8, and 26-9 for location restrictions. (*Delmar/Cengage Learning*)

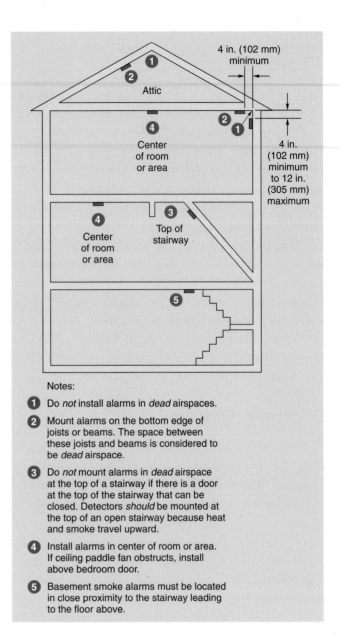

Notes:

1. Do *not* install alarms in *dead* airspaces.

2. Mount alarms on the bottom edge of joists or beams. The space between these joists and beams is considered to be *dead* airspace.

3. Do *not* mount alarms in *dead* airspace at the top of a stairway if there is a door at the top of the stairway that can be closed. Detectors *should* be mounted at the top of an open stairway because heat and smoke travel upward.

4. Install alarms in center of room or area. If ceiling paddle fan obstructs, install above bedroom door.

5. Basement smoke alarms must be located in close proximity to the stairway leading to the floor above.

FIGURE 26-6 Recommendations for the installation of heat and smoke alarms. Check manufacturers' installation instructions for their recommendations about installing smoke and heat alarms in attics. (*Delmar/Cengage Learning*)

smoke alarms. As you continue reading, study Figures 26-5, 26-6, 26-7, 26-8, and 26-9. Complete data are found in *NFPA Standard No. 72, Chapter 29,* and in the instructions furnished by the manufacturer of the equipment.

The Absolute Minimum Level of Protection

Here is a brief summary of the absolute minimum level of protection for smoke alarms for new residential construction in conformance to *NFPA 72, Chapter 29.* Some communities have adopted portions of *NFPA 72* for existing dwellings. You will have to check with your local authority having jurisdiction (AHJ). Because life safety is so important, attending an NFPA seminar on *NFPA 72* is highly recommended.

- Install multiple station smoke alarms:

  a. in all sleeping rooms and guest rooms.

  b. outside each separate sleeping area, within 21 ft (6.4 m) of any door to the sleeping area. This distance is measured along the path of travel.

  c. on every level of a dwelling unit. This includes basements but excludes crawl spaces and unfinished attics.

  d. in guest bedrooms and suites: More elaborate homes may perhaps have a guest bedroom(s) or suite, completely self-contained with bedroom(s), bathroom(s), and closet(s).

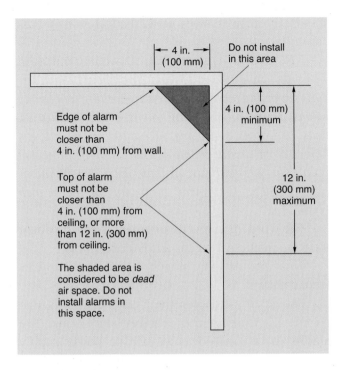

FIGURE 26-7   Do not mount smoke or heat alarms in the *dead* airspace where the ceiling meets the wall. (*Delmar/Cengage Learning*)

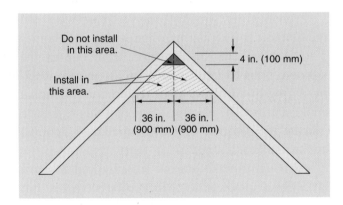

FIGURE 26-8   Installing smoke and heat detectors/alarms in peaked (cathedral) ceilings. (*Delmar/Cengage Learning*)

If these are separated from the rest of the house by a door, a smoke alarm shall be installed in the guest bedroom or suite, plus an additional smoke alarm shall be installed on the living area side of the door. In instances where a hallway is outside the sleeping area, a smoke alarm shall be installed in that hallway. If that hallway is

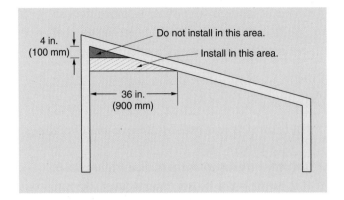

FIGURE 26-9   Installing smoke and heat detectors/alarms in sloped ceilings. (*Delmar/Cengage Learning*)

closed off by a door, an additional smoke alarm shall be installed on the living area side of the door. The bottom line is that three smoke alarms might be needed.

e. After installing smoke alarms required in (a), (b), (c), and (d), check one more thing for large homes. It might be necessary to install additional smoke alarms for a given floor area 1000 ft^2 (93 m^2) or greater. Don't include the garage area. To conform, either of the following is acceptable:

1. Check to verify that no point on the ceiling of the given floor area is more than 30 ft (9.1 m) from a smoke alarm.

2. Install a smoke alarm for every 500 ft^2 (46.5 m^2) in the given floor area.

f. for cathedral and/or vaulted ceilings extended over more than one floor. Meeting requirements (a), (b), (c), and (d) for the upper level will meet requirement (e).

g. For existing dwellings, listed battery-powered smoke alarms that are not interconnected are generally permitted. This provides a reasonable level of protection as opposed to having no smoke alarms at all. However, your local *Code* might require ac/dc dual-powered interconnected alarms.

• Smoke alarms must have at least two independent sources of power. The primary source is the 120-volt ac circuit, and the secondary (standby) source is the integral battery power supply.

- The ac source of power is permitted to be a dedicated branch circuit or an unswitched part of a conventional branch circuit.

Where to Install

In *bedrooms* and *halls outside of sleeping areas*. See Figure 26-5.

In *other rooms and areas*. See Figure 26-6.

On *walls*—not closer than 4 in. (100 mm) but not farther than 12 in. (300 mm) from the adjoining ceiling. Do not install in dead airspace. See Figures 26-6 and 26-7.

On *flat ceilings*—not closer than 4 in. (100 mm) from the adjoining wall. See Figures 26-6 and 26-7. A level ceiling is one that is level or has a slope of not more than 1 ft in 8 ft (1 m in 8 m).

On *peaked ceilings*—locate with 36 in. (900 mm) horizontally from the peak, but not closer than 4 in. (100 mm) vertically from the peak. A peaked ceiling might also be called a cathedral ceiling. Do not install in dead airspace. See Figure 26-8.

On *sloped ceilings*—a sloped ceiling is defined as having a slope greater than 1 ft in 8 ft (1 m in 8 m). Locate within 36 in. (900 mm) of the high side but not closer than 4 in. (100 mm) from the adjoining wall. Do not install in dead airspace. See Figure 26-9. A level ceiling is one that is level or has a slope of not more than 1 ft in 8 ft (1 m in 8 m).

How to Wire Smoke and Heat Alarms

In new construction, smoke and heat alarms require both primary (120-volt) and secondary (battery) sources of power. The *National Fire Alarm Code NFPA 72* does not specify where to pick up the 120-volt supply.

Smoke and Fire Alarms and AFCI Protected Branch Circuits?

An often asked question is: "Are home-type smoke and fire alarms permitted to be connected to an AFCI protected branch circuit? As it now stands, the answer is yes. Read the following words carefully.

NEC 210.12(A) states that in dwelling units, *All 120-volt, single-phase, 15- and 20-ampere branch circuits supplying outlets installed in dwelling unit family rooms, dining rooms, living rooms, parlors, libraries, dens, bedrooms, sunrooms, recreation rooms, closets, hallways, or similar rooms or areas shall be protected by a listed arc-fault circuit interrupter, combination-type, installed to provide protection of the branch circuit.**

The word *all* in this requirement leaves no doubt as to the intent. All means all! No exceptions!

This includes the branch circuit that supplies smoke alarms.

The justification for the "yes" answer is that *NFPA 72, Chapter 29*, requires smoke and fire alarms to be dual-powered—the ac power supply and the battery backup. The battery is required to last for 24 hours in normal conditions, followed by 4 minutes of alarm. In that amount of time, the homeowner, sleeping or awake, will become aware of a potential problem by the beeping or chirping. Some units have visual indication also.

Probably the simplest way to hard-wire and interconnect smoke alarms when wiring new homes is to pick up one smoke alarm from one of the bedroom branch circuits, then run the interconnecting wiring to all of the other required and optional smoke alarms. The electrical plans for this residence indicate the location of the required and optional smoke and heat alarms.

Although smoke alarms are required to have battery backup power, some local codes do not permit smoke alarms to be supplied by the same branch circuit as the bedroom lighting. They require that some other branch circuit supply the smoke alarms. Check this out with your electrical inspector.

Wireless systems are discussed earlier in this chapter.

Combination Smoke and Carbon Monoxide Detectors/Alarms

Combination smoke and carbon monoxide detectors/alarms are readily available. Some communities are requiring that this type be installed in

*Reprinted with permission from NFPA 70-2011.

new construction. Check with your local electrical inspector.

Some inspectors require a "lock-off" device over the handle of the circuit breaker that serves the alarms so that it will not unintentionally be turned off. This lock-off device is really not necessary. Should the breaker supplying the alarms get turned off intentionally or unintentionally, or if it trips due to trouble in the circuit, the battery backup would power the alarms. When the batteries get low, the alarms start "chirping."

The usual way to connect alarms is to run the 2-wire branch circuit to the first alarm, then run 3-wire nonmetallic-sheathed cable from the first alarm to all other alarms. The third conductor provides the interconnect feature that will sound all alarms should any one of the alarms activate. See Figure 26-4.

Summary of Dos and Do Nots

When installing detectors and alarms, ask yourself, "Where will the smoke and greatly heated air travel?" Because smoke and heated air rise, detectors and alarms must be in the path of the smoke and greatly heated air.

Figures 26-5, 26-6, 26-7, 26-8, and 26-9 illustrate some of the more important issues regarding smoke and heat alarms.

Here is a list of the *Dos* and *Do nots* in no particular order of importance.

The Dos

1. *Do* install listed smoke alarms, and install them according to the manufacturer's instructions.

2. *Do* install smoke alarms on the ceiling, as close as possible to the center of the room or hallway. Note: Because ceiling paddle fans are often installed in bedrooms, install the alarms on the wall space above the bedroom door, not closer than 4 in (102 mm) nor more than 12 in. (305 mm) from the ceiling.

3. *Do* make sure that the path of rising smoke will reach the alarm when the alarm is installed in a stairwell. This is usually at the top of the stairway. The alarm must not be located in a dead airspace created by a closed door at the head of the stairway.

4. *Do* space according to the instructions furnished with the devices.

5. *Do* install the type that has a "hush" button if located within 20 ft (6.1 m) of a cooking appliance, or install a photoelectric type.

6. *Do* install smoke alarms at the high end of a room that has a sloped, gabled, or peaked ceiling where the rise is greater than 1 ft (305 mm) per 8 ft (2.6 m). Do not install in the dead airspace.

7. *Do* mount on the bottom of open joists or beams.

8. *Do* install a smoke alarm on the basement ceiling close to the stairway to the first floor.

9. *Do* install smoke alarms in split-level homes. A smoke alarm installed on the ceiling of an upper level can suffice for the protection of an adjacent lower level if the two levels are not separated by a door. Better protection is to install alarms for each level.

10. *Do* install in new construction dual-powered smoke alarms that are hard-wired directly to a 120-volt ac source and also have a battery. In existing homes, battery-powered alarms are the most common, but 120-volt ac alarms or dual-powered alarms can be installed. Remember, battery-powered alarms will not operate with dead batteries. Ac-powered alarms will not operate when the power supply is off.

11. *Do* install interconnected smoke alarms in new construction so that the operation of any alarm will cause all other alarms that are interconnected to sound.

12. *Do* install smoke alarms so they are not the only load on the branch circuit. Connect smoke alarms to a branch circuit that supplies other lighting outlets in habitable spaces. Because smoke alarms draw such a miniscule amount of current, they can easily be connected to a general lighting branch circuit, such as a bedroom branch circuit.

13. *Do* always consider the fact that doors, beams, joists, walls, partitions, and similar obstructions

will interfere with the flow of smoke and heat, and, in most cases, create new areas needing additional smoke alarms and heat alarms.

14. *Do* make sure that the gap around the ceiling outlet box is sealed to prevent dust from entering the smoke chamber. This is a big problem for ceiling-mounted alarms.

15. *Do* clean smoke alarms according to the manufacturers' instructions.

16. *Do* consider that the maximum distance between heat alarms mounted on flat ceilings is 50 ft (15 m) and 25 ft (7.5 m) from the detector to the wall. This information is explained in detail by *NFPA 72*. Where obstructions such as beams and joists will interfere with the flow of heat, the 50-ft distance is reduced to 25 ft (7.5 m), and the 25-ft distance is reduced to 12½ ft (3.75 m).

17. *Do* install alarms after construction clean-up of all trades is complete and final.

18. *Do* install interconnected units of the same manufacturer. Different manufacturers' units may or may not be compatible.

19. *Do* connect all interconnected units to the same branch circuit. Different circuits cannot be shared.

The Do Nots

1. *Do not* install smoke alarms in the dead airspace at the top of a stairway that can be closed off by a door.

2. *Do not* install within 36 in. (914 mm) of a door to a kitchen or to a bathroom containing a shower or tub.

3. *Do not* locate where the smoke alarm will be subject to temperature and/or humidity that exceeds the limitations stated by the manufacturer.

4. *Do not* install within a 36-in. (914-mm) horizontal path from a supply hot air register. Install outside the direct flow of air from these registers.

5. *Do not* install within a horizontal path 36 in. (914 mm) from the tip of the blade of a ceiling-suspended paddle fan.

6. *Do not* install where smoke rising in a stairway could be blocked by a closed door or an obstruction.

7. *Do not* place the edge of a ceiling-mounted smoke alarm or heat alarm closer than 4 in. (102 mm) from the wall.

8. *Do not* place the top edge of a wall-mounted smoke alarm or heat alarm closer than 4 in. (102 mm) from the ceiling and farther than 12 in. (305 mm) down from the ceiling. Some manufacturers recommend placement not farther down than 6 in. (153 mm).

9. *Do not* install smoke alarms or heat detectors on an outside wall that is not insulated, or one that is poorly insulated. Instead mount the alarms on an inside wall.

10. *Do not* install smoke alarms or heat alarms on a ceiling where the ceiling will be excessively cold or hot. Smoke and heat will have difficulty reaching the alarms. This could be the case in older homes that are not insulated or are poorly insulated. Instead mount the alarms on an inside wall.

11. *Do not* install smoke alarms or heat alarms where the ceiling meets the wall because this is considered dead airspace where smoke and heat may not reach the detector.

12. *Do not* connect smoke alarms or heat alarms to wiring that is controlled by a wall switch.

13. *Do not* install smoke alarms or heat alarms where the relative humidity exceeds 85%, such as in bathrooms with showers, laundry areas, or other areas where large amounts of visible water vapor collect. Check the manufacturer's instructions.

14. *Do not* install smoke alarms or heat alarms in front of air ducts, air conditioners, or any high-draft areas where the moving air will keep the smoke or heat from entering the detector.

15. *Do not* install smoke alarms or heat alarms in kitchens where the accumulation of household smoke can result in setting off an alarm, even though there is no real hazard. The person in the kitchen will know why the alarm sounded, but other people in the house may panic. This problem exists in multifamily dwelling units, where

unwanted triggering of the alarm in one dwelling unit might cause people in the other units to panic. The photoelectric type may be installed in kitchens but must not be installed directly over the range or cooking appliance. A better choice in a kitchen is to install a heat alarm.

16. *Do not* install smoke alarms where the temperature can fall below 32°F (0°C) or rise above 120°F (49°C) unless the detector is specifically identified for this application.

17. *Do not* install smoke alarms in garages where vehicle exhaust might set off the detector. Instead of a smoke alarm, install a heat detector.

18. *Do not* install smoke alarms in airstreams that will pass air originating at the kitchen cooking appliances across the alarm. False alarms will result.

19. *Do not* install smoke alarms or heat detectors on ceilings that employ radiant heating.

20. *Do not* install smoke alarms or heat alarms in a recessed location.

21. *Do not* connect smoke alarms or heat alarms to a switched circuit or a circuit controlled by a dimmer.

22. *Do not* install smoke detectors in attics because of possible nuisance triggering. Check the manufacturers' instructions.

MAINTENANCE AND TESTING

Once installed, smoke, heat, and carbon monoxide alarms must be tested periodically and maintained (blow out accumulated dust) to make sure they are operating properly. This is a requirement of *NFPA 72*, which states, *Homeowners shall inspect and test smoke alarms and all connected appliances in accordance with the manufacturer's instructions at least monthly.*

Failure rate: Field studies indicate a probable failure rate of:

3% in the first year
30% in the first 10 years
50% in the first 15 years
Nearly 100% in 30 years

The need for periodic testing and replacement is obvious.

When to replace: In conformance to *10.4.7* and *29.8.1.4.(5)* in *NFPA 72*, smoke alarms *shall be replaced when they fail to respond to tests* and *shall not remain in service longer than 10 years from the date of manufacture.**

Note: The 10-year mandatory replacement requirement is often overlooked. This requirement is based on studies of tens of thousands of smoke alarms in operation, to determine the acceptable number of failures over many years. Code committees and manufacturers analyzed the data and agreed that 10 years in service provided a reasonable lifetime. The 10-year replacement program, along with regular testing per the manufacturer's instruction, results in very few homes going unprotected for any extended period of time. Hard-wired and battery-operated alarms are equally affected by age.

The expected life of lithium batteries used in some alarms is said to be 10 years.

At least one major manufacturer of smoke alarms has incorporated a feature that

- sets off a chirping sound after the alarm has been in service for 10 years. The chirping sound repeats every 30 seconds.
- sets off a chirping sound when the battery is low. The chirping sound repeats every minute.

Suggestion: It is difficult to know when a smoke, heat, or carbon monoxide alarm was installed. The instructions are usually filed away or were discarded or otherwise separated from these devices. It might not be a bad idea when installing the alarm to mark the installation date on the device.

After You Install Smoke and Heat Alarms, Then What?

After installing smoke and heat alarms, there is still more to do. *NFPA 72* requires that the installer of the fire warning system provide the homeowner with the following information:

1. An instruction booklet illustrating typical installation layouts
2. Instruction charts describing the operation, method, and frequency of testing and maintenance of fire warning equipment

---

*Reprinted with permission from NFPA 72-2010.

3. Printed information for establishing an emergency evacuation plan

4. Printed information to inform owners where they can obtain repair or replacement service, and where and how parts requiring regular replacement, such as batteries or bulbs, can be obtained within 2 weeks

5. Information that, unless otherwise recommended by the manufacturer, smoke alarms shall be replaced when they fail to respond to tests

6. Instruction that smoke alarms shall not remain in service longer than 10 years from the date of installation

This information is usually part of the instructions furnished by the manufacturer of the alarms.

Exceeding Minimum Levels of Protection

The following are recommendations for attaining levels of fire warning protection that exceed the minimum level stated previously and include guidelines for installing smoke and heat detectors in a home.

- Install smoke alarms in *all* rooms, basements, hallways, heated attached garages, and storage areas. Installing alarms in these locations will increase escape time, particularly if the room or area is separated by a door(s). In some instances, smoke alarms are installed in attics and crawl spaces.

- Install heat alarms in kitchens because conventional smoke alarms can nuisance sound an alarm.

- Install heat alarms in garages because gasoline fires give off little smoke.

- Consider special alarms that light up or vibrate for occupants who are hard of hearing.

- Consider smoke alarms that have an "escape" light.

- Consider low temperature alarms (e.g., 45°F [7°C]) that can detect low temperatures should the heating system fail. The damage caused by frozen water pipes bursting can be extremely costly.

Building Codes

In most instances the various building codes published by the International Code Council (ICC) (formerly BOCA, ICBO, and SBCCI) adopt *NFPA 72* and *NFPA 720* by reference, but it would be wise to check with your local AHJ to find out if there are any differences between these codes that might affect your installation. See Chapter 1 for information about these building code organizations. Also refer to the Web site list following the Appendix of this text.

Many companies specialize in installing complete fire and security systems. They also offer system monitoring at a central office and can notify the police or fire department when the system gives the alarm. How elaborate the system should be is up to the homeowner.

The ICC Code requires that smoke alarms be connected to a circuit that also supplies lighting outlets in habitable spaces. The logic is that should the alarms be connected to a separate circuit, that circuit could unknowingly be off.

Manufacturers' Requirements

Here are a few of the responsibilities of the manufacturers of smoke alarms and heat alarms. Complete data are found in *NFPA 72*.

- The power supply must be capable of operating the signal for at least 4 minutes continuously.

- Battery-powered units must be capable of a low-battery warning "chirp" of at least one chirp per minute for 7 consecutive days.

- Direct-connected 120-volt ac alarms must have a visible indicator that shows "power-on."

- Alarms must not signal when a power loss occurs or when power is restored.

  Note: In *4.3.1* and *2.3.1* in *NFPA 72*, we find that we must

- always install fire alarm equipment that is "Listed for the purpose."

- always install fire alarm equipment in accordance with the manufacturers' installation and maintenance instructions.

In this residence, the smoke alarms and heat detectors are connected to Circuit A17 located in Panel A in the Workshop. Run two conductors (the circuit) to the first alarm, then interconnect the other alarms by running three conductors between all other detectors.

According to the plans for this residence, smoke, heat, and carbon monoxide alarms (detectors) are located as follows:

Smoke: Bedrooms	3
Hall between bedrooms	1
Entry	1
Recreation Room	1
Rear Hall	1
Total	7
Heat: Kitchen	1
Laundry Room	1
Workshop	1
Garage	1
Total	4
Carbon monoxide:	
Recreation Room	1
Hall between bedrooms	1
Total	2

CARBON MONOXIDE ALARMS

Carbon monoxide alarms are installed in addition to smoke and heat alarms!

Carbon monoxide (CO) is referred to as the "silent killer"! It is responsible for more deaths than any other single poison.

When a heat or smoke alarm is triggered, it is generally not difficult to determine the source of the problem. But when a carbon monoxide alarm goes off, it is hard to determine the source.

Carbon monoxide is odorless, colorless, and tasteless—undetectable by any of a person's five senses, taste, smell, sight, touch, hearing—but is highly poisonous. Carbon monoxide replaces oxygen in the bloodstream, resulting in brain damage or total suffocation.

Carbon monoxide is produced by the incomplete burning of fuels. Common sources of carbon monoxide in a home are a malfunctioning furnace, gas appliances, kerosene heaters, automobile or other gas engines, charcoal grills, fireplaces, or a clogged chimney.

Symptoms of carbon monoxide poisoning are similar to those of flulike illnesses, such as dizziness, fatigue, headaches, nausea, diarrhea, stomach pains, irregular breathing, and erratic behavior.

Carbon monoxide has about the same characteristics as air. It moves about just like air.

Carbon monoxide alarms detect carbon monoxide from any source of combustion. They are not designed to detect smoke, heat, or gas. The Consumer Product Safety Commission (CPSC) recommends that at least one carbon monoxide alarm be installed in each home, and preferably one on each floor of a multilevel house. Some cities require this. If only one carbon monoxide alarm is installed, it should be in the area just outside individual bedrooms.

Carbon monoxide alarms monitor the air in the house and sound a loud alarm when carbon monoxide above a predetermined level is detected. They provide early warning before deadly gases build up to dangerous levels.

For residential applications, carbon monoxide alarms are available for hard-wiring (direct connection to the branch-circuit wiring), plug-in (the male attachment plug is built into the back of the detector for plugging directly into a 120-volt wall receptacle outlet), and combination 120-volt and battery-operated units. They are available with the interconnect feature for interconnecting with smoke and heat alarms. Some carbon monoxide alarms also have an explosive gas sensor.

NFPA 720 contains the installation recommendations for carbon monoxide alarms. Here is a summary of these recommendations.

Installation "Dos" for Carbon Monoxide Alarms

1. *Do* install carbon monoxide alarms that are tested and listed in conformance to UL Standard 2034.

2. *Do* install carbon monoxide alarms according to the manufacturer's instructions.

3. *Do* use household electricity as the primary source of power. In existing homes, monitored battery units are permitted.

4. *Do* install carbon monoxide alarms in or near bedrooms and living areas.

5. *Do* install carbon monoxide alarms in locations where smoke alarms are installed.

6. *Do* test carbon monoxide alarms once each month or as recommended by the manufacturer.

7. *Do* remove carbon monoxide alarms before painting, stripping, wallpapering, or using aerosol sprays. Store in a plastic bag until the project is completed.

8. *Do* carefully vacuum carbon monoxide alarms once each month or as recommended by the manufacturer.

9. *Do* interconnect alarms where multiple alarms are installed.

Installation "Do Nots" for Carbon Monoxide Alarms

1. *Do not* connect carbon monoxide alarms to a switched circuit or a circuit controlled by a dimmer.

2. *Do not* install carbon monoxide alarms closer than 6 in. (150 mm) from a ceiling. This is considered dead airspace.

3. *Do not* install carbon monoxide alarms in garages, kitchens, or furnace rooms. This could lead to nuisance alarms, may subject the detector to substances that can damage or contaminate the alarm, or the alarm might not be heard.

4. *Do not* install carbon monoxide alarms within 15 ft (4.5 m) of a cooking or heating gas appliance.

5. *Do not* install carbon monoxide alarms in dusty, dirty, or greasy areas. The sensor inside of the alarm can become coated or contaminated.

6. *Do not* install carbon monoxide alarms where the air will be blocked from reaching the alarm.

7. *Do not* install carbon monoxide alarms in dead airspaces such as in the peak of a vaulted ceiling or where a ceiling and wall meet.

8. *Do not* install carbon monoxide alarms where the detector will be in the direct airstream of a fan.

9. *Do not* install carbon monoxide alarms where temperatures are expected to drop below 40°F (4°C) or get hotter than 100°F (38°C). Their sensitivity will be affected.

10. *Do not* install carbon monoxide alarms in damp or wet locations such as showers and steamy bathrooms. Their sensitivity will be affected.

11. *Do not* mount carbon monoxide alarms directly above or near a diaper pail. The high methane gas will cause the detector to register, and this is not carbon monoxide.

12. *Do not* clean carbon monoxide alarms with detergents or solvents.

13. *Do not* spray carbon monoxide alarms with hair spray, paint, air fresheners, or other aerosol sprays.

FIRE ALARM SYSTEMS

These are the more comprehensive installations involving a central control panel, alarms, sensors, detectors as covered in the *National Fire Alarm Code, NFPA 72, Chapters 1* through *10*.

Fire alarm systems are also covered in *Article 760* of the *NEC*.

NEC 760.2 defines a fire alarm circuit as The portion of the wiring system between the load side of the overcurrent device or the power-limited supply and the connected equipment of all circuits powered and controlled by the fire alarm system. Fire alarm circuits are classified as either non power-limited or power-limited.*

Fire-protective signaling systems installed in homes have power-limited fire alarm (PLFA) circuits where the power output is limited by the listed power supply for the system. Here are the key requirements in *Article 760, Part III*:

- The power supply must have power output capabilities limited to the values specified in *Chapter 9, Table 12(A)* and *Table 12(B)*. Refer to *760.121*. Residential fire-protective signaling systems are generally "power-limited."

- Do not connect the equipment to a GFCI- or AFCI-protected branch circuit, *760.41(B)* and *760.121(B)*.

---

*Reprinted with permission from NFPA 70-2011.

- Wiring on the supply side of the equipment is installed according to the conventional wiring methods found in *Chapter 1* through *Chapter 4* of the *NEC*. See *760.127*.

- The branch-circuit overcurrent protection supplying the fire alarm system shall not exceed 20 amperes, *760.127*.

- Cables may be run on the surface or be concealed and shall be adequately supported and protected against physical damage, *760.130(B)(1)*.

- When run exposed within 7 ft (2.1 m) of the floor, the cable shall be securely fastened at intervals not over 18 in. (450 mm), *760.130(B)(1)*.

- Cables shall not be run in the same raceway, cable, outlet box, device box, or be in the same enclosure as light and power conductors unless separated by a barrier, or introduced solely to connect that particular piece of equipment. Refer to *760.136(A)* and *(B)*. This is also a ruling in *725.136(A)* and *(B)*.

- Cables shall not be supported by taping, strapping, or attaching by any means to any electrical conduits, or other raceways, *760.143*.

- Cables shall be supported by structural components, *760.24*.

- Wiring on the load side shall be insulated solid or stranded copper conductors of the types listed in *Table 760.179(A)*.

- *Table 760.179(I)* shows different types of cable that can be substituted for one another when necessary.

- If installed in a duct or plenum, the cable must be plenum rated, *760.179(D)*. The cable will be marked FPLP.

- Type FPL cable is permitted in one- and two-family dwellings, *760.179(F)*.

  - Conductors in cable shall not be smaller than 26 AWG. Single conductors shall not be smaller than 18 AWG. Refer to *760.179(B)*.

  - Cable minimum voltage rating is 300 volts, *760.179(C)*.

- Remove abandoned conductors and cables. They add fuel to a fire. Here is a list of where the removal requirements are found in the *NEC*:

372.13, 374.7, 390.7, 640.2, 640.6(C), 645.2, 645.5(F), 725.2, 725.3(B), 760.2, 760.3(A), 770.2, 770.25, 770.154(A), 800.2, 800.25, 800.154, 820.2, 820.25, 820.154, 830.2, 830.25.

It makes a lot of sense to contact your local inspection authority before installing a fire alarm system to be sure the system you intend to install meets the requirements of the local codes! Contractors who specialize in this kind of work are well aware of the *Code* requirements and will install the proper types of cables.

SECURITY SYSTEMS

It is beyond the scope of this text to cover all types of residential security systems. What follows is a typical system.

Professionally installed and homeowner installed security systems can range from a simple system to a very complex one. Features and options include master control panels, remote controls, perimeter sensors for doors and windows, motion sensors, passive infrared sensors, wireless devices, interior and exterior sirens, bells, electronic buzzers, strobe lights to provide audio (sound) as well as visual detection, heat sensors, carbon monoxide alarms, smoke alarms, glass break sensors, flood sensors, low temperature sensors, and lighting modules. Figure 26-10 shows some of these devices. Figure 26-11 shows a schematic one-line diagram for a typical security system.

Monitored security systems are connected by telephone lines to a central station, a 24-hour monitoring facility. If the alarm sounds, security professionals will contact the end user to verify the alarm before sending the appropriate authorities (fire, police, or ambulance) to the home.

The decision about how detailed an individual residential security system should be generally begins with a meeting between the homeowner and the security consultant or licensed electrical contractor before the actual installation. Most security system installers, consultants, and licensed electrical contractors are familiar with a particular manufacturer's system.

The systems are usually explained and even demonstrated during an in-home security presentation. They are then professionally installed by the

The wireless key allows the end user to disarm and arm the system with the press of a button. There is no need to remember a security code and it can also be used to operate lights and appliances.

Monitored smoke alarms are on 24 hours a day, even if the burglary protection is turned off.

Magnetic contacts are often located at vulnerable entry points such as doors.

Sirens, such as these placed inside the home, alert families to emergencies.

Motion detectors are interior security devices that sense motion inside the home.

A wireless panic button allows the user to press a button to signal for help.

The wireless keypad is portable so you can arm or disarm the system from the yard or driveway.

Glassbreak detectors, available in a variety of sizes, will sound the alarm when they detect the sound of breaking glass.

FIGURE 26-10 Types of security systems. (*Photos courtesy of BRK Brands*)

security system company or licensed electrical contractor.

Do-it-yourself kits that contain many of the features provided by the professionally installed

systems can be found at home centers, hardware stores, electrical distributors, and electronics stores. Depending on the system, it may or may not be possible to connect to a central monitoring facility.

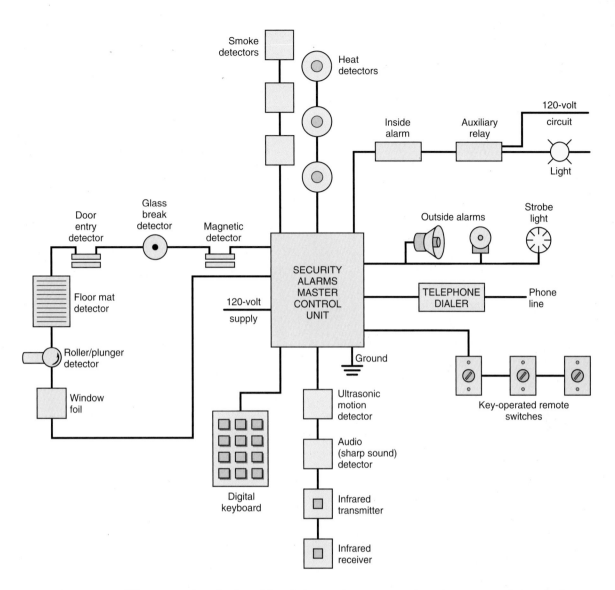

FIGURE 26-11   Diagram of typical residential security system showing some of the devices available. Complete wiring and installation instructions are included with these systems. Check local code requirements in addition to following the detailed instructions furnished with the system. Most of the interconnecting conductors are 18 AWG.
(*Delmar/Cengage Learning*)

The wiring of a security system consists of small, easy-to-install low-voltage, multiconductor cables made up of 18 AWG conductors. These conductors should be installed after the regular house wiring is completed to prevent damage to these smaller cables. Usually the wiring can be done at the same time as the chime wiring is being installed.

Security system wiring comes under the scope of *Article 725*.

When wiring detectors such as door entry, glass break, floor mat, and window foil detectors, circuits are electrically connected in series so that if any part of the circuit is opened, the security system will detect the open circuit. These circuits are generally referred to as "closed" or "closed loop." Alarms, horns, and other signaling devices are connected in parallel, because they will all signal at the same time when the security system is set off. Heat detectors and smoke detectors are generally connected in parallel because any of these devices will "close" the circuit to the security master control unit, setting off the alarms.

The instructions furnished with all security systems cover the installation requirements in detail, alerting the installer to *Code* regulations, clearances, suggested locations, and mounting heights of the systems' components.

Always check with the local electrical inspector to determine if there are any special requirements in your locality relative to the installation of security systems.

Sprinkler Systems Code

Home fire protection sprinkler systems are covered by *NFPA 13D*, which is the *Standard for the Installation of Sprinkler Systems in One- and Two-Family Dwellings and Manufactured Homes*.

▶The 2009 *International Residential Code* (*IRC*) in section *R313.2* will require, effective January 1, 2011, that new one and two-family dwellings have an automatic residential fire sprinkler system.

An exception provides that the automatic residential fire sprinkler system is not required for *additions* or *alterations* to existing buildings that are not already provided with an automatic residential sprinkler system.

Automatic residential fire sprinkler systems are required to be designed and installed in accordance with *IRC* Section *P2904* or *NFPA 13D*.◀

This rule will apply where the *International Residential Code* is adopted and enforced. A parallel requirement is in the 2009 *International Building Code*.

Standards Relating to Fire, Smoke, Carbon Monoxide, and Security Devices

UL 217—Standard for Single- and Multiple-Station Alarms

UL 268—Standard for Smoke Detectors for Fire Protective Signaling Systems

UL 365—Standard for Police Station Connected Burglar Alarm Units

UL 521—Standards for Heat Detectors for Fire Protection Signaling Systems

UL 539—Standard for Single- and Multiple-Station Heat Detectors

UL 609—Standard for Local Burglar Alarm Units and Systems

UL 827—Standard for Central Station Alarm Services

UL1023—Standard for Household Burglar Alarm System Units

UL 1610—Standard for Central Station Burglar Alarm Units

UL 1641—Standard for Installation and Classification of Residential Burglar Alarm Systems

UL 2034—Standard for Single- and Multiple-Station Carbon Monoxide Alarms

REVIEW

1. In your own words, explain the terms *alarm* and *detector*.

_____

_____

_____

_____

_____

2. The requirements for household fire alarm systems are found in what *NFPA Code* and in what chapter of this code?

_____

3. Do heat alarms and carbon monoxide alarms take the place of smoke alarms? Explain.

_____

_____

4. Name two basic ways that smoke alarms are powered.

   _____

   _____

5. Name the two types of smoke alarms.

   _____

   _____

6. List the absolute minimum level of smoke alarm protection in a new one-family dwelling.

   _____

   _____

7. Circle the correct answer from the following statements:

   a. Smoke alarms installed in new one- and two-family homes shall be battery operated only so as not to be affected by a power outage.

   b. Smoke alarms installed in new one- and two-family homes shall be dual powered by a 120-volt circuit and a battery.

8. Circle the correct answer from the following statements:

   a. Smoke alarms installed in new one- and two-family homes shall be interconnected so that if any one of them is triggered, all other alarms will also sound off.

   b. Smoke alarms installed in new one- and two-family homes shall not be interconnected so that each alarm will operate independently of all other alarms in the home.

9. Circle the correct answer from the following statements:

   a. Always install smoke and fire alarms in dead airspaces.

   b. Never install smoke and fire alarms in dead airspaces.

10. Circle the correct answer from the following statements:

    a. Mount wall-mounted smoke alarms so that the top edge is not closer than 4 in. (102 mm) from the ceiling and not more than 12 in. (305 mm) from the ceiling.

    b. Mount wall-mounted smoke alarms anywhere on the wall, but not lower than 36 in. (914 mm) from the ceiling.

11. Cooking and baking in the kitchen can produce quite a bit of smoke. The best choice for a smoke alarm in the kitchen is the photoelectric type. The better choice would be to install a (heat alarm) (carbon monoxide alarm). Circle the correct answer.

12. An important but often overlooked requirement in the *National Fire Alarm Code, NFPA 72* is that alarms must be replaced after being in service for: (Circle the correct answer.)

    a. not more than 5 years.

    b. not more than 10 years.

    c. not more than 15 years.

13. Carbon monoxide is _____, and is _____, _____, and _____.

14. Circle the correct answer from the following statements.

    a. Carbon monoxide is heavier than air.

    b. Carbon monoxide will always rise to the ceiling.

    c. Carbon monoxide will drop to the floor.

    d. Carbon monoxide is about the same weight as air.

15. The more complex household fire alarm systems are covered in what article of the *National Electrical Code*? Circle the correct answer.

    a. *Article 72*

    b. *Article 310*

    c. *Article 760*

16. Circle the correct answer from the following statements:

    a. Always connect a fire alarm circuit or system to a power supply that has ground-fault circuit interrupter (GFCI) protection.

    b. Never connect a fire alarm circuit or system to a power supply that has ground-fault circuit interrupter (GFCI) protection.

    c. It makes no difference whether or not the branch circuit is GFCI or AFCI protected because the alarms are required to have battery backup power. In the event of a loss of power for whatever reason, the alarms would still be operative.

17. Circle the correct answer from the following statements:

    a. Conductors for fire alarm systems shall not be installed in the same raceways, cables, or electrical boxes as the light and power conductors.

    b. It is permissible to install fire alarm conductors in the same raceways, cables, or electrical boxes as the light and power conductors because the conductors for fire alarm systems covered in *Article 760* are small and would easily fit into other electrical raceways and boxes.

18. Fire alarm cable generally used in more complex residential fire alarm and security systems is marked (FPL) (low-voltage bell wire) (THHN). Circle the correct answer.

19. When installing a fire alarm system or a complete security system package in a home, always follow the installation instructions from: (Circle the correct answer.)

    a. the manufacturer of the product and applicable codes.

    b. your neighbor, because he knows a lot about codes and standards.

    c. the man at the home center who sold you the product.

20. Residential sprinkler fire protection systems are required to be installed in compliance with the following (Circle the best answer):

    a. *NFPA 13D.*

    b. *IRC P2904.*

    c. either *NFPA 13D* or *IRC P2904.*

    d. neither *NFPA 13D* nor *IRC P2904.*

Service-Entrance Equipment

OBJECTIVES

After studying this chapter, you should be able to

- understand the *NEC* terminology and requirements for installing all types of residential electrical services.

- calculate the proper size of residential service conductors and raceways.

- understand and install the required grounding and bonding for residential electrical services.

- understand the meaning of the term *UFER* ground.

- know the *NEC* requirements for grounding electrical equipment in a separate building.

- calculate the cost of using electricity.

- understand the meaning of conductor *withstand rating*.

NEC 230.79(C) requires a minimum 100-ampere, 3-wire disconnect for a one-family dwelling. Let's continue with the full details regarding residential electrical services.

This chapter covers a lot of ground. Mastering the material in this chapter is quite an accomplishment. Be patient, understanding, and thorough as you tackle it. Congratulate yourself when you have completed this chapter.

An electric service is required for all buildings containing an electrical system and receiving electrical energy from a utility company. The *NEC* defines the term *service* as *the conductors and equipment for delivering electric energy from the serving utility to the wiring system of the premises served.** The point where the utility's supply ends and the customer's premises wiring begins is called the *Service Point*. The utility company generally must be contacted to determine where they want the meter to be located.

IMPORTANT DEFINITIONS

Several definitions relative to services have changed for the 2011 *NEC*. Many of these changes attempt to clarify the portion of the service supply that are the responsibility of the electric utility and which portion is the responsibility of the property owner or installer. The *Code* defines a *service drop* as ▶*The overhead conductors between the utility distribution system and the Service Point.** These conductors will be installed, owned, and maintained by the electric utility.

A new definition of *Overhead Service Conductors* was added that reads, ▶*Service Conductors, Overhead. The overhead conductors between the Service Point and the first point of connection to the service-entrance conductors at the building or other structure.** These conductors will be installed by the electrician rather than by the electric utility. However, this rarely happens in practice. The electric utility almost always installs a *service drop* to the building or structure.

The definition of *Service Point* in *NEC Article 100* is important to our discussion of these issues. It reads, *The point of connection between the facilities of the serving utility and the premises*

*Reprinted with permission from NFPA 70-2011.

*wiring.** This *service point* establishes the point or line of demarcation between the electric utility and its customer and is most often established by the electric utility. The electric utility is responsible for the conductors and equipment on the supply side of the *service point*, and the customer is responsible for the electrical conductors and equipment on the load side of the *service point*. See Figure 27-1 for an illustration of these definitions.

Furthermore, ▶*NEC 90.2(B)(5)* states that the electric utility's service drop is not covered by the *NEC*.◀ As a result, the title of *NEC 230.24* was revised to cover *Overhead Service Conductors* rather than *Service Drops*. This section clearly does not apply to the electric utility's service drop, but rather to the customer's overhead service conductors (which are rarely installed). However, *NEC 230.26* continues to require the point of attachment for the utility service drop conductors be located so as to provide for the requirements in *NEC 230.9* for building openings, and *NEC 230.24* for clearances above roofs and above the ground. The bottom line is the point of attachment for the utility's service drop must provide for the *NEC*-required clearances of overhead service conductors.

OVERHEAD SERVICE

An overhead service includes the service raceways, fittings, meter, meter socket, the main service disconnecting means, and the service conductors between the main service disconnecting means and the point of attachment to the utility's service-drop conductors. Usually, residential-type watt-hour meters are located on the exterior of a building. In some cases, the entire service-entrance equipment may be mounted outside the building. This includes the watt-hour meter and the disconnecting means. Figure 27-2 illustrates the *Code* terms for the various components of a service entrance.

Overhead service-drop conductors might be attached to a through-the-roof mast-type service or to the side of the house. In either case, proper clearance is required above the ground, roof, fences, windows, sidewalks, decks, and so on. Locate the mast insulator or screw-in knob so when the electric utility installs the service drop, the clearances above the ground or building features required in *NEC 230.24* are met.

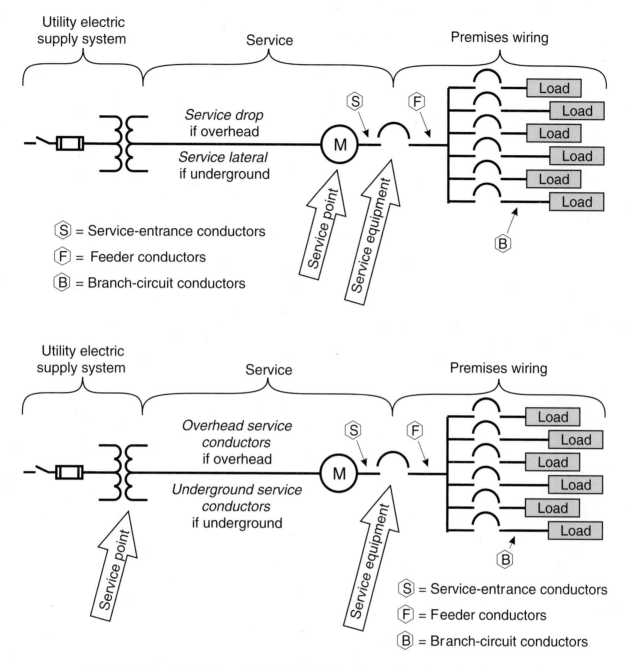

FIGURE 27-1   Definitions from *NEC* relative to overhead- and underground-
supplied services. (*Delmar/Cengage Learning*)

MAST-TYPE SERVICE

The mast service, Figures 27-3, and 27-4, is a commonly used method of installing an overhead service entrance. The overhead mast service is most commonly used in areas of the electric utility distribution area where the utilities have not been installed underground.

The service raceway is run through the roof, as shown in several figures including 27-2, 27-3 and 27-4, using a roof flashing and neoprene seal fitting, as illustrated in Figures 27-3(A) and 27-5. This fitting keeps water from seeping in and damaging the structure where the raceway penetrates the roof. The conduit is securely fastened to the building to comply with *Code* and electric utility requirements. Many

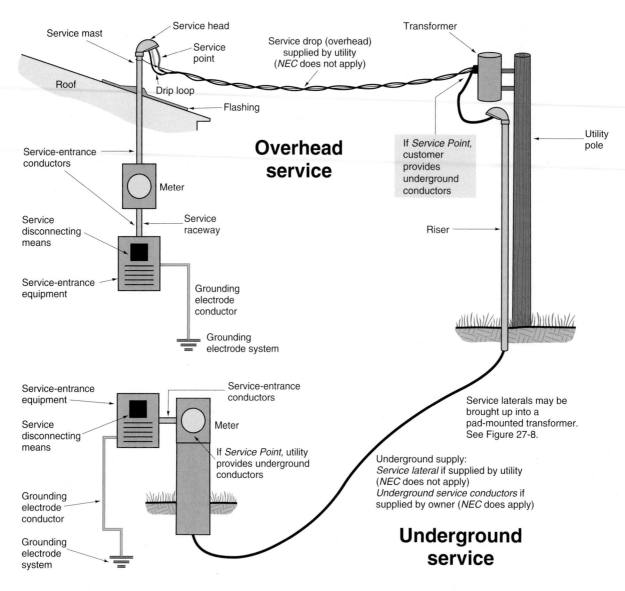

FIGURE 27-2 *Code* terms for services. (*Delmar/Cengage Learning*)

types of fittings are readily available at electrical supply houses for securely fastening raceways and other electrical equipment to wood, brick, masonry, and other surfaces. Figure 27-3(B) shows typical methods of securing the conduit mast to ensure it will safely carry the load of the service drop. Several conduit mast fittings are shown in Figure 27-5.

Service Mast as Support *(230.28)*

The bending force on the conduit increases with an increase in the distance between the roof support and the point where the service-drop conductors are attached. The pulling force of a service drop on a mast service conduit increases as the length of the service drop increases. As the length of the service drop decreases, the pulling force on the mast service conduit decreases.

If extra support is not to be provided, the mast service conduit must be at least 2 in. (50 mm) in diameter. This size prevents the conduit from bending due to the strain of the service-drop conductors.

If extra support is needed, it is usually in the form of guy wires attached to a mast fitting and to the roof in a "Y" configuration. For fairly flat roofs, "stiff-legs" made out of galvanized pipe or conduit are often installed from the mast to the roof. Figure 27-3 shows several methods for securing conduit masts that are used to support the utility service drop. Some of the hardware that may be used for terminating the

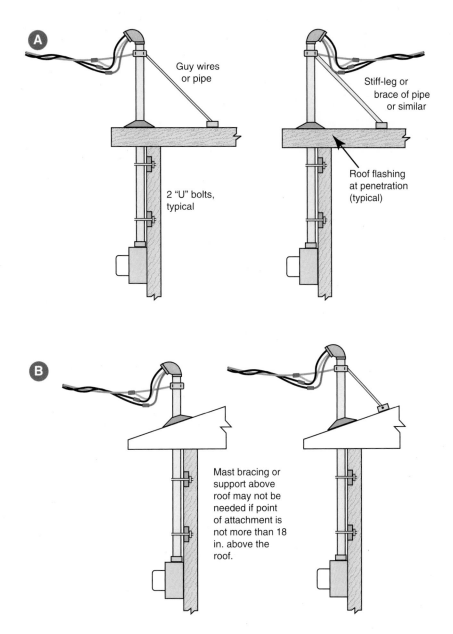

Guy wires
or pipe

Stiff-leg or
brace of pipe
or similar

Roof flashing
at penetration
(typical)

2 "U" bolts,
typical

Mast bracing or
support above
roof may not be
needed if point
of attachment is
not more than 18
in. above the
roof.

FIGURE 27-3   Typical methods of securing and supporting a mast-type service. Mast clamp positioned to provide for clearances in *NEC 230.24(A)* and *(B)*. Follow electric utility service requirements for all clearances. (*Delmar/Cengage Learning*)

service drop is shown in Figure 27-5(B), 27-5(C), 27-5(E). Again, many support fittings and devices are available at electrical supply houses.

When used as a through-the-roof service mast, rigid metal conduit or intermediate metal conduit is not required to be supported within 3 ft (900 mm) of the service head, *344.30(A)* and *342.30(A)*. Verify the application of this provision with the local electrical inspector and electrical utility, as many utilities have specific installation requirements for securing masts to which their service drop will be

attached. The roof mast kit provides adequate support when properly installed.

Clearance Requirements for Service Drop and Overhead Service Conductors for Mast Installations

As stated above, the 2011 *NEC* significantly changed the requirements for *service drops* and introduced a new definition for *overhead service conductors*. Several factors determine the maximum

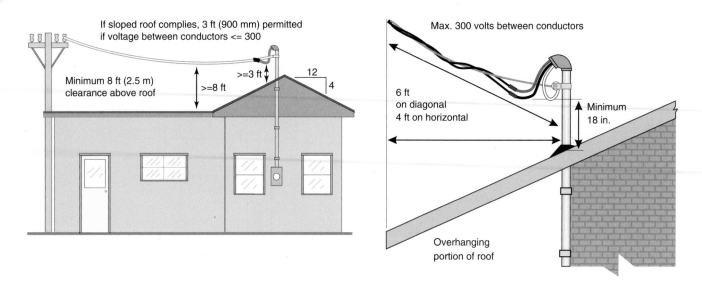

FIGURE 27-4 Clearances of overhead service conductors over roofs. (*Delmar/Cengage Learning*)

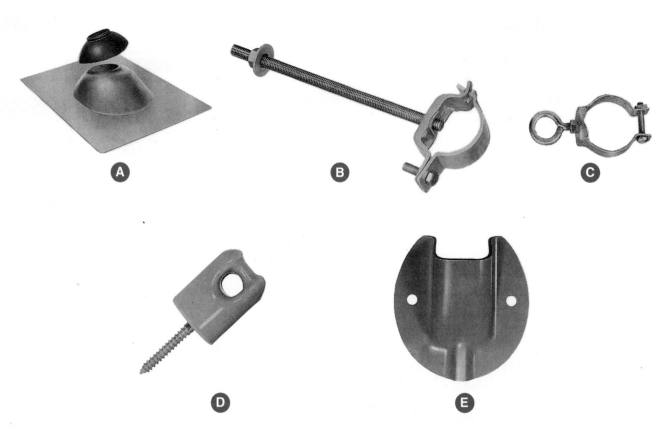

FIGURE 27-5 Fittings for conduit mast include (A) neoprene seal with metal flashing to use where the service raceway comes through the roof; (B) through-the-wall support for securing the service raceway; (C) clamp that fastens to the service raceway to which a guy wires can be attached for support of the service riser; (D) insulator that can be screwed into a barge rafter or at the gable end of the building; (E) is an insulating clamp that is fastened to a conduit mast for terminating the service drop. (*Delmar/Cengage Learning*)

length of conduit that can be installed between the roof support and the point where the service-drop conductors are attached. Verify these factors with the local electric serving utility. Typical requirements are shown in Figures 27-3 and 27-4.

The *NEC* rules for insulation and clearances apply to the service-drop and service-entrance conductors. For example, the service conductors must be insulated, except where the voltage to ground does not exceed 300 volts. In this case, the grounded neutral conductors are not required to have insulation.

Clearance above Roofs: Consult the electric utility that will serve the residence. *NEC 230.26* requires that the point of attachment is to provide for the clearances in *NEC 230.24(A)* for overhead service conductors. The utility service drop is treated as an equivalency to the customer's overhead service conductors. See Figure 27-4.

Only power service-drop conductors are permitted to be attached to and supported by the service mast, *230.28*. Refer to Figure 27-6.

All fittings used with raceway-type service masts must be identified for use with service masts, *230.28*.

Consult the utility company and electrical inspection authority for information relating to their specific requirements for clearances, support, and so on, for service masts.

Clearance from Ground: Once again, consult the electric utility that will serve the residence. *NEC 230.26* requires the point of attachment to provide for the clearances in *NEC 230.24(B)* for the customer's overhead service conductors. These rules give the required clearances for overhead service conductors above ground. See Figure 27-7 for clearances.

Clearance from Windows: *NEC 230.24(C)* refers back to 230.9 to determine the required clearances for service conductors from windows and other building openings. See Figure 27-7.

The installation requirements for a typical service entrance are shown in Figures 27-2, 27-3, 27-4, 27-7, and 27-8.

Installing Service-Entrance Cable

The installation of service-entrance cable is covered in *NEC Article 338*. Service-entrance cable for aboveground use is supplied in one of two varieties. A "U" style consists of two insulated conductors with a bare neutral wrapped around the insulated conductors in a spiral configuration. At terminations, the spiral neutral conductor is unwound and twisted together to form the third or neutral conductor. A 4-wire variety of Type SE cable is manufactured and is suitable for installation where a separate neutral and equipment grounding conductor are required, such as for wiring ranges, dryers, and some panels.

As shown in Figure 27-9, fittings are specifically designed to facilitate installation of service-entrance cable outside the building. These fittings include a weatherhead and raintight and non-raintight cable clamps. Sill plates are available to protect the cable and weatherproof the installation where the cable is routed inside the building.

The cable must be protected against physical damage in accordance with *NEC 230.50*.

Be sure to check with the local electrical inspector, as some jurisdictions limit the length of service-entrance cables inside a building. It is considered more fragile than a metal raceway such as EMT, IMC,

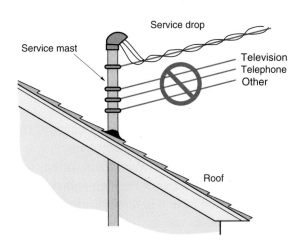

FIGURE 27-6 Only power service-drop conductors are permitted to be attached to and supported by the service mast. *Do not* attach or support television cables, telephone cables, or anything else to the mast, *230.28*.
(*Delmar/Cengage Learning*)

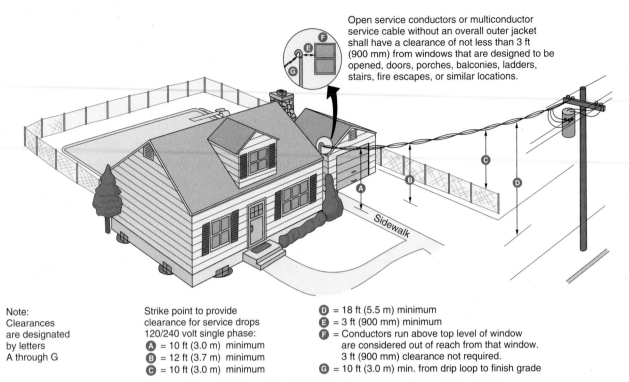

Open service conductors or multiconductor service cable without an overall outer jacket shall have a clearance of not less than 3 ft (900 mm) from windows that are designed to be opened, doors, porches, balconies, ladders, stairs, fire escapes, or similar locations.

Note:
Clearances
are designated
by letters
A through G

Strike point to provide
clearance for service drops
120/240 volt single phase:
- Ⓐ = 10 ft (3.0 m) minimum
- Ⓑ = 12 ft (3.7 m) minimum
- Ⓒ = 10 ft (3.0 m) minimum

- Ⓓ = 18 ft (5.5 m) minimum
- Ⓔ = 3 ft (900 mm) minimum
- Ⓕ = Conductors run above top level of window are considered out of reach from that window. 3 ft (900 mm) clearance not required.
- Ⓖ = 10 ft (3.0 m) min. from drip loop to finish grade

Note: Electric utilities follow the National Electrical Safety Code (NESC). The clearance requirements in the NESC may differ from those in the *National Electrical Code* (*NEC*). The deciding factor: Is the installation customer installed, owned, and maintained by the customer, or is the installation installed, owned, and maintained by the utility?

FIGURE 27-7 Clearances from ground for a residential service in accordance with typical utility requirements. Clearances (A) and (C) of 10 ft (3.0 m) permitted only if the service-entrance drop conductors are supported on and cabled together with a grounded bare messenger wire where the voltage to ground *does not* exceed 150 volts. This 10-ft (3.0-m) clearance must also be maintained from the lowest point of the drip loop. Clearance (B) of 12 ft (3.7 m) is permitted over residential property such as lawns and driveways where the voltage to ground *does not* exceed 300 volts. The circled insert shows clearance from building openings. (*Delmar/Cengage Learning*)

or rigid steel conduit and requires protection against physical damage in accordance with *NEC 300.4*.

Verify Which Overhead Clearances Apply

Electricians must abide by the *NEC*. Utilities must go along with the *National Electrical Safety Code*. These codes may have different clearance requirements for service-drop conductors and drip loops. When involved with installing an overhead service, check with your electrical inspector to determine which code is enforced in your area.

The installation requirements of the serving electric utility must always be complied with so long as they do not reduce the requirements of the *NEC*.

UNDERGROUND SERVICE

Underground service means the cable installed underground from the point of connection to the system provided by the utility company.

New residential developments often include underground installations of the high-voltage electrical systems. The conductors in these distribution systems end in the bases of pad-mounted transformers, Figure 27-10. These transformers are placed at the rear lot line or in other inconspicuous locations in the development. The transformer and primary high-voltage conductors are generally installed by, and are the responsibility of, the utility company.

If installed by the electric utility, the conductors installed between the pad-mounted transformer

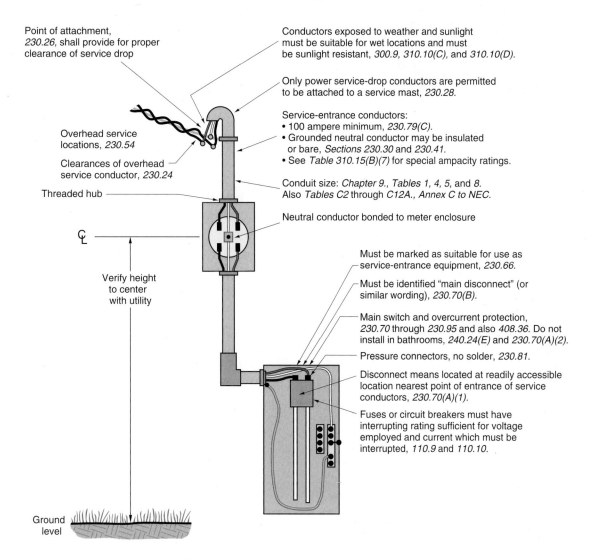

Point of attachment, *230.26*, shall provide for proper clearance of service drop

Conductors exposed to weather and sunlight must be suitable for wet locations and must be sunlight resistant, *300.9*, *310.10(C)*, and *310.10(D)*.

Only power service-drop conductors are permitted to be attached to a service mast, *230.28*.

Service-entrance conductors:
• 100 ampere minimum, *230.79(C)*.
• Grounded neutral conductor may be insulated or bare, *Sections 230.30* and *230.41*.
• See *Table 310.15(B)(7)* for special ampacity ratings.

Overhead service locations, *230.54*

Clearances of overhead service conductor, *230.24*

Threaded hub

Conduit size: *Chapter 9.*, *Tables 1, 4, 5,* and *8.* Also *Tables C2* through *C12A.*, *Annex C to NEC*.

Neutral conductor bonded to meter enclosure

Ꮯ

Verify height to center with utility

Must be marked as suitable for use as service-entrance equipment, *230.66*.

Must be identified "main disconnect" (or similar wording), *230.70(B)*.

Main switch and overcurrent protection, *230.70* through *230.95* and also *408.36*. Do not install in bathrooms, *240.24(E)* and *230.70(A)(2)*.

Pressure connectors, no solder, *230.81*.

Disconnect means located at readily accessible location nearest point of entrance of service conductors, *230.70(A)(1)*.

Fuses or circuit breakers must have interrupting rating sufficient for voltage employed and current which must be interrupted, *110.9* and *110.10*.

Ground level

FIGURE 27-8   The wiring of a typical service-entrance installation. Today, most services are supplied underground, as shown in Figure 27-10. (*Delmar/Cengage Learning*)

and the meter are called *service lateral conductors*. Normally, the electric utility furnishes and installs them. In many areas, the utility will install both the service lateral conductors and the communication cables in the same trench. In some areas, but not too commonly, service laterals, communication cables, and gas lines are run in the same trench. Figure 27-10 shows a typical underground installation.

The wiring from the external meter to the main service equipment is the same as the wiring for a service connected from overhead lines, as in Figure 27-8. Some local codes may require conduit to be installed underground from the pole to the service-entrance equipment. *NEC* requirements for underground services are given in *Article 230, Part III*. If installed by

the owner or electrician, the underground conductors must be suitable for direct burial in the earth, Type USE single-conductor or Type USE cable containing more than one conductor, *NEC Article 338*.

If the electric utility installs the underground service conductors, the work must comply with the rules established by the utility. Utilities follow the National Electrical Safety Code, NESC. These rules may not be the same as those given in *NEC 90.2(B)(5)*.

As stated in *NEC 90.2(B)(5)*, the *NEC* does not cover installations under the exclusive control of an electric utility where such installations:

▶a. consist of service drops or service laterals and associated metering, or

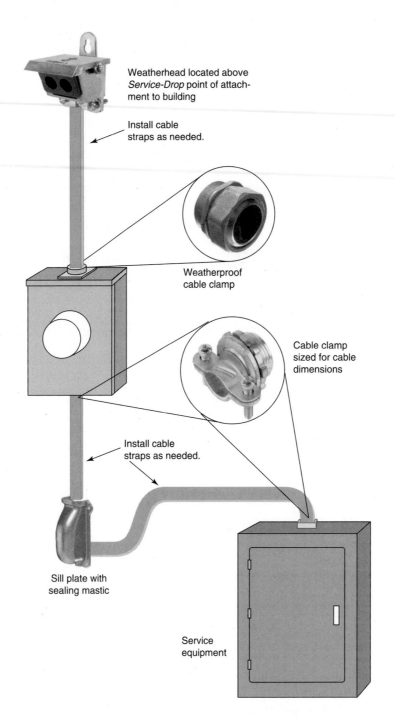

Weatherhead located above *Service-Drop* point of attachment to building

Install cable straps as needed.

Weatherproof cable clamp

Cable clamp sized for cable dimensions

Install cable straps as needed.

Sill plate with sealing mastic

Service equipment

FIGURE 27-9 Typical hardware used for installation of service-entrance cable for services. (*Delmar/Cengage Learning*)

b. are on property owned or leased by the electric utility for the purpose of communications, metering, generation, control, transformation, transmission, or distribution of electric energy, or

c. are located in legally established easements, rights-of-way, or

d. are located by other written agreements either designated by or recognized by public service commissions, utility commissions, or other regulatory agencies having jurisdiction for such installations. These written agreements shall be limited to installations for the

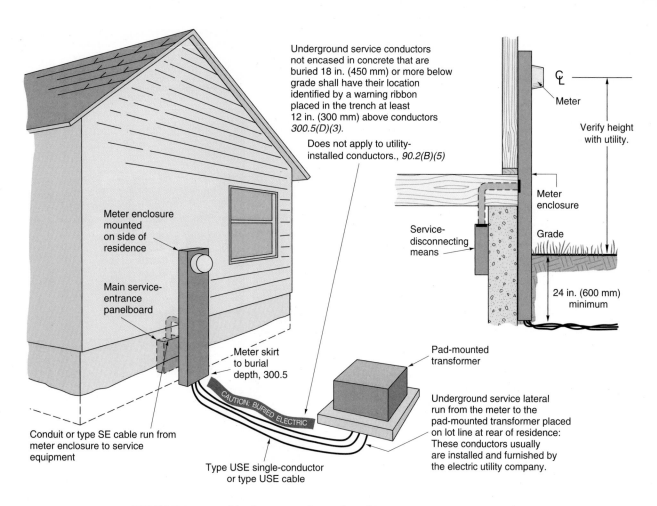

Underground service conductors not encased in concrete that are buried 18 in. (450 mm) or more below grade shall have their location identified by a warning ribbon placed in the trench at least 12 in. (300 mm) above conductors *300.5(D)(3)*.

Does not apply to utility-installed conductors., *90.2(B)(5)*

Meter

Verify height with utility.

Meter enclosure

Service-disconnecting means

Grade

24 in. (600 mm) minimum

Meter enclosure mounted on side of residence

Main service-entrance panelboard

Meter skirt to burial depth, 300.5

Pad-mounted transformer

Underground service lateral run from the meter to the pad-mounted transformer placed on lot line at rear of residence: These conductors usually are installed and furnished by the electric utility company.

Conduit or type SE cable run from meter enclosure to service equipment

CAUTION: BURIED ELECTRIC

Type USE single-conductor or type USE cable

FIGURE 27-10   Underground service. (*Delmar/Cengage Learning*)

purpose of communications, metering, generation, control, transformation, transmission, or distribution of electric energy where legally established easements or rights-of-way cannot be obtained. These installations shall be limited to Federal Lands, Native American Reservations through the U.S. Department of the Interior Bureau of Indian Affairs, military bases, lands controlled by port authorities and State agencies and departments, and lands owned by railroads. ◀

When underground service conductors or service-entrance cables are installed by the electrician, *230.50* applies. This section deals with the protection of service-entrance conductors and service-entrance cables against physical damage. *NEC 230.50* also refers to *300.5*, which covers all situations involving underground wiring, such as the sealing of raceways where they enter a building. Sealing

raceways is covered in "Meter/Meter Base Location" later on in this chapter.

MAIN SERVICE DISCONNECT LOCATION

In conformance to *230.70(A)(1)* and *230.70(A)(2)*, the main service disconnecting means shall

- be located inside the building at a readily accessible location nearest the point of entrance of the service conductors, Figure 27-11, or

- be located outside the building at a readily accessible location, and

- not be installed in a bathroom.

If for some reason it is necessary to locate service-entrance equipment some distance inside the building, *NEC 230.6* considers service-entrance conductors to be outside of a building if the

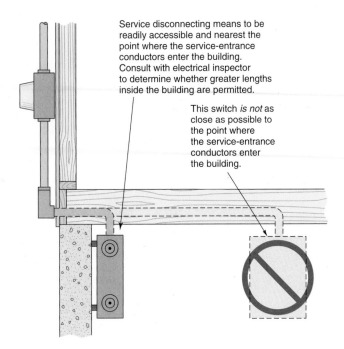

Service disconnecting means to be readily accessible and nearest the point where the service-entrance conductors enter the building. Consult with electrical inspector to determine whether greater lengths inside the building are permitted.

This switch *is not* as close as possible to the point where the service-entrance conductors enter the building.

FIGURE 27-11 Main service disconnect location, *230.70*. Do not install in a bathroom. Refer to *230.70(A)(2)* and *240.24(E)*. (*Delmar/Cengage Learning*)

service-entrance conductors are one of the following:

- installed under not less than 2 in. (50 mm) of concrete beneath the building, or

- installed within the building in a raceway encased in at least 2 in. (50 mm) of concrete or brick, or

- installed in conduit, and covered by at least 18 in. (450 mm) of earth beneath the building.

Circuit Directory/Circuit Identification

NEC 408.4 is very clear as to its requirements on the subject. It requires that *Every circuit and circuit modification shall be legibly identified as to its clear, evident, and specific purpose or use. The identification shall include sufficient detail to allow each circuit to be distinguished from all others. Spare positions that contain unused overcurrent devices or switches shall be described accordingly. The identification shall be included in a circuit directory that is located on the face or inside of the panel door in the case of a panel board, and located at each switch on a switchboard.*

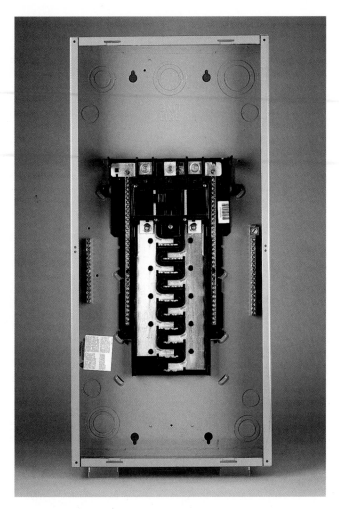

FIGURE 27-12 A typical 120/240-volt, single-phase, 3-wire load center. This load center has a main circuit breaker. Note the neutral terminal bars for termination of the white branch-circuit grounded conductors, and the equipment grounding terminal bars for termination of the equipment grounding conductors of nonmetallic-sheathed cable. See *230.66*. (*Courtesy of Schneider Electric*)

*No circuit shall be described in a manner that depends on transient conditions of occupancy.**

Figure 27-12 shows a typical 240/120-volt, single-phase, 3-wire panelboard.

Figure 27-13 shows the circuit identification for Main Service Panel A. Figure 27-14 shows the circuit identification for sub-Panel B.

NEC 110.22 also required that disconnect switches be "legibly marked," describing what the disconnect is for.

---

*Reprinted with permission from NFPA 70-2011.

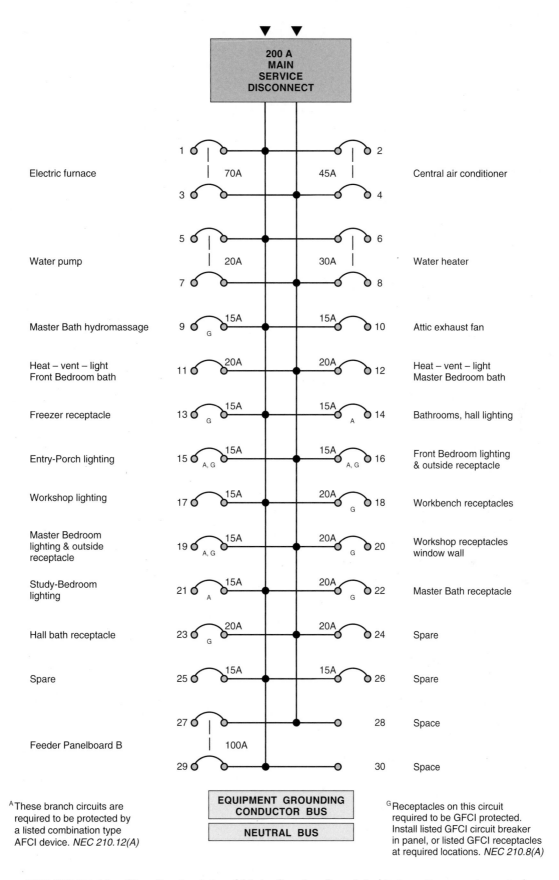

Electric furnace — 70A — 45A — Central air conditioner

Water pump — 20A — 30A — Water heater

Master Bath hydromassage — 15A — 15A — Attic exhaust fan

Heat – vent – light Front Bedroom bath — 20A — 20A — Heat – vent – light Master Bedroom bath

Freezer receptacle — 15A — 15A — Bathrooms, hall lighting

Entry-Porch lighting — 15A — 15A — Front Bedroom lighting & outside receptacle

Workshop lighting — 15A — 20A — Workbench receptacles

Master Bedroom lighting & outside receptacle — 15A — 20A — Workshop receptacles window wall

Study-Bedroom lighting — 15A — 20A — Master Bath receptacle

Hall bath receptacle — 20A — 20A — Spare

Spare — 15A — 15A — Spare

Feeder Panelboard B — 100A — Space

Space

EQUIPMENT GROUNDING CONDUCTOR BUS

NEUTRAL BUS

^A These branch circuits are required to be protected by a listed combination type AFCI device. *NEC 210.12(A)*

^G Receptacles on this circuit required to be GFCI protected. Install listed GFCI circuit breaker in panel, or listed GFCI receptacles at required locations. *NEC 210.8(A)*

FIGURE 27-13   Circuit schedule of Main Service Panel A. (*Delmar/Cengage Learning*)

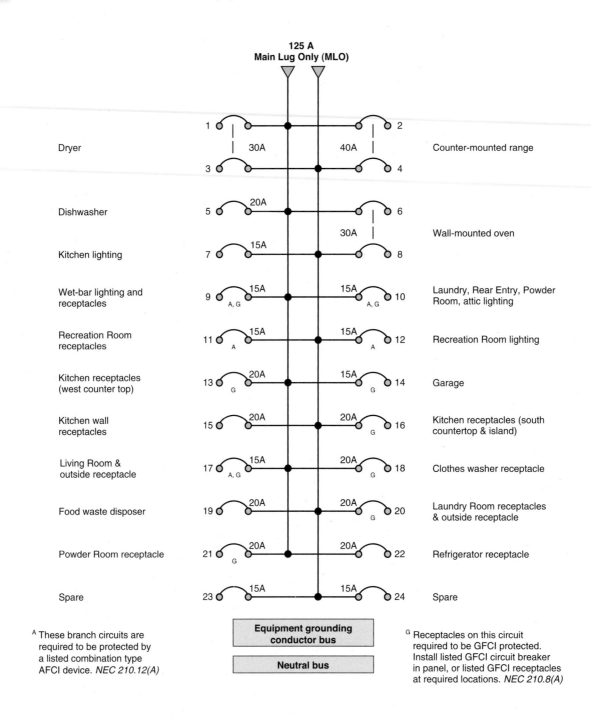

125 A
Main Lug Only (MLO)

Dryer	1	30A	40A	2 Counter-mounted range
	3			4
Dishwasher	5	20A		6
Kitchen lighting	7	15A	30A	8 Wall-mounted oven
Wet-bar lighting and receptacles	9 A, G	15A	15A A, G	10 Laundry, Rear Entry, Powder Room, attic lighting
Recreation Room receptacles	11 A	15A	15A A	12 Recreation Room lighting
Kitchen receptacles (west counter top)	13 G	20A	15A G	14 Garage
Kitchen wall receptacles	15	20A	20A G	16 Kitchen receptacles (south countertop & island)
Living Room & outside receptacle	17 A, G	15A	20A G	18 Clothes washer receptacle
Food waste disposer	19	20A	20A G	20 Laundry Room receptacles & outside receptacle
Powder Room receptacle	21 G	20A	20A	22 Refrigerator receptacle
Spare	23	15A	15A	24 Spare

Equipment grounding conductor bus

Neutral bus

A These branch circuits are required to be protected by a listed combination type AFCI device. *NEC 210.12(A)*

G Receptacles on this circuit required to be GFCI protected. Install listed GFCI circuit breaker in panel, or listed GFCI receptacles at required locations. *NEC 210.8(A)*

FIGURE 27-14   Circuit schedule of Panelboard B. (*Delmar/Cengage Learning*)

Service Disconnecting Means (Panel A)

It is extremely important that a disconnect or panelboard used as service equipment be marked "Suitable for Use as Service Equipment." This is a requirement of *230.66*. Some disconnects and panelboards are so marked—others are not. Such marking ensures that proper bonding together of the neutral bus, equipment ground bus, and metal enclosure is provided, as well as a termination (lug) for the grounding electrode conductor.

More requirements for service disconnect means are found in *230.70* through *230.82*. *NEC 230.71(A)* requires that the service disconnecting means consist of not more than six switches or six circuit breakers mounted in a single enclosure, in a group of separate enclosures in the same location, or in or on a switchboard. This permits the disconnection of all electrical equipment in the house with not more than six hand operations. Some local codes take exception to the six-disconnect rule, and require a single main disconnect. Some localities require the main disconnecting means to be located outside. If located outside, *230.70(A)(1)* requires that the service disconnect must be located on the building, and must be readily accessible.

You need to check this out with the local authority that has jurisdiction in your area.

In damp and wet locations, to prevent the accumulation of moisture that could lead to rusting of the enclosure and damage to the equipment inside the enclosure, *312.2* specifies that there be at least a ¼-in. (6-mm) air space between the wall and a surface-mounted enclosure. Most disconnect switches, panelboards, meter sockets, and similar equipment have raised mounting holes that provide the necessary clearance. This is clearly visible in the open-meter socket in Figure 27-15 of the "Meter/Meter Base Location" section in this chapter. The *NEC* (in Definitions) considers masonry in direct contact with the earth to be a wet location. The *Exception* to *312.2* allows nonmetallic enclosures to be mounted without an air space on concrete, masonry, tile, and similar surfaces.

Panelboards shall be installed in accordance with the listing of the panelboard. A panelboard used as service equipment that has multiple overcurrent devices is not required to have overcurrent protection on the line side of the panelboard, *408.36, Exception No. 1*. When used as other than service equipment, overcurrent protection for the panelboard shall not exceed the rating of the panelboard.

NEC 230.79 tells us that the service disconnect shall have a rating not less than the calculated load to be carried, determined in accordance with *Parts III, IV, or V* of *Article 220*. We show how these calculations are made in Chapter 29.

For a one-family residence, the minimum size main service disconnecting means is 100 amperes, 3-wire, *NEC 230.79(C)*.

NEC 230.70(B) requires that the main service disconnect must be permanently marked to identify it as service disconnect.

When nonmetallic-sheathed cable is used, a panelboard must have a terminal bar for attaching the equipment grounding conductors, *408.20*. See Figure 27-16.

FIGURE 27-15 Typical residential self-contained meter socket. Combination meter socket/main breaker in one enclosure is also available. (*Courtesy of Schneider Electric*)

For this residence, panelboard A serves as the main disconnecting means for the service. This panelboard consists of a main circuit breaker and many branch-circuit breakers. Figure 27-12 is a photo of this type of panelboard. Figure 27-13 is the schedule of the circuits connected to panelboard A. Panelboard A is located in the Workshop.

Working Space. To provide for safe working conditions around electrical equipment, *110.26* contains a number of rules that must be followed. For residential installations, there are some guidelines for working space.

- A working space not less than 30 in. (762 mm) wide and 3 ft (914 mm) deep must be provided in front of electrical equipment, such as panelboards A and B in this residence.

- This working space must be kept clear and not used for storage.

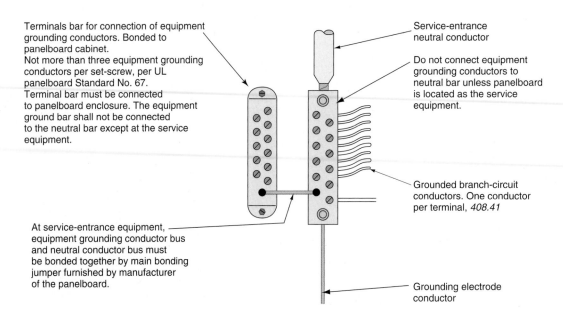

Terminals bar for connection of equipment grounding conductors. Bonded to panelboard cabinet.
Not more than three equipment grounding conductors per set-screw, per UL panelboard Standard No. 67.
Terminal bar must be connected to panelboard enclosure. The equipment ground bar shall not be connected to the neutral bar except at the service equipment.

Service-entrance neutral conductor

Do not connect equipment grounding conductors to neutral bar unless panelboard is located as the service equipment.

Grounded branch-circuit conductors. One conductor per terminal, *408.41*

At service-entrance equipment, equipment grounding conductor bus and neutral conductor bus must be bonded together by main bonding jumper furnished by manufacturer of the panelboard.

Grounding electrode conductor

FIGURE 27-16   Connections of service neutral conductor, branch-circuit neutral conductors, and equipment grounding conductors at panelboards, *408.40* and *408.41*. (*Delmar/Cengage Learning*)

- Do not install electrical panelboards inside of cabinets, nor above shelving, washers, dryers, freezers, work benches, and so forth.

- Do not install electrical panels in bathrooms.

- Do not install electrical panels above or close to sump pump holes.

- The hinged cover (door) on the panel shall be able to be opened to a full 90°.

- A "zone" equal to the width and depth of the electrical equipment, from the floor to a height of 6 ft (1.8 m) above the equipment or to the structural ceiling, whichever is lower, shall be dedicated to the electrical installation. No piping, ducts, or equipment foreign to the electrical installation shall be located in this zone. This zone is dedicated space intended for electrical equipment only! See *110.26(E)*.

Mounting Height.   Disconnects and panelboards must be mounted so the center of the main disconnect handle in its highest position is not higher than 6 ft 7 in. (2.0 m) above the floor, *404.8*.

Headroom.   To provide for safe adequate working space and easy access to service equipment and electrical panels, the *NEC* in *110.26(A)(3)* requires that there be at least 6½ ft (2.0 m) of headroom. If

you are installing service equipment in an existing dwelling, services of 200 amperes or less are permitted to be located in areas where the headroom is less than 6½ ft (2.0 m).

Lighting Required for Service Equipment and Panels.   Working on electrical equipment with inadequate lighting can result in injury or death. In *110.26(D)*, there is a requirement that illumination be provided for service equipment and panelboards. But the *Code* does not spell out how much illumination is required. It becomes a judgment call on the part of the electrician and/or electrical inspector. Note that the electrical plans for the workshop show a porcelain pull-chain lampholder to the front of and off to one side of the main service panelboard A. In many cases, adjacent lighting, such as the fluorescent luminaires in the recreation room, is considered to be the required lighting for equipment such as panelboard B.

Panelboard Setback in Walls

Panelboards are sometimes installed within a wall for flush mounting. The setback and repair opening rules for panelboards are found in *NEC 312.3* and *312.4*:

- In plaster, drywall, or plasterboard: repair wall so no gap or opening is greater than ⅛ in. (3 mm).

- In combustible walls: flush.
- In wall of concrete, tile, or other noncombustible material: set back not more than ¼ in. (6 mm).

Close Openings in Panelboards

Inside a panelboard, there are live parts! There is a real fire and shock hazard when there are openings (knockouts) on the sides, top, or bottom of a panelboard or when circuit breaker openings in the cover are inadvertently broken out.

NEC 408.7 requires that *Unused opening for circuit breakers and switches shall be closed using identified closures, or other approved means that provide protection substantially equivalent to the wall of the enclosure.**

Identified means *Recognizable as suitable for the specific purpose, function, use, environment, application, and so forth, where described in a particular Code requirement.**

Approved means *Acceptable to the authority having jurisdiction.**

Panelboard B

When a second panelboard is installed, it is oftentimes called a subpanel.

For installation where the main service panelboard is a long distance from areas that have many circuits and/or heavy load concentration, as in the case of the kitchen and laundry of this residence, it is recommended that at least one additional panel be installed near these loads. This results in the branch-circuit wiring being short. Line losses (voltage and wattage) are considerably less than if the branch circuits had been run all the way back to the main panelboard.

The cost of material and labor to install an extra panelboard should be compared to the cost of running all of the branch-circuit "home runs" back to the main service panelboard.

Panelboard B in this residence is located in the Recreation Room. It is fed by three 3 AWG THHN or THWN conductors run in a trade size 1 EMT originating at the service (panelboard A). This feeder is protected by a 100-ampere, 2-pole circuit breaker located in panelboard A.

Do not install panelboards in clothes closets, *240.24(D)*. There is a fire hazard because of the presence of ignitable materials.

Do not install panelboards in bathrooms, *240.24(E)*. Damaging moisture problems may result from the presence of water, and shock hazard may result from both the presence of water and the close proximity of metal faucets, and other plumbing fixtures.

Panelboard B is shown in Figure 27-17. The circuit schedule for panelboard B is found in Figure 27-14.

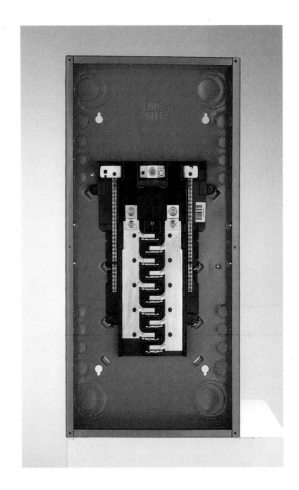

FIGURE 27-17 Typical 240/120-volt, single-phase, 3-wire panelboard, sometimes called a load center. This panelboard does not have a main disconnect and is referred to as a *main lug only (MLO)* panel. The branch-circuit breakers are not shown. This is the type of panelboard that could be used as panelboard B in the residence discussed in this text. Most panelboards have a separate terminal bar on which to terminate the equipment grounding conductors of nonmetallic-sheathed cable, as in Figure 27-12. (*Courtesy of Schneider Electric*)

*Reprinted with permission from NFPA 70-2011.

SERVICE-ENTRANCE CONDUCTOR SIZING

NEC 230.42(A) states that the minimum size service-entrance conductors must be of sufficient size to carry the load as calculated according to *Article 220*.

NEC 230.79(C) calls for a minimum service of 100 amperes for a one-family residence.

The standard and optional methods for calculating the minimum size of service-entrance conductors are discussed in detail in Chapter 29.

RUNNING CABLES INTO TOP OF SERVICE PANEL

See Figure 4-21 for running nonmetallic-sheathed cables into the top of a panelboard.

SERVICE-ENTRANCE OVERCURRENT PROTECTION

Overcurrent protection (fuses and circuit breakers) is discussed in Chapter 28.

SERVICE-ENTRANCE RACEWAY SIZING

This residence is served with an underground service. The conduit running from the meter to Main panelboard A must be sized correctly for the conductors it will contain. In the case of a through-the-roof mast service, the conduit might be sized for both mechanical strength reasons and conductor fill. All utilities have requirements for size and guying of mast services. Table 27-1 shows the calculation for sizing the service-entrance conduit.

METER/METER BASE LOCATION

The electric utility must be contacted before the installation of the service equipment begins. The utility will determine when, where, and how the connection will be made to the homeowner's service conductors. Because of the ever-increasing number of wood decks and concrete patios being added to homes, many utilities prefer that the meter be mounted on the side of the house rather than in the rear. If there is a raised deck, the meter could be only a short distance above the deck, making it both subject to physical damage and hard to read. If concrete patios are poured over underground utility cables, it is very difficult to repair or replace these cables should there be problems. Utilities furnish manuals and brochures detailing their requirements for services that are not found in the *NEC*.

The electric utility furnishes and installs the watt-hour meter. The meter base, also referred to as a meter socket, is usually furnished and installed by the electrical contractor, although some utilities furnish the meter base to the electrical contractor for installation.

Figure 27-7 shows a service supplied from an underground service. Figure 27-11 discusses the requirement that the main disconnect must be as close as possible to the point of entrance of the

---

TABLE 27-1

Calculations for sizing the EMT between Main Panelboard A and the meter base. (A) uses the smaller size conductors permitted by *310.15(B)(7)*. (B) uses conductors sized per the standard calculations. Dimensional data for the conductors and circular raceways are found in *Chapter 9, Tables 1, 4, 5, and 8 of the NEC*.

A Conductor Size Based on Ampacity per *310.15(B)(7)*		**B** Conductor Size Based on Ampacity per *Table 310.15(B)(16)*	
Two 2/0 AWG THWN Copper	0.2223 in²	Two 3/0 AWG THWN Copper	0.2679 in²
	0.2223 in²		0.2679 in²
One 1 AWG Bare Copper	0.0870 in²	One 1 AWG Bare Copper	0.0870 in²
TOTAL	0.5316 in²	TOTAL	0.6228 in²
EMT	Trade Size 1¼	EMT	Trade Size 1½

NEC 300.5(G) and 300.7 state that when raceways pass through areas having great temperature differences, some means must be provided to prevent passage of air back and forth through the raceway. Note that outside air is drawn in through the conduit whenever a door opens. Cold outside air meeting warm inside air causes the condensation of moisture. This can result in rusting and corrosion of vital electrical components. Equipment having moving parts, such as circuit breakers, switches, and controllers, is especially affected by moisture. The sluggish action of the moving parts in this equipment is undesirable.

Sealant shall be identified for such use, and shall not have an adverse effect on the conductor insulation, 230.8.

NEC 230.8 requires seals where service raceways enter from an underground distribution system.

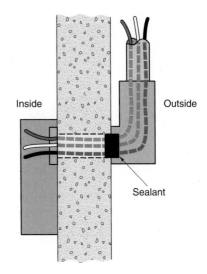

Inside   Outside

Sealant

FIGURE 27-18   Installation of conduit through a basement wall. (*Delmar/Cengage Learning*)

service-entrance conductors. Figure 27-15 shows a typical meter socket for an overhead service, usually mounted at eye level on the outside wall of the house. The service raceway or service-entrance cable is connected to the threaded hub (boss) on top of the meter socket. Figure 27-18 shows how to seal the raceway where it enters the building, to keep out moisture.

A combination meter socket/main breaker in one enclosure is also available.

Raintight and Draining

Outdoor locations are generally wet locations. However, under an overhang may be considered a damp rather than a wet location. If you are using EMT outdoors in a wet location, the fittings (couplings and connectors) must be listed as "Raintight." If the fittings and the container they come in are not marked "Raintight," then the fittings have not been listed for raintight applications. Raintight fittings are generally of the compression type that has a special sealing ring.

NEC 225.22 states that *Raceways on exteriors of buildings or other structures shall be arranged to drain and shall be raintight in wet locations.**

NEC 230.53 states that *Where exposed to the weather, raceways enclosing service-entrance conductors shall be raintight and arranged to drain. Where embedded in masonry, raceways shall be arranged to drain.**

---

*Reprinted with permission from NFPA 70-2011

Pedestals

Meter "pedestals," as shown in Figure 27-19, are sometimes used for residential services. Usually mounted on the outside wall of the house, a meter "pedestal" could also be installed on the lot line between residential properties. Contact the electric utility on this issue.

For underground services, the utility generally runs the underground lateral service-entrance conductors in a trench from a pad-mount transformer to the line-side terminals of the meter base. The electrician then takes over to complete the service installation by running service-entrance conductors from the load side of the meter to the line side of the main service disconnect.

Quite often, the electric utility and the telephone company have agreements whereby the electric utility will lay both power cables and telephone cables in the same trench. The electric utility carries both types of cables on its underground installation vehicles. This is a much more cost-effective way to do this than having each utility dig and then backfill its own trenches.

Often overlooked is a requirement in *NFPA Standard No. 54 (Fuel Gas Code), Section 2.7.2(c)*, requiring that gas meters be located at least 3 ft (900 mm) from sources of ignition. Examples of sources of ignition might be an air-conditioner disconnect where an arc can be produced when the disconnect is opened under load. Furthermore,

The electric utility provides the underground service lateral.

FIGURE 27-19 Installing a metal "pedestal" allows for ease of installation of the underground service-entrance conductors by the utility. (*Photo courtesy of Milbank Manufacturing Co.*)

some electric utilities require a minimum of 3 ft (900 mm) clearance between electric metering equipment and gas meters and gas regulating equipment. It is better to be safe than sorry. Check this issue with the local electrical inspector if there might be a problem with the installation you are working on.

COST OF USING ELECTRICAL ENERGY

A watt-hour meter is always connected into some part of the service-entrance equipment to record the amount of energy used. Billing by the utility is generally done on a monthly basis. The meter might be mounted on the side of the house or on a pedestal somewhere on the premise on the lot line. The utility makes this decision. The residence discussed in

this text has one meter, mounted on the back of the house near the sliding doors of the Master Bedroom.

The kilowatt (kW) is a convenient unit of electrical power. One thousand watts (w) is equal to one kilowatt. The watt-hour meter measures and records both wattage and time.

For residential metering, most utilities have rate schedules based on "cents per kilowatt-hour." Stated another way: How much wattage is being used and for how long?

Burning a 100-watt light bulb for 10 hours is the same as using a 1000-watt electric heater for 1 hour. Both equal 1 kilowatt-hour.

$$\text{kWh} = \frac{\text{watts} \times \text{hours}}{1000} = \frac{100 \times 10}{1000} = 1 \text{ kWh}$$

$$\text{kWh} = \frac{\text{watts} \times \text{hours}}{1000} = \frac{1000 \times 1}{1000} = 1 \text{ kWh}$$

In these examples, if the electric rate is $0.08 cents per kilowatt-hour, the cost to operate the 100-watt light bulb for 10 hours and the cost to operate the electric heater for 1 hour are the same—$0.08. Both loads use 1 kilowatt-hour of electricity.

Simple Formula for Calculating the Cost of Using Electricity

The cost of electrical energy used can be calculated as follows:

$$\text{Cost} = \frac{\text{watts} \times \text{hours used} \times \text{cost per kWh}}{1000}$$

 EXAMPLE

Find the cost of operating a color television set for 8 hours. The label on the back of the television set indicates 175 watts. The electric rate is $0.10494 per kilowatt-hour.

$$\text{Cost} = \frac{175 \times 8 \times \$0.10494}{1000} = \frac{\$0.1469}{\text{(approx. 15¢)}}$$

 EXAMPLE

Find the approximate cost per day of operating a central air conditioner that on average runs 50% of the time during a 24-hour period on a typical hot summer day. The unit's nameplate is marked 240 volts, single-phase, 23 amperes. The electric rate is $0.09 per kilowatt-hour. The steps are as follows:

1. The time the air conditioner operates each day is: $24 \times 0.50 = 12$ hours.

2. Convert the nameplate data to use in the calculations:

$$\text{Watts} = \text{volts} \times \text{amperes} = 240 \times 23$$
$$= 5520 \text{ watts}$$

3. $\text{Cost} = \dfrac{\text{watts} \times \text{hours used} \times \text{cost per kWh}}{1000}$

$$= \dfrac{5520 \times 12 \times \$0.09}{1000} = \$5.96$$

Note: The answers to these examples are approximate because they were based on watts. Power factor and efficiency factors were not included as they generally are unknown when making rough estimates of an electric bill. For all practical purposes, the answers are acceptable.

Table 27-2 is an example of what a typical monthly electric bill might look like.

Some utilities increase their rates during the hot summer months, when the air-conditioning load is high. Other utilities provide a second meter for specific loads such as electric water heaters, air conditioners, heat pumps, or total electric heat. Other utilities use electronic watt-hour meters that have the capability of registering kilowatt-hour consumption during a specific "time of use." These electronic watt-hour meters might have up to four different "time-of-use" periods, each period having a different "cents per kilowatt-hour" rate.

Other charges that might appear on a "light bill" might be a fuel adjustment charge based on a "per kWh" basis. Such charges enable a utility to recover from the consumer extra expenses it might incur for fuel costs used in generating electricity. Fuel charges can vary with each monthly bill, without the utility having to apply to the regulatory agency for a rate change.

GROUNDING/ BONDING *(ARTICLE 250)*

All residential electrical systems are grounded systems. Within the system, some things are *grounded* and some things are *bonded*. The terms *grounding* and *bonding* are used throughout the *NEC*. There is a distinct difference between

---

TABLE 27-2

A typical monthly electric bill.

GENERIC ELECTRIC COMPANY Anyplace, USA

Days of Service	From 06-01-20XX	To 06-28-20XX	Due Date 07-25-20XX
Present reading			84,980
Previous reading			83,655
Kilowatt-hours used			1,325
Rate/kWh 1st 400 kWh @ $0.10494			$41.98
remaining 925 kWh @ $0.06168			57.05
Energy charge			99.03
Basic service charge (single-dwelling)			8.91
State tax			4.24
Total current charges			$112.18
Total amount due by 07-25-20XX			$112.18
Total amount due after 07-25-20XX			$117.79

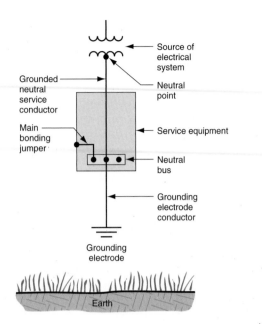

FIGURE 27-20 Grounding of an electrical system at the utility transformer and at the service connects the electrical system to earth. (*Delmar/Cengage Learning*)

grounding and bonding. Each serves a different purpose. Let's take a look at these two very important terms.

What Does Grounded (Grounding) Mean?

Grounded (grounding) is defined in *NEC Article 100* as *Connected to ground or to a conductive body that extends the ground connection.* * See Figure 27-20. As shown, the electrical system is grounded by the electric utility at the transformer and again at the service equipment.

What Does Bonded (Bonding) Mean?

Bonded (bonding) is defined in *NEC Article 100* as *Connected to establish electrical continuity and conductivity.* * See Figure 27-21. In its simplest form, bonding means "connected together."

Now that we have talked about the meaning of the words "grounding" and "bonding," here are more detailed definitions you should be familiar with.

*Reprinted with permission from NFPA 70-2011

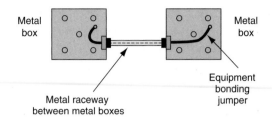

FIGURE 27-21 Two methods for bonding the metal boxes or other metal parts together. One method uses the metal raceway. The second method is to install a separate equipment grounding conductor between the two metal boxes, as would be required if a nonmetallic raceway had been used. (*Delmar/Cengage Learning*)

⎍⎍⎍ GROUNDING

The following definitions related to grounding are found in *NEC Article 100, 250.2,* and *250.4(A).*

Ground: *The Earth.* * Throughout this country and, for that matter, throughout the world, the earth is the common reference point for all electrical systems.

Grounded Conductor: *A system or circuit conductor that is intentionally grounded.* * In residential wiring, this is the white conductor. It is also referred to as the neutral conductor.

Grounding Conductor, Equipment (EGC): *The conductive path(s) installed to connect normally non-current-carrying metal parts of equipment together and to the system grounded conductor or to the grounding electrode conductor, or both.* * The definition speaks of connecting equipment together. When you think about it, it is logical to conclude that the equipment grounding conductor also functions as a bonding conductor. In residential wiring, the equipment grounding conductor is the bare conductor in nonmetallic-sheathed cable, the metal jacket plus the bonding strip in armored cable, metal raceways, green insulated conductors, and so on. Different acceptable EGCs are listed in *250.118.* Size the EGC per *Table 250.122.*

EGCs are illustrated in Figures 5-30, 5-32, 5-33, 5-34, 27-16, and 30-4.

Ground Fault: *An unintentional, electrically conducting connection between a normally current-carrying conductor of an electrical circuit, and the normally non-current-carrying conductors, metallic enclosures, metallic raceways, metallic equipment, or earth.* * Ground faults occur when an ungrounded "hot" conductor comes in contact with a grounded surface or grounded conductor. This could be a result of insulation failure or an ungrounded conductor connection coming loose.

Ground-Fault Current Path: *An electrically conductive path from the point of a ground fault on a wiring system through normally non-current-carrying conductors, equipment, or the earth to the electrical supply source.** This is the path that the flow of ground-fault current will take. Whatever fault current flows, that fault current must return to its source. What goes out must come back! The current might return through connectors, couplings, bonding jumpers, grounding conductors, ground clamps, and other components that make up the ground-fault return path.

Ground-Fault Current Path, Effective: As defined in *250.2, An intentionally constructed, reliable, low-impedance electrically conductive path designed and intended to carry current under ground-fault conditions from the point of a ground fault on a wiring system to the electrical supply source and that facilitates the operation of the overcurrent protective device or ground fault detectors on high-impedance grounded systems.** If and when a ground fault occurs, we want the overcurrent device ahead of the ground fault to open as fast as possible to clear the fault. To accomplish this, the integrity of the ground-fault current path must be unquestionable.

We sometimes think that electricity follows the path of least resistance. That is not totally correct! Electricity follows all paths. As stated earlier, ground-fault current might return through connectors, couplings, bonding jumpers, grounding conductors, ground clamps, and any other components that make up the ground-fault return path. Some of the fault current might flow on sheet metal ductwork and on metal water piping or metal gas piping. That is why the *NEC* is so strict about keeping the impedance of the ground-fault current path as low as possible. We want the ground-fault current to return on the electrical system—not on other nonelectrical parts in the building. Think about it this way—the lower the impedance, the higher the current flow. The higher the current flow, the faster the overcurrent device will clear the fault.

As required by *250.4(A)(5), Electrical equipment and wiring and other electrically conductive material*

likely to become energized shall be installed in a manner that creates a low-impedance circuit facilitating the operation of the overcurrent device or ground detector for high-impedance grounded systems. It shall be capable of safely carrying the maximum ground-fault current likely to be imposed on it from any point on the wiring system where a ground fault may occur to the electrical supply source. The Earth shall not be considered as an effective ground-fault current path.**

Grounding Electrical System: *Electrical systems that are grounded shall be connected to earth in a manner that will limit the voltage imposed by lightning, line surges, or unintentional contact with higher voltage lines and that will stabilize the voltage to earth during normal operation.** See Figure 27-20.

Grounding of Electrical Equipment: *Normally non-current-carrying conductive materials enclosing electrical conductors or equipment, or forming part of such equipment, shall be connected to earth so as to limit the voltage to ground on these materials.**

Grounding Electrode: *A conducting object through which a direct connection to earth is established.** In residential wiring, the grounding electrode might be an underground metal water piping supply, a concrete-encased reinforcing steel bar or bare copper conductor (UFER ground), or it might be a ground rod(s). Acceptable grounding electrodes are listed in *250.52(A)(1)* through *(7)*. See *250.52*.

Grounding Electrode Conductor: *A conductor used to connect the system grounded conductor or the equipment to a grounding electrode or to a point on the grounding electrode system.** See *250.62*. For typical residential wiring, the grounding electrode conductor is run between the main service panel neutral/ground terminal bar and the grounding electrode system.

Grounding electrode conductors

- are permitted to be aluminum or copper [see the limitation on the installation of aluminum conductors in *250.64(A)*].

- shall be protected against physical damage, *250.64(B)*.

- ▶If required, splices or connections shall be made as permitted in parts (1) through (4) of *250.64(C)*.

- *(1) Splicing of the wire-type grounding electrode conductor shall be permitted only by*

---

*Reprinted with permission from NFPA 70-2011

irreversible compression-type connectors listed as grounding and bonding equipment or by the exothermic welding process.

- *(2) Sections of busbars shall be permitted to be connected together to form a grounding electrode conductor.*

- *(3) Bolted, riveted, or welded connections of structural metal frames of buildings or structures*

- *(4) Threaded, welded, brazed, soldered or bolted-flange connections of metal water piping are sized according to Table 250.66.* *◄

In this residence, a 4 AWG armored copper grounding electrode conductor (GEC) runs from Main Panel A, across the workshop ceiling to the water pump area. The GEC is terminated with a listed ground clamp to the metal water piping where the piping comes through the basement wall. Depending on the size of the metal water pipe, ground clamps of the types illustrated in Figures 27-22, 27-23,

27-24, and 27-25 are used. As previously mentioned, because of the ever-increasing use of nonmetallic water pipe, there is a corresponding decrease in the use of a water pipe grounding electrode and more use of the concrete-encased grounding electrode, referred to as the UFER ground. To comply with the requirements in *NEC 250.50* to create a grounding electrode system, a 4 AWG copper bonding conductor was connected to a 20-ft (6.0 m) section of the reinforcing steel bar in the foundation footing before the concrete was poured. The end of the bonding conductor was coiled and left in the area of the water pipe so the two grounding electrodes could be bonded together.

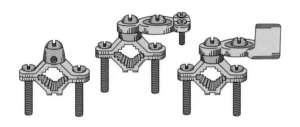

FIGURE 27-22 Typical ground clamps used in residential systems. (*Delmar/Cengage Learning*)

FIGURE 27-24 Ground clamp of the type used to bond (jumper) around water meter. (*Courtesy of Thomas & Betts*)

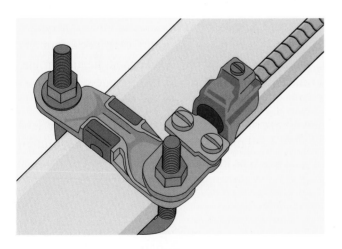

FIGURE 27-23 Armored grounding conductor connected with ground clamp to water pipe. (*Courtesy of Thomas & Betts*)

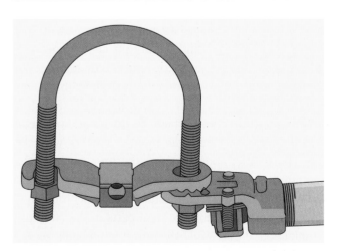

FIGURE 27-25 Ground clamp of the type used to attach ground wire to well casings. (*Courtesy of Thomas & Betts*)

*Reprinted with permission from NFPA 70-2011

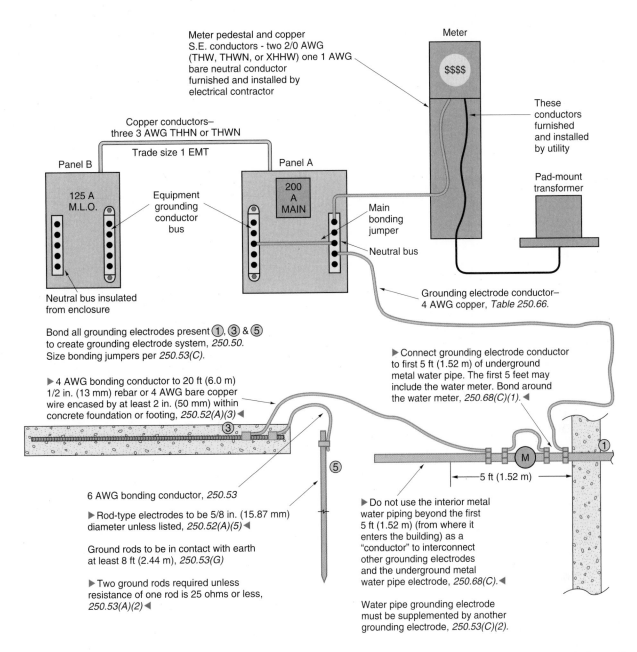

Meter pedestal and copper
S.E. conductors - two 2/0 AWG
(THW, THWN, or XHHW) one 1 AWG
bare neutral conductor
furnished and installed by
electrical contractor

Meter

$$$$

These
conductors
furnished
and installed
by utility

Pad-mount
transformer

Copper conductors–
three 3 AWG THHN or THWN

Trade size 1 EMT

Panel B

Panel A

125 A
M.L.O.

200
A
MAIN

Equipment
grounding
conductor
bus

Main
bonding
jumper

Neutral bus

Neutral bus insulated
from enclosure

Grounding electrode conductor–
4 AWG copper, *Table 250.66*.

Bond all grounding electrodes present ①, ③ & ⑤
to create grounding electrode system, *250.50*.
Size bonding jumpers per *250.53(C)*.

▶ Connect grounding electrode conductor
to first 5 ft (1.52 m) of underground
metal water pipe. The first 5 feet may
include the water meter. Bond around
the water meter, *250.68(C)(1)*. ◀

▶ 4 AWG bonding conductor to 20 ft (6.0 m)
1/2 in. (13 mm) rebar or 4 AWG bare copper
wire encased by at least 2 in. (50 mm) within
concrete foundation or footing, *250.52(A)(3)* ◀

③

⑤

M

①

5 ft (1.52 m)

6 AWG bonding conductor, *250.53*

▶ Rod-type electrodes to be 5/8 in. (15.87 mm)
diameter unless listed, *250.52(A)(5)* ◀

Ground rods to be in contact with earth
at least 8 ft (2.44 m), *250.53(G)*

▶ Two ground rods required unless
resistance of one rod is 25 ohms or less,
250.53(A)(2) ◀

▶ Do not use the interior metal
water piping beyond the first
5 ft (1.52 m) (from where it
enters the building) as a
"conductor" to interconnect
other grounding electrodes
and the underground metal
water pipe electrode, *250.68(C)*. ◀

Water pipe grounding electrode
must be supplemented by another
grounding electrode, *250.53(C)(2)*.

FIGURE 27-26 A typical electrical service grounded to a grounding electrode system consisting of
an underground metal water piping supply, concrete-encased electrode, and ground rod.
(*Delmar/Cengage Learning*)

Reference to grounding electrode conductors is
found in Figures 27-2, 27-16, 27-20, 27-26, 27-27,
27-28, and 27-29 through 27-34.

Ungrounded: *Not connected to ground or a conductive body that extends the ground connection.**

⎍⌁ BONDING

The following definitions related to bonding are
found in *NEC Article 100*.

Bonded (Bonding): *Connected to establish
electrical continuity and conductivity.** Figure 27-21
shows two metal boxes bonded together with the
metal raceway installed between the two boxes. The
bonding jumper could also have been a conductor.
Requirements for the size of a conductor of the wire
type are found at various locations in *Article 250*.

▶**Bonding Conductor or Jumper:** *A conductor to ensure the required electrical conductivity*

*Reprinted with permission from NFPA 70-2011

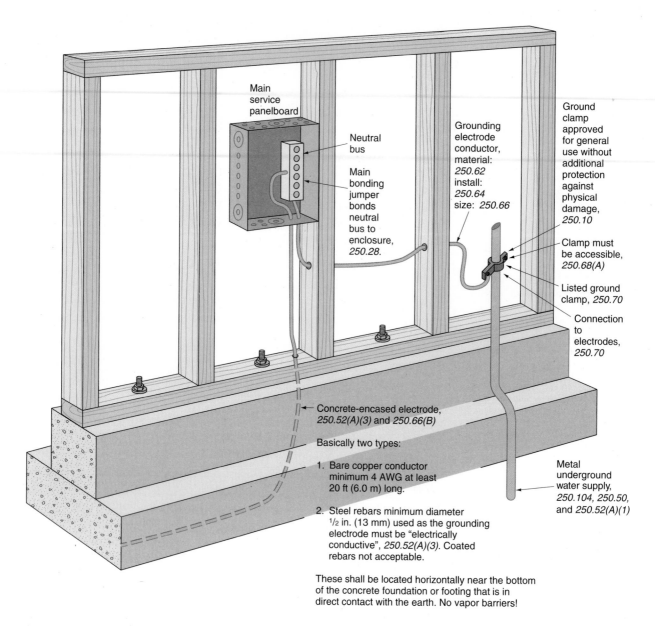

Main service panelboard

Neutral bus

Main bonding jumper bonds neutral bus to enclosure, 250.28.

Grounding electrode conductor, material: 250.62 install: 250.64 size: 250.66

Ground clamp approved for general use without additional protection against physical damage, 250.10

Clamp must be accessible, 250.68(A)

Listed ground clamp, 250.70

Connection to electrodes, 250.70

Concrete-encased electrode, 250.52(A)(3) and 250.66(B)

Basically two types:

1. Bare copper conductor minimum 4 AWG at least 20 ft (6.0 m) long.

2. Steel rebars minimum diameter ½ in. (13 mm) used as the grounding electrode must be "electrically conductive", 250.52(A)(3). Coated rebars not acceptable.

Metal underground water supply, 250.104, 250.50, and 250.52(A)(1)

These shall be located horizontally near the bottom of the concrete foundation or footing that is in direct contact with the earth. No vapor barriers!

FIGURE 27-27 Connection of concrete-encased grounding electrode and underground metal water pipe. (*Delmar/Cengage Learning*)

between metal parts required to be electrically connected.* ◀ A conductor of the wire type is required to be installed in a nonmetallic raceway to provide continuity between metal parts. With a metal raceway, the bonding would be accomplished by the metal raceway.

Bonding Jumper, Equipment: *The connection between two or more portions of the equipment grounding conductor.** Bonding equipment together

does not mean that the equipment is grounded. However, typical electrical systems in residential occupancies are grounded. The equipment grounding conductor then serves a dual function of bonding metal enclosures together and providing a ground-fault return path back to the source.

Bonding Jumper, Main: *The connection between the grounded circuit conductor and the equipment grounding conductor at the service.* See *250.28.* Main bonding jumpers are clearly illustrated in Figures 27-8, 27-16, 27-20, 27-26, and 27-28.

*Reprinted with permission from NFPA 70-2011

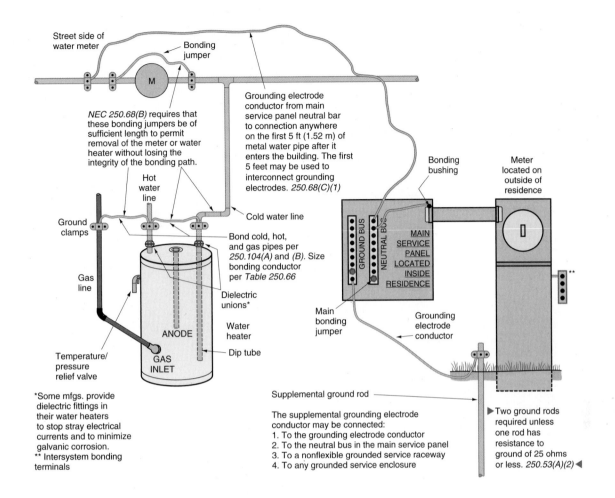

FIGURE 27-28 One method that may be used to provide proper electrical system grounding of service-entrance equipment, and bonding of the cold and hot water piping and gas piping. **Provide a listed intersystem bonding terminal strip at the meter enclosure, service equipment enclosure, or on the grounding electrode conductor. (*Delmar/Cengage Learning*)

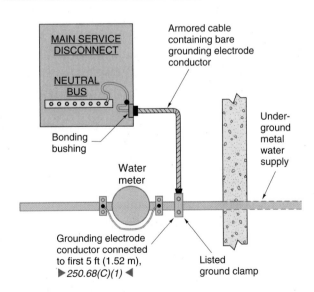

FIGURE 27-29 This installation "Meets *Code*." (*Delmar/Cengage Learning*)

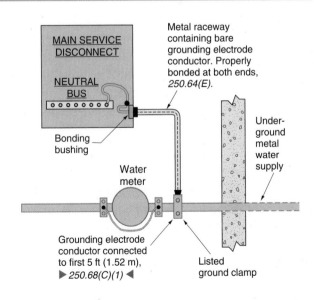

FIGURE 27-30 This installation "Meets *Code*." (*Delmar/Cengage Learning*)

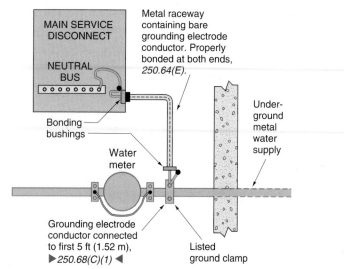

FIGURE 27-31 This installation "Meets *Code*." (*Delmar/Cengage Learning*)

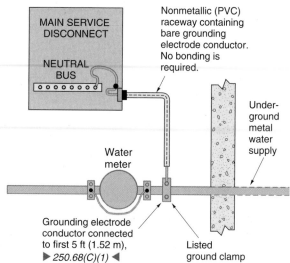

FIGURE 27-33 This installation "Meets *Code*." (*Delmar/Cengage Learning*)

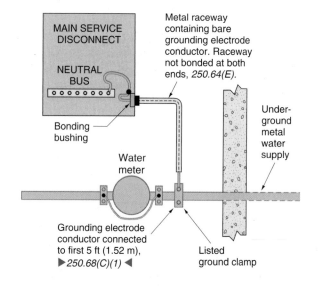

FIGURE 27-32 This installation *does not* "Meet *Code*." (*Delmar/Cengage Learning*)

Bonding of Electrically Conductive Materials and Other Equipment: *Normally non-current-carrying electrically conductive materials that are likely to become energized shall be connected together and to the electrical supply source in a manner that establishes an effective ground-fault current path. NEC 250.4(A)(4).* *

---

*Reprinted with permission from NFPA 70-2011

Bonding of Electrical Equipment: *Normally non-current-carrying conductive materials enclosing electrical conductors or equipment, or forming part of such equipment, shall be connected together and to the electrical supply source in a manner that establishes an effective ground-fault current path. NEC 250.4(A)(3).* *

These are a few other key *Code* sections relating to system grounding:

250.20(B)(1): This section requires that an electrical system be grounded where the maximum voltage to ground on the ungrounded "hot" conductors does not exceed 150 volts. The serving electric utility grounds the midpoint of their transformer. This midpoint becomes the system *neutral point*.

The electrical system in a typical home is single-phase, 3-wire, 120/240 volts. The voltage between the ungrounded "hot" phase conductors is 240 volts, and the voltage between the ungrounded "hot" phase conductors and the neutral (the grounded conductor) is 120 volts. The grounded neutral service conductor is again grounded at the main service.

The electrical system in some large multifamily buildings (condos, apartments, etc.) is derived from a 3-phase, 4-wire, 208Y/120-volt system. Here, the voltage between the ungrounded "hot" phase conductors is 208 volts, and the voltage between the ungrounded "hot" phase conductors and the neutral (the grounded conductor) is 120 volts. The supply to each dwelling

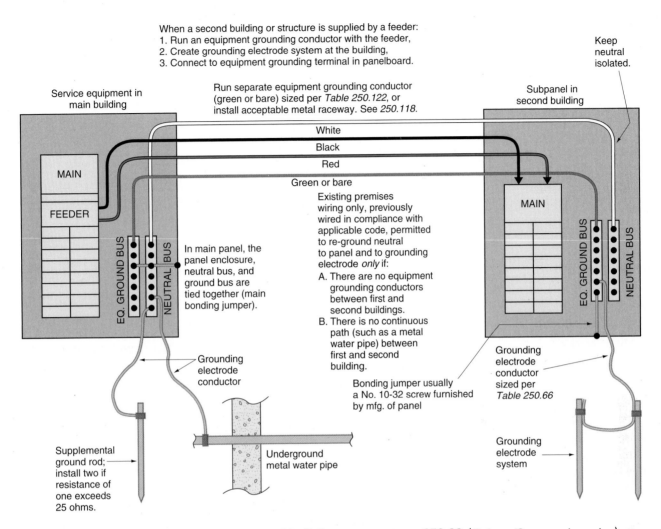

When a second building or structure is supplied by a feeder:
1. Run an equipment grounding conductor with the feeder,
2. Create grounding electrode system at the building,
3. Connect to equipment grounding terminal in panelboard.

Keep neutral isolated.

Service equipment in main building

Run separate equipment grounding conductor (green or bare) sized per *Table 250.122*, or install acceptable metal raceway. See *250.118*.

Subpanel in second building

White

Black

Red

Green or bare

MAIN

FEEDER

EQ. GROUND BUS

NEUTRAL BUS

In main panel, the panel enclosure, neutral bus, and ground bus are tied together (main bonding jumper).

Existing premises wiring only, previously wired in compliance with applicable code, permitted to re-ground neutral to panel and to grounding electrode *only* if:
A. There are no equipment grounding conductors between first and second buildings.
B. There is no continuous path (such as a metal water pipe) between first and second building.

Bonding jumper usually a No. 10-32 screw furnished by mfg. of panel

MAIN

EQ. GROUND BUS

NEUTRAL BUS

Grounding electrode conductor

Grounding electrode conductor sized per *Table 250.66*

Supplemental ground rod; install two if resistance of one exceeds 25 ohms.

Underground metal water pipe

Grounding electrode system

FIGURE 27-34 Grounding at a second building or structure, *250.32*. (*Delmar/Cengage Learning*)

unit is referred to as a single-phase, 3-wire, 120/208-volt system.

250.26(2): This section requires that the *neutral conductor* be grounded for single-phase, 3-wire systems. As mentioned earlier, the transformer midpoint (neutral point) is grounded at the utility transformer and again at the main service.

Providing a proper ground-fault current path helps ensure that overcurrent protective devices will operate fast when responding to ground faults. One essential component of an effective ground-fault current path consists of a low-impedance (ac-resistance) ground path. Ohm's law verifies that, in a given circuit, "the lower the impedance, the higher the value of current." As the value of ground-fault current increases, there is an increase

in the speed with which a fuse will open or a circuit breaker will trip. This is called an *inverse time* relationship.

Clearing Ground Faults

Clearing a ground fault or short circuit is important because arcing damage to electrical equipment as well as conductor insulation damage is closely related to a value called *ampere-squared-seconds* (I^2t), where

I = the current in amperes flowing phase to ground, or phase to phase.

t = the time in seconds that the current is flowing.

Thus, there will be less equipment and/or conductor damage when the fault current is kept to a low value and when the time that the fault current is allowed to flow is kept to a minimum. The impedance of the circuit determines the *amount* of fault current that will flow. The speed of operation of a fuse or circuit breaker determines the amount of *time* the fault current will flow.

Grounding electrode conductors and equipment grounding conductors carry an insignificant amount of current under normal conditions. However, when a ground fault occurs, the equipment grounding conductor as well as the ungrounded ("hot") circuit conductors must be capable of carrying whatever value of fault current might flow (*how much?*) for the time (*how long?*) it takes for the overcurrent protective device to clear the fault and reduce the fault current to zero.

This potential hazard is recognized in the *Note* below *Table 250.122*, which states that *Where necessary to comply with 250.4(A)(5) or 250.4(B)(4), the equipment grounding conductor shall be sized larger than given this table.* This note calls attention to the fact that equipment grounding conductors may have to be sized larger than indicated in the table when high-level ground-fault currents are possible.

What is a high-level ground fault? To determine the available short-circuit and ground-fault current, a short-circuit calculation is necessary. It is also necessary to know the time/current characteristic of the overcurrent protective device. This is discussed in *Electrical Wiring—Commercial* (© Cengage Delmar Learning).

This is referred to as the conductor's "withstand rating."

Table 250.66 and *Table 250.122* are based on the fact that copper conductors and their bolted connections can withstand

- one ampere
- for 5 seconds
- for every 30 circular mils.

Thermoplastic insulation on a copper conductor rated 75°C can safely withstand

- one ampere
- for 5 seconds
- for every 42.25 circular mils.

To exceed these values will result in damage to the conductors, with possible burn-off and loss of the grounding or bonding path.

When bare equipment grounding conductors are in the same raceway or cable as insulated conductors, always apply the insulated conductor withstand rating limitation.

Properly selected fuses and circuit breakers for normal residential installations generally will protect conductors and other electrical equipment against the types of ground faults and short circuits to be expected in homes. Fault currents can be quite high in single-family homes where the pad-mounted transformer is located close to the service equipment, and in multi-family dwellings (apartments, condos, town houses), making it necessary to take equipment short-circuit ratings and conductor withstand rating into consideration.

The subject of conductor and equipment withstand rating is covered in greater depth in *Electrical Wiring—Commercial* (© Delmar Cengage Learning).

Grounding Electrode System

Article 250, Part III, covers the requirements for establishing a grounding electrode system.

NEC 250.50 requires that metal underground water piping, the metal frame of a building, a concrete-encased electrode, a ground ring, and rod-pipe-plate electrodes be bonded together to create a grounding electrode system if any or all of the grounding electrodes *are present* in a new installation.

Metal gas piping shall never be used as a grounding electrode, *250.52(B)(1)*. Past experience has shown that because of galvanic action, gas pipes have deteriorated, resulting in serious incidents. However, metal gas piping is required to be bonded. Normally, the equipment grounding conductor for the branch circuit supplying a gas furnace serves as the required bond. *NEC 250.104(B)* covers the rules for bonding metal gas piping.

Should any one of these become disconnected, the integrity of the grounding system is maintained through the other paths. Here are a few key points:

- *250.90* states that *bonding shall be provided where necessary to ensure electrical continuity and the capacity to conduct safely any fault current likely to be imposed.**

*Reprinted with permission from NFPA 70-2011.

- *250.92(A)* explains what parts of a service must be bonded together.

- *250.94* explains what is acceptable as a bonding means.

- *250.96(A)* states in part that bonding of metal raceways, cable armor, enclosures, frames, fittings, and so on, that serve as the grounding path *shall be effectively bonded where necessary to ensure electrical continuity and the capacity to conduct safely any fault current likely to be imposed on them.**

By having all metal parts bonded together for a grounding electrode system, potential differences between non-current-carrying metal parts is virtually eliminated. In addition, the grounding electrode system serves as a means to bleed off lightning, stabilize the system voltage, and ensure that the overcurrent protective devices will operate.

Hazard of Improper Bonding

Figure 27-35 and the following steps illustrate what can happen if the electrical system is not properly grounded and bonded:

1. A "live" wire contacts the gas pipe. The bonding jumper "A" has not been installed.

2. The gas pipe now has 120 volts on it. The pipe is energized. It is "hot."

*Reprinted with permission from NFPA 70-2011.

3. The insulating joint in the gas pipe at the gas meter results in no current flow as the circuit is open.

4. The 20-ampere overcurrent device does not open, but the gas pipe remains energized.

5. If a person touches the "live" gas pipe and the water pipe at the same time, current flows through the person's body. If the hand-to-hand body resistance is 1100 ohms, the current is

$$I = \frac{E}{R} = \frac{120}{1,100} = 0.11 \text{ amperes}$$

This amount of current passing through a human body can cause death. See Chapter 6 for a discussion regarding electric shock.

6. The overcurrent device does not open.

7. If the bonding jumper "A" had been installed, it would have kept the voltage difference between the water pipe and the gas pipe at or near zero. The overcurrent device would have opened. Checking *Table 8, Chapter 9, NEC*, we find the dc resistance of 1000 feet (305 m) of an uncoated 4 AWG copper wire is 0.308 ohms. The resistance of 100 ft (30 m) is 0.0308 ohms. The resistance of 10 ft (3.0 m) is 0.00308 ohms. The current would be

$$I = \frac{E}{R} = \frac{120}{0.00308} = 38,961 \text{ amperes}$$

In an actual installation, the total impedance of *all* parts of the circuit would perhaps be much higher

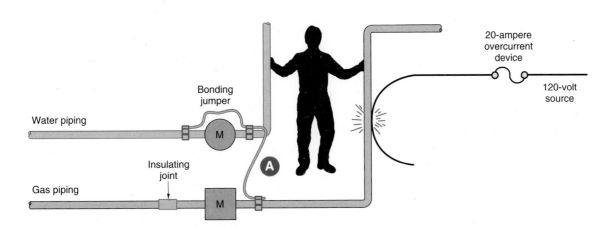

FIGURE 27-35 Proper bonding of gas piping system eliminates shock hazard. (*Delmar/Cengage Learning*)

than these simple calculations. A lower current would result. The value of current would be enough to cause the overcurrent device to open.

Where to Connect the Grounding Electrode Conductor

NEC 250.24(A) tells us that the grounding electrode conductor shall be connected to the grounded (neutral) service conductor. See Figure 27-26. *NEC 250.24(A)(1)* goes on to tell us that the connection is permitted to be made at any accessible point from the load end of the service drop or service lateral up to, and including, the terminal or bus to which the grounded (neutral) service conductor is connected at the service disconnecting means.

All residential panelboards have

- a neutral bus for the white grounded circuit conductors, and

- an equipment grounding bus for the bare equipment grounding conductors when nonmetallic-sheathed cable is used as the wiring method.

The neutral conductors and equipment grounding conductors are often connected to the same terminal bar in the service equipment as permitted by *NEC 408.40*. Only the equipment grounding conductors are connected to the equipment grounding conductor terminal bar in subpanels.

For just about all residential services, the connection of the grounding electrode conductor is made to the neutral bus. *NEC 250.24(A)(1)*. Many residential panelboards have a green hexagon-shaped No.10-32 screw that becomes the main bonding jumper between the neutral bus and the enclosure when properly installed.

Figures 27-16, 27-20 and 27-29 through 27-34 show the grounding electrode conductor connected to the neutral bus in the main service panelboard.

Grounding and Bonding the Electrical System in a Typical Residence

A metal underground water piping system 10 ft (3.0 m) or longer, in direct contact with the earth, is acceptable as a grounding electrode, *250.52(A)(1)*. The connection of the grounding electrode conductor must be made on the first 5 ft (1.52 m) of where the metal underground water piping enters the building, *250.68(C)(1)*.

When metal underground water piping is used as the only grounding electrode, it must be supplemented by at least one additional grounding electrode, *250.53(D)(2)*. In this residence, the concrete-encased electrode and a driven ground rod are all connected together to create the grounding electrode system and automatically satisfy the requirement for the supplemental grounding electrode, as permitted by *NEC 250.50* and *250.52(A)(5)*.

▶As shown in Figure 27-26, the 20 ft (6.0 m) length of steel reinforcing bar that is encased in 2 in. (50 mm) of concrete near the bottom of the foundation footing is recognized in *NEC 250.52(A)(3)* as a grounding electrode. A 4 AWG bare copper conductor at least 20 ft (6.0 m) long installed identically to the steel reinforcing rod is also permitted as a concrete-encased grounding electrode. This is often referred to as a UFER ground, named after Herbert G. Ufer, who worked for Underwriters Laboratories. Connecting the grounding electrode conductor to a concrete-encased electrode (a minimum 4 AWG bare copper conductor or reinforcing bars (rebars) are required by *250.66(B)*. A concrete encased electrode does not require a supplemental electrode as does underground metal water piping, *250.53(D)(2)*.◀

▶A concrete-encased electrode could also be installed vertically in a foundation wall. The key is that the foundation wall be in direct contact with the earth, *250.52(A)(3)*.◀

Watch out when using ground rods, pipes, or plates as grounding electrodes! ▶*NEC 250.53(A)(2)* states that two rods, pipes, or plates be installed unless it can be shown that a single grounding electrode has a resistance to ground of 25 ohms or less.◀ There are a number of manufacturers of testers for measuring ground–earth resistance.

What's So Good about a Concrete-Encased Electrode?

Unquestioned reliability!

The increasing use of nonmetallic water mains brought about the need to use concrete-encased electrodes. In addition, a concrete-encased electrode is recognized as a grounding electrode that provides a good connection to the earth.

When used as a grounding electrode, a concrete-encased electrode does not have to be supplemented as does a metal underground water piping system. There is no need to check for the maximum 25-ohm requirement as there is for ground rods. Many communities have mandated using a concrete-encased electrode as the primary electric service electrode because of its proven performance record of providing an excellent connection to earth.

The permanent moisture under a concrete foundation or footing ensures a low-impedance direct connection to earth. When using a concrete-encased grounding electrode, be sure that the footing or foundation is in direct contact with the earth. Make sure that there is no vapor barrier underneath the footing or foundation.

The electrician must work closely with the concrete and rebar contractor. It is necessary to bring one end (stub-up) of a reinforcing bar (called "rebar") or the bare 4 AWG copper conductor upward out of the concrete slab or footing at a location near the likely location of the electrical main service. This makes for a rather easy connection point for the grounding electrode conductor. See *250.52(A)(3)*. If the rebar is brought up close to the metal water service, the bonding together of the rebar and the metal water service is easily accomplished.

If a grounding electrode conductor connection is made underground or is imbedded in concrete, the connectors must be listed for direct burial. See Figure 27-26.

Coated rebars are not acceptable.

More Ground Rod Rules for Residential Application

- The most common ground rods are copper-clad steel. Copper makes for an excellent connection between the rod and the ground clamp. The steel gives it strength to withstand being driven into the ground. Galvanized and stainless steel rods are also available. Aluminum rods are not permitted.

- Ground rods must be at least ⅝ in. (15.87 mm) in diameter unless listed by a qualified electrical testing laboratory.
- They must be installed below the permanent moisture level if possible.
- They must not be less than 8 ft (2.5 m) in length.
- They must be driven to a depth so that at least 8 ft (2.5 m) is in contact with the soil.
- If solid rock is encountered so the rod cannot be driven vertically to the proper depth, the rod must be driven at an angle not greater than 45° from vertical.
- If the ground rod cannot be driven at an oblique angle not greater than 45° from the vertical, the rod is permitted to be installed in a trench that is at least 2½ ft (750 mm) deep.
- The rod should be driven so the upper end is flush with or just below ground level. If the upper end is exposed, the ground rod, the ground clamp, and the grounding electrode conductor must be protected from physical damage.
- If more than one rod is needed, they must be kept at least 6 ft (1.8 m) apart. Driving the rods close together reduces their effectiveness because their sphere of influence overlaps. Actually, it is better to space multiple rods twice the length of the longest rod. For example, when driving two 8-ft (2.5-m) ground rods, space them 16 ft (4.9 m) apart.

Table 27-3 shows accurate multipliers for multiple ground rods spaced one rod-length apart. To use this chart, divide the resistance value of one rod by the number of rods used, then apply the multiplier.

TABLE 27-3

Multipliers for multiple ground rods.

Number of Rods	Multiplier
2	1.16
3	1.29
4	1.36

EXAMPLE

The ground–earth resistance reading of one 8 ft (2.5 m) ground rod is 30 ohms. This exceeds the *NEC* maximum of 25 ohms. A second ground rod is driven and connected in parallel to the first ground rod. What is the approximate ground–earth resistance of these two ground rods when spaced 8 ft (2.5 m) apart?

$$\frac{30}{2} \times 1.16 = 17.4 \text{ ohms}$$

- A grounding electrode conductor that is the sole (only) connection to a driven ground rod need not be larger than 6 AWG copper or 4 AWG aluminum, *250.53(E), 250.66(A)*.

In some parts of the country, a water pipe ground is considered unreliable. The concern is the prolific use of insulating fittings and nonmetallic water piping services. These parts of the country have found that a concrete-encased electrode provides excellent grounding capabilities. Refer to *250.52(A)(3)*.

Other Acceptable Grounding Electrodes

- The metal frame of a building (this residence is constructed of wood), *250.50(A)(2)*.

- A ground ring that encircles the building consisting of not less than 20 ft (6.0 m) of bare copper wire, minimum size 2 AWG, buried directly in the earth at least 2½ ft (750 mm) deep. See *250.52(A)(4)* and *250.52(G)*.

- Ground plates must be at least 2 ft² (0.186 m²), *250.52(A)(7)*. Two ground plates are required unless the resistance of one plate is found to be 25 ohms or less, *250.53(A)(2)*. The ground plates must be buried at least 2½ ft (750 mm) deep.

- "Other Listed Electrodes," *NEC 250.52(A)(6)*. This "catch-all" title was added specifically to accommodate chemical ground rods. These consist of copper tubing that is filled with a natural earth salt chemical. Because they are quite expensive, it is uncommon to use them for other than commercial installations.

- *250.53(D)(2)* tells us that a metal underground water piping system must have a supplemental electrode installed. Ground rods, ground plates, and underground metal water piping systems are required to have a supplemental grounding electrode. Other types of grounding electrodes, such as structural steel electrodes, concrete-encased electrodes, and ground rings, do not need a supplemental electrode.

Bond All Grounding Electrodes Together

NEC 250.50 requires that if any or all of the following are present in a new installation, they must be bonded together: metal underground water pipe, metal frame of the building, concrete-encased electrode, ground ring, rod and pipe electrodes, and plate electrodes. This creates the grounding electrode system.

Some Issues Regarding Bonding of Metal Pipes Inside or Attached to the Outside of the Home

Metal water piping and metal gas piping must be properly bonded, *250.104(A)* and *(B)*.

Over the years, underground natural gas services installed from the street to the home might have been black iron pipe, copper tubing, or plastic piping designed for gas underground installations. You might even find lead water and gas pipes in homes built prior to 1963, but not likely after 1963. Today, metal gas piping installed underground has a factory-applied corrosion protection coating and/or wrapping as required by *NFPA 54, Natural Fuel Gas Code*. Most underground gas pipes today are plastic.

Homes have been blown off their foundations as a result of ignition and explosion of leaking gas from underground piping. The culprit causing the leak usually can be traced to corrosion. Where gas is found to be leaking from an underground copper pipe, the gas company workers pull the copper pipe out of the ground, pulling in its place an approved plastic pipe. A 14 AWG yellow tracer wire is pulled in (buried) with the plastic pipe so as to be able to locate the underground nonmetallic piping at some later date.

Underground gas line piping is not a major concern for electricians! What is of concern to the electrician is that the metal water and gas piping within or on the home must be bonded together.

Here are the *NEC* requirements that deal with bonding metal water and gas piping:

- *250.52(B)(1):* Metal underground gas piping is not permitted to be used as a grounding electrode.

    Similarly, the *National Fuel Gas Code, NFPA 54, Section 7.13.3* states that *Gas piping shall not be used as a grounding conductor or electrode.*

    To prevent galvanic corrosion of underground metal gas piping, gas utilities install dielectric (insulating) fittings at the line side port of the gas meter. A dielectric fitting "isolates" the underground metal gas piping from the interior metal piping and prevents the corrosion problem. Today, gas companies generally install plastic piping underground. This solves the corrosion problem present when metal piping is installed underground.

- *250.104(A):* Requires that interior metal water piping be bonded, and that the bonding conductor be sized according to *Table 250.66*. Figure 27-28 clearly shows this bond.

    To control galvanic corrosion of the water heater tank and any directly connected steel, galvanized, and/or copper piping, you will usually find dielectric fittings (unions) in the water heater's cold and hot water lines. Always install the bonding jumper *above* the dielectric fittings.

- *250.104(B):* All metal piping, including gas piping, shall be bonded. Figure 27-28 clearly shows this bond.

- *250.104(B):* The bonding conductor for metal gas piping shall be sized according to *Table 250.122* for the ampere rating of the circuit that might energize the metal piping.

    Similarly, the *National Fuel Gas Code, NFPA 54,* in *Section 7.13.1* requires that "*Each above ground portion of gas piping other than Corrugated Stainless Steel Tubing (CSST) that is likely to become energized shall be electrically continuous and bonded to an effective ground-fault current path. Gas piping, other than CSST, shall be considered to be bonded when it is connected to gas utilization equipment that is*

connected to the appliance grounding conductor of the circuit supplying the appliance."*

In *NFPA 54 7.13.2,* CSST gas piping systems shall be bonded to the electrical service grounding electrode system at the point where the gas service enters the building. The bonding jumper shall not be smaller than 6 AWG copper wire or equivalent.

The logic to this requirement is that if the "hot" conductor of the electrical circuit supplying a gas utilization equipment such as a gas furnace comes in contact with the metal frame of the appliance, the ground-fault current path through the equipment grounding conductor will cause the branch-circuit overcurrent device to clear the fault. In the meantime, bonding maintains an equal voltage potential between the metal frame of the gas appliance and its metal gas supply piping. Equal voltage potential is also maintained between the faulted appliance and other nearby equipment.

If there is no likelihood of the gas pipe becoming energized (no gas appliance that is also served by electricity), then no bonding of the gas piping is required.

- *250.104(B), Informational Note: "Bonding all piping and metal air ducts within the premises will provide additional safety."* Informational Notes are not mandatory but certainly provide excellent recommendations to improve safety.

In a typical home, there is usually a large number of bonds between the hot and cold water pipes.

There are interconnections at metallic mixer faucets, water heaters, clothes washers, and similar plumbing connections. Many plumbers use short lengths of copper tubing to bridge a stud space or joist space to support and keep water lines in place. They solder (tack) the water lines to these cross pieces. This bonds the hot and cold water pipes in a number of places inside the walls.

Figure 27-28 illustrates the bonding together of the cold and hot water metal pipes and the metal gas piping. This bonding jumper is sized per *Table 250.66*. This installation is easy and eliminates the questionable bonding/grounding of the

---

*Reprinted with permission from NFPA 54-2009.

many interconnections of the metal water and gas piping through various removable gas appliances. This bonding above the water heater *does not* mean that the gas pipe is serving as a grounding electrode.

Some electrical inspectors will consider the hot/cold mixing faucets and the water heater as an acceptable means of bonding the hot and cold water metal pipes together. Most, however, will require a bonding jumper as illustrated in the figures, consistent with the thinking that electrical grounding and bonding shall not be dependent on the plumbing trade.

Don't depend on the plumber to do your job!

Do it right! Follow Figure 27-28 for reliably bonding together all metal water piping and metal gas piping.

The ground-fault current return path in Figure 27-36 is the EMT. Where nonmetallic-sheathed cable is used, the equipment grounding conductor in the cable serves as the ground-fault current return path. Where armored cable is used, the armor plus the bonding strip is the ground-fault current return

path. See *NEC 250.118* for other accepted equipment grounding conductors.

NEC 300.6 addresses corrosion that an electrician needs to be concerned with when working in wet locations, underground, in concrete, and similar installations.

Ground Clamps

Ground clamps used for bonding and grounding must be listed for the purpose. The use of solder to make up bonding and grounding connections is not acceptable because under high levels of fault current, the solder would probably melt, resulting in loss of the integrity of the bonding and/or grounding path. See *250.8*.

Various types of ground clamps are shown in Figures 27-22, 27-23, 27-24, and 27-25. These clamps and their attachment to the grounding electrode must conform to *250.70*.

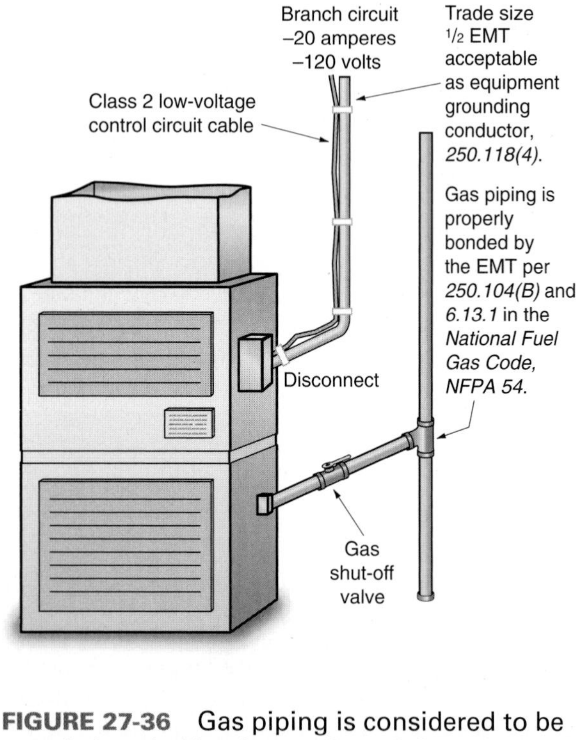

FIGURE 27-36 Gas piping is considered to be properly bonded by the appliance's branch-circuit equipment grounding conductor. The equipment grounding conductor in this illustration is the EMT, which provides an excellent ground-fault current return path.
(*Delmar/Cengage Learning*)

Table 250.66 Grounding Electrode Conductor for Alternating-Current Systems

Size of Largest Ungrounded Service-Entrance Conductor or Equivalent Area for Parallel Conductors[a] (AWG/kcmil)		Size of Grounding Electrode Conductor (AWG/kcmil)	
Copper	Aluminum or Copper-Clad Aluminum	Copper	Aluminum or Copper-Clad Aluminum[b]
2 or smaller	1/0 or smaller	8	6
1 or 1/0	2/0 or 3/0	6	4
2/0 or 3/0	4/0 or 250	4	2
Over 3/0 through 350	Over 250 through 500	2	1/0
Over 350 through 600	Over 500 through 900	1/0	3/0
Over 600 through 1100	Over 900 through 1750	2/0	4/0
Over 1100	Over 1750	3/0	250

Notes:
1. Where multiple sets of service-entrance conductors are used as permitted in 230.40, Exception No. 2, the equivalent size of the largest service-entrance conductor shall be determined by the largest sum of the areas of the corresponding conductors of each set.
2. Where there are no service-entrance conductors, the grounding electrode conductor size shall be determined by the equivalent size of the largest service-entrance conductor required for the load to be served.
[a]This table also applies to the derived conductors of separately derived ac systems.
[b]See installation restrictions in 250.64(A).

Reprinted with permission from NFPA 70-2011

Connection of Equipment Grounding and Grounded Conductors in Main Panel and Subpanel

According to the second paragraph of *408.20*, equipment *grounding* conductors (the bare copper equipment grounding conductors found in nonmetallic-sheathed cable) shall not be connected to the *grounded* conductor (the neutral conductor) terminal bar (neutral bar) unless the bar is identified for the purpose, and is located where the *grounded* conductor is connected to the *grounding* electrode conductor. In this residence, this occurs in Main Panel A. The green main bonding jumper screw furnished with Panel A is installed in Panel A. This bonds the neutral bar, the ground bus, the grounded neutral conductor, the grounding electrode conductor, and the panel enclosure together, as in Figure 27-16. *Grounding* conductors and *grounded* conductors are not to be connected together anywhere on the load side of the main service disconnect, *250.24(5)* and *250.142(B)*. The green bonding screw furnished with Panel B will *not* be installed in Panel B. More on this a little later in this chapter.

Sheet Metal Screws Not Permitted for Connecting Grounding Conductors

Sheet metal screws shall not be used to connect grounding conductors or connection devices to enclosures, *250.8*. Sheet metal screws do not have the same fine thread that No. 10-32 machine screws have, which match the tapped No. 10-32 threaded holes in outlet boxes, device boxes, and other enclosures. Sheet metal screws "force" themselves into a hole instead of nicely threading themselves into a pretapped matching hole. Sheet metal screws have *not* been tested for their ability to safely carry ground-fault currents as required by *250.4(A)(5)*.

A typical bonding jumper (equipment) is illustrated in Figure 27-37.

Bonding Service Equipment

At the main service-entrance equipment, the grounded neutral conductor must be bonded to the metal enclosure. For most residential panels, this main bonding jumper is a bonding screw that is furnished with the panel. This bonding screw is inserted through the neutral bar into a threaded hole in the back of the panel. This bonding screw must be green and must be clearly visible after it is in place, *250.28(B)*.

The required main bonding jumpers for services must not be smaller than the grounding electrode conductor, *250.28(D)*. The lugs on bonding bushings are based on the ampacity of the conductors that would normally be installed in that particular size of raceway. The lugs become larger as the trade size of the bushing increases. However, there is no need to calculate the adequacy of the device furnished by the panelboard manufacturer that is intended to function as the main bonding jumper so long as the panelboard is listed by a qualified electrical testing laboratory such as UL.

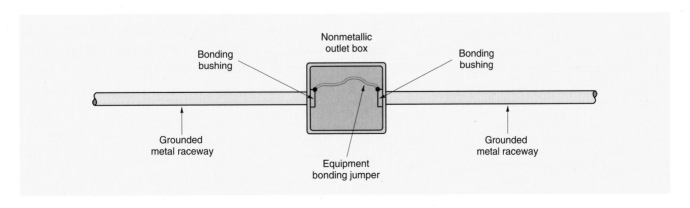

FIGURE 27-37 Bonding jumper (equipment). (*Delmar/Cengage Learning*)

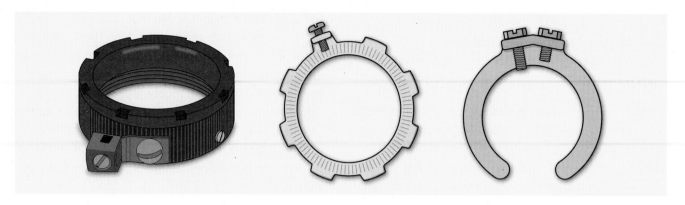

FIGURE 27-38 An insulating grounding/bonding bushing with grounding lug, a bonding locknut, and a bonding wedge. (*Delmar/Cengage Learning*)

Bonding at service-entrance equipment is very important because service-entrance conductors do not have overcurrent protection at their line side, other than the electric utility company's primary transformer fuses. Overload protection for service-entrance conductors is at their load end. High available fault currents can result in severe arcing, which is a fire hazard. For all practical purposes, available short-circuit current is limited only by (1) the kVA rating and impedance of the transformer that supplies the service equipment, and (2) the size, type, and length of the service conductors between the transformer and service equipment. Fault currents can easily reach 20,000 amperes or more at the main service in residential installations. Much higher fault current is available in multifamily dwellings such as apartments and condominiums.

Fault current calculations are presented later in this chapter. The text *Electrical Wiring— Commercial* (© Delmar Cengage Learning) covers fault current calculations in greater depth.

NEC 250.92(A) and *(B)* lists the parts of the service-entrance equipment that must be bonded. These include the meter base, service raceways, service cable armor, and main disconnect. *NEC 250.94* lists the methods acceptable for bonding all of the previous equipment.

Grounding/Bonding Bushings

Grounding bushings have a means (lug) for connecting a grounding or bonding conductor to it. A grounding bushing might also have a means (one or

more set screws) that ensure reliable bonding of the bushing to a metal raceway, in which case the bushing serves as a grounding and bonding bushing. These bushings are available with an insulated throat. See Figure 27-38. These are the most commonly used for bonding residential electrical services.

Bonding Bushings, Bonding Locknuts, Bonding Wedges

Bonding bushings, bonding locknuts, and bonding wedges have one or more set screws that "dig in" and secure the bushing to the electrical equipment enclosure. These products do not have a means for connecting a grounding or bonding conductor.

Why Are Grounding/Bonding/ Insulating Bushings Installed?

Grounding/bonding bushings, bonding wedges, grounding conductors, and bonding jumpers are installed to ensure a low-impedance path if a fault occurs. Bonding jumpers are required where there are concentric or eccentric knockouts, or where reducing washers are installed. *NEC 300.4(F)* states that if 4 AWG or larger ungrounded circuit conductors enter a cabinet, box enclosure or raceway, an insulating bushing or equivalent must be used, shown in Figures 27-38, 27-39, and 27-40. Similar insulating bushing requirements are found in *312.6(C)*, *314.17(D)*, *314.28(A)*, *354.46*, and *362.46*. Insulating bushings protect the conductors from abrasion where they pass through the bushing. Combination metal/insulating

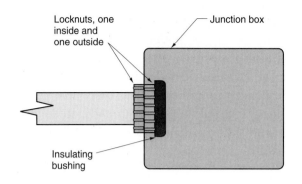

FIGURE 27-40   The use of double locknuts plus an insulating bushing. See *300.4(F)* and *312.6(C)*. (*Delmar/Cengage Learning*)

FIGURE 27-39   Insulating bushings. (*Delmar/Cengage Learning*)

bushings can be used. Some EMT connectors have an insulated throat. Insulating sleeves are also available that slide into a bushing between the conductors and the throat of the bushing.

If the bushing is made of insulating material only, as in Figure 27-39, then two locknuts must be used, as shown in Figure 27-40.

Installing a Grounding Electrode Conductor from Main Service Disconnect to Metal Underground Water Pipe that Will Serve as the Grounding Electrode

The *Code* specifies in *NEC 250.64* how a grounding electrode conductor is to be installed, as shown in Table 27-4.

In order to carry high values of ground-fault current, the ground-fault current path must have as low an impedance as possible. This means that the path must be continuous, must have the capability of carrying any value of fault current that it might be called on to carry, and shall have an impedance low enough to limit the voltage to ground and to ensure that the protective overcurrent device will operate. *NEC 250.90* and *250.96* have similar requirements. For example, the metal raceway enclosing a grounding electrode conductor must be continuous from

the main disconnect to the ground clamp, or be made continuous by proper bonding.

NEC 250.64(E) requires that the metal raceway that encloses a grounding electrode conductor be bonded to the conductor at both ends. At first thought, it might appear that simply installing a grounding electrode conductor in a metal raceway makes for a neat and workmanlike installation.

TABLE 27-4		
Installing a grounding electrode conductor.		
4 AWG or larger	**6 AWG**	**Smaller than 6 AWG**
May be attached to the surface on which it is carried. Does not need additional mechanical protection unless exposed to physical damage, which is highly unlikely in the case of residential installations.	If not subject to physical damage, may be run along the surface of the building structure without additional protection. If subject to physical damage, install in RMC, EMT, IMC, or Schedule 80 PVC, or cable armor.	Must be installed in rigid metal conduit, electrical metallic tubing, intermediate metal conduit, rigid PVC conduit, or cable armor.

A typical armored grounding electrode conductor.

However, unless the enclosing metal raceway is properly bonded on both ends, the ground path will be a high-impedance path. Bonding the enclosing metal raceway at one end only will result in a ground path impedance approximately twice that of bonding the metal raceway on both ends. Remember—the higher the impedance, the lower the current flow.

A TYPICAL ARMORED GROUNDING ELECTRODE CONDUCTOR

Past testing (See *Some Fundamentals of Equipment-Grounding Circuit Design* included in *Electrical Grounding and Bonding* published by Delmar Cengage Learning) has shown that under certain conditions for a 300-ampere ground fault, approximately 295 amperes flowed in the metal raceway, and 5 amperes flowed in the enclosed grounding electrode conductor. Another test revealed that for a 590-ampere ground fault, 585 amperes flowed in the metal raceway, and 5 amperes flowed in the enclosed grounding electrode conductor. These tests showed that under the conditions of the test, the conduit provided a lower impedance path than the contained conductor. Proper bonding at both ends of a metal raceway is necessary to ensure a low impedance path that will safely transfer the high magnitudes of ground-fault current flowing in the metal conduit to the ground bus in the main panel, and/or to the ground clamp.

Figures 27-29, 27-30, 27-31, 27-32, and 27-33 illustrate accepted methods of installing a grounding electrode conductor between the main panel and the ground clamp. Additional information concerning grounding electrode conductors can be found in Chapter 4.

GROUNDING ELECTRICAL EQUIPMENT AT A SECOND BUILDING

Detached Garage—Grounding

A detached garage or other second building on the same residential property is almost always served by the same electrical service as the house.

For dwellings, a detached garage is not required to have any electrical supply. However, if you do run electric power to a detached garage, specific *Code* requirements must be complied with.

We discussed the receptacle and lighting requirements for garages in Chapter 3. We will now discuss the grounding requirements for a second building, *250.32*. See Figure 27-34.

Branch Circuit to Detached Garage

If a single or multiwire branch circuit is run to a detached garage or other second building, the simplest way to provide proper and adequate grounding for the equipment in that detached garage is to install an underground cable Type UF that contains an equipment grounding conductor. A grounded metal raceway between the house and the outbuilding may also serve as the equipment grounding conductor.

A separate grounding electrode at the second building is not required if only a single branch circuit supplies the building, *250.32(A), Exception*.

Feeder to Detached Garage

When a feeder supplies a panel in a detached garage or other second building that has more than one branch circuit, proper grounding at the second building is accomplished by:

- installing a metal raceway, which is an acceptable equipment grounding conductor in *NEC 250.118*, between the main building panel and the panel in the second building, or

- installing a separate equipment grounding conductor of the wire type (in a cable or in a nonmetallic raceway) between the main building panel and the panel in the second building.

In either case, a grounding electrode system is required at the second building. A grounding electrode conductor is run between the grounding electrode and the equipment grounding bus in the panelboard in the second building.

The grounding electrode conductor is sized according to *Table 250.66*, but it does not have to be larger than the largest ungrounded supply conductor, *250.32(E)*.

The neutral bus and the equipment grounding bus in the second building are not connected together. If these two buses were tied together in the second building, there would be more than one path (a parallel path) for neutral current to flow back on. Take a look at Figure 27-34. Imagine a tie between the neutral bus and the equipment grounding bus in the second building. Now imagine that there is some value of neutral current. You can readily see that this neutral current would have many ways to return to the source at the main building panel. Some current would return on the neutral conductor, some on the equipment grounding conductor, some on the grounding electrode conductor—to the grounding electrode—and back through the earth, and some through a metal water pipe if one is present. These multiple paths of neutral return current are not permitted.

All of the above requirements are found in *230.32*, and are illustrated in Figure 27-33.

REVIEW

Note: Refer to the *NEC,* the text, and/or the plans where necessary.

1. Define the *Service Point.* _____

   _____

   _____

   _____

   _____

2. Define *service-drop conductors.* _____

   _____

   _____

   _____

   _____

3. Who is responsible for determining the service location? _____

4. a. The service head must be located (above) (below) the point where the service-drop conductors are spliced to the service-entrance conductors. (Circle the correct answer.)

   b. What *Code* section provides the answer to part (a)? _____

5. What is a mast-type service entrance? _____

   _____

   _____

6. When a conduit is extended through a roof, must it be guyed? _____

   _____

7. a. What size and type of conductors are installed for this service? _____

   _____

   b. What size conduit is installed? _____

   c. What size grounding electrode conductor is installed? (not neutral) _____

   d. Is the grounding electrode conductor insulated, armored, or bare? _____

8. How and where is the grounding electrode conductor attached to the water pipe?

_____

_____

9. What are the minimum distances or clearances for the following?

 a. Overhead service conductor clearance over private driveway _____

 b. Overhead service conductor clearance over private sidewalks _____

 c. Overhead service conductor clearance over alleys _____

 d. Overhead service conductor clearance over a roof having a roof pitch of not less than 4/12. (Voltage between conductors does not exceed 300 volts.) _____

 e. Overhead service conductor horizontal clearance from a porch _____

 f. Overhead service conductor clearance from a fence that can be climbed on _____

10. What are the minimum size ungrounded conductors using Type THW copper for the following residential electrical services? The terminals in the panelboard are marked 75°C.

Ampere Rating of Service	Service-Entrance Conductors Sized per *Table 310.15(B)(16)*	Service-Entrance Conductors Sized per *Table 310.15(B)(7)*
100		
200		
400		

11. What is the minimum size copper grounding electrode conductor for each of the following residential electrical services? Refer to *Table 250.66*. The ungrounded conductors are Type THW copper.

Ampere Rating of Service	Grounding Electrode Conductor When the Service-Entrance Conductors Are Sized per *Table 310.15(B)(16)*	Grounding Electrode Conductor When the Service-Entrance Conductors Are Sized per *Table 310.15(B)(7)*
100		
200		
400		

* *Table 310.15(B)(7)* is only for 120/240-volt, single-phase residential services and feeders. This table does not apply to services and feeders other than residential.

12. What is the recommended height of a meter socket from the ground? _____

13. a. May the bare grounded neutral conductor of a service be buried directly in the ground? _____

 b. What section of the *Code* covers this? _____

14. How far must mechanical protection be provided when underground service conductors are carried up a pole? _____

15. a. A service disconnect may consist of not more than how many switches or circuit breakers? _____

    b. Must these devices be in one enclosure? _____

    c. What type of main disconnect is provided in this residence? _____

    d. Does your city permit more than one service disconnect? _____

16. Complete the following table by filling in the columns with the appropriate information.

	Circuit Number	Ampere Rating	Poles	Volts	Conductor Size
A. Living room receptacle outlets					
B. Workbench receptacle outlets					
C. Water pump					
D. Attic exhaust fan					
E. Kitchen lighting					
F. Hydromassage tub					
G. Attic lighting					
H. Counter-mounted cooking unit					
I. Electric furnace					

17. a. What size conductors supply Panel B? _____

    b. What size raceway? _____

    c. Is this raceway run in the form of electrical metallic tubing or rigid conduit?

    d. What size overcurrent device protects the feeders to Panel B? _____

18. How many electric meters are provided for this residence? _____

19. a. According to the *NEC*, is it permissible to ground a residential rural electrical service to driven ground rods only when a metallic water system is available?

    _____

    b. What sections of the *Code* apply? _____

20. What table in the *NEC* lists the sizes of grounding electrode conductors required for electrical services of various sizes? *Table* _____

21. *NEC 250.53(D)(2)* requires that a supplemental ground be provided if the available grounding electrode is: (Circle the correct answer.)

    a. metal underground water pipe.

    b. building steel.

    c. concrete encased ground.

22. Do the following conductors require mechanical protection?

    a. 8 AWG grounding electrode conductor _____

    b. 6 AWG grounding electrode conductor _____

    c. 4 AWG grounding electrode conductor _____

23. Why is bonding of service-entrance equipment necessary? _____
    _____
    _____

24. What special types of bushings are required on service entrances? _____
    _____

25. When insulated conductors 4 AWG or larger that are required to be insulated are installed in conduit, what additional provision is required on the conduit ends? _____
    _____

26. What minimum size copper bonding jumpers must be installed to bond properly the electrical service for the residence discussed in this text? _____
    _____

27. a. Where is Panel A located? _____
    b. On what type of wall is Panel A fastened? _____
    c. Where is Panel B located? _____
    d. On what type of wall is Panel B fastened? _____

28. When conduits pass through the wall from outside to inside, the conduit must be _____ to prevent air circulation through the conduit.

29. Briefly explain why electrical systems and equipment are grounded. _____
    _____
    _____
    _____
    _____

30. In general, systems are required to be grounded if the maximum voltage to ground does not exceed: (Circle the correct response.)
    a. 120 volts.
    b. 150 volts.
    c. 300 volts.

31. To ensure a complete grounding electrode system: (Circle the correct response.)
    a. everything must be bonded together.
    b. all metal pipes and conduits must be isolated from one another.
    c. the service neutral is grounded to the water pipe only.

32. An electric clothes dryer is rated at 5700 watts. The electric rate is 10.091 cents per kilowatt-hour. The dryer is used continuously for 3 hours. Find the cost of operation, assuming the heating element is on continuously. $_____

33. A heating cable rated at 750 watts is used continuously for 72 hours to prevent snow from freezing in the gutters of the house. The electric rate is 8.907 cents per kilowatt-hour. Find the cost of operation. $_____

34. When used as service equipment, a panelboard (load center) must be _____ that is suitable for use as service equipment, *230.66*. If the panelboard (load center) contains the main service disconnect, it must be clearly marked _____ _____, *230.70(B)*.

35. Here are five commonly used terms in the electrical industry. Enter the letter of the term that corresponds to its definition.

    a. Grounding electrode conductor

    b. Main bonding jumper

    c. Grounded circuit conductor

    d. Equipment grounding conductor

    e. Underground service conductors (Service Point at transformer)

    _____ The neutral conductor.

    _____ The term used to define underground service-entrance conductors that run between the meter and the utilities connection.

    _____ The conductor (sometimes a large threaded screw) that connects the neutral bar in the service equipment to the service-entrance enclosure.

    _____ The conductor that runs between the neutral bar in the main service equipment to the grounding electrode (water pipe, ground rod, etc.).

    _____ The bare copper conductor found in nonmetallic-sheathed cable.

36. The electrician mounted the disconnect switch for a central air-conditioning unit 8 ft (2.5 m) above the ground. This was easy to do because all he had to do was run the conduit across the ceiling joists in the basement, through the outside wall, and directly into the back of the disconnect switch. The electrical inspector turned the job down, citing *NEC* _____ that requires that the disconnecting operating handle must not be higher than _____ feet above the ground.

37. a. According to the *NEC* definition of a wet location, a basement cement wall that is in direct contact with the earth is considered to be a wet location. A panelboard mounted on this wall must have at least [¼ in. (6 mm)], [½ in. (13 mm)] [1 in. (25 mm)] space between the wall and the panel. Circle the correct answer.

    b. How is this required spacing accomplished? _____

    _____

38. When using rebars as the concrete-encased electrode, the rebars must be (a) insulated with plastic material so they will not rust, or (b) bonded together by the usual steel tie wires or other effective means. Underline the correct answer.

39. The circuit directory in a panelboard (may) (shall) be filled out according to *NEC* _____. Circle the correct answer, and enter the correct *NEC* section number.

40. A main service panel is located in a dark corner of a basement, far from the basement light. In your opinion, does this installation meet the requirements of *110.26(D)*? Explain your answer. _____

_____

41. Does the electric utility in your area allow the location of the meter to be on the back side of a residence? _____

42. The term used when the utility charges different rates during different periods during the day is _____.

43. A ground rod is driven below the meter outside of the house. A grounding electrode conductor connects between the meter base and the ground rod. The ground clamp is buried under the surface of the soil. This is permitted if the ground clamp is _____ for direct burial according to *NEC* _____.

44. Copper-coated steel ground rods are the most commonly used grounding electrodes. These rods shall:

    a. be at least _____ in. (_____ mm) in diameter, *250.52(A)(5)*.

    b. be driven to a depth of at least _____ ft (_____ m) unless solid rock is encountered, in which case the rod may be driven at a _____-° angle, or it may be laid in a trench that is at least _____ ft (_____ mm) deep, *250.53(G)*.

    c. be separated by at least _____ ft (_____ m) when more than one rod is driven, *250.53(A)(3)*.

    d. have a ground resistance of not over _____ ohms for one rod, *250.53(A)(2) Exception*.

45. What section of the *Code* prohibits the use of sheet metal screws as a means of attaching grounding conductors to enclosures? *NEC* _____

46. Which section of the *NEC* prohibits using the space below and in front of an electrical panel as storage space? *NEC* _____.

47. What section of the *NEC* prohibits using an underground metal gas pipe as the grounding electrode for an electrical service? _____

48. When wiring a gas furnace, what additional steps, if any, are necessary in order to make sure that the gas piping supplying the furnace is adequately bonded? Explain.

_____

_____

_____

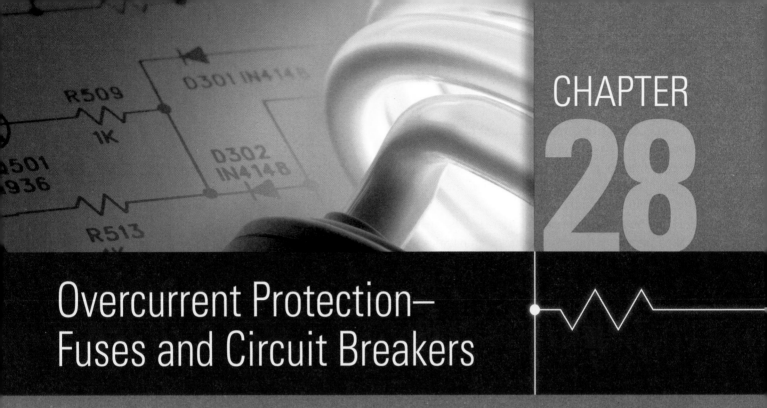

Overcurrent Protection–
Fuses and Circuit Breakers

OBJECTIVES

After studying this chapter, you should be able to

- understand the important *NEC* requirements for fuses and circuit breakers.

- discuss the five possible circuit conditions.

- understand the various types and operation of fuses and circuit breakers.

- know when to use single-pole and 2-pole circuit breakers.

- understand the term *interrupting rating* for fuses and circuit breakers.

- calculate available short-circuit current using a simple formula.

- understand *series-rated* panelboards.

- understand the meaning of *selective coordination* and *nonselective coordination*.

THE BASICS

The *NEC* covers overcurrent protection of conductors in *Article 240*. Overcurrent protection for residential services, branch circuits, and feeders is provided by circuit breakers or fuses. These are the "safety valves" of an electrical circuit.

Fuses and circuit breakers are sized by matching their ampere ratings to conductor ampacities and connected load currents. They sense overloads, short circuits, and ground faults, and protect the wiring and equipment from reaching dangerous temperatures.

A fuse will function only one time. It does its job of protecting the circuit. When a fuse opens, find out what caused it to open, fix the problem, and replace it with the size and type that will provide proper overcurrent protection. Do not keep replacing a fuse without finding out why the fuse blew in the first place.

When a circuit breaker trips, find out what caused it to trip, fix the problem, then reset it. Do not repeatedly reset the breaker again and again into a fault. This is looking for trouble.

KEY *NEC* REQUIREMENTS FOR OVERCURRENT PROTECTION

- *NEC Table 210.24:* This table shows the maximum overcurrent protection for branch-circuit conductors.
- *NEC 240.4:* Overcurrent protection is sized according to the ampacity of a conductor.
- *NEC 240.4(B):* Where the standard ampere rating of fuses or circuit breaker does not match the ampacity of the conductor, the next higher ampere rating is permitted. This is permitted for 800 amperes or less if the branch circuit does not supply cord-and-plug-connected loads.
- *NEC 240.4(D):* The maximum overcurrent protection of small conductors is

Conductor Size	Maximum Overcurrent Protection
14 AWG copper	15 amperes
12 AWG copper	20 amperes
10 AWG copper	30 amperes

- *NEC 240.6:* Lists the standard ampere ratings for fuses and circuit breakers.
- *NEC 240.20(A):* Overcurrent protection shall be provided for each ungrounded ("hot") conductor.
- *NEC 240.21:* Overcurrent protection shall be at the point where a conductor receives its supply. Refer to this section for exceptions.
- *NEC 240.22:* Overcurrent devices are generally not permitted in the grounded conductor. Exceptions are: (1) if the overcurrent device opens *all* conductors of the circuit at the same time, and (2) where the overcurrent devices are used for motor overload protection.
- *NEC 240.24(A):* Overcurrent protective devices shall be accessible.
- *NEC 240.24(D):* Overcurrent protective devices shall not be near easily ignitable material, such as in clothes closets.
- *NEC 240.24(E):* Overcurrent devices shall not be located in bathrooms.
- *NEC 240.24(F):* Overcurrent devices shall not be located over steps of a stairway.
- *NEC 230.70(A)(2):* Service disconnect(s) shall not be located in bathrooms.
- *NEC 230.79(C):* 100-ampere minimum-size service for a one-family dwelling.
- *NEC 230.90(A):* Overload protection shall be provided for each ungrounded "hot" service-entrance conductor. This is accomplished by the overcurrent protective device(s) in the service-entrance main panel.
- *NEC 230.91:* The main service overcurrent device is usually an integral part of the service disconnecting means. For large services, the service disconnecting means could be a separate disconnect switch.
- *NEC Table 310.15(B)(7):* This table shows special ampacities permitted for service-entrance conductors for three-wire, single-phase, 120/240-volt services for dwellings. Overcurrent protection for these conductors is not to exceed that specified in the table.

FIVE CIRCUIT CONDITIONS

For the following five conditions, each conductor is 5 feet (1.52 m) long and is solid copper. From *Table 8* in *Chapter 9* in the *NEC*, we find the resistance of a 14 AWG solid copper conductor to be 3.07 ohms per 1000 feet (304.8 m). This equates to 0.307 ohms per 100 feet (30.48 m), 0.0307 ohms per 10 feet (3.05 m), and 0.01535 ohms per 5 feet (1.52 m). To keep the calculation simple, we of rounded off the resistance values for each 5-foot (1.52-m) conductor length to 0.015 ohms. We are not trying to be rocket scientists. We are merely pointing out the fundamentals of different circuit conditions.

Normal: Normal loading of a circuit is when the current flowing is within the capability of the circuit and/or the connected equipment. In Figure 28-1, we have a 15-ampere circuit carrying approximately 10 amperes.

An *overload* is a condition where the current flowing is more than the circuit and/or connected equipment is designed to safely carry. Figure 28-2 shows a 15-ampere circuit carrying 20 amperes. A momentary overload is harmless, but a continuous overload condition will cause the conductors and/or equipment to overheat—a potential cause for fire. The current flows through the "intended path"—the conductors and/or equipment.

A *short circuit* is a condition when two or more normally insulated circuit conductors come in contact with one another, resulting in a current flow that bypasses the connected load. A short circuit might be two "hot" conductors coming together, or it might be a "hot" conductor and a "grounded" conductor coming together. In either case, the current flows outside of the "intended path." The only resistance (impedance) is that of the conductors, the source, and the arc. This low resistance results in high levels of short-circuit current. The heat generated at the point of the arc can result in a fire, as shown in Figure 28-3.

A *ground fault* is a condition when a "hot" or ungrounded conductor comes in contact with a grounded surface, such as a grounded metal raceway, metal water pipe, sheet metal, and so on, as shown in Figure 28-4. The current flows outside of the "intended path." A ground fault can result in a flow of current greater than the circuit rating, in which case the overcurrent device will open. A ground fault can also result in a current flow less than the circuit rating, in which case the overcurrent device will not open.

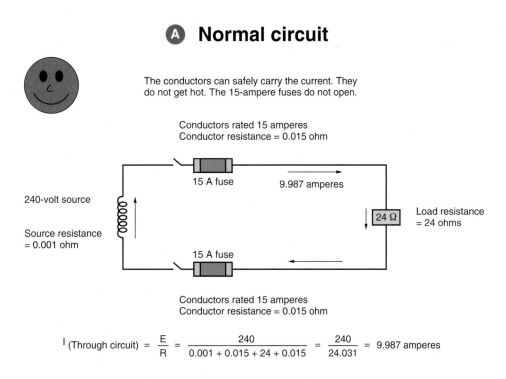

A Normal circuit

The conductors can safely carry the current. They do not get hot. The 15-ampere fuses do not open.

Conductors rated 15 amperes
Conductor resistance = 0.015 ohm

15 A fuse   9.987 amperes

240-volt source

Source resistance = 0.001 ohm

24 Ω   Load resistance = 24 ohms

15 A fuse

Conductors rated 15 amperes
Conductor resistance = 0.015 ohm

$$I \text{ (Through circuit)} = \frac{E}{R} = \frac{240}{0.001 + 0.015 + 24 + 0.015} = \frac{240}{24.031} = 9.987 \text{ amperes}$$

FIGURE 28-1   A normally loaded circuit. (*Delmar/Cengage Learning*)

B Overloaded circuit

The conductors begin to get hot, but the 15-ampere fuses will open before the conductors are damaged.

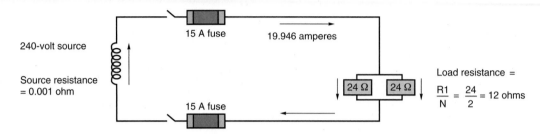

Conductors rated 15 amperes
Conductor resistance = 0.015 ohm

240-volt source

15 A fuse   19.946 amperes

Source resistance = 0.001 ohm

24 Ω   24 Ω

Load resistance =

$\dfrac{R1}{N} = \dfrac{24}{2} = 12$ ohms

15 A fuse

Conductors rated 15 amperes
Conductor resistance = 0.015 ohm

$$I \text{ (Through circuit)} = \frac{E}{R} = \frac{240}{0.001 + 0.015 + 12 + 0.001} = \frac{240}{12.031} = 19.946 \text{ amperes}$$

FIGURE 28-2 An overloaded circuit. (*Delmar/Cengage Learning*)

C Short circuit

The conductors get extremely hot. The insulation will melt off and the conductors will melt unless the fuses open in a very short (fast) period of time. Current-limiting overcurrent devices will limit the amount of "let-through" current by opening so fast (fraction of a cycle) that the full value of fault current will not be reached.

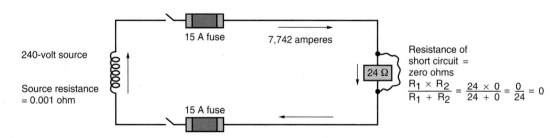

Conductors rated 15 amperes
Conductor resistance = 0.015 ohm

240-volt source

15 A fuse   7,742 amperes

Source resistance = 0.001 ohm

24 Ω

Resistance of short circuit = zero ohms

$\dfrac{R_1 \times R_2}{R_1 + R_2} = \dfrac{24 \times 0}{24 + 0} = \dfrac{0}{24} = 0$

15 A fuse

Conductors rated 15 amperes
Conductor resistance = 0.015 ohm

$$I \text{ (Through circuit)} = \frac{E}{R} = \frac{240}{0.001 + 0.015 + 0.015} = \frac{240}{0.031} = 7,742 \text{ amperes}$$

FIGURE 28-3 Note that the connected load is short-circuited. (*Delmar/Cengage Learning*)

Ⓓ Ground fault

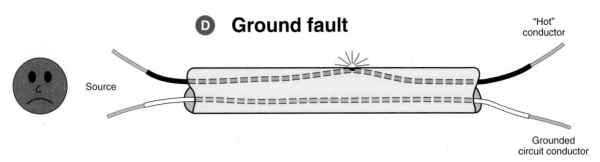

"Hot" conductor comes in contact with metal raceway or other metal object. If the return ground path has low resistance (impedance), the overcurrent device protecting the circuit will clear the fault. If the return ground path has high resistance (impedance), the overcurrent device will not clear the fault. The metal object will then have a voltage to ground the same as the "hot" conductor has to ground. In house wiring, this voltage to ground is 120 volts. Proper grounding and ground-fault circuit interrupter protection is discussed elsewhere in this text. The calculation procedure for a ground fault is the same as for a short circuit; however, the values of "R" can vary greatly because of the unknown impedance of the ground return path. Loose locknuts, bushings, set screws on connectors and couplings, poor terminations, rust, etc., all contribute to the resistance of the return ground path, making it extremely difficult to determine the actual ground-fault current values.

FIGURE 28-4 The insulation on the "hot" conductor has come in contact with the metal conduit. This is termed a "ground fault". (*Delmar/Cengage Learning*)

Ⓔ Open circuit

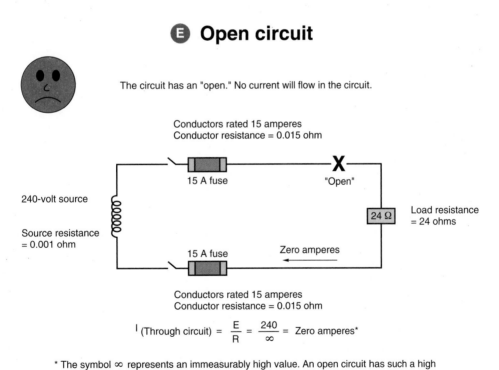

The circuit has an "open." No current will flow in the circuit.

$$I \text{ (Through circuit)} = \frac{E}{R} = \frac{240}{\infty} = \text{Zero amperes*}$$

* The symbol ∞ represents an immeasurably high value. An open circuit has such a high resistance value that ordinary ohmmeters cannot read it. We can ignore the resistance values of the other components of the circuit and use infinity for the "R" value.

Caution: Line voltage (in this case 240 volts) appears across the "open."

FIGURE 28-5 The circuit is "open" where marked "X." (*Delmar/Cengage Learning*)

A *GFCI* is an example of a device that protects against very low levels of ground-fault current passing through the human body when a person touches a "live" wire and a grounded surface. See Chapter 6.

An *open circuit* is a condition where the circuit is not closed somewhere in the circuit, as in Figure 28-5, and no current can flow.

FUSES

Fuses are a reliable and economical form of overcurrent protection. *NEC 240.50* through *240.54* provides the requirements for plug fuses, fuseholders, and adapters. Here is a brief recap:

Edison-Base Plug Fuses

- Plug fuses may be used (1) in circuits that do not exceed 125 volts between conductors, and (2) in circuits having a grounded neutral where no conductor operates at over 150 volts to ground. In a residential 120/240-volt electrical system, the voltage between the two "hot" conductors is 240 volts, and the voltage from either "hot" conductor to ground is 120 volts.

- The fuses shall have ampere ratings of 0 through 30 amperes.

- Plug fuses shall have a hexagonal configuration somewhere on the fuse when rated at 15 amperes or less.

- The screw shell of the fuseholder must be connected to the load side of the circuit.

- Edison-base plug fuses may be used only to replace fuses in existing installations where there is no sign of overfusing or tampering.

- Typical applications for plug fuses are using box-cover units where you need to provide close overcurrent protection, such as for an attic fan, a gas furnace, an appliance, and similar loads. This already has been discussed elsewhere in this text.

Type S Fuses

All new plug fuse installations shall be Type S fuses, *240.52*. That is because the ampere ratings of conventional Edison-base plug fuses are all interchangeable—any ampere rating up to and including 30 amperes.

- Type S fuses are classified at 0 through 15 amperes, 16 through 20 amperes, and 21 through 30 amperes. The reason for this classification is given in the following paragraph.

TABLE 28-1

This table shows the many ampere rating combinations for Type S fuses and adapters.

TYPE S FUSE INFORMATION

Type S Fuse Ampere Ratings	Type S Adapter Ratings	Type S Fuse Ampere Rating That Fit Into This Adapter
$3/10$, $1/2$, $8/10$, $4/10$, $6/10$, 1	1	1 ampere & smaller
$1^1/8$, $1^1/4$	$1^1/4$	All smaller
$1^4/10$, $1^6/10$	$1^6/10$	All smaller
$1^8/10$, 2	2	$1^8/10$, 2
$2^1/4$, $3^1/2$	$2^1/2$	$1^8/10$, 2, $2^1/4$, $2^1/2$
$2^8/10$, $3^8/10$	$3^2/10$	$1^8/10$, 2, $2^1/4$, $2^1/2$, $2^8/10$, $3^2/10$
$3^1/2$, 4	4	$3^1/2$, 4
$4^1/2$, 5	5	$3^1/2$, 4, $4^1/2$, 5
$5^6/10$, $6^1/4$	$6^1/4$	$3^1/2$, 4, $4^1/2$, 5, $5^6/10$, $6^1/4$
7, 8	8	7, 8
9, 10	10	7, 8, 9, 10
12, 14	14	7, 8, 9, 10, 12, 14
15	15	15
20	20	20
25	30	20, 25, 30
30	30	20, 25, 30

In the 0- to 15-ampere range, there are many ampere ratings to choose from, excellent for protecting motors. See Table 28-1. Type S fuses are used with a corresponding size adapter.

When the electrician installs fusible equipment, the ampere rating of the circuit(s) must be determined. After determining the ampere rating of the Type S fuse to be used, an adapter of the proper size is inserted into the Edison-base fuseholder. The proper Type S fuse is then screwed into the adapter. Because of the adapter, the fuseholder is nontamperable and noninterchangeable. For example, assume that a 15-ampere adapter is inserted for a 15-ampere branch circuits; it is impossible to substitute a Type S fuse with a higher ampere rating without removing the 15-ampere adapter.

A Type S fuse and adapter is shown in Figure 28-6.

FIGURE 28-6 Type S fuse and adapter.
(*Courtesy of Cooper Bussmann*)

Fuse Characteristics

Fuses have different time-current characteristics. The term *time-current* refers to how long it will take a fuse or circuit breaker to open under different current values.

- Nontime-delay: This type of fuse has one link (fusible element). One part of the link is "necked down" so when excessive current flows, it will open in the weakest part of the link—the necked down portion. Nontime-delay fuses are not the best choice for motor circuits because of the high starting inrush current of motors. See the "W" fuse, Figure 28-7.

FIGURE 28-7 Three types of plug fuses: single-element, nontime-delay (W); dual-element, time-delay (T); loaded link, time-delay (TL). (*Courtesy of Cooper Bussmann*)

- Time-delay, loaded link: This type of fuse has one link (fusible element) that is "loaded" with a heat sink next to the "necked down" portion of the link. This "load" acts as a heat sink that absorbs a considerable amount of heat before the "necked down" portion of the link melts open. This heat sink provides the fuse with time delay. See the "TL" fuse, Figure 28-7.

- Time-delay, dual-element: One fuse element opens quickly when a short circuit, heavy overload, or ground fault occurs. The other element in series with the first element opens slowly on overload conditions. Dual-element, time-delay fuses are an excellent choice for motor circuits because they will not open needlessly on momentary overloads. See the "T" fuse, Figure 28-7.

Cartridge Fuses

Cartridge fuses are covered in *NEC 240.60* and *240.61*.

Cartridge fuses are available with the same three basic types of time-current characteristics as plug fuses.

The most common type of dual-element cartridge fuse is shown in Figure 28-8. They are available in 250-volt and 600-volt sizes with ratings from 0 through 600 amperes.

Large ampere rating cartridge fuses may have more than one fuse element.

Time-delay, dual-element fuses provide a time-delay of about 10 seconds at a current of 500% of the fuse rating before it opens. This provides accurate protection for prolonged overloads.

Dual-element fuses are used on motor and appliance circuits where the long time-delay characteristic is required. Single-element fuses are more suitable for circuits that are not expected to have high inrush currents.

Cartridge fuses are available in Class H and Class R types. Refer to Table 28-2.

General-Purpose Cartridge Fuses

These are single-element cartridge fuses that have no intentional time-delay designed into them, as are dual-element, time-delay fuses. These are the type of fuses used in conventional disconnect switches.

FIGURE 28-8 Cartridge-type dual-element fuse (A) is a 250-volt, 100-ampere fuse. The cutaway view in (B) shows the internal parts of the fuse. When this type of fuse has a rejection slot in the blades of 70–600-ampere sizes, or a rejection ring in the end ferrules of the 0–60-ampere sizes, it is a UL Class RK1 or RK5 fuse. (*Courtesy of Cooper Bussmann*)

They are available in ampere ratings from ⅛ through 600 amperes, in both 250-volt and 600-volt ratings.

Those rated through 60 amperes are classified as K5 and have an interrupting rating of 50,000 amperes. Sizes 0 through 60 have a brass ferrule on each end. See Figure 28-9.

Those rated 65–600 are classified as class H and have an interrupting rating of 10,000 amperes. These sizes have a blade coming out of each end and are therefore referred to as "knife blade" fuses. See Figure 28-9.

Other Classes of Fuses

Class CC, Class G, and Class T fuses are special-purpose fuses that are found in original equipment, or in commercial and industrial panelboards, multimetering equipment, and motor controllers. They are used in original equipment because of their small physical size. If you need more information about these types of fuses, consult the manufacturer's technical literature and/or Web sites. These fuses are rarely used in residential applications but are mentioned here to make you aware of them.

TABLE 28-2

The interrupting ratings of the more common fuses and circuit breakers.

Type of Overcurrent Device	Interrupting Rating
Circuit breakers	Are marked with their interrupting rating if other than 5000 amperes.
Plug fuses (Edison base and Type S)	10,000 amperes ac RMS symmetrical.
	Plug fuses 15 amperes and less will have a hexagon shape in the window—or the fuse body of the fuse will have the hexagon shape.
Class H cartridge fuses	10,000 amperes ac RMS symmetrical.
	This is the most common type of low-cost cartridge fuse.
	Renewable link Class H cartridge fuses are permitted only on existing installations where there is no evidence of overfusing or tampering.
	Rarely, if ever, will these be found in residential electrical systems.
	The interrupting rating is marked when other than 10,000 amperes.
Class R cartridge fuses	The letter "R" stands for "Rejection." Knife blade types have a notch in the blade. Ferrule types have an annular ring in one of the ferrules. This means that if the switch or fuseholder is designed for Class R fuses, then ordinary Class H fuses will not fit.
	These are further broken down into RK1 and RK5 categories.
	RK fuses are available with interrupting ratings of 50,000, 100,000, and 200,000 amperes ac RMS symmetrical.
	The interrupting rating is marked on the fuse.

Note: Using a fuse or circuit breaker having a high interrupting rating does not necessarily constitute a safe installation. The entire assembly (panelboard, switch, controller) is marked with its short-circuit rating when used with a specific type of overcurrent device. That is why it is so important to "read the label."

FIGURE 28-9 A 60-ampere and a 100-ampere general purpose, one-time fuse. (*Courtesy of Cooper Bussmann*)

FIGURE 28-10 Typical molded-case circuit breakers. (*Courtesy of Schneider Electric*)

Sometimes poor connections at the terminals or loose fuse clips can be found because the fuse tubing will appear to be charred. This problem can be dangerous and should be corrected as soon as possible. Quite often, the heat resulting from poor connections in the fuse clips or fuseholder will cause the brass or copper fuse clips or fuseholders to turn from a bright, shiny appearance to that of antique brass or copper.

Fuse Ampere Ratings

The standard ampere ratings for fuses are given in *240.6(A)*. In addition to the standard ratings listed, there are many other "in-between" ampere ratings available to match specific load requirements. Check a fuse manufacturer's catalog for other available ampere ratings.

CIRCUIT BREAKERS

Installations in dwellings normally use thermal-magnetic circuit breakers. On a continuous overload, a bimetallic element in such a breaker moves until it unlatches the inner tripping mechanism of the breaker. Momentary small overloads do not cause the element to trip the breaker. If the overload is heavy or if there is a short circuit, a magnetic

coil in the breaker causes it to interrupt the branch-circuit almost instantly. Figure 28-10 illustrates a typical single-pole and a 2-pole circuit breaker.

NEC 240.80 through *240.86* give the requirements for circuit breakers. The following points are taken from these sections.

- Circuit breakers shall be trip-free so that even if the handle is held in the "ON" position, the internal mechanism will trip to the "OFF" position.

- Breakers shall indicate clearly whether they are on or off.

- A breaker shall be nontamperable so that it cannot be readjusted (trip point changed) without dismantling the breaker or breaking the seal.

- The rating shall be durably marked on the breaker. For small breakers rated at 100 amperes or less and 600 volts or less, the rating must be molded, stamped, or etched on the handle or on another part of the breaker that will be visible after the cover of the panel is installed.

- Every breaker with an interrupting rating other than 5000 amperes shall have this rating marked on the breaker. Circuit breakers used in dwellings typically are rated to interrupt 10,000 amperes during a short circuit or a ground-fault condition.

- Circuit breakers rated at 120 volts and 277 volts and used for fluorescent loads shall not be used as switches unless marked "SWD."

Thermal-magnetic circuit breakers are temperature sensitive. Some circuit breakers are ambient (surrounding temperature) compensated. This means that the tripping characteristic of the breaker is partially or completely neutralized by the effect of ambient temperature. Underwriters Laboratories (UL) Standard 489 specifies calibration testing at various loads and ambient temperatures. The continuous test is run at 104°F (40°C). One manufacturer states that no rerating is necessary when the circuit breaker is installed in an ambient temperature in the range of 14°F to 140°F (–10°C to 60°C). If you are installing circuit breakers in extremely hot or extremely cold temperatures, consult the manufacturer's literature.

It is a good practice to turn the breaker on and off periodically to "exercise" its moving parts.

NEC 240.6(A) gives the standard ampere ratings of circuit breakers.

How Much Load Can a Circuit Breaker Safely Carry?

The *NEC* permits 100% loading on an overcurrent device *only* if that overcurrent device is listed for 100% loading. See *NEC 210.20* and *215.3*. At the time of this writing, UL has no listing of residential-type molded-case circuit breakers that are suitable for 100% loading. A *maximum* loading of 80% is good practice—less is even better!

One 2-Pole or Two Single-Pole Circuit Breakers

Use 2-pole circuit breakers on 3-wire 120/240-volt branch circuits (two "hots" and one "neutral") for loads such as electric ranges, electric ovens, and electric clothes dryers.

Use 2-pole circuit breakers for straight 240-volt branch circuits (two "hots") for such loads as electric water heaters, electric furnaces, electric baseboard heaters, air conditioners, and heat pumps.

The use of "handle ties," which connect the handles of two single-pole circuit breakers together so both poles trip together, is permitted for a few very specific applications.

Handle ties *do* allow two single-pole circuit breakers to be switched simultaneously to the "Off" or "On" positions.

Handle ties generally *do not* simultaneously trip both handle-tied single-pole circuit breakers on overloads, short circuits, or ground faults on one of the circuit breakers. This could result in a false sense of security for someone working on an appliance where two single-pole circuit breakers with a handle tie has tripped. This person may think that the branch circuit is totally in the "Off" position, when in fact it is quite possible that one of the circuit breakers may still be "On."

Identified Handle Ties and the *Code*

First appearing in the 2005 *NEC* is that handle ties for circuit breakers must be *identified*. *Identified* in the *NEC* means *Recognizable as suitable for the specific purpose, function, use, environment, application, and so forth, where described in a particular Code requirement.** In addition, they may have been tested and listed by a nationally recognized testing laboratory (NRTL) for use on a specific manufacturer's circuit breaker.

- *210.4(B):* Where multiwire branch circuits are used, a means must be provided to simultaneously disconnect all ungrounded conductors. Identified handle ties on a circuit breaker meet this requirement. A disconnect switch also meets this requirement. We discussed the pros and cons of multiwire branch circuits earlier in this text.

- *210.4(C):* A multiwire branch circuit is generally only permitted to serve line-to-neutral loads. There are two exceptions to this rule:

  1. where the branch-circuit supplies only one piece of equipment such as an electric range, in which case the branch circuit is actually supplying both line-to-neutral and line-to-line loads, or

  2. where the overcurrent device simultaneously disconnects all ungrounded conductors of the multiwire branch-circuit.

- *225.33(B):* Identified handle ties are permitted to disconnect all conductors of an outside multiwire branch circuit or feeder with no more than six operations of the hand.

- *230.71(B):* Identified handle ties are permitted for circuit breakers to disconnect all conductors

---

*Reprinted with permission from NFPA 70-2011.

of a service as long as it takes no more than six operations of the hand to shut everything off.

- *240.15:* Requires that a circuit breaker open all ungrounded conductors of a circuit.

- *240.16(A):* Individual single-pole circuit breakers, with or without handle ties, are permitted to protect the ungrounded conductors of a multiwire branch circuit only when serving single-phase, line-to-neutral loads. See *210.4(B).*

- *240.16(B):* Individual single-pole circuit breakers with identified handle ties permitted to protect the ungrounded conductors for line-to-line loads for single-phase circuits.

With all of the above restrictions on the use of handle ties, they should only be used as "the last resort." If you have to use handle ties, use only those that are *identified.*

INTERRUPTING RATINGS FOR FUSES AND CIRCUIT BREAKERS

Here are some very important *NEC* sections regarding interrupting ratings.

- *NEC 110.9:* **Interrupting Rating.** *Equipment intended to interrupt current at fault levels shall have an interrupting rating not less than the nominal circuit voltage and the current that is available at the line terminals of the equipment. Equipment intended to interrupt current at other than fault levels shall have an interrupting rating at nominal circuit voltage not less than the current that must be interrupted.**

- *NEC 110.10:* **Circuit Impedance, Short-Circuit Current Ratings, and Other Characteristics.** *The overcurrent protective devices, the total impedance, the component short-circuit current ratings, and other characteristics of the circuit to be protected shall be selected and coordinated to permit the circuit protective devices used to clear a fault to do so without extensive damage to the electrical components of the circuit. This fault shall be assumed to be either*

between two or more of the circuit conductors, or between any circuit conductor and the grounding conductor or enclosing metal raceway. Listed products applied in accordance with their listing shall be considered to meet the requirements of this section.*

- UL Standard No. 67 requires that panelboards be marked with their short-circuit current rating.

The overcurrent protective device must be able to interrupt the current that may flow under any condition (overload or short circuit). Such interruption must be made with complete safety to personnel and without damage to the panel or switch in which the overcurrent device is installed.

CAUTION: Overcurrent devices with inadequate interrupting ratings are, in effect, bombs waiting for a short circuit to trigger them into an explosion. Personal injury may result and serious damage will be done to the electrical equipment.

Series Rated versus Fully Rated Panelboards

Panelboards are available in two types: fully rated and series rated. Figures 28-11, 28-12, 28-13, and 28-14 explain the meaning of these terms.

When you hear and see the term *series rated,* the term means exactly what it says. The main overcurrent device and the branch-circuit overcurrent devices are connected in series.

The unique characteristic of a series-rated combination is that the upstream circuit breaker has a high interrupting rating and the downstream circuit breakers have a low interrupting rating. Series-rated systems are less costly than fully rated systems. It is safe to say that high interrupting rated circuit breakers cost more than low interrupting rated circuit breakers. What is compromised is selectivity, discussed a little later.

The upstream high interrupting rated overcurrent device is permitted to be in the same enclosure as the lower interrupting rated devices—or is permitted to be remote, such as at a main switchboard panelboard—and the lower interrupting rated circuit breakers are located in a subpanel.

*Reprinted with permission from NFPA 70-2011.

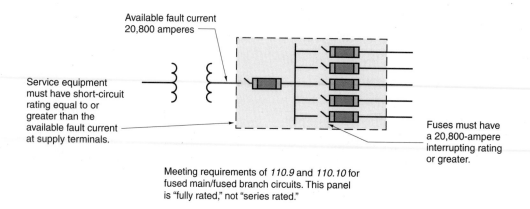

Available fault current
20,800 amperes

Service equipment
must have short-circuit
rating equal to or
greater than the
available fault current
at supply terminals.

Fuses must have
a 20,800-ampere
interrupting rating
or greater.

Meeting requirements of *110.9* and *110.10* for
fused main/fused branch circuits. This panel
is "fully rated," not "series rated."

FIGURE 28-11 A *fully rated* system using main fuses and branch-circuit fuses. The entire assembly (fuses and panelboard) is tested, listed, and marked with a maximum short-circuit rating. Look for the marking on the panelboard. This type of combination can be found in commercial and industrial facilities. (*Delmar/Cengage Learning*)

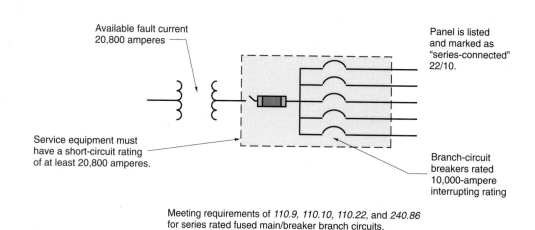

Available fault current
20,800 amperes

Panel is listed
and marked as
"series-connected"
22/10.

Service equipment must
have a short-circuit rating
of at least 20,800 amperes.

Branch-circuit
breakers rated
10,000-ampere
interrupting rating

Meeting requirements of *110.9*, *110.10*, *110.22*, and *240.86*
for series rated fused main/breaker branch circuits.
This panel is "series-rated," not "fully rated."

FIGURE 28-12 A *series-rated* system using main fuses and branch-circuit breakers. The entire assembly (main fuses, branch-circuit breakers, and panelboard) is tested, listed, and marked with a maximum short-circuit rating. Look for the marking on the panelboard. The lower interrupting rating of the branch-circuit breakers is acceptable because the combination has been tested and listed as a series-rated system. This type of combination can be found in light commercial installations such as multimetering service equipment. (*Delmar/Cengage Learning*)

Interrupting Ratings

Interrupting ratings of overcurrent devices, as shown in Table 28-2, are given in root-mean-square (RMS) values.

The interrupting rating of a fuse or circuit breaker does not indicate the short-circuit rating of a panelboard. These are stand-alone ratings. The short-circuit rating of a panelboard is derived by testing the panel as an assembly with breakers and/or fuses installed in the panel. The short-circuit rating of a panelboard is marked on its label.

SHORT-CIRCUIT CURRENTS

This text does not cover in detail the methods of calculating short-circuit currents. For a detailed discussion, see *Electrical Wiring—Commercial*

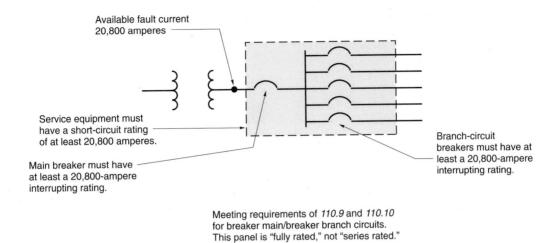

FIGURE 28-13   A *fully rated* system using a main circuit breaker circuit and branch-circuit breakers. The entire assembly (main breaker, branch-circuit breakers, and panelboard) is tested, listed, and marked with a maximum short-circuit rating. Look for the marking on the panelboard. This type of combination can be found in commercial and industrial installations. (*Delmar/Cengage Learning*)

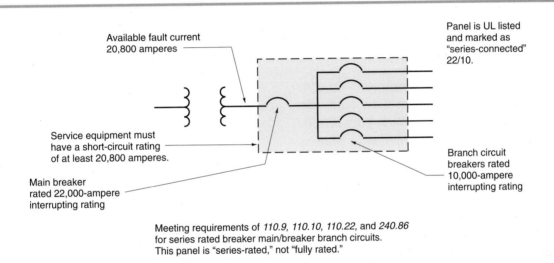

FIGURE 28-14   A *series-rated* system using a main circuit breaker and branch-circuit breakers. The entire assembly (main breaker, branch-circuit breakers, and panelboard) is tested, listed, and marked with a maximum short-circuit rating. Look for the marking on the panelboard. The lower interrupting rating of the branch-circuit breakers is acceptable because the assembly has been listed as a series–rated system. This type of combination is most commonly found in typical residential and light commercial applications. (*Delmar/Cengage Learning*)

(© Delmar Cengage Learning). The ratings required to determine the maximum available short-circuit current delivered by a transformer are the kVA and impedance values of the transformer. The size and length of wire installed between the transformer and the overcurrent device must be considered as well.

The transformers used in modern electrical installations are efficient and have very low imped-ance values. A low-impedance transformer having a given kVA rating delivers more short-circuit cur-rent than a transformer with the same kVA rating and a higher impedance. When an electrical service is connected to a low-impedance transformer, the problem of available short-circuit current is very serious.

The examples given in Figures 28-11, 28-12, 28-13, and 28-14 show that the available short-circuit current at the transformer secondary terminals is 20,800 amperes.

HOW TO CALCULATE SHORT-CIRCUIT CURRENT

Short-circuit current is also referred to as fault current.

Calculating fault currents is just as important as calculating load currents. Overloads will cause conductors and equipment to run hot, will shorten their life, and will eventually destroy them. Fault currents that exceed the interrupting rating of the overcurrent protective devices and panel rating can cause violent electrical explosions of equipment, with the potential of serious injury to people standing near it. Fire hazard is also present.

The local electric utility and the electrical inspector are good sources of information when short-circuit current is to be determined.

A simplified method is given as follows to determine the approximate available short-circuit current at the terminals of a transformer. This method assumes that there is an infinite amount of primary short-circuit current available (infinite bus), and a bolted fault at the secondary transformer terminals.

1. Determine the normal full-load secondary current delivered by the transformer.

   For single-phase transformers:

   $$I = \frac{kVA \times 1000}{E}$$

   For 3-phase transformers:

   $$I = \frac{kVA \times 1000}{E \times 1.73}$$

   where

   I = current, in amperes
   kVA = kilovolt-amperes (transformer name-plate rating)
   E = secondary line-to-line voltage (transformer nameplate rating)

2. Using the impedance value given on the transformer nameplate, find the multiplier to determine the short-circuit current.

   $$multiplier = \frac{100}{percent\ impedance}$$

3. The short-circuit current = normal full-load secondary current × multiplier.

 EXAMPLE

A transformer is rated at 100 kVA and 120/240 volts. It is a single-phase transformer with an impedance of 1% (from the transformer nameplate). Find the short-circuit current.

For a single-phase transformer:

$$I = \frac{kVA \times 1000}{E}$$

$$I = \frac{100 \times 1000}{240}$$

$$= 416\ amperes,\ full\text{-}load\ current$$

The multiplier for a transformer impedance of 1% is

$$multiplier = \frac{100}{percent\ impedance}$$

$$= \frac{100}{1}$$

$$multiplier = 100$$

The short-circuit current = I × multiplier

$$= 416\ amperes \times 100$$

$$= 41,600\ amperes$$

Thus, the available short-circuit current at the terminals of the transformer is 41,600 amperes.

This value decreases as the distance from the transformer increases.

If the transformer impedance is 1.5%, the multiplier is

$$multiplier = \frac{100}{1.5} = 66.6$$

The short-circuit current $= 416 \times 66.6$
$$= 27,706\ amperes$$

If the transformer impedance is 2%,

$$multiplier = \frac{100}{2} = 50$$

The short-circuit current $= 416 \times 50$
$$= 20,800\ amperes$$

(**Note:** A short-circuit current of 20,800 amperes is used in Figures 28-11, 28-12, 28-13, and 28-14.)

CAUTION: The line-to-neutral short-circuit current at the secondary of a single-phase transformer is approximately 1½ times greater than the line-to-line short-circuit current.

For the previous example,

Line-to-line short-circuit current = 20,800 amperes

Line-to-neutral short-circuit current
= 20,800 × 1.5 = 31,200 amperes

It is significant to note that UL Standard 1561 allows the marked impedance on the nameplate of a transformer to vary plus or minus (±) 10% from the actual impedance value of the transformer. Thus, for a transformer marked "2%Z," the actual impedance could be as low as 1.8%Z to as much as 2.2%Z. Therefore, the available short-circuit current at the secondary of a transformer as calculated previously should be increased by 10% if one wishes to be on the safe side when considering the interrupting rating of the fuses and breakers to be installed.

An excellent point-to-point method for calculating fault currents at various distances from the secondary of a transformer that uses different size conductors is covered in detail in *Electrical Wiring—Commercial* (© Delmar Cengage Learning). It is simple and accurate. The point-to-point method can be used to calculate both single-phase and 3-phase faults currents.

For an easy and fun computer program for making short-circuit calculations, visit the Bussmann Web site at http://www.bussmann.com. Look for "Electrical UL/CSA," then "Solution Center," then "Software," then "Short-Circuit Current Calculator."

PANELBOARDS . . . WHAT ARE THEY?

The terms *panel*, *panelboard*, and *load center* are all used, and for all practical purposes mean the same thing. The *NEC* and the UL Standards use the term *panelboard*. Neither the *NEC* nor the UL Standards contain the term *load center*. A load center is really a panelboard with less gutter (wiring) space and less depth and width, and generally does not have features such as integral relays, remote-controlled circuit breakers, load-shedding devices, and other

similar optional features. However, some load centers are available for home automation applications that do have certain remote control features. Circuit breakers used in load centers are usually of the "plug-on" type as opposed to the "bolt-on" type found in commercial and industrial panelboards. Load centers generally are for use in a shallow wall (2 × 4 studs) such as found in residential and light commercial construction. They are available with a main circuit breaker or main fuses, or with main lugs only (MLO).

Panelboards are tested and listed under UL Standard 67.

The most popular residential panelboard installed nowadays are *series-rated*, as shown in Figure 28-14. The most common rating is the 22/10 panelboard. The main circuit breaker has an interrupting rating of 22,000 amperes. The branch-circuit breakers have an interrupting rating of 10,000 amperes. The series rating of 22/10 is adequate for most typical residential main panelboards. If you are installing a large electrical service where a large kVA-rated transformer is involved, don't take a chance! You should have real concern for services over 200 amperes. Obtain the available fault current on the line side of the service equipment from the local electric utility; then select panels, fuses, and circuit breakers suitable for the available fault current.

Table 28-3 compares the requirements for different types of panelboards.

Selective Coordination/ Nonselective Coordination

These are terms used by the real pros!

Selective Coordination: Under overload, short-circuit, or ground-fault conditions, only the overcurrent device nearest the fault opens. The main fuses or circuit breaker remain closed. Refer to *240.2*. See the definition in *NEC Article 100*.

Nonselective Coordination: Under overload, short-circuit, or ground-fault conditions, the branch breaker or fuses might open, the main breaker or fuses might open, or both might open. For example, a short circuit in a branch circuit could also cause the main breaker to trip, resulting in a total power outage.

TABLE 28-3

Comparison of panelboard types.

	Fused Main/ Fused Branches (Fully-Rated) Figure 28-11	Fused Main/ Fused Branches (Series-Rated) Figure 28-12	Breaker Main/ Breaker Branches (Fully Rated) Figure 28-13	Breaker Main/ Breaker Branches (Series Rated) Figure 28-14
Must be listed and marked as suitable for use as service equipment when used as service equipment. Does not need this marking when used as a subpanel.	Yes	Yes	Yes	Yes
Must have short-circuit current rating adequate for the available fault current at the line side of the panelboard.	Yes	Yes	Yes	Yes
Must be marked with its short-circuit current rating.*	Yes	Yes	Yes	Yes
Must be marked "Series-Rated."	No	Yes Marking: Caution—Series Combination System Rated _____ Amperes. Identified Replacement Components Required.	No	Yes Marking: Caution—Series Combination System Rated _____ Amperes. Identified Replacement Components Required.
Main fuses or circuit breaker must have interrupting rating adequate for the available fault current on the line side of the panelboard.	Yes	Yes	Yes	Yes
Branch fuses or breakers must have interrupting rating adequate for the available fault current on the line side of the panelboard.	Yes	No Main fuses might have 200,000-ampere interrupting rating. Branch breakers might have 10,000-ampere interrupting rating. Other combinations available. Check manufacturers' data.	Yes	No Main breaker might have 22,000-ampere interrupting rating. Branch breakers might have 10,000-ampere interrupting rating. Other combinations available. Check manufacturers' data.
Main fuses are current limiting. See *240.2*.	Yes	Yes	Not applicable	Not applicable
Must be sized as determined by service-entrance calculations and/or local codes.	Yes	Yes	Yes	Yes
Check manufacturer's data for time-current characteristic curves for fuses and breakers, and for "unlatching times" of breakers.	Always a good idea.	Always a good idea.	Always a good idea.	Always a good idea.

*If a panelboard has no marked short-circuit rating, then its short-circuit rating is based on the lowest interrupting-rated device installed.

Series-rated systems are susceptible to nonse-lectivity when available fault currents are high.

The theory behind selective and nonselective systems is covered in detail in *Electrical Wiring—Commercial* (© Delmar Cengage Learning). Time-current curves for fuses and circuit breakers, circuit-breaker "unlatching time" data, and fuse melting/total clearing data are discussed.

REVIEW

1. What is the minimum size service for a one-family dwelling? _____

2. a. What is a Type S fuse? _____

   _____

   b. Where must Type S fuses be installed? _____

   _____

   _____

3. a. What is the maximum voltage permitted between conductors when using plug fuses?

   _____

   _____

   b. May plug fuses (Type S) be installed in a switch that disconnects a 120/240-volt clothes dryer? _____

   _____

   c. Give a reason for the answer to (b). _____

   _____

   _____

4. Will a 20-ampere, Type S fuse fit properly into a 15-ampere adapter? _____

5. A time-delay, dual-element fuse will hold 500% of its rating for approximately (5 seconds) (10 seconds) (20 seconds). Circle the correct answer.

6. What part of a circuit breaker causes the breaker to trip

   a. on an overload? _____ b. on a short circuit? _____

7. What is meant by an ambient-compensated circuit breaker? _____

   _____

   _____

   _____

8. List the standard sizes of fuses and circuit breakers up to and including 100 amperes.

   _____

   _____

9. When hooking up a 240-volt electric baseboard heater, you should use (circle the correct answer)

   a. a 2-pole circuit breaker.

   b. two single-pole breakers and a handle tie.

10. Using the method shown in this unit, what is the approximate short-circuit current available at the terminals of a 50-kVA single-phase transformer rated 120/240 volts? The transformer impedance is 1%.

    a. Line-to-line

    b. Line-to-neutral

11. State four possible combinations of service equipment that meet the requirements of *110.9* and *110.10* of the *Code*.

   a. _____   c. _____

   b. _____   d. _____

12. Which *Code* section states that all overcurrent devices must have adequate interrupting ratings for the available fault current to be interrupted? _____

   _____

13. All electrical components have some sort of "withstand rating." The *NEC* and UL Standards refer to this as the component's short-circuit current rating. This rating indicates the ability of the component to withstand a specified amount of fault current for the time it takes the overcurrent device upstream of the component to open the circuit. *NEC* _____ refers to short-circuit current rating.

14. Arc-fault damage is closely related to the value of _____.

15. The utility company has provided a letter to the contractor stating that the available fault current at the line side of the main service-entrance equipment in a residence is 17,000 amperes RMS symmetrical, line to line. In the space following each statement, write in "Meets *Code*" or "Violation" of *110.9* and *110.10* of the *Code*.

   a. Main breaker has a 10,000-ampere interrupting rating; branch breakers have a 10,000-ampere interrupting rating. _____

   b. Main current-limiting fuse has a 200,000-ampere interrupting rating; branch breakers have a 10,000-ampere interrupting rating. The panel is marked "Series-Rated."_____

   c. Main breaker has a 22,000-ampere interrupting rating; branch breakers have a 10,000-ampere interrupting rating. The panel is marked "Series-Connected."

   _____

16. While working on the main panel with the panel energized, the electrician inadvertently causes a direct short circuit (line to ground) on one of the branch-circuit breakers. The available fault current at the main service equipment is rather high. The panel is labeled "Series-Connected." The main breaker is rated 100 amperes. Both branch-circuit breaker and main breaker trip "Off." This condition is referred to as being (selective) (nonselective). Circle the correct answer.

17. Repeat Question 13 for a main service panel that consists of 100-ampere main current-limiting fuses and breakers for the branch circuits. The branch circuit trips; the 100-ampere fuses do not open. This condition is referred to as being (selective) (nonselective). Circle the correct answer.

18. *NEC* _____ prohibits the connecting of fuses and/or circuit breakers into the grounded circuit conductor.

19. In general, overcurrent protective devices are inserted where the conductor receives its _____.

20. a. Does the *NEC* permit panelboards to be installed in clothes closets? _____

   b. Does the *NEC* permit panelboards to be installed in bathrooms? _____

21. Overcurrent devices must be accessible. (True) (False). Circle the correct response.

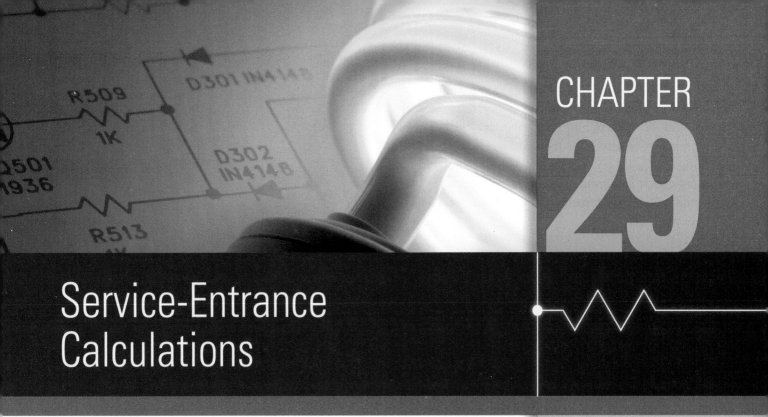

CHAPTER

29

Service-Entrance Calculations

OBJECTIVES

After studying this chapter, you should be able to

- calculate the size of the service entrance, including the size of the neutral conductors, using conventional and optional methods.

- understand why the neutral conductor is permitted to be smaller than the ungrounded conductors.

- understand special *Code* rules that permit smaller size service-entrance conductors for single-family dwellings.

- realize that derating service-entrance conductors may be necessary if the installation is located in a hot climate.

- understand how to read a watt-hour meter.

The various load values determined in earlier chapters of this text are now used to illustrate the proper method of determining the size of the service-entrance conductors for the residence. The calculations are based on *NEC* requirements. The student must check local and state electrical codes for any variations in requirements that may take precedence over the *NEC*.

SIZING OF SERVICE-ENTRANCE CONDUCTORS AND SERVICE DISCONNECTING MEANS

NEC 230.42 tells us that service-entrance conductors must be sufficient to carry the load as calculated according to *Article 220* and shall have an ampacity not less than the rating of the service disconnect.

For a one-family residence, the minimum-size main service disconnecting means is 100 amperes, 3 wire, *230.79(C)*.

There are two methods for calculating service size for homes:

1. The conventional method is found in *Article 220, Part III*.

2. The optional method is found in *Article 220, Part IV*.

Method 1 (*Article 220, Part III*)

All of the volt-ampere values in the following calculations are taken from previous chapters of this text.

Single-Family Dwelling Service-Entrance Calculations

1. General Lighting Load, *220.12*

   3232 ft² @ 3 VA per ft² = 9696 VA

Note: Included in this floor area calculation are all lighting outlets and general-use receptacles. Do not include open porches, garages, or unused or unfinished spaces not adaptable for future use. See *NEC 220.12, Table 220.12,* and *220.14(J)*.

2. Minimum Number of 15-Ampere Lighting Branch Circuits

   9696 VA ÷ 120 volts = 80.8 amperes

   then, $\dfrac{\text{amperes}}{15} = \dfrac{80.8}{15} = 5.387$

   (round up to 6) 15-ampere branch circuits

3. Small-Appliance Load, *210.11(C)(1)*, *210.52(B)*, and *220.52(A)*

   (minimum of two 20-ampere branch circuits)

   Kitchen 3 small-appliance branch circuits installed

   3 branch circuits @ 1500 VA each = 4500 VA

4. Laundry Branch Circuit, *210.11(C)(2)*, *210.52(F)*, and *220.52(B)*

   (Minimum of one 20-ampere branch circuit)

   1 branch circuit(s) @ 1500 VA = 1500 VA
   1* Clothes washer @ 1500 VA = 1500 VA
                     Total = 3000 VA

5. Total General Lighting, Small Appliance, and Laundry Load

   Lines 1 + 3 + 4   = 17,196 VA

6. Net Calculated General Lighting, Small Appliance, and Laundry Loads (less ranges, ovens, and "fastened-in-place" appliances). Apply demand factors from *Table 220.42*.

   a. First 3000 VA @ 100%   = 3000 VA
   b. Line 5   17,196 − 3000
             = 14,196 @ 35% = 4969 VA
             Total a + b = 7969 VA

7. Electric Range, Wall-Mounted Ovens, Counter-Mounted Cooking Units (*Table 220.55*)

   Wall-mounted oven   7450 VA
   Counter-mounted
     cooking unit   6600 VA
           Total   14,050 VA (14 kW)

   14 kW exceeds 12 kW by 2 kW
   2 kW × 5% = 10% increase; therefore:
   8 kW + 0.8 kW = 8.8 kW   = 8,800 VA

8. Electric Dryer, *220.54* = 5700 VA

9. Electric Furnace, *220.51*, Air Conditioner, Heat Pump, *Article 440*

   Air conditioner: 26.4 × 240   = 6336 VA
   Electric furnace: 13,000 VA
   (Enter largest value, *220.60*)   = 13,000 VA

*This is in addition to the required minimum of one laundry branch circuit. Thus, the laundry is served by a 20-ampere branch circuit B18 for the automatic washer and another 20-ampere branch circuit B20 supplying two receptacles for plugging in an electric iron, steamer, or clothes press.

10. Net Calculated General Lighting, Small Appliance, Laundry, Ranges, Ovens, Cooktop Units, HVAC

    Lines 6 + 7 + 8 + 9   = 35,469 VA

11. List "Fastened-in-Place" Appliances *in Addition* to Electric Ranges, Air Conditioners, Clothes Dryers, Space Heaters

Appliance		VA Load
Water Heater	=	4500 VA
Dishwasher 9.2 × 120	=	1104 VA
Garage Door Opener 5.8 × 120	=	696 VA
Food Waste Disposer 7.2 × 120	=	864 VA
Water Pump 8 × 240	=	1920 VA
Hydromassage Tub 10 × 120	=	1200 VA
Attic Exhaust Fan 5.8 × 120	=	696 VA
Heat/Vent/Lights 1500 × 2	=	3000 VA
Freezer 5.8 × 120	=	696 VA
	Total	14,676 VA

12. Apply 75% Demand Factor, *220.53*, if Four or More "Fastened-in-Place" Appliances. If Less than Four, Figure at 100%. Do not include electric ranges, clothes dryers, electric space heaters, or air-conditioning equipment.

    Line 11   14,676 × 0.75   = 11,007 VA

13. Total Calculated Load (Lighting, Small Appliance, Ranges, Dryers, HVAC, "Fastened-in-Place" Appliances)

    Line 10   35,469
    + Line 12   11,007   = 46,476 VA

14. Add 25% of Largest Motor (*220.50* and *430.24*)

    This is the water pump motor:

    8 × 240 × 0.25 =   480 VA

    Note: The largest motor can be difficult to determine because nothing is in place when service-entrance load calculations are made. It might be an air-conditioning unit or a heat pump. If the dwelling is cooled by an evaporative cooler, the largest motor might be a water pump motor, a large attic exhaust fan, a large food waste disposer, or a sump pump. For simplicity in this example, the water pump motor was chosen for this example calculation.

    Had we used the air-conditioning unit as the largest motor, we would use the Branch-Circuit Selection Current value of 19.9 amperes. See

Chapter 23 for complete explanation of how air-conditioning current ratings are derived. This would result in 19.9 × 240 = 4,776 VA. Next: 4,776 × 0.25 = 1,194 VA. Next: 46,401 + 1,194 = 47,595 VA. Next: 47,595 ÷ 240 = 198.3125 amperes. The service-entrance size would be 200 amperes.

15. Total of Line 13 + Line 14 = 46,956 VA

16. Minimum Ampacity for Ungrounded Service-Entrance Conductors

$$\text{Amperes} = \frac{\text{Line 15}}{240} = \frac{46,956}{240}$$
$$= 195.6 \text{ amperes}$$

17. Ungrounded Conductor Size (copper)
    2/0 AWG THWN

    Note: This could be a 3/0 AWG THW, THHW, THWN, XHHW, or THHN per *Table 310.15(B)(16)*, or a AWG 2/0 (same types) using *Table 310.15(B)(7)*. *Table 310.15(B)(7)* may only be used for 120/240-volt, 3-wire, residential single-phase service-entrance conductors, underground service conductors, and feeder conductors that serve as the main power feeder to a dwelling unit. This table *cannot* be applied to the feeder to Panel B because that feeder carries only part of the load in the residence. We have assumed that the terminals in the equipment are marked 75°C. We read the ampacity from the 75°C column in *Table 310.15(B)(16)* as stated in *110.14(C)*.

18. Minimum Ampacity for Neutral Service-Entrance Conductor, *220.61*, and *310.15(B)(7)*. Do not include loads connected only phase-to-phase such as the water heater (240-volt loads).
    a. Line 6   = 7969 VA
    b. Line 7, Oven and
       Cooktop 8800 × 0.70   = 6160 VA
    c. Line 8 Dryer 5700 × 0.70 = 3990 VA
    d. Line 11 (include only 120-volt loads)

Freezer	= 696 VA
Food Waste Disposer	= 864 VA
Garage Door Opener	= 696 VA
Heat/Vent/Light	= 3000 VA
Attic Fan	= 696 VA
Hydromassage tub	= 1200 VA
Dishwasher	= 1104 VA
Total	8256 VA

e. Line d Total @ 75% Demand Factor
per 220.53: 8256 × 0.75 = 6192 VA

f. Add 25% of Largest 120-volt Motor.
This is the hydromassage pump motor:
10 × 120 × 0.25 = 300 VA

g. Total a + b + c + e + f = 25,492 VA

$$Amperes = \frac{volt-amperes}{volts} = \frac{24,611}{240}$$
$$= 102.5\ amperes$$

19. Neutral Conductor Size (copper), *220.61*, 1 AWG

Note: The calculations indicate that a 2 AWG copper neutral conductor would be adequate based on the unbalanced load calculations. *NEC 310.15(B)(7)* states that the neutral conductor is permitted to be smaller than the ungrounded "hot" conductors if the requirements of *215.2, 220.61,* and *230.42* are met. *NEC 220.61* states that a feeder or service neutral load shall be the maximum unbalance of the load determined by *Article 220*. The specifications for this residence specify that a 1 AWG neutral conductor be installed. The neutral conductor shall not be smaller than the grounding electrode conductor, *250.24(C)(1)*. Where bare conductors are used with insulated conductors, their ampacity is based on the ampacity of the other insulated conductors in the raceway, *310.15(B)(4)*.

20. Grounding Electrode Conductor Size (copper), *Table 250.66*, 4 AWG

21. Conduit (see Table 27-1 for calculations) Trade Size 1¼

Note: A one-page short version of the above service-entrance calculation form has been added to the Appendix of this text to encourage the student to perform additional residential load calculations. Practice makes perfect!

Table 310.15(B)(7) provides special consideration for 120/240-volt, 3-wire, single-phase service entrance, service lateral, and feeder conductors that serve as the main power supply to a dwelling unit. Table 29-1 in this text replicates *Table 310.15(B)(7)* from the *NEC*. The size of conductor permitted for various services is typically smaller than the

allowable ampacities found in *Table 310.15(B)(16)* because of the tremendous diversity of electricity use in homes.

Table 310.15(B)(7) cannot be used to determine the feeder size to Panel B, because this feeder does not carry the main power to the home.

Be careful! Underground wiring, raceways in direct contact with the earth, installations subject to saturation of water or other liquids, and unprotected locations exposed to the weather are wet locations per the definition in the *NEC*. Insulated conductors and cables in these locations shall be listed for wet locations, *300.5(B)*.

A similar wet location requirement found in *NEC 300.9* requires, *Where raceways are installed in wet locations above grade, the interior of these raceways shall be considered to be a wet location. Insulated conductors and cables installed in raceways in wet locations above grade shall comply with*

TABLE 29-1

This table is similar to *Table 310.15(B)(7)*, and is *only* for residential 120/240-volt, 3-wire, single-phase service-entrance conductors, service-lateral conductors, and residential feeder conductors that carry the main power to the dwelling.

CONDUCTORS PERMITTED FOR RESIDENTIAL 120/240-VOLT, 3-WIRE, SINGLE-PHASE SERVICE-ENTRANCE CONDUCTORS, SERVICE-LATERAL CONDUCTORS, FEEDER CONDUCTORS, AND CABLES THAT SERVE AS THE MAIN POWER FEEDER TO A DWELLING UNIT. (AWG) SIZE FOR RHH, RHW-2, THHN, THHW, THW, THW-2, THWN, THWN-2, XHHW, XHHW-2, SE, USE, USE-2

Copper Conductor	Aluminum or Copper-Clad Aluminum Conductors	Service or Feeder Rating (Amperes)
4	2	100
3	1	110
2	1/0	125
1	2/0	150
1/0	3/0	175
2/0	4/0	200
3/0	250 kcmil	225
4/0	300 kcmil	250
250 kcmil	350 kcmil	300
350 kcmil	500 kcmil	350
400 kcmil	600 kcmil	400

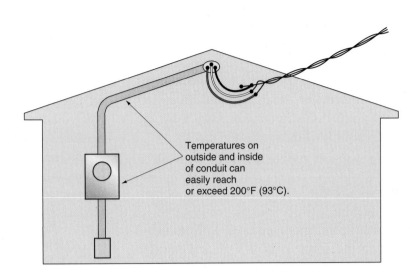

Temperatures on outside and inside of conduit can easily reach or exceed 200°F (93°C).

FIGURE 29-1   Example of a high-temperature location. (*Delmar/Cengage Learning*)

310.8(C).* Note the importance of the words "interior of these raceways."

Suffice it to say that insulated conductors and cables in wet locations must have a "W" designation in their type marking. See *310.10(C)*.

High Temperatures

In certain parts of the country, such as the southwestern desert climates where extremely hot temperatures are common, the AHJ will probably enforce *310.10(D)* and *310.15(A)(3)* regarding temperature limitations of insulated conductors. This means that conductors in raceways or cables exposed to direct sunlight (on a roof or side of a building) be corrected (derated) according to the "Correction Factors" found in *Table 310.15(B)(2)(a)*. See Figure 29-1.

*Reprinted with permission from NFPA 70-2011.

EXAMPLE

he U.S. Weather Bureau lists the average summer temperature as 113°F (45°C), then a correction factor of 0.82 must be applied. For instance, a 3/0 XHHW copper conductor per *Table 310.15(B)(16)* is 200 amperes at 86°F (30°C). At 113°F (45°C), the conductor's ampacity is

$$200 \times 0.82 = 164 \text{ amperes}$$

A properly sized conductor capable of carrying 200 amperes safely requires an ampacity before application of the temperature correction factor of

$$\frac{200}{0.82} = 243.9 \text{ amperes}$$

Therefore, according to *Table 310.15(B)(16)* and the applied correction factor, a 250-kcmil Type XHHW copper conductor is required.

Should you wish to learn more about the derating of conductors installed in raceways in direct exposure to sunlight on or above a roof, *NEC 310.15(B)(3)(c)* is covered in detail in the *Electrical Wiring—Commercial* text, a companion to *Electrical Wiring—Residential*.

Neutral Conductor

Why is the neutral conductor permitted to be smaller than the "hot" ungrounded conductors?

The neutral conductor must be able to carry the calculated maximum neutral current, *NEC 220.61*. Loads connected only phase to phase or at 240-volts are not included in the calculation of neutral conductors, as no load from these circuits is placed on the neutral. The calculations resulted in a calculated load of approximately 195 amperes for the ungrounded conductors and approximately 103 amperes for the neutral conductor. Referring to *Table 310.15(B) (16)*, this could be two 3/0 AWG THW, THHW,

THWN, XHHW, or THHN copper conductors and one 2 AWG neutral conductor. The neutral conductor may be insulated or bare per *230.30* and *230.41*.

NEC 310.15(B)(7) also permits the neutral conductor to be smaller than the "hot" ungrounded conductors, but only if it can be verified by calculations that the requirements of *215.2, 220.61,* and *230.42* are met. We verified this in our calculations for the residence discussed in this text.

In general, service-entrance conductors must be insulated, *230.30, 230.41,* and *310.2*. However, the exceptions to these rules permit the use of bare neutral conductors.

Where bare conductors are used with insulated conductors in raceways or cables, their allowable ampacity is limited to the allowable ampacity for the adjacent insulated conductors, *310.15(B)(4)*.

For example, for a service consisting of two 3/0 AWG Type THWN conductors and one 1/0 AWG bare neutral conductor, the allowable ampacity of the bare neutral conductor is found in *Table 310.15(B)(16)* in the THWN column.

3/0 AWG THWN phase conductors	200 amperes
1/0 AWG bare neutral conductor	150 amperes

The specifications for this residence call for two 2/0 AWG ungrounded "hot" conductors and one 1 AWG neutral conductor.

The service neutral conductor serves another purpose. In addition to carrying the unbalanced current, the neutral conductor must also be capable of safely carrying any fault current that it might be called on to carry, such as a line-to-ground fault in the service equipment back to the utility transformer. Sizing the neutral conductor only for the normal neutral unbalance current could result in a neutral conductor too small to safely carry fault currents. The neutral conductor might burn off, causing serious voltage problems, damage to equipment, and the creation of fire and shock hazards.

In sizing neutral conductors for dwellings, this generally is not a problem. But installing reduced size neutrals on commercial and industrial services can present a major problem. The grounded conductor of *any* service must never be smaller than the required grounding electrode conductor for that particular service. Refer to *230.23, 230.31, 230.42(C), 250.2, 250.4,* and *250.24(B)*.

Subpanel B Calculations. We will next calculate the load on the feeder to panelboard B. This will determine the minimum ampacity of the feeder conductors and the rating of the overcurrent device protecting the feeder. We will include in the calculation only the loads that will be supplied from this panelboard and will apply appropriate demand factors. Note that some demand factors are not permitted, because a reduced number of appliances are supplied. The concept for application of demand factors is that all connected loads will not be operating at the same time, and reduced loading of the service or feeder occurs.

1. General Lighting Load (*220.12, 220.43*)
   1420 sq. ft. @ 3 volt-amperes per ft^2 (Kitchen, Living Room, Laundry, Rear Entry Hall, Powder Room, Recreation Room) = 4260 VA

2. Small-Appliance Loads [*220.52(A)*]

Kitchen	3	
Laundry	1	[*220.52(B)*]
Automatic washer	1	[*220.52(B)*]
Total	5	@ 1500 VA per circuit = 7500 VA

3. Total General Lighting and Small-Appliance Load = 11,760 VA

4. Application of Demand Factors (*Table 220.42*)
   3000 volt-amperes @100% = 3000 VA
   11,760 − 3000 = 8760 @ 35% = 3066 VA

5. Net Calculated Load (Less Range and "Fastened-in-Place" Appliances) 6066 VA

6. Wall-Mounted Oven and Counter-Mounted Range: *Table 220.55, Note 4:*

Wall-mounted oven	7450 VA
Counter-mounted range	6600 VA
Total	14,050 VA

14 kW exceeds 12 kW by 2 kW:

   $2 \text{ kW} \times 5\% = 10\%$
   increase
   $8 \text{ kW} \times 0.10 = 0.8 \text{ kW}$
   (*Column C, Table 220.55*)

Cooking appliances
   $8 + 0.8 = 8.8 \text{ kW}$ = 8800 VA

7. Dryer (*220.54*)   5700 VA

8. Net Calculated Load
   = Lines 5 + 6 + 7   20,566 VA

9. List "Fastened-in-Place" Appliances *in Addition* to Electric Ranges, Air Conditioners, Clothes Dryers, Space Heaters

Appliance	VA Load
Dishwasher Motor:	
9.2×120 =	1104 VA
Food waste disposer	
7.2×120 =	864 VA
Garage door opener	
5.8×120 =	696 VA
Total	2664 VA

Note that a 75% demand factor cannot be applied as fewer than four of these appliances are supplied by the feeder.

10. Calculated load
    (lines 8 and 9)   = 23,230 VA

11. Add 25% of Largest Motor
    $7.2 \times 120 \times 0.25$ = 216 VA

12. Net Calculated Load (Lighting, Small Appliance Circuits, Ranges, Dryer, Dishwasher, Food Waste Disposer, Garage Door Opener)   23,446 VA

$$\text{Amperes} = \frac{\text{volt–amperes}}{\text{volts}} = \frac{23,446}{240}$$
$$= 97.7 \text{ amperes}$$

13. The Feeder Conductors Supplying Panel B Could Be 3 AWG THHW, THW, THHN, THWN, or XHHW per *Table 310.15(B)(16)*. We have not shown the calculations for the neutral conductor to Panel B. The specifications and also Figure 27-2 require that three 3 AWG THHN feeder conductors supply Panel B.

Method 2 (Optional Calculations) (*Article 220, Part IV*)

A second method for determining the load for a one-family dwelling is given in *Article 220, Part IV*. This method simplifies the calculations. But remember, the minimum-size service for a one-family dwelling is 100 amperes, *230.42(B)* and *230.79(C)*.

Let's take a look at *Part IV* of *Article 220*. This is an alternative method of calculating service loads and feeder loads. It is referred to as the *optional method*.

NEC 220.80 permits us to calculate service-entrance conductors and feeder conductors (both phase conductors and the neutral conductor) using an optional method. *NEC 220.82* addresses single-family dwellings. Note that several of the loads require nameplate information. For example, you are not permitted to use the general loads for electric ranges or dryers from *Part III* of *Article 220*. If the nameplate information is not available at the time the load calculation is made, a standard, rather than optional, load calculation must be performed. *NEC 220.84* addresses multifamily dwellings.

NEC 220.82(B) tells us to

1. include 1500 volt-amperes for each 20-ampere small-appliance and laundry branch circuit.

2. include 3 volt-amperes per ft^2 for lighting and general-use receptacles.

3. include the nameplate rating of appliances
   - that are fastened in place, or
   - that are permanently connected, or
   - that are located to be connected to a specific circuit, such as ranges, wall-mounted ovens, counter-mounted cooking units, clothes dryers, and water heaters.

4. include nameplate ampere rating or kVA for motors and all low-power-factor loads. The intent of the reference to low-power factor is to address such loads as low-cost, low-power-factor fluorescent ballasts. It is always recommended that high-power-factor ballasts be installed.

If there is heating and air conditioning, *NEC 220.82(C)* tells us to make the load calculation by selecting the largest load in kVA from a list of six types of loads:

1. 100% of nameplate rating of air conditioning and cooling.

2. 100% of nameplate rating of heat pump that does not have supplemental electric heating.

3. 100% of nameplate rating of heat pump compressor and 65% of supplemental electric heat. The supplemental electric heating load does not have to be added in if it cannot come on at the same time as the heat pump compressor.

4. 65% of electric space heating if less than four separately controlled units.

5. 40% of electric space heating if four or more separately controlled units.

6. 100% of electric thermal storage heating if expected to be continuous at full nameplate value.

So let's begin our optional calculation for the residence discussed in this text. The residence has an air conditioner and an electric furnace.

Air conditioner
$$26.4 \times 240 = 6336 \text{ volt-amperes}$$

Electric furnace
$$13,000 \times 0.65 = 8450 \text{ volt-amperes}$$

Therefore, we will select the electric furnace load for our calculations because it is the largest load. It is also a noncoincidental load, as defined in *220.60*. We can omit the air-conditioner load from our calculations from here on.

We now add up all of the other loads:

General lighting load 3232 ft² @ 3 volt-amperes per ft² =	9696 volt-amperes
Small-appliance and laundry circuits (5) @ 1500 volt-amperes each	7500 volt-amperes
Wall-mounted oven (nameplate rating)	6600 volt-amperes
Counter-mounted cooking unit (nameplate rating)	7450 volt-amperes
Water heater (nameplate rating)	4500 volt-amperes
Clothes dryer (nameplate rating)	5700 volt-amperes
Dishwasher 9.2 × 120	1104 volt-amperes
Food waste disposer 7.2 × 120	864 volt-amperes
Water pump 8 × 240	1920 volt-amperes
Garage door opener 5.8 × 120	696 volt-amperes
Heat-vent-lights (2) 1500 × 2	3000 volt-amperes
Attic exhaust fan 5.8 × 120	696 volt-amperes
Hydromassage tub 10 × 20	1200 volt-amperes
Total other loads	50,926 volt-amperes

We can now complete our optional calculation:
Enter electric furnace load
$$13,000 \times 0.65 = 8450 \text{ volt-amperes}$$

Plus first 10 kVA of all
other loads at 100% 10,000 volt-amperes
Plus remainder all other loads at 40%:
$$50,926 - 10,000 = 40,926 \times 0.4 = 34,820$$

$$\text{Amperes} = \frac{\text{volt-amperes}}{\text{volts}} = \frac{34,820}{240}$$

$$= 145.1 \text{ amperes}$$

Using the optional calculation method, and referring to *Table 310.15(B)(7)*, we find that for the 145+ amperes calculated load, a 1 AWG copper conductor with any of the insulation types listed is permitted to be used for a service with a rating of 150 amperes and could be installed for this service entrance. This seems a bit small, and that is why some local electrical codes do not permit the use of this optional calculation with the smaller conductor sizes permitted in *Table 310.15(B)(7)*.

The specifications for the residence discussed in this text call for a full 200-ampere service consisting of two AWG 2/0 THW, THHN, THWN, or XHHW copper phase conductors and one 1 AWG bare copper neutral conductor.

Existing Homes

When adding new loads in an existing home, *220.83* provides an optional calculation method that may be used to confirm that the existing service is large enough to handle the additional load. The procedure is a mirror image of the optional calculation method we discussed above.

Service-Entrance Conductor Size Table

Table 29-2 has been taken from one major city that prefers to show minimum service-entrance conductor size requirements rather than having electrical contractors make calculations each and every time they install a service.

It is within the realm of local authorities to publish code requirements specifically for their communities.

TABLE 29-2

Service-entrance conductor sizing used in some cities so that service-entrance calculations do not have to be made for each service-entrance installation.

SERVICE-ENTRANCE CONDUCTOR SIZE

Size (amperes)	COPPER			ALUMINUM		
	Phase (Hot) Conductors	Neutral Conductors	Conduit Trade Size	Phase (Hot) Conductors	Neutral Conductors	Conduit Trade Size
100	2 AWG	4 AWG	1½	1 AWG	2 AWG	1½
125	1 AWG	4 AWG	1½	00 AWG	1 AWG	1½
150	0 AWG	4 AWG	1½	000 AWG	0 AWG	2
175	00 AWG	2 AWG	2	0000 AWG	00 AWG	2
200	000 AWG	2 AWG	2	250 kcmil	000 AWG	2

Main Disconnect

The main service disconnecting means in this residence is rated 200 amperes, discussed in Chapter 27.

Grounding Electrode Conductor

Grounding electrode conductors are sized according to *Table 250.66* of the *Code*.

As an example, this residence is supplied by 2/0 AWG copper service-entrance conductors. Checking *Table 250.66*, we find that the minimum grounding electrode conductor must be a 4 AWG copper conductor.

The grounding electrode conductor for the service in this residence is a 4 AWG copper armored ground cable, discussed in Chapter 27.

TYPES OF WATT-HOUR METERS

Figure 29-2 shows different styles of watt-hour meters. The first, (A), is a typical electronic residential kilowatt-hour-only meter. The second, (B), is a single-phase watt-hour meter that has two sets of registers: the electromechanical register records the total kilowatt-hours, and the electronic register records and displays kilowatt-hour consumption for up to four different rate schedules.

Table 250.66 Grounding Electrode Conductor for Alternating-Current Systems

Size of Largest Ungrounded Service-Entrance Conductor or Equivalent Area for Parallel Conductors[a] (AWG/kcmil)		Size of Grounding Electrode Conductor (AWG/kcmil)	
Copper	Aluminum or Copper-Clad Aluminum	Copper	Aluminum or Copper-Clad Aluminum[b]
2 or smaller	1/0 or smaller	8	6
1 or 1/0	2/0 or 3/0	6	4
2/0 or 3/0	4/0 or 250	4	2
Over 3/0 through 350	Over 250 through 500	2	1/0
Over 350 through 600	Over 500 through 900	1/0	3/0
Over 600 through 1100	Over 900 through 1750	2/0	4/0
Over 1100	Over 1750	3/0	250

Notes:
1. Where multiple sets of service-entrance conductors are used as permitted in 230.40, Exception No. 2, the equivalent size of the largest service-entrance conductor shall be determined by the largest sum of the areas of the corresponding conductors of each set.
2. Where there are no service-entrance conductors, the grounding electrode conductor size shall be determined by the equivalent size of the largest service-entrance conductor required for the load to be served.
[a]This table also applies to the derived conductors of separately derived ac systems.
[b]See installation restrictions in 250.64(A).

(Reprinted with permission from NFPA 70-2011)

The third, (C), is a solid-state programmable polyphase watt-hour meter that can register kilowatt-hour consumption for up to five different rate schedules, demand, reactive measurements, and load profile if

FIGURE 29-2 Photo A shows an electronic watt-hour meter. Photo B shows a single-phase watt-hour meter that has two sets of registers: one registers the total kilowatt-hours, the second records and displays kilowatt-hour consumption for up to four different rate schedules. Photo C shows an electronic programmable polyphase watt-hour meter that is capable of registering five different time-of-use rates, demand, reactive measurement, and load profile if so equipped. (*Courtesy of Landis + Gyr*)

so equipped. Utilities that offer special "time-of-use" rates install meters of the second and third types.

Wiring diagrams for the metering and control of electric water heaters, air conditioners, and heat pumps are discussed in Chapter 19. Electric utilities across the country have different requirements for hooking up these loads. Always consult your local utility for their latest information regarding electric service.

Time-Based Metering

A rather recent innovation for residential metering is "time-based" metering, also referred to as "real-time" or "smart" metering. These electronic solid-state meters are capable of reading and calculating electric usage by the hour. Utilities offering this type of "time-based" metering have rate pricing that changes by the hour. You can plan ahead and use high-energy-consumption appliances (dishwasher, washer, clothes dryer, clothes iron, ranges, ovens, cooktops, air conditioning, etc.) when the rate is low.

Low rates generally are early in the morning and late at night. Typically, midnight is the lowest rate, which slowly increases until noon; then a steeper increase occurs in the afternoon, peaking between 3 and 6 PM, and then into the night rates drop off.

Some utilities have a Web site where you can check their rates for all times of the day. You can keep track of your actual usage. This puts the burden on the consumer to choose when to do household tasks to reduce the electric bill.

The world of electronics is amazing. The transition from "off-peak" metering to "time-base" metering will no doubt become very popular.

Commercial and industrial customers have had this type of "time-based" metering for years.

- Does your local electric utility offer this type of metering?

- Does it charge a monthly premium for this type of metering?

- It just might pay to check it out!

READING WATT-HOUR METERS

Reading a digital watt-hour meter is easy. Just read the numbers.

For dial-type meters, starting with the first dial, record the last number the pointer has just passed. Continue doing this with each dial until all dial readings have been recorded. The reading on the five-dial watt-hour meter in Figure 29-3 is 18,672 kilowatt-hours.

FIGURE 29-3   The reading of this five-dial meter is 18,672 kilowatt-hour. (*Delmar/Cengage Learning*)

FIGURE 29-4   One month later, the meter reads 18,975 kilowatt-hours, indicating that 303 kilowatt-hours were used during the month. (*Delmar/Cengage Learning*)

If the meter reads 18,975 one month later, as indicated in Figure 29-4, by subtracting the previous reading of 18,672, it is found that 303 kilowatt-hours of electricity were used during the month. The 303 kilowatt-hours of electricity used is multiplied by the rate per kilowatt-hour, and the power company bills the consumer for the energy used. The utility may also add a fuel adjustment charge or other charges.

REVIEW

Note: Refer to the *Code* or the plans where necessary.

1. According to *NEC* _____, the minimum-size service required for a one-family dwelling is _____ amperes.

2. a. What is the unit load per ft² for the general lighting load of a residence? _____
   _____

   b. What are the demand factors for the general lighting load in dwellings? _____
   _____
   _____
   _____

3. a. What is the ampere rating of the circuits that are provided for the small-appliance loads? _____

b. What is the minimum number of small-appliance circuits permitted by the *Code*?

_____

c. How many small-appliance circuits are included in this residence?_____

4. Why is the air-conditioning load for this residence omitted in the service calculations?

_____

_____

_____

_____

_____

5. What demand factor may be applied when four or more fixed appliances are connected to a service, in addition to an electric range, air conditioner, clothes dryer, or space-heating equipment? *(220.53)* _____

_____

6. What load may be used for an electric range rated at not over 12 kW? *(Table 220.55)*

_____

7. What is the calculated load for an electric range rated at 16 kW? *(Table 220.55)* Show calculations.

8. What is the calculated load when fixed electric heating is used in a residence? *(220.51)*

_____

9. On what basis is the neutral conductor of a service entrance determined? _____

_____

_____

_____

_____

10. Why is it permissible to omit an electric space heater, water heater, and certain other 240-volt equipment when calculating the neutral service-entrance conductor for a residence? _____

_____

_____

11. a. What section of the *Code* contains an optional method for determining residential service-entrance loads? _____

b. Is this section applicable to a two-family residence? _____

12. Calculate the minimum size of copper service-entrance conductors and grounding electrode conductor required for a residence containing the following: floor area is 24 ft × 38 ft (7.3 m × 11.6 m). The dwelling will have:

    • a 12-kW, 120/240-volt electric range
    • a 5-kW, 120/240-volt electric clothes dryer consisting of a 4-kW, 240-volt heating element, a 120-volt motor, and 120-volt lamp that have a combined load of 1 kW
    • a 2200-watt, 120-volt sauna heater
    • six individually controlled 2-kW, 240-volt baseboard electric heaters
    • a 12-ampere, 240-volt air conditioner
    • a 3-kW, 240-volt electric water heater
    • a 1000-watt, 120-volt dishwasher
    • a 7.2-ampere, 120-volt food waste disposer
    • a 5.8-ampere, 120-volt attic exhaust fan

    Determine the sizes of the ungrounded "hot" conductors and the neutral conductor. Use Type THWN. Use *Table 310.15(B)(7)*. Also determine the correct size grounding electrode conductor.

    Do not forget to include the required two 20-ampere small-appliance circuits and the laundry circuit.

    The panelboard is marked for 75°C terminations.

    You may use the blank Single-Family Dwelling Service-Entrance Calculation form in the Appendix of this text, or use the form as a guide to make your calculations in the proper steps.

    Two_____AWG THWN ungrounded "hot" conductors
    One_____AWG THWN or bare neutral conductor
    One_____AWG grounding electrode conductor

STUDENT CALCULATIONS

13. Read the meter shown. Last month's reading was 22,796. How many kilowatt-hours of electricity were used for the current month?

14. After studying this chapter, we realize that electric utilities across the country have many ways to meter residential customers. They may have lower rates during certain hours of the day. A common term used to describe these programs is _____.

15. a. Does your electric utility offer "time-based" metering or some form of it?

    _____

    b. Do you use it to reduce your electric bills? _____

Swimming Pools, Spas, Hot Tubs, and Hydromassage Baths

OBJECTIVES

After studying this chapter, you should be able to

- recognize the importance of proper swimming pool wiring with regard to human safety.

- discuss the hazards of electrical shock associated with faulty wiring in, on, or near pools.

- understand and apply the basic *NEC* requirements for the wiring of swimming pools, spas, hot tubs, and hydromassage bathtubs.

People, water, and electricity do not mix!

To protect people against the hazards of electric shock associated with swimming pools, spas, hot tubs, and hydromassage bathtubs, special *Code* requirements are necessary.

A picture is worth a thousand words. A detailed drawing of the *NEC* requirements for swimming pools appears on Sheet 10 of 10 in the Plans found in the back of this text. Refer to this drawing often as you study this chapter.

ELECTRICAL HAZARDS

A person can suffer an immobilizing or lethal shock in a residential-type pool in either of two ways.

1. *Direct contact:* An electrical shock can be deadly to someone in a pool who touches a "live" wire or touches an object such as a metal ladder, metal fence, metal parts of a luminaire, metal enclosures of electrical equipment, or appliances that have become "live" by an ungrounded conductor coming in contact with the metal enclosure or exposed metal parts of an appliance. Figure 30-1 shows a person in the water touching a faulty appliance.

2. *Indirect contact:* An electrical shock can be lethal to someone in a pool if the person is merely in the water when voltage gradients in the water are present. This is illustrated in Figure 30-2.

As shown in Figure 30-2, "rings" of voltage radiate outward from the radio to the pool walls and bottom of the pool. These rings can be likened to the rings that form when a rock is thrown into the water. The voltage rings or gradients range from 120 volts at the radio to zero volts at the pool walls and bottom. The pool walls and bottom are assumed to be at ground or zero potential. The gradients, in varying degrees, are found throughout the body of water. Figure 30-2 shows voltage gradients in the pool of 90 volts and 60 volts. This figure is a simplification of the actual situation in which there are many voltage gradients. In this case, the voltage differential, 30 volts, is an extremely hazardous value. The person in the pool immersed in these voltage gradients is subject to severe shock, immobilization (which can result in drowning), or actual electrocution. Tests conducted over the years have shown that a voltage gradient of 1½ volts per foot can cause paralysis.

The shock hazard to a person in a pool is quite different from that of the normal "touch" shock hazard to a person not submersed in water. The water makes contact with the entire skin surface of the body rather than just at one "touch" point. Skin wounds, such as cuts and scratches, reduce the body's resistance to shock to a much lower value than that of the skin alone. Body openings such as ears, nose, and mouth further reduce the body resistance. As Ohm's law states, for a given voltage, the lower the resistance, the higher the current.

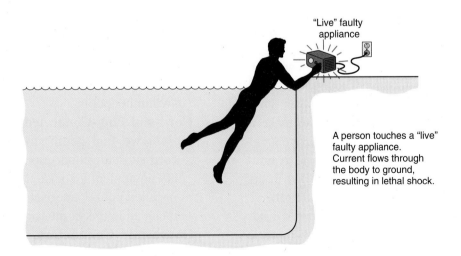

"Live" faulty appliance

A person touches a "live" faulty appliance. Current flows through the body to ground, resulting in lethal shock.

FIGURE 30-1   Touching a "live" faulty appliance can cause lethal shock. (*Delmar/Cengage Learning*)

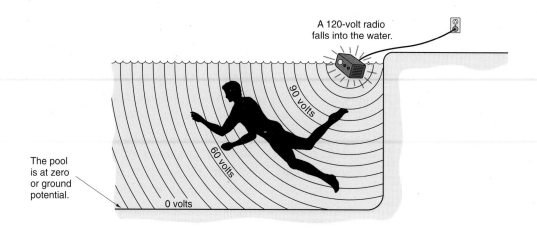

FIGURE 30-2 Voltage gradients surrounding a person in the pool can cause severe shock, drowning, or electrocution. (*Delmar/Cengage Learning*)

KEY *NEC* REQUIREMENTS—WIRING FOR SWIMMING POOLS

Article 680 covers swimming pools, wading pools, hydromassage bathtubs, fountains, therapeutic and decorative pools, hot tubs, and spas.

Here are some of the general *Code* rules. Comprehensive details for swimming pool requirements are found on Sheet 10 of 10 of the Plans at the back of this text.

General, Part I. Contains general requirements that apply to all types of swimming pools, spas, hot tubs, hydromassage tubs, and similar water related equipment. *Parts II, III, IV, V, VI,* and *VII* discuss the wiring requirements for specific equipment.

Scope, *680.1:* The scope tells us exactly what is covered by *Article 680*.

Definitions, *680.2:* There are many one-of-a-kind terms listed in *680.2*. These terms might seem like a foreign language. It makes sense to read these definitions carefully to make it easier to understand the rest of *Article 680*.

Other Articles, *680.3:* Many of the basic requirements of *Chapter 1* through *Chapter 4* in the *NEC* are modified by special rules in *Article 680* that are unique to swimming pools, wading pools, hydromassage bathtubs, therapeutic pools, decorative pools, hot tubs, and spas. This section of the *NEC* contains a convenient table that lists other articles and chapters that apply.

Ground-Fault Circuit Interrupters, *680.5:* GFCIs shall be listed circuit breakers or receptacles, or other listed types. GFCIs are discussed in Chapter 6. Throughout *Article 680*, you will find the need for GFCIs.

Be careful when connecting underwater luminaires. Once you run conductors beyond (on the load side of) a GFCI device protecting that wiring, you are not permitted to have other conductors in the same raceway, box, or enclosure. The exceptions to this rule are when (1) the other conductors are protected by GFCIs; (2) the other conductors are equipment grounding conductors; (3) the other conductors are supply conductors to a feed-through type GFCI; or (4) the GFCI is installed in a panelboard, where obviously there will be a "mix" of other circuit conductors that are not GFCI protected. This is covered in *680.23(F)(3)*.

Grounding, *680.6:* In general, grounding shall be done as required in *Article 250*. But there are modifications that specifically address swimming pools.

Cord-and-Plug-Connected Equipment, *680.7:* Here we find permission to use a flexible cord-and-plug connection for fixed or stationary equipment associated with a permanently installed pool. Underwater luminaires are not included in this permission.

Overhead Conductors, *680.8:* Overhead open conductors must be kept out of reach. Clearances are shown in Plan 10 of 10 in the back of this text.

Electric Pool Water Heaters, *680.9:* The branch-circuit conductor ampacity and the rating of the branch-circuit overcurrent devices must not be less than 125% of the total nameplate rated-load ampere rating of the electric heater. Figure 30-3 shows the requirements for unit heaters, radiant heaters, and embedded heating cables.

Underground Wiring Location, *680.10:* Underground wiring is not permitted under pools or within 5 ft (1.5 m) horizontally from the inside wall of the pool. There are a few exceptions.

Maintenance Disconnecting Means, *680.12:* A means shall be provided to simultaneously disconnect all ungrounded conductors of a circuit supplying pool-associated equipment. The disconnecting means shall be readily accessible, within sight, and located at least 5 ft (1.5 m) horizontally from the inside walls of the pool, spa, or hot tub. Pool lighting is not covered by *680.12.*

▶**GFCI Protection for Pump Motors, *680.22(B):*** All 15- and 20-ampere outlets supplied by branch circuits protected by an overcurrent device rated 120 volts through 240 volts, single phase, that supply pump motors whether by receptacle or direct connection shall be GFCI protected.◀

Permanently Installed Pools, Part II. Here we find the rules governing permanently installed pools such as for motors; clearances for overhead lighting; underwater lighting; existing installations; receptacles; requirements for switching; equipment associated with the pool; bonding, grounding, and audio equipment; and electric heaters. Electric deck heater *Code* requirements are shown in Figure 30-3.

Grounding and bonding together all metal parts in and around pools ensures that these metal parts are at the same voltage potential to ground, greatly reducing the shock hazard. Stray voltage and voltage gradient (see Figure 30-2) problems are kept to a minimum. Proper grounding and bonding also facilitates the opening of the overcurrent protection devices should a fault occur in the circuit(s).

You will see on Sheet 10 of 10 of the Plans that a metal conduit by itself is not considered an adequate equipment grounding means for equipment in and around pools.

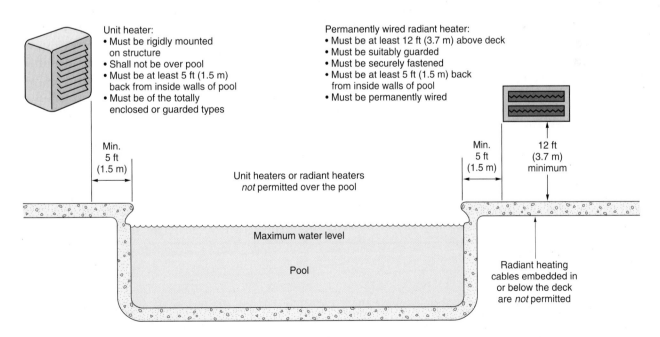

Unit heater:
• Must be rigidly mounted on structure
• Shall not be over pool
• Must be at least 5 ft (1.5 m) back from inside walls of pool
• Must be of the totally enclosed or guarded types

Permanently wired radiant heater:
• Must be at least 12 ft (3.7 m) above deck
• Must be suitably guarded
• Must be securely fastened
• Must be at least 5 ft (1.5 m) back from inside walls of pool
• Must be permanently wired

Min. 5 ft (1.5 m)

Unit heaters or radiant heaters *not* permitted over the pool

Min. 5 ft (1.5 m)

12 ft (3.7 m) minimum

Maximum water level

Pool

Radiant heating cables embedded in or below the deck are *not* permitted

FIGURE 30-3 Deck-area electric heating within 20 ft (6.0 m) from inside edge of swimming pools, *680.27(C)*. (*Delmar/Cengage Learning*)

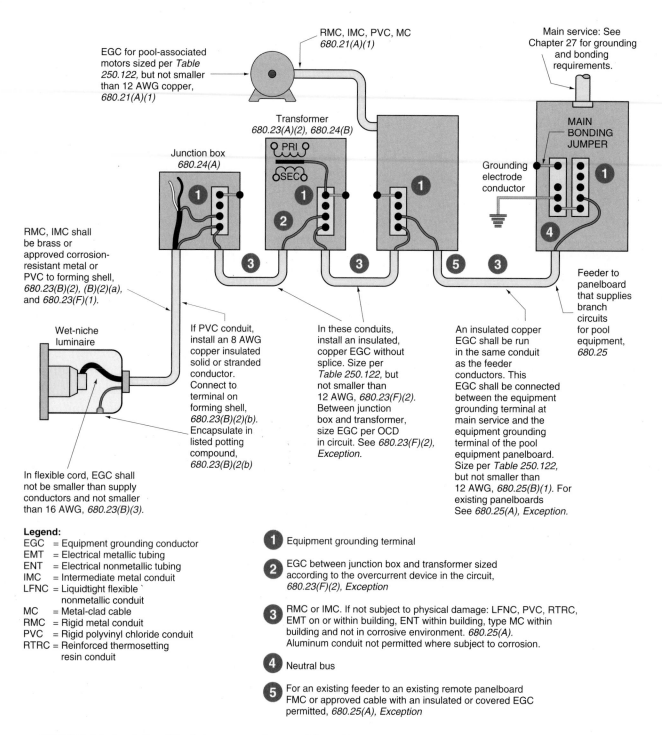

FIGURE 30-4 A "trail" of the grounding and bonding requirements from a wet-niche luminaire back to the main service panelboard. (*Delmar/Cengage Learning*)

Figure 30-4 is a detailed illustration showing the grounding of important metal parts of a permanently installed pool. As can be seen, the equipment grounding conductors also bond or connect the enclosures together, as stated in the *Informational Note*

following the definition of Equipment Grounding Conductor in *NEC Article 100*.

Figure 30-5 shows an equipotential bonding grid that consists of the conductive reinforcing steel bars of a permanently installed swimming pool.

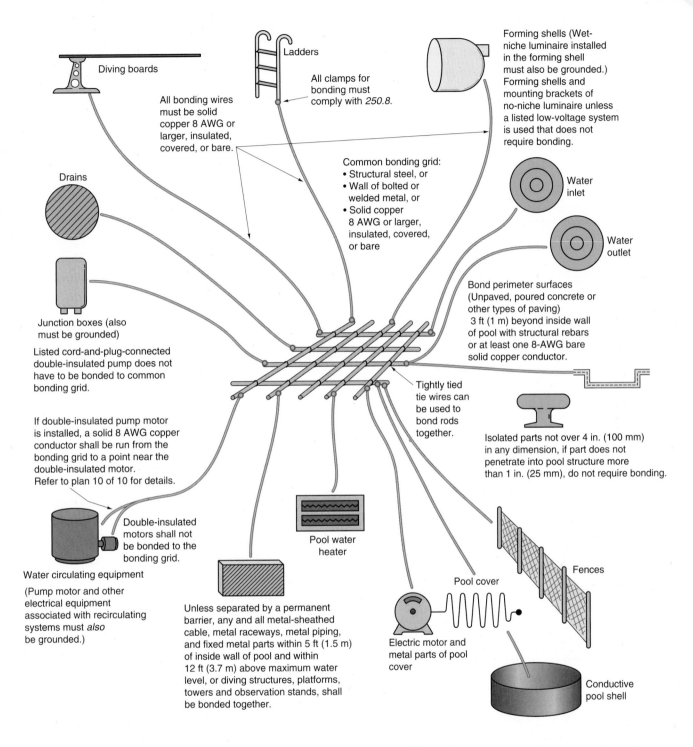

Diving boards

Ladders

All clamps for bonding must comply with *250.8*.

All bonding wires must be solid copper 8 AWG or larger, insulated, covered, or bare.

Forming shells (Wet-niche luminaire installed in the forming shell must also be grounded.) Forming shells and mounting brackets of no-niche luminaire unless a listed low-voltage system is used that does not require bonding.

Drains

Common bonding grid:
• Structural steel, or
• Wall of bolted or welded metal, or
• Solid copper 8 AWG or larger, insulated, covered, or bare

Water inlet

Water outlet

Junction boxes (also must be grounded)

Listed cord-and-plug-connected double-insulated pump does not have to be bonded to common bonding grid.

If double-insulated pump motor is installed, a solid 8 AWG copper conductor shall be run from the bonding grid to a point near the double-insulated motor. Refer to plan 10 of 10 for details.

Bond perimeter surfaces (Unpaved, poured concrete or other types of paving) 3 ft (1 m) beyond inside wall of pool with structural rebars or at least one 8-AWG bare solid copper conductor.

Tightly tied tie wires can be used to bond rods together.

Isolated parts not over 4 in. (100 mm) in any dimension, if part does not penetrate into pool structure more than 1 in. (25 mm), do not require bonding.

Double-insulated motors shall not be bonded to the bonding grid.

Water circulating equipment

(Pump motor and other electrical equipment associated with recirculating systems must *also* be grounded.)

Unless separated by a permanent barrier, any and all metal-sheathed cable, metal raceways, metal piping, and fixed metal parts within 5 ft (1.5 m) of inside wall of pool and within 12 ft (3.7 m) above maximum water level, or diving structures, platforms, towers and observation stands, shall be bonded together.

Pool water heater

Electric motor and metal parts of pool cover

Pool cover

Fences

Conductive pool shell

FIGURE 30-5 Equipotential Bonding Grid: The purpose of bonding together all metal parts and other conductive surfaces that are part of, in, and around a permanently installed swimming pool is to reduce voltage gradients in the pool and adjacent areas around the pool. Bonding is done with solid 8 AWG copper conductors (insulated, covered, or bare) or with rigid metal conduit made of brass or other identified corrosion-resistant metal. See the list of items required to be included in the equipment bonding grid in *NEC 680.26* and Plan 10 of 10. In installations where the structural steel is encapsulated with nonconductive material, you will have to install a copper conductor grid. See *680.26(B)(1)(a)* and *(b)*. (*Delmar/Cengage Learning*)

The conductor that connects various components together must not be smaller than 8 AWG and must be a solid copper conductor. If the pool shell does not have conductive steel reinforcing bars, a copper bonding grid is required. See *NEC 680.26(B)* for details about the following. Components that must be bonded include these:

- Conductive pool shells
- Perimeter surfaces within 3 ft (1 m)
- Metallic components
- Underwater lighting
- Metal fittings
- Electrical equipment associated with the water circulating system
- Fixed metal parts within 5 ft (1.5 m).

Note that the bonding conductor does not have to be connected to any service equipment or any grounding electrode. It merely ties all metal parts together.

Pool water contains dissolved chemicals that are conductive to some extent. Some pools are made of fiberglass (nonconductive) and some have insulating liners (nonconductive). Where there are no metal parts in contact with the pool water, there still is a need to bond the water to the pool deck to eliminate voltage gradients between the pool water and the deck. For these situations, *An intentional bond of a minimum conductive surface area of 9 in.² (5806 mm²) shall be installed in contact with the pool water. See 680.26(C).*

Storable Pools, Part III. Storable pools are just that: they are used on or above the ground and can be dismantled and stored.

The wiring for storable pools must follow the basic requirements found in *Article 680, Part 1*, plus the additional requirements found in *Part III*. The additional requirements for storable pools are as follows:

- Cord-connected pool filter pumps are double-insulated, and have provisions for grounding only internal and nonaccessible noncurrent-carrying metal parts. This is the equipment grounding conductor in the cord that has a grounding-type attachment plug cap.
- All electrical equipment used with storable pools shall be GFCI protected. For cord-

connected pool filter pumps, the GFCI device is required to be part of the attachment plug cap or located in the power supply cord not more than 12 in. (300 mm) from the attachment plug cap, *680.31*.

- All 125-volt, 15- and 20-ampere receptacles within 20 ft (6.0 m) of inside walls of a storable pool must be GFCI protected, *680.32*.
- Receptacles are not permitted less than 6 ft (1.83 m) from the inside walls of a storable pool, *680.22(A)(1)*.
- Receptacles rated 15 or 20 amperes, 125 through 250 volts, that supply pool pump motors shall be GFCI protected.

Spas and Hot Tubs, Part IV. What's the difference? In reality, there is no difference. Many years ago, hot tubs were constructed of wood such as redwood, teak, cypress, or oak. Along came acrylics, plastics, fiberglass, and other synthetic products. The name "spa" was created. Today, both names are used interchangeably depending on where you live. They have therapeutic jets that create whirling currents of hot water, a filtering system, and some have mood lighting, sound systems, and waterfalls. Spas and hot tubs are not designed or intended to have the water drained or discharged after each use. Here are the code rules for spas and hot tubs. (See Figure 30-6.)

Outdoor Installations, *680.42:* Shall be wired according to *Parts I and II of Article 680*, except for the following:

a. Flexible connections using flexible conduit and cord-and-plug connections are permitted.

b. Different metal components where there is metal-to-metal mounting of the same metal frame are acceptable as the required bond.

c. For one-family dwellings, the interior wiring to a self-contained spa or hot tub or a packaged spa or hot tub assembly is permitted to be conventional wiring methods that contain a copper equipment grounding conductor not smaller than 12 AWG. Conventional wiring methods are found in *Chapter 3* of the *NEC*. However, outdoor wiring to the spa or hot tub must be those wiring methods required in *Parts I and II* of *Article 680* for an in-ground swimming pool.

FIGURE 30-6 A typical spa.
(*Courtesy of Hot Spring Spas*)

Nonmetallic-sheathed cable with 14 AWG conductors has a 14 AWG equipment grounding conductor. Nonmetallic-sheathed cable with 12 AWG conductors has a 12 AWG equipment grounding conductor.

Indoor Installations, *680.43*: Shall be wired according to *Parts I* and *II* of *Article 680*, except for the following:

a. Receptacles:

- At least one 125-volt, 15- or 20-ampere receptacle connected to a general-purpose branch circuit must be installed at least 6 ft (1.83 m) but not more than 10 ft (3.0 m) from the inside wall of the spa or hot tub.

- All other 125-volt, 15- or 20-ampere receptacles must be located at least 6 ft (1.83 m), measured horizontally, from the inside wall of the spa or hot tub.

- All 125-volt receptacles rated 30 amperes or less located within 10 ft (3.0 m) of the inside walls of a spa or hot tub must be GFCI protected.

- A receptacle that supplies power to a spa or hot tub must be GFCI protected.

b. Luminaires, lighting outlets, and ceiling-suspended paddle fans above spa or hot tub and within 5 ft (1.5 m) from the inside walls of the spa or hot tub:

- to be not less than 12 ft (3.7 m) above if not GFCI protected.

- to be not less than 7 ft 6 in. (2.3 m) if GFCI protected.

- if less than 7 ft 6 in. (2.3 m), must be GFCI protected and be suitable for damp locations and be:

  – recessed luminaires with a glass or plastic lens and nonmetallic or isolated metal trim.

  – surface-mounted luminaires with a glass or plastic globe, a nonmetallic body, or a metallic body isolated from contact.

- if underwater lighting is installed, the rules for regular swimming pools apply, in which case refer to Sheet 10 of 10 of the Plans.

c. Switches:

- wall switches must be kept at least 5 ft (1.5 m), measured horizontally, from the inside edge of the spa or hot tub.

d. Bonding: Bonding requirements are similar to those of conventional swimming pools. Bonding keeps everything at the same potential. Excluded are small conductive surfaces not likely to become energized such as drains and air and water jets. Bond together:

- all metal fittings within or attached to the spa or hot tub structure.

- metal parts of associated electrical equipment including pump motors.

- metal raceways, metal piping, and other metal surfaces within 5 ft (1.5 m) of the inside walls of the spa or hot tub, unless separated by a permanent barrier, such as a wall or building.

- electrical devices and controls not associated with the pool or hot tub and located not less than 5 ft (1.5 m) from the spa or hot tub. Bonding to the spa or hot tub system also is acceptable.

e. Bonding all of the above things together can be accomplished by means of threaded piping and fittings, by metal-to-metal mounting on a common base or frame, or by a solid copper 8 AWG or larger solid, insulated, covered or bare bonding jumper.

f. Bond all electrical equipment within 5 ft (1.5 m) of the inside wall of a spa or hot tub. Also bond all electrical equipment associated with the circulating system of the spa or hot tub.

Protection, 680.44: The word *protection* refers to GFCI protection. Outlets for spas and hot tubs shall be GFCI protected, unless the unit has integral GFCI protection. A few other exceptions are mentioned in *680.44.*

Fountains, Part V. The residence discussed in this text does not have a fountain. Permanently installed fountains come under the requirements of *Part V* of *Article 680.* For fountains that share water with a regular swimming pool, the wiring must conform to the requirements found in *Part II* of *Article 680,* which covers permanently installed swimming pools.

Small self-contained portable fountains that have no dimension over 5 ft (1.5 m) do not come under the requirements of *Part V.*

Pools and Tubs for Therapeutic Use, Part VI. The residence discussed in this text does not have any therapeutic equipment.

Hydromassage Bathtubs, Part VII. A hydromassage tub is intended to be filled, used, then drained after each use. These are also known as whirlpool bathtubs. Figure 22-12 illustrates one manufacturer's hydromassage bathtub.

- Protection, *680.71:* The word *protection* refers to GFCI protection. Hydromassage bathtubs, together with their associated electrical components, must be GFCI protected, and must be supplied by an individual branch circuit.

- Protection, *680.71:* All 125-volt, single-phase receptacles that are located within 6 ft (1.83 m), measured horizontally, from the inside walls of the hydromassage bathtub must be GFCI protected.

- Other Electrical Equipment, *680.72:* A hydromassage bathtub does not constitute any more of a shock hazard than a regular bathtub. All wiring (fixtures, switches, receptacles, and other equipment) in the same room but not directly associated with the hydromassage bathtub is installed according to all the normal *Code* requirements covering installation of that equipment in bathrooms.

- Accessibility, *680.73:* Electrical equipment associated with a hydromassage bathtub must be accessible without damaging the building structure or finish.

  ▶If the hydromassage bathtub is cord-and-plug-connected with the supply receptacle accessible only through a service access opening, the receptacle is required to be installed so that its face is within direct view and not more than 1 ft (300 mm) of the opening.◀

- Bonding, *680.74:* All metal piping systems and all grounded metal parts in contact with the circulating water shall be bonded together using a copper 8 AWG or larger solid, insulated, covered, or bare.

  ▶The 8 AWG or larger solid copper bonding jumper is required to be long enough to terminate on a replacement non-double-insulated pump motor and shall be terminated to the equipment grounding conductor of the branch circuit of the motor when a double-insulated circulating pump motor is used.◀

This residence has a hydromassage tub in the master bathroom. The wiring of it is covered in Chapter 22.

⊣⋀⊢ GETTING TRAPPED UNDER WATER

There is another real hazard, not truly electrical in nature, but solvable electrically.

Drownings have occurred when a person is trapped under water by the tremendous suction at an unprotected drain. The grate over the drain may have been removed or is broken. One report indicated that three adult males could not pull a basketball from a drain because of the suction!

To address this hazard, *680.41* requires that a clearly labeled emergency shutoff switch for the control of the recirculation system and jet system for spas and hot tubs shall be installed not less than 5 ft (1.5 m) away, adjacent to, and within sight of the spa or hot tub. Although this requirement does not apply to single-family dwellings, some states require this disconnect for single-family dwelling pools.

Another hazard associated with spas and hot tubs is prolonged immersion in hot water. If the water is too hot and/or the immersion too long, lethargy (drowsiness, hyperthermia) can set in. This can increase the risk of drowning. UL Standard 1563 establishes the maximum water temperature at 104°F (40°C). The suggested maximum time of immersion is generally 15 minutes. Instructions furnished with spas and hot tubs specify the maximum temperature and time permitted for using the spa or hot tub. Hot water, in combination with drugs and alcohol, presents a real hazard to life. Caution must be observed at all times.

Although not an electrical issue, it is interesting to note that Underwriters' Laboratories has a category referred to as *Suction Fittings for Swimming Pools, Wading Pools, Spas, and Hot Tubs*. For listing these fittings, UL uses the American Society of Mechanical Engineers Standard ASME A112.19.8-2007 that addresses the safety issues regarding hair, body, finger, and limb entrapment in suction fittings that are designed to be totally submerged in swimming pools, wading pools, spas, and hot tubs.

UNDERWRITERS LABORATORIES STANDARDS

UL standards of interest are these:

UL 676	Underwater Lighting Fixtures
UL 943	Ground-Fault Circuit Interrupters
UL 1081	Swimming Pool Pumps, Filters, and Chlorinators
UL 1241	Junction Boxes for Underwater Pool Luminaires
UL 1563	Electric Spas, Equipment Assemblies, and Associated Equipment
UL 1795	Hydromassage Bathtubs

REVIEW

Note: Refer to the *Code* or the plans where necessary.

1. Most of the requirements for the wiring of swimming pools are covered in *NEC Article* _____.

2. Name the two ways in which a person may sustain an electrical shock when in a pool.

   a. _____

   b. _____

3. Using the text, the Plans in the back of this text, and the *NEC*, determine whether the following items must be grounded. Place an "X" in the correct space.

	True	False
a. Wet-and-dry-niche luminaires	_____	_____
b. Electrical equipment located within 5 ft (1.52 m) of inside walls of pool	_____	_____
c. Electrical equipment located within 10 ft (3.0 m) of inside walls of pool	_____	_____

   d. Recirculating equipment and pumps   _____   _____

   e. Luminaires and ceiling fans installed more than 15 ft
      (4.5 m) from inside walls of pool   _____   _____

   f. Junction boxes, transformers, and GFCI enclosures that
     contain conductors that supply wet-niche luminaires   _____   _____

   g. Panelboards that supply the electrical equipment for
      the pool   _____   _____

   h. Panelboards 20 ft (6.0 m) from the pool that do not
      supply the electrical equipment for the pool   _____   _____

4. Is it permitted to run other conductors in the same conduit as the GFCI-protected conductors that run to underwater luminaires? _____

5. *NEC* _____ prohibits running electrical conduits under a swimming pool.

6. What is the purpose of grounding and bonding? _____

   _____

7. Does the *NEC* require that, after all of the metal parts of a swimming pool are bonded together to form a common grounding grid, all of this grid must be connected to a grounding electrode? _____

8. May conductors be run above the pool? Explain. _____

   _____

9. What is the closest distance from the inside wall of a pool that a receptacle may be installed? _____

10. Receptacles located within 20 ft (6.0 m) from the inside wall of a pool must be _____ protected.

11. Luminaires installed above a new outdoor swimming pool must be mounted not less than (10 ft. [3.0 m]) (12 ft. [3.7 m]) above the maximum water level. Circle the correct answer.

12. Junction boxes and enclosures for transformers and GFCIs have one thing in common. They all (are made of brass) (have threaded hubs) (must be mounted at least 8 in. [200 mm] above the deck). Circle the correct answer.

13. For indoor spas and hot tubs, the following statements are either true or false. Check one. Refer to *680.43(A), (B),* and *(C)*.

	True	False
a. Receptacles may be installed within 5 ft (1.5 m) from the edge of the spa or hot tub.	_____	_____
b. All receptacles within 10 ft (3.0 m) of the spa or hot tub must be GFCI protected.	_____	_____
c. Any receptacles that supply power to pool equipment must be GFCI protected.	_____	_____

    d. Wall switches must be located at least 5 ft (1.5 m) from the pool. _____ _____

    e. In general, luminaires above a spa shall not be less than 7 ft 6 in. (2.3 m) above the maximum water level if the circuit is GFCI protected. _____ _____

14. Bonding and grounding of electrical equipment in and around spas and hot tubs (is required by the *NEC*) (is decided by the electrician). Circle the correct answer.

15. Where a spa or hot tub is installed in an existing bathroom or other suitable location, an existing receptacle outlet within 5 ft (1.5 m) of the tub is permitted to remain, but only if the circuit feeding the receptacle has _____ protection.

16. Underwater pool luminaires shall be installed so that the top of the lens is not less than _____ in. below the normal water line, unless they are listed for a lesser depth.

17. a. Hydromassage bathtubs and their associated electrical components (shall) (shall not) be protected by a GFCI. Circle the correct answer.

    b. All receptacles located within 6 ft (1.83 m) of a hydromassage tub (shall) (shall not) be protected by a GFCI. Circle the correct answer.

Wiring for the Future: Home Automation Systems

OBJECTIVES

After studying this chapter, you should be able to

- understand the basics of Insteon, ZigBee, Z-Wave, and X10 systems.
- understand the basics of "wireless" and "structured wiring" systems.
- understand some of the terminology used in the home automation industry.
- understand the types of category-rated cables.

Before the ink on this page is dry, companies and products for home automation technology will have come and gone. Some systems are wireless, some systems are hard-wired, and some systems are hybrid, using both wireless and hard-wiring. Without question, home automation is a moving target. It is almost impossible to stay on top of this fast-changing technology.

Wiring for the future comes under many names: *home automation*, *advanced home automation*, *structured wiring*, *voice-data-video (VDV)*, and *smart house*, to name a few.

Will home automation (wireless, hard-wired, or hybrid) become the norm? Will conventional premises wiring become a thing of the past? Will homeowners really want to have elaborate auto-mated control of things? Will everyone want a home theater? The answer to these and similar questions could be yes, no, or maybe. Only time will tell.

There is an increasing trend in upper-end, upscale residential wiring in the installation of advanced home automation systems. These systems are generally installed in larger, more expensive homes but are also suitable for the typical home on a limited basis.

This chapter is but a mere introduction to home automation wiring. It is beyond the scope of this book to cover the subject in detail because of the many manufacturers' products available today. Some of these products can be interchanged; others cannot. There are texts, videos, and Web sites devoted specifically to advanced home automation. Visit the many Web sites listed after the Appendix of this text. Those identified with an asterisk (*) are related to home automation. You will want to browse these Web sites and download specific information that fits your needs. Check the Delmar Cengage Learning catalog of available texts on this subject.

We hear about, read about, and might even attend seminars about home automation systems. The list of companies in manufacturing and installing home auto-mation systems is endless! Some of the names associ-ated with home automation systems are ActiveHome, Avaya, BlueTooth, CEBus, Cisco, Decora Home Controls (Leviton), Eaton, Echelon, Elan, Enerlogic, Greyfox, Honeywell, IBM, IES Technologies, Insteon, Intermatic, Logitech, LightTouch, LonWorks, Lutron, Motorola Home Sight, On-Q/Legrand, Perceptive

Automation, Radio Shack's "Plug 'N Power," PowerMark, SmartHome, SmartHouse, SmartLinc, Lightminder, Wi-Fi, X10 Powerhouse, ZigBee, and Z-Wave. At the time of this writing, the key systems in the wireless arena seem to be Insteon, ZigBee, and Z-Wave. However, the X10 technology has been around a long time. Manufacturers develop and design their devices around one of these technologies. In most cases, one does not mix different manufacturers' products.

Home automation components are available at electronic retail stores and home centers as well as online from many sources.

Everyone is familiar with individual control devices. Examples are wall switches, thermostats, motion detectors, occupancy sensors, timers, TV remote controllers, and similar devices. They control one "thing."

Home automation gives us the ability to control lighting (on, off, and dimming), heating, cooling, ventilation, appliances, home theater, audio/video entertainment, phone systems, irrigation control, and security in a home. Control of single or multiple loads from any number of locations is possible. Modules are available that can be controlled through a personal computer. Quite a few components merely plug in (plug-and-play). Others need to be wired:

Those devices using infrared function when the receiver is in sight of the sender.

Those devices using radio frequency (RF) function through walls.

Today, there are different types of advanced home automation systems for homes.

THE X10 SYSTEM

X10 has been around the longest. X10 technology was invented in Scotland in 1975. The original X10 patent expired in 1997, so it is now an open standard for any manufacturer to apply to their products.

X10 technology is simple and is the least costly home automation system, especially for retrofit wiring in existing homes. For the most part, special wiring is not required for basic functions. For many applications, all that is necessary is to plug in the various components as needed.

This system transmits carrier wave (120 kilohertz [kHz]) signals that are superimposed on the regular 120/240-volt, 60-cycle branch-circuit wiring in the home. Components communicate with each other over the existing electrical wiring without a central controller.

Transmitting devices take a small amount of power from the 120-volt power line, modulate or change it to a higher frequency, and then superimpose this high-frequency signal back onto the ac circuit. At the other end, a receiving device responds to the high-frequency burst. Xl0 is basically a one-way communications system. Some devices "talk" and others "listen."

The control of appliances, audio/video equipment, outdoor and indoor lighting, and receptacles, as well as the arming of a security system, is possible from just about any place in the home. You might call it a "plug-n-play" system. Components are easily added as needed.

An X10 system might include programmers; receiving modules (Figure 31-1); transmitters; keypad wall or tabletop controller (Figure 31-2);

FIGURE 31-2 A typical X10 tabletop keypad controller makes it easy to transmit preprogrammed signals to receiver modules. (*Courtesy Leviton*)

dimmers; single-pole, 3-way, 4-way, and double-pole switches; on/off/dim lamp modules; on/off appliance modules; receptacle modules; thermostats; timers; surge protection devices; burglar alarm devices; motion sensors (turn on lights, sound an alarm); wireless controllers; handheld remotes; photocells for light/darkness sensing; drapery controllers; telephone responders; tie-in with a personal computer; voice activation; and so on. A typical system of this type will have the capability of 256 easily adjusted different "addresses," more than adequate for just about any size home. Receivers and switches generally mount in standard wall boxes. X10 wall switches and receptacles can replace conventional switches and receptacles. Make sure that the wall box has ample space for the larger X10 devices so the wires in the wall box will not be jammed.

To ensure trouble-free operation, a phase-coupler (Figure 31-3) connected at the main panel is used to provide proper strength command signals to both sides of the ac wiring.

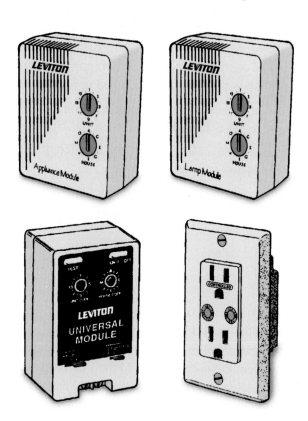

FIGURE 31-1 Various types of typical receiver modules used with X10 systems. (*Courtesy Leviton*)

Limitations of X10 Technology

X10 modules are susceptible to interference from an ac power line. Many things can generate "noise" on 120-volt power lines, including fluorescent luminaire ballasts, appliances, and wireless intercoms. And because standard home wiring (most often Type NM cable) is unshielded, it can pick up

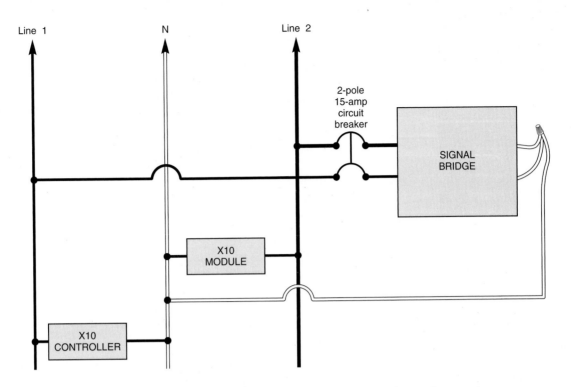

FIGURE 31-3 A signal bridge phase coupler connected to the main panel to ensure proper signals on both sides of an ac line. (*Delmar/Cengage Learning*)

unintended induced signals in the same way that an antenna does, even from devices that are not attached to the same circuit.

In fact, where two or more houses are connected to the secondary of the same utility transformer, signals could be transmitted from one house to another. X10 filters are available that install at the service panel to block interference from outside the house.

X10 signals are relatively slow. The maximum data rate is about 60 bits per second (bps) as compared to systems that handle 10,000 bps. This means there can be a noticeable delay between the time you activate a control and the time a controlled action takes place.

ZIGBEE

ZigBee is a wireless communication system that uses low-power radio signals. The system is aimed at the wireless personal area networks (WPANs), typical for home automation systems. Over 200 manufacturers around the world have formed an alliance to design devices and equipment that operate on the system specifications found in IEEE Standard 802.15.4. The system does not require hard-wiring.

The ZigBee system uses three different types of devices:

- Coordinator
- Router
- End Device

In the United States, the ZigBee system primarily operates on a 2.4 gigahertz (GHz) band. It also transmits on 915 megahertz (MHz).

Z-WAVE

Z-Wave is a low-power wireless communication system. More than 100 manufacturers have formed an alliance and are designing home automation devices based on the Z-Wave standard. These devices work fine within a range of 100 ft (30 m). The devices are readily available at electronic and home center retail centers. Z-Wave devices are "plug-in" of the type illustrated in Figure 31-1.

In the United States, the Z-Wave system operates on 868.42 megahertz (MHz).

Z-Wave devices are generally not compatible with X10 devices. The manufacturers' instruction manuals provide the necessary information relative to compatibility with other systems.

INSTEON

Insteon systems use both wireless and hard-wired technology. The system is compatible with X10 devices. As with most typical home automation systems, the Insteon system incorporates remote control (on/off, dimming) lighting, security sensors and alarms, heating, cooling, humidity, door locking, and control of appliances, as well as audio and video control.

Insteon devices have the ability to send and receive X10 commands. Thus, they are said to be compatible with X10 devices. If X10 devices are already in use in a home, it will not be necessary to discard the X10 devices if newer Insteon devices are used. Insteon's power line frequency is 131.65 kilohertz (kHz). Its wireless radio frequency is 902 to 924 megahertz (MHz). Insteon systems work through walls.

STRUCTURED RESIDENTIAL WIRING SYSTEMS

Another type of system is referred to as "structured wiring" or "infrastructure wiring," and is completely separate from the conventional branch-circuit wiring in the home. It is nicely suited for new construction. Some electricians refer to this concept as "future proofing," which means wiring the new home during the original construction (rough-in) stage for all the applications that the homeowner may want now or at some later date.

These systems are wired with special cables such as unshielded twisted pair (UTP), shielded twisted pair (STP), speaker wire, coaxial, fiber optic (FO), and cables that contain multiple types of conductors, as shown in Figure 31-4. The UL Standard refers to FO cables as *optical fiber cables*. A single pair of FO cables can handle the same amount of voice traffic as 1400 pairs of copper conductors.

FIGURE 31-4 A typical cable consisting of ac power, telephone, coaxial cable, and low-voltage control conductors. (*Delmar/Cengage Learning*)

Structured premises wiring systems, sometimes called "home LANs" (for local area networks), are intended to allow distribution of high-speed data, audio, and video signals throughout the home. They offer such advantages as faster Internet access for home computers and convenient networking of data devices (computers, printers, and fax machines) in home offices. Structured premises wiring systems get their name from ANSI/TIA/EIA standards 568 and 570, which describe "structured" or standardized cabling systems for businesses and homes (see box on home wiring standards).

Although different systems may not be interchangeable, they all provide the same wiring infrastructure and similar product interfaces such as cables, FO cables, modules, receptacles, switches, outlets, connectors, jacks, keypads, and so forth.

Here are a few of the common devices used in a "structured" wiring installation.

Combination Outlets

Telecommunications, coaxial, and sometimes FO connectors can be combined in the same wall plates, with multiple cables to each outlet location, as shown in Figure 31-5.

Central Distribution

Cables from room outlets are run to a central cabinet, sometimes referred to as a "service-center." Except for outlets, most system components are located at this cabinet, which also functions as a patch panel for making most wiring changes at a central location, as shown in Figure 31-6.

You will need a 120-volt receptacle near the location where the central distribution panel is located.

STANDARDS

Home Wiring Standards

The EIA/TIA 570-A *Telecommunications Cabling Standard* is available from Global Engineering, 15 Inverness Way East, Englewood, CO 80112-5776 (phone numbers are 1-800-854-7179 and 1-303-397-7956). This standard was jointly developed by the Electronic Industries Association (EIA) and the Telecommunications Industry Association (TIA). It describes a wiring system that uses twisted 4-pair telecommunications cabling, modular jack outlets, and star wiring (home runs) to each room from the distribution panel.

The NEMA Standard WC 63.1-2000 entitled *Performance Standard for Twisted Pair Premise Voice and Data Communications Cable* contains the minimum electrical performance as well as material and mechanical specifications, definitions, and test methods for these types of cables.

FIGURE 31-5 A single-gang combination wall outlet with two telecommunications outlets and two coaxial cable outlets. Many other combinations are available. (*Courtesy Leviton*)

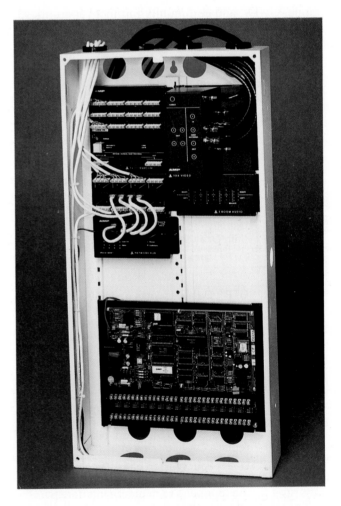

FIGURE 31-6 A "service-center" is the hub of a structured wiring system with telecommunications, audio, video, and home automation controllers installed. (*Courtesy Leviton*)

CABLE TYPES AND INSTALLATION RECOMMENDATIONS

Communication cable is tested and listed in accordance with UL Standard 444. Optical fiber cable is tested and listed in accordance to UL Standard 1651. These standards are mirror images of the EIA and TIA standards. Cables that have passed the UL testing will bear a "listing" marking.

If the cable has also passed the data transmission performance category program, the cable will be surface-marked and tag-marked, "Verified to UL Performance Category Program."

Some communities are installing underground hybrid FO/coaxial cable systems that combine telephone, CATV, and community intranet capabilities.

The categories are shown in Table 31-1.

TABLE 31-1

Types of category-rated cables.

Category 1:	Four-pair nontwisted cable. Referred to as *quad wire*. Okay for audio and low-speed data. Not suitable for modern audio, data, and imaging applications. This is the old fashioned Plain Old Telephone Service (POTS) cable. If you have any, get rid of it!
Category 2:	Usually 4-pair with slight twist to each pair. Okay for audio and low-speed data.
Category 3:	UTP*. Data networks up to 16 MHz.
Category 4:	UTP*. Data networks up to 20 MHz.
Category 5:	UTP*. Data networks up to 100 MHz. By far the most popular, having four pairs with 24 AWG copper conductors, solid (for structured wiring) and stranded (for patch cords). Cable has an outer PVC jacket.
Category 5e:	UTP*. Basically an enhanced Cat 5 manufactured to tighter tolerances for higher performance. Data networks up to 350 MHz. This is detailed in the new TIA/EIA 568A-5 Standard. Will eventually replace Cat 5 for new installations.
Category 6:	UTP*. Four twisted pairs. Data networks up to 250 MHz. Has a spline between the four pairs to minimize interference between the pairs.
Category 7:	Four twisted pairs surrounded by a metal shield. Data networks up to 750 MHz. Specifications not yet standardized at the time of publishing this text.

* UTP means unshielded twisted-pair.

Proper selection and installation of any type of audio/video/data cable is extremely important. The final performance will be that of the weakest link in the system. Improper installation, connection hardware, outlets, and other components can reduce the performance to that of an old-fashioned phone system.

Although you may not experience problems with audio transmission, using the proper cable for data transmission is critical. Losing your online Internet connection for no apparent reason could very well be pinpointed to older style untwisted-pair connecting cables that are unable to handle high-speed data transmission. Using high-quality cables is extremely important.

For residential applications, a minimum of one Category 5 or 5e (preferred) unshielded twisted-pair (UTP) cable with four twisted pairs of 24 AWG copper conductors or two Category 5 (or 5e) UTP cables with two twisted pairs of 24 AWG copper conductors should be installed for the "home runs" between each outlet location and a central distribution panel where all of the audio/video/data cables are run. Also install one or two RG-6 quad-shielded coax cable to each outlet location. Figure 31-7 illustrates a typical bundled cable with two coaxial cables and two Category 5E cables.

There are still widespread problems with existing telephone lines outside of the home causing havoc with online Internet connections. There is not much you can do about this, other than complain to the telephone company.

Installation Recommendations

Because it is so easy to damage the cable, thus affecting its transmission properties, here are some cautions for installing any audio/video/data cables:

- Keep the runs as short as possible.
- Run the cables in one continuous length from Point A to Point B without splices.
- Do not to make sharp bends in the cable. Make slow sweeping bends: radius 4D for Category 5 cable, radius 10D for coaxial cable.
- Do not pull so hard as to damage the conductor insulation and the cable jacket (25 lb is the maximum it can stand).

FIGURE 31-7 Typical bundled cables showing both coaxial and Category 5E cables.
(*Courtesy of Honeywell International Inc.*)

- Do not kink or knot the cable.

- Use rounded or depth-stop plastic staples.

- Where a number of cables are bundled, use tie-wraps loosely around the bundle, then nail the tie-wrap to the framing member.

- Do not use nail guns or metal staples. This can only lead to trouble. Use cable-tie guns.

- Always untwist the least amount, about ½ in. (13 mm) of the conductors at point of termination. The tighter the twist in a cable, the less distortion and interference.

- Do not run these cables through the same holes as power wiring, conduit, pipes, and ducts. There is too much possibility of damage to the cables.

- Keep cables away from hot water pipes, hot air ducts, and other sources of high heat.

- Keep cables at least 12 in. (300 mm) from power wiring where the cables are run parallel to the power wiring. This caution was written before twisted pairs and category-rated cables entered the scene. With the advent of properly installed twisted and shielded Category 5 and 5e cables, this recommendation from the past is probably not an issue. Category 5 and 5e cables are highly immune to interference.

- Run a dedicated 15-ampere, 120-volt branch circuit to the central distribution panel that will be hub of the audio/video/data system.

- Always check the specifications and recommendations of the cable you are installing.

- Consider installing two or three trade size 2 PVC empty conduits from the basement to the attic. Additional cables can be installed at a later date without time-consuming fishing of cables or damaging walls and ceilings.

Many types of bundled cables are available that contain a multitude of cables. Examples are one RG-6 (75-ohm) coaxial cable and one Cat 5 cable; two RG-6 (75-ohm) coaxial cables, two Cat 5 cables, and one ground; two FO cables, two RG-6 (75-ohm) coaxial cables, two Cat 5 cables, and one ground; or two RG-6 (75-ohm) coaxial cables, and one Cat 5 cable. The combinations are endless, and you need to check out manufacturers' catalogs to find the cable that meets your needs.

TERMINOLOGY

The letters "RG" stand for *Radio Grade*.

The jacks (connectors) for Cat 5 cables are generally the RJ-11 6-pin jack and the RJ-45 8-pin jack. The letters "RJ" stand for *Registered Jack*.

Connectors for coaxial cable are referred to as *F* connectors. The threaded type is preferred over the push-on type.

The letter "M" means *mega*, which is 1 million (for short: *meg*).

The letter "G" means *giga*, which is 1 billion (for short: *gig*).

The letters "bps" mean *bits per second*.

The letters "Hz" mean *hertz*, which are cycles per second.

The letters "MHz" mean *megahertz*, which are millions of cycles per second.

A *bit* is the smallest unit of measurement in the binary system. The expression is derived from the term *binary digit*.

It takes 8 bits to make 1 byte. A *byte* is the amount of data required to describe a single character of text.

The letters "Mbps" mean *megabits per second*. For example, 20 Mbps means that 20 million (20,000,000) bits of data are being transmitted per second over copper or FO cable.

There is no direct correlation between *megahertz* and *megabits*. Different encoding schemes represent information in different ways.

WIRELESS

Structured wiring involves a separate system of conductors and cables that run between controllers, sensors, and all of the other equipment. X10 technology uses the existing wiring in your home. The devices are electronically coupled to the premises wiring in your house. Both systems use wireless remote control devices using infrared technology. Infrared (IR) technology is the heart of most wireless systems, although some operate on RF. "RF" stands for radio frequency. The main drawback of IR is that you must have a straight unobstructed line of sight between the remote and the equipment being controlled. With RF devices, you do not need a straight, unobstructed line of sight. RF sends out a radio signal that goes around corners and through walls and doors. Both systems can control TVs, radios, sound systems, VCR players, CD players, DVD players, lighting, ceiling paddle fans, garage door openers, small relays that plug into a wall receptacle, remote car locking systems, and so on. Sound familiar? You may not have thought about it, but you are already in the world of home automation.

SUMMARY

Wireless or Wired? The choice is yours!

Since home automation systems vary, visit the many Web sites marked with an asterisk (*) listed after the Appendix of this text.

An excellent text on the subject is *Premises Cabling*, available from Delmar Cengage Learning. See Data and Voice Communications Cabling and Fiber Optics in the front material of this text for other related books.

REVIEW

1. One type of home automation system is referred to as
   a. X10.
   b. X20.
   c. HAS.

2. When using an X10 system, which of the following statements is most correct?
   a. A totally independent wiring system must be installed.
   b. The devices generally are "plug-and-play" (PnP), requiring no additional wiring.
   c. An X10 system is ideally suited for installations in new construction.

3. A home automation system that requires a totally separate wiring system is referred to as
   a. an X10 system.
   b. a structured system.
   c. a DC-powered system.

4. The most recommended "category-rated" cable for wiring from various outlets back to the main "service-center" for the interconnection of computers is
   a. Category 1.
   b. Category 2.
   c. Category 3.
   d. Category 5 or 5e.

5. When installing Category 5 rated cable, the minimum bending radius generally recommended is
   a. 4 times the diameter of the cable.
   b. 6 times the diameter of the cable.
   c. 10 times the diameter of the cable.

6. When installing coaxial cable, the minimum bending radius generally recommended is
   a. 4 times the diameter of the cable.
   b. 6 times the diameter of the cable.
   c. 10 times the diameter of the cable.

7. When selecting cables for a structured wiring system or computer hookups, the best choice is to select a cable that has
   a. untwisted pairs of conductors.
   b. twisted pairs of conductors.

8. Match each of the following terms with its description by drawing a line between them.
   a. RG   cycles per second
   b. RJ   the smallest unit of measurement in the binary system
   c. bps   registered jack
   d. mbps   radio grade
   e. Hz   bits per second
   f. bit   megabits per second

32

Standby Power Systems

OBJECTIVES

After studying this chapter, the student should be able to

- understand some of the safety issues concerning optional standby power systems.

- understand the basics of standby power.

- understand the types of standby power systems.

- understand wiring diagrams for portable and standby power systems.

- understand transfer switches, disconnecting means, and sizing recommendations.

- understand the *National Electric Code* requirements for standby power systems.

The text that follows is a general discussion and overview of home-type temporary generator systems. It is not possible in this chapter to cover all the variables and details involved with the many types of standby power systems available in the marketplace today.

WARNING: Exhaust gases contain deadly carbon monoxide, the same as an automobile engine. One 5.5-kilowatt generator produces as much carbon monoxide as six idling automobiles. Carbon monoxide can cause severe nausea, fainting, or death. It is particularly dangerous because it is an odorless, colorless, tasteless, nonirritating gas. Typical symptoms of carbon monoxide poisoning are dizziness, headache, light-headedness, vomiting, stomachache, blurred vision, and inability to speak clearly. If you suspect carbon monoxide poisoning, remain active. Do not sit down, lie down, or fall asleep. Breathe fresh air—fast!

Precautions to Follow When Operating a Generator

- Always install equipment such as generators, power inlets, transfer switches, and panelboards that are listed by a nationally recognized testing laboratory (NRTL).

- Always carefully read, understand, and follow the manufacturer's installation instructions.

- Always turn off the power when hooking up standby systems. Do not work on "live" equipment.

- Do not stand in water or otherwise work on electrical equipment with wet hands.

- Do not touch bare wires.

- Store gasoline only in approved red containers clearly marked "GASOLINE."

- Do not operate indoors.

- Do not operate in the garage.

- Operate a portable generator only outside in open air, not close to windows, doors, or other vents. Exhaust fumes from the generator can infiltrate the house or a neighbor's house.

- Death is just around the corner when the above warnings are not heeded!

WHY STANDBY (TEMPORARY) POWER?

Everyone has experienced a power outage, sometimes more often than one would like. You are left in the dark. Your furnace does not operate. You are concerned about water pipes freezing in the winter. Your refrigerator and freezer are not running. Your sump pump is not operating, and your basement is flooding. Your overhead garage door does not operate. You want the peace of mind that the essential equipment and appliances will continue to function while you are on an extended trip. Or, possibly worst of all, you may miss the most important sports game of the year on TV!

What can you do? The Y2K crisis made everyone aware of the consequences of a major power outage. Generators were carried out of home centers and other dealers at a furious pace. Now that the supposed Y2K crisis is history, it still makes sense to consider a standby power system.

A few questions have to be answered. In your home, which loads are critical and must continue to operate in the event of a utility power outage? You need to know this so you can select a generator panelboard with adequate space for branch circuits. Only you can make this decision. Another question that must be answered is, how large must the generator be to serve the loads that are considered critical in your home? Still another question is, how simple or complicated a standby system do you want?

WHAT TYPES OF STANDBY POWER SYSTEMS ARE AVAILABLE?

The Simplest

Home centers usually carry the simplest and most economical portable generators, as shown in Figure 32-1. These consist of a gasoline-driven motor/generator set that must be started manually. Some models have the recoil "pull-to-start" feature; others have battery electric start. These are the types construction workers use for temporary power on job sites. Depending on the size of the fuel tank and the load being supplied, these smaller generators might be capable of

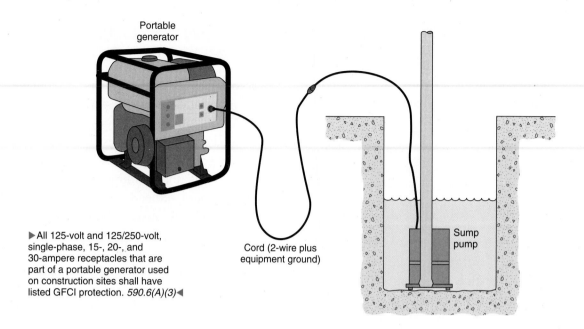

Portable generator

▶All 125-volt and 125/250-volt, single-phase, 15-, 20-, and 30-ampere receptacles that are part of a portable generator used on construction sites shall have listed GFCI protection. *590.6(A)(3)*◀

Cord (2-wire plus equipment ground)

Sump pump

FIGURE 32-1 A portable generator serving standby power to a 120-volt sump pump. (*Delmar/Cengage Learning*)

running for 4 to 8 hours. This type of portable generator is commonly referred to as "backup" power.

These small portable generators consist of a gasoline engine driving a small electrical generator, just like a gasoline engine drives the blades of a lawn mower or snow thrower. Some generator sets have one or more standard 15- or 20-ampere single or duplex grounding-type receptacles, some have GFCI receptacles, and others have 20- and 30-ampere twistlock-type receptacles. Some are rated 120 volts, whereas others are rated 120/240 volts.

These generators are available in sizes up to about 7000 watts. Also available are models with more "bells and whistles" such as oil alerts, adjustable output voltage, larger fuel tanks, and so on.

The procedure for this type is to start up the generator, then plug an extension cord into the receptacle and run it to whatever critical cord-and-plug-connected load needs to operate.

If you are not home when a power outage occurs, you are out of luck!

Receptacles on Portable Generator Sets

▶Because of the extreme hazards always present for workers on construction sites, particularly when using portable generators for temporary power, *NEC 590.6(A)(3)* requires that *All 125 volt and*

*125/250 volt, single-phase, 15-, 20-, and 30 ampere receptacle outlets that are a part of a 15 kW or smaller portable generator shall have listed ground-fault circuit interrupter protection for personnel.** The requirements of *Article 590* apply to *Temporary Wiring* and clearly cover construction sites.◀

CAUTION: Electricity from generators can also cause electric shock when used at dwellings. Purchase a generator that has GFCI protection. For existing generators without GFCI protection, use a listed portable cord that has GFCI protection.

The Next Step Up

The next step up involves both permanent wiring and cord-and-plug-connected wiring.

This is by far the most popular when it comes to standby power for homes. In the event of a power outage, you must manually start the generator, flip the transfer switching device over from normal power to standby power, and plug in the power "patch" cord (Figure 32-2).

The permanent wiring part of the installation involves connecting specific branch circuits to a separate generator panelboard (usually located right next to the main panelboard), installing and properly

---

*Reprinted with permission from NFPA 70-2011.

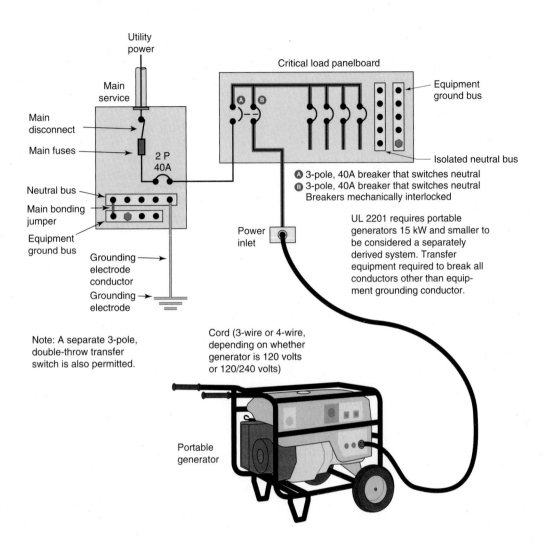

Utility power

Main service

Main disconnect

Main fuses

2 P 40A

Neutral bus

Main bonding jumper

Equipment ground bus

Grounding electrode conductor

Grounding electrode

Critical load panelboard

Equipment ground bus

Ⓐ Ⓑ

Isolated neutral bus

Ⓐ 3-pole, 40A breaker that switches neutral
Ⓑ 3-pole, 40A breaker that switches neutral
Breakers mechanically interlocked

Power inlet

UL 2201 requires portable generators 15 kW and smaller to be considered a separately derived system. Transfer equipment required to break all conductors other than equipment grounding conductor.

Note: A separate 3-pole, double-throw transfer switch is also permitted.

Cord (3-wire or 4-wire, depending on whether generator is 120 volts or 120/240 volts)

Portable generator

FIGURE 32-2 A portable generator supplies standby power to a critical load panelboard. The circuit breakers in the critical load panelboard must be manually turned to the temporary power position. These breakers are equipped with a mechanical interlock so there will be no feedback of electricity between the generator power and the utility normal power. (*Delmar/Cengage Learning*)

connecting a transfer switch (discussed later), and installing a power inlet receptacle (discussed later). The generator panelboard serves those circuits that you have selected as critical to operate in the event of a power outage. This panelboard is sometimes referred to as a generator panelboard, a selected load panelboard, an emergency panelboard (these systems in dwellings almost never meet the definition of an emergency system in *NEC 700.2*), and sometimes a critical load panelboard (Figure 32-3).

The cord-and-plug-connected part of the installation consists of merely a flexible power patch cord that runs between the generator and the power inlet receptacle. This cord and the generator are set up as needed.

These are gasoline-powered generators that use a flexible 4-wire rubber cord (12 AWG for 5000-watt generators, 10 AWG for 7500-watt generators) that plugs into a female polarized twistlock receptacle on the generator. The other end of the cord plugs into a polarized twistlock receptacle mounted on the outside of the house in a weatherproof power inlet box. This polarized twistlock receptacle is permanently connected to transfer equipment or to a special electrical generator panelboard in which the critical branch circuits are connected. The generator panelboard will have a transfer mechanism plus from 4 to 20 branch-circuit breakers; it all depends on the wattage rating of the generator. Some generator panelboards have the polarized twistlock receptacle

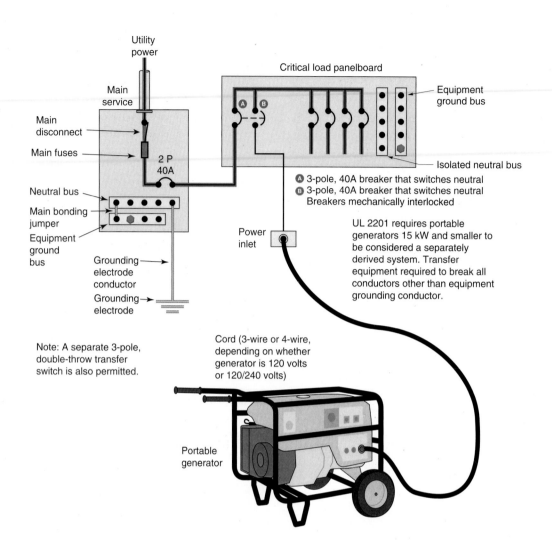

FIGURE 32-3 A critical load panelboard served by normal power from the main service panelboard. A portable generator is shown plugged into the power inlet. To switch to standby power, the circuit breakers in the critical load panelboard must be manually turned to the standby position. These breakers are equipped with a mechanical interlock so there will be no feedback of electricity between the generator power and the utility normal power. (*Delmar/Cengage Learning*)

mounted as part of and just below the generator panelboard.

During normal operation, this generator panelboard is fed from a 2-pole circuit breaker in the main panelboard. This 2-pole breaker might be rated 30, 40, 50, or 60 amperes, depending on the number of critical branch circuits to be supplied. When utility power is lost, this generator panelboard is fed from the generator. When the electric utility loses power, you must *manually* turn the "transfer switch" from its normal power position to its temporary power position. One manufacturer of electrical equipment furnishes a panelboard that contains two circuit breakers that are mechanically

interlocked. These are required to be 3-pole breakers because the neutral from a portable generator must be switched. These breakers serve as the transfer switching means. Only one breaker can be in the "On" position at the same time. One of these breakers brings the normal power to the generator panelboard. The second breaker brings the power from the generator to the generator panelboard. When one breaker is turned "Off," the other is turned "On." This panelboard will have four or more branch-circuit breakers.

With this system, the procedure generally is to first plug the 4-wire cord into the polarized twistlock receptacle on the generator; plug the other end into the power inlet receptacle; and then start the generator, which is always located outdoors. The transfer switch in the generator panelboard is then switched to the "GEN" position.

When normal power is restored, turn the transfer switch in the generator panelboard back to its normal position; turn off the generator according to the manufacturer's instructions; unplug the power cord from the generator; and finally, unplug the other end of the cord from the power inlet receptacle.

UL Standard Requirements

The UL Safety Standard UL 2201 is titled, "Portable Engine-Generator Assemblies." This standard covers portable generators that are rated up to 15 kW. The guide card information for the standard in the UL *White Book* in category "Engine Generators for Portable Use (FTCN)" reads,

"This category covers internal-combustion-engine-driven generators rated 15 kW or less, 250 V or less, which are provided only with receptacle outlets for the ac output circuits. The generators may incorporate alternating- or direct-current generator sections for supplying energy to battery-charging circuits.

When a portable generator is used to supply a building wiring system:

1. The generator is considered a separately derived system in accordance with ANSI/NFPA 70, National Electrical Code (NEC).

2. The generator is intended to be connected through permanently installed Listed transfer equipment that switches all conductors other than the equipment grounding conductor.

3. The frame of a listed generator is connected to the equipment grounding conductor and the grounded (neutral) conductor of the generator. When properly connected to a premises or structure, the portable generator will be connected to the premises or structure grounding electrode for its ground reference.

4. Portable generators used other than to power building structures are intended to be connected to ground in accordance with the NEC.*"

The safety standard adds to or supplements requirements in the *NEC*. Let's look at these rules one by one.

1. "The generator is considered a separately derived system in accordance with ANSI/NFPA 70, *National Electrical Code (NEC).*" The term *separately derived system* as defined in *NEC Article 100* means that the transfer equipment must break all system conductors from the generator, including the neutral. This rule is in place to prevent the equipment grounding conductor from being in parallel with the neutral and carrying neutral current. Normally, this rule would require the generator to be grounded at the source. The *NEC* has special rules for grounding portable generator electrical systems in *NEC 250.34*. See our discussion following list item 4.

2. "The generator is intended to be connected through permanently installed Listed transfer equipment that switches all conductors other than the equipment grounding conductor." The installation of transfer equipment that switches all conductors, including the grounded or neutral conductor, is the deciding factor in determining that the system is in fact a *separately derived system*. Once again, separately derived systems are generally required to have the neutral grounded (connected to earth) at the source. This is not the case for portable generators. See our discussion following list item 4.

3. "The frame of a Listed generator is connected to the equipment-grounding conductor and the grounded (neutral) conductor of the generator. When properly connected to a premises or structure, the portable generator will be connected to the premises or structure grounding electrode for its ground reference." These connections ensure that a ground-fault return

---

*Reprinted from the *White Book* with permission from Underwriters Laboratories Inc.®. Copyright © 2010 Underwriters Laboratories Inc.®.

path is established back to the source, which is the generator winding. Be very cautious when selecting a generator. Some manufacturers produce generators with a floating neutral. As a result, it is impossible for an overcurrent device to operate on a ground fault because no circuit for fault current can be established. The connection of the neutral and equipment grounding conductor to the frame of the generator is required in *NEC 250.34(A)(2)* and *(C)*.

4. "Portable generators used other than to power building structures are intended to be connected to ground in accordance with the *NEC*." This sentence might better be read "Portable generators used other than to power building structures are intended to be connected to ground *(grounded)* if required by the *NEC*." The *NEC* contains special requirements for grounding of portable generators in *250.34*. So long as the generator supplies loads only through receptacles mounted on the generator and the equipment grounding conductor is connected to the generator frame, the frame of the portable generator is not required to be connected to a grounding electrode.

When connected to the service-supplied electrical system by cord-and-plug connection, the equipment grounding conductor serves as a ground or earth reference for the electrical system because the conductor is extended to the windings of the generator.

Important: If you intend to have a GFCI breaker-protected branch circuit transferred to the generator panelboard, be sure to connect this circuit to a GFCI breaker in the generator panelboard.

Because GFCIs and AFCIs require a neutral conductor, make sure the neutral conductor is properly connected between the main panelboard neutral bus and the neutral bus in the critical panelboard.

The neutral bus in the generator panelboard is insulated from the generator panelboard enclosure.

The ground bus in the generator panelboard is bonded to the generator panelboard enclosure.

Standby generators of this type can have run times as long as 16 hours. Again, it depends on the loading. The manufacturer's installation and operating instructions will provide this information.

If you are not home when the power outage occurs, you are out of luck!

The Top of The Line

Standby power served by a permanently installed generator is truly standby power. These types of generator systems are covered under UL Standard 2200, *Stationary Engine Generator Assemblies*.

Standby power is a system that allows you to safely provide electrical power to your home in the event of a power outage. Instead of running extension cords from a portable generator located outside of the house to whatever critical loads (lights, sump pumps, refrigerators, etc.), you select the branch-circuits you feel are critical or essential to satisfactorily operate your home electrical system. These branch circuits are placed in a panelboard that will be transferred to the generator in the event of a power outage from the electric utility. The diagrams in this chapter show this.

The generator for this type of system is permanently installed on a concrete pad at a suitable location outside of the home, convenient to making up all of the electrical connections to serve the selected critical loads. These systems generally run on natural gas (Figure 32-4) or propane (liquid petroleum gas—LPG), and in rare cases, on diesel fuel. All the conductors between the generator, the transfer switch, and the generator panelboard installed for the selected critical loads are permanently wired. Note than the power conductors and control conductors are installed in separate raceways to satisfy the requirements in *NEC 725.136*.

The automatic transfer switch monitors the incoming electric utility's voltage. Should a power outage occur, the standby generator automatically starts, the transfer switch "transfers," and the selected loads continue to function. When normal power is restored, the transfer switch automatically "transfers" back to the normal power source, and the generator shuts down.

Because the fuel supply is a permanently installed natural gas line, these systems can run indefinitely in conformance to the manufacturer's specifications.

To be totally automatic, an electronic controller that is part of the transfer switch immediately senses

FIGURE 32-4 A natural gas-fueled generator. (*Courtesy of Kohler Power Systems*)

the loss of normal power and "tells" the standby generator to start, and to "tell" the transfer switch to transfer from normal power to standby power. This controller senses when normal power is restored, allowing the transfer switch to return to normal power and to shut off the standby generator. All of the circuitry is reset, and it is now ready for the next power interruption.

WIRING DIAGRAMS FOR A TYPICAL STANDBY GENERATOR

For simplicity, the diagrams in Figures 32-5 and 32-6 are one-line diagrams. Actual wiring consists of two ungrounded conductors and one grounded "neutral" conductor. Equipment grounding of all of the components is accomplished through the metal raceways that interconnect the components. See *250.118*.

The neutral bus in a panelboard that serves selected loads must *not* be connected to the metal panelboard enclosure. Connecting the neutral conductor, the grounding electrode conductor, and the equipment grounding conductors together is permitted only in the main service panelboard.

If you were to bond the neutral conductor and the metal enclosures of the equipment (panelboard, transfer switch, and generator) together beyond the main service panelboard, you would create a parallel path. A parallel path means that some of the normal return current and fault current will flow on the grounded neutral conductor and some will flow on the metal raceways and other enclosures. This is not a good situation! See *NEC 250.6* and *250.24(A)(5)*.

Manufacturers of generators in most cases do not connect the generator neutral conductor lead to the metal frame of the generator. Instead, they connect it to an isolated terminal. Then it is up to you to determine how to connect the generator in compliance with the *NEC* and/or local electrical codes.

Most inspectors will permit the internal neutral bond in a portable generator to remain in place. In

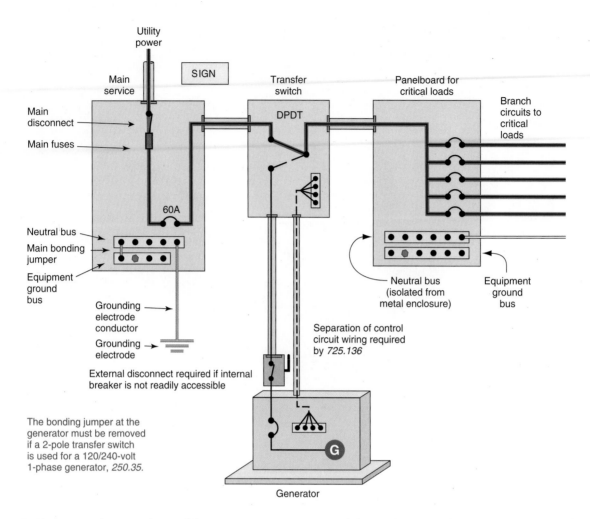

FIGURE 32-5 A natural gas-driven generator connected through a transfer switch. The transfer switch is in the normal power position. Power to the critical load panelboard is through the main service panelboard, then through the transfer switch. (*Delmar/Cengage Learning*)

fact, when you purchase a portable generator, you should insist the neutral-to-case bond is in place. Without the bond, an overcurrent device cannot function on a ground fault because a return path does not exist. For permanently installed generators, inspectors will generally not permit a bond between the generator neutral and the metal frame of the generator because of the explicit requirements in the *NEC*.

A sign must be placed at the service-entrance main panelboard indicating where the standby generator is located and what type of standby power it is. See *702.8(A)*.

The total time for complete transfer to standby power is approximately 45–60 seconds.

To further give the homeowner assurance that the standby generator will operate when called

upon, some systems provide automatic "exercising" of the system periodically, such as once every 7 or 14 days for a run time of 7 to 15 minutes.

WARNING: When an automatic type of standby power system is in place and set in the automatic mode, the engine may crank and start at any time without warning. This would occur when the utility power supply is lost. To prevent possible injury, be alert, aware, and very careful when working on the standby generator equipment or on the transfer switch. Always turn the generator disconnect to the "Off" position, then lock out and tag out the switch, warning others not to turn the switch back "On." In the main panelboard, locate the circuit breaker that supplies the transfer equipment, turn it "Off," then

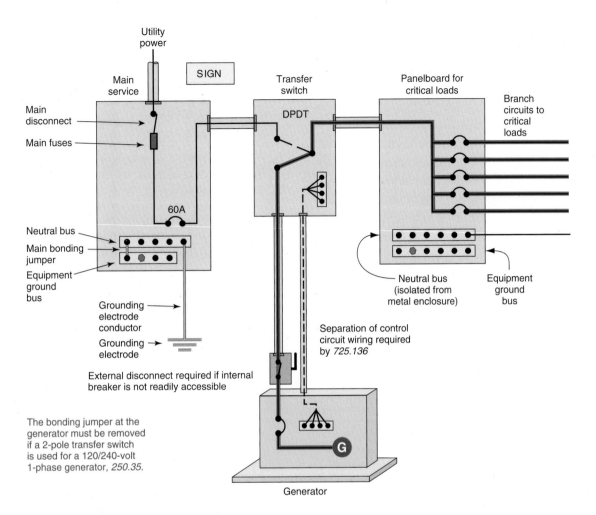

FIGURE 32-6 A natural gas-driven generator connected through a transfer switch. The transfer switch is in the standby position, feeding power to the critical load panelboard. (*Delmar/Cengage Learning*)

lock out and tag out the circuit breaker feeding the transfer switch.

TRANSFER SWITCHES OR EQUIPMENT

A transfer switch or equipment shifts electrical power from the normal utility power source to the standby power source. A transfer switch isolates the utility source from the standby source in such a way that there is no feedback from the generator to the utility's system, or vice versa. When normal power is restored, the automatic transfer switch resets itself and is ready for the next power outage.

Transfer switches or equipment for typical residential applications might have ratings of 40 to 200 amperes. Some have wattmeters for balancing generator loads.

For home installations, transfer switches may be 3-pole double throw (TPDT) or double-pole, double-throw (DPDT), depending on how the type of system is being installed. As stated above for listed portable generators, a 3-pole, double-throw transfer switch or equipment is required.

For permanently installed generators, either a 2-pole or 3-pole transfer switch or equipment is permitted so long as the generator produced electrical system is coordinated with the transfer equipment.

Two-pole transfer switches or equipment do not break the "neutral" conductor of a 120/240-volt single-phase system. Technically speaking, the *NEC* refers to this type of system as a *nonseparately derived system*. A nonseparately derived system is properly grounded through the grounding electrode system of the normal premises wiring at the service equipment. A system where the neutral is disconnected by the transfer switch is referred to as a *separately derived system* and must be regrounded according to *250.30*. Separately derived systems are common in commercial and industrial installations and are required for connection of portable generators.

If a transfer switch is connected on the line side of the main service disconnect, it must be listed as being suitable for use as service equipment. In Figures 32-5 and 32-6, the transfer switch is not on the line side of the main service disconnect and therefore does not have to be listed for use as service equipment.

A transfer switch must "break-before-make." Without the "break-before-make" feature, a dangerous situation is present and could lead to destruction of the generator, personal injury, or death. If the transfer switch does not separate the utility line from the standby power line, utility workers working on the line could be seriously injured. For low-cost, nonautomatic transfer systems, the transferring means might consist of two 2-pole or 3-pole circuit breakers that are mechanically interlocked so both cannot be "On" at the same time.

Capacity and Ratings of Transfer Equipment (*NEC 702.6*)

The rating of the transfer switch must be capable of safely handling the load to be served. Otherwise, the transfer switch could get so hot as to cause a fire. This is nothing new; it is the same hazard as overloading a conductor. Specific rules on the capacity of the generator are dependent on the type of the transfer switch used, as follows:

1. If manual transfer equipment is used, an optional standby system is required to have adequate capacity and rating for the supply of all equipment intended to be operated at one time. The user of the optional standby system is permitted to select the load that is connected to the system.

2. If automatic transfer equipment is used, an optional standby system must comply with parts (a) or (b).

   (a) Full Load. The standby source shall be capable of supplying the full load that is transferred by the automatic transfer equipment.

   (b) Load Management. If a system is employed that will automatically manage the connected load, the standby source must have a capacity sufficient to supply the maximum load that will be connected by the load management system.

A transfer switch must also be capable of safely interrupting the available faultcurrent that the generator or utility is capable of delivering. Listed transfer switches or equipment may be provided with or without integral overcurrent protection. The suitability of listed transfer equipment for interrupting or withstanding short-circuit current is marked on the transfer equipment.

A typical transfer switch might take 10 to 15 seconds to start the generator. This eliminates nuisance start-ups during momentary utility power outages. After start-up, another 10- to 15-second delay is provided to allow the voltage to stabilize. After start-up and warm-up, full transfer takes place. Similar time delay features are used for the return to normal power. Some systems can do the full transfer in a few seconds.

The UL Product Standard is 1008, *Transfer Switch Equipment*.

See the book *Electrical Grounding and Bonding* published by Cengage Delmar Learning for additional information on grounding and bonding of separately derived and nonseparately derived alternate power systems supplied by generators.

DISCONNECTING MEANS

A disconnecting means is required for a generator. *NEC 445.18* states that *Generators shall be equipped with disconnect(s), lockable in the open position, by means of which the generator and all protective devices and control apparatus are able to be disconnected entirely from the circuits supplied by the generator except where both of the following conditions apply:*

1. *The driving means for the generator can be readily shut down*

2. *The generator is not arranged to operate in parallel with another generator or other source of voltage.**

NEC 225.34 requires that all disconnecting means shall be grouped.

If an outdoor housed generator set is equipped with a readily accessible disconnecting means that is located within sight of the building or structure supplied, and additional disconnecting means is not required where ungrounded ("hot") conductors serve or pass through the building or structure, *NEC 702.12.* Most electrical inspectors will allow the cord-and-plug connection for a portable generator to serve as the disconnecting means.

Where the normal power main service disconnecting means is located inside of the home, some electrical inspectors will require that a disconnect for the standby power from the generator be installed near the main service equipment.

NEC 225.34(B) states that the required disconnect for the standby generator be located remote from the normal service disconnect. This is to lessen the possibility of accidentally shutting off both disconnects at the same time.

Having one disconnecting means inside the home and another outside of the home presents quite a challenge for firefighters or others wanting to totally shut off the power to the home. This is why *NEC 702.9(A)* requires that a sign be placed at the service-entrance equipment that indicates the type and location of on-site optional standby power sources.

GROUNDING

Grounding the metal frame of a portable generator is through the equipment grounding conductor in the power cord. Grounding of a hard-wired generator is accomplished by means of a metallic wiring method or other means acceptable by *250.118.*

---

*Reprinted with permission from NFPA 70-2011.

It is not necessary to connect the frame of a portable generator to a grounding electrode, *250.34(A).*

CONDUCTOR SIZE FROM STANDBY GENERATOR

The conductors that run from the standby generator to the transfer switch and to the generator panelboard that contains the branch-circuit overcurrent devices shall be sized not less than 115 percent of the generator's nameplate current rating. See *NEC 445.13.* This applies if overcurrent protection of the conductors is not provided at the generator source. If the generator is equipped with an overcurrent device, the standard rules for sizing conductors in *NEC 240.4* apply.

EXAMPLE

A 7500-watt, single-phase, 120/240-volt generator.

Step 1: Calculate the generator's current rating.

$$\text{Amperes} = \frac{\text{Watts}}{\text{Volts}} = \frac{7500}{240} = 31.25 \text{ Amperes}$$

Step 2: $31.25 \times 1.15 = 35.94$ amperes.

Step 3: Refer to *NEC Table 310.15(B)(16)* to determine the minimum size conductors.

In the 60°C column of *Table 310.15(B)(16)*, we find that an 8 AWG copper conductor has an allowable ampacity of 40 amperes. The discussion on when to use the 60°C and when to use the 75°C columns of *NEC Table 310.15(B)(16)* is covered in Chapter 4.

The neutral conductor is permitted to be sized according to *220.61*, which under certain conditions allows the neutral conductor to be smaller than the phase conductors. This information is found in *NEC 445.13.*

GENERATOR SIZING RECOMMENDATIONS

No matter what type or brand of generator you purchase, you will have to size it properly. There is no better way to do this than to follow the manufacturer's recommendations. Generators for home use are generally rated in watts. The manufacturer of the generator has taken into consideration that watts = volts × amperes. Some manufacturers suggest that after adding up all of the loads to be picked up by the generator, add another 20% capacity for future loads.

As stated above, the type of transfer switch installed determines the minimum capacity required for the generator. If a manual transfer switch or equipment is installed, a generator is required to have capacity for all of the equipment intended to be operated at one time. The user of the optional standby system is permitted to select the load connected to the system. See *NEC 702.6(B)(1)*.

If the generator is supplied with an automatic transfer switch or equipment, it is required to be capable of supplying the full load that is transferred by the automatic transfer equipment. A load management system is permitted to be installed, in which case the generator must be sized not smaller than required to supply the maximum load that the load management system will allow for it to be connected to the equipment. See *NEC 702.6(B)(2)*.

Wattage ratings of appliances vary greatly, as shown in Table 32-1. Resistive loads such as toasters, heaters, and lighting do not have an initial high

TABLE 32-1

Wattage ratings of appliances.

	Typical Operating Wattage Requirements	Starting Wattage Requirements
Blender	600	700
Broiler	1600	1600
Central Air Conditioner		
10,000 BTU	1500	4500
20,000 BTU	2500	7500
24,000 BTU	3800	11,000
32,000 BTU	5000	15,000
40,000 BTU	6000	18,000
Coffee Maker	900–1750	900–1750
CD Player	50–100	50–100
Clothes Dryer		
Electric	5000 @ 240 volts	6500
Gas	700	2200
Computer	300–800	300–800
Curlers	50	50
Dehumidifier	250	350
Dishwasher	1500	2500
Electric		
Drill	250–750	300–900
Fry Pan	1300	1300
Blanket	50–200	50–200
Space Heater	1650	1650
Water Heater	4500–8000 @ 240 volts	4500–8000 @ 240 volts
Electric Range		
6-in. surface unit	1500	1500
8-in. surface unit	2100	2100
Oven	4500	4500
Lights	Add wattage of bulbs	Add wattage of bulbs

(Continued)

TABLE 32-1

(*Continued*)

	Typical Operating Wattage Requirements	Starting Wattage Requirements
Furnace Fan		
⅛ Horsepower Motor	300	900
⅙ Horsepower Motor	500	1500
¼ Horsepower Motor	600	1800
⅓ Horsepower Motor	700	2100
½ Horsepower Motor	875	2650
Garage Door Opener		
⅓ Horsepower	725	1400
¼ Horsepower	550	1100
Hair Dryers	300–1650	300–1650
Iron	1200	1200
Microwave Oven	700–1500	1000–2300
Radio	15 to 500 (with components)	15 to 500
Refrigerator	700–1000	2200
Security, Home	25–100	25–100
Sump Pump		
⅓ Horsepower	800	2400
½ Horsepower	1050	3150
Television	300	300
Toaster		
4-Slice	1500	1500
2-Slice	950	950
Automatic Washer	1150	3400
Well Pump Motor		
⅓ Horsepower	800	2400
½ Horsepower	1050	3150
1 Horsepower	1920 @ 240 volts	5500
Vacuum Cleaner	1100	1600
VCR	150–250	150–250
Window Fan	200	300

inrush surge of current. Motors, on the other hand, do have a high inrush surge of starting current, which lasts for only a few seconds until the motor gets up to speed. This inrush must be taken into consideration when selecting a generator. Here are some typical *approximate* values for household loads. For heating-type appliances, the typical operating wattage values are used. For motor-operated appliances, the starting wattage values should be used. Verify actual wattage ratings by checking the nameplate on the appliance. Verify that the connected loads do not exceed the generator's marked capacity.

Most generators are capable of handling a momentary "inrush" of an extra 50% of their rating. Here again, check the manufacturer's specifications.

THE *NATIONAL ELECTRICAL CODE* REQUIREMENTS

The *NEC* addresses optional standby systems in *Article 702*. This includes those systems that are *permanently installed in their entirety*, and those systems that are intended for *connection to a premises wiring system from a portable alternate power supply.*

- *NEC 702.6*: All of the equipment must have the capacity and rating for the loads that will be supplied by the equipment. The equipment shall be suitable for the maximum available fault current at its terminals.

- *NEC 702.7*: Transfer equipment shall be suitable for the intended use and designed to prevent the interconnection of the generator and the normal utility supply.

- *NEC 702.8*: Some systems, where practical, require audible and visual signal devices to indicate when the normal power system has been transferred to the standby system. The manufacturer of the equipment provides these features. Signal devices are not required for portable standby generators, *702.8, Exception.*

- *NEC 702.9(A)*: A sign is required at the service-entrance equipment to indicate the type and location of the standby generator.

- *NEC 702.9(B):* Where the grounded circuit conductor connected to the optional standby generator is connected to a grounding electrode conductor at a location remote from the standby source, there shall be a sign at the grounding location that shall identify all standby and normal sources connected at that location. For most residential standby systems, this connection is made on the neutral bus in the generator panelboard. This neutral bus in turn has a conductor run to the neutral bus in the main panelboard, where the neutral bus and equipment grounding bus are connected together, and a grounding electrode conductor is connected and run to the grounding electrode of the system. The neutral bus in the generator panelboard is isolated from the panelboard enclosure. The equipment grounding bus in the generator panelboard is bonded to the panelboard enclosure.

- *NEC 702.9:* It is permissible to run standby power conductors and normal power conductors in the same raceway or enclosure.

- *NEC 702.11(B):* Where a portable optional standby source is used as a nonseparately derived system, the equipment grounding conductor shall be bonded to the system grounding electrode. The equipment grounding conductor (green or bare) from the standby generator to the generator panelboard is connected to the equipment grounding bus in the generator panelboard. The equipment grounding bus in the generator panelboard is bonded to the generator panelboard enclosure. In turn, the equipment grounding bus in the generator panelboard is bonded back to the main panelboard either by a metallic wiring method, or by a separate equipment grounding conductor. At the main panelboard, the grounding electrode conductor is connected to both the equipment ground bus and the neutral bus.

- *NEC 702.12* tells us that if an optional outdoor housed standby system generator is
  - equipped with a readily accessible disconnecting means, and
  - located within sight of the building it serves, an additional disconnecting means is not required where the ungrounded conductors from the generator serve or pass through the building it serves. See *225.31* and *225.32.*

Permits

Permanently installed generators require a considerable amount of electrical work and gas line piping. The installation will probably require applying for a permit so that proper inspection by the authorities can be made. Check with your local building official.

Sound Level

Because generators produce a certain level of "noise" (decibels) when running, you will want to check with your building authorities to make sure the generator you choose is in compliance with local codes relative to any sound ordinance that might be applicable. Sound-level information is provided by manufacturers in their descriptive literature.

REVIEW

Note: Refer to the *Code* or the plans where necessary.

1. The basic safety rule when working with electricity is to _____
   _____.

2. Where would the logical location be for running a portable generator?

3. The best advice to follow is to always use (listed) (cheapest) (smallest) equipment.
   Circle the correct answer.

4. How would you define the term *standby power*? _____
   _____

5. Describe in simple terms the three types of standby power systems.
   _____
   _____
   _____
   _____
   _____

6. Is it permitted to ground the neutral conductor of the standby generator to the metal
   enclosure of the generator, transfer switch, or critical (generator) panelboard?
   Explain.
   _____
   _____
   _____
   _____

7. Briefly explain the function of a transfer switch.
   _____
   _____
   _____
   _____

8. When a transfer switch transfers to standby power, the electrical connection inside the
   switch (circle the correct answer):
   a. maintains connection to the normal power supply as it makes connection to the
      standby power supply.
   b. breaks the connection to the normal power supply before it makes connection to the
      standby power supply.

9. A typical transfer switch for residential application is (circle the correct answer):
   a. a three-pole, double pole switch (TPDT).
   b. a double-pole, double-throw switch (DPDT).
   c. a single-pole, double-throw switch (SPDT).
   d. a double-pole, single-throw switch (DPST).

10. The technical *NEC* definition of a system in which the neutral conductor is not switched is referred to as (circle the correct answer):

    a. a separately derived system.

    b. a nonseparately derived system.

11. The *NEC* requires that a _____ means be provided for standby power systems. In the case of portable gen sets, this might be as simple as pulling out the plug on the extension cord that is plugged into the receptacle on the gen set. For permanently installed standby power generators, this might be on the gen set and/or separately provided inside or outside of the home.

12. The conductors running from a permanently installed standby power generator to the transfer switch shall not be less than (100%) (115%) (125%) of the generator's output rating if overcurrent protection of the conductors is not provided at the generator. (Circle the correct answer.)

13. *NEC 590.6(C)* requires that all 125-volt and 125/250-volt, single-phase, 15-, 20-, and 30-ampere receptacle outlets used on a construction site that are a part of a portable generator shall have listed (circle the correct answer):

    a. ground-fault circuit interrupter protection for personnel.

    b. grounding-type receptacles.

    c. twist-lock receptacles.

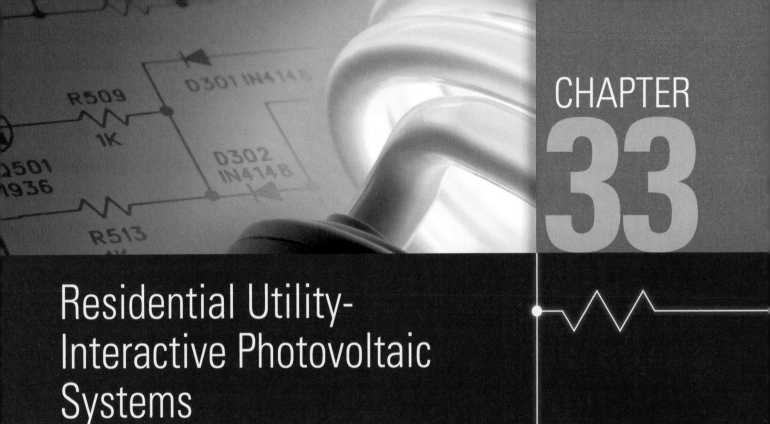

Residential Utility-Interactive Photovoltaic Systems

OBJECTIVES

After studying this chapter, you should be able to

- identify the components of a residential utility-interactive solar photovoltaic system.

- recognize the electrical hazards unique to solar photovoltaic systems.

- apply *National Electrical Code* requirements to the installation of a residential utility-interactive solar photovoltaic system.

Electricity from sunlight? Is this possible? This is not only possible, but it has become very practical through the installation of utility-interactive photovoltaic systems. A combination of factors recently has made photovoltaic (PV) systems very popular:

- Increased use of electricity throughout the United States

- Increased costs for electricity production

- Environmental concerns with the use of fossil fuels

- Concern over dependence on foreign sources of energy

- Increased efficiency of photovoltaic systems

Installation of photovoltaic systems is being encouraged by government and electrical utilities through tax incentives and rebates. Some states require utilities to produce defined amounts of electricity through use of renewable resources. Electrical utilities can meet the requirement by construction of large central photovoltaic generating plants or through distributed generation. Generation of electricity by the consumer at the point of use (distributed generation) decreases the need for utility-generated power. Excess electricity will be supplied to the utility grid for use by other customers. Electrical utility companies are required to purchase customer-generated electricity at predetermined rates. Photovoltaic systems such as these are known as utility interactive. Depending on size, a utility-interactive PV system can supply most or all of a home's electrical load. Photovoltaic systems are being retrofitted to existing homes and even installed on new homes as they are constructed. Battery storage of the generated electricity may also be included, but is not as common.

Electrical Hazards

Electrical work for the installation of photovoltaic systems is not complex, but there are significant differences when compared to typical residential wiring. First and foremost, a utility-interactive PV system is a supply of electricity for the home, not another load. Photovoltaic modules on the roof are generating electricity when exposed to sunlight. Although the utility disconnect (main circuit breaker) for a house can be turned off, dangerous levels of electricity will still be present on the dwelling as long as the sun is shining. Modules on the roof of a house are connected in series in strings that operate at up to 600 volts. Electricity generated by the photovoltaic modules is direct current (dc). All conductors and components must be listed for use with dc voltage. The sun shines for more than three hours at a time, so all conductors/components must be sized for continuous currents (three hours or longer of operation). Because string conductors of the correct type are permitted to be exposed on the roof of a dwelling, good workmanship is critical. There are open conductors, exposed to the elements, operating at up to 600 volts of continuous current (Vdc) that cannot be turned off! If a short circuit does occur, it is not always obvious. Module short-circuit current is only slightly higher than normal operating current. Contrast this with the short-circuit current from sources such as a utility or even 12-Vdc vehicle batteries. Shorting out these sources results in extremely high fault currents. High-fault currents are easier to detect. High-fault currents will also cause an overcurrent device (fuse or circuit breaker) to open quickly. The connection of the utility-interactive PV system to the existing service panel is of concern. Service disconnecting means, grounding, and proper labeling must be accomplished to maintain a safe electrical installation. It is obvious that there are some special considerations for the installation of residential photovoltaic systems. *Article 690* was added to the *National Electrical Code (NEC)* in 1984 to establish minimum electrical standards for the installation of photovoltaic systems. *Chapters 1* through *4* of the *NEC*, along with the requirements of *Article 690*, will apply to residential PV installations. Local jurisdictions may have amendments to the *NEC* for PV installations. Always check with the local Authority Having Jurisdiction (AHJ) before starting the installation in a new area.

THE BASIC UTILITY-INTERACTIVE PV SYSTEM

Several components are required in order to convert sunlight into useful amounts of electricity. A basic utility-interactive system will consist of modules, mounting racks, combiner/transition boxes, inverter(s), and several disconnects. A grounding electrode system and connection to the existing service panel will be required. See Figures 33-1 and 33-2 for examples of basic PV system components and arrangements.

Solar Cells, Modules, and Arrays

A photovoltaic module is the basic unit of power production in the system. A module is a manufactured unit made up of many semi-conductor PV cells encased in a protective covering and mounted to an aluminum frame. All modules are required by the *NEC* to be listed to nationally recognized standards (*ANSI/UL 1703*). Individual photovoltaic modules are mounted to a support rack which is connected to roof members, or in

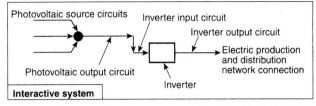

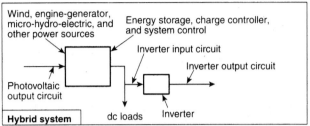

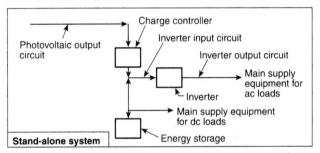

Notes:
1. These diagrams are intended to be a means of identification for photovoltaic system components, circuits, and connections.
2. Disconnecting means and overcurrent protection required by *Article 690* are not shown.
3. System grounding and equipment grounding are not shown. See *Article 690, Part V.*
4. Custom designs occur in each configuration, and some components are optional.

FIGURE 33-2 Identification of solar photovoltaic system components in common system configurations. (*Figure 690.1(B) reprinted with permission of NFPA 70-2011.*)

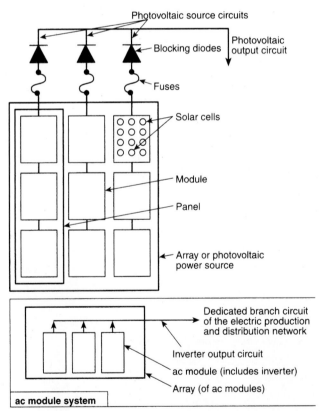

Notes:
1. These diagrams are intended to be a means of identification for photovoltaic system components, circuits, and connections.
2. Disconnecting means required by *Article 690, Part III*, are not shown.
3. System grounding and equipment grounding are not shown. See *Article 690, Part V.*

FIGURE 33-1 Identification of solar photovoltaic system components. (*Figure 690.1(A) reprinted with permission of NFPA 70-2011.*)

some cases, to ground mounted structural supports. Individual modules are wired together in a series circuit. Factory installed leads are provided by the manufacturer for this purpose. Type *USE-2* or *PV* conductors can be spliced to module leads to facilitate circuiting. A separate equipment grounding conductor is used to ground each module. Most strings will consist of eight to fifteen modules. The quantity of modules in a series circuit will be limited by the combined open circuit voltage of the string. A typical residential array of roof mounted modules is shown in Figure 33-3.

Multiple strings are combined together (in parallel) at the combiner box. This allows a single pair of conductors to deliver current to an inverter. Combiner boxes may be fused or non-fused. Most

FIGURE 33-3 Rooftop array of modules. (*Delmar/Cengage Learning*)

residential PV systems are made up of three to six strings that can be combined in a single box. String fusing is generally not required for three strings or fewer. Module ratings will determine this, but some designers specify fused combiner boxes even when not required. Many inverters have the ability to combine several strings at the input terminals. With an inverter like this, the string conductors are routed to the inverter without being combined first. The combiner box would not be required. A junction box, known as a transition box, is used to splice the open string conductors to a wiring method (usually conduit and THWN conductors) for connection of the inverter. The entire assembly of racks, modules and combiner/junction boxes is known as a photovoltaic array. Available space, sunlight, and the ultimate amount of electricity desired will determine the number of modules in the array. Obviously, the larger the system, the more expensive the installation, so the budget must be considered.

A disconnect for the ungrounded dc string conductor(s) is required before the conductors enter the dwelling unless a metallic raceway is used as the wiring method. Use of a metallic wiring method such as EMT or FMC is common so that string conductors can be routed through a dwelling unit attic. Metallic raceways provide more protection for the 400–600 Vdc string conductors, which remain energized until the sun goes down. Firefighters responding to a house fire will open the service

disconnecting means on arrival. This does not de-energize the PV string conductors if the sun is still shining. Metallic raceways will provide a level of protection for the firefighters who may encounter the dc wiring method when responding to a fire. Use of a metallic raceway eliminates the requirement for an array disconnect on the roof but not at the inverter. The inverter will be energized from two different sources: dc from the array and ac from the utility connection. Disconnecting means must be provided for both sources. Some inverters are manufactured with integral disconnects. A circuit breaker in an adjacent electrical panel can serve as the ac disconnect, but all safety disconnects must be within sight of the inverter and may need to be grouped. The dc grounded conductor is not permitted to be opened by any disconnecting means. The inverter in Figure 33-4 has an integrated dc disconnect.

The Inverter

The inverter will need to be installed in a location with sufficient working space per *NEC 110.26*. Larger systems may use two or more inverters with output combined in a separate ac panelboard before supplying the service. Grounding/bonding of the utility-interactive PV system is required and is usually accomplished at the inverter location. Both the ac and the dc systems contain grounded current carrying conductors. Grounding of the dc conductor

FIGURE 33-4 A residential utility-interactive inverter. (*Delmar/Cengage Learning*)

is accomplished at the inverter. A new grounding electrode must be installed (and bonded to the existing premises' grounding electrode) or the existing grounding electrode for the dwelling must be accessed for connection. Correct termination of the grounding electrode conductor, equipment grounding conductors, and grounded ac/dc conductors at the inverter is critical. Utility-interactive inverters are required to provide ground fault protection for the array. Proper operation of the ground fault protection system is dependent on the correct terminations of the grounding and grounded conductors. Inverter design will dictate whether the positive or the negative dc conductor is the grounded current carrying conductor for the array. Inverter size is based on the capacity of the array. Most residential inverters are in the 2 kW to 10 kW range. There are advantages to installing an inverter indoors. Cooler temperatures result in better operating efficiency, but outdoor installation is also common.

Safety Features of Inverters

Just as with modules and combiner boxes, inverters are required to be manufactured to nationally recognized standards. Inverters that are listed by a Nationally Recognized Testing Laboratory (NRTL) to ANSI/UL 1741 have met this standard. One requirement of UL 1741 is known as anti-islanding. Anti-islanding means that an inverter must automatically turn off ac output when utility power is lost. This feature prevents an inverter from supplying electricity into the utility grid when there is an outage. Electrical utility workers risk being electrocuted by inverter-supplied electricity without this.

Connection of Inverters

The two options for connection of the inverter to the electrical service panel are known as a supply-side or a load-side connection. A connection on the utility side of the main disconnect is known as a supply-side point of connection. A load-side connection will supply electricity on the customer side of the main disconnect. Supply-side connection size is not limited by the *NEC*, but a second electrical service is created so the requirements of *NEC Article 230 (Services)* will apply. A load-side

connection will have limitations. A dedicated circuit breaker (suitable for backfeed) or fusible disconnect is required. The total supply of current to a panelboard is limited to 120% of the busbar rating. The service panel is supplied by both the utility (through the main circuit breaker) and the inverter (through a back-fed circuit breaker). The sum of the ampere ratings of the main circuit breaker and the inverter circuit breaker cannot exceed the rating of the panelboard bus multiplied by 1.2 (120%). A circuit breaker used for connection of the inverter will have to be located at the opposite end of the bus from the main circuit breaker, to avoid overloading of the bus.

Building-Integrated Photovoltaic Modules

Photovoltaic modules that also serve as an outer protective finish for a building are known as building-integrated photovoltaic (BIPV) modules. This type of module is often installed in the form of roofing tiles intended to blend in with surrounding non-PV roof tiles. See Figure 33-5. BIPV modules are investigated through the listing process for conformance to appropriate fire-resistance and waterproofing standards, which apply to roofing tiles, along with the electrical standards of *UL/ANSI 1703*. A module the size of a roofing tile is obviously much

FIGURE 33-5 Building-integrated photovoltaic roof tiles. (*Delmar/Cengage Learning*)

smaller than the standard photovoltaic module. Many more of the smaller modules are required to create strings.

Micro-Inverters and ac Photovoltaic Modules

A single small inverter connected to each photovoltaic module is known as a micro-inverter. Instead of connecting multiple modules to a single inverter, each module will have its own attached inverter. The output of each module is connected directly to the micro-inverter with the existing module leads. Multiple micro-inverters are connected in parallel on a single circuit, which then supplies the service panel of the home. Micro-inverters are required to be listed to ANSI/UL 1741 just as with the larger string inverters. Ground-fault protection, anti-islanding, and the other requirements of utility-interactive inverters are applicable. Output current for a single micro-inverter is approximately 0.8 amps. This would permit up to 15 inverters on a single 15-amp circuit (allowing for the continuous current multiplier). The inverter shown in Figure 33-6 is a micro-inverter.

An ac photovoltaic module is essentially a normal dc module with the micro-inverter installed at the factory. The dc conductors are covered and inaccessible on an ac module. The absence of field-installed dc string conductors is a big advantage with both micro-inverters and ac modules. Hazards associated with the dc string conductors are minimized with the use of micro-inverters and eliminated with ac modules. *NEC* requirements for ac circuits connecting micro-inverters/ac modules to service panels are similar to normal branch-circuit rules.

FIGURE 33-6 A utility-interactive micro-inverter.
(*Courtesy Enphase Energy*)

NATIONAL ELECTRICAL CODE REQUIREMENTS

Article 690, Solar Photovoltaic Systems, of the *NEC* provides the specific requirements for installation of utility-interactive PV systems. In addition, the general requirements of *NEC Chapters 1* through *4* apply, except as modified by *Article 690*. A utility-interactive PV system operates in parallel with the utility (primary source); portions of *Article 705, Interconnected Electric Power Production Sources*, will apply as well. We will look at some of the requirements of *Article 690* in the following section.

Part I. General

Article 690 begins by providing the scope or what is covered by the article and defining many terms used in the article and PV industry. Specific applicable sections of *Article 705* are referenced in *NEC 690.3*. An exception to *NEC 690.3* makes it clear that PV systems installed in hazardous (classified) locations must comply with the applicable portions of *Articles 500* through *516*. Many of the basic rules are covered in *690.4*. Separation of PV system conductors from other systems, arrangement of module connections, and the requirement for listing of equipment are all found in *NEC 690.4*. Ground-fault protection for grounded dc photovoltaic arrays is required, *NEC 690.5*. *Part I* ends with *NEC 690.6*, the rules for the installation of alternating current modules.

Part II. Circuit Requirements

Requirements for PV source and output circuits are found in this section. Voltage and current parameters for design of PV circuits are defined. Maximum dc voltage for a series string of modules is defined as the sum of the rated open-circuit voltage (V_{oc}) of the modules, corrected for the lowest expected ambient temperature. *Part I* must be used if the manufacturer does not provide temperature correction coefficients. This table provides multipliers that correspond to the lowest expected ambient temperature. Multiply the sum of the module V_{oc} by the factor from *Table 690.7* to find maximum voltage.

The size of PV system conductors and components is determined by the amount of current that flows through them. *NEC 690.8* provides the method for determining the circuit size and current. This is a two-step process. First, a multiplier is used to calculate the maximum source circuit current, *690.8(A)*; then a second multiplier is applied to determine the size of system conductors and overcurrent devices, *690.8(B)*. The rated module short-circuit current (I_{sc}) is the starting point for both of these calculations. Sunlight may be more intense in the field than at the testing lab, so *690.8(A)* requires the module short-circuit current to be multiplied by 1.25 (125%) to determine maximum circuit current. Module current is continuous (3 hours or more), which is why the maximum circuit current is required to be multiplied by 1.25 (125%) again for sizing of conductors and overcurrent devices, *690.8(B)*. An *Informational Note* states that applying both *690.8(A)(1)* and *(B)(1)* results in a multiplication factor of 1.56 (1.25 × 1.25 = 1.56). Multiplying module short-circuit current by 1.56 will provide the current value required to size the dc system conductors and overcurrent devices. Inverter current output is continuous, so the rated output of the inverter must be multiplied by 1.25 (125%) for sizing of conductors and overcurrent devices. See *690.8(A)(3)* and *690.8(B)(1)*.

Requirements for photovoltaic system overcurrent protection are found in *690.9*. The general requirement is that all conductors and equipment are to be protected in accordance with *Article 240, Overcurrent Protection*. There is an exception to this general requirement, which is often used. If circuit conductors were sized using the requirements of *690.8(B)* and there are no external sources of parallel currents that exceed the ampacity of the conductors, then overcurrent protection is not required. This exception permits parallel strings of modules to be combined without fuse protection of the individual strings. As long as the sum of the short-circuit currents from parallel strings does not exceed the rating of a faulted module or conductor, no overcurrent protection is required. Consider the following example: an array contains three parallel strings

of modules, and short-circuit current from each module/string is 7 amperes. If a short circuit occurs in one of the modules/strings, the maximum fault current would be the sum of the short-circuit currents from the other two strings. In this example the sum is 14 amperes (7 + 7 = 14). As long as the modules have a rating greater than 14 amperes, the exception to *690.9* permits combination of the strings without fuse protection. A fused combiner box would still be allowed, of course. Many designers specify fused combiner boxes even if not required.

Part III. Disconnecting Means

Means shall be provided to disconnect all current-carrying conductors of a PV source from all other conductors in a building or structure per *NEC 690.13*. This does not apply to grounded dc conductors. Energized grounded current-carrying conductors are not to be interrupted by the disconnect(s), just as with grounded ac conductors (neutrals). Several disconnects may be required for a typical residential installation. The direct-current circuit generated by the array must have a disconnecting means installed at a readily accessible location outside the building, or if inside, closest to the point of entrance. The disconnect is not permitted to be located in bathrooms. An exception to *690.14(C)(1)* permits the direct-current photovoltaic conductors to enter the building without a disconnecting means if contained in metal raceways or enclosures. See *690.31(E)*.

Photovoltaic equipment, such as an inverter, is required to have a disconnecting means. The disconnects are required to be grouped and identified if the equipment is energized from more than one source. A utility-interactive inverter is connected to two sources of electricity. The array on the roof generates direct current anytime the sun is shining, and the conductors that connect the inverter to the service are energized by the electric utility. Interruption of the array circuit to the inverter will prevent the inverter from supplying alternating current to the service panel, but the conductors are still energized by the utility supply. If the inverter is located within sight of the circuit breaker connection in the service panel, then the circuit breaker can serve as the disconnecting means. Always check with the local

Authority Having Jurisdiction (AHJ) and electrical utility company for policies regarding disconnecting means. Some utility companies will require a dedicated utility disconnect.

Part IV. Wiring Methods

Exposed single-conductor cable is permitted to be installed for array interconnection. Only types *USE-2* and listed *PV* wire are permitted, and there are other limitations. If the circuit operates at over 30 volts and is in a readily accessible location, a raceway must be used. Direct-current array conductors that enter the structure are required to be installed in metal raceways or enclosures from the point of penetration to the first readily accessible disconnecting means, per *690.31(E)*. Wiring methods for the inverter output circuit are only limited by the general requirements of *Chapters 1* through *4* of the *NEC*. Inverter wiring to the service is essentially a branch circuit if a load-side connection is used. A supply-side connection will involve the service wiring method limitations of *Article 230*. Junction, pull, and outlet boxes are permitted to be located behind removable modules that are connected with flexible wiring methods. See *690.34*. Ungrounded photovoltaic power systems are permitted if the installation complies with *690.35(A)* through *(G)*.

Part V. Grounding

According to *690.41*, one conductor of a 2-wire PV system over 50 volts shall be grounded. The exception to *690.42* allows the grounded conductor bond to be made by the ground-fault detection device required by *690.5*. The grounded conductor of the dc system can be either the positive or the negative conductor. Inverter design will determine which conductor is grounded, but any PV system over 50 volts is required to be a grounded system unless it complies with *690.35*.

Exposed non-current-carrying metal parts of module frames and other equipment are required to be grounded. Methods of grounding module frames will vary among different manufacturers. Installation instructions for the module must be referenced. The size of the equipment grounding conductor is determined by *NEC Table 250.122*.

When there is no overcurrent device in the circuit, the module/string short-circuit current shall be the assumed overcurrent device size for reference to *Table 250.122* per *690.45(A)*.

A grounding electrode system is required for both dc and ac photovoltaic systems. Systems with ac and dc grounding requirements, such as utility-interactive systems, must meet the requirements of *690.47(C)*. Optional methods of installing a grounding electrode system are permitted. All methods include a bond to the existing premises' grounding system. A conductor that serves as an equipment grounding conductor and the bond between ac and dc systems is permitted by *690.47(C)(3)* for an inverter with ground-fault protection. Ground-mounted arrays require a local grounding electrode. The structure of a ground-mounted array is permitted to serve as the grounding electrode if the requirements of *250.52(A)* are met. An array installed within 6 feet (1.8 m) of the premises' wiring grounding electrode will not require an additional grounding electrode according to *Exception No. 2* of *690.47(D)*.

Part VI. Marking

Photovoltaic modules will have labels marked with specific information as required by *690.51* (dc modules) and *690.52* (ac modules). This information is required so that system designers and installers are able to size the balance of system components. The completed system will have operating parameters (voltage and current) unique to the design. Field labels marked with the specific dc- and ac-operating parameters must be installed per the rules of *690.53* and *690.54*. Buildings with stand-alone or utility-interactive PV systems are required to have a plaque or directory indicating the presence of the system and location(s) of disconnecting means. See *690.56*. Installers of PV systems should be aware that there are marking requirements for specific components throughout *Article 690*.

Part VII. Connection to Other Sources

The rules for connection of the utility-interactive PV system to the electrical service are found in this part. An inverter or ac module is required to automatically de-energize its output when utility power is lost. An inverter must stay in this de-energized state until utility power is restored. This is known as "anti-islanding" and is a requirement of *690.61* as well as ANSI/UL 1741. A neutral conductor smaller than the ungrounded (phase) conductors is permitted per *690.62*. A smaller neutral for the inverter is only allowed when used for instrumentation or detection purposes. Installation instructions for an inverter will provide the minimum neutral conductor size.

The *National Electrical Code* permits two methods for connection of a utility-interactive PV system to the electrical service. Requirements for both methods are found in *690.64 Point of Connection*. A connection on the line side of the service disconnecting means is known as a supply-side connection, *690.64(A)*. Size of the photovoltaic system is virtually unlimited, but the rules of *NEC Article 230* will apply to these service conductors. Requirements for load-side connections are found in *690.64(B)*. The photovoltaic system size is limited by the panel bus rating and main circuit breaker size. See *690.64(B)(2)*. A dedicated PV system circuit breaker, suitable for backfeed and positioned at the opposite end of the bus from the main circuit breaker, is a requirement of *690.64(B)*. See Figures 33-7 and 33-8 for a representation of the two connection methods.

Part VIII. Storage Batteries and Part IX. Systems over 600 Volts

Requirements for photovoltaic storage battery systems are found in *Part VIII* of *Article 690*. The provisions of *Article 480, Storage Batteries,* will apply as well. Specific rules for dwelling units are found in *690.71(B)*.

Part IX of *Article 690* requires photovoltaic systems with a maximum dc voltage over 600 volts dc to comply with *Article 490, Equipment, over 600 Volts, Nominal*. Note that *690.7(C)* limits photovoltaic source and output circuits to 600 volts or less for one- and two-family dwellings.

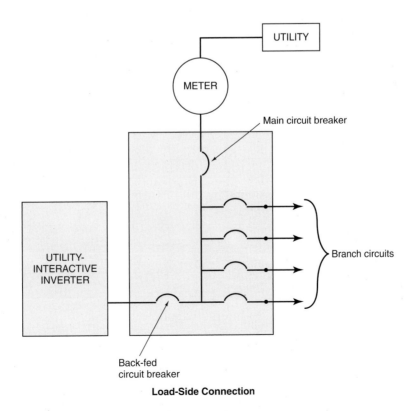

FIGURE 33-7 Load-side connection. (*Delmar/Cengage Learning*)

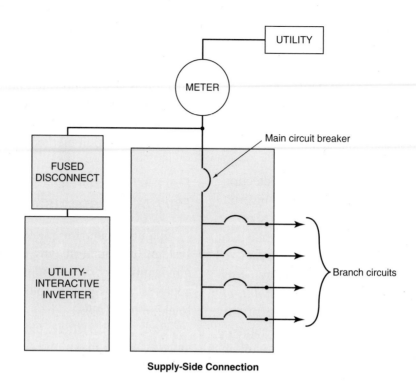

FIGURE 33-8 Supply-side connection. (*Delmar/Cengage Learning*)

REVIEW

Refer to the *National Electrical Code* when necessary. Where applicable, responses should be written in complete sentences.

1. Article _____ of the *National Electrical Code* contains most of the requirements for installation of photovoltaic systems.

2. *NEC Chapters* _____ through _____ will apply to PV installations except as modified by *Article* _____.

3. Name four electrical hazards associated with photovoltaic systems:

   _____

   _____

   _____

   _____

4. Name five components of a utility-interactive photovoltaic system:

   _____

   _____

   _____

   _____

   _____

5. Which conductor types are permitted to be installed exposed for array circuits?

   _____.

6. To size conductors and overcurrent devices for source circuits, you must multiply the module string short-circuit current by _____.

7. When is a metallic raceway required for photovoltaic source circuits?

   _____

8. Photovoltaic modules that also serve as an outer protective finish for a building are known as _____.

9. Grounding electrode system requirements for a utility-interactive photovoltaic system are found in section _____ of the *NEC*.

10. The two methods permitted for connection of a utility-interactive photovoltaic system to a service are known as _____ and _____.

Specifications for Electrical Work— Single-Family Dwelling

1. **GENERAL:** The "General Clause and Conditions" shall be and are hereby made a part of this division.

2. **SCOPE:** The electrical contractor shall furnish and install a complete electrical system as shown on the drawings and/or in the specifications. Where there is no mention of the responsible party to furnish, install, or wire for a specific item on the electrical drawings, the electrical contractor will be responsible completely for all purchases and labor for a complete operating system for this item.

3. **WORKMANSHIP:** All work shall be executed in a neat and workmanlike manner. All exposed conduits shall be routed parallel or perpendicular to walls and structural members. Junction boxes shall be securely fastened, set true and plumb, and flush with finished surface when wiring method is concealed.

4. **LOCATION OF OUTLETS:** The electrical contractor shall verify location, heights, outlet and switch arrangements, and equipment prior to rough-in. No additions to the contract sum will be permitted for outlets in wrong locations, in conflict with other work, and so on. The owner reserves the right to relocate any device up to 10 ft (3.0 m) prior to rough-in, without any charge by the electrical contractor.

5. **CODES:** The electrical installation is to be in accordance with the latest edition of the *National Electrical Code,* all local electrical codes, and the utility company's requirements.

6. **MATERIALS:** All materials shall be new and shall be listed and bear the appropriate label of Underwriters Laboratories, Inc., or another nationally recognized testing laboratory for the specific purpose. The material shall be of the size and type specified on the drawings and/or in the specifications.

7. **WIRING METHOD:** Wiring, unless otherwise specified, shall be nonmetallic-sheathed cable, armored cable, or electrical metallic tubing (EMT), adequately sized and installed according to the latest edition of the *National Electrical Code* and local ordinances.

8. **PERMITS AND INSPECTION FEES:** The electrical contractor shall pay for all permit fees, plan review fees, license fees, inspection fees, and taxes applicable to the electrical installation and shall be included in the base bid as part of this contract.

9. **TEMPORARY WIRING:** The electrical contractor shall furnish and install all temporary wiring for handheld tools and construction lighting per latest OSHA standards and *Article 590, NEC,* and include all costs in base bid.

10. **WORKSHOP:**   Workshop wiring is to be installed in EMT using steel compression gland fittings. Wiring with Type NM cable protected by EMT from the joist space to the outlet, device, or junction box is acceptable as is installation in Type MC cable.

11. **NUMBER OF OUTLETS PER CIRCUIT:**   In general, not more than ten (10) lighting and/or receptacle outlets shall be connected to any one lighting branch circuit. Exceptions may be made in the case of low-current-consuming outlets.

12. **CONDUCTOR SIZE:**   General lighting branch circuits shall be not smaller than 14 AWG copper protected by 15-ampere overcurrent devices. Small-appliance branch circuits and branch circuits serving bathroom receptacles shall be 12 AWG copper protected by 20-ampere overcurrent devices. Conductors shall be Type THHN. For service-entrance conductors, see Clause 28.

    All other circuits: conductors and overcurrent devices as required by the *Code*.

13. **LOAD BALANCING:**   The electrical contractor shall connect all loads, branch circuits, and feeders per the Panelboard Schedule, but shall verify and modify these connections as required to balance connected and calculated loads to within 10% variation.

14. **SPARE CONDUITS:**   Furnish and install two empty trade size 1 EMT conduits between workshop and attic for future use.

15. **GUARANTEE OF INSTALLATION:**   The electrical contractor shall guarantee all work and materials for a period of one full year after final acceptance by the architect/engineer, electrical inspector, and owner.

16. **APPLIANCE CONNECTIONS:**   The electrical contractor shall furnish all wiring materials and make all final electrical connections for all permanently installed appliances such as, but not limited to, the furnace, water heater, water pump, built-in ovens and ranges, food waste disposer, dishwasher, and clothes dryer.

    These appliances are to be furnished by the owner.

17. **CHIMES:**   Furnish and install two (2) two-tone door chimes where indicated on the plans, complete with two (2) push buttons and a suitable chime transformer. Allow $150.00 for above items. Chimes and buttons to be selected by owner.

18. **DIMMERS:**   Furnish and install dimmer switches where indicated.

19. **EXHAUST FANS:**   Furnish, install, and provide connections for all exhaust fans indicated on the plans, including, but not limited to, ducts, louvers, trims, speed controls, and lamps. Included are the recreation room, laundry, rear entry powder room, range hood, and bedroom hall ceiling fans. Allow a sum of $700.00 in the base bid for this. This allowance does not include the two bathroom heat/vent/light units.

20. **LUMINAIRES:**   A luminaire allowance of $1500.00 shall be included in the electrical contractor's bid for all surface-mounted luminaires and post lanterns, not including the ceiling paddle fans. Luminaires to be selected by owner. This allowance includes the three (3) medicine cabinets.

    The electrical contractor shall
    a. furnish and install all incandescent and fluorescent recessed luminaires.
    b. install all surface, recessed, track, strip, pendant, and hanging luminaires, and post lanterns.
    c. furnish and install two 2-lamp fluorescent luminaires above the workbench.
    d. furnish and install all lamps except as noted. Fluorescent lamps and incandescent lamps shall be energy-efficient type. Lamps for ceiling paddle fans to be furnished by owner.
    e. furnish and install all porcelain pull-chain and keyless lampholders.

    This luminaire allowance does not include the two bathroom ceiling heat/vent/light units. See Clause 21.

21. **HEAT/VENT/LIGHT CEILING UNITS:**   Furnish and install two heat/vent/light units where indicated on the plans, complete with switch assembly, ducts, and louvers required to perform the heating, venting, and lighting operations as recommended by the manufacturer.

22. **PLUG-IN STRIP:**   Where noted in the workshop, furnish and install a multioutlet assembly with six outlets.

23. **SWITCHES, RECEPTACLES, AND FACEPLATES:**   All flush switches shall be of the quiet ac-rated toggle type. They shall be mounted 46 in. (1.15 m) to center above the finished floor unless otherwise noted.

    Receptacle outlets shall be mounted 12 in. (300 mm) to center above the finished floor unless otherwise noted. All convenience receptacles shall be of the grounding type. Furnish and install, where indicated, ground-fault circuit interrupter (GFCI) receptacles to provide ground-fault circuit protection as required by the *National Electrical Code*. All wiring devices are to be provided with ivory handles or faces and shall be trimmed with ivory faceplates except in the kitchen, where chrome-plated steel faceplates shall be used.

    Receptacle outlets, where indicated, shall be of the split-circuit design.

24. **TELEVISION OUTLETS:**   Furnish and install 4-in. square, 1½-in.-deep outlet boxes with single-gang raised plaster covers at each television outlet where noted on the plans. Mount at the same height as receptacle outlets. Furnish and install 75-ohm coaxial cable to each television outlet from a point in the workshop near the main service-entrance switch. Allow 6 ft (1.8 m) of cable in workshop. Furnish and install television plug-in jacks at each location. Faceplates are to match other faceplates in home. All remaining work done by others.

25. **TELEPHONES:**   Furnish and install a 3-in.-deep device box or 4-in. square box, 1½-in. deep, with a suitable single-gang raised plaster cover at each telephone location, as indicated on the plans.

    Furnish and install four-conductor 18 AWG copper telephone cables from the telephone company's point of demarcation near the service-entrance panelboard to each designated telephone location. Terminate in a proper modular jack, complete with faceplates that match the electrical device faceplates. Allow 6 ft (1.8 m) of cable to hang below ceiling joists. Telephone company to furnish, install, and connect any and all equipment (including grounding connection) up to and including their Standard Network Interface (SNI) device.

    Installation shall be in accordance with any and all applicable *National Electrical Code* and local code regulations.

26. **MAIN SERVICE PANELBOARD:**   Furnish and install in the Workshop where indicated on the plans one (1) 200-ampere, 120/240-volt, single-phase, 3-wire panelboard with a 200-ampere main circuit breaker. Panelboard shall be series-rated for 22,000 amperes available fault current.

    Furnish and install all active and spare breakers as indicated on the Panelboard Schedule.

    Circuit breakers shall be AFCI, GFCI, or dual-function AFCI/GFCI as required by the *NEC*.

27. **SERVICE-ENTRANCE UNDERGROUND LATERAL CONDUCTORS:**   To be furnished and installed by the utility. Meter enclosure (pedestal type) to be furnished by the utility and installed by the electrical contractor where indicated on plans. Electrical contractor to furnish and install all panelboards, conduits, fittings, conductors, and other materials required to complete the service-entrance installation from the demarcation point of the utility's equipment to and including the Main Panelboard.

28. **SERVICE-ENTRANCE CONDUCTORS:**   Service-entrance conductors supplied by the electrical contractor shall be two 2/0 AWG THHN/THWN phase conductors and one 1 AWG bare neutral conductor. Install trade size 1½ EMT from Main Panelboard A to the meter pedestal.

29. **BONDING AND GROUNDING:**   Bond and ground service-entrance equipment in accordance with the latest edition of the *National Electrical Code*, local, and utility code requirements. Install one 4 AWG copper grounding electrode conductor.

30. **PANELBOARD B:** Furnish and install in the Recreation Room where indicated on the plans one (1) 20-circuit, 125-ampere, 120/240-volt, single-phase, 3-wire MLO load center. Panelboard shall be series-rated for 22,000 amperes available fault current.

> Furnish and install all active and spare breakers as indicated on the Panelboard Schedule.
> Circuit breakers shall be AFCI, GFCI, or dual-function AFCI/GFCI as required by the *NEC*.
> Feed panelboard with three 3 AWG THHN or THWN conductors from a 100-ampere, two-pole breaker in Main Panelboard A. Install conductors in trade size 1 EMT.

31. **CIRCUIT IDENTIFICATION:** All panelboards shall be furnished with typed-card directories with proper designation of the branch-circuit loads, feeder loads, and equipment served. The directories shall be located on the inside of the panelboard cover door in a holder for clear viewing.

32. **SEALING PENETRATIONS:** The electrical contractor shall seal and weatherproof all penetrations through foundations, exterior walls, and roofs.

33. **FINAL SITE CLEAN-UP:** Upon completion of the installation, the electrical contractor shall review and check the entire installation, clean equipment and devices, and remove surplus materials and rubbish from the owner's property, leaving the work in neat and clean order and in complete working condition. The electrical contractor shall be responsible for the removal of any cartons, debris, and rubbish for equipment installed by the electrical contractor, including equipment furnished by the owner or others and removed from the carton by the electrical contractor.

34. **SMOKE DETECTORS:** Furnish and install all smoke detectors and associated wiring per manufacturer's instructions and all codes. See Clause 5. Detectors to be of the ac/dc type. Interconnect to provide simultaneous signaling.

35. **SPECIAL-PURPOSE OUTLETS:** Install, provide, and connect all wiring for all special-purpose outlets. Upon completion of the job, all luminaires and appliances shall be operating properly. See plans and other sections of the specifications for information as to who is to furnish the luminaires and appliances.

36. **AFCI AND GFCI:** Furnish and install ground-fault circuit interrupter (GFCI) and arc-fault circuit interrupter (AFCI) protection as required by the latest edition of the *National Electrical Code*.

37. **CEILING PADDLE FANS:** The electrical contractor shall furnish and install outlet boxes identified as suitable for fan support at all locations where a ceiling suspended paddle fan is likely to be installed. The ceiling-suspended paddle fans, controls, timers, and lamps for same to be furnished by the owner and installed by the electrical contractor.

Schedule of Special-Purpose Outlets

Symbol	Description	Volts	Horse-power	Appliance Ampere Rating	Total Appliance Wattage Rating (or VA)	Circuit Ampere Rating	Poles	Wire Size THHN	Circuit Number	Comments
▲A	Hydromassage tub, Master Bedroom	120	½	10	1200	15	1	14	A9	Connect to separate circuit. Class "A" GFCI circuit breaker.
▲B	Water pump	240	1	8	1920	20	2	12	A(5−7)	Run circuit to disconnect switch on wall adjacent to pump. Install 10-ampere dual-element fuses as "back-up" protection. Pump controller has integral overload protection.
▲C	Water heater: top element 4500 W. Bottom element 4500 W.	240	−0−	18.8	4500	30	2	10	A(6−8)	Connected for limited demand. Both elements cannot operate at the same time.
▲D	Dryer	120/240	120 V ⅙ Motor Only	23.75	5700 Total	30	2	10	B(1−3)	Provide flush-mounted 30-A dryer receptacle.
▲E	Overhead garage door opener	120	See Unit 16 for explanation of overhead door operator horse-power rating	5.8	696	15	1	14	B14	Unit comes with 3-W cord. Provide box-cover unit (fuse/switch). Install Fustat Type S fuse, 8 amperes. Unit has integral protection. Fustat fuses are additional "back-up" protection. Connect to garage lighting circuit. Shall be GFCI protected
▲F	Wall-mounted oven	120/240	−0−	27.5	6600	30	2	10	B(6−8)	
▲G	Countertop range	120/240	−0−	31	7450	40	2	8	B(2−4)	
▲H	Food waste disposer	120	⅓	7.2	864	20	1	12	B19	Controlled by S. P. switch on wall to the right of sink.
▲I	Dishwasher	120	⅓	9.2 Includes 875-W heater	1104	20	1	12	B5	
▲J	Heat/vent/light Master Bedroom bath	120	−0−	12.5	1500	20	1	12	A12	
▲K	Heat/vent/light Front Bedroom bath	120	−0−	12.5	1500	20	1	12	A11	
▲L	Attic exhaust fan	120	¼	5.8	696	15	1	14	A10	Run circuit to 4-in. square box. Locate near fan in attic. Provide box-cover unit. Install Fustat Type S fuse, 8 amperes. Unit has integral protection. Fustat fuses provide additional back-up protection.
▲M	Electric furnace	240	⅓ Motor Only	Motor 3.5 heater 50.7 Total 54.2	13000	70	2	4	A(1−3)	The overcurrent device and branch-circuit conductors shall not be less than 125% of the total load of the heaters and motor. 54.2 × 1.25 = 67.75 (424.3(B)).
▲N	Air conditioner	240	−0−	26.4	6336	45	2	10	A(2−4)	
▲O	Freezer	120	¼	5.8	696	15	1	14	A13	Install single receptacle outlet.

Appendix

Useful Formulas			
To Find	**Single-Phase**	**Three-Phase**	**Direct Current**
AMPERES when kVA is known	$\dfrac{kVA \times 1000}{E}$	$\dfrac{kVA \times 1000}{E \times 1.73}$	Not applicable
AMPERES when horsepower is known	$\dfrac{HP \times 746}{E \times \% \text{ eff.} \times pf}$	$\dfrac{HP \times 746}{E \times 1.73 \times \% \text{ eff.} \times pf}$	$\dfrac{HP \times 746}{E \times \% \text{ eff.}}$
AMPERES when kilowatts is known	$\dfrac{kW \times 1000}{E \times pf}$	$\dfrac{kW \times 1000}{E \times 1.73 \times pf}$	$\dfrac{kW \times 1000}{E}$
HORSEPOWER	$\dfrac{I \times E \times \% \text{ eff.} \times pf}{746}$	$\dfrac{I \times E \, 1.73 \times \% \text{ eff.} \times pf}{746}$	$\dfrac{I \times E \times \% \text{ eff.}}{746}$
KILOVOLT AMPERES	$\dfrac{I \times E}{1000}$	$\dfrac{I \times E \times 1.73}{1000}$	Not applicable
KILOWATTS	$\dfrac{I \times E \times pf}{1000}$	$\dfrac{I \times E \times 1.73 \times pf}{1000}$	$\dfrac{I \times E}{1000}$
VOLT-AMPERES	$I \times E$	$I \times E \times 1.37 \times pf$	$I \times E$
WATTS	$I \times E \times pf$	$E \times I \times 1.37 \times pf$	$E \times I$

$$\text{ENERGY EFFICIENCY} = \frac{\text{Load Horsepower} \times 746}{\text{Load Input kVA} \times 1000}$$

$$\text{POWER FACTOR (pf)} = \frac{\text{Power Consumed}}{\text{Apparent power}} = \frac{W}{VA} = \frac{kW}{kVA} = \cos\varnothing$$

I = Amperes   E = Volts   kW = Kilowatts   kVA = Kilovolt-amperes

HP = Horsepower   % eff. = Percent Efficiency e.g., 90% eff. is 0.90   pf = Power Factor e.g., 95% pf is 0.95

EQUATIONS BASED ON OHM'S LAW:

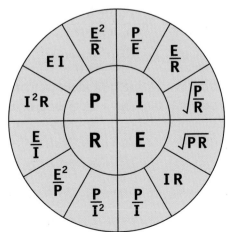

P = POWER, IN WATTS
I = CURRENT, IN AMPERES
R = RESISTANCE, IN OHMS
E = ELECTROMOTIVE FORCE, IN VOLTS

METRIC SYSTEM OF MEASUREMENT

The metric system is known as the International System of Units (SI), taken from the French Le Système International d'Unités. The following conversion table was developed from the latest National Institute of Standards and Technology (NIST) publications. In the table, whenever the slant line (/) is found, say it as "per."

The NIST Web site is http://www.nist.gov.

The *NEC* shows both inch-pound and metric units. The term *inch-pound* refers to the way dimensions (i.e., inches, feet, yards), weights (i.e., ounces, pounds), and volume (i.e., cubic inches, cubic feet) are given in the United States. Inch-pound units are also called U.S. Customary units. Whenever possible, use the measurements found in the *NEC* instead of making your own conversions.

Throughout the *NEC*, inch-pound values were converted to metric units in accordance to specific rules established by the National Fire Protection Association.

Soft Metric Conversion

A *soft conversion* is a direct mathematical conversion. A soft conversion is the result of taking a product already designed, manufactured to inch-pound values, then converting those values to metric units. The products dimension does not change in any way. Only the measurement units change.

Converting an inch-pound measurement to a metric measurement always results in bizarre, fractional values. The same holds true when converting a metric measurement to an inch-pound measurement.

In the *NEC*, to obtain a more convenient numerical expression and eliminate awkward values, an "exact metric conversion" is often rounded (up or down) to some value within acceptable tolerances of the original value, making sure that the final measurement does not violate a "maximum" or "minimum" *Code* requirement. This is called a *soft conversion*.

Here is an example of a soft conversion where there is a real safety issue.

 EXAMPLE ───────────

The minimum clearance from combustible material for a NON-IC recessed luminaire is ½ in. (13 mm). One-half inch is exactly 12.70 mm. The *NEC* rounded this measurement up to 13 mm to obtain a more convenient, easy-to-use numerical expression.

Here is an example of a soft conversion where safety is not an issue.

 EXAMPLE ───────────

Type NM cable shall be secured at intervals of 4½ ft (1.4 m) and not more than 12 in. (300 mm) from an electrical box. The exact conversion of 4½ ft is 1.3716 m and of 12 in. is 304.8 mm. These values would be cumbersome to use, so the *NEC* rounded off the measurements to obtain convenient numerical expressions.

Hard Metric Conversion

A *hard metric conversion* involves a change in physical size of a product or measurement. An inch-pound standard size is replaced with an accepted metric standard size for a particular purpose.

A hard metric dimension is attained by designing a product to metric dimensions. No conversion from inch-pound units is involved. For example, a box is designed to measure $50 \times 100 \times 100$ mm. It's a done deal.

Multiply This Unit(s)	By This Factor	To Obtain This Unit(s)
acre	4 046.9	square meters (m^2)
acre	43 560	square feet (ft^2)
ampere hour	3 600	coulombs (C)
angstrom	0.1	nanometer (nm)
atmosphere	101.325	kilopascals (kPa)
atmosphere	33.9	feet of water (at 4°C)
atmosphere	29.92	inches of mercury (at 0°C)
atmosphere	0.76	meter of mercury (at 0°C)
atmosphere	0.007 348	ton per square inch
atmosphere	1.058	tons per square foot
atmosphere	1.0333	kilograms per square centimeter
atmosphere	10 333	kilograms per square meter
atmosphere	14.7	pounds per square inch
Bar	100	kilopascals (kPa)
barrel (oil, 42 U.S. gallons)	0.158 987 3	cubic meter (m^3)
barrel (oil, 42 U.S. gallons)	158.987 3	liters (L)
board foot	0.002 359 737	cubic meter (m^3)
bushel	0.035 239 07	cubic meter (m^3)
Btu	778.16	foot-pounds
Btu	252	gram-calories
Btu	0.000 393 1	horsepower-hour
Btu	1 054.8	joules (J)
Btu	1.055 056	kilojoules (kJ)
Btu	0.000 293 1	kilowatt-hour (kWh)
Btu per hour	0.000 393 1	horsepower (hp)
Btu per hour	0.293 071 1	watt (W)
Btu per degree Fahrenheit	1.899 108	kilojoules per kelvin (kJ/K)
Btu per pound	2.326	kilojoules per kilogram (kJ/kg)
Btu per second	1.055 056	kilowatts (kW)
calorie	4.184	joules (J)
calorie, gram	0.003 968 3	Btu
candela per feet squared (cd/ft^2)	10.763 9	candelas per meter squared (cd/m^2)

Note: The former term *candlepower* has been replaced with the term *candela*.

candela per meter squared (cd/m^2)	0.092 903	candela per feet squared (cd/ft^2)
candela per meter squared (cd/m^2)	0.291 864	footlambert*
candela per square inch (cd/in^2)	1 550.003	candelas per square meter (cd/m^2)

Celsius = (Fahrenheit − 32) × 5/9		
Celsius = (Fahrenheit − 32) × 0.555555		
Celsius = (0.556 × Fahrenheit) − 17.8		

Note: The term *centigrade* was officially discontinued in 1948, and was replaced by the term *Celsius*. The term *centigrade* may still be found in some publications.

centimeter (cm)	0.032 81	foot (ft)
centimeter (cm)	0.393 7	inch (in.)
centimeter (cm)	0.01	meter (m)
centimeter (cm)	10	millimeters (mm)

(*continued*)

(*continued*)

Multiply This Unit(s)	By This Factor	To Obtain This Unit(s)
centimeter (cm)	393.7	mils
centimeter (cm)	0.010 94	yard
circular mil	0.000 005 067	square centimeter (cm.²)
circular mil	0.785 4	square mil
circular mil	0.000 000 785 4	square inch (in.²)
cubic centimeter (cm³)	0.061 02	cubic inch (in.³)
cubic foot per second	0.028 316 85	cubic meter per second (m³/s)
cubic foot per minute	0.000 471 947	cubic meter per second (m³/s)
cubic foot per minute	0.471 947	liter per second (L/s)
cubic inch (in.³)	16.39	cubic centimeters (cm³)
cubic inch (in.³)	0.000 578 7	cubic foot (ft³)
cubic meter (m³)	35.31	cubic feet (ft³)
cubic yard per minute	12.742 58	liters per second (L/s)
cup (c)	0.236 56	liter (L)
decimeter	0.1	meter (m)
decameter	10	meters (m)
Fahrenheit = (9/5 Celsius) + 32		
Fahrenheit = (Celsius × 1.8) + 32		
fathom	1.828 804	meters (m)
fathom	6.0	feet
foot	30.48	centimeters (cm)
foot	12	inches
foot	0.000 304 8	kilometer (km)
foot	0.304 8	meter (m)
foot	0.000 189 4	Mile (statute)
foot	304.8	millimeters (mm)
foot	12 000	mils
foot	0.333 33	yard
foot, cubic (ft³)	0.028 316 85	cubic meter (m³)
foot, cubic (ft³)	28.316 85	liters (L)
foot, board	0.002 359 737	cubic meter (m³)
cubic feet per second (ft³/s)	0.028 316 85	cubic meter per second (m³/s)
cubic feet per minute (ft³/min)	0.000 471 947	cubic meter per second (m³/s)
cubic feet per minute (ft³/min)	0.471 947	liter per second (L/s)
foot, square (ft²)	0.092 903	square meter (m²)
footcandle	10.763 91	lux (lx)
footlambert*	3.426 259	candelas per square meter (cd/m²)
foot of water	2.988 98	kilopascals (kPa)
foot-pound	0.001 286	Btu
foot-pound-force	1 055.06	joules (J)
foot-pound-force per second	1.355 818	joules (J)
foot-pound-force per second	1.355 818	watts (W)
foot per second	0.304 8	meter per second (m/s)
foot per second squared	0.304 8	meter per second squared (m/s²)

Multiply This Unit(s)	By This Factor	To Obtain This Unit(s)
gallon (U.S. liquid)	3.785 412	liters (L)
gallons per day	3.785 412	liters per day (L/d)
gallons per hour	1.051 50	milliliters per second (mL/s)
gallons per minute	0.063 090 2	liter per second (L/s)
gauss	6.452	lines per square inch
gauss	0.1	millitesla (mT)
gauss	0.000 000 064 52	weber per square inch
grain	64.798 91	milligrams (mg)
gram (g) (a little more than the weight of a paper clip)	0.035 274	ounce (avoirdupois)
gram (g)	0.002 204 6	pound (avoirdupois)
gram per meter (g/m)	3.547 99	pounds per mile (lb/mile)
grams per square meter (g/m^2)	0.003 277 06	ounces per square foot (oz/ft^2)
grams per square meter (g/m^2)	0.029 494	ounces per square yard (oz/yd^2)
gravity (standard acceleration)	9.806 65	meters per second squared (m/s^2)
quart (U.S. liquid)	0.946 352 9	liter (L)
horsepower (550 ft • lbf/s)	0.745 7	kilowatt (kW)
horsepower	745.7	watts (W)
horsepower hours	2.684 520	megajoules (MJ)
inch per second squared (in./s^2)	0.025 4	meter per second squared (m/s^2)
inch	2.54	centimeters (cm)
inch	0.254	decimeter (dm)
inch	0.083 3	feet
inch	0.025 4	meter (m)
inch	25.4	millimeters (mm)
inch	1 000	mils
inch	0.027 78	yard
inch, cubic (in.^{3})	16 387.1	cubic millimeters (mm^3)
inch, cubic (in.^{3})	16.387 06	cubic centimeters (cm^3)
inch, cubic (in.^{3})	645.16	square millimeters (mm^2)
inches of mercury	3.386 38	kilopascals (kPa)
inches of mercury	0.033 42	atmosphere
inches of mercury	1.133	feet of water
inches of water	0.248 84	kilopascal (kPa)
inches of water	0.073 55	inch of mercury
joule (J)	0.737 562	foot-pound-force (ft•lbf)
kilocandela per meter squared (kcd/m^2)	0.314 159	lambert*
kilogram (kg)	2.204 62	pounds (avoirdupois)
kilogram (kg)	35.274	ounces (avoirdupois)
kilogram per meter (kg/m)	0.671 969	pound per foot (lb/ft)
kilogram per square meter (kg/m^2)	0.204 816	pound per square foot (lb/ft^2)
kilogram-meter-squared (kg•m^2)	23.730 4	pounds-foot squared (lb•ft^2)
kilogram-meter-squared (kg•m^2)	3 417.17	pounds-inch squared (lb•in.^{2})
kilogram per cubic meter (kg/m^3)	0.062 428	pound per cubic foot (lb/ft^3)
kilogram per cubic meter (kg/m^3)	1.685 56	pound per cubic yard (lb/yd^3)

(*continued*)

(*continued*)

Multiply This Unit(s)	By This Factor	To Obtain This Unit(s)
kilogram per second (kg/s)	2.204 62	pounds per second (lb/s)
kilojoule (kJ)	0.947 817	Btu
kilometer (km)	1 000	meters (m)
kilometer (km)	0.621 371	mile (statute)
kilometer (km)	1,000,000	millimeters (mm)
kilometer (km)	1 093.6	yards
kilometer per hour (km/h)	0.621 371	mile per hour (mph)
kilometer squared (km^2)	0.386 101	square mile (mile^2)
kilopound-force per square inch	6.894 757	megapascals (MPa)
kilowatts (kW)	56.921	Btus per minute
kilowatts (kW)	1.341 02	horsepower (hp)
kilowatts (kW)	1 000	watts (W)
kilowatt-hour (kWh)	3 413	Btus
kilowatt-hour (kWh)	3.6	megajoules (MJ)
knots	1.852	kilometers per hour (km/h)
lamberts*	3 183.099	candelas per square meter (cd/m^2)
lamberts*	3.183 01	kilocandelas per square meter (kcd/m^2)
liter (L)	0.035 314 7	cubic foot (ft^3)
liter (L)	0.264 172	gallon (U.S. liquid)
liter (L)	2.113	pints (U.S. liquid)
liter (L)	1.056 69	quarts (U.S. liquid)
liter per second (L/s)	2.118 88	cubic feet per minute (ft^3/min)
liter per second (L/s)	15.850 3	gallons per minute (gal/min)
liter per second (L/s)	951.022	gallons per hour (gal/hr)
lumen per square foot (lm/ft^2)	10.763 9	lux (lx)
lumen per square foot (lm/ft^2)	1.0	footcandles
lumen per square foot (lm/ft^2)	10.763	lumens per square meter (lm/m^2)
lumen per square meter (lm/m^2)	1.0	lux (lx)
lux (lx)	0.092 903	lumen per square foot (foot-candle)
maxwell	10	nanowebers (nWb)
megajoule (MJ)	0.277 778	kilowatt-hours (kWh)
meter (m)	100	centimeters (cm)
meter (m)	0.546 81	fathom
meter (m)	3.2809	feet
meter (m)	39.37	Inches
meter (m)	0.001	kilometer (km)
meter (m)	0.000 621 4	mile (statute)
meter (m)	1 000	millimeters (mm)
meter (m)	1.093 61	yards
meter, cubic (m^3)	1.307 95	cubic yards (yd^3)
meter, cubic (m^3)	35.314 7	cubic feet (ft^3)
meter, cubic (m^3)	423.776	board feet
meter per second (m/s)	3.280 84	feet per second (ft/s)
meter per second (m/s)	2.236 94	miles per hour (mph)

Multiply This Unit(s)	By This Factor	To Obtain This Unit(s)
meter, squared (m^2)	1.195 99	square yards (yd^2)
meter, squared (m^2)	10.763 9	square feet (ft^2)
mho per centimeter (mho/cm)	100	siemens per meter (S/m)

Note: The older term *mho* has been replaced with *siemens*. The term *mho* may still be found in some publications.

micro inch	0.025 4	micrometer (μm)
mil	0.002 54	centimeters
mil	0.000 083 33	feet
mil	0.001	Inches
mil	0.000 000 025 40	kilometers
mil	0.000 027 78	yards
mil	25.4	micrometers (μm)
mil	0.025 4	millimeter (mm)
miles per hour	1.609 344	kilometers per hour (km/h)
miles per hour	0.447 04	meter per second (m/s)
miles per gallon	0.425 143 7	kilometer per liter (km/L)
miles	1.609 344	kilometers (km)
miles	5 280	feet
miles	1 609	meters (m)
miles	1 760	yards
miles (nautical)	1.852	kilometers (km)
miles squared	2.590 000	kilometers squared (km^2)
millibar	0.1	kilopascal (kPa)
milliliter (mL)	0.061 023 7	cubic inch ($in.^3$)
milliliter (mL)	0.033 814	fluid ounce (U.S.)
millimeter (about the thickness of a dime)	0.1	centimeter (cm)
millimeter (mm)	0.003 280 8	foot
millimeter (mm)	0.039 370 1	Inch
millimeter (mm)	0.001	meter (m)
millimeter (mm)	39.37	mils
millimeter (mm)	0.001 094	yard
millimeter squared (mm^2)	0.001 550	square inch ($in.^2$)
millimeter cubed (mm^3)	0.000 061 023 7	cubic inches ($in.^3$)
millimeter of mercury	0.133 322 4	kilopascal (kPa)
ohm	0.000 001	megohm
ohm	1,000,000	micro ohms
ohm circular mil per foot	1.662 426	nano ohms meter ($n\Omega \cdot m$)
oersted	79.577 47	amperes per meter (A/m)
ounce (avoirdupois)	28.349 52	grams (g)
ounce (avoirdupois)	0.062 5	pound (avoirdupois)
ounce, fluid	29.573 53	milliliters (mL)
ounce (troy)	31.103 48	grams (g)
ounce per foot squared (oz/ft^2)	305.152	grams per meter squared (g/m^2)
ounces per gallon (U.S. liquid)	7.489 152	grams per liter (g/L)
ounce per yard squared (oz/yd^2)	33.905 7	grams per meter squared (g/m^2)

(*continued*)

(continued)

Multiply This Unit(s)	By This Factor	To Obtain This Unit(s)
pica	4.217 5	millimeters (mm)
pint (U.S. liquid)	0.473 176 5	liter (L)
pint (U.S. liquid)	473.177	milliliters (mL)
pound (avoirdupois)	453.592	grams (g)
pound (avoirdupois)	0.453 592	kilograms (kg)
pound (avoirdupois)	16	ounces (avoirdupois)
poundal	0.138 255	newton (N)
pound-foot (lb•ft)	0.138 255	kilogram-meter (kg•m)
pound-foot per second	0.138 255	kilogram-meter per second (kg•m/s)
pound-foot squared (lb•ft^2)	0.042 140 1	kilogram-meter squared (kg•m^2)
pound-force	4.448 222	newtons (N)
pound-force foot	1.355 818	newton-meters (N•m)
pound-force inch	0.112 984 8	newton-meter (N•m)
pound-force per square inch	6.894 757	kilopascals (kPA)
pound-force per square foot	0.047 880 26	kilopascal (kPa)
pound per cubic foot (lb/ft^3)	16.018 46	kilograms per cubic meter (kg/m^3)
pound per foot (lb/ft)	1.488 16	kilograms per meter (kg/m)
pound per foot squared (lb/ft^2)	4.882 43	kilograms per meter squared (kg/m^2)
pound per foot cubed (lb/ft^3)	16.018 5	kilograms per meter cubed (kg/m^3)
pound per gallon (U.S. liquid)	119.826 4	grams per liter (g/L)
pound per second (lb/s)	0.453 592	kilogram per second (kg/s)
pound-inch squared (lb•in.^{2})	292.640	kilograms-millimeter squared (kg•mm^2)
pound per mile	0.281 849	gram per meter (g/m)
pound-square foot (lb•ft^2)	0.042 140 11	kilogram square meter (kg/m^2)
pound per cubic yard (lb/yd^3)	0.593 276	kilogram per cubic meter (kg/m^3)
quart (U.S. liquid)	946.353	milliliters (mL)
square centimeter (cm^2)	197 300	circular mils
square centimeter (cm^2)	0.001 076	square foot (ft^2)
square centimeter (cm^2)	0.155	square inch (in.^{2})
square centimeter (cm^2)	0.000 1	square meter (m^2)
square centimeter (cm^2)	0.000 119 6	square yard (yd^2)
square foot (ft^2)	144	square inches (in.^{2})
square foot (ft^2)	0.092 903 04	square meter (m^2)
square foot (ft^2)	0.111 1	square yard (yd^2)
square inch (in.^{2})	1 273 000	circular mils
square inch (in.^{2})	6.4516	square centimeters (cm^2)
square inch (in.^{2})	0.006 944	square foot (ft^2)
square inch (in.^{2})	645.16	square millimeters (mm^2)
square inch (in.^{2})	1,000,000	square mils (m^2)
square meter (m^2)	10.764	square feet (ft^2)
square meter (m^2)	1 550	square inches (in.^{2})
square meter (m^2)	0.000 000 386 1	square mile
square meter (m^2)	1.196	square yards (yd^2)
square mil	1.273	circular mils

Multiply This Unit(s)	By This Factor	To Obtain This Unit(s)
square mil	0.000 001	square inch (in.^{2})
square mile	2.589 988	square kilometers (km^2)
square millimeter (mm^2)	1 973	circular mils
square yard (yd^2)	0.836 127 4	square meter (m^2)
tablespoon (tbsp)	14.786 75	milliliters (mL)
teaspoon (tsp)	4.928 916 7	milliliters (mL)
therm	105.480 4	megajoules (MJ)
ton (long) (2 240 lb)	1 016.047	kilograms (kg)
ton (long) (2 240 lb)	1.016 047	metric tons (t)
ton, metric	2 204.62	pounds (avoirdupois)
ton, metric	1.102 31	tons, short (2 000 lb)
ton, refrigeration	12 000	Btus per hour
ton, refrigeration	4.716 095 9	horsepower-hours
ton, refrigeration	3.516 85	kilowatts (kW)
ton (short) (2 000 lb)	907.185	kilograms (kg)
ton (short) (2 000 lb)	0.907 185	metric ton (t)
ton per cubic meter (t/m^3)	0.842 778	ton per cubic yard (ton/yd^3)
ton per cubic yard (ton/yd^3)	1.186 55	tons per cubic meter (ton/m^3)
torr	133.322 4	pascals (Pa)
watt (W)	3.412 14	Btus per hour (Btu/hr)
watt (W)	0.001 341	horsepower
watt (W)	0.001	kilowatt (kW)
watt-hour (Wh)	3.413	Btus
watt-hour (Wh)	0.001 341	horsepower-hour
watt-hour (Wh)	0.001	kilowatt-hour (kW/h)
yard	91.44	centimeters (cm)
yard	3	feet
yard	36	Inches
yard	0.000 914 4	kilometer (km)
yard	0.914 4	meter (m)
yard	914.4	millimeters
yard, cubic (yd^3)	0.764 55	cubic meter (m^3)
yard, squared (yd^2)	0.836 127	meter squared (m^2)

* These terms are no longer used, but may still be found in some publications.

ARCHITECTURAL DRAFTING SYMBOLS

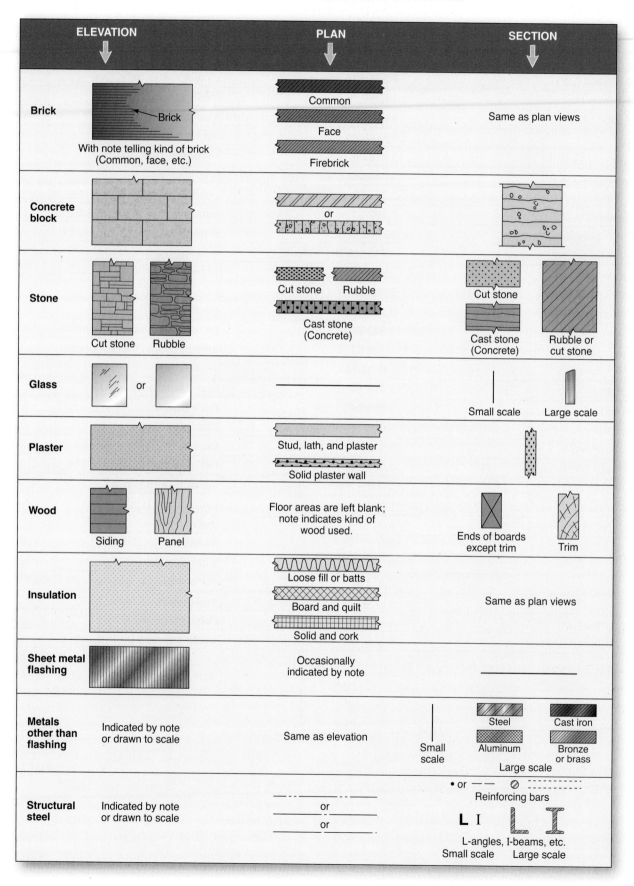

ARCHITECTURAL DRAFTING SYMBOLS (*CONTINUED*)

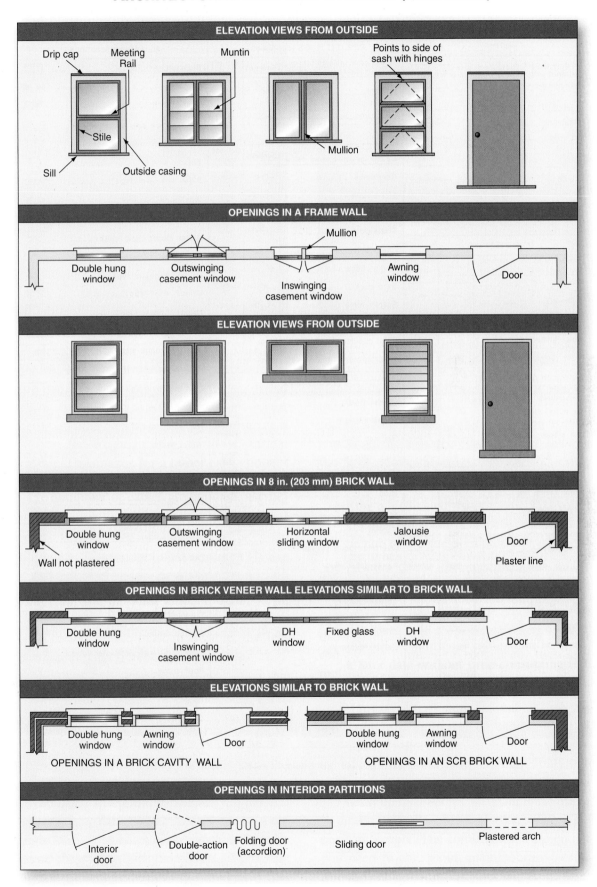

STANDARD SYMBOLS FOR PLUMBING, PIPING, AND VALVES

PIPING

Piping, in general _____
(Lettered with name of material conveyed)
Non-intersecting pipes _____
Steam _____
Condensate _____
Cold water _____
Hot water _____
Air _____
Vacuum _____
Gas _____
Refrigerant _____
Oil _____

PIPE FITTINGS

For welded or soldered fittings, use joint indication shown below	Screwed	Bell and spigot
Joint		
Elbow—90°		
Elbow—45°		
Elbow—turned up		
Elbow—turned down		
Elbow—long radius		
Side outlet elbow—outlet down		
Side outlet elbow—outlet up		
Base elbow		
Double-branch elbow		
Single-sweep tee		
Double-sweep tee		
Reducing elbow		
Tee		
Tee—outlet up		
Tee—outlet down		
Side outlet tee outlet up		
Side outlet tee outlet down		
Cross		
Reducer		
Eccentric reducer		

PIPE FITTINGS (continued)

For welded or soldered fittings, use joint indication shown below	Screwed	Bell and spigot
Lateral		
Expansion joint flanged		

VALVES

For welded or soldered fittings, use joint indication shown below	Screwed	Bell and spigot
Gate valve		
Globe valve		
Angle globe valve		
Angle gate valve		
Check valve		
Angle check valve		
Stop cock		
Safety valve		
Quick opening valve		
Float opening valve		
Motor operated gate valve		

PLUMBING

Corner bath _____
Recessed bath _____
Roll rim bath _____
Sitz bath _____ (SS)
Foot bath _____ (FB)
Bidet _____ (B)
Shower stall _____
Shower head _____ (Plan) (Elev)
Overhead gang shower _____ (Plan) (Elev)
Pedestal lavatory _____ (PL)
Wall lavatory _____ (WL)
Corner lavatory _____ (Lav)
Manicure lavatory Medical lavatory _____ (ML)
Dental lavatory _____ (Dental lav)

PLUMBING (continued)

Plain kitchen sink _____ (S)
Kitchen sink, R & L drain board _____
Kitchen sink, L H drain board _____
Combination sink & dishwasher _____
Combination sink & laundry tray _____ (S&T)
Service sink _____ (SS)
Wash sink (Wall type) _____
Wash sink _____
Laundry tray _____ (LT)
Water closet (Low tank) _____
Water closet (No tank) _____
Urinal (Pedestal type) _____
Urinal (Wall type) _____
Urinal (Corner type) _____
Urinal (Stall type) _____
Urinal (Trough type) _____ (TU)
Drinking fountain (Pedestal type) _____ (DF)
Drinking fountain (Wall type) _____ (DF)
Drinking fountain (Trough type) _____ (DF)
Hot water tank _____ (HW)
Water heater _____ (WH)
Meter _____ (M)
Hose rack _____ (HR)
Hose bibb _____ (HB)
Gas outlet _____ (G)
Vacuum outlet _____
Drain _____ (D)
Grease separator _____
Oil separator _____
Cleanout _____
Garage drain _____
Floor drain with backwater valve _____
Roof sump _____

Types of joints

Flanged	Screwed	Bell & spigot	Welded	Soldered

SHEET METAL DUCTWORK SYMBOLS

Blank Off. adjustable	Damper automatic	Damper deflecting	Damper deflecting up

Damper deflecting down	Damper volume	Duct flow direction	Duct inclined drop
	PLAN		
	ELEVATION		

Duct inclined rise	Duct section exhaust, return	Duct section supply	Duct section notation
	riser to 2nd floor	riser to 2nd floor	type exhaust
	riser to 1st floor	riser to 1st floor	place kitchen

Duct connection below joist	Fan flexible connection	Vanes	Louver & screen air intake

Ventilator, cowl	Ventilator, gooseneck	Ventilator, rainproof	Ventilator, standard
PLAN	PLAN	PLAN	PLAN
ELEVATION	ELEVATION	ELEVATION	ELEVATION

SINGLE LINE REPRESENTATION

Supply	Return	Damper & retractor	Anchor
S	R		PA.

Hanger	Expansion joint	Louver opening	Register or grille
H		L	

SINGLE-FAMILY DWELLING SERVICE-ENTRANCE CALCULATIONS

1. **General Lighting Load (*220.12*).**

   _____ ft² @ 3 VA per ft²   = _____ VA

   Note: Included in this floor area calculation are all lighting outlets and general-use receptacles. Do not include open porches, garages, or unused or unfinished spaces not adaptable for future use. See *NEC 220.12, Table 220.12,* and *220.14(J).*

2. **Minimum Number of 15-ampere Lighting Branch Circuits.**

   $$\frac{\text{Line 1}}{120} = \frac{}{120} \qquad = \text{_____amperes}$$

   then, $\dfrac{\text{amperes}}{15}$ $\dfrac{}{15}$ = _____15-ampere branch circuits

3. **Small-Appliance Load [*210.11(C)(1), 220.52(A),* and *210.52(B)*].**

   (Minimum of two 20-ampere branch circuits)

   _____ branch circuits @ 1500 VA each   = _____ VA

4. **Laundry Branch Circuit [*210.11(C)(2), 220.52(B),* and *210.52(F)*].**

   (Minimum of one 20-ampere branch circuit)

   _____ branch circuit(s) @ 1500 VA each   = _____ VA

5. **Total General Lighting, Small-Appliance, and Laundry Load.**

   Lines 1 + 3 + 4   = _____ VA

6. **Net Calculated General Lighting, Small-Appliance, and Laundry Loads (less ranges, ovens, and "fastened-in-place" appliances). Apply demand factors from *Table 220.42.***

   a. First 3000 VA @ 100%   = ___3000___ VA

   b. Line 5 _____ − 3000 = _____ @ 35%   = _____ VA

   Total a + b   = _____ VA

7. **Electric Range, Wall-Mounted Ovens, Counter-Mounted Cooking Units (*Table 220.55*).**   = _____ **VA**

8. **Electric Clothes Dryer (*Table 220.51*).**   = _____ **VA**

9. **Electric Furnace (*220.54*).**

   Air Conditioner, Heat Pump (*Article 440*).

   (Enter largest value, *220.60*)   = _____ **VA**

10. **Net Calculated General Lighting, Small-Appliance, Laundry, Ranges, Ovens, Cooktop Units, HVAC.**

    Lines 6 + 7 + 8 + 9   = _____ VA

11. **List "Fastened-in-Place" Appliances *in addition* to Electric Ranges, Electric Clothes Dryers, Electric Space Heating, and Air-Conditioning Equipment.**

Appliance	VA Load
Water heater:	= _____ VA
Dishwasher:	= _____ VA
Garage door opener:	= _____ VA
Food waste disposer:	= _____ VA
Water pump:	= _____ VA
Gas-fired furnace:	= _____ VA
Sump pump:	= _____ VA
Other: _____	= _____ VA
_____	= _____ VA
_____	= _____ VA
_____	= _____ VA
Total	= _____ VA

12. **Apply 75% Demand Factor (*220.53*) if Four or More "Fastened-in-Place" Appliances. If Less Than Four, Figure @ 100%.** Do not include electric ranges, electric clothes dryers, electric space heating, or air-conditioning equipment.

    Line 11 Total: _____ × 0.75   = _____ VA

13. **Total Calculated Load (Lighting, Small-Appliance, Ranges, Dryer, HVAC, "Fastened-in-Place" Appliances).**

    Line 10 _____ + Line 12 _____   = _____ VA

14. **Add 25% of Largest Motor (*220.50* and *430.24*).**

    _____ × 0.25   = _____ VA

    Note: The largest motor can be difficult to determine because nothing is in place when service-entrance load calculations are made. It might be an air-conditioning unit or a heat pump. If the dwelling is cooled by an evaporative cooler, the largest motor might be a water pump, a large attic exhaust fan, a large food waste disposer, or a sump pump. For simplicity in this example, the water pump was chosen. The additional 25% of the largest motor is a small portion of the total service-entrance load calculation.

15. **Grand Total Line 13 + Line 14.**   = _____ VA

16. **Minimum Ampacity for Ungrounded Service-Entrance Conductors.**

    $\text{Amperes} = \dfrac{\text{Line 15}}{240} = \dfrac{\text{_____}}{240}$ = _____ amperes

17. **Ungrounded Conductor Size (copper).**   _____ AWG

    Note: Table 310.15(B)(7) may be used only for 120/240-volt, 3-wire, residential single-phase service-entrance conductors, service lateral conductors, and feeder conductors that serve as the main power feeder to a dwelling unit.

18. **Minimum Ampacity for Neutral Service-Entrance Conductor, *220.61* and *310.15(B)(7)*. Do Not Include Straight 240-Volt Loads.**

    a. Line 6:   = _____ VA

    b. Line 7: _____ @ 0.70   = _____ VA

    c. Line 8: _____ @ 0.70   = _____ VA

    d. Line 11: (Include only 120-volt loads.)

        _____   _____ VA

        _____   _____ VA

        _____   _____ VA

        _____   _____ VA

        _____   _____ VA

            Total   _____ VA

    e. Line d total @ 75% demand factor if four or more per *220.53*, otherwise use 100%.

        _____ × 0.75   = _____ VA

    f. Add 25% of largest 120-volt motor.

        _____ × 0.25   = _____ VA

                                Total = _____ VA

    g. Total a + b + c + e + f.   = _____ VA

$$\text{Amperes} = \frac{\text{Line g}}{240} = \frac{}{240} \qquad = \underline{\hspace{2cm}} \text{ amperes}$$

19. **Neutral Conductor Size (copper)(*220.61*).**   _____ AWG

    Note: NEC *310.15(B)(7)* permits the neutral conductor to be smaller than the ungrounded "hot" conductors if the requirements of *215.2, 220.61*, and *230.42* are met. NEC *220.61* states that a feeder or service neutral load shall be the maximum unbalance of the load determined by *Article 220*. When bare conductors are used with insulated conductors, the conductors' ampacity is based on the lowest temperature rating of the insulated conductors in the raceway, *310.15(B)(4)*. The neutral conductor shall not be smaller than the grounding electrode conductor, *250.24(C)(1)*.

20. **Grounding Electrode Conductor Size (copper) (*Table 250.66*).**   _____ AWG

21. **Raceway Size.**   _____ Trade Size

    Obtain dimensional data from *Table 1, Table 4, Table 5*, and *Table 8, Chapter 9, NEC*.

SINGLE-FAMILY DWELLING SERVICE-ENTRANCE CALCULATIONS

JOB _____

Note: See back of page for line item instructions.

1. **General Lighting Load** _____ ft² @ 3 VA per ft² = _____ VA

2. **Minimum Number of 15-Ampere Lighting Branch Circuits.**

$$\frac{\text{Line 1}}{120} = \frac{}{120} = \text{amperes, then } \frac{\text{amperes}}{15} = \frac{}{15} = 15 \text{ ampere lighting branch circuits}$$

3. **Small-Appliance Load** _____ branch circuits @ 1,500 VA each = _____ VA

4. **Laundry Branch Circuit** ____branch circuit(s) @ 1,500 VA each = _____ VA

5. **Total General Lighting, Small-Appliance, and Laundry Load.**
   Lines 1 + 3 + 4 = _____ VA

6. **General Lighting, Small-Appliance, Laundry.**

         a. First 3000 VA @ 100% = _____ VA

         b. Line 5 _____ − 3000 = _____ @ 35% = _____ VA

         Total a + b = _____ VA

7. **Electric Range, Ovens, Counter-Mounted Cooking Units** = _____ VA

8. **Electric Clothes Dryer** = _____ VA

9. **Electric Furnace, Air Conditioner, Heat Pump (Enter largest)** = _____ VA

10. **Add Lines 6 + 7 + 8 + 9** = _____ VA

11. **List "Fastened-in-Place" Appliances *in addition* to Electric Ranges, Electric Clothes Dryers, Electric Space Heating, and Air-Conditioning Equipment.**

Appliance	VA Load	
Water heater:	=	_____ VA
Dishwasher:	=	_____ VA
Garage door opener:	=	_____ VA
Food waste disposer:	=	_____ VA
Water pump:	=	_____ VA
Gas-fired furnace:	=	_____ VA
Sump pump:	=	_____ VA
Other: _____	=	_____ VA
_____	=	_____ VA
_____	=	_____ VA
Total	=	_____ VA

12. **Apply 75% Demand Factor (*220.53*) if Four or More "Fastened-in-Place" Appliances. If Less Than Four, Figure @ 100%.**

        Line 11 Total: _____ × 0.75 = _____ VA

13. **Total Calculated Load (Lighting, Small-Appliance, Ranges, Dryer, HVAC, "Fastened-in-Place" Appliances).**

        Line 10 _____ + Line 12 _____ = _____ VA

14. **Add 25% of Largest Motor (*220.50* and *430.24*).**

        _____ × 0.25 = _____ VA

15. Grand Total Line 13 + Line 14.   = _____ VA

16. Minimum Ampacity for Ungrounded Service-Entrance Conductors.

$$\text{Amperes} = \frac{\text{Line 15}}{240} = \frac{\underline{\hspace{2cm}}}{120} = \underline{\hspace{3cm}} \text{ amperes}$$

17. Ungrounded Conductor Size (copper). _____ **AWG**

18. Minimum Ampacity for Neutral Service-Entrance Conductor

    a. Line 6 total: = _____ VA

    b. Line 7: _____ @ 0.70   = _____ VA

    c. Line 8: _____ @ 0.70   = _____ VA

    d. Line 11: (Include only 120-volt loads.)

          _____ VA   _____ VA

          _____ VA   _____ VA

          Total VA   _____ VA

    e. Line d total @ 75% demand factor if four or more per 220.53; otherwise use 100%.

          _____ × 0.75 = _____ VA

    f. Add 25% of largest 120-volt motor. _____ × 0.25 = _____ VA

    g. Total a + b + c + e + f   Total = _____ VA

19. Neutral Conductor Size (copper)   _____ AWG

20. Grounding Electrode Conductor Size (copper)   _____ AWG

21. Raceway Size.   _____ Trade Size (inch)

SINGLE-FAMILY DWELLING SERVICE-ENTRANCE CALCULATIONS
Line item instructions

1. Included in this floor area calculation are all lighting outlets and general-use receptacles. Do not include open porches, garages, or unused or unfinished spaces not adaptable for future use. See *NEC 220.12, Table 220.12*, and *220.14(J)*.

2. No further explanation needed.

3. Minimum of two 20-ampere branch-circuits. See *210.11(C)(1)*, *210.52(B)*, and *220.52(A)*.

4. Minimum of one 20-ampere branch-circuit. See *210.11(C)(2)*, *210.52(F)*, and *220.52(B)*.

5. No further explanation needed.

6. Apply the 100% and 35% demand factors from *Table 220.42*.

7. See *Table 220.55* for load values. Typical household electric ranges come under the 8 kW level.

8. See *Table 220.54* for load values. 5000 VA minimum or nameplate rating, whichever is larger.

9. Enter largest value, *220.60*. Also see *220.51* and *Article 440*.

10. This is the total for General Lighting, Small-Appliance, Laundry, Ranges, Ovens, Cooktop Units, and HVAC.

11. List fastened-in-place appliances. See *220.53*.

12. Apply 75% demand factor. Do not include electric ranges, electric clothes dryers, electric space heating, or air-conditioning equipment. See *220.53*.

13. No further explanation needed.

14. What is the largest motor? The largest motor can be difficult to determine because nothing is in place when service-entrance load calculations are made. It might be an air-conditioning unit or a heat pump. If the dwelling is cooled by an evaporative cooler, the largest motor might be the evaporative cooler motor. It could also be a furnace fan motor, a water pump, a large attic exhaust fan, a sump pump, or a large food waste disposer. Don't waste valuable time quibbling. Enter a large volt-ampere value to be on the safe side.

15. No further explanation needed.

16. No further explanation needed.

17. Shall not be smaller than 100 amperes, *230.42(B)* and *230.79*. See *Table 310.15(B)(16)*. Special conductor sizing might be permitted. *Table 310.15(B)(7)* may be used only for 120/240-volt, 3-wire, residential single-phase service-entrance conductors, service lateral conductors, and feeder conductors that serve as the main power feeder to a dwelling unit. Check whether this is permitted in your locality.

18. See *220.61* and *310.15(B)(7)*. Do Not Include Straight 240-Volt Loads.

19. See *220.61*. *NEC 310.15(B)(7)* permits the neutral conductor to be smaller than the ungrounded "hot" conductors if the requirement of *215.2*, *220.61*, and *230.42* are met. *NEC 220.61* states that a feeder or service neutral load shall be the maximum unbalance of the load determined by *Article 220*. When bare conductors are used with insulated conductors, the conductors' ampacity is based on the ampacity of the other insulated conductors in the raceway, *310.15(B)(4)*. The neutral conductor must not be smaller than the grounding electrode conductor, *250.24(B)(1)*.

20. See *Table 250.66*. Grounding electrode conductor based on size of service-entrance conductors.

21. Obtain dimensional data from *Table 1*, *Table 4*, *Table 5*, and *Table 8*, *Chapter 9, NEC*.

MEMBERSHIP APPLICATION
For Inspector Member/Associate Member

Annual Membership Dues $102
Multi-year Dues (3-year) $286

Services from the IAEI for you and your organization include all this and more:

- *IAEI News* magazine subscription
- Educational Opportunities
- Networking
- Career Advancement
- Great Discount Savings

PLEASE PRINT

Name - Last	First	M.I.	Chapter, where you live or work, if known

Title	(Division, where appropriate)

E-mail Address	

Employer	**For Office Use**	Section No.	Ch. No.	Div. No.

Address of Applicant	If previous member, give last membership number

City	State or Province	Zip or Postal Code	and last year of membership

Endorsed by RAY C. MULLIN	Endorser's Membership Number **533800**

Applicant's Signature	(Area Code) Telephone Number - -

☐ Check ☐ Money Order ☐ Diners Club
☐ MasterCard ☐ Visa ☐ Amex ☐ Discover

Name on Card	Charge Card Number	Exp. Date

☐ Inspector $102   ☐ Associate $102

☐ Inspector/Multi-year (3-year) $286   ☐ Associate/Multi-year (3-year) $286

Inspector Members MUST sign below:

☐ I,_____, meet the qualifications for inspector member type as described below.

Inspector members must regularly make electrical inspections for preventing injury to persons or damage to property on behalf of a governmental agency, insurance agency, rating bureau, recognized testing laboratory or electric light and power company.

Contact the IAEI Customer Service Department for information on our special membership categories - Section, National, and International Member; Sustaining Member (Bronze, Silver, Gold, or Platinum); and Inspection Agency Member.

Email addresses and telephone numbers are not required but are requested to facilitate IAEI's communicating with members and conducting IAEI/member business.

Mail to: IAEI, P. O. Box 830848 Richardson, TX 75083-0848   For information call: (972) 235-1455 (8-5 CT)
www.iaei.org • email: iaei@iaei.org

MA2010

Key Terms

The following is a list of terms used extensively in the electrical trades. Many of these terms are unique to the electrical industry. Words marked with an * are defined in the *National Electrical Code*. Definitions are found in *Article 100, NEC*.

Accent Lighting: Directional lighting to emphasize a particular object or to draw attention to a part of the field in view.

Accessible (equipment)*: Admitting close approach: not guarded by locked doors, elevation, or other effective means.

Accessible, Readily*: Capable of being reached quickly for operation, renewal, or inspections, without requiring those to whom ready access is requisite to climb over or remove obstacles or to resort to portable ladders, chairs, and so forth.

Accessible (wiring methods)*: Capable of being removed or exposed without damaging the building structure or finish, or not permanently closed in by the structure or finish of the building.

Addendum: Modification (change) made to the construction documents (plans and specifications) during the bidding period.

Adjustment Factor: A multiplier (penalty) that is applied to conductors when there are more than three current-carrying conductors in a raceway or cable. These multipliers are found in *Table 310.15(B)(3) (a)*. An *adjustment factor* is also referred to as a *derating factor*.

AL/CU: Terminal marking on switches and receptacles rated 30 amperes and greater, suitable for use with aluminum, copper, and copper-clad aluminum conductors. If not marked, suitable for copper conductors only.

Alternate Bid: Amount stated in the bid to be added or deducted from the base bid amount proposed for alternate materials and/or methods of construction.

Ambient Lighting: Lighting throughout an area that produces general illumination.

Ambient Temperature: The environmental temperature surrounding the object under consideration.

American National Standards Institute (ANSI): An organization that identifies industrial and public requirements for national consensus standards and coordinates and manages their development, resolves national standards problems, and ensures effective participation in international standardization. ANSI does not itself develop standards. Rather, it facilitates development by establishing consensus among qualified groups. ANSI ensures that the guiding principles—consensus, due process, and openness—are followed.

Ampacity*: ▶The maximum current, in amperes, that a conductor can carry continuously under the conditions of use without exceeding its temperature rating.◀

Ampere: The measurement of intensity of rate of flow of electrons in an electric circuit. An ampere is the amount of current that will flow through a resistance of 1 ohm under a pressure of 1 volt.

Appliance*: Utilization equipment, generally other than industrial, normally built in standardized sizes or types, that is installed or connected as a unit to perform one or more functions such as clothes washing, air conditioning, food mixing, deep frying, and so forth.

Approved*: Acceptable to the authority having jurisdiction (AHJ).

Arc-Fault Circuit-Interrupter (AFCI)*: A device intended to provide protection from the effects of arc faults by recognizing characteristics unique to arcing and by functioning to de-energize a circuit when an arcing fault is detected.

Architect: One who designs and supervises the construction of buildings or other structures.

Architectural Drawing: A line drawing showing plan and/or elevation views of the proposed building for the purpose of showing the overall appearance of the building.

As-Built Drawings: Contract drawings that reflect changes made during the construction process.

As-Built Plans: When required, a modified set of working drawings that are prepared for a construction project that includes all variances from the original working drawings that occurred during the project construction.

Authority Having Jurisdiction (AHJ): The organization, office, or individual responsible for approving equipment, materials, an installation, or a procedure. An AHJ is usually a governmental body that has legal jurisdiction over electrical installations. The AHJ has the responsibility for making interpretations of the rules, for deciding upon the approval of equipment and materials, and for granting special permission if required. Where a specific interpretation or deviation from the *NEC* is given, get it in writing. See *90.4* of the *NEC*.

Average Rated Life (lamp): How long it takes to burn out the lamp; for example, a 60-watt lamp is rated 1000 hours. The 1000-hour rating is based on the point in time when 50% of a test batch of lamps burn out and 50% are still burning.

AWG: The American Wire Gauge. Previously known as the Brown & Sharpe Gauge. The smaller the AWG number, the larger the conductor. There are 40 electrical conductor sizes from 36 AWG through 0000 AWG. Starting from the smallest size, 36 AWG, each successive size is approximately 1.26 times larger than the previous AWG size. *Table 8* of *Chapter 9* in the *NEC* lists conductor sizes commonly found in the electrical industry. Conductor sizes through 0000 AWG are shown using the AWG designation. Conductors larger than 0000 AWG are shown in circular mil area. The 1.26 relationship between sizes is easily confirmed in *Table 8* by checking conductor sizes 4 AWG through 0000 AWG.

Backlight: Illumination from behind a subject directed substantially parallel to a vertical plane through the optical axis of the camera.

Ballast: Used for energizing fluorescent lamps. It is constructed of a laminated core and coil windings or electronic components.

Ballast Factor: A measure of the light output (lumens) of a ballast and lamp combination in comparison to an ANSI Standard "reference" ballast operated with the same lamp. It is a measure of how well the ballast performs when compared to the "reference ballast."

Bare Lamp: A light source with no shielding.

Bid: An offer or proposal of a price.

Bid Bond: A written form of security executed by the bidder as principal and by a surety for the purpose of guaranteeing that the bidder will sign the contract, if awarded the contract, for the stated bid amount.

Bid Price: The stipulated sum stated in the bidder's bid.

Black Line Print: Another term for a photocopy or computer-aided drawing (CAD).

Blueprints: A term used to refer to a plan or plans. A photographic print in white on a bright blue ground or blue on a white ground, used especially for copying maps, mechanical drawings, and architects' plans. See text for further explanation.

Bonded (bonding)*: Connected to establish electrical continuity and conductivity.

▶**Bonding Conductor or Jumper*:** A reliable conductor to ensure the required electrical conductivity between metal parts required to be electrically connected.◀

Bonding Jumper, Equipment*: The connection between two or more portions of the equipment grounding conductor.

Bonding Jumper, Main*: The connection between the grounded circuit conductor and the equipment grounding conductor at the service.

Branch Circuit*: The circuit conductors between the final overcurrent device protecting the circuit and the outlet(s).

Branch Circuit, Appliance*: A branch circuit supplying energy to one or more outlets to which appliances are to be connected; such circuits are to have no permanently connected luminaires not a part of an appliance.

Branch Circuit, General Purpose*: A branch circuit that supplies a number of outlets for lighting and appliances.

Branch Circuit, Individual*: A branch circuit that supplies only one utilization equipment.

Branch Circuit—Multiwire*: A branch circuit consisting of two or more ungrounded conductors having a voltage between them, and a grounded conductor having equal voltage between it and each ungrounded conductor of the circuit, and that is connected to the neutral or grounded conductor of the system.**

Branch-Circuit Overcurrent Protective Device*: A device capable of providing protection for service, feeder, and branch circuits and equipment over the full range of overcurrent between its rated current and the interrupting rating. Branch-circuit overcurrent protective devices are provided with interrupting ratings appropriate for the intended use but no less than 5000 amperes.

Branch Circuit Selection Current*: Branch circuit selection current is the value in amperes to be used instead of the rated-load current in determining the ratings of motor branch-circuit conductors, disconnecting means, controllers, and branch-circuit short-circuit and ground-fault protective devices wherever the running overload protective device permits a sustained current greater than the specified percentage of the rated-load current. The value of branch-circuit selection current will always be equal to or greater than the marked rated-load current. This definition is found in *Article 440.*

Brightness: In common usage, the term "brightness" usually refers to the strength of sensation that results from viewing surfaces or spaces from which light comes to the eye.

Building Code: The legal requirements set up by the prevailing various governing agencies covering the minimum acceptable requirements for all types of construction.

Building Inspector/Official: A qualified government representative authorized to inspect construction for compliance with applicable building codes, regulations, and ordinances.

Building Permit: A written document issued by the appropriate governmental authority permitting construction to begin on a specific project in accordance with drawings and specifications approved by the governmental authority.

Candela: The international (SI) unit of luminous intensity. Formerly referred to as a "candle," as in "candlepower."

Candlepower: A measure of intensity mathematically related to lumens.

Carrier: A wave having at least one characteristic that may be varied from a known reference value by modulation.

Carrier Current: The current associated with a carrier wave.

CC: A marking on wire connectors and soldering lugs indicating they are suitable for use with copper-clad aluminum conductors only.

CC/CU: A marking on wire connectors and soldering lugs indicating they are suitable for use with copper or copper-clad aluminum conductors only.

Change Order: A written document signed by the owner and the contractor, authorizing a change in the work or an adjustment in the contract sum or the contract time. A contract sum and the contract time may be changed only by a change order.

Circuit Breaker*: A device designed to open and close a circuit by nonautomatic means and to open the circuit automatically on a predetermined overcurrent without damage to itself when properly applied within its rating.

Clothes Closet*: A nonhabitable room or space intended primarily for the storage of garments and apparel.

CO/ALR: Terminal marking on switches and receptacles rated 15 and 20 amperes that are suitable for use with aluminum, copper, and copper-clad aluminum conductors. If not marked, they are suitable for copper or copper-clad conductors only.

Conductor*:

Bare: A conductor having no covering or electrical insulation whatsoever. (See *Conductor, Covered.*)

Covered: A conductor encased within material of a composition or thickness that is not recognized by this *Code* as electrical insulation. (See *Conductor, Bare.*)

Insulated: A conductor encased within material of a composition and thickness that is recognized by this *Code* as electrical insulation.

Construction Documents: A term used to represent all drawings, specifications, addenda, and other pertinent construction information associated with the construction of a specific project.

Construction Drawing: See *Drawings.*

Construction Types: (Also see *Annex E, NEC.*)

Type I: Fire-resistive construction. A building constructed of noncombustible materials such as reinforced concrete, brick, stone, and so on, and having any metal members properly "fireproofed" with major structural members designed to withstand collapse and to prevent the spread of fire. Type I construction is subdivided into Types IA and IB.

Type II: Noncombustible construction. A building having all structural members, including walls, floors, and roofs, of noncombustible materials such as reinforced concrete, brick, stone, and so on, and not qualifying as fire-resistive construction. Type II construction is divided into Types IIA, IIB, and IIC.

Type III: Construction where the exterior walls are of concrete, masonry, or other noncombustible material. The interior structural members are constructed of any approved materials such as wood or other combustible materials. Type III construction is subdivided into Types IIIA and IIIB.

Type IV: Construction where the exterior walls are constructed of approved noncombustible materials. The interior structural members are of solid or laminated wood.

Type V: Construction where the exterior walls, load-bearing walls, partitions, floors, and roofs are constructed of any approved material. This is wood construction as found in typical residential construction. Type V construction is divided into Types VA and VB.

Continuous Load*:   A load where the maximum current is expected to continue for 3 hours or more.

Contract: An agreement between two or more parties, especially one that is written and enforceable by law.

Contract Modifications: After the agreement has been signed, any additions, deletions, or modifications of the work to be done are accomplished by change order, supplemental instruction, and field order. They can be issued at any time during the contract period.

Contractor: A properly licensed individual or company that agrees to furnish labor, materials, equipment, and associated services to perform the work as specified for a specified price.

Correction Factor: A multiplier (penalty) that is applied to conductors when the ambient temperature is greater than 86°F (30°C). These multipliers are found in *Table 310.15(B)(2)(a)* of the *NEC*.

CSA: This is the Canadian Standards Association that develops safety and performance standards in Canada for electrical products, similar to but not always identical to those of (Underwriters Laboratories) UL in the United States.

CU: A marking on wire connectors and soldering lugs indicating they are suitable for use with copper conductors only.

Current: The flow of electrons through an electrical circuit, measured in amperes.

Derating Factor: A trade or industry jargon term for *Ambient Temperature Correction Factors [Table 310.15(B)(2)(a)]* for ambient temperatures exceeding 86°F (30°C), *Adjustment Factors [Table 310.15(B)(3)(a)]* that is applied to conductors when there are more than three current-carrying conductors in a raceway or cable.

Details: Plans, elevations, or sections that provide more specific information about a portion of a project component or element than smaller scale drawings.

Device*:   A unit of an electrical system that carries or controls electric energy as its principal function.

Diagrams: Nonscaled views showing arrangements of special system components and connections not possible to clearly show in scaled views. A *schematic diagram* shows circuit components and their electrical connections without regard to actual physical locations. A *wiring diagram* shows circuit components and the actual electrical connections.

Dimmer: A switch with components that permits variable control of lighting intensity. Some dimmers have electronic components; others have core and coil (transformer) components.

Dimming Ballast: Controls light output of fluorescent lamps.

Direct Glare: Glare resulting from high luminances or insufficiently shielded light sources in the field of view. Usually associated with bright areas, such as luminaires, ceilings, and windows that are outside the visual task of the region being viewed.

Disconnecting Means*:   A device, or group of devices, or other means by which the conductors of a circuit can be disconnected from their source of supply.

Downlight: A small, direct lighting unit that guides the light downward. It can be recessed, surface-mounted, or suspended.

Drawings: (1) A term used to represent that portion of the contract documents that graphically illustrates the design, location, and dimensions of the components and elements contained in a specific project. (2) A line drawing.

Dry Niche Luminaire*:   A luminaire intended for installation in the wall of a pool or fountain in a niche that is sealed against the entry of pool water. This definition is found in *Article 680*.

Dwelling Unit: A single unit, providing complete and independent living facilities for one or more persons, including permanent provisions for living, sleeping, cooking, and sanitation.

Effective Ground-Fault Current Path: See *Ground-Fault Current Path, Effective*.

Efficacy: The total amount of light energy (lumens) emitted from a light source divided by the total lamp and ballast power (watts) input, expressed in lumens per watt.

Elevations: Views of vertical planes, showing components in their vertical relationship, viewed perpendicularly from a selected vertical plane.

Equipment*: A general term including material, fittings, devices, appliances, luminaires, apparatus, machinery, and the like used as a part of, or in connection with, an electrical installation.

Feeder*: All circuit conductors between the service equipment, the source of a separately derived system, or other power-supply source and the final branch-circuit overcurrent device.

Fill Light: Supplementary illumination to reduce shadow or contract range.

Fire Rating: The classification indicating in time (hours) the ability of a structure or component to withstand fire conditions.

Flame Detector: A radiant energy-sensing fire detector that detects the radiant energy emitted by a flame.

Fluorescent Lamp: A lamp in which electric discharge of ultraviolet energy excites a fluorescing coating (phosphor) and transforms some of that energy to visible light.

Footcandle: The unit used to measure how much total light is reaching a surface. One lumen falling on 1 ft^2 of surface produces an illumination of 1 footcandle.

Fully Rated System: All devices installed shall have an interrupting rating greater than or equal to the specified available fault-current. In a fully rated system, the panelboard short-circuit current rating will be equal to the lowest interrupting rating of any branch-circuit breaker or fuse installed. (See *Series-Rated System.*)

Fuse: An overcurrent protective device with a fusible link that operates to open the circuit on an overcurrent condition.

General Conditions: A written portion of the contract documents set forth by the owner, stipulating the contractor's minimum acceptable performance requirements including the rights, responsibilities, and relationships of the parties involved in the performance of the contract. General conditions are usually included in the book of specifications but are sometimes found in the architectural drawings.

General Lighting: Lighting designed to provide a substantially uniform level of illumination throughout an area, exclusive of any provision for special lighting.

Glare: The sensation produced by luminance within the visual field that is sufficiently greater than the luminance to which the eyes are adapted to cause annoyance, discomfort, or loss of visual performance and visibility.

Ground*: The earth.

Grounded (Grounding)*: Connected (connecting) to ground or to a conductive body that extends the ground connection.

Grounded, Solidly*: Connected to ground without inserting any resistor or impedance device.

Grounded Conductor: A system or circuit conductor that is intentionally grounded.

Ground Fault*: An unintentional, electrically conducting connection between a normally current-carrying conductor of an electrical circuit, and the normally non-current-carrying conductors, metallic enclosures, metallic raceways, metallic equipment, or earth.

Ground-Fault Circuit Interrupter (GFCI)*: A device intended for the protection of personnel that de-energizes a circuit or portion thereof within an established period of time when a current to ground exceeds the values established for a Class A device.

Informational Note*: Class A ground-fault circuit interrupters trip when the current to ground is 6 mA or higher and do not trip when the current to ground is less than 4 mA. For further information, see UL 943, *Standard for Ground-Fault Circuit Interrupters.*

Ground-Fault Current Path*: An electrically conductive path from the point of a ground fault on a wiring system through normally non-current-carrying conductors, equipment, or the earth to the electrical supply source. This definition found in *NEC 250.2.*

Ground-Fault Current Path, Effective*: An intentionally constructed, permanent, low-impedance electrically conductive path designed and intended to carry current under ground-fault conditions from the point of a ground fault on a wiring system to the electrical supply source and that facilitates the operation of the overcurrent protective device or ground fault detectors on high-impedance grounded systems. See *NEC 250.2* and *250.4(A)(5).*

Ground-Fault Protection of Equipment*: A system intended to provide protection of equipment from damaging line-to-ground fault currents by operating to cause a disconnecting means to open all ungrounded conductors of the faulted circuit. This protection is provided at current levels less than those required to protect conductors from damage through the operation of a supply circuit overcurrent device.

▶ **Grounding Conductor, Equipment (EGC)*:** The conductive path(s) installed to connect normally non-current carrying metal parts of equipment together and to the system grounded conductor or to the grounding electrode conductor, or both.

Grounding Electrode*: A conducting object through which a direct connection to earth is established.

Grounding Electrode Conductor*: A conductor used to connect the system grounded conductor or the equipment to a grounding electrode or to a point on the grounding electrode system.

HACR: Circuit breakers subjected to additional tests unique to HVAC equipment, have been marked with the letters HACR. The letters stood for Heating, Air Conditioning, and Refrigeration. The HVAC equipment would also be marked with the letters HACR if that is the type of overcurrent protection required by the manufacturer of the HVAC equipment.

Today, all standard molded case circuit breakers tested and listed to UL 489 are suitable for use on HVAC equipment. The letters *HACR* will no doubt disappear from the scene as time goes on.

Habitable Room*: A room in a residential occupancy used for living, sleeping, cooking, and eating, but excluding bath, storage and service area, and corridors. (NFPA definition)

Halogen Lamp: An incandescent lamp containing a halogen gas that recycles tungsten back onto the tungsten filament surface. Without the halogen gas, the tungsten would normally be deposited onto the bulb wall.

Heat Alarm: A single or multiple station alarm responsive to heat.

Hermetic Refrigerant Motor-Compressor: A combination consisting of a compressor and motor, both of which are enclosed in the same housing, with no external shaft or shaft seals; the motor operating in the refrigerant. This definition is found in *Article 440.*

High-Intensity Discharge Lamp (HID): A general term for a mercury, metal-halide, or high-pressure sodium lamp.

High-Pressure Sodium Lamp: A high-intensity discharge light source in which the light is primarily produced by the radiation from sodium vapor.

Horsepower (tools): Horsepower is a measurement of motor torque multiplied by speed. Horsepower also refers to the rate of work (power) an electric motor is capable of delivering. See *Torque, Rated Horsepower,* and *Maximum Developed Horsepower.*

Horsepower, Rated: Rated horsepower is a motor's running torque at its rated running speed. The motor can be run continuously at its rated horsepower without overheating. If the motor is required to give an extra spurt of effort while running, the motor is overloaded and develops extra horsepower to compensate. The most horsepower that can be drawn from a motor to handle this extra effort is its maximum developed horsepower. See *Torque, Horsepower,* and *Maximum Developed Horsepower.*

Horsepower, Maximum Developed: If a motor is required to work harder than its idle speed, it will be overloaded and must develop extra horsepower. The most horsepower that can be expected from a motor to handle this extra effort is referred to as its maximum developed horsepower. See *Horsepower, Torque,* and *Rated Horsepower.*

Household Fire Alarm System: A system of devices that produces an alarm signal in the household for the purpose of notifying the occupants of the presence of a fire so that they will evacuate the premises.

ICC Building Code: One of the families of codes and related publications published by the International Code Council. See www.iccsafe.org for a complete listing of codes produced for the built environment. The International Code Council is an amalgamation of the former International Conference of Building Officials (ICBO), the Building Officials Code Administration (BOCA), and Southern Building Codes Congress International.

Identified* (conductor): The identified conductor is the insulated grounded conductor.

- For sizes 6 AWG or smaller, the insulated grounded conductor shall be identified by a continuous white or gray outer finish or by three continuous white stripes on other than green insulation along its entire length.

- For sizes larger than 6 AWG, the insulated grounded conductor shall be identified either by a continuous white or gray outer finish or by three continuous white stripes on other than green insulation along its entire length, or at the time of installation by a distinctive white or gray marking at its terminations that encircles the insulation. See *NEC 200.6.*

Identified* (equipment): Recognizable as suitable for the specific purpose, function, use, environment, application, and so forth, where described in a particular *Code* requirement.

Identified* (terminal): The identification of terminals to which a grounded conductor is to be connected shall be substantially white in color. The identification of other terminals shall be of a readily distinguishable different color. See *NEC 200.10.*

IEC: The International Electrotechnical Commission is a worldwide standards organization. These standards differ from those of Underwriters Laboratories. Some electrical equipment might conform to a specific IEC Standard but may or may not conform to the UL Standard for the same item.

Illuminance: The amount of light energy (lumens) distributed over a specific area expressed as footcandles (lumens/ft^2) or lux (lumens/m^2).

Illumination: The act of illuminating or the state of being illuminated.

Immersion Detection Circuit Interrupter (IDCI): A device integral with grooming appliances that will shut off the appliance when the appliance is dropped in water.

Incandescent Filament Lamp: A lamp that provides light when a filament is heated to incandescence by an electric current. Incandescent lamps are the oldest form of electric lighting technology.

Indirect Lighting: Lighting by luminaires that distribute 90% to 100% of the emitted light upward.

Inductive Load: A load that is made up of coiled or wound wire that creates a magnetic field when energized. Transformers, core and coil ballasts, motors, and solenoids are examples of inductive loads.

Informational Note*: Explanatory material, such as references to other standards, references to related sections of the *NEC,* or information related to a *NEC* rule, is included in the *NEC* in the form of *Informational Notes.* Such notes are informational only and are not enforceable as requirements of the *NEC.* See *NEC 90.4.*

In Sight*: Where this *Code* specifies that one equipment shall be "in sight from," "within sight from," or "within sight," and so forth, of another equipment, the specified equipment is to be visible and not more than 50 ft (15 m) from the other.

Instant Start: A circuit used to start specially designed fluorescent lamps without the aid of a starter. To strike the arc instantly, the circuit utilizes higher open-circuit voltage than is required for the same length preheat lamp.

International Association of Electrical Inspectors (IAEI): A not-for-profit and educational organization cooperating in the formulation and uniform application of standards for the safe installation and use of electricity, and collecting and disseminating information relative thereto. The IAEI is made up of electrical inspectors, electrical contractors, electrical apprentices, manufacturers, electrical testing laboratories and governmental agencies.

Interrupting Rating*: The highest current at rated voltage that a device is identified to interrupt under standard test conditions.

Intersystem Bonding Termination*: A device that provides a means for connecting bonding conductors for communications to the grounding electrode system.

Ionization Detector: This type of detector triggers an alarm when oxygen and nitrogen particles are ionized in the ionization chamber. The internal circuitry measures a minute amount of electrical current between two plates. When smoke enters the ionization chamber, the current is reduced, triggering the alarm. This type of detector works well for detecting small amounts of smoke such as that resulting from gasoline fires. Consult the manufacturer's literature for more information.

Isolated Ground Receptacle: A grounding-type device in which the equipment ground contact and terminal are electrically isolated from the receptacle mounting means.

Kilowatt (kW): One thousand watts equals 1 kilowatt.

Kilowatt-hour (kWh): One thousand watts of power in 1 hour. One 100-watt lamp burning for 10 hours is 1 kilowatt-hour. Two 500-watt electric heaters operated for 1 hour is 1 kilowatt-hour.

▶ **Kitchen*:** An area with a sink and permanent provisions for food preparation and cooking.◀

Labeled*: Equipment or materials to which has been attached a label, symbol, or other identifying mark of an organization that is acceptable to the authority having jurisdiction (AHJ) and concerned with product evaluation that maintains periodic inspection of production of labeled equipment or materials and by whose labeling the manufacturer indicates compliance with appropriate standards or performance in a specified manner.

Labor and Material Payment Bond: A written form of security from a surety (bonding) company to the owner, on behalf of an acceptable prime or main contractor or subcontractor, guaranteeing payment to the owner in the event the contractor fails to pay for all labor, materials, equipment, or services in accordance with the contract. (See *Performance Bond* and *Surety Bond.*)

Lamp: A generic term for an artificial source of light.

Light: The term generally applied to the visible energy from a source. Light is usually measured in lumens or candlepower. When light strikes a surface, it is either absorbed, reflected, or transmitted.

Lighting Outlet*: An outlet intended for the direct connection of a lampholder or luminaire.

Listed*: Equipment, materials, or services included in a list published by an organization that is acceptable to the authority having jurisdiction (AHJ) and concerned with evaluation of products or services, that maintains periodic inspection of production of listed equipment or materials or periodic evaluation of services, and whose listing states that either the equipment, material, or services meets appropriate designated standards or has been tested and found suitable for a specified purpose.

Load: The electric power used by devices connected to an electrical system. Loads can be figured in amperes, volt-amperes, kilovolt-amperes, or kilowatts. Loads can be intermittent, continuous intermittent, periodic, short-time, or varying. See the definition of "Duty" in the *NEC*.

Load Center: A common name for residential panelboards. A load center may not be as deep as a panelboard and generally does not contain relays or other accessories as are available for panelboards. Circuit breakers "plug in" as opposed to the "bolt-in" types used in panelboards. Manufacturers' catalogs will show both load centers and panelboards. The UL standards do not differentiate.

Location, Damp*: Locations protected from weather and not subject to saturation with water or other liquids but subject to moderate degrees of moisture. Examples of such locations include partially protected locations under canopies, marquees, roofed open porches, and like locations, and interior locations subject to moderate degrees of moisture, such as some basements, some barns, and some cold-storage warehouses.

Location, Dry*: A location not normally subject to dampness or wetness. A location classified as dry may be temporarily subject to dampness or wetness, as in the case of a building under construction.

Location, Wet*: Installations underground or in concrete slabs or masonry in direct contact with the earth; in locations subject to saturation with water or other liquids, such as vehicle washing areas; and in unprotected locations exposed to weather.

Locked Rotor Current (LRC): The steady-state current taken from the line with the rotor locked and with rated voltage and frequency applied to the motor.

Low-Pressure Sodium Lamp: A discharge lamp in which light is produced from sodium gas operating at a partial pressure.

Lumens: The SI unit of luminous flux. The units of light energy emitted from the light source.

Luminaire*: A complete lighting unit consisting of a light source such as a lamp or lamps, together with the parts designed to position the light source and connect it to the power supply. It may also include parts to protect the light source, ballast, or distribute the light. A lampholder itself is not a luminaire. (Prior to the *National Electrical Code* adopting the International System (SI) definition of *luminaire*, the commonly used term in the United States was and in most instances still is "lighting fixture." It will take years for the electrical industry to totally change and feel comfortable with the term *luminaire*.)

Luminaire Efficiency: The total lumen output of the luminaire divided by the rated lumens of the lamps inside the luminaire.

Lux: The SI (International System) unit of illumination. One lumen uniformly distributed over an area 1 square meter in size.

Mandatory Rules: Required. The terms *shall* or *shall not* are used when a *Code* statement is mandatory.

Maximum Rating of Branch-Circuit Short-Circuit and Ground-Fault Protection: A term used with equipment that has a hermetic motor compressor(s). This is the maximum ampere rating for the equipment's branch-circuit overcurrent protective device. The ampere rating is determined by the manufacturer and is marked on the nameplate of the equipment. The nameplate will also indicate if the overcurrent device can be a circuit breaker, a fuse, or either.

Mercury Lamp: A high-intensity discharge light source in which radiation from the mercury vapor produces visible light.

Metal-Halide Lamp: A high-intensity discharge light source in which the light is produced by the radiation from mercury together with the halides of metals such as sodium and candium.

Minimum Supply Circuit Conductor Ampacity: A term used with equipment that has a hermetic motor compressor(s). This is the minimum ampere rating value used to determine the proper size conductors, disconnecting means, and controllers. It is determined by the manufacturer and is marked on the nameplate of the equipment.

National Electrical Code (NEC): The electrical code published by the National Fire Protection Association.

This *Code* provides for practical safeguarding of persons and property from hazards arising from the use of electricity. It does not become law until adopted by federal, state, or local laws and regulations.

National Electrical Manufacturers Association (NEMA): NEMA is a trade organization made up of many manufacturers of electrical equipment. They develop and promote standards for electrical equipment.

National Fire Protection Association (NFPA): Located in Quincy, MA. The NFPA is an international Standards Making Organization dedicated to the protection of people from the ravages of fire and electric shock. The NFPA is responsible for developing and writing the *National Electrical Code*, the *Installation of Sprinkler Systems in One- and Two-Family Dwellings and Manufactured Homes*, the *Life Safety Code, the National Fire Alarm Code*, and over 300 other codes, standards, and recommended practices. The NFPA phone number is (800) 344-3555. Its Web site is http://www.nfpa.org.

Nationally Recognized Testing Laboratory (NRTL): The term used to define a testing laboratory that has been recognized by OSHA; for example, Underwriters Laboratories (UL), Intertek Testing, and MET Laboratories.

Neutral Conductor*: The conductor connected to the neutral point of a system that is intended to carry current under normal conditions.

In residential wiring, the neutral conductor is the grounded conductor in a circuit. A neutral conductor is always a grounded conductor. A grounded conductor is not always a neutral conductor, such as a "grounded B phase" system discussed in *Electrical Wiring—Commercial*.

Neutral point*: The common point on a wye-connection in a polyphase system or midpoint on a single-phase, 3-wire system, or midpoint of a single-phase portion of a 3-phase delta system, or a midpoint of a 3-wire, direct current system.

No-Niche Luminaire*: A luminaire intended for installation above or below the water without a niche. The definition is found in *Article 680*.

Noncoincidental Loads: Loads that are not likely to be on at the same time. Heating and cooling loads would not operate at the same time. See *220.60* of the *NEC*.

Notations: Words found on plans to describe something.

Occupational Safety and Health Act (OSHA): This is the code of federal regulations developed by the Occupational Safety and Health Administration, U.S. Department of Labor. The electrical regulations are covered in *Part 1910, Subpart S*. The *NEC* must still be referred to in conjunction with OSHA regulations.

Ohm: A unit of measure for electric resistance. An ohm is the amount of resistance that will allow 1 ampere to flow under a pressure of 1 volt.

Outlet*: A point on the wiring system at which current is taken to supply utilization equipment.

Overcurrent*: Any current in excess of the rated current of equipment or the ampacity of a conductor. It may result from overload, short circuit, or ground fault.

Overcurrent Device: Also referred to as an overcurrent protection device. A form of protection that operates when current exceeds a predetermined value. Common forms of overcurrent devices are circuit breakers, fuses, and thermal overload elements found in motor controllers.

Overload*: Operation of equipment in excess of the normal, full-load rating, or of a conductor in excess of rated ampacity that, when it persists for a sufficient length of time, would cause damage or dangerous overheating. The current flow is contained in its normal intended path. A fault, such as a short circuit or a ground fault, is not an overload.

Owner: An individual or corporation that owns a real property.

Panelboard*: A single panel or group of panel units designed for assembly in the form of a single panel, including buses and automatic overcurrent devices; equipped with or without switches for the control of light, heat, or power circuits; designed to be placed in a cabinet or cutout box placed in or against a wall or partition and accessible only from the front.

Performance Bond: (1) A written form of security from a surety (bonding) company to the owner, on behalf of an acceptable prime or main contractor or subcontractor, guaranteeing payment to the owner in the event the contractor fails to perform all labor, materials, equipment, or services in accordance with the contract. (2) The surety companies generally reserve the right to have the original prime or main or subcontractor remedy any claims before paying on the bond or hiring other contractors.

Permissive Rules: Allowed but not required. Terms such as *shall be permitted or shall not be required* are used when a *Code* statement is permissive.

Photoelectric Detector: This type of detector triggers an alarm when light is blocked between its internal light source and sensor. Typically, heavy smoke will block the light. The action of blocking light is similar to that of the safety feature on a garage door operator. Consult the manufacturer's literature for more information.

Photometry: The pattern and amount of light that is emitted from a luminaire, normally represented as a cross-section through the luminaire distribution pattern.

Plan: (1) A line drawing (by floor) representing the horizontal geometrical section of the walls of a building. The section (a horizontal plane) is taken at an elevation to include the relative positions of the walls, partitions, windows, doors, chimneys, columns, pilasters, and so on. (2) A plan can be thought of as cutting a horizontal section through a building at an eye level elevation.

Plan Checker: A term sometimes used to describe a building department official who examines the building permit documents.

Plans: A term used to represent all drawings, including sections and details and any supplemental drawings, for complete execution of a specific project.

Power Factor: A ratio of actual power (W or kW) being used to apparent power (VA or kVA) being drawn from the power source.

Power Supply: A source of electrical operating power including the circuits and terminations connecting it to the dependent system components.

Preheat: A type of ballast that is easily identified because it has a starter. One type of starter is automatic with two buttons on one end; the other type is a manual On/Off switch that has a momentary "make" position just beyond the "On" position.

Preheat Fluorescent Lamp Circuit: A circuit used on fluorescent lamps wherein the electrodes are heated or warmed to a glow stage by an auxiliary switch or starter before the lamps are lighted.

Preliminary Drawings: The drawings that precede the final approved drawings. Usually stamped "PRELIMINARY."

Prime Contractor: Any contractor having a contract directly with the owner. Usually the main (general) contractor for a specific project.

Prints: A term used to refer to a plan or plans.

Project: A word used to represent the overall scope of work being performed to complete a specific construction job.

Proposal: A written offer from a bidder to the owner, preferably on a prescribed proposal form, to perform the work and to furnish all labor, materials, equipment and/or service for the prices and terms quoted by the bidder.

Punch List (Inspection List): A list prepared by the owner or his or her authorized representative of items of work requiring immediate corrective or completion action by the contractor.

Qualified Person*: One who has skills and knowledge related to the construction and operation of the electrical equipment and installations and has received safety training to recognize and avoid the hazards involved.

Raceway*: An enclosed channel of metallic or nonmetallic materials designed expressly for holding wires, cables, or busbars, with additional functions as permitted in this *Code*. Raceways include, but are not limited to, rigid metal conduit, rigid PVC conduit, intermediate metal conduit, liquidtight flexible conduit, flexible metallic tubing, flexible metal conduit, electrical nonmetallic tubing, electrical metallic tubing, underfloor raceways, cellular concrete floor raceways, cellular metal floor raceways, surface raceways, wireways, and busways.

Rapid Start: The most common type of ballast used today that does not require a starter.

Rapid Start Fluorescent Lamp Circuit: A circuit designed to start lamps by continuously heating or preheating the electrodes. Lamps must be designed for this type of circuit. This is the modern version of the "trigger start" system. In a rapid-start, 2-lamp circuit, one end of each lamp is connected to a separate starting winding.

Rate of Rise Detector: A device that responds when the temperature rises at a rate exceeding a predetermined value.

Rated-Load Current: The rated-load current for a hermetic refrigerant motor-compressor is the current resulting when the motor-compressor is operated at the rated load, rated voltage, and rated frequency of the equipment it serves. See *NEC Article 440*.

Receptacle*: A receptacle is a contact device installed at the outlet for the connection of an attachment plug. A single receptacle is a single contact device with no other contact device on the same yoke. A multiple receptacle is two or more contact devices on the same yoke.

Receptacle Outlet*: An outlet where one or more receptacles are installed.

Resistive Load: An electric load that opposes the flow of current. A resistive load does not contain cores or coils of wire. Some examples of resistive loads are the electric heating elements in an electric range, ceiling heat cables, and electric baseboard heaters.

Root-Mean-Square (RMS): The square root of the average of the square of the instantaneous values of current or voltage. For example, the RMS value of voltage, line to neutral, in a home is 120 volts. During each electrical cycle, the voltage rises from zero to a peak value ($120 \times 1.4142 = 169.7$ volts), back through zero to a negative peak value, then back to zero. The RMS value is 0.707 of the peak value ($169.7 \times 0.707 = 120$ volts). The root-mean-square value is what an electrician reads on an ammeter or voltmeter.

Schedules: Tables or charts that include data about materials, products, and equipment.

Scope: A written range of view or action; outlook; hence, room for the exercise of faculties or function, capacity for achievement; all in connection with a designated project.

Sections: Views of vertical cuts through and perpendicular to components showing their detailed arrangement.

Series-Rated System: Panelboards are marked with their short-circuit rating in RMS symmetrical amperes. A series-rated panelboard will be determined by the main circuit breaker or fuse, and branch-circuit breaker combination tested in accordance to UL Standard 489. The series rating will be less than or equal to the interrupting rating of the main overcurrent device, and greater than the interrupting rating of the branch-circuit overcurrent devices. (See "Fully Rated System.")

Service*: The conductors and equipment for delivering energy from the serving utility to the wiring system of the premises served.

Service Conductors*: The conductors from the service point to the service disconnecting means.

▶ **Service Conductors, Overhead*:** The overhead conductors between the service point and the first point of connection to the service-entrance conductors at the building or other structure. ◀

▶ **Service Conductors, Underground*:** The underground conductors between the service point and the first point of connection to the service-entrance conductors in a terminal box, meter, or other enclosure, inside or outside the building wall.

Informational Note*: Where there is no terminal box, meter, or other enclosure with adequate space, the point of connection shall be considered to be the point of entrance of the service conductors into the building. ◀

▶ **Service Drop*:** The overhead service conductors between the utility electric supply system and the service point. ◀

Service Equipment*: The necessary equipment, usually consisting of a circuit breaker or switch and fuses and their accessories, located near the point of entrance of supply conductors to a building or other structure, or an otherwise defined area, and intended to constitute the main control and means of cutoff of the supply.

▶ **Service Lateral*:** The underground conductors between the utility electric supply system and the service point. ◀

Service Point*: The point of connection between the facilities of the serving utility and the premises wiring.

▶ **Informational Note*:** The service point can be described as the point of demarcation between where the serving utility ends and the premises wiring begins. The serving utility generally specifies the location of the service point based on the conditions. of service. ◀

Shall: Indicates a mandatory requirement.

Short Circuit: A connection between any two or more conductors of an electrical system in such a way as to significantly reduce the impedance of the circuit. The current flow is outside of its intended path, thus the term *short circuit*. A short circuit is also referred to as a fault.

Short-Circuit Current Rating: The prospective symmetrical fault current at a nominal voltage to which an apparatus or system is able to be connected without sustaining damage exceeding defined acceptance criteria.

Should: Indicates a recommendation or that which is advised but not required.

Site: The place where a structure or group of structures was, or is, to be located (as in construction site).

Smoke Alarm: A single or multiple station alarm responsive to smoke.

Smoke Detector: A device that detects visible or invisible particles of combustion.

Sone: A unit of loudness equal to the loudness of a sound of 1 kilohertz at 40 decibels above the threshold of hearing of a given listener.

Specifications: Text setting forth details such as description, size, quality, performance, workmanship, and so forth. Specifications that pertain to all of the construction trades involved might be subdivided into "General Conditions" and "Supplemental

General Conditions." Further subdividing the specifications might be specific requirements for the various contractors such as electrical, plumbing, heating, masonry, and so forth. Typically, the electrical specifications are found in Division 16.

Split-Wired Receptacle: A receptacle that can be connected to two branch circuits or a multiwire branch circuit. These receptacles may also be used so that one receptacle is live at all times, and the other receptacle is controlled by a switch. The terminals on these receptacles usually have breakaway tabs so the receptacle can be used either as a split-wired receptacle or as a standard receptacle.

Standard Network Interface (SNI):   A device usually installed by the telephone company at the demarcation point where their service leaves off and the customer's service takes over. This is similar to the service point for electrical systems.

Starter: A device used in conjunction with a ballast for the purpose of starting an electric discharge lamp.

Structure*:   That which is built or constructed.

Subcontractor: A qualified subordinate contractor to the prime or main contractor.

Surety Bond: A legal document issued to ensure the completion of an act by another person.

Contractors usually are required to purchase surety bonds, if they are working on public projects. A surety bond guarantees to one party that another (the contractor) will perform specified acts, usually within a stated period of time. The surety company typically becomes responsible for fulfillment of a contract if the contractor defaults. In the case of a public works project, such as a road, that means that the surety bond protects taxpayers should a contractor go out of business.

Surface-Mounted Luminaire: A luminaire mounted directly on the ceiling.

Surge-Protective Device (SPD):   A protective device for limiting transient voltages by diverting or limiting surge current; it also prevents continued flow of follow current while remaining capable of repeating these functions and is designated as follows:

Type 1:   Permanently connected SPDs intended for installation between the secondary of the service transformer and the line side of the service disconnect overcurrent device.

Type 2:   Permanently connected SPDs intended for installation on the load side of the service disconnect overcurrent device, including SPDs located at the branch panel.

Type 3:   Point of utilization SPDs.

Type 4:   Component SPDs, including discrete components, as well as assemblies.

Informational Note: For further information on Type 1, Type 2, Type 3, and Type 4 SPDs, see UL 1449, Standard for Surge Protective Devices.

Suspended (pendant) Luminaire: A luminaire hung from a ceiling by supports.

Switches*:

General-Use Snap Switch: A form of general-use switch constructed so that it can be installed in device boxes or on box covers, or otherwise used in conjunction with wiring systems recognized by this *Code*.

General-Use Switch: A switch intended for use in general distribution and branch circuits. It is rated in amperes, and it is capable of interrupting its rated current at its rated voltage.

Motor-Circuit Switch: A switch, rated in horsepower, capable of interrupting the maximum operating overload current of a motor of the same horsepower rating as the switch at the rated voltage.

Symbols: Graphic representations that stand for or represent other things. A symbol is a simple way to show such things as lighting outlets, switches, and receptacles on an electrical plan. The American Institute of Architects has developed a very comprehensive set of symbols that represent just about everything used by all building trades. When an item cannot be shown using a symbol, then a more detailed explanation using a notation or inclusion in the specifications is necessary.

T & M: An abbreviation for a contracting method called Time and Material. A written agreement between the owner and the contractor wherein payment is based on the contractor's actual cost for labor equipment, materials, and services plus a fixed add-on amount to cover the contractor's overhead and profit.

Task Lighting: Lighting directed to a specific surface or area that provides illumination for visual tasks.

Terminal: A screw or a quick-connect device where a conductor(s) is intended to be connected.

Thermocouple: A pair of dissimilar conductors so joined at two points that an electromotive force is developed by the thermoelectric effects when the junctions are at different temperatures.

Thermopile: More than one thermocouple connected together. The connections may be series or parallel, or both.

Torque: A measurement of rotation or turning force. Torque is measured in ounce-inches (oz.-in.), ounce-feet (oz.-ft) and pound-feet (lb.-ft). See *Horsepower, Rated Horsepower,* and *Maximum Developed Horsepower.*

Troffer: A recessed lighting unit, usually long and installed with the opening flush with the ceiling. The term is derived from "trough" and "coffer."

UL: Underwriters Laboratories (UL) is an independent not-for-profit organization that develops standards and tests electrical equipment to these standards.

UL-Listed: Indicates that an item has been tested and approved to the standards established by UL for that particular item. The UL Listing Mark may appear in various forms, such as the letters UL in a circle. If the product is too small for the marking to be applied to the product, the marking must appear on the smallest unit container in which the product is packaged.

UL-Recognized: Refers to a product that is incomplete in construction features or limited in performance capabilities. A "Recognized" product is intended to be used as a component part of equipment that has been "listed." A "Recognized" product must not be used by itself. A UL product may contain a number of components that have been "Recognized".

Ungrounded: Not connected to ground or a conductive body that extends the ground connection.

Ungrounded Conductor: The conductor of an electrical system that is not intentionally connected to ground. This conductor is often referred to as the "hot" or "live" conductor.

Volt: The difference of electric potential between two points of a conductor carrying a constant current of 1 ampere, when the power dissipated between these points is equal to 1 watt. A voltage of 1 volt can push 1 ampere through a resistance of 1 ohm.

Voltage (of a circuit)*: The greatest root-mean-square (effective) difference of potential between any two conductors of the circuit concerned.

Voltage (nominal)*: A nominal value assigned to a circuit or system for the purpose of conveniently designating its voltage class (e.g., 120/240 volts, 480Y/277 volts, 600 volts).

The actual voltage at which a circuit operates can vary from the nominal within a range that permits satisfactory operation of equipment.

Voltage to Ground*: For grounded circuits, the voltage between the given conductor and that point or conductor of the circuit that is grounded; for ungrounded circuits, the greatest voltage between the given conductor and any other conductor of the circuit.

Voltage Drop: Also referred to as IR drop. Voltage drop is most commonly associated with conductors. A conductor has resistance; when current is flowing through the conductor, a voltage drop will be experienced across the conductor. Voltage drop across a conductor can be calculated using Ohm's law $E = IR$.

Volt-ampere: A unit of power determined by multiplying the voltage and current in a circuit. A 120-volt circuit carrying 1 ampere is 120 volt-amperes.

Watt: A measure of true power. A watt is the power required to do work at the rate of 1 joule per second. Wattage is determined by multiplying voltage times amperes times the power factor of the circuit: $W = E \times I \times PF$.

Watertight: Constructed so that moisture will not enter the enclosure under specified test conditions.

Weatherproof*: Constructed or protected so that exposure to the weather will not interfere with successful operation.

> **Informational Note*:** Rainproof, raintight, or watertight equipment can fulfill the requirements for weatherproof where varying weather conditions other than wetness, such as snow, ice, dust, or temperature extremes, are not a factor.

Wet Niche Luminaire*: A luminaire intended for installation in a forming shell mounted in a pool. This definition is found in *Article 680.*

Working Drawing(s): A drawing sufficiently complete with plan and section views, dimensions, details, and notes, so that whatever is shown can be constructed and/or replicated without instructions but subject to clarifications, see drawings.

Work Order: A written order, signed by the owner or his or her representative, of a contractual status requiring performance by the contractor without negotiation of any sort.

Zoning: Restrictions of areas or regions of land within specific geographical areas based on permitted building size, character, and uses as established by governing urban authorities.

---

*Reprinted with permission from NFPA 70-2011.

Web Sites

The following list of World Wide Web sites has been included for your convenience. These are current as of the time of printing. Web sites are a "moving target" with continual changes. We do our utmost to keep these Web sites current. If you are aware of Web sites that should be added, deleted, or that are different from those shown, please let us know. Most manufacturers will provide catalogs and technical literature upon request either electronically or by mail. You will be amazed at the amount of electrical and electronic equipment information that is available on these Web sites.

You can search the Web using a search engine such as http://www.google.com, http://www.dogpile.com, http://www.yahoo.com, or http://www.ask.com. Just type in a key word (for example, fuses) and you will be led to many Web sites relating to fuses. It's easy!

Web sites marked with an asterisk * are closely related to the "home automation" market.

ACCUBID
http://www.accubid.com

Acme Electric Corporation
http://www.acmepowerdist.com

Active Home*
http://www.X10-beta.com/activehome/

Active Power
http://www.activepower.com

Adalet
http://www.adalet.com

Advanced Cable Ties
http://www.actfs.com/

AEMC Instruments
http://www.aemc.com

AFC Cable Systems
http://www.afcweb.com

Air-Conditioning & Refrigeration Association
http://www.ahrinet.org/

Alcan Cable
http://www.cable.alcan.com

Allen-Bradley
http://www.ab.com

Allied Moulded Products
http://www.alliedmoulded.com

Allied Tube & Conduit
http://www.alliedtube.com

Alpha Wire
http://www.alphawire.com

The Aluminum Association, Inc.
http://www.aluminum.org

American Council for an Energy Efficient Economy
http://www.aceee.org

American Gas Association
http://www.aga.org

American Heart Association
http://www.americanheart.org

American Institute of Architects
http://www.aia.org

American Lighting Association
http://www.americanlightingassoc.com

American National Standards Institute
http://www.ansi.org

American Pipe & Plastics, Inc.
http://www.ampipe.com

American Power Conversion
http://www.apcc.com

American Public Power Association
http://www.appanet.org

American Society for Testing and Materials
http://www.astm.org

American Society of Heating, Refrigerating, and
Air-Conditioning Engineers
http://www.ashrae.org

American Society of Mechanical Engineers
http://www.asme.org

American Society of Safety Engineers
http://www.asse.org

Amprobe
http://www.amprobe.com

Anixter (data & telecommunications cabling)*
http://www.anixter.com

Appleton Electric Co.
http://www.appletonelec.com

ArcWear
http://www.arcwear.com

Arlington Industries, Inc.
http://www.aifittings.com

Arrow Fasteners
http://www.arrowfastener.com

Arrow Hart Wiring Devices
http://www.arrowhart.com

Assisted Living: The World of Assisted Technology
http://www.abledata.com

Associated Builders & Contractors
http://www.abc.org

Association of Cabling Professions
http://www.wireville.com

Association of Home Appliance Manufacturers
http://www.aham.org

Automated Home Technologies*
http://www.automatedhomenashville.com/

Automatic Switch Co. (ASCO)
http://www.asco.com

AVO Training Institute
http://www.avotraining.com

Baldor Motors & Drives
http://www.baldor.com

R. W. Beckett Corporation
(oil burner technical info.)
http://www.beckettcorp.com

Belden Wire & Cable Co.
http://www.belden.com

Bell, Hubbell Inc.
http://www.hubbell-bell.com

Bender
http://www.bender.org

Best Power
http://www.bestpower.com

BICC General
http://www.generalcable.com

BICSI (Building Industry Consulting Services
International, a telecommunications assoc.)*
http://www.bicsi.org

BidStreet
http://www.BidStreetUSA.com

Black and Decker
http://www.blackanddecker.com

The Blue Book (construction)
http://www.thebluebook.com

Bodine Company
http://www.bodine.com

boltswitch, inc.
http://www.boltswitch.com

Brady Worldwide, Inc.
http://www.whbrady.com

Bridgeport Fittings Inc.
http://www.bptfittings.com

Broan
http://www.broan.com

Bryant–Hubbell Inc.
http://www.hubbell-bryant.com

Burndy Electrical Products
http://www.fciconnect.com

Cable Telecommunications Engineers*
http://www.scte.org

Caddy
http://www.erico.com

Canadian Standards Association
http://www.csa-international.org

Cantex
http://www.cantexinc.com

Carlon, A Lamson & Sessions Co.
http://www.carlon.com

Carol Cable
http://www.generalcable.com

Casablanca Fan Company
http://www.casablancafanco.com

Caterpillar
http://www.cat.com

Caterpillar
http://www.cat-engines.com

CEBus Industry Council*
http://www.cebus.org

CEE News magazine
http://www.ceenews.com

Cengage Delmar Learning
http://www.delmarlearning.com

Cengage Delmar Learning (electrical)
http://www.delmarelectric.com

Center for Disease Control (CDC)
http://www.cdc.gov

Centralite*
http://www.centralite.com

CEPro*
http://www.ce-pro.com

Channellock Inc
http://www.channellock.com

Chromalox
http://www.chromalox.com

Circon Systems Corporation*
http://www.circon.com

Coleman Cable
http://www.colemancable.com

Columbia Lighting
http://www.columbia-ltg.com

CommScope* (coax cable mfg.)
http://www.commscope.com

Communications Systems, Inc.
http://www.commsystems.com

Computer and Telecommunications Page*
http://www.cmpcmm.com/cc/

ConnectHome*
http://www.connecthome.com

Construction Specifications Institute
http://www.csinet.org

Construction Terms/Glossary
http://www.constructionplace.com/glossary.asp

Consumer Electronic Association (CEA)*
http://www.ce.org

Consumer Product Safety Commission (CPSC)
http://www.cpsc.gov

Continental Automated Building Association (CABA)*
http://www.caba.org

Contrast Lighting (recessed)
http://www.contrastlighting.com

Cooper Bussmann, Inc.
http://www.cooperbussman.com

Cooper Industries
Here you will find links to all Cooper Divisions
for electrical products, lighting, wiring devices,
power and hand tools.
http://www.cooperindustries.com

Cooper Lighting
http://www.cooperlighting.com

Copper (technical information)
http://energy.copper.org

Copper Development Association
http://www.copper.org

Craftsman Tools
http://www.craftsman.com

Crouse-Hinds
http://www.crouse-hinds.com

Cummins Power Generation*
http://www.cumminspower.com

Custom Electronic Design & Installation Association*
http://www.cedia.net

Custom Electronic Professional*
http://www.ce-pro.com

Eaton
http://www.eaton.com/EatonCom/Markets/
Electrical/SelectRegion/index.htm

DABMAR
http://www.dabmar.com

Danaher Power Solutions
http://www.danaherpowersolutions.com

Daniel Woodhead Company
http://www.danielwoodhead.com

Day-Brite Lighting
http://www.daybrite.com

Department of Energy
http://www.energy.gov

Dimension Express (to obtain roughing-in dimensions
from all appliance manufacturers)
http://www.dexpress.com

F. W. Dodge
http://www.fwdodge.com

DOMOSYS Corporation*
http://www.domosys.com

Douglas Lighting Control
http://www.douglaslightingcontrol.com

Dremel
http://www.dremel.com

Dual-Lite
http://www.dual-lite.com

Dynacom Corporation
http://www.dynacomcorp.com

Eaton (links to all divisions)
http://www.eatonelectrical.com

EC Online
http://www.econline.com

EC & M magazine
http://www.ecmweb.com

Edison Electric Institute
http://www.eei.org

Edwards Signaling and Security Products
http://www.edwards-signals.com

ELAN Home Systems*
http://www.elanhomesystems.com

Electric-Find
http://www.electric-find.com

Electric Smarts
http://www.electricsmarts.com

Electri-flex
http://www.electriflex.com

Electrical Apparatus Service Association
http://www.easa.com

Electrical Contracting & Engineering News
http://www.ecpzone.com

Electrical Contractor magazine
http://www.ecmag.com

The Electrical Contractor Network
http://www.electrical-contractor.net

Electrical Designers Reference
http://www.edreference.com

Electrical Distributor magazine
http://www.tedmag.com

Electrical Reliability Services, Inc.
(formerly Electro-Test)
http://www.electro-test.com

Electrical Safety
http://www.electrical-safety.com

Electrical Safety Foundation International
http://www.esfi.org

Electrician.com
http://www.electrician.com

Electro-Test Inc.
http://www.electro-test.com

Electronic House*
http://www.electronichouse.com

Electronic Industry Alliance*
http://www.eia.org

Emerson
http://www.emersonfans.com

Encore Wire Limited
http://www.encorewire.com

Energy Efficiency & Renewable Energy Network
(U.S. Dept. of Energy)
http://www.eere.energy.gov

Energy Star
http://www.energystar.gov

Environmental Protection Agency
http://www.epa.gov

Erico
http://www.erico.com

Essex Group, Inc.*
http://www.essexgroup.com

Estimation Inc.
http://www.estimation.com

ETL Semko
http://www.intertek-etlsemko.com/

Exide Technologies
http://www.exide.com

Factory Mutual (FM Global)
http://www.fmglobal.com

Faraday
http://www.faradayfirealarms.com

FCI/Burndy (telecom connectors)
http://www.fciconnect.com

Federal Communications Commission*
http://www.fcc.gov

Federal Pacific
http://www.federalpacific.com

Federal Signal Corp.
http://www.federalsignal.com

Ferraz-Shawmut
http://www.ferrazshawmut.com

Fiber Optic Association, Inc.*
http://www.thefoa.org

Fibertray
http://www.ditel.net

First Alert
http://www.firstalert.com

Flex-Core
http://www.flex-core.com

Fluke Corporation
http://www.fluke.com

Fluke Networks*
http://www.flukenetworks.com

Fox Meter Inc.
http://www.arcfaulttester.com

Fox Meter Inc.
http://www.foxmeter.com

Gardner Bender
http://www.gardnerbender.com

GE Appliances
http://www.geappliances.com

GE Electrical Distribution & Control
http://www.ge.com/edc

GE Industrial Systems
http://www.geindustrial.com

GE Lighting
http://www.gelighting.com

General Cable
http://www.generalcable.com

Generators:

 Coleman Powermate
  http://www.powermate.com/

 Cummins Power Generation
  http://www.cumminspower.com

 Generac Power Systems
  http://www.generac.com

 Gen-X
  http://www.genxnow.com

 Gillette Generators, Inc.
  http://www.gillettegenerators.com

 Kohler Power Systems
  http://www.kohlergenerators.com

 Onan Corporation
  http://www.onan.com

 Reliance Electric
  http://www.reliance.com

Genesis Cable Systems*
http://www.genesiscable.com

Global Engineering Documents
http://www.global.ihs.com

Grayline, Inc. (tubing)
http://www.graylineinc.com

Greenlee Inc.
http://www.greenlee.com

Greyfox Systems*
http://www.legrand.us/OnQ.aspx?NewLookRedir=yes

Guth Lighting
http://www.guth.com

The Halex Company
http://www.halexco.com

Hand Tools Institute
http://www.hti.org.

Handicapped Suggestions
The World of Assisted Technology
http://www.abledata.com

Hatch Transformers, Inc
http://www.hatchtransformers.com

Heavy Duty
http://www.milwaukeetool.com

Heyco Products, Inc.
http://www.heyco.com

Hilti Corporation
http://www.hilti.com

Hinkley Lighting
http://www.hinkleylighting.com

HIOKI (measuring instruments)
http://www.hioki.co.jp/

Hitachi Cable Manchester, Inc.*
http://www.hcm.hitachi.com

Hoffmann
http://www.hoffmanonline.com

Holophane
http://www.holophane.com

Holt, Mike Enterprises
http://www.mikeholt.com

Home Automation and Networking Association*
http://www.homeautomation.org

Home Automation, Inc.*
http://www.homeauto.com

Home Automation Links*
http://www.asihome.com

Home Automation Systems*
http://www.smarthome.com

Home Controls, Inc.*
http://www.homecontrols.com

Home Director, Inc.*
http://www.homedirector.net

HomeStar*
http://www.wellhome.com/home

Homestore*
http://www.homestore.com

Home Touch*
http://www.home-touch.com

Home Toys*
http://www.hometoys.com

Honeywell
http://www.honeywell.com

Honeywell*
http://www.honeywell.com/yourhome

Housing & Urban Development
http://www.hud.gov

Houston Wire & Cable Company
http://www.houwire.com

Hubbell Incorporated (links to all divisions)
http://www.hubbell.com

Hubbell Incorporated
http://www.hubbell-premise.com

Hubbell Lighting Incorporated
http://www.hubbell-ltg.com

Hubbell Wiring Devices
http://www.hubbell-wiring.com

Hunt Control Systems, Inc.
http://www.huntdimming.com

Hunter Fans
http://www.hunterfan.com

Husky
http://www.mphusky.com

ICC
http://www.iccsafe.org/Pages/default.aspx

Ideal Industries, Inc.
http://www.idealindustries.com

Illuminating Engineering Society of North America
http://www.iesna.org

ILSCO
http://www.ilsco.com

Independent Electrical Contractors
http://www.ieci.org

Institute of Electrical and Electronic Engineers, Inc.
http://www.ieee.org

Instrumentation, Systems, and Automation Society
http://www.isa.org

Insulated Cable Engineers Association (ICEA)
http://www.icea.net

Insulated tools (certified insulated products)
http://www.insulatedtools.com

Intel's Home Network
http://www.intel.com/

Intermatic. Inc.
http://www.intermatic.com

International Association of Electrical
Inspectors (IAEI)
http://www.iaei.org

International Association of Plumbing &
Mechanical Officials
http://www.iapmo.org

International Brotherhood of Electrical Workers
http://www.ibew.org

International Code Council (ICC)
http://www.iccsafe.org

International Dark-Sky Association
http://www.darksky.org

International Electrotechnical Commission (IEC)
http://www.iec.ch/

International Organization for Standardization (ISO)
http://www.iso.org

iSqFt*
http://www.isqft.com/

Jacuzzi
http://www.jacuzzi.com

Jefferson Electric
http://www.jeffersonelectric.com

Johnson Controls, Inc.
http://www.johnsoncontrols.com

Joslyn
http://www.joslynmfg.com

Juno Lighting, Inc.
http://www.junolighting.com

Kenall Lighting
http://www.kenall.com

Kichler Lighting
http://www.kichler.com

Kidde (home safety)
http://www.kiddeus.com

King Safety Products
http://www.kingsafety.com

King Wire Inc.
http://www.kingwire.com

Klein Tools, Inc.
http://www.kleintools.com

Kohler Generators
http://www.kohler.com

Lamp recycling information (sponsored by NEMA)
http://www.lamprecycle.org

LEDs Magazine
http://www.ledsmagazine.com

Leviton Mfg. Co., Inc (links to all divisions)
http://www.leviton.com

Lew Electric
http://www.lewelectric.com

Liebert
http://www.liebert.com

Lighting
http://www.lightsearch.com

Lighting Research Center
http://www.lrc.rpi.edu

Lightning Protection Institute
http://www.lightning.org

Lightolier
http://www.lightolier.com

Lightolier Controls*
http://www.lolcontrols.com

Litetouch*
http://www.litetouch.com

Littelfuse, Inc.
http://www.littelfuse.com

LonWorks*
http://www.lonworks.echelon.com

Lucent Technologies*
http://www.lucent.com

Lumisistemas, S.A. de C.V
http://www.lumisistemas.com

Lutron Electronics Co., Inc.
http://www.lutron.com

Lutron Home Theater*
http://www.classic.lutron.com

MagneTek (links to all divisions)
http://www.magnetek.com

Makita Tools
http://www.makita.com

Manhattan/CDT
http://www.manhattancdt.com

McGill
http://www.mcgillelectrical.com

Megger
http://www.megger.com/us/index.asp

MET Laboratories, Inc.
http://www.metlabs.com

MI Cable Company
http://www.micable.com

Midwest Electric Products, Inc.
http://www.midwestelectric.com

Milbank Manufacturing Co.
http://www.milbankmfg.com

Milwaukee Electric Tool
http://www.milwaukeetool.com

Minerallac
http://www.minerallac.com

Molex Inc.
http://www.molex.com

Motor Control
http://www.motorcontrol.com

Motorola Lighting Inc.
http://www.motorola.com

MTU Onsite Energy
http://www.mtuonsiteenergy.com/mtuonline/

Multiplex Technology
http://www.multiplextechnology.com/

National Association of Electrical Distributors
http://www.naed.org

National Association of Home Builders
http://www.nahb.org

National Association of Manufacturers
http://www.nam.org

National Association of Radio and Telecommunications
Engineers, Inc.
http://www.narte.com

National Conference of States on Building
Codes and Standards
http://www.ncsbcs.org

National Electrical Contractors Association (NECA)
http://www.necanet.org

National Electrical Manufacturers Association (NEMA)
http://www.nema.org

National Fire Protection Association
http://www.nfpa.org

National Fire Protection Association
National Electrical Code® information
http://www.nfpa.org

National Institute for Occupational Safety & Health
http://www.cdc.gov/niosh

National Institute of Standards and Technology (NIST)
http://www.nist.gov

National Joint Apprenticeship and
Training Committee (NJATC)
http://www.njatc.org

National Lighting Bureau
http://www.nlb.org

National Resource for Global Standards
http://www.nssn.org

National Safety Council
http://www.nsc.org

National Spa and Pool Institute
http://www.nspi.org

National Systems Contractors Association*
http://www.nsca.org

NECDirect
http://www.necdirect.org

Newton's Electrician
http://www.electrician.com

Nora Lighting
http://www.noralighting.com

NuTone
http://www.nutone.com

Occupation Health & Safety
http://www.ohsonline.com

Occupational Safety & Health Administration
http://www.osha.gov

Okonite Co.
http://www.okonite.com

Olflex Cable
http://www.olflex.com

Onan Corporation
http://www.onan.com

ONEAC Corporation
http://www.oneac.com

OnQ Technologies*
http://www.onqtech.com

Optical Cable Corporation*
http://www.occfiber.com

Ortronics Inc.
http://www.ortronics.com

Osram Sylvania Products, Inc.
http://www.sylvania.com

Overhead door information: Door & Access
System Manufacturers Association
http://www.dasma.com

OZ/Gedney
http://www.o-zgedney.com

Panasonic Lighting
http://www.panasonic.com/MHCC/pl/technoi.htm

Panasonic Technologies
http://www.panasonic.com

Panduit Corporation
http://www.panduit.com

Pass and Seymour, Legrand
http://www.legrand.us/PassAndSeymour.
aspx?NewLookRedir=yes

Pelican Rope Works
http://www.pelicanrope.com

Penn-Union Corp.
http://www.penn-union.com

Phase-A-Matic, Inc.
http://www.phase-a-matic.com

Philips
http://www.philips.com

Philips Advance
http://www.advance.philips.com/default.aspx

Philips Lamps
http://www.ALTOlamp.com

Philips Lighting Company
http://www.philipslighting.com

Philtek Power Corporation
http://www.philtek.com

Plant Engineering On-Line
http://www.plantengineering.com

Popular Home Automation*
http://www.phoenixcontact.com

Porter Cable
http://www.porter-cable.com

Power & Tel.
http://www.ptsupply.com/enterprise.asp

Power Quality Monitoring
http://www.pqmonitoring.com

Power-Sonic Corporation
http://www.power-sonic.com

Premise Wiring Products
http://www.unicomlink.com

Prescolite
http://www.prescolite.com

Progress Lighting
http://www.progresslighting.com

RACO
http://www.hubbell-raco.com

Radix Wire Company
http://www.radix-wire.com

Rayovac Corporation
http://www.rayovac.com

RCA*
http://www.rca.com

Refrigeration Service Engineering Society
http://www.rses.org

Regal Fittings
http://www.regalfittings.com

Reliable Power Meters
http://www.reliablemeters.com

Remke Industries, Inc.
http://www.remke.com

Rhodes, M. H. (timers)
http://www.mhrhodes.com

Ridgid Tool Company
http://www.ridgid.com

Robertson Transformer Co.
http://www.robertsontransformer.com

Robicon
http://www.robicon.com

Robroy Industries – Conduit Div.
http://www.robroy.com

Rockwell Automation
http://www.automation.rockwell.com

RotoZip Tool Corp.
http://www.rotozip.com

Russelectric
http://www.russelectric.com

Safety Technology International, Inc.
http://www.sti-usa.com

S&C Electric Company
http://www.sandc.com

SBCCI
http://www.sbcci.org

Sea Gull Lighting
http://www.seagulllighting.com

Seatek, Inc.
http://www.SeatekCo.com

SECURITY*

Caddx*
http://www.caddx.com

Digital Monitoring Products*
http://www.dmpnet.com

Home Automation*
http://www.homeauto.com

Honeywell Security*
http://www.security.honeywell.com/sce

NAPCO Security Systems*
http://www.napcosecurity.com

Siemens (main site)
http://www.siemens.com

Siemens Energy & Automation
http://www.sea.siemens.com

The Siemon Company*
http://www.siemon.com

Sensor Switch Inc.
http://www.sensorswitchinc.com

Shat-R-Shield
http://www.shat-r-shield.com

Silent Knight
http://www.silentknight.com

Simplex Grinnell
http://www.simplexgrinnell.com

Simpson Electric Co.
http://www.simpsonelectric.com

Skil Tools
http://www.skil.com

Smart Home*
http://www.smarthome.com

Smart Home Pro*
http://www.smarthomepro.com

Smart House, Inc.*
http://www.smart-house.com

Smoke & Fire Extinguisher Signs
http://www.smokesign.com

Sola/Hevi Duty
http://www.sola-hevi-duty.com

Southwire
http://www.southwire.com

SP Products, Inc.
http://www.spproducts.com

A.W. Sperry Instruments, Inc.
http://www.awsperry.com

Stanley Tools
http://www.stanleyworks.com

Starfield Controls*
http://www.starfieldcontrols.com

Steel Tube Institute
http://www.steeltubeinstitute.org

Straight Wire*
http://www.straightwire.com

Square D, Groupe Schneider
http://www.squared.com

Square D (Homeline)
http://www.squared.com/retail

Square D, Power Logic
http://www.powerlogic.com

Steel Tube Institute, Conduit Committee
http://www.steelconduit.org

Sunbelt Transformers
http://www.sunbeltusa.com

Superior Electric
http://www.superiorelectric.com

Superior Essex Group
http://www.superioressex.com

Suttle
http://www.suttleonline.com

Sylvania
http://www.sylvania.com

Sylvania Lighting International
http://www.sli-lighting.com

Telecommunications Industry Association (TIA)*
http://www.tiaonline.org

Terms /Construction /Glossary
http://www.constructionplace.com/glossary.html

Thermostats, Communicating*
http://www.homeauto.com

This Old House
http://www.thisoldhouse.org

Thomas & Betts Corporation
http://www.tnb.com

Thomas Lighting
http://www.thomaslighting.com

Thomas Register
http://www.thomasregister.com

3M Electrical Products
http://www.3m.com/elpd

Tork
http://www.tork.com

Touch-Plate Lighting Controls
http://www.touchplate.com

Trade Service Corporation
http://www.tradeservice.com

Trade Service Systems
http://www.tradepower.com

Triplett Corporation
http://www.triplett.com

TYCO (links to all divisions)
http://www.tyco.com

Tyco Electronics
http://www.tycoelectronics.com/components/default.aspx

Tyton Hellermann
http://www.hellermanntyton.us

UEI Test Equipment
http://www.ueitest.com

Underwriters Laboratories
http://www.ul.com

UL Standards Department
http://ulstandardsinfonet.ul.com

Unistrut (Tyco)
http://www.unistrut.com

Unity Manufacturing
http://www.unitymfg.com

Universal Lighting Technologies
http://www.universalballast.com

USA Wire and Cable, Inc.
http://www.usawire-cable.com

U.S. Department of Energy (DOE)
http://www.doe.gov

U.S. Department of Labor
http://www.dol.gov

U.S. Department of Labor, Employment & Training Administration
http://www.doleta.gov

U.S. Environmental Protection Agency (EPA)
http://www.epa.gov

Ustec* (structured wiring)
http://www.legrand.us/OnQ.aspx?NewLookRedir=yes

Vantage Automation & Lighting Control*
http://www.vantagecontrols.com

Watt Stopper
http://www.wattstopper.com

Waukesha Electric Systems
http://www.waukeshaelectric.com

Werner Ladder Company
http://www.wernerladder.com

Westinghouse
http://www.westinghouselighting.com/

Westinghouse Lighting
http://www.westinghouselighting.com

Wheatland Tube Company
http://www.wheatland.com

Wireless Industry*
http://www.compnetworking.about.com/

Wireless Thermostats*
http://www.wirelessthermostats.com

Wiremol
http://www.wiremold.com

Woodhead Industries
http://www.danielwoodhead.com

X-10*
http://www.x10.com

X-10 Pro Automation Products*
http://www.x10pro.com

Zenith Controls, Inc.*
http://www.geindustrialsystems.com

Zircon
http://www.zircon.com

Code Index

Note: Bold indicates material is in figures.

Index

Note: Page numbers in **bold** reference non-text material.

circuit number, 454
circuit requirement for, 449–451
connection methods, 449–450
disconnecting means, 451, 454
grounding, 454
grounding frames, 451–453
load calculations, 450, 454
nameplates on, 450, 454
overcurrent protection of, 451, 454
receptacles for, 451–452
wire size for, 450–451, 453, 454
COPALUM, 102
Cord AFCI, 206–207
Cord-and-plug connections
appliances, 348
water heater, 441
Cord-connections
dishwashers, 468
electric furnaces, 491
food waste disposer, 468
pools, 650
Cords, terminal identification, 453
Correction factors
defined, 71
due to high temperatures, 107, 410
Corrosion, water heater tanks, 430
Counterfeit products, 22
Countertops
GFCI's and, 79
kitchens, 78–81, **80**
CPSC (Consumer Product Safety Commission), 101
Crawl spaces, 192
Creep, aluminum wire, 100
Cross-talk problems, telephones, 535–536
CSA (Canadian Standards Association International), 23, **23**
Current, water pumps, 424–425

D

Damp locations, defined, 105
Data outlets, symbol for, **34**
Decks, receptacles for, 84–85
Decorative pools, code requirements for, 650
Demarcation point, telephone, 533
Derating correcting examples, 412–413

Derating factor
conductors, in conduits/cables, 409–410
defined, 71
high-temperature and, 107
Detector
defined, 551, 552
types of, 553
Device
boxes, molded fiberglass, 50
codes for, 36–42
described, 30
determining size of, 263
grounding clip, attaching, **182**
Diameters, inside, vs. trade size, **18**
Digital multimeters, 7–8
Dimmer controls, for homes, 321–335
Dimming ballasts, 235
Dinette tables, luminaries for, **310**
Dining room
receptacle outlets, **313**
tables, luminaries for, **311**
Direct-current, PV systems
disconnecting means, 693
general discussion, 688
wiring, 694
Disconnecting means
accessibility of, 502–503
appliances, 481–482
attic exhaust fan, 478
baseboard heaters, 492
central heating systems, 511
clothes dryer, 348
cooking unit, counter-mounted, 451, 454
dishwashers, 469
food waste disposers, 469
furnaces, 29, 511
generators, 680–681
pools, 651
PV systems, 693–694
ranges
electric, 451
freestanding, 455
rating, 503
service-entrance, 636–643
special-purpose outlets, 451
wall-mounted ovens/cooking units, 451, 453
water heaters, 439
water pumps, 426–427

Disconnect switch
symbol for
fused, **34**
unfused, **34**
temperature ratings, 106
Dish (satellite antennas), 528–533
Dishes, washing, 434
Dishwashers
code rules for, 468–469
disconnecting means, 469
general discussion, 467–468
grounding, 469
overcurrent protection, 469
portable, 469–470
small-appliance circuit for, 76
water temperature, 470
Door chimes
general discussion, 539–540
for hearing impaired, 540, **541, 542**
multiple locations, 543
power consumption, **543**
symbol for, **34**, 539, **540**
transformers, 540–543
wiring, 543–544
Doorjamb switch, 304, **305**
Doors, sliding glass, receptacle outlets and, 276
Double-pole switches, 175–177, **176, 177**
Draft regulator, described, 514
Draft stopping, 49
Drawings, 11
Dropped ceilings, 386
Dryers, clothes
cord sets for, 452
grounding frames of, 349–350
receptacles for, 451–452
Dry locations, defined, 105
Duplex receptacle outlet, symbol for, **37**
Dwellings
existing, electrical inspection, 15
lighting outlets in, **87**
multifamily, common areas in, 89, **90**
one- and two-family, 15
unit
defined, 17, 552
tamper-resistant receptacles in, **164, 197**, 208–209, **210, 211**